Global
Perspectives

Global Perspectives

A World Regional Geography

Second Edition

Charles A. Stansfield, Jr.
Chester E. Zimolzak

Glassboro State College

Merrill Publishing Company
A Bell & Howell Information Company
Columbus | Toronto | London | Melbourne

Cover Photo: Rio de Janeiro, by Ann Purcell.

Published by
Merrill Publishing Company
A Bell & Howell Information Company
Columbus, Ohio 43216

This book was set in Garamond.

Administrative Editor: Stephen Helba
Production Editor: Constantina Geldis
Art Coordinator: Ruth Kimpel
Cover Designer: Brian Deep
Text Designer: Cynthia Brunk
Photo Editor: Terry L. Tietz

This book was previously published under the title
World Regions: Changing Interactions.

Acknowledgments for photographs appear on page 660.

The Goode's Homolosine Equal-Area Projection base
maps in this text are used by permission of The
University of Chicago Committee on Geographical
Studies. Goode Base Map Series copyright © The
University of Chicago.

Library of Congress Catalog Card Number: 89–63248
International Standard Book Number: 0–675–20719–3
Printed in the United States of America
1 2 3 4 5 6 7 8 9—94 93 92 91 90

To our children,
our reach toward the future.
May the world they share be one of
justice and peace.

Preface

We all live in a time and a place. To fully comprehend our own time and place, we must understand both historical and geographic linkages. Where we live has a history and is part of a web of interconnections with other places. An educated person should be able to place events and prospects in the perspectives of time and place and in the context of history and geography.

On a daily basis contemporary Americans watch televised events and interviews from around the world. Newsworthy happenings in the Soviet Union, South Africa, Brazil, or Japan share the six o'clock news with more localized concerns, not only because we're curious about the rest of the world, but because we know that what goes on in faraway places has an impact on our lives as well. It is a world of increasing emphasis upon interregional, intercontinental connections—economic, cultural, and political. The decisions and actions of political leaders, business executives, voters, and consumers in each of the world's nations influence our future just as

inevitably as our choices affect their lives and fortunes.

This second edition of *Global Perspectives: A World Regional Geography* is about geographic perspectives and the complex interactions of the world's regions and peoples. To understand the present social and economic developmental characteristics and qualities of each region some historical background is provided.

This book divides the world into ten regions, starting with the United States and Canada, the region most familiar to American and Canadian students. The regional organization basically moves from the technologically advanced, highly industrialized countries to the newer and then the less industrialized countries. This system supports a developed–less developed world organizational approach, although the text is written so that the instructor can design any sequence of chapters following Chapter 1, "An Introduction to the World and Its Geography." While many smaller national territories are classed as subregions or even several states grouped as a sub-region, the larger nations (Australia, Brazil, Canada, China, India, Mexico, the United States, and the Soviet Union) are subdivided into subregions in recognition of their vast areas and internal diversity.

USING THIS BOOK

While reading each chapter, students should be instructed to keep these fundamental, organizational questions in mind: Physically and culturally, what is this region like? Why is this region like this? What is the nature of this region's connections with other regions and with the world's future?

Each chapter opens with an introduction highlighting special characteristics of the region. Salient cultural, economic, and historical geographic factors are outlined, followed by a survey of the physical frame. Subsequently, a series of subregions is presented in a more detailed survey. Finally, concluding statements relate this region to others and to the United States and Canada in particular. The specific topics selected for more thorough examination vary according to the nature of the region's trends and current problems.

Global Perspectives contains the following features designed to enhance students' comprehension of geographic concepts and regional characteristics:

1. Two hundred and thirty-five full-color pieces of artwork, mostly maps, are produced to the highest standards of professional cartography; they are designed to incorporate the latest boundary changes and place-name spellings.
2. Eighty-three full-color photos, especially selected to illustrate the unique physical, economic, and cultural qualities of each region, have been integrated carefully with text material and have been linked to major geographic concepts.
3. Statistical tables, graphs, and charts provide the factual information to substantiate text generalizations.
4. Several vignettes per chapter provide more detailed insights into some interesting aspects of the culture, economy, or political life and essence of a region.

5. Key terms are printed in boldface to alert students to geographic concepts and terminology. These selected terms are defined when first introduced, and examples are provided.
6. Review questions at the end of each chapter help students evaluate their comprehension of material and organize their studies.
7. For those students who wish to explore a region in greater detail, a compact bibliography at the end of each chapter lists selections that are both interesting and generally available in undergraduate libraries.
8. A complete glossary at the end of the book organizes definitions of all geographic terms and concepts mentioned in the text, and it serves as an important study guide for objective test questions which frequently involve definitions of terms.

A separate student study guide is available to assist students in focusing on the major points of each chapter and to increase efficient use of study time. An outline of text material is provided, along with sample essay topics and objective questions.

For the instructor, an instructor's resource manual has been compiled. This manual includes additional vignettes for each chapter that provide the basis for lectures that complement, but do not repeat, themes and concepts presented in the student text. A test bank of over 700 multiple choice questions is provided in the instructor's resource manual and is also available on computer disk. Additionally, slides and transparencies will be supplied to instructors adopting *Global Perspectives* for class use. (Please contact the publisher for more information.)

ACKNOWLEDGMENTS

In writing this book, we have benefited greatly from the research, teaching, assistance, and advice of many people. First, we must recognize a long line of teachers; among our mentors we thank Peirce Lewis, E. Willard Miller, Arthur Robinson, Jean Gottman, and Richard Hartshorne for inspiring our curiosity and guiding our quests for geographic comprehension.

The original manuscript was rigorously evaluated by a team of fellow geographers; these perceptive, thorough, and enthusiastic reviewers were: Crystal Boyd, Schoolcraft College; Clyde Browning, University of North Carolina—Chapel Hill; Charles Buddenhagen, Anne Arundel Community College; David L. Clawson, University of New Orleans; Jane Ehemann, Shippensburg State College; Jeff Gordon, Bowling Green State University; Peter L. Halvorson, University of Connecticut; Edward Karabenick, California State University—Long Beach; Steven Kimbrel, University of North Carolina; David R. Lee, Florida Atlantic University; Gordon R. Lewthwaite, California State University—Northridge; Jerry N. McDonald, Radford University; Edwin Moreland, Emporia State University; John Patterson, University of Wisconsin—Whitewater; Paul E. Phillips, Fort Hays State University; Peter Poletti, University of Missouri—St. Louis; Milton Rafferty, Southwest Missouri State University; Bruce W. Smith, Bowling Green State University; Joe Spinelli, Bowling Green State University; Bob J. Walter, Ohio University; William Wyckoff, Montana State University; Leon Yacher, Southern Connecticut State University; and Charles C. Yahr, San Diego State University.

Our friends and colleagues at Glassboro State College kindly shared their research, personal libraries, and time with us; we are grateful to Edward Behm, Wade Currier, Robert Edwards, David Kasserman, Jerry Lint, and Dick Scott for their friendship and support. Dr. Takashi Yamaguchi of Tokyo University, our valued friend and a onetime fellow graduate student, made several suggestions for the East Asia chapter and helped obtain photographs of that key region. William Top, a professional planner and longtime friend in Hove, England, supplied many insights on urban and economic trends in the United Kingdom and the Common Market.

Karen Haynes, Patt Martinelli, and Diane Stansfield typed the original manuscript; their patience and friendship were as important to us as their typing skills. They have our deepest thanks and appreciation.

Many talents beyond those of the authors are required to produce a book. When we first started writing textbooks, we were fortunate to establish a good relationship with the professional staff of Merrill Publishing. We continue to be impressed by the high quality of work produced by this talented, dedicated, and hard-working team: Stephen Helba, Connie Geldis, Cindy Brunk, Ruth Kimpel, Linda Johnstone, Marilyn Phelps, Anne Vega, Terry Tietz, Brian Deep, Jan Wagner, and Bruce Johnson.

Finally, we appreciate our families' tolerance of our seemingly permanent preoccupation with this book. We owe our wives and children many weekends and a good vacation; we know that Diane and Chesha share our pleasure in finishing this book.

Charles A. Stansfield, Jr.
Chester E. Zimolzak

Contents

CHAPTER FIVE

Australia-Oceania   253

CHAPTER SIX

East Asia   281

CHAPTER SEVEN

South Asia   341

Global
Perspectives

This political map reveals a highly compartmentalized world. The numerous political entities range in size from the colossal USSR to minute but significant countries such as Singapore, Malta, or Grenada. The names of these political entities evoke images of different environments, peoples, cultures, and levels of well-being. However, the political boundaries that segregate five billion inhabitants do not clearly reveal the underlying geographic complexities of our world. Numerous other sets of boundaries could be imposed: boundaries which delineate multinational alliances, boundaries which classify economic and agricultural environments, and boundaries which outline the myriad of peoples, languages, and ideologies that affect our daily lives. Unraveling the complexities of our world requires intellectual attention to many questions relating to geography. This map and the chapters that follow are intended to begin the student on a journey toward understanding our exciting and complex world.

NOVA TOTIVS TERRARVM SIVE NOVI OR

CHAPTER ONE

An Introduction to the World and Its Geography

Map of the World by Willem Blaeu

Geography is an extremely broad discipline; the literal meaning is "description of the earth." The field's earliest practitioners, the Greeks, set about doing just that—describing the earth, in all its aspects. Modern geography has moved beyond description, adding explanation, interpretation, and forecasting to its tasks. However, location—describing where things are (or are not)—is still the first step in the process of areal analysis.

There is a pattern to the distribution of people across the world, just as there are distributional patterns for everything else with a location. Population is concentrated in some areas, absent in others. Why is India so densely populated? Why is Canada so empty over most of its extent? Geography explores the reasons why. Certainly, India has been settled longer than Canada, and the two countries are strikingly different physically. India is much warmer and generally wetter, and upon investigation of such physical factors as soil fertility and depth, degree of slope, and length of growing season, it becomes obvious that India has a higher degree of utility for farming and, therefore, a greater overall ability to support dense populations than is the case in Canada.

Is variable population distribution of any significance? India's apparently crowded circumstances may have something to do with its relatively poor dietary standards. Geographers cannot assume that high population density and poverty go hand in hand. If they did, Japan and West Germany would be among the poorest nations of the world. Likewise, relatively low densities do not guarantee a Canadian degree of prosperity and abundance. The poor, food-deficient nations of Africa are characterized by relatively low population densities. The poorest nations of the world include both densely and lightly populated units.

Geography allows us to know the world, to assess the productive potential of its parts, and to anticipate, but not always solve, many of its problems. It utilizes all the data and materials available in its tasks of description, analysis, interpretation, and forecasting.

Geography is also concerned with the organization of space. Over much of the midwestern United States, fields and farms are laid out in a rectangular pattern. Local roads, following property lines, feature right-angle turns, difficult to negotiate at high speed. Ours is a society where virtually everyone drives, where time and speed are deemed both valuable and critical. The Midwest is an area with few physical obstacles to discourage road construction from the straightest possible path, yet rural roads exhibit this peculiar pattern. Even midwestern counties are generally rectangular in shape. Farm families live in dwellings widely dispersed across this landscape of lines and right angles. In contrast, fields in tropical Africa are often nearly circular islands scattered amid dense forest, and the traditional European farm is often a collection of strips (rather than a compact ownership unit) in a densely farmed landscape almost completely cleared of forest.

Landscapes differ, worldwide, in response to cultural preferences, available materials, historical evolution, physical obstacles, and government regulations, reflecting the wide range of factors that impinge on an area's geography. Yet, despite the seemingly endless diversity of the world, broad similarities are readily observable over large areas. There is a tendency for landscape and land use types to appear quite similar over vast reaches. Organizing these similar parts of the world into units based on this kind of relative homogeneity is one of geography's tasks—the creation of units called **regions**.

Location and distribution, together with interpretation and analysis, give us the **geographical perspective**—the primary tool that enables us to view the world in an orderly fashion. The regional concept allows us to organize our knowledge further by developing areal units within which a variety of distributions have a similar extent or exhibit similar patterns of occurrence.

The study of the world becomes easier when the planet is broken down into manageable spatial units, each exhibiting a reasonable degree of similarity across the space within its borders. It is also easier to learn about the world when an organizational framework for study—the geographical perspective—provides a specific set of guidelines for understanding the complexities.

Geography, then, provides a key to world understanding. However, it is not sufficient to learn one set of factors or one group of generalizations. The field is as dynamic as the world itself. Patterns change over time. New resources are discovered, new technologies are developed, tastes change, and cultures may become more receptive to change with increased education or a shift in governmental structure or policy. Regional borders may also change, and certain aspects of a region's geography may be altered. Once an understanding of geographic method is mastered, analysis can continue, and new conclusions can be reached. In a period characterized by rapid technological change, radical innovation, shifting political alliances, and repeated social upheaval, the geographic perspective is more viable and valuable than ever before. We live in such an era. Individual and national survival and economic prosperity are made easier with a solid base of geographic knowledge.

WHY STUDY GEOGRAPHY?

Recent surveys have shown that most Americans know little about the field of geography and exhibit a relatively low degree of familiarity with the concepts of location and distance, or even with place names and locations. Increasingly, public figures find their geographic blunders receiving widespread notice in the media. Even the huge American foreign trade deficit has been blamed on prevalent geographic ignorance as the press relates such marketing errors as attempts by U.S. corporations to sell pork products in Islamic countries (where pork consumption is forbidden by religion) and red hair coloring in societies where cultural values deem that hair color to be ugly or an omen of bad luck. These examples illustrate the need to understand better the world we live in and cultures other than our own.

Geography is a bottom-line discipline in the sense that it is fundamental to all learning, because all things have location. Modern geography provides a basic framework for understanding the world and the processes of change. Geography can help to make you more competitive in today's world by supplying knowledge applicable in such job market areas as land use planning, environmental conservation, and marketing. It contributes to a personal understanding of world economics, current affairs, and the potential implications of any country's national or international policies.

On a more mundane, but practical level, geography can teach you what clothes to take and what to expect on the menu when you travel abroad. At home, it can provide useful information that will aid house buyers in making a wise decision. It can provide a reserve of information that is useful in the study of a variety of disciplines. Geography affects us at every turn, and a firm geographic base is a crucial part of any individual's knowledge.

There is also a certain amount of pleasure associated with geographic study. Each of us enjoys solving a mystery or formulating the solution to a problem. Most of us like travel, and an increasing number of jobs require it. We all have a taste for the different and exotic, at least occasionally. Everyone likes to appear knowledgeable, and society demands that we remain "current." Geography supplies something of each of these.

People are naturally curious. *Why?* and *why not?* are among our first spoken words. Knowing what makes things work (or not work) is crucial to daily living and a source of self-satisfaction. Geography supplies background material and areal perspective, allowing each individual to be an expert.

THE FIELD OF GEOGRAPHY: ORIGINS AND EARLY EVOLUTION

Geography is an old discipline, as old as civilization itself. Perhaps the first geographers were the tax agents of ancient times, who measured land, recorded deeds, estimated crop yields, and levied taxes. Even nomads knew the locations of water holes and where grass was likely to be found in abundance at different times of the year. Knowing the earth, in either case, was necessary to prosperity or survival. In both cases, it combined a knowledge of the physical and the human (modified) environments.

The Greeks gave the field of geography its name and were its earliest academic practitioners. Aristotle, writing in the fourth century B.C., described it as the "study of natural processes [operating] in and near the surface of the earth."[1] Other Greek scholars wrestled with the size, dimensions, and shape of the earth and developed theories of earth-sun relationships, establishing the physical environment as a part of the discipline of geography.

Itinerant Greek merchants described the inhabitants of foreign lands, noted cultural practices, and observed differences in the cities and settlements of non-Greeks. The human contributions to geography were noted almost as early as the physical contributions. All of these observations were collected and published in the second century A.D. by Ptolemy in his writings entitled *Geographia*. Within the first 400 years of geography's formal existence, the groundwork for modern geography had been laid, including the duality of the field in dealing with both the physical and cultural environments.

Modern geography derives from the efforts of German scholars in the seventeenth, eighteenth, and nineteenth centuries. The first academic chair in geography per se was established at Berlin University in 1820,[2] and the study of geography diffused from there throughout Europe, and, ultimately, the rest of the world. By 1904, the Association of American Geographers, the discipline's first professional organization in the United States, was established in Philadelphia, one year after the establishment of the first graduate program in geography at the University of Chicago.

Much of the early work of American geographers focused on the physical environment. The cultural landscape was viewed as being determined by physical environmental factors. A school of thought that emerged in Europe soon came to dominate the field. Known as **environmental determinism**, that viewpoint claimed that the physical environment, particularly climate, exerted a controlling influence on people and culture. As in the case of many other disciplines, the search for cause-and-effect relationships erred through oversimplification.

Environmental determinism's succeeding antithesis, **cultural determinism**, or **possibilism**, has fared little better. A world poised at the brink of a scientific revolution in the late nineteenth century was only too willing to reject ideas of environmental controls on human behavior in the face of a rapidly expanding technology. Rails had conquered distance by reducing travel time; ocean liners had conquered stormy seas. Expositions and world's fairs in turn-of-the-century Europe and America were geared to themes of human conquest over the forces of nature. The age, and the public, lionized explorers—conquerors of space and the elements. It took 60 years or more for the public to reexamine the physical environment and its influence on humans.

By the mid-1920s, geographers abandoned both of these simplistic philosophies and began to search for answers that encompassed a variety of cultural factors—economic, political, and technological—never forgetting, though perhaps at times minimizing, the role of physical factors in the total environment.

WHAT IS GEOGRAPHY NOW?

Geography remains an integrative discipline. By its very nature, it attempts to comprehend the whole world or the whole of a region—a segment of that world. Commonly, this search for an explanation of distributions, or for a thorough regional comprehension, requires geographers to search out data from a wide range of disciplines (Figure 1–1). There are no solid answers without all the facts.

When two or more geographic variables tend to be concentrated (or absent) together in roughly the same pattern, one can speak of **spatial correlation**. Is a causal relationship implied? Perhaps, but not necessarily. Modern geography seeks answers cautiously. Important, viable spatial concepts have emerged over time to clarify relationships among various cultural and physical factors. Spatial relationships help bring order to the chaos implicit in a lengthy list of individual facts.

What, then, is the core of modern geography? It is the study of the earth, in its entirety or in part. The attributes of and explanations for all locations, distributions, and spatial correlations that occur on the earth form its locus of study. The definition of regions and subregions, the making of maps, and the

FIGURE 1–1
Geography as an Integrative Discipline. The search for a thorough comprehension of the world in all its complexity requires that geographers gather data from a wide range of disciplines. The field integrates knowledge obtained from all these sources into a comprehensive picture of all or a portion of the planet. In an earlier era, geography was devoted primarily to location and description. Modern geographers emphasize analysis, interpretation, and forecasting, though location is still a central theme.

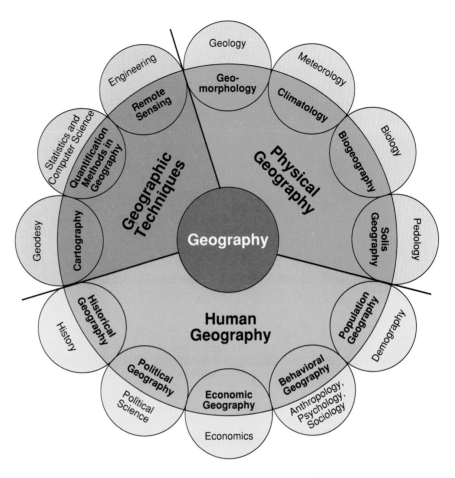

interpretation of data are the tools with which analysis is organized and accomplished; they constitute the proofs displayed in support of theories and concepts developed by geographers. Implicit in the entire process are the interrelationships between both halves of the environment—humans and their constructs and the various forms and forces of nature. Such traditional terms as *human-land relationships, human-milieu interrelationships, areal distribution,* and *spatial correlation* still apply.

The breadth of the field is both intimidating and exciting. Because of this breadth, any definition of geography must be extremely broad and therefore is subject to both disagreement among geographers and territorial dispute with other disciplines. In the search for an ideal, wholly acceptable definition, a committee of geographers issued, in 1954, a summary document called *American Geography: Inventory and Prospect.*[3] It was just that—an inventory of

the past and a prospect for the field in the then future. Like committee work in general, it reflected a series of compromises and yielded a definition that did not satisfy all geographers.

In his 1939 historical analyses of the field, *The Nature of Geography,* Richard Hartshorne defined geography as "the study of the areal distribution of phenomena on the face of the earth as the home of man."[4] This definition clearly encompasses area, space, and anything with a location or distribution attribute; it fits the objectives of this book and those of much of geographic research.

Geography is a field with many traditions, befitting a discipline that has a broad base, a long history, and a variety of interests. In a 1963 speech before the National Conference for Geographic Education (another of the field's professional organizations), William Pattison, a geographer from California, outlined "four traditions of geography":[5]

THE PHILOSOPHIES OF ENVIRONMENTAL DETERMINISM AND POSSIBILISM

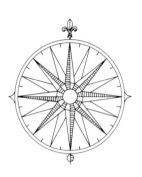

Geography seeks to explain locational patterns and explores spatial associations, along with possible cause-and-effect relationships. Periodically through the more than 2,000-year history of the discipline, a misdirected passion for precise and uncomplicated answers has resulted in a deterministic view of the complex relationships among physical and cultural geographic phenomena.

Determinism implies that people do not have freedom of decision making, that their actions and decisions are determined by some force or influence other than human will. American geography evolved as an offshoot of geology, and so the first generation of American geographers were trained mainly as physical geographers. Their interests lay in the study of landforms, climate, and natural vegetation. Ultimately, they began to study cultural geography—the distribution patterns of the human population, settlement, land use, resource exploitation, and levels of economic development. While seeking explanations for cultural geographic patterns strictly within the physical world, they attempted to show a direct causal connection between physical environmental factors and human activities. The first president of the Association of American Geographers, William Morris Davis, attempted to explain the organizational viewpoint of geography in 1906 by stating that it showed the relationship between "an element of inorganic control and one of organic response."* Control from the nonliving elements of the natural world was assumed to be imposed on all living organisms, including people.

The central concepts of environmental determinists were forcefully, if poetically, stated by geographer Ellen Churchill Semple. Her 1911 book, *The Influences of Geographic Environment,* opens:

> Man is a product of the earth's surface. This means not merely that he is a child of the earth, dust of her dust, but that the earth has mothered him, fed

1. **spatial tradition**—analysis of such aspects of occurrence as distance, form, direction, and position, essentially encompassed in the geographic essentials of geometry (form, pattern, distribution, movement)
2. **area studies tradition**—organization of space into regions (areas) and intensive study of those areas; organizing knowledge about an area contained within constructed boundaries
3. **man-land tradition**—analysis of the interplay of the physical and cultural halves of our environment—the way in which humans organize space, encompassing such diverse topics as conservation, resource utilization, and planning
4. **earth science tradition**—study emphasizing the physical environment—the natural processes, geologic evolution, weather occur-

him, set him tasks, directed his thoughts, confronted him with difficulties that have strengthened his body and sharpened his wits, given him his problems of navigation or irrigation, and at the same time whispered hints for their solution. She has entered into his bone and tissue, into his mind and soul.**

All those active verbs! In the determinists' view, people respond to the actions and initiatives of nature. Many environmental determinists identified climate as the most powerful factor of the physical world. As Ellsworth Huntington asserted in his conclusions to *Civilization and Climate:*

We are slowly realizing that character in the broad sense of all that pertains to industry, honesty, purity, intelligence, and strength of will is closely dependent upon the condition of the body. Each influences the other. Neither can be at its best while its companion is dragged down. The climate of many countries seems to be one of the great reasons why idleness, dishonesty, immorality, stupidity, and weakness of will prevail.***

At the other extreme from environmental determinism is cultural determinism, or possibilism. In this view, people can do anything, anywhere that they choose. Humans do things not because they are restrained by the environment, but *despite* the physical environment. The role of people is to conquer the environment, changing wilderness into productive landscape. Technology level is the main consideration, not the environment itself.

Contemporary geographers would avoid subscribing to either of these two extreme viewpoints. Physical environment is seen as passive; it does not dictate responses from people. People do indeed consider physical factors in making their decisions about land use and resource development, but they make these decisions within the framework of a culture. They consider available technology; prevailing costs of labor, energy, and capital; government regulations; preferences; and other cultural factors.

*William Morris Davis, "An Inductive Study of the Content of Geography," *Bulletin of the American Geographical Society* 38 (1) (1906): 71.
**Ellen Churchill Semple, *The Influences of Geographic Environment* (New York: Henry Holt & Co., 1911), p. 1.
***Ellsworth Huntington, *Civilization and Climate,* 3rd ed. (New Haven, CT: Yale University Press, 1924), p. 411.

rences, and zones of climate as they interact together, and with humans, on our habitable planet

Geography, as Pattison noted, pursues all four traditions concurrently. The planet, in its exceptional diversity, is what gives unity to the field. Geography, like the planet it studies, reflects both that unity and diversity as well.

THE BASIC TOOLS OF GEOGRAPHY

Each field of study is characterized by a set of research methods and techniques used to gather, organize, and interpret data. These tools and techniques generally derive from what is termed standard scientific method. Most are not unique to any one field but are shared with other research disciplines.

DEFINITIONS OF GEOGRAPHY

Most geographers agree with the statement, "No person is educated who lacks a sense of time and place." Geography stands beside history as a fundamental social science because it provides a sense of place. Whereas history places human knowledge in the context of time, geography provides the place, or locational, context.

Geography seeks to understand the character of places and the relationships among them. The study of area—its location, its special qualities, and its relationships with other areas—is central to geography. Geography is also concerned with distributional patterns and with understanding the reasons for those patterns.

Definitions as stated by noted geographers include:

Geography is the science of places. (La Blache)
Geography is the science of distribution. (Marthe)
Geography is human ecology. (Barrows)
Geography is the correlative science. (Taylor)
Geography tells what is where, why and what of it. (Bowman)
Geography seeks to interpret the significance of likenesses and differences among places in terms of cause and effect. (James)
Geography is that discipline that seeks to describe and interpret the variable character from place to place of the earth as the home of man. (Hartshorne)

Field Observation

Field observation, perhaps the most basic technique in the study of geography, studies an area with a trained eye, sorting data, observing patterns, looking for correlations, and noting detail. It may involve running tests on soil or air, measuring physical features or phenomena, taking censuses, administering interviews or surveys, photographing features in the landscape, and numerous other tasks. Field research seeks to document impressions with solid, corroborating data. This method is most applicable to small areas, pilot projects, or highly specialized topics. The trained eye seeks geographic order in every segment of the landscape traveled. Field observation is a sophisticated descendant of the Greek merchants' pen-

chant for note taking and a modest, if more technically exact, continuation of the explorers' tradition.

Maps

Maps are an inseparable part of geography. Though the entire world uses maps for a variety of purposes, this intimate popular association is quite accurate because cartographers—the people who design and make most maps—are trained in the discipline of geography. Because spatial patterns and associations are basic to the field, the use of maps is as critical as the written word in most geographic studies. Maps represent a kind of shorthand; they depict a selected portion of reality at a convenient scale and in se-

READING MAPS

Cartographers have developed a series of conventions to aid the uninitiated in reading maps. On most maps, the orientation of direction follows a simple pattern: up is north, down is south, to the right is east, and to the left is west. Such maps are called **right-oriented**. Where this convention is violated for some practical reason, a directional arrow indicates north. Maps that use color follow basic color schemes. Urban areas are red, forested areas are green, water is colored blue, topographic lines are in brown, and place names are printed in black. On special-purpose maps, known generally as **thematic maps**, colors are divided into hot and cold categories. Reds, oranges, and browns depict a heavy density or great frequency of occurrence of whatever is being mapped. Colder colors, particularly yellows and greens, indicate a lesser value.

Map symbols are fairly standard. Some variation of a circle indicates a settlement, and linear features such as roads and railroads are simply shown as lines. Many symbols are so generally understood as to leave little doubt in the reader's mind; others require explanation. Crucial clues to reading any map appear in a box called the **map legend**, where symbols are explained with appropriate captions.

A seasoned map reader or geographer can go far beyond the obvious and clearly stated, seeing a great deal more than what initially meets the eye on any map. This analysis beyond the obvious is known as **map interpretation**.

lected detail. A map is like a scale model, a faithful depiction of the original produced in smaller size and showing only those details deemed essential.

The **scale** of a map depicts the ratio of distance and area on the earth to the distance and area on the map itself. It is usually expressed as a proportional fraction, a ratio, a line showing distance, or sometimes a verbal statement. A ratio scale of 1:1,000,000, a common scale for wall maps, indicates that 1 unit (an inch or a centimeter) on the map is equivalent to 1 million units (inches or centimeters) on the surface of the earth. Various depictions of scale are shown in Figure 1–2. Large-scale maps show small areas in a high degree of detail; the reverse is true of small-scale maps. The larger the number in the scale, the smaller the scale of the map.

Transferring an area from the spherical earth to a two-dimensional map surface always involves some distortion. In a simple analogy, one cannot flatten the skin of an orange without considerable distortion. Cartographers have devised ways to control the distortion that occurs during the transfer from three to two dimensions (Figure 1–3a). The result is a series of **map projections**. Each projection is a systematic way of constructing a map to control for the amount, type, and location of distortion, while accurately preserving one or more of the original relationships. It is possible to preserve true (equal) area, the crossing of grid lines at right angles, true distance, directional relationships, and shape, among other characteristics. However, it is not possible to maintain all such characteristics at once. Each projection, then, is a

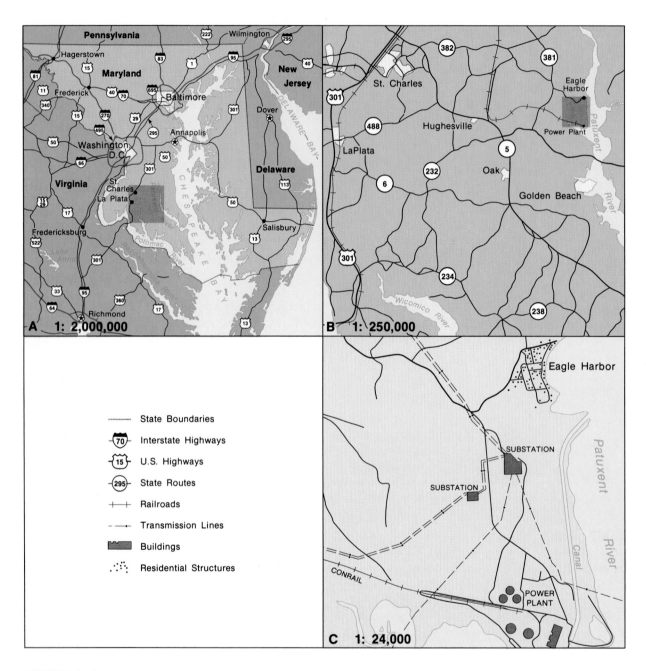

FIGURE 1–2
Meaning of Scale. These three maps illustrate the concept of scale—the ratio of map size to earth size. A portion of Maryland is shown here at three, progressively larger scales; each portion shows a smaller area in greater detail. Scale is chosen in accordance with the level of detail appropriate to the area studied or the purpose of the map.

FIGURE 1–3
Projections. In the process of creating maps, cartographers reduce the actual size and increase the degree of generalization. There is unavoidable distortion as a three-dimensional earth is depicted on a two-dimensional surface. (a) Process of converting the earth to a map. (b) Robinson projection, which attempts to reduce many negative aspects of distortion without interruption. (c) Goode's Homolosine projection, which maintains the property of equal area by utilizing interruption. (d) World scale projection used in this text, a modified Goode's Homolosine designed to maximize land area.

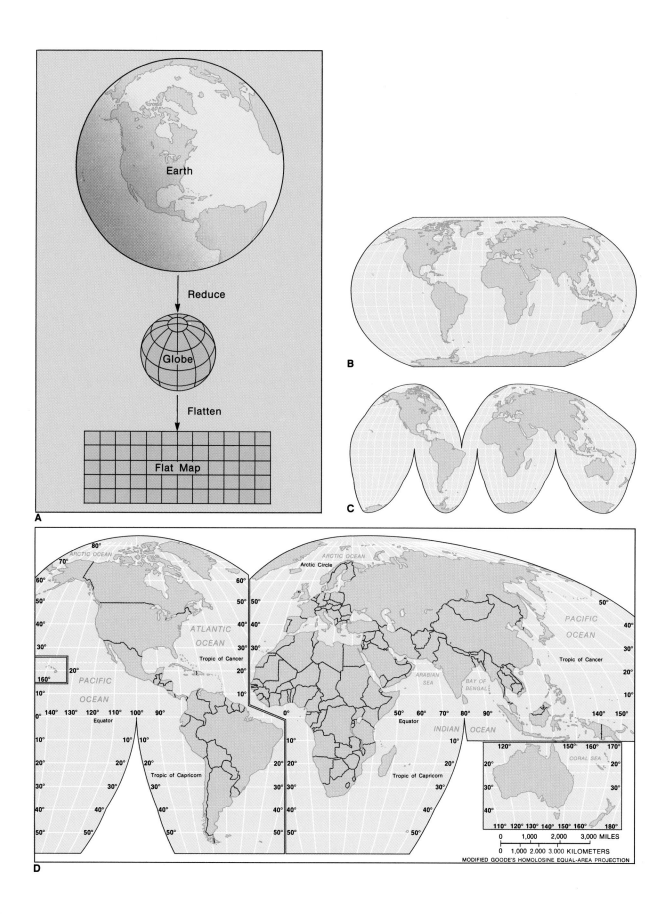

A

Earth

Reduce

Globe

Flatten

Flat Map

B

C

D

ARCTIC OCEAN

80°
70°
60°
50°
40°
30°
20°
160°
10°
0° 140° 130° 120° 110° 100° 90°
10°
20°
30°
40°
50°

ATLANTIC
OCEAN

Tropic of Cancer

PACIFIC
OCEAN

Equator

Tropic of Capricorn

60°
50°
40°
30°
20°
10°
0°
10°
20°
30°
40°
50°

ARCTIC OCEAN

Arctic Circle

50°
40°
30°
20°
10°
0°
10°
20°
30°
40°
50°

ARABIAN
SEA

BAY OF
BENGAL

50° 60° 70° 80° 90°

Equator

INDIAN OCEAN

Tropic of Cancer

PACIFIC
OCEAN

50°
40°
30°
20°
10°

140° 150°

CORAL SEA

120° 150° 160° 170°

20°
30°
40°

110° 120° 130° 140° 150° 160° 180°

0   1,000   2,000   3,000 MILES

0   1,000 2,000 3,000 KILOMETERS

MODIFIED GOODE'S HOMOLOSINE EQUAL-AREA PROJECTION

THE MAP GRID

The grid system on a map is a handy system for pinpointing the location of any place or feature. The Greeks originally designated a sphere as being composed of 360 degrees (°). Once the earth (a sphere) was given a finite number that did not vary, it became possible to divide it into parts on a systematic basis by devising a grid system. The line known as the **equator** (designated as 0°) was the beginning point of the system. Located midway between the northern and southern extremes of the earth, it bisects the earth into two equal parts. From the equator, lines of **latitude** run east and west as a series of parallel, concentric circles which are used to measure distance north and south of the equator (Figure Aa). Lines of **longitude** are superimposed at right angles to lines of latitude. Longitudinal lines run north-south and measure distance east and west of a central line called the **prime meridian**, or 0° longitude. Longitude lines meet at the poles, whereas lines of latitude cannot cross or converge (Figure Ab).

Although the location of the equator is unequivocal, there has been much disagreement, historically, over the location of the prime meridian. Various nations have, at times, chosen their capital city or chief observatory as the location of the prime meridian, renumbering lines of longitude accordingly. Washington, DC, Paris, Berlin, Moscow, and Beijing (Peking) have all been designated the central meridian by nationalistic governments at some time. The resulting confusion was intolerable. The location of the prime meridian now used in all navigation and on virtually all maps is that of the Royal Observatory at Greenwich, a London suburb. Britain's supremacy on the seas during the nineteenth century made this a widely adopted world standard.

Not all maps have a grid, nor do all maps need one, depending on the central purpose of the map. On many maps of relatively small areas, such as states or cities, lines of latitude and longitude are replaced with a simple, more dense grid using letters and numbers; location is often more important than navigation on a road or street map. On even smaller areas (larger-scale maps), as on topographic sheets, latitude and longitude are used for reasons of compass reference in field work and as base lines for surveying.

FIGURE A

Grid System. A series of imaginary lines are used to provide a uniform frame of reference for locations. (a) Meridians (lines of longitude) extend north-south and give east-west references. Parallels (lines of latitude) cross meridians at right angles and provide north-south references. (b) Any place on the earth can be pinpointed with the aid of a reading in longitude and latitude. For example, New Orleans is located at 30° north (latitude) and 90° west (longitude).

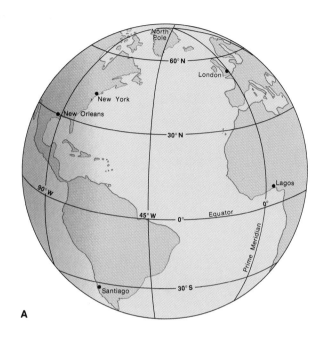

A

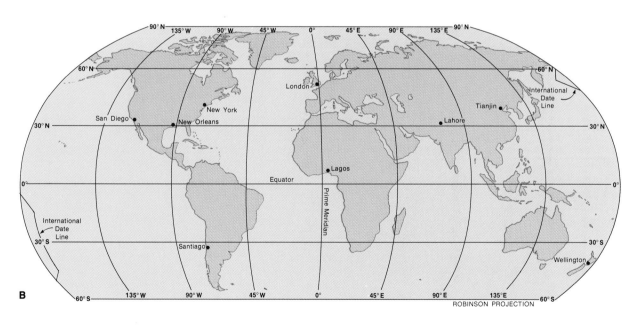

B

ROBINSON PROJECTION

carefully controlled distortion of reality. The distortion problem is extremely difficult to solve in situations involving a map of the entire world. The smaller the area depicted, the fewer the problems of distortion.

Arthur Robinson, a noted cartographer, has spent years developing a world projection that minimizes distortions in area, direction, and shape without interrupting the projection (breaking it into pieces) (Figure 1–3b). **Interruption**, however, is a valid means of maximizing the land area and minimizing the areas of sea, where distributions do not usually occur (Figure 1–3c).

The concept of equal area is extremely important on maps showing distribution. The visual value of such things as density can be badly distorted if equal area is not maintained. For that reason, world maps in the text use an equal-area projection (Figure 1–3d).

Aerial Photography and Remote Sensing

The Apollo Moon Program produced the most famous and familiar photograph of the earth taken from out in space. Only a handful of astronauts have actually seen the earth from deep in space, yet virtually all of us have seen the photograph—thus engaging in a form of **remote sensing**. Any technique used to "see" objects without being in direct contact with (or in close proximity to) them is an exercise in remote sensing. A camera is the most common remote sensing device. Aerial photography is among the oldest and still most useful remote sensing tool in geography.

The vast potential of aerial photography has expanded in three general ways: (1) photos may be taken of larger earth areas from progressively farther away as the camera mount has passed from balloon, to plane, to satellite; (2) the art and science of inter-

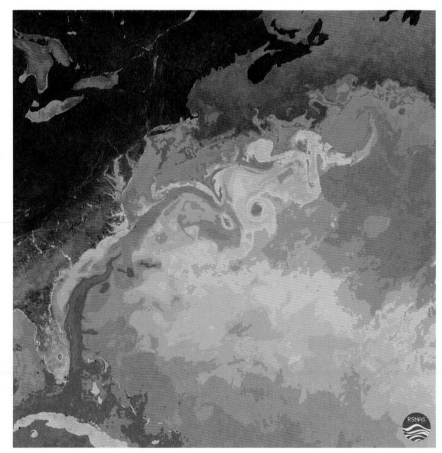

Satellite Imagery. *This photograph was taken from miles above the earth with temperature-sensitive infrared film. Technically, it is not a photograph, but rather the image of heat reflected from various surfaces. Land appears in brown, and the ocean in a variety of colors. Rivers appear as yellow. The warmer waters of the Gulf Stream current appear as red and orange, in contrast to the colder surrounding waters, in shades of green. Cooler waters appear in blue, while cold currents and waters are shown in shades of purple. Such products of modern technology yield detailed information that would be difficult or impossible to obtain by traditional research methods and techniques.*

preting aerial photography has advanced enormously; and (3) the qualities and capabilities of both cameras and film types (including electronic transmission of images) have improved, resulting in a broadening spectrum of the types of photography that can be done.

False-color or infrared photography can be used to identify patterns not discernible in ordinary photographs. Photographs, unlike maps, include everything present in an area. Except for small parcels of land, topographic surveys are now done from photo overflights rather than through laborious manual surveying. Air photos have proven useful in detecting forest diseases, in discovering groundwater resources, in prospecting for minerals, and in tracing subsurface systems of faults. In all cases, aerial photos can provide new information, simplify data updating, and make possible more detailed and exact interpretation. As a technological tool, remote sensing increases accuracy, speeds data collection, and increases the scope of work that can be accomplished in geographic research.

Statistical Research Tools

Traditionally, geographers have relied on their national censuses of population, manufacturing, and agriculture to supply much of the data for domestic research. United Nations agencies, the World Bank, and individual national statistical yearbooks are the counterparts for research done in foreign areas. The number (and accuracy) of statistical sources has increased rapidly in the last 40 years. The computer has been even more instrumental in advancing this statistical revolution, improving both speed and accuracy and enabling the development of increasingly complex systems of statistical analysis. Ease of storage on disks has enabled the retention of data for longer periods and in greater detail. (Census statistics are available down to the level of census district and block data for the United States and many developed countries.) With the aid of the computer, correlations among spatially related statistical patterns can be assessed for accuracy and tested for validity. There is an increasing capability for producing accurate computer graphics, including maps.

Beyond data analysis, and analysis of geographers' inferences about those data, is the rapidly expanding field of computer simulation. For example, historical trend data and information on relationships among different factors can be entered into a computer, along with a program, to depict current spatial relationships, to project future spatial relationships, or to re-create the historical development of these relationships. With a program involving location references, computers can be instructed to design and produce maps of many spatial variables and correlations entered into its data banks (Figure 1–4, p. 18).

Applying Geographic Knowledge and Insights to Location

What do geographers do with this potentially enormous mass of data after they have gathered it and organized it in some meaningful way? The two key questions of geography—*where?* and *why there?*—provide some guidance. Location is a deceptively simple concept. Actually, it should be divided into two conceptual types: location-description (what it is, and where it is in terms of a location grid—north/south and east/west of fixed points or lines) and relative location (location in relation to other geographic phenomena and to the larger region and the world).*

Location can be studied at either, or both, of two levels or scales. **Site** is the most local scale of location: it refers to the immediate surroundings. It can include a description of the boundaries, an assessment of the physical characteristics of the local landscape, the grid location, and accessibility by water or from neighboring settlements. **Situation**, a broader definition of location, is relative location with reference to facts and features farther removed from the site. It gives the location in perspective to a larger region, area, or country. Generally, situation refers to the productive advantages of a given location. Situational advantages can change over time and can be enhanced by constructions that improve accessibility.

Spatial Correlations in Geography: Looking for the Explanation

Once location has been determined, questions of spatial associations and possible correlations arise.

THE SITE AND SITUATION OF PHILADELPHIA

Philadelphia's site was not arbitrarily selected; William Penn chose to locate the city on a peninsula between two rivers, the Delaware and the Schuylkill. The banks of both rivers afforded a sheltering harbor distant from open seas, yet tidewater reached to the site of Market Street, the principal thoroughfare across the peninsula both then and now. Because the river junction itself was marshy, Penn chose a site on higher, flood-free ground, where the peninsula was well drained. The head of tidewater on Delaware Bay, at the junction of the two rivers and with a geodesic reference location of 40° north and 75°13′ west, is the site of Philadelphia (Figure Aa).

The situational advantages of the city were probably not immediately evident. Falls on both rivers limited ship navigation to the city itself on the Schuylkill side and to only 25 miles above the city on the Delaware (Figure Ab). Nonetheless, the rivers offered navigational possibilities by small boat and portage to an unknown interior, as well as routes for easier foot travel along their banks on existing Indian paths. The falls were to become the city's original source of energy as direct water power.

The situational advantages evolved over time. Some, however, were inherent in the site. A narrow valley to the west offered an easy overland route to what was to become the rich agricultural region of the Lancaster plain. First a road, then a canal, and finally a railroad occupied this level route which eventually became known as the Main Line (after the canal). Thus connected, Philadelphia became the chief colonial-era market and port; it was assured of a stable food supply and a farm product surplus.

The anthracite coalfields, the country's earliest major fossil fuel resource, were first reached by the Schuylkill and the Lehigh, a tributary of the Delaware (Figure Ac). Philadelphia technology adapted this fuel resource to the industrial needs of the day, and city capital canalized the rivers that reached the coalfields in their very headwaters. Both rivers tapped into heavily forested regions of prime timber; iron deposits occurred at dozens of locations along the Schuylkill. The natural wealth of the hinterland contributed to the situational advantages of Philadelphia.

Roughly midway between New York and Baltimore, central within the East Coast Megalopolis, Philadelphia's location was also enhanced by the growth of its neighbors (Figure Ad). Innovations in transportation (canal, rail, and highway) continued to foster growth and enhance the city's situational advantages. Competition forced Philadelphia to improve its rail connections, providing it with the best possible access to the increasingly important Appalachian bituminous fields, the steel of Pittsburgh, and the farm produce of the Midwest.

Thus, it was Philadelphia's situation, more than the physical site, that contributed to the city's ultimate economic success, and only with the enhancement of that situation by human endeavor has Philadelphia maintained its competitive position in an ever-changing economy.

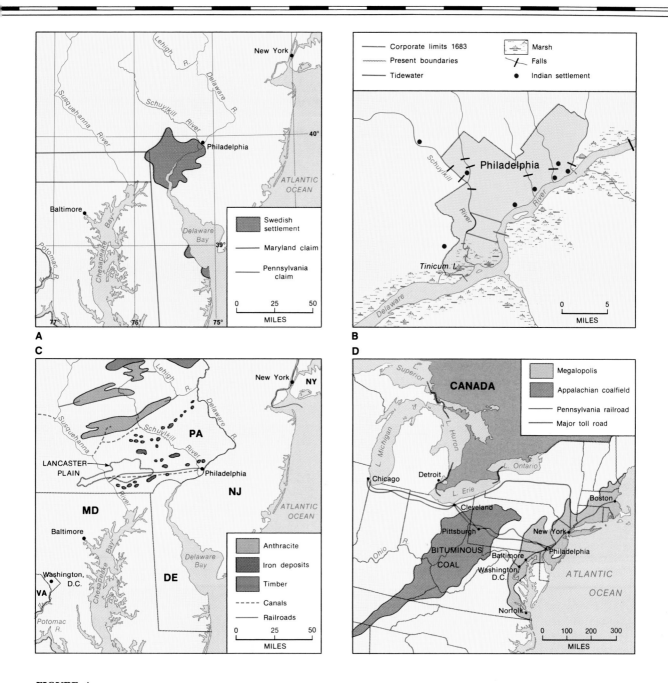

A

B

C

D

Corporate limits 1683
Present boundaries
Tidewater
Marsh
Falls
Indian settlement

New York
40°
Philadelphia
ATLANTIC OCEAN
Baltimore
Delaware Bay
39°
Chesapeake Bay
Potomac R.
77°   76°   75°

Swedish settlement
Maryland claim
Pennsylvania claim
0   25   50
MILES

Schuylkill River
Philadelphia
Tinicum I.
Delaware
0   5
MILES

Lehigh R.
New York   NY
Schuylkill River   PA
LANCASTER PLAIN
Philadelphia   NJ
MD
Baltimore
Washington, D.C.   DE
VA
Potomac R.
Delaware Bay
ATLANTIC OCEAN

Anthracite
Iron deposits
Timber
Canals
Railroads
0   25   50
MILES

L. Superior
CANADA
L. Michigan   L. Huron   L. Ontario
Chicago   Detroit   L. Erie   Boston
Cleveland
Pittsburgh   New York
BITUMINOUS COAL   Baltimore   Philadelphia
Washington, D.C.   ATLANTIC OCEAN
Ohio R.
Norfolk
Megalopolis
Appalachian coalfield
Pennsylvania railroad
Major toll road
0   100   200   300
MILES

FIGURE A

Site and Situation of Philadelphia. The site characteristics of Philadelphia are shown at two different scales in (a) geodesic location (latitude and longitude) and (b) site accessibility and physical controls. These illustrations explain the reasons behind the choice of a particular parcel of land for the siting of the settlement. The relative location within the larger regional context (situation) is shown in (c) regional resource base and (d) market centrality and connections to the interior. The situational location helps explain the reasons for the city's growth.

FIGURE 1–4

Computers in Geography.
Computers aid not only in
statistical analysis but also in the
production of maps. These two
maps were produced as part of
a study on geographic trends in
the Philadelphia metropolitan
area. (a) There is a strong
correlation between location
and income. Levels of income
are much lower in most of the
central city and in older
industrial suburbs and satellites.
(b) Population growth reflects
the same general patterns, with
a few exceptions. This situation
is typical of U.S. metropolitan
areas. In this case, the
exceptions rather than the
general correlation would be of
greater interest to geographers.

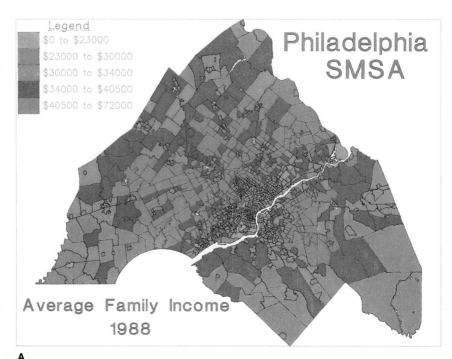

Legend
$0 to $23000
$23000 to $30000
$30000 to $34000
$34000 to $40500
$40500 to $72000

Philadelphia
SMSA

Average Family Income
1988

A

Percent
-20% to -5%
-5% to 0%
0% to 5%
5% to 15%
15% to 30%
30% to 35%

Philadelphia
SMSA

Percent Change in
Population: 1980-88

B

Does what we have looked at show a pattern that seems to be associated with the spatial pattern of something else? Any apparently consistent spatial association may reflect a **correlation**—a relationship that *may* be one of cause and effect. Possible correlations are often suggested by comparing two maps. If, for example, we compare two maps of the world, one showing average income, the other showing calorie consumption, we see an apparent correlation of the two patterns (Figure 1–5). Where food consumption levels are high, income levels generally are also high. This relationship indicates a **positive correlation**, because the two factors exhibit the same pattern. A **negative correlation** is reflected in a situation where one factor, say income, varies inversely with another, such as infant mortality. Where income is highest, it is likely that infant deaths will be lowest. Of course, such spatial correlations do not prove any specific cause-and-effect relationship, nor can they be isolated from many other social and economic factors. High rates of food consumption may *result* from high income levels, but income does not *produce* the food. The value of correlation, as for statistical models, is that it suggests a general relationship. When researchers find a case that does not "fit" the general situation, they recognize it as an exception, study it in more detail, and look for alternative explanations.

Geography and Time

Spatial correlations and the analyses of locations are seldom complete without some sense of the time dimension. Just as any map, photograph, or statistical table captures a moment, spatial patterns of distribution or attributes of location exist at a particular, static time. A complete understanding of any part of the earth requires some knowledge of how the present situation came about. Geographic patterns, in particular, often must be placed in historical context to be understood fully. All geographic occurrences evolve and change over time; without the time context, comprehension is limited and often distorted.

THE METHODOLOGICAL DIVISION: TWO SYSTEMS OF APPROACH

All geography can be categorized as either systematic or regional in terms of the form of analytical treatment. A **systematic geography** zeroes in on a set of distributions and occurrences that share a categorical commonality, such as farming, climates, or transportation. The subject group of the study is plotted on a map, discussed, and analyzed wherever it occurs.

Regional geography focuses on an area, defined by political boundaries, academic endeavor, or both. All distributions—all categorical types found within the area—are fair game for study. Of course, an individual investigation can be both systematic and regional, if both methodologies are used (Figure 1–6a). But a course, or a text, is usually predominantly one or the other.

The world maps that appear throughout this chapter are essentially systematic: they most frequently deal with one topic in its worldwide distribution and are intended for repeated reference. Each chapter contains several regional maps indicating the occurrence, location, and distribution of a wide variety of relevant geographic phenomena (Figure 1–6b). This book and the course of study are dominated by the regional treatment.

The Region as a Geographic Concept

The region is a geographic area, and the regional approach to the study of an area involves defining the area's extent and describing and analyzing the phenomena found within its boundaries. The concept is universal. As children, individuals recognize the neighborhoods they live in, whether they be in urban or rural surroundings. This neighborhood constitutes a region at the local scale. At a larger scale, in most countries there is a sense of sectional differentiation, such as the highland versus lowland environments of Scotland or the lifestyle and dialectical differences between North and South in the United States.

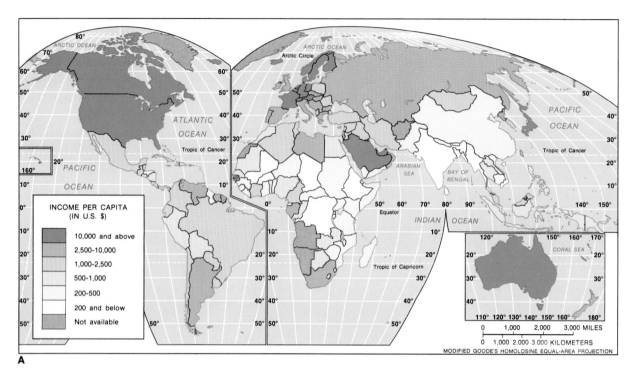

A

INCOME PER CAPITA
(IN U.S. $)

- 10,000 and above
- 2,500-10,000
- 1,000-2,500
- 500-1,000
- 200-500
- 200 and below
- Not available

MODIFIED GOODE'S HOMOLOSINE EQUAL-AREA PROJECTION

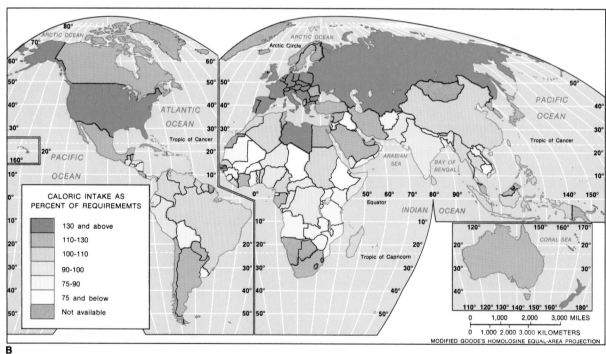

B

CALORIC INTAKE AS
PERCENT OF REQUIREMEMTS

- 130 and above
- 110-130
- 100-110
- 90-100
- 75-90
- 75 and below
- Not available

MODIFIED GOODE'S HOMOLOSINE EQUAL-AREA PROJECTION

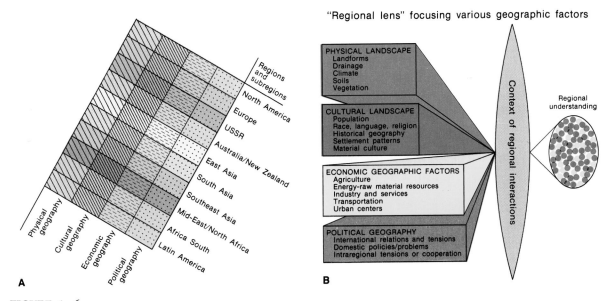

A

B

"Regional lens" focusing various geographic factors

Regions and subregions
North America
Europe
USSR
Australia/New Zealand
East Asia
South Asia
Southeast Asia
Mid-East/North Africa
Africa South
Latin America

Physical geography
Cultural geography
Economic geography
Political geography

PHYSICAL LANDSCAPE
Landforms
Drainage
Climate
Soils
Vegetation

CULTURAL LANDSCAPE
Population
Race, language, religion
Historical geography
Settlement patterns
Material culture

ECONOMIC GEOGRAPHIC FACTORS
Agriculture
Energy-raw material resources
Industry and services
Transportation
Urban centers

POLITICAL GEOGRAPHY
International relations and tensions
Domestic policies/problems
Intraregional tensions or cooperation

Context of regional interactions

Regional understanding

FIGURE 1–6
Systematic and Regional Geography. (a) The challenge of geography is the
need to organize an encyclopedic breadth of information into a comprehensive, yet
digestible form. Two methodologies are available: systematic (organized by subject
matter) and regional (organized by area). All geographic studies emphasize one or
the other, but most studies use elements of both methodologies. (b) The region, an
area designated and delineated by the geographer, provides the organizational
framework for research and analysis. Inputs are drawn from many subcategories of
four basic subject matter groupings: the physical environment, the cultural attributes
of the region's residents, the economic organization of the area, and relevant
political factors.

The region is a familiar construct shared by all as
part of a common experience. The word itself derives from the Latin *rex,* or "ruler," and, inherently,
the concept of political rulership, including boundaries, is a part of it. The word *foreign,* "ruled by

← **FIGURE 1–5**
Spatial Association of Geographic Variables. The
apparent correlation between the two sets of data
displayed on these maps may suggest a cause-and-effect
relationship. The map of income (a) correlates fairly
strongly with that of calorie intake (b), but the level of
coincidence is only general. Poor countries tend to
exhibit poor diets as well. The highest levels of calorie
consumption, however, occur over nations with a broad
range of incomes. Other variables must be considered
before reaching sound conclusions.

others," stems from the same root. The Russian word
for foreigner, *inostraniets*— literally, "those who live
on the other side"— gives an even better geographic
sense to the difference in people and cultures anticipated over distance and space. Regions automatically imply some degree of similarity to things and
people found *within,* and a significant degree of differentiation from those found *without.*

Perspective plays a role in regional recognition.
St. Louis and Chicago are the East for a Californian,
but not for a New Yorker. Texas is perceived as both
South and West; indeed, it is properly both, depending on the criteria chosen for inclusion and differentiation. The regional concept appears to be deceptively simple. In reality, regions are complex because
they are composed of an aggregate of cultural and
physical phenomena; they are rarely based on one
criterion.

A region is a subsection of the earth with some degree of internal homogeneity. All physical and cultural phenomena within the region can be the object of study. The purpose of defining regions is to simplify geographic analysis—to comprehend the whole by understanding the individual parts. All regions have location, areal extent, boundaries, and a degree of internal similarity. Most regions have a central **core**, but there can be much disagreement as to their actual extent and ultimate boundaries.

Are regions of significance? In a word, yes, and at virtually all levels of size and importance. Inclusion within a "wetlands region" as defined by U.S. federal law defines what types of development are permitted or excluded from the area. At the most local level, municipalities develop land use regions through zoning laws that accomplish the segregation of function. European countries have designated depressed economic and developmental regions wherein taxes are lower and industries locating there are given special privileges or incentives. There are international regions that have an enormous impact on our thinking, even though they are not official or backed by the force of government. Such concepts as the Free World and the Communist World—a two-region division of the planet—long influenced American foreign policy. Then-Soviet Premier Khrushchev challenged this view by introducing the concept of the Third World in 1956, indicating a third region composed of nonaligned, underdeveloped countries. Thus, a regional construct has become the standard term used to refer to the less developed nations of the world.

Many regions are devised as simply an areal framework for academic research, and some are nothing more than a convenient designation. At times, regions begun as academic constructs have been converted to administrative function. O.E. Baker of the U.S. Department of Agriculture devised a series of crop production regions in the 1920s. These regions, still often used as convenient designations, for a time served as institutionalized regions for the administration of federal agricultural assistance programs, federal crop subsidies, and price supports. In many cases, crop patterns have altered radically in recent decades, and the remaining agricultural programs are administered with a greater degree of sophistication.

Types of Regions. An important distinction is made between formal and functional regions. A **formal region** is relatively uniform throughout. A **functional region**, sometimes called a **nodal region**, operates out of a definite center, or node. It is from this center that an interaction system or network emanates. Nodal regions imply movement to and from all parts of the region, into and out of a central node (Figure 1–7).

The homogeneity of a formal region may be based on physical or cultural factors. Any political unit, from township through country, would constitute such a formal region because the jurisdiction is uniform throughout. The Appalachian Mountains constitute a uniform physical region of the formal type regardless of how many different state jurisdictions are involved over their length and breadth (Figure 1–7a). The Riviera of France is both a physical and a cultural region; it has a high degree of uniformity in both cases. New England is relatively uniform culturally, if not physically; it is widely acknowledged as a cultural region, as are Cajun Louisiana, French Canada, and Scandinavia. All of these units are character-

FIGURE 1–7
Types and Interpretations of Regions. (a) Uniform (formal) regions are based on homogeneity or the strong similarity of characteristics found within the area. The cultural region of Appalachia depicted here is based on a legal definition, an official recognition of widespread unemployment, and other social problems prevalent in much of the area. It is also a developmental region slated for government assistance in promoting economic change. (b) Any function administered to or consumed by the inhabitants of an area from a central point can be the basis for demarcating a functional (nodal) region. In this case, it is the viewing area of a particular television station or stations. Functional regions may overlap boundaries at local, state, or national levels. (c) Several different functional regions, often with quite different boundaries, may operate from the same nodal center. Characteristically, they occupy portions of the same space, but only rarely is there complete coincidence between boundaries.

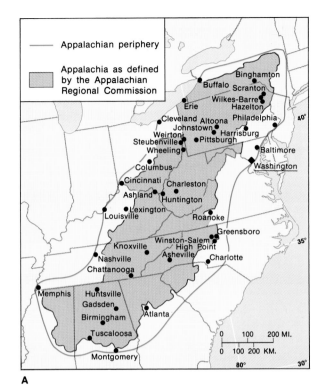

A

Legend (Map A):
- Appalachian periphery
- Appalachia as defined by the Appalachian Regional Commission

Cities shown: Binghamton, Buffalo, Scranton, Wilkes-Barre, Hazelton, Erie, Cleveland, Altoona, Philadelphia, Johnstown, Harrisburg, Weirton, Pittsburgh, Steubenville, Baltimore, Wheeling, Washington, Columbus, Cincinnati, Charleston, Ashland, Huntington, Lexington, Louisville, Roanoke, Greensboro, Winston-Salem, High Point, Knoxville, Asheville, Nashville, Charlotte, Chattanooga, Memphis, Huntsville, Gadsden, Atlanta, Birmingham, Tuscaloosa, Montgomery

Scale: 0 100 200 MI. / 0 100 200 KM.

40°, 35°, 30°, 80°

C

Ohio, Pennsylvania, Maryland, West Virginia

ALLEGHENY COUNTY, Pittsburgh

Legend (Map C):
- County boundary
- State boundary
- Pittsburgh industrial district
- Pittsburgh Diocese, Roman Catholic Church
- Major urban area

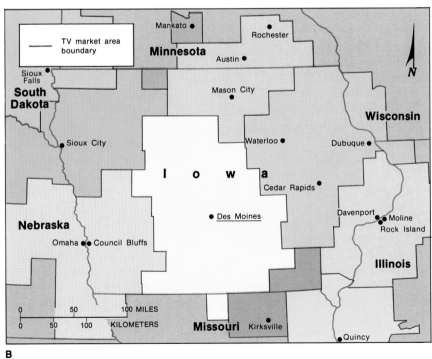

B

Legend (Map B):
- TV market area boundary

States and cities: Minnesota (Mankato, Rochester, Austin, Mason City), South Dakota (Sioux Falls), Wisconsin, Iowa (Sioux City, Waterloo, Dubuque, Cedar Rapids, Des Moines, Davenport, Moline, Rock Island), Nebraska (Omaha, Council Bluffs), Illinois, Missouri (Kirksville, Quincy)

Scale: 0 50 100 MILES / 0 50 100 KILOMETERS

N

ized by a high degree of uniformity—physical, cultural, or both—and are varied examples of formal regions.

All movements, flows, interactions, and networks that characterize functional regions emanate from the central core, or node. The node may be a settlement, a control point communications center, or even a shopping mall (Figure 1–7b). In all cases, people, goods, and services move along a network to and from the central point. The commuting area for a major metropolitan center is an example. Because the relationships involved are dynamic, the precise boundaries drawn by the regionalization can differ significantly (Figure 1–7c).

Regional Hierarchies. In this text, the world has been divided into 10 major cultural regions, with one region covered in each chapter. The basis for their construction is the relatively uniform cultures that prevail within them. In turn, each region is divided into countries or groupings of countries. Larger or more complex countries are divided into subregions as needed. The sheer size and diversity of some political units such as Brazil and the USSR require division into subregions. Some regions, such as the Amazon Basin, overlap political boundaries. Although there is no satisfactory answer that applies in all cases, political boundaries are generally used because statistics conform to them. The framework of national units is a handy and familiar collection of subunits that can be used as building blocks for regions or as subregional units.

Regions as a Concept for Organizing Space.
Regions are structures for organizing and presenting geographic generalizations. As in all generalizations, there are exceptions. The existence of exceptional or contradictory facts presents some problems in the generalization process, but users recognize this situation. Nevertheless, generalizations are a necessity. Because world regional geography seeks to explain the spatial variation of a great number of physical and cultural phenomena, geographers need a system for ordering and classifying information. Regions are created for the purpose of rational organization.

Except for gross physical units such as continents, regions are seldom defined by purely physical circumstances. They are defined most often by some specific, critical aspect of the physical environment, or by some outstanding cultural factor, or by a combination of both. Identifying (and bounding) criteria can be chosen from an almost infinite number of possibilities. Regions can be defined at virtually any scale of territory or space. To identify a region, it is necessary only to select some overall characteristics or functions that can be used to establish a regional boundary, differentiating that region from others and from the rest of the world in the process. The intellectual challenge is the choice of significant criteria from among an abundance of facts. Because regions are identified by individuals for their own purposes, they are intellectual constructs—products of human imagination, values, and perceptions.

For example, distinctive graffiti or other deliberate markings may delineate gang territory within cities. Children may be acutely aware of social boundaries in urban areas, learning early what is "their" area as opposed to that of others. Urban regions are often defined by criteria acceptable to the researcher, even when defined by the residents themselves. Like all regions, urban neighborhoods are places with special identities, special functions, and special relationships with other regions.

Significance of National Boundaries in a Regionalization Scheme. National boundaries are a vitally important factor throughout the world. They may affect economic development and the standard of living; they may also reflect differences in opportunities and in such cultural factors as race, language, and religion. Political boundaries can be far more significant than physical factors in the resulting cultural landscape. Anyone who has crossed the U.S.-Mexican border would agree. Whether one lives in Corpus Christi, Texas, or Monterrey, Mexico, can have a far greater significance than whether one lives in Anchorage or Boston, or in Mexico City or Vera Cruz. Because national states *are* regions, in the economic and cultural senses as well as the political sense, national boundaries are used to define portions of major world regional boundaries, even though political factors are not the sole criterion for establishing the regions themselves.

Urban Graffiti.

Neighborhoods are a form of urban region. At times, graffiti can yield clues to the extent, ethnic makeup, problems, and even attitudes of the neighborhood and its residents. This scene in Montreal reveals that the neighborhood is bilingual (French and English) and that its residents are upset with the changes in neighborhood status.

GEOGRAPHY AS THE STUDY OF BOTH HALVES OF THE ENVIRONMENT

To all of us, the forces of both the physical and the cultural environment are evident at every turn. Cars that don't start on wintry mornings, increased peach prices after late Georgia frosts, and news of dwellings destroyed by forest fires, mudslides, or earthquakes are all reminders of the presence and continued importance of the forces of nature in the human environment. Polluted water supplies, smog, the "hole" in the ozone layer of the atmosphere, and acid rain all bear testimony to human misuse or abuse of the natural environment. Carcinogens in our food, fish kills, and the problems associated with trash disposal have all sobered society on the once-thought unmitigated blessings of technology. Surely a look at both halves of the environment is necessary. Geography, one of the disciplines equally at home in both the physical and cultural environments, has taken on a new significance in an age of environmental awareness.

Reflecting this traditional dichotomy, geography courses and texts are usually categorized as dominantly **physical geography** or dominantly **cultural (human) geography**, yet neither perspective can reject the influence of, or the consequences for, the other half of the environment. Because the regions used in this text are based primarily on cultural similarities, this is dominantly a cultural geography text. Nonetheless, the physical parameters operative in each region are presented in each chapter, and the implications of that physical geography for each region are noted throughout.

CULTURAL LANDSCAPE DYNAMICS

Different people, from different cultures (or even different time periods in the same culture), would likely perceive the landscape differently and establish a set of regions on the basis of quite different criteria. Their respective available technologies would result in widely varying identifications of what was important. Even the names given to regions would reflect this difference. Any cultural region unavoidably reflects **ethnocentrism**, the tendency to

COMMON REGIONAL TYPES AT THE MACRO SCALE

There are as many different types, scales, and concepts of regions as there are human motivations and needs for their creation. Regions and their boundaries are not necessarily fixed forever. The conceptualization, identification, and bounding of regions reflect changing concerns, perceptions, and realities.

Continental landmasses have traditionally been the largest subdivisions and most generalized regions of the earth. In any organized study of geography, or even history, the continents are common subdivisions.

Physical regions are most commonly based on rainfall, climate, landforms, vegetation, soil, or some combination of these physical attributes. Although the data are measures of physical variables, the choice of significant or bounding data is a personal decision. The world's rainforests form a physical region based on vegetation (Figure A). They can be viewed as a single region, even though they are dispersed across parts of several continents. The recently understood effects of the rainforests on our weather and atmosphere, combined with the threat of their demise, are what give this region currency in the public mind.

Economic regions concern human activities, systems, and structures. They may focus on any economic activity with spatial distribution. Geographers have devised regions such as "rice bowls" and manufacturing regions. On a world scale, the level of development, industrialization, or technology can be used to identify gross-scale economic regions, such as the Third World (Figure B).

On a smaller scale, economic regions can be depressed economic areas such as Appalachia or Britain's Lancashire, singled out for government investment and subsidy to hasten economic change. Frontier areas or other units associated with colonization schemes are also economic regions, based on the idea of ultimate development.

The New World is a good example of a **cultural-historical region** that is widely accepted by scholars of all disciplines. Other examples are the Islamic world and the Orient.

Political regions are sharply defined, often because they frequently correspond to international boundaries. Any individual country is a political region. Political regions at the most generalized world scale include such concepts as the Soviet bloc and the NATO alliance. The member countries of the Organization of African Unity constitute a region of states with similar problems and something of a unified political agenda.

FIGURE B
First, Second, and Third Worlds. The designation Third World is widely used to refer collectively to a group of less developed national economies. Originally, it was a political term, referring to the group of nonaligned countries within the larger context of Soviet-Western international tensions.

→

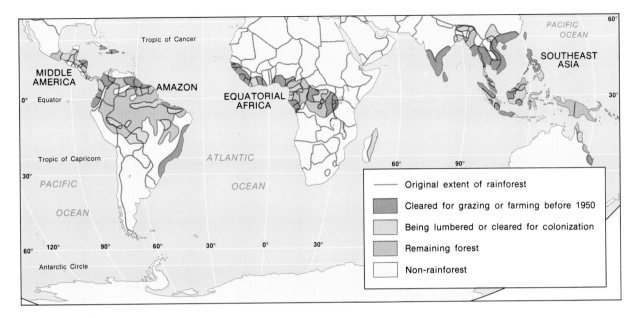

FIGURE A

World's Rainforests. Physical regions are based on elements of the natural environment; in this case, the regional core element is a type of vegetation. This region is not contiguous because of the interposition of land and water. The destruction of the world's rainforests through lumbering and the extension of agriculture underscores the changeable nature of regional boundaries as well as the larger environmental problem associated with that change.

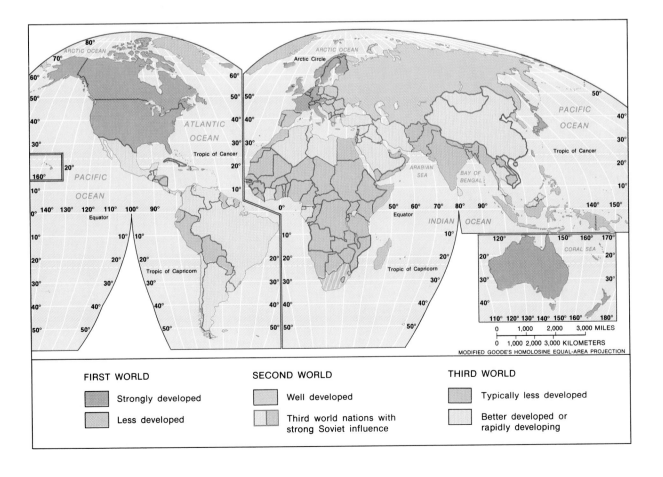

FIRST WORLD
- Strongly developed
- Less developed

SECOND WORLD
- Well developed
- Third world nations with strong Soviet influence

THIRD WORLD
- Typically less developed
- Better developed or rapidly developing

view the world from an individual's personal viewpoint and geographic location. Nonetheless, from any viewpoint, cultural characteristics and practices are a valid basis for developing regions, because the backgrounds, experiences, and values of people can be broadly similar over large areas. To develop regions based on *culture,* it is necessary to define this widely used, often misunderstood term.

What Is Culture?

Culture refers to the total way of life of a group of people. Each societal group develops a lifestyle that is characteristic, underlining that group's identity. The fact that groups of people do things differently contributes to their distinction. Each individual is born into a particular culture and, over the maturation process, learns certain forms of behavior that are deemed acceptable. The individual learns to follow a pattern of behavior, acquires knowledge of the use of certain tools, and develops habits and skills that conform to the demands of that lifestyle. Certain beliefs, fundamental ideas, and values are implicit in any culture; they operate as guides to the individual, offering help in understanding and utilizing the surrounding world. Cultures are created and shaped by the people themselves.

Culture is dynamic. Each generation modifies and adds to its inherited culture. Culture, then, is historically derived and, simultaneously, in the process of further evolution. Cultures set a standard of behavior that is taught to all their members, who in turn accept most of its elements as part of that acquired base of knowledge. Cultures are handed down through the generations.

Culture and Perception

Perception is the ability to observe the surrounding world and to make certain judgments about it. Perception is strongly influenced by culture. Colonists of European cultural origins perceived the forests of eastern North America quite differently than the indigenous inhabitants did. Europeans were culturally influenced to think that the proper use of land was agriculture; they viewed the forests as something to be cleared. Native Americans perceived them as a rich natural environment full of game and food for the gathering. They perceived the forests as something to be conserved. In both cases, these perceptions were culturally derived behavior based on each group's learned behaviors and values. What the world "is" to any group, what is expected from that world, and how that world is to be used, will differ because of differences in learned behavior and remembered data. What is selectively "seen," interpreted, and understood from the surrounding environment is the product of cultural perception.

Elements of Culture and the Cultural Landscape

Some cultural differences are readily apparent; the reasons for these differences are less apparent and, therefore, the focus of geographic study and research. Geographers emphasize the spatial extent of different elements of culture and their relationship to various physical and cultural environments. The spatial distribution of *individual* elements of culture is related to other geographic patterns.

Language, religion, political organization, population distribution, economic endeavors, settlement patterns, and the systems that connect them are the most commonly studied cultural elements. In its geographic sense, culture includes the actual physical manifestations of a people on the landscape, not just their behavior. The imprint of a culture on the land in all its physical forms is known as the **cultural landscape**. Early German geographers viewed *landschaft* ("landscape") as a composite of all the physical and cultural aspects of a region that gave it character; they emphasized the physical while attempting to understand the interrelationships between the two. French practitioners of the discipline also viewed geographic regions as representing this duality of environment, but tended to emphasize the cultural aspects in their geographic studies.

Culture, Crisis, and Change

A world regional geography is, at least in part, a geography of world crises. The energy crisis of the 1970s, the starvation in parts of Africa in the 1980s, and the ongoing population time bomb are a few of the alarming situations familiar to most people. Each

Rothenburg, West Germany. *This city appears as if lifted from a fairy tale of the brothers Grimm. Its turreted buildings and steeply pitched roofs of red tile within stout defensive walls suggest not only its age but its German cultural origins. Today's television antennae, intermingled with the nests of storks, detract only minimally from this cityscape so German in mood and spirit.*

of these crises has a spatial or geographic aspect, and each is related to population. How well people live on various parts of this planet is closely related to the population growth rate, the current ratio of population to resource base, and the level of technology available in that area. Not just fuels and metals, it must be remembered, but systems of government and educational levels of a population can be viewed as resources, too. The most basic resource of any nation is its people; their energies, talents, and motivation are essential to the use of all other resources. A nation able to devise political and social systems that liberate and enhance its citizens' talents will achieve the one absolutely vital resource.

Most of the cultural factors in any landscape are dynamic; change is both probable and ongoing. For example, the quantity and variety of known natural resources change constantly. Technological advances continually redefine "resources," since these are simply naturally available materials for which people have use and that can be extracted, processed, and marketed economically. The oil and gas under northwest Europe's North Sea was not a resource until undersea drilling technology and high prices for oil combined to make exploitation both possible and profitable.

The natural resource "inventory" of any state or region, then, is not fixed. It contracts and expands

GEOGRAPHY AND THE AMERICAN CULTURAL LANDSCAPE

American interest in the cultural landscape developed relatively late, a logical consequence in an area preoccupied with completing settlement and securing full development until about 1920. Our own culture, however indebted to its European origins, soon came to be engrossed with technological feats and the development of new technologies. Our national consensus was new, consciously based on cultural aspects that differed from those of European nations and, like our people, derived from myriad ethnic inputs. Much time was necessary before these widely differing cultural elements became welded into a recognizable unity, and even more time before this unity was mirrored in an identifiable and characteristic cultural landscape. Europeans asked to characterize America perceive it as a landscape of skyscrapers, sprawling suburbs, automobiles, open space, heavily mechanized, giant farms, enormous factories, and endless superhighways that is peopled with informal, ambitious, materially oriented citizens leading fast-paced lives. Although not fully accurate, nor always uniformly applicable, this is not a totally incorrect characterization of what has evolved in North America. During the 1920s, American geographers began to recognize a distinctly American cultural landscape. They proceeded to write about it, to interpret it, and to study the cultural processes that shaped the particular cultural patterns that characterized North America.

This "noticing" of a North American culture has not been exclusively a domestic endeavor. A Swedish geographer, Sten de Geer, visiting America in the early twentieth century, first observed that there was something different about the pattern of manufacturing that had evolved in North America. The concept of the North American Manufacturing Belt was born, perhaps the original article on North American cultural landscapes that evoked American geographers' interest in their own cultural surroundings. Shortly thereafter, geographers at the Universities of Chicago and Wisconsin began to pursue research on American cities and rural landscapes in search of geographic understanding. Carl Sauer developed cultural geography at the University of

through technological advances and change. New mineral resources will continue to be discovered, especially in remote and developing areas. The level of technology available is another variable; some nations can be categorized as technology innovators, developers, and exporters of technology, whereas others are deficient in technology and therefore recipients. The production of new technology and the purchase of existing technology are costly. The sale of resources is the normal way of paying such costs.

Innovation and Diffusion

The geographic aspect of the processes by which ideas and technology spread across space and time is called **diffusion**. It is possible for the same idea or invention to occur more or less simultaneously at two or more locations. In most cases, however, it occurs *once*, at a particular time and place. Any new concept, process, idea, or invention, or the significant modification of an existing one, is called an in-

California at Berkeley during the 1920s, popularizing interest in cultural geography and cultural landscapes through his extensive writings and broadening the horizons of American cultural geographers to include the entire world.

A second generation of cultural landscape research in America was also inspired by the perceptions of a European geographer, when Jean Gottmann "discovered" megalopolis, the American supercity (see Chapter 2, Figure 2–8). This time, the focus shifted toward service functions and systems of regional interaction. The new emphasis was on urban landscapes.

The cultural landscape is all-encompassing. It contains a great deal that is tangible and easily mappable: buildings (at times of identifiable architectural style, reflecting age and cultural origin), transportation facilities and systems, patterns of survey and settlement, and even decorative arts and crafts. Combined, these elements constitute what has come to be known as **material culture**, which reflects mode and style of living and is mirrored in the societal organization of space. Equally important to a culture is the way in which people make a living, in the sense that culture and its accompanying technology reflect and determine a culture's standard of living. Technology and employment are also influential in developing settlement patterns and other aspects of the organization of space.

There are other less mappable, less tangible elements of culture: the smell of an oil refinery, the characteristic taste of an urban water supply, the friendliness of a people, the sound of traffic, the degree of cleanliness of surroundings, and the sights of a city. They, too, are reflections of the cultural character of an area and a part of its geographic image. New Orleans and San Francisco have distinctly different flavors, in both cases, no small part of their ability to attract both foreign and domestic tourists. Geographers use words to convey their impressions of the environment, including not only the cultural feel of an area, but how it developed.

novation. The outward spread of an innovation to other people or places is called the **diffusion of innovation**. It is the geographic phenomenon that we can follow outward from the center of innovation, or **hearth**, to the places where it is received and adapted. An innovation does not diffuse evenly from its center in relation to distance alone. Rarely would a diffusion pattern resemble a series of expanding concentric circles (Figure 1–8). The barriers to diffusion can be physical, cultural, or both. A great desert, a vast ocean, or some other feature that eliminates or minimizes contacts would be a physical barrier to diffusion, if only temporarily. The hearth may simply refuse to accept or transmit an innovation. In that case, it may also deny (temporarily) the innovation's availability to other peoples. The rejecting group would function as a cultural barrier to diffusion. The study of innovation and diffusion is an important part of the study of changing interactions among and within the world's regions.

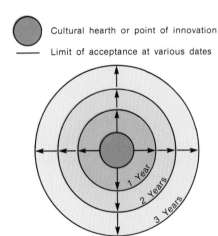

○ Cultural hearth or point of innovation

— Limit of acceptance at various dates

IDEAL PATTERN OF DIFFUSION

Under ideal conditions, without physical or cultural obstacles, outward diffusion would be influenced only by distance and time, yielding uniform speeds of acceptance.

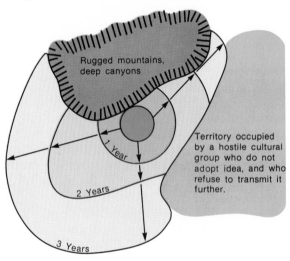

A MORE TYPICAL PATTERN

Diffusion pattern distorted by physical and cultural barriers to its adaptation by others.

FIGURE 1–8

Process of Diffusion. Ideas, pieces of new technology, or new applications of existing tools or concepts diffuse, or travel outward from their hearth or point of innovation. Ideally, such diffusion would take place at an even rate and distance over time, but physical or cultural barriers, or both, can alter an otherwise even, wavelike diffusion outward from the hearth.

PHYSICAL LANDSCAPE DYNAMICS

Changes in the cultural landscape are the more obvious because they commonly happen within brief time periods. Yet, change is universal and ongoing in the physical landscape as well, if sometimes on a much slower timetable. Details of the physical world change quickly enough to be casually noticed by people—sandbars appear in rivers, coastal erosion destroys houses, and earthquakes may be accompanied by sudden shifts in the land itself. Less obvious is the fact that very large scale changes are occurring, if at imperceptible rates in the context of a lifespan. The "eternal hills" are anything but; the mightiest mountains *will* be worn down. The continents themselves move, changing their relationships with one another and even their shapes and masses, however slowly (Figure 1–9). The great Sahara once had a humid climate, and tropical forests flourished, eons ago, in Antarctica and Alaska. The familiar assumption that the physical landscape is inert, changing only in superficial details, is simply untrue.

Tectonic Plate Theory

Tectonic activity encompasses the processes involved in the building and changing of the earth's landforms. The complexity of the earth's topography results from the contact of large, rigid rock plates. The crust of the earth—its brittle, outermost layer—together with the upper mantle which underlies it, form a zone called the **lithosphere** ("zone of rock"). The lithosphere is composed of a series of

FIGURE 1–9

→

Plate Tectonics and Continental Drift. The earth's crust and upper mantle are composed of a series of huge tectonic plates in slow, almost imperceptible motion, buoyed along by convectional cells deep in the earth's molten interior. Continents are the lighter, higher-riding portions of these plates. Where plates collide or override one another, crustal materials are deformed into mountain ranges in processes accompanied at times by earthquakes and volcanic action. (a) Theorized extent of continental plates. (b) Ancient continental mass of Pangaea as it existed 300 million years ago. (c) Continents as they are now. Shaded areas show the evidence of ancient continental glaciation that took place when Pangaea existed.

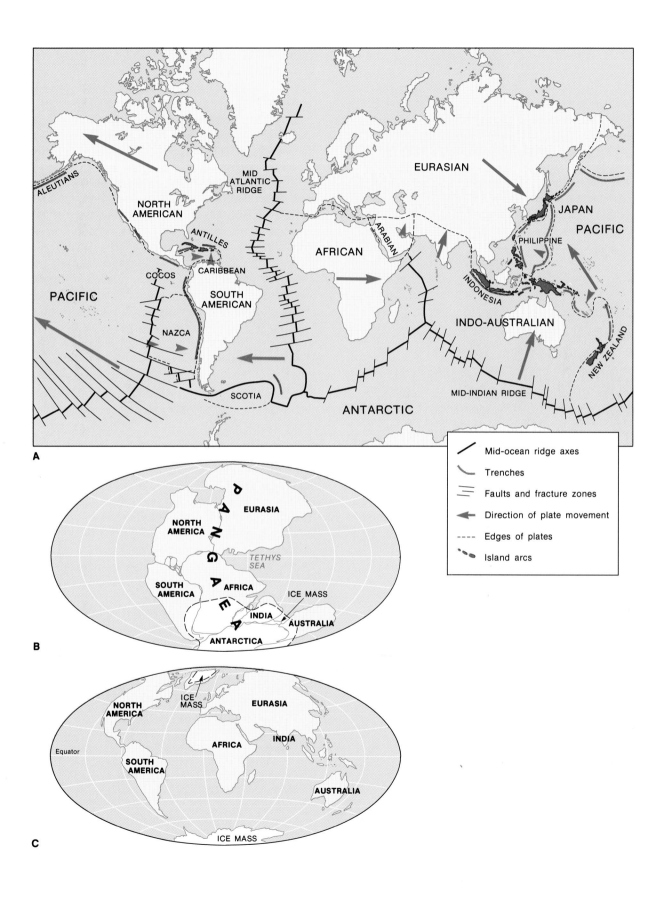

A

ALEUTIANS

NORTH
AMERICAN

MID
ATLANTIC
RIDGE

EURASIAN

JAPAN

PACIFIC

ANTILLES

PHILIPPINE

COCOS

CARIBBEAN

ARABIAN

AFRICAN

PACIFIC

NAZCA

SOUTH
AMERICAN

INDONESIA

INDO-AUSTRALIAN

NEW ZEALAND

SCOTIA

MID-INDIAN RIDGE

ANTARCTIC

╱	Mid-ocean ridge axes
⌣	Trenches
≡	Faults and fracture zones
←	Direction of plate movement
- - -	Edges of plates
•◦•	Island arcs

B

PANGAEA

EURASIA

NORTH
AMERICA

*TETHYS
SEA*

SOUTH
AMERICA

AFRICA

ICE MASS

INDIA

AUSTRALIA

ANTARCTICA

C

NORTH
AMERICA

ICE
MASS

EURASIA

INDIA

Equator

AFRICA

SOUTH
AMERICA

AUSTRALIA

ICE MASS

plates about 60 miles thick. These plates literally move over the surface of the earth at an exceedingly slow pace. Some are ocean bottom, while others carry continents above them. Where these plates collide, there is much deformation of the crust through warping, folding, and faulting, as one plate tends to buckle and slide under the other. Earthquakes and volcanic eruptions accompany these collisions and the splitting apart of some plates (Figure 1–9).

The plates are set in motion by convection cells within the earth. Enormously hot, molten material rises toward the surface much as hot water rises in a boiling pot. As it rises, cooling takes place; and when the temperature approximates that of the surrounding material, the cooling mass moves laterally and then sinks. The movement of these giant convection cells carries plates with it. Thus, the continents and oceans are in a state of motion, even at the current time.

Diastrophism is the name given to a series of mountain-building, elevation-causing processes and forces caused by such plate movement. Most experts agree that one large continent, called **Pangaea** once existed and has since broken up into the individual continents we know today. Vulcanism is associated with the movements of, and stresses within, tectonic plates as volcanoes occur most likely where rifts and cracks in the crust allow molten material to break through to the surface. We can theorize the number and location of these plates by mapping activity such as vulcanism and earthquakes (Figure 1–10).

Erosion and Deposition

Opposed to tectonic processes, which are constructional forces creating relief differences on the surface, is a set of abrading and smoothing forces that attempt to gradually level and obliterate differences in relief. These forces of reduction and transport of earth materials are **weathering** and **erosion**, known together as **degradation**. Weathering refers to the in-place disintegration of earth materials; erosion is the movement of these disintegrated materials through the agencies of gravity, running water, moving ice, wind, or ocean currents. Erosion eventually results in **deposition**, the creation of new physical features through the deposit of erosion-transported materials. Landforms associated with weathering, erosion,

and deposition are seldom as spectacular as the grand-scale forms that can result from vulcanism and diastrophism. However, depositional landforms such as river deltas and barrier beaches can be of great size, and the residual landforms of erosion such as the ice-sharpened peaks of the Rockies or the Grand Canyon can be quite dramatic.

The importance of these concepts to a study of world regional geography is that virtually any sizable area of the world will include representative landforms created by the continuing interactions of tectonic and weathering-erosional-depositional forces. The very surface of the earth is a reflection of these forces on the grandest of scales (Figure 1–11). Societies are sensitive to the physical landscape; people and cultures learn to make careful evaluations of the problems and potentials of their physical surroundings. To some extent, they adapt to, use, and even modify the physical landscape. Individually and collectively, people are both users and shapers of the physical landscape. Frequently, and unfortunately, they accelerate erosion and otherwise participate in natural physical processes.

Continental Glaciation

Glaciers are a dramatic agent of erosion and deposition, capable of stripping a land surface to bare, solid bedrock and gouging out huge chunks of that as well. A **glacier** is a moving sheet of ice. Glaciers can transport a wide range of particle sizes, from finely ground rock flour to boulders as big as a small house. Because the actions and effects of many relatively small glaciers in high mountains can be observed, it is possible to theorize the existence of mammoth, continental-scale glaciers that once operated on the surface of the earth in the recent geological past (1 to 3 million years ago). The evidence

FIGURE 1–10 →
Vulcanism and Earthquakes. Earthquakes and active volcanoes outline the extent of tectonic plates. Exceptions occur, but the coincidence of such tectonic activity with the theorized boundaries of plates is truly remarkable. (a) Some more recently active volcanoes. (b) Earthquake activity over a recent 10-year period. (Data from NOAA)

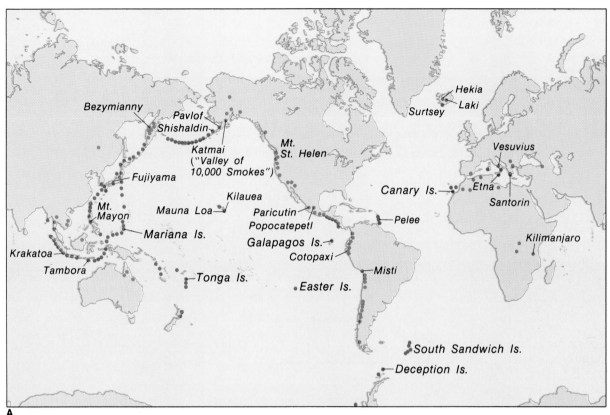

Bezymianny
Pavlof
Shishaldin
Katmai
("Valley of
10,000 Smokes")
Mt.
St. Helen
Hekia
Laki
Surtsey
Vesuvius
Fujiyama
Canary Is.
Etna
Kilauea
Santorin
Mauna Loa
Mt.
Mayon
Paricutin
Popocatepetl
Mariana Is.
Galapagos Is.
Pelee
Krakatoa
Cotopaxi
Kilimanjaro
Tambora
Tonga Is.
Misti
Easter Is.
South Sandwich Is.
Deception Is.

A

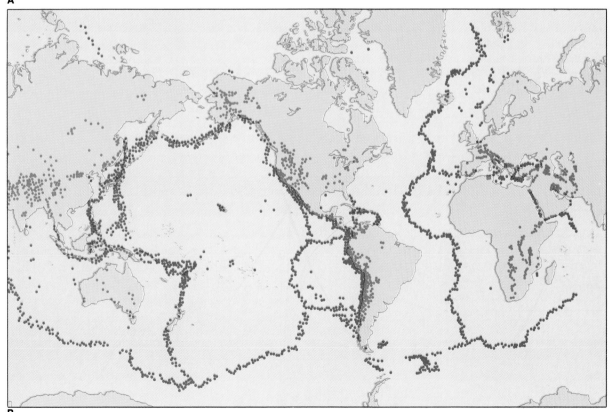

B

FIGURE 1–11

World Landforms. The world is composed of a series of landform regions that contribute to (or detract from) the agricultural utility of the continents. Ancient continental remnants are preserved in plateaus, highlands, and shields in all the continents. Mountain ranges rise along the contact zones of tectonic plates, blocking or channeling climatic influences and, at times, presenting physical obstacles to transportation. Vast lowlands are often home to dense populations. Topography is one of the important physical variables that influences the utility of the planet.

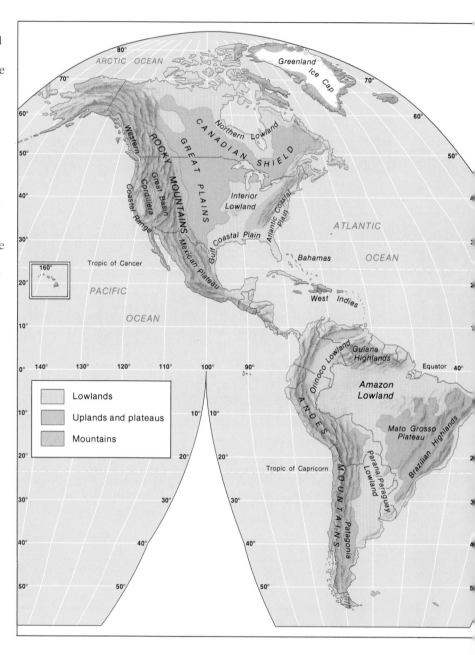

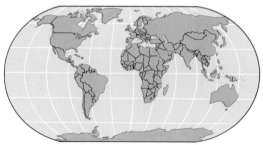

to support this theory is clearly written in the landscape in the form of grand-scale landforms identical to smaller ones found in the vicinity of today's mountain glaciers. The results of continental glaciation were both positive and negative, depending on what processes were dominant in a given area. The type of terrain and the quality of soil left behind after the glaciers receded are of particular interest to geographers. The forces of continental glaciation have had

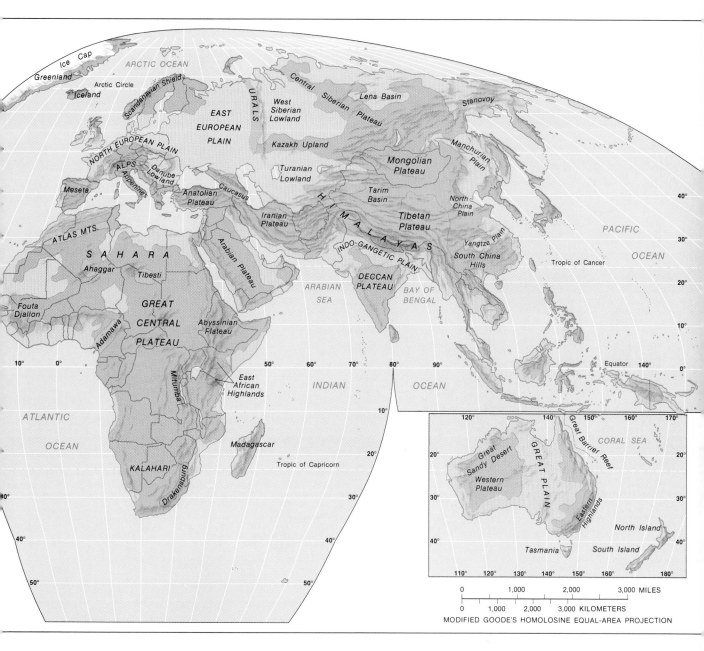

great impact on the landscape and physical environments of parts of Europe, North America, and the USSR (Figure 1–12).

Climate

Climate is the long-term average of temperature and precipitation conditions found in an area. Not just the amount or intensity of these two variables is considered; seasonality is also important. Although the human species probably originated in the tropics, it has since spread to areas with various combinations of temperature and precipitation regimes. In the process of moving to new environments, humans developed cultural adaptations to minimize the discomforts and problems to human survival experienced in less ideal climates. Climate is one of the most important variables in the physical environment and is best

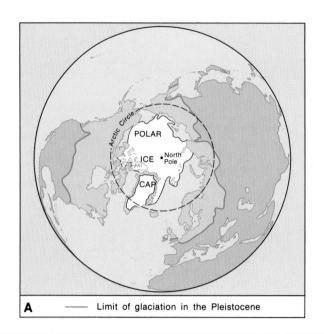

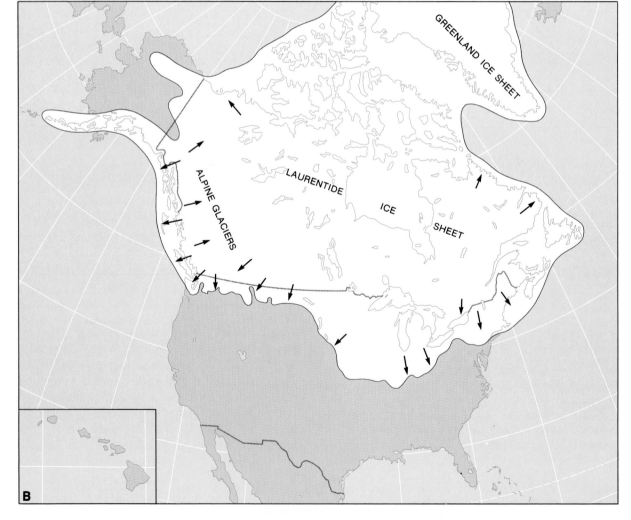

A ————— Limit of glaciation in the Pleistocene

POLAR

ICE

CAP

North
Pole

Arctic Circle

GREENLAND ICE SHEET

LAURENTIDE

ICE

SHEET

ALPINE GLACIERS

B

understood in the context of broad, general patterns across the planet.

Earth-Sun Relations and Temperature Variations.
The nature of the relationship of the earth to the sun is the fundamental control on climate. Temperature variations at the earth's surface are strongly related, as both cause and effect, to the atmospheric pressure and wind systems and, through them, also to precipitation patterns. The more concentrated, more "directly overhead" rays of the sun deliver more incoming solar energy per areal unit in the tropics than do the more oblique rays delivering energy to the higher latitudes (Figure 1–13). This situation produces an equator-straddling belt of "heat surplus," where the heating effects of the incoming solar rays outweigh the outgoing heat radiation. Flanking this heat energy surplus belt are two areas of net heat loss, the high-latitude polar and subpolar zones. If the earth's atmosphere and oceans did not function as temperature equalizers by continuously transferring heat poleward and cold equatorward, the tropics would be even hotter, and the polar-subpolar regions even colder than they are now. Just as heat energy can be moved vertically and horizontally within the ocean, heat energy is transferred vertically and horizontally through the earth's envelope of relatively dense atmosphere, the **troposphere**.

Air near the relatively hot surface of the equatorial region has a strong tendency to rise vertically, whereas air near the cold surfaces of the polar regions has the least tendency to be heated, to expand, and to rise. There are huge convection cells developed within the troposphere associated with temperature variations on the surface (Figure 1–14a and b). The equatorial low pressure, or **intertropical convergence**, is an updraft of warm air that rises, cools, and recycles down to the surface in the vicinity of 30° north and south, where it produces subtropical high pressures (Figure 1–14b). Air moving parallel to the surface either heads equatorward, to complete the convection cell, or moves toward the poles, where this relatively warm air collides with the cooler air of the subpolar regions and forms a zone of subpolar low pressure (Figure 1–14c). This contact zone of air masses of sharply different temperature characteristics produces low-pressure cells or cyclonic storms —the purveyors of much of the precipitation to the midlatitudes of the world.

Precipitation Variations and the Hydrologic Cycle.
Low atmospheric pressures, with their rising, cooling air currents, commonly produce precipitation; moisture condenses into clouds when an air mass is cooled (Figure 1–14c). High pressures, with their tendency to produce descending, stagnant, or warming air, seldom induce condensation and thus produce relatively cloudless skies and dry surface conditions (Figure 1–14c). There are two general areas of heavier precipitation in each (Northern and Southern) hemisphere, the equatorial zone and the subpolar zone. There are also two general areas of lower precipitation in each hemisphere, at the subtropical and polar highs (Figure 1–15, pp. 42–43).

Water, the absolute essential to humans and every other form of life, is in motion from atmosphere to surface (land or water) and back to the atmosphere in an endless cycle called the **hydrologic cycle** (Figure 1–16, p. 44). The condensation of water vapor into tiny droplets of water or crystals of ice, their precipitation as liquid or solid to the surface, and the eventual evaporation of water or sublimation of ice back into gas in the atmosphere is one of the great energy-recycling systems of the planet. Heat energy is absorbed in evaporation or sublimation and released in condensation. Because the hydrologic cycle produces precipitation, it also powers agents of erosion and deposition.

← **FIGURE 1–12**
Northern Hemisphere Pleistocene Glaciers.
(a) Although the cultural factors in the landscape are obviously dynamic, the physical world is not unchanging. Within the most recent geologic era, the Pleistocene (the last 1 to 2 million years), climatic changes produced immense ice sheets in the area around the Arctic basin. (b) At their maximum extent, glaciers occupied 4 million square miles, over half the North American continent. Vast areas of the Canadian Shield were eroded down to barren rock by the forces of moving ice, while portions of the northern United States and southern Canada were mantled in till, the finer soil and rock particles deposited by retreating glaciers. Continental glaciers severely modified local landscapes and even rearranged the continent's drainage patterns.

FIGURE 1–13
Earth-Sun Relationships.
(a) The sun's rays strike the earth at different angles because of the curvature of the earth's surface. (b) Rays striking at a high angle (left) exhibit a high degree of solar intensity. Intense solar radiation characterizes the tropics. More oblique (and less intense) radiation characterizes the midlatitudes (middle), and higher latitudes (right) experience an even lower intensity. (c) The earth is rendered more livable by the tilt on its axis, 23½° from the vertical. The belt of most intense solar radiation moves with the seasons, spreading warmth to other parts of the planet. Longer summer days in polar latitudes intensify the benefits of the limited solar radiation received in those regions. (From *Earth Science,* 5th ed., by Edward J. Tarbuck and Frederick K. Lutgens, copyright © 1988 Merrill Publishing Co., Columbus, Ohio. Illustrations by Dennis Tasa, Tasa Graphic Arts, Inc.)

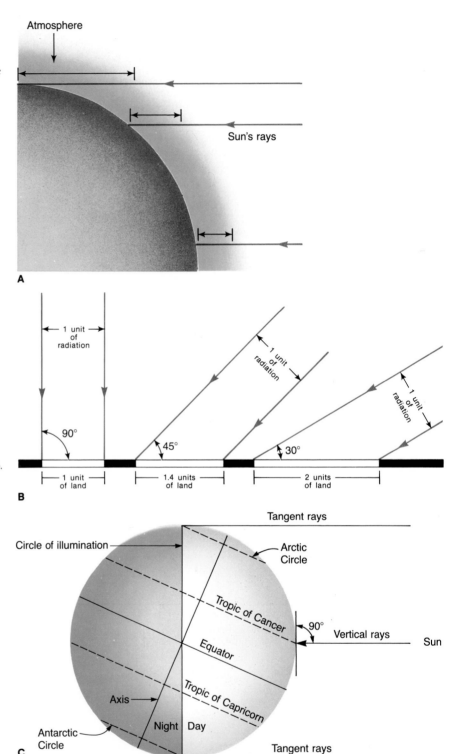

FIGURE 1–14
Continental Air Masses and Wind Systems. Heat is transferred across the planet by the movement of air. Surface winds are only the lower portion of giant, heat-transferring convectional cells that move heat and cold between major portions of the earth. (a) Winds circulate from high- to low-pressure areas as if filling a vacuum. Air currents rise from the heated surface (the low) to higher altitudes, where cooling takes place and condensation results. Clouds and rain are associated with low pressures. (b) In high-pressure areas, associated with relatively cooler surface temperatures, air currents descend from higher altitudes, diverging outward. Descending air warms, reducing the chances of condensation; high-pressure areas are characterized by clear skies. (c) The relatively stationary, semipermanent system of highs and lows above the earth's surface induces air movement and acts as a climatic control. Differential solar radiation creates a planetary weather machine that gives the earth its habitability by transferring heat and water. (From *Earth Science,* 5th ed., by Edward J. Tarbuck and Frederick K. Lutgens, copyright ©1988 Merrill Publishing Co., Columbus, Ohio. Illustrations by Dennis Tasa, Tasa Graphic Arts, Inc.)

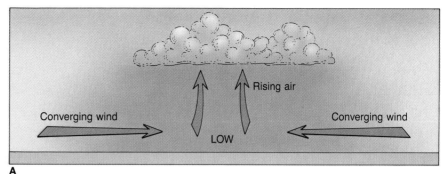

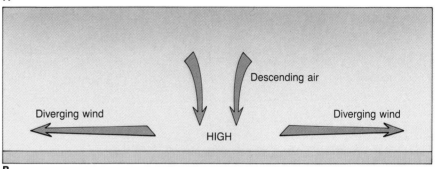

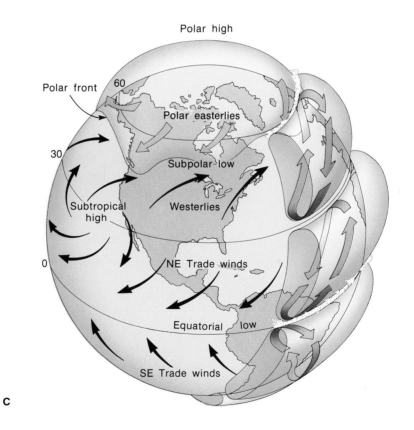

FIGURE 1-15
World Mean Annual Precipitation. Precipitation varies greatly over the planet. Vast regions receive excessive rain, while other areas receive only limited and occasional precipitation. The tropics and certain highland areas receive the greatest amounts, while the polar reaches and portions of the subtropics often receive very little. Rainfall varies seasonally in response to changes in the level of solar radiation.

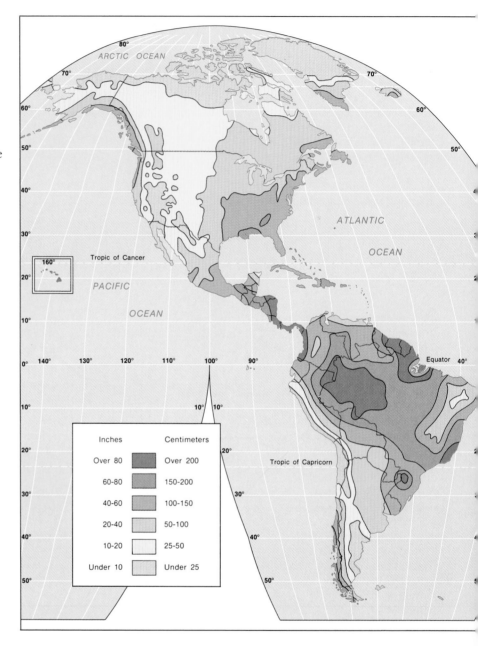

Inches		Centimeters
Over 80		Over 200
60-80		150-200
40-60		100-150
20-40		50-100
10-20		25-50
Under 10		Under 25

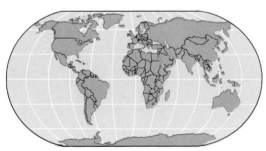

The Oceans and Climate. Latitude and associated atmospheric circulation patterns (see Figures 1-13a and b and 1-14c), although basic, are not the sole controls or influences on climates. Water, like air, circulates warmth around the planet. As a basic rule of thumb, land heats and cools faster than water. Therefore, land surfaces experience greater temperature differences, both daily and seasonally, than do

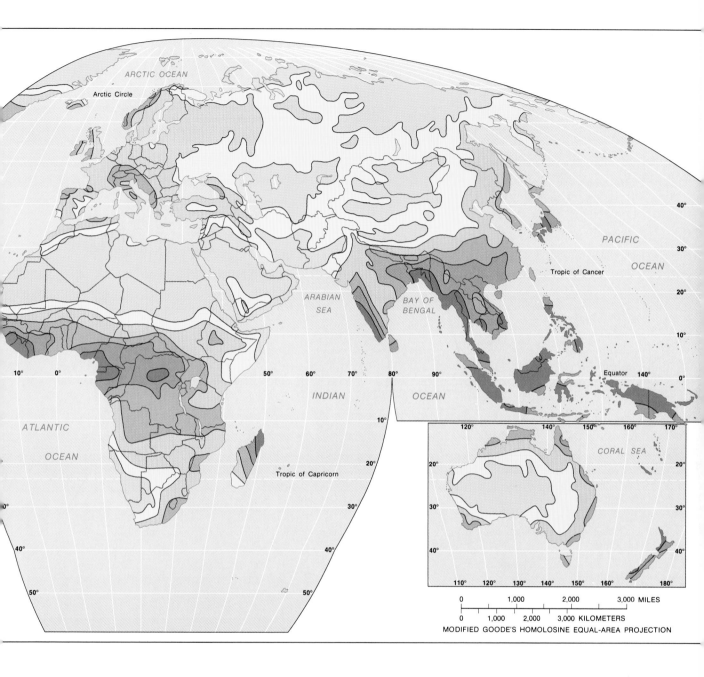

ARCTIC OCEAN

Arctic Circle

PACIFIC

OCEAN

40°

30°

Tropic of Cancer

20°

ARABIAN
SEA

BAY OF
BENGAL

10°

INDIAN   OCEAN

Equator   140°

0°

ATLANTIC

OCEAN

Tropic of Capricorn

CORAL SEA

MODIFIED GOODE'S HOMOLOSINE EQUAL-AREA PROJECTION

water surfaces. The deeper and larger the body of water, the greater its influence on weather and climate, in both the area and the degree of the influence.

Solar energy is in effect stored in ocean waters. Like thermal energy in the atmosphere, it is transferred around the planet in huge cells, vertically between the ocean depths and surface and laterally across the surface of the waters. This transfer of surface waters is accomplished by currents (Figure 1–17).

All currents are classed as either warm or cold in relation to what the expected temperature would be at a particular latitude. Thus, the Alaska current is warm, and the California current is cold. Where the wind blows from sea to land, warm currents tend to

PHYSICAL LANDSCAPE DYNAMICS | **43**

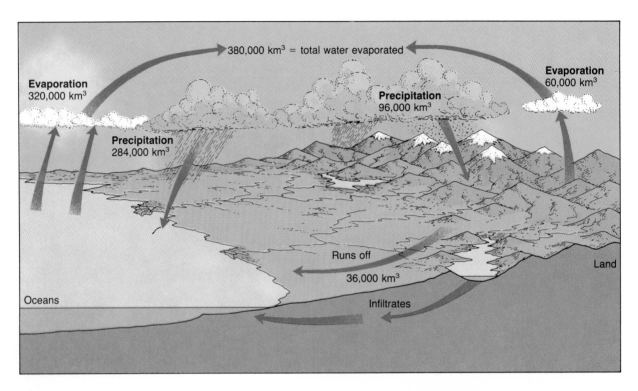

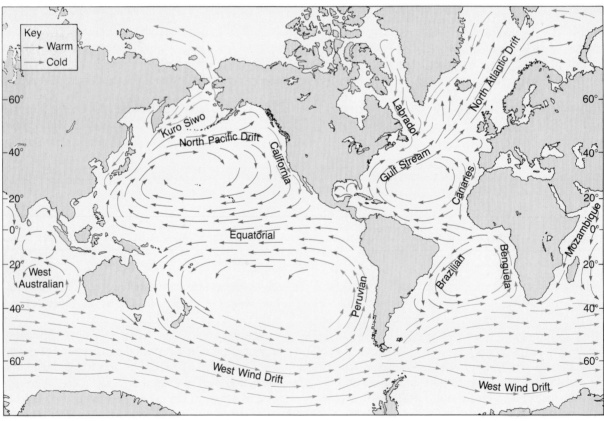

← FIGURE 1−16

FIGURE 1−16

The Hydrologic Cycle. Water is transferred across the planet (in both its vapor and liquid forms) in a natural series of mechanisms that keep water supplies in relative balance. Evaporation occurs over land and sea surfaces, but the oceans contribute the overwhelming share of atmospheric moisture. Most of this water falls back into the oceans, but about 30 percent falls on the land. Precipitation over the land surface is stored, in part, as groundwater but ultimately finds its way back to the oceans through surface and subsurface drainage. (From *Earth Science,* 5th ed., by Edward J. Tarbuck and Frederick K. Lutgens, copyright ©1988 Merrill Publishing Co., Columbus, Ohio. Illustrations by Dennis Tasa, Tasa Graphic Arts, Inc.)

warm the climate of adjacent land, whereas cold currents have the opposite effect. Again, where winds blow over warm currents toward the land, they also tend to increase the chances of precipitation, whereas cold currents decrease those chances. These generalizations apply only to limited areas where mountains block the inland penetration of these maritime (oceanic) influences. Where the areas are relatively level, open and well positioned to receive these maritime influences, these influences can penetrate inland over great distances. Geographers frequently refer to climates as either **maritime** (strongly modified by ocean influences) or its opposite, **continental**. So-called maritime climates have lower variations in seasonal and daily temperatures than do continental climates. Continental climates are characterized by temperature and precipitation extremes.

The prevailing wind direction heightens or lessens this maritime effect. Onshore winds facilitate the spread of maritime effects, and offshore winds effec-

← FIGURE 1−17

FIGURE 1−17

Ocean Currents. Currents may be viewed as rivers traveling across the oceans, pushed along by prevailing surface winds. Because the oceans absorb and store the warmth of solar radiation, currents also transfer heat and atmospheric warmth. Weather systems passing over warm currents can be rendered unstable, leading to precipitation. The reverse situation occurs where weather systems traverse cold currents. (From *Earth Science,* 5th ed., by Edward J. Tarbuck and Frederick K. Lutgens, copyright ©1988 Merrill Publishing Co., Columbus, Ohio. Illustrations by Dennis Tasa, Tasa Graphics Arts, Inc.)

tively spread the influence of continental conditions. The Atlantic coast of the United States, for example, receives little maritime influence from the warm Gulf Stream current off its shores. Winds move dominantly from land to sea in that area. The same current, however, bathes the shores of Europe with its warm waters. There the winds move from sea toward land, making the area both warmer and wetter than might be expected.

Climate and Altitude. Generally, the higher the area, the cooler the temperature, which is basically why mountain vacations are popular in summer, and why skiing conditions are better, longer, in areas of higher altitude. Because temperature affects livability and land use, so, indirectly, does altitude. Select parts of the tropics have almost midlatitude temperature conditions because of altitude. This factor has had a serious influence on the distribution of people in Latin America, where highlands are often crowded and some tropical lowlands are nearly population voids.

Slopes that face prevailing winds (and weather systems) are wetter; slopes that face away from them are drier. The tendency for altitude to increase precipitation on their windward (weather-facing) side is called the **orographic effect**. The air rises, cools, and releases its moisture. The tendency for highland areas to have a dry side (resulting from descending, warming air) is called the **rainshadow effect**. The effects of altitude may be highly localized or widespread, depending on the areal extent of high-altitude physical features (see Figure 1−15).

Generalized World Climate Patterns. As shown in Figure 1−18 (pp. 46−47) and Table 1−1 (p. 48), there are fairly few representative climates of the world. Temperatures are either hot all year, cold all year, or vary with the seasons. Each part of the earth's surface is either wetter than the potential for evaporation, or drier. Beyond these simple facts, there are the questions associated with seasonality. How long and intense is the season of warmth associated with plant growth? How long is the dry season, and does it coincide with a season warm enough for plant growth? The emphasis in a world geography course must be on the patterns and results of weather and climate, rather than on the causal mechanisms.

FIGURE 1–18

World Climate. Scientists have combined and averaged weather conditions into generalized regions of similar climate. Climate is the long-term average of temperature and precipitation that characterize the individual region. Climatic regions take into account the amount and seasonality of precipitation, the length of the frost-free season, and the intensity of seasonal warmth and cold. The relative surplus or deficit of moisture is deemed one of the most critical factors.

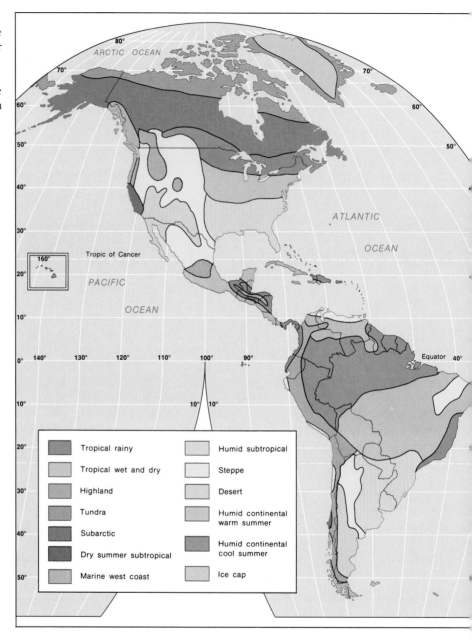

■ Tropical rainy		■ Humid subtropical	
■ Tropical wet and dry		■ Steppe	
■ Highland		■ Desert	
■ Tundra		■ Humid continental warm summer	
■ Subarctic		■ Humid continental cool summer	
■ Dry summer subtropical		■ Ice cap	
■ Marine west coast			

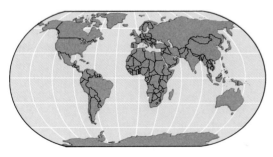

Soils: The Medium of Growth

Soils are a crucial factor in the production of food, and food, of course, is critical to the support of human population. Such factors as drainage, chemical composition, texture, and acidity are the bases on which soil fertility is assessed. Just as all plant life has

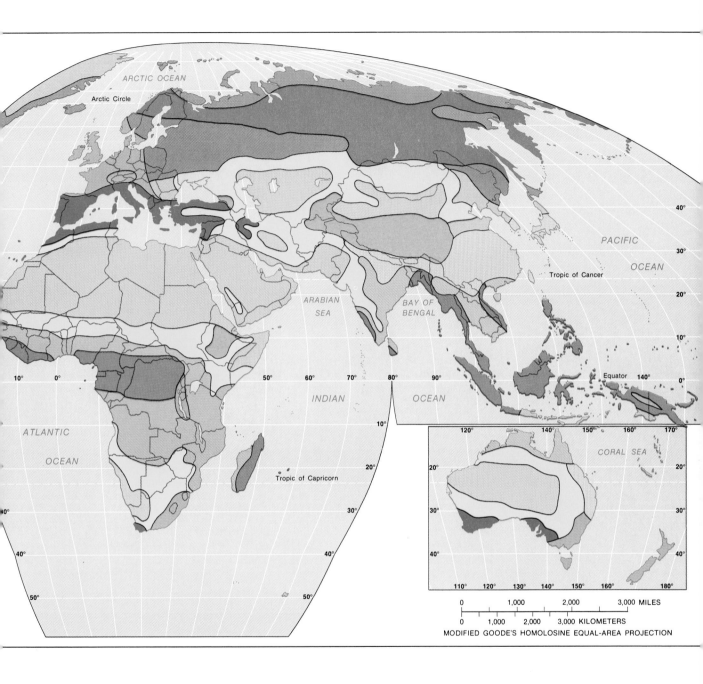

a certain range of climatic tolerance, requiring certain temperature and water conditions in order to grow and yield, all plants have a certain range of soil conditions they will tolerate. Within that total range is a set of soil parameters that are ideal for each plant. When particular plants are paired with ideal soil and climatic conditions, the yields are tremendous. Where conditions are less than ideal, yields are poorer. Beyond certain limits of tolerance of either soil or climate, crops will fail.

Soils are the product of various physical factors: parent material (the rock from which soil has decomposed), slope, climatic conditions, and the vegetational cover that was in place before the landscape

TABLE 1–1
Characteristics of world climate types.

Climate	Location by Latitude (continental position if distinctive)	Temperature	Precipitation (inches/year)
Tropical rainy	Equatorial	Warm, range* less than 5° F; no winter	60+; no distinct dry season
Tropical wet and dry	5°–20°	Warm, range* 5°–15° F; no winter	25–60; summer rainy, low-sun period dry
Steppe	Subtropics and middle latitudes (sheltered and interior continental positions)	Hot and cold seasons; dependent on latitude	Normally less than 20, much less in middle latitudes
Desert	Subtropics and middle latitudes (sheltered and interior continental positions)	Hot and cold seasons; dependent on latitude	Normally less than 10, much less in middle latitudes
Dry summer subtropical	30°–40° (western and subtropical coastal portions of continents)	Warm to hot summers; mild but distinct winters	20–30; dry summer; maximum precipitation in winter
Humid subtropical	0°–40° (eastern and southeastern subtropical portions of continents)	Hot summers; mild but distinct winters	30–65; rainy throughout the year; occasional dry winter (Asia)
Marine west coast	40°–60° (west coasts of midlatitude continents)	Mild summers and winters	Highly variable, 20–100; rainfall throughout the year; tendency toward winter maximum
Humid continental hot summer	35°–45° (continental interiors and east coasts, Northern Hemisphere only)	Warm to hot summers; cold winters	20–45; summer concentration; no distinct dry season
Humid continental cool summer	35°–45° (continental interiors and east coasts, Northern Hemipshere only)	Short, mild summers; severe winters	20–45; summer concentration; no distinct dry season
Subarctic	46°–70° (Northern Hemisphere only)	Short, mild summers; long, severe winters	20–45; summer concentration; no distinct dry season
Tundra	60° and poleward	Frost anytime; short growing season, vegetation limited	Limited moisture, 5–10, except at exposed locations
Ice cap	Polar areas	Constant winter	Limited precipitation, but surface accumulation
Highland	Variable	Variable	Heavier on the windward side, lighter on the rainshadow side

*The difference in temperature between the warmest months and the coldest months.
SOURCE: James S. Fisher, *Geography and Development: A World Regional Approach.* 2d ed. (Columbus, OH: Merrill Publishing, 1989).

was cleared and plowed for farming. Because all these factors vary widely across the earth, so do soil types.

There are literally tens of thousands of types of soils categorized as **soil series**. A system called the **seventh approximation** (so named because it was the seventh try) simplifies this to 10 large soil orders (Table 1–2 and Figure 1–19). However, a system that lumps together the rich soils of the Canadian prairies with the soils of the troubled Sahel and tropical India leaves something to be desired. The oldest widely accepted classification is that of the Russian V.V. Dokuchaev. With 16 soil types, it is as easily manageable and mapped as the seventh approximation and is still widely used. Highly adaptable in the developed world, it is of lower utility in tropical ar-

TABLE 1–2
World soil types.

Soil Group	7th Approximation	Location	Features
Tundra soils	Inceptisols	High latitudes	Often poorly drained permafrost
Podzolic; podzols	Spodosols and histosols	High latitudes	Severely leached, acidic, infertile, poorly decomposed organic material; requires lime and fertilizer if it is to be used
Gray-brown podzolic	Alfisols	Middle latitudes	Slightly acidic to basic, good humus accumulation; useful if well managed and fertilized
Red-yellow podzolic	Utisols	Subtropical latitudes	Severe leaching, low in nutrients, less-than-moderate fertility; requires careful management
Latosolic; laterites	Oxisols	Tropical latitudes	Severe leaching, low in nutrients, less-than-moderate fertility; requires careful management
Chernozemic	Mollisols	Middle latitudes	Good organic content, good nutrient supply; highly fertile soils
Grumusolic (savanna soils)	Vertisols	Tropical and subtropical latitudes	Moderate humus and nutrient supply; difficult to manage because of wet/dry climatic conditions; prone to formation of hardpan
Desertic	Aridisols	Subtropical and middle latitudes	Low organic content but high nutrient content except for nitrogen; fertile soils; possible salt accumulation problem when irrigated
Alluvial	Entisols	Variable	Deposited by water; often good nutrient content; fertile in most cases
Bog	Histosols	Variable	Poorly drained, poor material; difficult to use; can yield well with artificial drainage
Mountain	Mountain soils	Variable	Highly varied types

NOTE: The correspondence of the two classification systems is valid only at a highly generalized level. A more detailed description of the soil groups is found in the U.S. Department of Agriculture, *Soils and Men: Yearbook of Agriculture, 1938* (Washington, DC: Government Printing Office, 1938). A detailed description of the comprehensive soil classification system is found in the U.S. Department of Agriculture, Soil Conservation Service, *Soil Taxonomy: A Basic System of Soil Classification for Making and Interpreting Soil Surveys*, Agriculture Handbook No. 436.

FIGURE 1–19

FIGURE 1–19
World Soil Regions. World soil regions are more highly generalized than climatic regions. The system of climatic classification is based on two prime variables, whereas soils are affected by a dozen or more. Although there are literally tens of thousands of individual soil types, a general correspondence of soil regions to climatic and vegetational regions exists. The system shown here is the seventh approximation, a recent scholarly attempt to organize soils into a few meaningful categories. The older system, developed by the Russian V.V. Dokuchaev, is still widely used, and its descriptive terminology is still relevant.

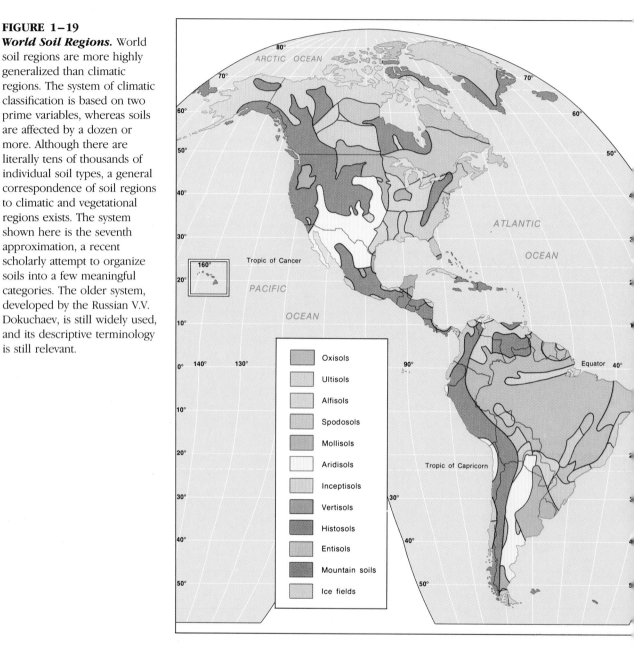

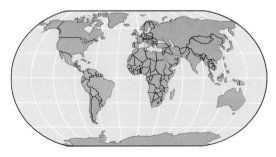

eas. Both these systems, when displayed on a map, show strong correlations to the climatic map (see Figure 1–18) and a map of original vegetation (see Figure 1–20). Local deviations can often be explained by local conditions.

Plants have the ability to manufacture food through photosynthesis. Warmth, carbon dioxide,

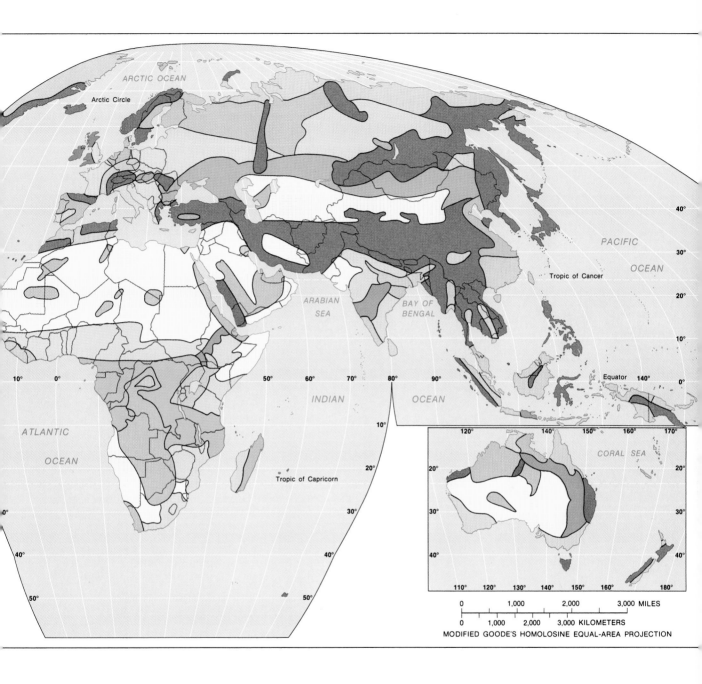

sunlight, water, and certain chemical elements are necessary to the process. Soils provide some of the chemical elements and a certain amount of the water. Of all the chemical elements present in the soil, three are crucial to growth: nitrogen, phosphorus, and potassium. These are the three "numbers" that appear on a bag of garden fertilizer. Certain other elements, called trace elements, are needed in tiny (but often crucial) amounts. All plants have a range of soil acidity that they tolerate, and within that range, a preferred condition; there is a similar range of ideal soil drainage conditions.

Agriculture is an attempt to maximize food production through providing the best possible soil and climatic conditions for plant growth. Farming is essentially human improvement of natural conditions to secure a greater amount of food than would result under purely natural conditions. For example, calcium (lime) is added to reduce acidity, and sulfur or iron is added to increase acidity to desirable levels. Water is added through irrigation or subtracted through drainage systems. Plowing increases the oxygen supply, improves drainage, kills weeds (the competition for the food supply), and breaks up soil into a finer texture.

On a grand and imperfect scale of generalization, the following soil concepts are useful. Areas that are climatically too wet (precipitation greatly exceeds evaporation) have lower fertility levels than areas that are dry. Soils of rainy tropical areas have a generally low fertility level because of this, and so do soils in areas of excessively heavy rain in cooler, seasonal climates. This low fertility results from an important natural process called **leaching**. In effect, an excess of precipitation rinses the upper layers of the soil of its critical chemical elements. Crop plants are generally shallow rooted, and because of leaching, they cannot "reach" the nutrients they need. These nutrients don't just disappear. They collect at deeper levels in the soil. In some cases, they mix with fine clays to form a "waterproof" layer called a **hardpan**, which can interrupt soil drainage, impeding the storage of water. Soils with hardpans have poor drainage and water storage characteristics. In areas of extreme cold, soils are generally thin (an undesirable condition overall) and poorly drained. Where climates are very dry, harmful concentrations of chemical salts tend to occur in the soil.

Where original vegetation was thick and lush, there is a high content of humus (decomposed plant matter) and, because of it, a good supply of nitrogen (Figure 1–20). Where vegetation was scanty, as in deserts, nitrogen is lacking—if other minerals are present. A thick cover of tall grass yields the greatest amount of humus (and therefore nitrogen) in the shallow layers of soil that humans farm. A vegetational cover of trees yields less. Burning of the forest cover to replace it with grass is an old, primitive means of increasing fertility. Soils formed under needle leaf trees tend to be acid and have little humus.

Those developed under broad-leafed trees tend to be less acid and somewhat richer in humus.

If the rock from which the soil was formed was rich in desirable chemical elements, the soil generally is also rich in those minerals. If the rock material was hard to erode, soils will be thin and stony; if the rock eroded easily, the soil will often be thicker.

Where vulcanism occurs, the ash that falls on the land can be an enricher of soils. Streams carry dissolved chemicals and suspended soil particles as a part of their load, not just water. When streams flood, they deposit these soil-enriching materials, called alluvium, on surrounding land. There are examples of good volcanic or alluvial soils in areas that otherwise have poor soils. There are also barren volcanic plains where the rock has not yet weathered into soil, and there are areas of worthless sand and stones that rivers deposited as alluvium. Generally, however, such adverse conditions are the exception.

Although the intent of agriculture is to improve conditions, it may also contribute to their decay. Removing the natural vegetational cover increases soil erosion by the forces of wind and water. With all its benefits, plowing is a cause of increased erosion as well. Irrigation can contribute to soil **salination**, the accumulation of harmful salts in the surface layers of soil.

As noted earlier, there are strong correlations between the maps of soil and those of climate and vegetation; however, none of them correlates particularly well with the map of population (Figure 1–21, pp. 56–57). The distribution of the earth's population is a reflection of the *potential* for agricultural production. The spread of human population from the area of origin to the rest of the world coincides with changes in technology and knowledge. Humans progressively adapted to and learned how to use a greater variety of environments. Improving soils has been one of those adaptations. The soils of China may bear little resemblance to what nature originally provided as thousands of years of human activity have undoubtedly modified them considerably. To a lesser degree, this is true of most of the developed world and some parts of the less developed world. Technological advances in agriculture could further alter population distributions. Viewed in that light, soil fertility is definitive only within the framework of human utility.

POPULATION GROWTH, DENSITY, AND CHARACTERISTICS

Population is the fastest changing variable over large areas of the earth. World maps of population distribution and growth rates portray this dynamic nature of the world's human population (Figures 1–21 and 1–22, pp. 58–59). The popular term *population explosion* refers to the recent spectacular change in the rate of growth of the population. A chart of world population increase shows a long period of relatively slow growth up until about A.D. 1500 (Figure 1–23, p. 60).

Population growth is the result of a simple relationship between two vital rates—births and deaths—both expressed per 1,000 people. A surplus of births over deaths—additions versus subtractions—results in growth. Migration is also a part of the process. A surplus of immigrants over emigrants will result in growth. Historically, births barely exceeded deaths in the period before A.D. 1500. Since that date, the rate of increase has grown rapidly. Not only has the population increased, but it has grown at an ever-increasing rate.

Humans have proven highly adaptable to the great range of climates, soils, landform types, and living conditions found over the earth. Their density varies from up to 100,000 per square mile in urban concentrations to fewer than 1 per square mile in parts of the Arctic, parts of the rainforests, and the world's great deserts. One whole continent, icebound Antarctica, has no permanent human residents at all. Population densities, growth rates, and the reasons for them, form an important basis for analyzing regional differences.

Important characteristics of populations include age and sex compositions. A commonly used graph showing age and sex components is called a **population pyramid** because of its geometrical similarity to that familiar structure (Figure 1–24a, p. 61). In a typical population, approximately equal numbers of males and females are born each year. Attrition in the form of disease and accident gradually reduces the number of people in older age groupings, or **cohorts**. Males generally have a higher death rate, especially in late maturity, and so the female side of each cohort surviving becomes larger than the male side. Significant variations from this average pyramid can present social concerns and other special problems to a population. The continued high birthrates and falling death rates of most developing nations' populations can produce a broader-based age-sex profile with large numbers of young and dependent people and few elderly (Figure 1–24b [left]). Such a population will have to spend a disproportionate share of national income on education and on creating jobs fast enough to absorb large numbers of new arrivals in job markets.

On the other hand, a mature industrial state will have an almost rectangularly shaped graph, reflecting low birthrates *and* low death rates (Figure 1–24b [middle]). A birthrate lower than the death rate, unless balanced by in-migration, could threaten the future of a nation. In-migration of a different ethnic or national group would, of course, result in sweeping cultural changes in a very low birthrate situation. Some states such as West Germany and France are actively trying to increase their birthrates in order to maintain stable numbers of their own nationals. A high proportion of dependent elderly will result from long life expectancies, with a smaller proportion of economically active population to support the retired cohorts (Figure 1–24b [right]). Medical costs will be disproportionately high. Recent changes in Social Security levies and payments in the United States reflect such concerns as the population ages. Normal age and sex ratios can be affected by war, high migration rates due to economic problems or opportunities, or inadequate medical care. For example, East Germany has a distorted pyramid in which females greatly outnumber males. This situation was caused by high death rates among males in the age cohorts that were of military service age during World War II and was further complicated by the fact that many East Germans migrated to West Germany. Young adult cohorts, wherever heavy out-migration occurs, show a surplus of females left at home. In societies where primitive sanitation and health care result in high numbers of deaths in child-

FIGURE 1–20
Original Extent of World Vegetational Regions. The distribution of original vegetation corresponds closely to climatic conditions. In turn, the original vegetation has had great influence on the composition and fertility of soils. Together, climate, soils, and vegetation reflect the agricultural potential of an area since the success or failure of crops depends on those environmental elements.

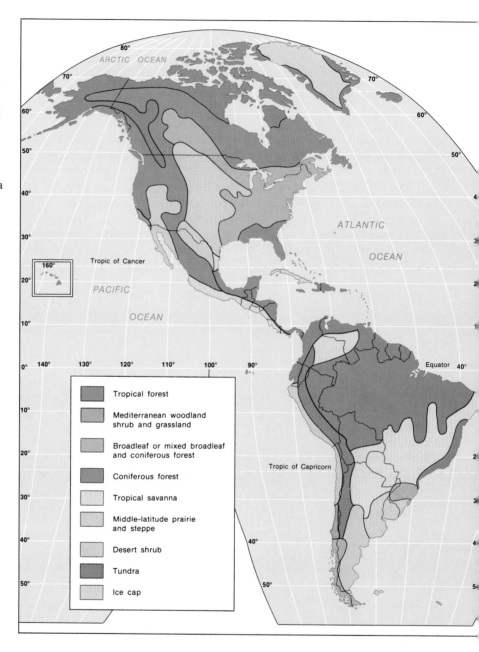

birth, young adult males will outnumber young adult females. These various characteristics of population can be as significant as total numbers or densities.

Population and Resource Pressures

Population and resource pressures have been major topics of study for centuries. The almost infinite capacity of populations to expand, compared to the

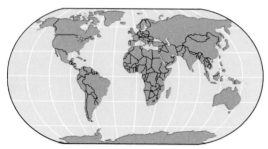

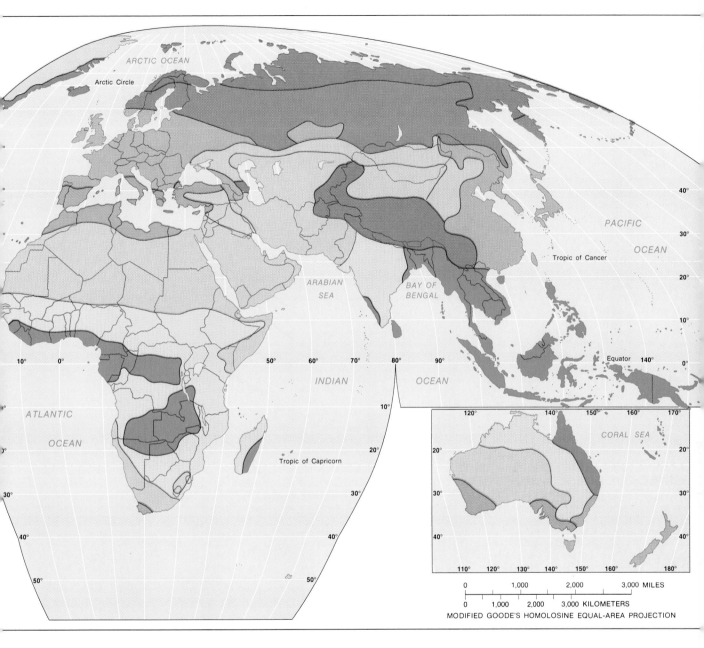

much slower "expansion" of the resource base through redefinition of resources, has attracted pessimistic attention since at least 1798. In that year, Thomas Malthus's *Essay on Population* expressed the view that populations tend to increase faster than the means of supporting them, leading to disaster. Unchecked population growth, he felt, would lead to famine, which in turn would reduce the population. When numbers were reduced, population would be-

gin to grow again until the next famine. Malthus's dire predictions for Britain never materialized because agricultural productivity increased, industrialization provided more jobs (which produced products for export to pay for imported food), and heavy out-migration commenced to North America and Oceania. Neo-Malthusians contend that his point was valid and assert that his timetable for disaster luckily has been delayed, but not canceled. Less pessimistic

FIGURE 1–21
World Population. Population
is not distributed uniformly
across the planet. For the most
part, the patterns of
concentration and void are in
equilibrium with the
environmental potential for
crop production.

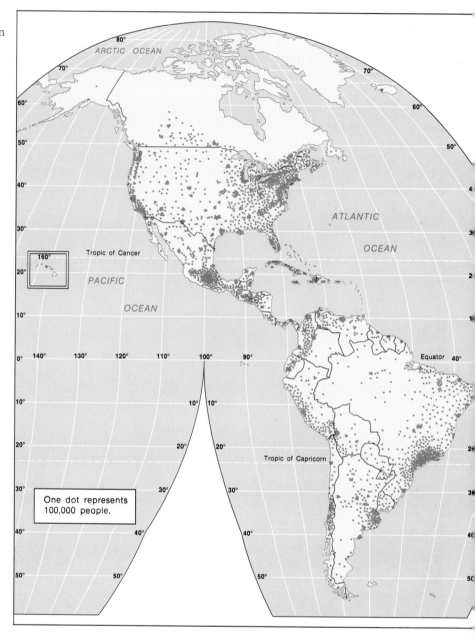

One dot represents
100,000 people.

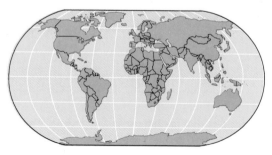

analysts assume that what occurred in Britain will
occur elsewhere. Development will result in a reduction of births, as well as deaths, and a low growth rate
will result. This theory of reduced growth through
development is called the **demographic transition.**

Many of the world's less developed countries are
in a crucial race between expanding population and

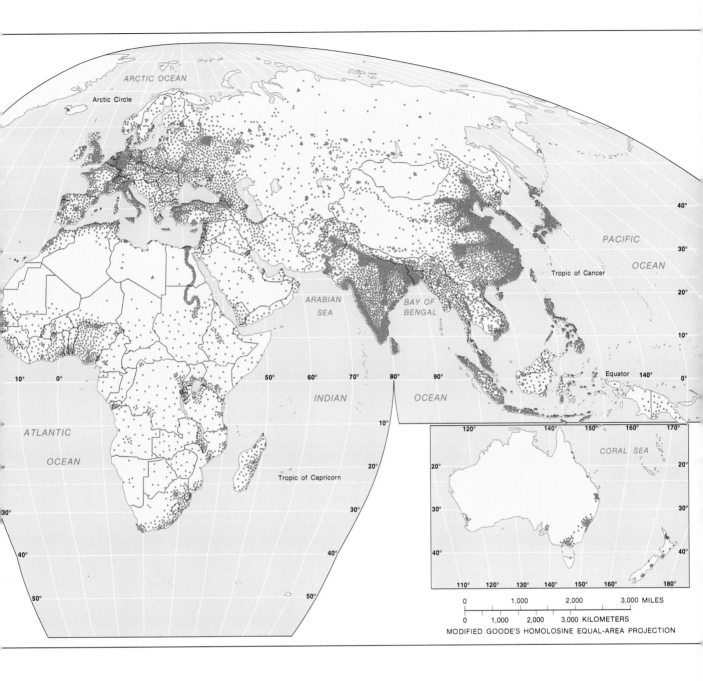

ARCTIC OCEAN

Arctic Circle

PACIFIC

OCEAN

40°

30°

Tropic of Cancer

20°

ARABIAN
SEA

BAY OF
BENGAL

10°

10°   0°   50°   60°   70°   80°   90°

Equator   140°   0°

INDIAN   OCEAN

ATLANTIC

10°

OCEAN

CORAL SEA

20°

120°   140°   150°   160°   170°

20°

30°

Tropic of Capricorn

30°

110°   120°   130°   140°   150°   160°   180°

0   1,000   2,000   3,000 MILES

0   1,000   2,000   3,000 KILOMETERS

MODIFIED GOODE'S HOMOLOSINE EQUAL-AREA PROJECTION

expanding food supplies and resource bases. The demographic transition is depicted in graph form in Figure 1–25, p. 62. It shows the theorized relationship between industrialization-modernization and rates of population increase. Based on historical trends, it offers some hope that world population growth rates will continue to drop.

Development and the Demographic Transition

During industrialization, the vital characteristics of the human population also go through a sequence of changes (Figure 1–25). There are two ways in which the surplus of births over deaths needed for a gain in

FIGURE 1–22
World Population Growth Rates. With the large and significant exception of China, development and growth rates vary almost inversely. Areas that are highly developed are also highly urbanized, and highly urbanized areas tend to have lower birth- and growth rates.

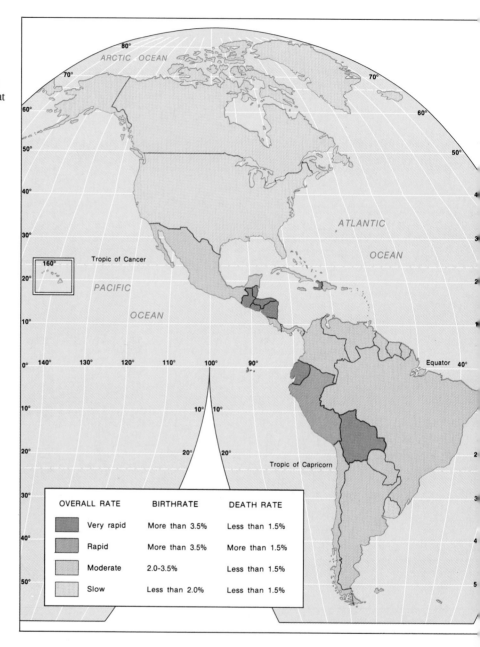

OVERALL RATE	BIRTHRATE	DEATH RATE
Very rapid	More than 3.5%	Less than 1.5%
Rapid	More than 3.5%	More than 1.5%
Moderate	2.0-3.5%	Less than 1.5%
Slow	Less than 2.0%	Less than 1.5%

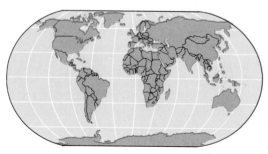

population can be achieved: the birthrate can go higher, or the death rate can drop lower. In stage 1 of the demographic transition, the growth rate is relatively low because there is only a small surplus of births over deaths. However, in this preindustrial stage, both birth- and death rates are high by industrial nation standards. High death rates at this time reflect a lack of modern sanitation and medicine, a

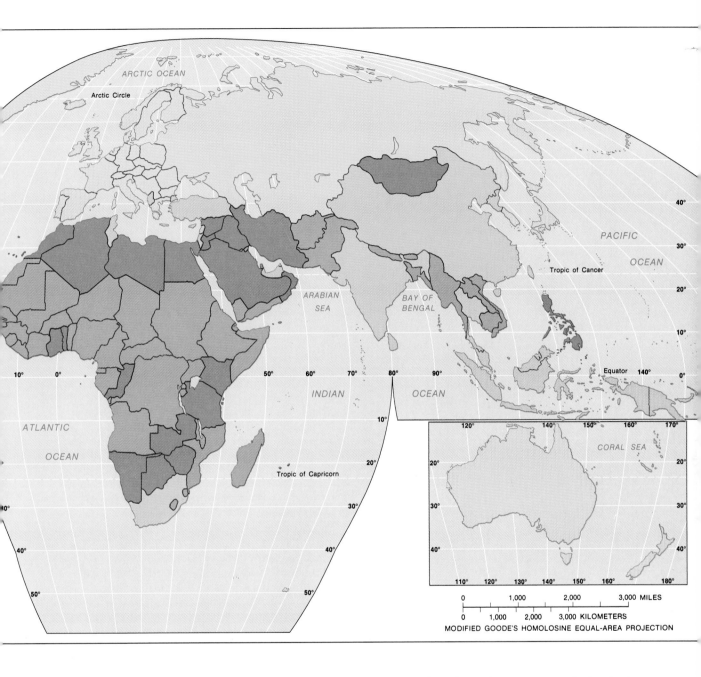

local food supply highly vulnerable to natural hazards, and a probability of less stable government, leading to civil unrest or wars. The smooth trend lines as generalized on the chart would be, in reality, continually fluctuating from year to year, but the long-time trend would be as shown. Only high birthrates counterbalance high death rates and ensure the continuity of the population.

During early industrialization (stage 2), the death rate begins to fall. Although the early industrial cities were notoriously unhealthy environments, their poor sanitation, fluctuating food supply, cramped housing conditions, and inadequate medical care were slowly improved. Similar improvements are anticipated with development wherever it occurs. Cheap, efficient long-distance transport of food and

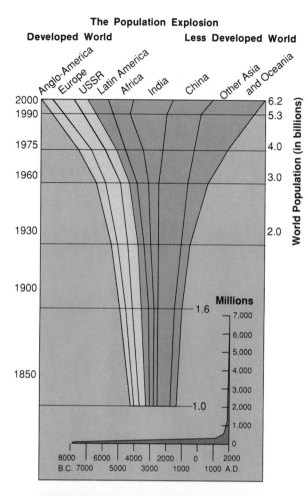

The Population Explosion

Region	Year 2000 Population (in millions)	World Share
Anglo-America	296	4.8%
Europe	507	8.3%
USSR	312	5.1%
Latin America	537	8.7%
Africa	880	14.3%
India	1013	16.5%
China	1200	19.4%
Other Asia and Oceania	1414	23.0%

FIGURE 1–23

World Population Expansion. For most of early recorded history, the world's population was rather sparse, and growth occurred at a slow and sporadic rate. Over the last 450 years, however, population has increased rapidly; growth has proceeded at an ever-increasing rate. This intensification of growth has been termed the *population explosion*. Only during the last decade has there been a noticeable decline in the rate of growth. Numbers continue to increase, but at a slowly diminishing rate.

FIGURE 1–24

(a) *Population Pyramids for the United States.* Population pyramids are a graphic means for displaying the composition of a population. Populations are divided into five-year age segments called cohorts, and the pyramid is divided centrally into male and female portions. In the early stages of development (1900), the pyramid has a broad base and the population is of a young average age. With increased development (1980), birthrates decline and lower-age cohorts become proportionately smaller. By the year 2000, it is theorized that slow growth and improved survival rates will further modify the U.S. pyramid to an increasingly columnar form as population stabilizes. A fully stabilized population is the theoretical norm for a fully matured economy. (b) *Population Pyramids for Selected Countries, 1987.* Chile has reached the stages of early population and economic maturity. Death rates declined sharply after the decade of the 1930s, as reflected in the bulge of growth observed in the cohorts under age 50. Denmark is approaching population stability; the population has remained virtually unchanged since the mid-1970s. The pyramid for Cape Verde reflects some gross abnormalities that occurred due to heavy out-migration over decades and is reflected in the small size of older cohorts. Decreased possibilities for migration since 1965, coupled with rapidly declining death rates since 1950, have resulted in an abnormally broad base to the pyramid. ([a] Adapted from Charles F. Westoff, "The Population of the Developed Countries," *Scientific American* 231 [1974] and U.S. Bureau of the Census, *Population Profile of the United States: 1980* [Washington, DC: U.S. Government Printing Office, 1981])

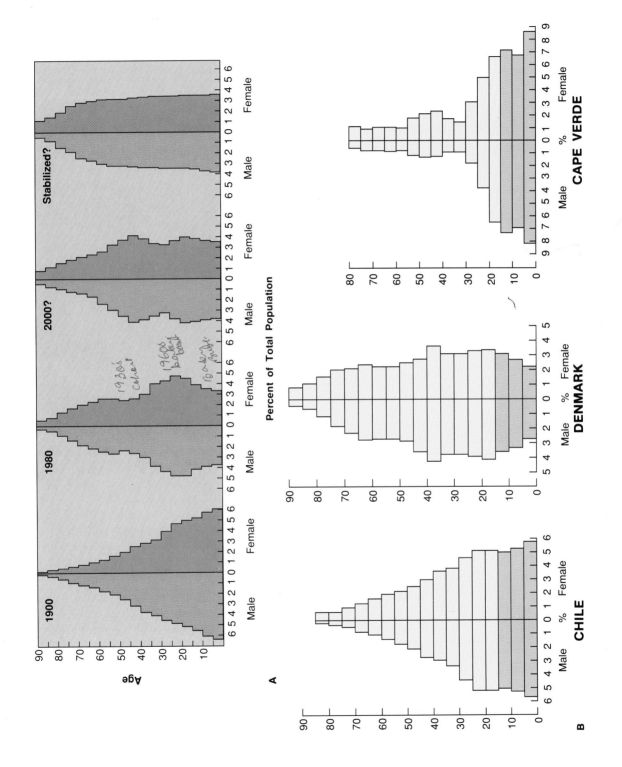

Percent of Total Population

Age

1900   1980   2000?   Stabilized?

Male   Female   Male   Female   Male   Female   Male   Female

1930s
cohort

1960s
babent
boom

Working
people

A

DENMARK

% Female

Male   %

CHILE

Male   %   Female

CAPE VERDE

Male   Female

B

FIGURE 1–25
Demographic Transition.
(a) There are four stages in the process that demographers call the *demographic transition:* *stage 1*—very high birth- and death rates with little net growth; *stage 2*—rapid population increase as a result of a radical decline in death rates and continued high birthrates; *stage 3*—moderating growth as birthrates decline and the decline of death rates slows; and *stage 4*—a return to little net growth as both birth- and death rates remain low. Migration rates are excluded from these theoretical models. (b) Note the strong differences in the demographic experiences of developed countries (red) and countries not yet fully developed (blue). (Adapted by permission from *The World Development Report 1982* [New York: Oxford University Press, 1982], p. 26)

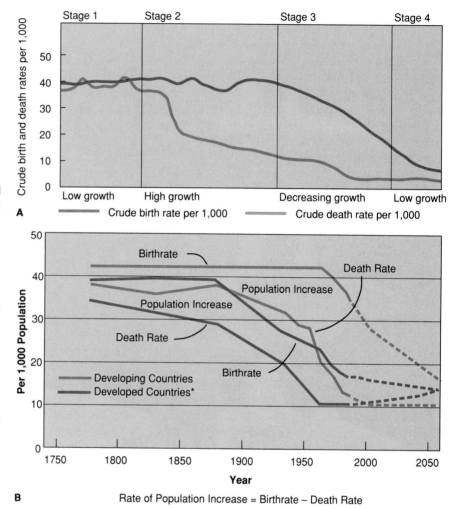

development of food preservation technologies improve the quantity and quality of diet. Because industrialization is most likely in a stable political situation, deaths from war and civil unrest also decrease. Medical and sanitation technologies advance. As the birthrate at first remains high, a much larger surplus of births over deaths develops. This is the critical stage in an industrializing society, the explosive growth rate sometimes termed the **population explosion**. The western European states that industrialized during the nineteenth century were able to send out large numbers of migrants to the New World, Oceania, and even Africa. The United States could absorb a high birthrate during industrialization, together with many immigrants, because it had sparsely settled frontier lands awaiting settlement

and rapidly developing industrial cities. Contemporary countries now in this first phase of industrialization do not have the option of large-scale outmigration to relieve population pressures.

During the maturation phase (stage 3), the birthrate begins to drop. Precisely why is open to discussion. It happens in societies of different religious beliefs and different races, and in different parts of the world. It seems to be related to the urbanization of the population. Children may be viewed as economic assets in a farming society, but in an urban-industrial one, they do not become economically productive until much later. In a high-technology society such as the United States (stage 4), children may be economically dependent for several decades after birth. Education stretches out over more years and is correspondingly expensive. Also, there is a greater likelihood that each child born will survive to maturity; there is no longer such a grim race between high birthrate and high death rate to ensure family survival into the next generation. Then, too, having many children used to guarantee support for aged parents; now this "social security" is institutionalized rather than dependent on children. Also, as more and more women join the labor force, their changing role favors later marriage and fewer children.

Thus, the motivation shifts from having many children to having fewer. Demographers have concluded that motivation to limit the number of children is far more important than any official policies or the availability of any specific birth control technology.

Without unduly minimizing the unfortunate, even catastrophic, implications of unchecked population expansion, it must be remembered that human population is the ultimate resource of any society or area. Although the threats of overpopulation are a clear and imminent danger in some areas, any other resource is useless without human enterprise, effort, and consumption. While many developing nations, particularly those of Latin America, Southeast Asia,

and Africa, are deeply concerned about the future implications of rapid population growth, others, including the Germanies and the Scandinavian states, are just as concerned about birthrates *below* zero population growth (in zero population growth, births just "replace" deaths) (Figure 1–26).

The vigor, innovation, and productivity of its human population is the most valuable resource of any state. People, not minerals, are the key to economic development. Resource-rich Zaire has a per capita income of $160, whereas resource-poor Japan has one of the highest average incomes in the world, at well over $10,000.

World Population Patterns

The world's population, estimated to be 5.25 billion people in 1990, is not equally distributed across the planet (see Figure 1–21). As pointed out in the sections on the physical environment, many parts of the land are too dry, too infertile, too cold, too barren of soil, or too rugged to support life. It has been estimated that only a sixth of the land surface can support crop farming or the more intensive types of

Nutrition Clinic. There is a tendency for death rates to decline earlier and more rapidly than death rates in the demographic transition. This clinic in Mexico provides information on hygiene, nutrition, child care, and birth control in addition to its functions as a medical dispensary and school for teaching modern agricultural practices. Improved health practices, medical care, and food supply—all primary goals of this government clinic—will tend to reduce death rates by increasing the rates of infant survival and reducing deaths due to childbirth complications.

livestock raising. Perhaps an equal proportion can support less intensive grazing. As a result, almost all of the earth's population is accounted for in one tenth of the surface area of the planet and one third of the land area.

The small portion of the earth that is readily usable and habitable is called the **ecumene**. Crop farming can support some enormous densities of population. Parts of China and Bangladesh support upwards of 2,000 and 3,000 persons to the square mile, but in quite productive parts of the Dakotas or the plains of Argentina, populations of a little more than 5 people to the square mile are common. Areas where the only ways to support life are hunting, fishing, and gathering are in the non-ecumene. That lower category of usability is also associated with **nomadic herding**, a lifestyle in which humans guide animals in foraging for food across vast empty areas with little vegetation.

Most of the earth's population is concentrated in the few highly favored areas where farming yields abundant food. This fact is a sobering reminder of our close ties to the physical environment: without food, we cannot exist. Four large concentrations of population house over 70 percent of the world's people. It helps to put the world in perspective when it is realized that three of these four great concentrations are entirely within the continent of Asia, and the fourth is primarily confined to Europe (see Figure 1–21). The Old World, not the New World, houses most of the people.

The greatest of these areas of population concentration is in East Asia. It includes China, Japan, both Koreas, and Taiwan. Most of China is classed as non-ecumene; its people are grouped in the eastern, wetter one third of the country. The Japanese portion is one of the most densely settled areas of the world. When measured as a ratio of people to usable farmland, mountainous Japan supports over 6,000 people to a square mile. This measure (people per square mile of farmland) is called **physiological density** (Figure 1–27). In the case of Japan, it shows just how much food humans are able to produce from a piece of ground.

Not all farmland can support such numbers, but the major clusters do correlate remarkably well with favorable environmental conditions for farming. The Asian clusters also correlate with cultures that learned to maximize production from the land in the earliest of times. Japan and the Koreas are highly urbanized, and China has perhaps 40 cities with over a million people, but this East Asian concentration is still basically a reflection of enormous rural densities in which an intensive set of agricultural techniques has been able to produce food for a quarter of the world's population.

The second largest concentration consists of India and its neighbors in the subcontinent of South Asia; it accounts for over one fifth of the world's total. The population of India alone is so large that every seventh person in the world is a resident of that country. The wetlands of the Indus and Ganges rivers support the densest populations; only the desert northwest and a few mountain areas are zones of low density. Less urbanized than East Asia, this South Asian cluster is tied more directly to earning a living from the land.

The third concentration is metro Europe, one of the most highly usable parts of the earth. It is not confined to the traditional continent of Europe, but includes the neighboring fringes of North Africa and Turkey and even the north of Iran. It also contains the more heavily settled western parts of the USSR, including and slightly beyond the Urals. Even though its southern fringes lag somewhat in development, this area is dominated by urban life and pursuits. The development of industry has caused the most recent growth of this huge concentration, though it had already grown to considerable size on its still-important agricultural riches. The greatest densities within it are collected along two axes that cross in the vicinity of the lower Rhine Valley. These were the trade routes of historical times. Formerly, the more important of the two connected the Mediterranean with the North Sea by way of the Rhine. The other axis interconnected the rich soil areas of northern central Europe which stretched from the Paris Basin to the steppes of southern Russia.

The fourth concentration is considerably smaller and much less compact. Southeast Asia is composed in part of islands, making it difficult to focus on any center there. Dense populations in the Philippines, Vietnam, and the islands of Java and Sumatra in Indonesia circle an almost empty center of lightly peopled islands and empty sea. On the rest of the mainland, there are "islands" of high density in the irrigated river valleys and the south of the Malay Pen-

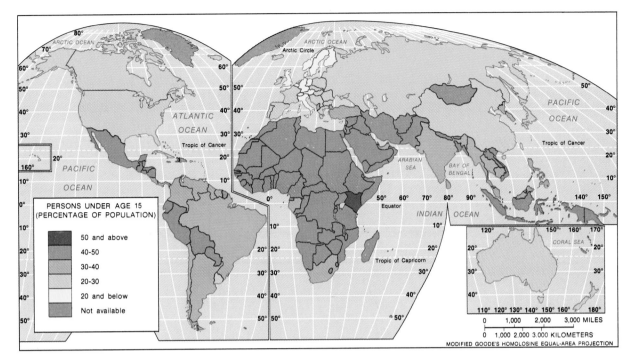

FIGURE 1–26

Percentage of Population under Age 15. Over one-third of the world's population is under age 15. That figure will increase to 40 percent by the year 2000. The percentage of young under working age varies from nearly 50 percent over most of Africa and the Middle East to a mere 15 or 20 percent in parts of Europe. Societies with a high percentage of young people experience high educational and child care costs. They also will experience difficulty in providing jobs for these large numbers of citizens when they reach working age. The reproductive potential of disproportionately young populations is staggering.

insula. They are discontinuous centers of density because of the alternation of land and water, lowland and mountain, habitable and uninhabitable surfaces. Collectively, they form a major concentration.

There are six secondary concentrations, and all are considerably smaller than the four concentrations we have looked at. Central Mexico; the highlands of Central America, Colombia, and Venezuela; the islands of the Caribbean and the Florida Gulf Coast form a circle of continuous dense population around the Caribbean Sea. Five major metropolitan centers dominate its northern portion, and Mexico City dominates it in the west. Highly urbanized Colombia and the city of Caracas (Venezuela) vie for urban-industrial supremacy in the south, while its islands remain dominantly rural. This area repre-

sents the full range of development possibilities, from least to most developed. Rampant growth is characteristic, even in a few parts of the South (U.S.). Culturally, there is much interchange between its Latin and North American portions. The fusing of this concentration with the neighboring concentration to the north is imminent.

The bulk of North America's population concentrates in the eastern United States and adjacent portions of Canada. This concentration surrounds the great navigable inland waters of the continent much as the Caribbean concentration encircles that sea. This eastern U.S.-southeastern Canada population concentration is still fairly lightly settled. It is supremely urban-industrial, though historically its densities were often built on agricultural wealth.

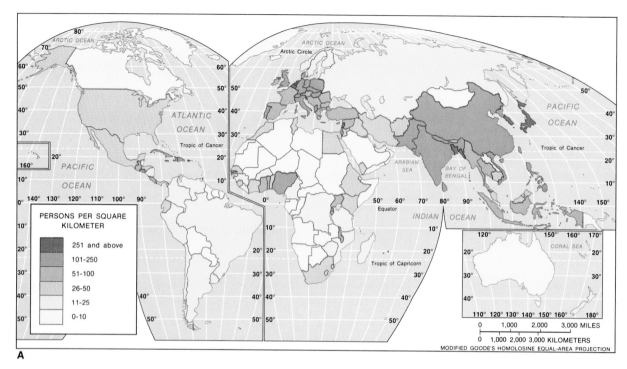

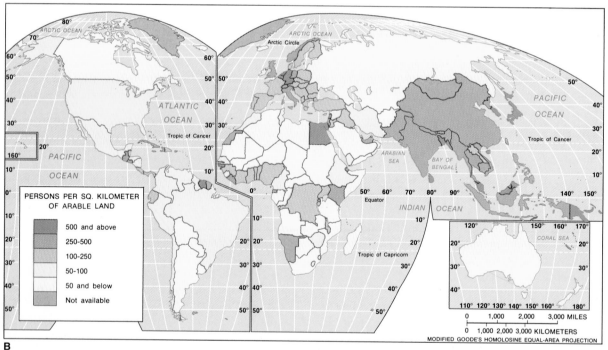

FIGURE 1–27

(a) *Arithmetic Population Density.* The figure normally given for population density is derived by dividing total population by total land area. This rather crude statistic does not take into account the utility of the land nor does it measure the difficulty of providing a decent standard of living. (b) *Physiological Density*. This figure measures the number of people per unit of land suitable for agriculture. It provides a more sophisticated measure of density by eliminating vast areas of essentially unusable land from consideration. Although a better measure of a country's ability to feed itself, it is still no measure of well-being. To some extent, it measures pressure on the agricultural resource.

Its core is a matter of dispute. Traditionally, it has been considered to be a megalopolis, the urban concentration on its Atlantic coast that stretches from north of Boston to south of Washington, DC. No one questions the importance, however, of a similar urban concentration that roughly parallels the southern Great Lakes from Pittsburgh through Chicago, or a third urban concentration that stretches from Montreal through Toronto to Detroit and Buffalo. This North American Manufacturing Belt has lost some of its preeminence in recent years but is still, collectively, the core of the concentration. Viewed another way, it has enlarged to encompass parts of the southeastern United States and the Mid-South. Atlanta, Dallas, Kansas City, and Minneapolis are its southern and western urban anchors. It is smaller by far than metro Europe, with power and production that greatly exceed its small population size.

Atlantic South America only now emerges as a population concentration. Nearly 20 cities of a million or more give it a basically urban character. Three of these cities, Rio de Janeiro, São Paulo, and Buenos Aires are among the 10 largest in the world. Formerly, it was a series of isolated urban clusters in a loose zone of important agricultural development; these clusters have recently coalesced. Rapid growth here, much as in the Caribbean concentration, will result in its steadily increasing importance.

The remaining three concentrations are in the Eastern Hemisphere. Those along the western and eastern margins of the continent of Africa are basically agricultural, though minerals play an important role in isolated cases. These two concentrations currently are experiencing the most rapid growth in the world. Although representing the best natural environmental conditions in that continent, they are not areas easily used for farming. It is questionable, at this time, whether this rapid growth can be handled in a somewhat capricious environment.

The last concentration is the Fertile Crescent, nearly as ancient as humankind. Here two ribbons of agriculture and population follow rivers in an otherwise basically desert environment; they are connected by some wetter, densely settled coastal and mountain tracts. Nowhere in the world is the transition from dense population to uninhabited void so abrupt. Oil, more than agriculture, has been responsible for its recent growth. As in the African cases, its ability to sustain continued rapid growth is questionable.

A mere 7.5 percent of the world's population occupies the rest of the land surface, and much of that (4 percent of the world total) is concentrated along the Pacific coasts of North and South America, collects along the lines of the Trans-Siberian Railroad in the USSR, or clusters in the expanding oases of Soviet Central Asia.

The empty areas cover more total territory than the ecumene. The world's population, then, is largely peripheral to the landmass, concentrating at the contact of land and sea. The edges of the Eurasian landmass house the bulk of the world's population in four major and one secondary concentrations. The New World, with less than 15 percent of the total, is far outweighed by the Old World in numbers. Recent rapid growth in the less developed world areas has resulted in a decrease in the relative importance of the U.S.-Canadian population concentration. The population of metropolitan Europe remains stable.

Overall, there is still a strong correlation between the distribution of the world's population and ideal conditions for crop agriculture. Within that persistent framework, there has been much change, as people increasingly live in urban circumstances and increasingly pursue occupations outside agriculture. Europe and the Americas are the most highly urbanized areas in the densely populated portions of the world. The secondary concentrations of Africa and the major concentrations of East Asia and South Asia remain heavily rural. East Asia, however, seems to be following the evolutionary pattern of Europe—toward a dominance of urban-industrial concentration. The rates of growth in East Asia are leveling off. Urbanization, with or without its traditional association with industry, is proceeding rapidly in the less developed countries of the world. Growth in that less developed world threatens to outpace the ability of the land to produce.

These population patterns are critical to understanding world geography. All questions and answers involving the world and its parts ultimately relate back to this distribution of people. There are reasons behind it, explanations of why. Additionally, it is a critical factor in answering the questions of how, where, and why for the world's future.

FIGURE 1–28
World GNP per Capita. Per capita GNP is widely used to measure economic well-being. Differences in official and actual currency values, subsidized social benefits, and controlled prices make per capita income figures less comparable. High per capita GNPs have traditionally coincided with areas of high levels of development. The rapid rise of oil prices in the 1970s led to inclusion of the oil-exporting states in the higher categories of GNP. The lack of an *official profit* for many product categories and the low value assigned to medical and social services in Communist countries make comparability difficult. GNP does not measure average living standards, since the distribution of income among the citizenry varies greatly from country to country. *Change* in the GNP, not the figure itself, is a measure of economic progress or contraction. Lacking better measures, it suffices as a crude index for economic comparisons. (From *Population Data Sheet 1988* [Washington, DC: Population Reference Bureau, 1989])

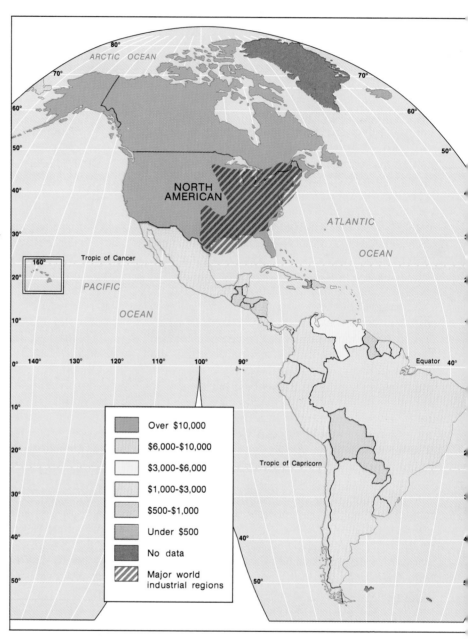

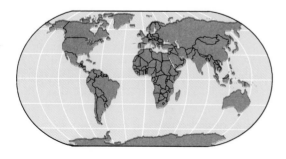

QUESTIONS OF DEVELOPMENT

The worldwide significance of regional variations in development levels is so great that some aspects of economic development must be considered before we progress through the world's regions. Once the general structure of industrialization and related changes in the human population's characteristics

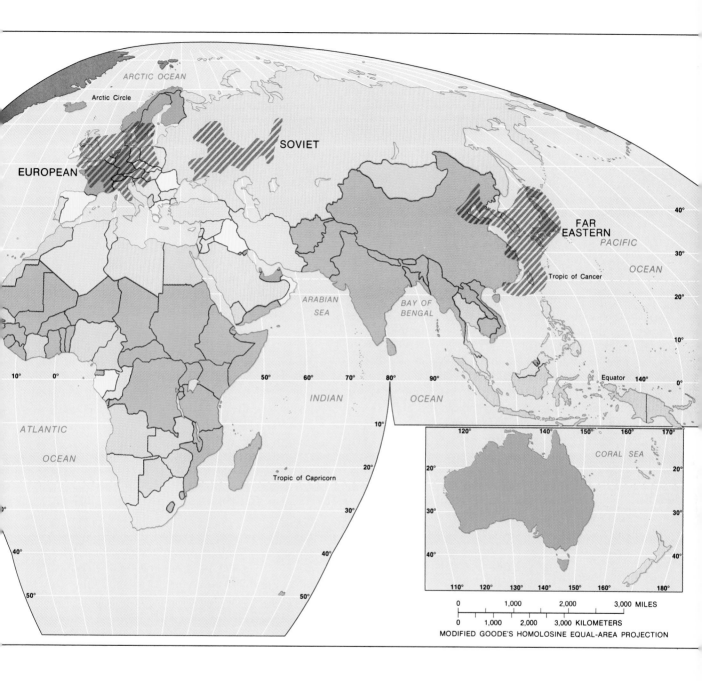

are understood, each region and subregion can be understood in developmental terms.

People generally see rising standards of living and an improved quality of life as the end products of economic development. Certainly, everyone has a personal interest in income levels, longevity, and general quality of life. A critical and dominant aspect of development is the current **economic stage** of any region. There is not an absolute relationship between level of industrialization and standard of living. Some preindustrial but resource-rich areas enjoy high living standards, and individual poverty in wealthy industrial nations is unfortunately common (Figure 1–28). In general terms, however, highly industrialized countries have higher average standards of living than do preindustrial ones.

Resources, Technology, and Industrialization

Industrialization is a process that occurs over a varying time span. It appears to have an internal sequence of development that persists independently of regional, cultural, political, or other variables. The best-known, succinct statement of the industrialization sequence is that of American economic historian Walt Rostow. Though Rostow's theories are not without problems, and other theories can be equally helpful and convincing, the geographic relationship among industrial development, natural resources, population characteristics, and other relevant variables are most easily understood with reference to the Rostow model.[6]

The Rostow thesis posits a predictable series of stages of economic growth. The sequence progresses in the same manner, apparently independent of the country involved. Even capitalism and communism appear to make little difference in either sequence or outcome. Once industrialization is under way, cultural variables seem to be of limited importance; equally significant is the fact that physical geographic variables (including natural resource base) have no clear relationship to industrialization, at least in a direct cause-and-effect sense.

Social, economic, and political conditions, however, *are* critical to the beginning phase of industrialization. The right economic-political-social climate must precede industrial development. In traditional societies, political power and wealth are concentrated in the hands of a landed aristocracy. Land, the source of food and raw materials for clothes and crafts, is the key to wealth and power. Land ownership confers prestige and income. The predominantly farming economy produces little surplus; what surplus exists is concentrated in the hands of the few at the top. They tend to spend this wealth on luxuries rather than investing it in expanding the economy. Few aristocrats apply much effort to increasing productivity, because such concerns are not proper to their social rank. Should the work force work harder and produce more, the results will be another palace, not reinvestment in expansion of new enterprises. Rostow emphasizes that a high rate of reinvestment in the economy is essential for an economy to begin to develop. Clearly, if the aristoc-

racy does not reinvest in new economic activity, it must be removed from power or otherwise restricted if the economy is to move forward. Not all industrial states have first had an upheaval on the scale of the Russian Revolution, but all have limited the power of feudal aristocracies.

The British achievement of a gradual political evolution rather than a thorough-going upheaval on the French model made it easier for the political "offspring" of the United Kingdom. The United States, after independence, and some of the British Commonwealth nations such as Australia and Canada were, in Rostow's words, "born free." They were born free of the deadening effect on economic development of a powerful feudal aristocracy.

Following this revolution or evolutionary change in society, the country must have political stability. A high rate of reinvestment is likely only when the decision to postpone consumption or enjoyment of the surplus in favor of reinvestment is a logical one. Reinvestment postpones enjoyment and consumption, but its attraction is that an expanding enterprise will eventually generate still larger surpluses for enjoyment in the future. If one cannot reasonably expect to collect the fruits of reinvestment eventually, there is little incentive to reinvest.

During the second stage, called **preconditions for takeoff**, government policy must favor the expansion of the economy through tax, tariff, labor, and regulatory policies. The critical stage, the **takeoff**, follows establishment of the preconditions. During takeoff, the sluggishly expanding agricultural economy is replaced by a rapid growth of industrial and transportation facilities and output. Economic growth is not only desirable; it is accepted as normal. Reinvestment takes place at a high rate, fueling this rapid growth. The takeoff at first may be limited to one or two industries (textile factories are a common beginning). Later, all sectors of the economy—agriculture, mining, construction, transportation, and communications, along with manufacturing—expand and modernize. Practical science and technology are thus applied, with inanimate power and machines, throughout the entire economy in this stage, known as the **drive to maturity**.

The concluding stage of the sequence is that of **high mass consumption**. The country's mature industrial plant no longer requires the very high rates of reinvestment that characterized the drive to maturity. A smaller portion of a larger gross product is sufficient reinvestment. A large consumable surplus can now be used to raise standards of living. The coincidence of a high gross national product with an already accomplished industrial development seems to confirm Rostow's thesis (see Figure 1–28).

Rostow was intrigued by the apparent association in time of two historical events. As some countries neared the end of their drive to maturity, they greatly expanded their colonial possessions, or expanded their home territories in wars. Some countries may choose to invest in military forces. Still other societies will choose to spend some surplus in funding social welfare programs—socialized education, housing, and medicine. Geographers can better understand a country's economic position, domestic policies, and international objectives if they view that country in the context of the stages of development.

THE SYSTEM OF WORLD REGIONS

The regions used in this text are based on a combination of economic and cultural factors (Figure 1–29, p. 74). As the interdependence of the world's economies becomes more of a reality and the diffusion of culture and technology proceeds more rapidly, the traditional cultural elements become less useful as a means of differentiating regions. Five major religious traditions account for most of the world's population, and religion, above all other factors, seems to overlap regional boundaries in the broadest possible way. Language is a more reasonable criterion for categorization, though only if the widest language grouping is used in some areas, and the narrowest in others. Most regions based on language uniformity would also contain some major exceptions. Cultural landscapes differ widely in detail, particularly in less developed economies, in a few highly individualistic areas, and in places and aspects hallowed by tradition. Traditional dress is disappearing as a uniform style of clothing seems to have been popularized throughout most countries of the world. Although there are notable exceptions to each of these statements, it is easy to see the tendency for traditional cultural elements to become submerged

in, or isolated from, an emerging worldwide cultural blending.

Perhaps the most critical differentiating element in today's world is development. The division of the world into developed and less developed portions is both obvious and exceedingly important. Once the division has been accomplished, aspects of culture can be used to differentiate each of these essentially economic designations into smaller, more nearly uniform regions.

The more developed portion of the world can be divided into four parts: The United States and Canada, Europe, the USSR, and Australia-Oceania. These four regions all have a European cultural background but differ somewhat in their political systems and often even more widely in their history. Generally, there is a greater degree of cultural similarity within these highly developed areas than is commonly found within less developed regions. The region of East Asia contains elements and areas of both a developed and an underdeveloped nature. In that sense, it bridges both economic extremes. Its ancient and well-developed culture is a source of unity, yet there is an increasing tendency for that culture to become modified along with development. The remainder of the world is clearly less developed, yet its parts are perhaps more distinctly differentiated into individual components along cultural and national lines.

The world has been divided into 10 regions, each based on a slightly different set of criteria. Most regions contain all or a portion of a major or a secondary population concentration. In each region, there is a strong tendency to share the same difficulties and to exhibit a high degree of economic and developmental uniformity regionwide (except in transitional East Asia). Each contains exceptions, yet in all, the similarity always outweighs the unique.

The United States and Canada

The North American region consists of two countries, each rich in mineral resources, highly urbanized, and heavily industrialized. Both enjoy a high standard of living and are peopled overwhelmingly by the descendants of migrants from other regions.

CONTEMPORARY DEVELOPMENT PROBLEMS

Many Third World countries have invested in the necessary infrastructure and have encouraged (or decreed) the necessary social change to establish the preconditions for takeoff. A smaller number of Third World countries (e.g., China, Brazil) have already arrived at takeoff and are proceeding into increasingly advanced stages of industrialization. Progress toward development, however, is highly uneven. Many industrializing states (e.g., Mexico, Egypt) remain short of food, and diversion of funds and energies to industrialization efforts leave food deficits unaddressed. The race to develop commercial, currency-earning agricultural exports often results in the need to increase basic food imports, negating some of the anticipated real economic growth.

The answer does not lie in a simple shift of food from areas of surplus to those of need. Two decades ago, there was a genuine fear that the world could not generate enough food for all its inhabitants. In the economic order of these times, a handful of later-settled, developed countries were the dominant surplus food generators. They had two potential markets for their overproduction: food-short, developed nations with the capital necessary to purchase any food needed (e.g., Japan and much of Western Europe) and food-short, underdeveloped nations without the available capital to purchase food. In the latter situation, profits made on sales of food to a developed Europe could be used to subsidize food shipments to capital-short, underdeveloped nations. Growing food self-sufficiency in Europe has changed that basic formula. For a time, Soviet and Chinese markets in the process of increasing their capital wealth and developmental infrastructure took up the slack in world markets. Today, China is relatively self-sufficient in food. The Soviets, still short of grain, can pick and choose among exporters, seeking the lowest price. Without a ready market for surplus food among developed nations, there is no surplus of agricultural profits with which to continue direct gifts, emergency aid, or food purchase subsidies to poor nations.

Traditionally, profitable industrial crops such as sugar, coffee, cotton, tropical fibers, tobacco, and vegetable oils were earners of hard currency that could be used to defray the costs of importing cheaper food and feed crops. A general glut of production of these industrial and specialty crops has resulted from attempts to develop as more nations add or emphasize these crops in their agriculture. Commodity prices have become seriously depressed, and there is no great profit from them with which to purchase basic food.

An alternative, where the resource base is present, was the development of mineral production. Again, the race toward development has encouraged the overproduction of 20 or more major minerals. In a glut situation, only the best-quality, cheapest-to-produce mineral deposits remain profitable. Therefore, only most-favored producers earn sufficient income to pay for imported food.

Increasingly, because basic commodities no longer provide solid profits, Third World countries turn to industrial production as the acceptable, and profitable, route to development. Where surplus unskilled labor is present and raw materials are scarce, such light industries as textiles, clothing, shoes, electronics, and component assembly are being developed. Where superior resources are present, basic heavy industry, such as metallurgy, chemicals, synthetics, cement, and fertilizer, are emphasized. These, too, are industries requiring little skill, and they can be constructed and operated within a framework of low overall technological development in a society.

Technology is fairly cheap and readily accessible because patents and licenses are easily procured. Industrial development can proceed without any "from scratch" development of expensive technology. Cheap labor (e.g., Singapore, South Korea) and cheap fuel (e.g., Mexico, Libya) can counter traditional productive advantages such as skilled labor. Capital is expensive, yet it has been readily borrowable; at times, cheap labor can be substituted. The rush to industrialize has resulted in massive Third World indebtedness as technology is purchased and infrastructure is built on borrowed capital.

Within already developed areas, long-established industries are threatened by competition from industrializing Third World producers. Can the developed world survive on lending, services, and the development of increasingly sophisticated technology? Can everyone, or even most, adjust to this changing world economy?

Development is relative; it is dynamic, not stable. Undoubtedly, new members will be added to the growing number of successful developed nations. The Soviet Union, and even the United States, were not fully developed a century ago, and Japan's degree of economic success was certainly not foreseen even 30 years ago. The stages of Rostow's theory may be no more accurate than the stages of history suggested by Hegel and Marx, because both sequences are interpretations of the future based on the historical past. New roads to development may open, and new factors may assume a critical role in development, just as new resources and new technologies are apparently revolutionizing, or at least dramatically changing, the world's economy at the current time.

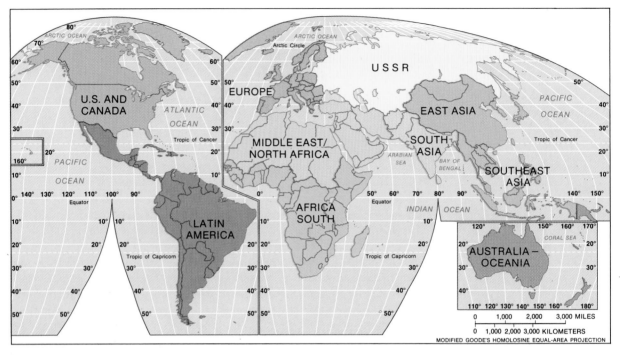

FIGURE 1–29

The World's Regions. These 10 regions are the organizational framework for the remainder of the text. Cultural similarities are the primary basis for regionalization; each region exhibits a reasonable degree of cultural homogeneity. To some extent, the regions are developmental regions as well. Four—the United States and Canada, Europe, the USSR, and Australia-Oceania—are characterized by highly developed economies. The remaining 6 are less developed by most standards of measure. There are exceptions, however. In the East Asia region, the highly developed Japanese economy contrasts with the lower degree of development in China. Highly developed Europe contains some weaker economic units. In South Africa, the level of development contrasts sharply along racial lines. There are regional developmental differences within virtually every country, and certainly within every region. Development is secondary to culture in this system of regionalization.

Their heritage is dominantly European, but their evolution and environments are clearly different. They have developed a separate culture from that of Europe. Their low population densities and spacious surroundings have played a large part in that differentiation. Despite the fact that by world standards these are "young" cultures and national units, they are highly developed and economically advanced.

Europe

Europe is perhaps the most fully usable of all the world's regions. There are few areas unsuited to ag-

riculture. It was the first region to develop modern industrialization and, despite the emergence of competitors, is still enormously important in the total world economic picture. Europe is, and has always been, a great center of innovation. Widely receptive to the technical (and some cultural) innovations of other regions, it continues to be deeply involved in commercial exchange.

Europe's ethnic and cultural diversity has proven to be both its greatest strength and its greatest vulnerability. Despite a long series of destructive wars accompanied by lasting animosities and long-standing rivalries, there is an increasing tendency toward more cooperation and a greater regional unity despite the East-West difference.

The peak of European power is past, but the loss of overseas territories did not diminish living standards or Europe's ability to compete. The production of its industrial establishments is still renowned for precision and quality. Sustained productivity, increasing efficiency, and an incredible resiliency keep Europe powerful, influential, and competitive.

The USSR

The Soviet Union is of continental scale; it is sufficiently important and large enough to be considered as a separate region. Culturally, it is European, but unique enough in its history and political philosophy to be considered separately from that region. It is planned, controlled, and operated as a unit, making it a functional region, with its central node at Moscow, as well as a formal region.

Although the government has literally created an industrial power, it has failed to create a unified nationality. Fully half of the country's population is not Russian. The Islamic minorities, particularly, have rejected assimilation and continue to grow at triple the Russian rate. Ethnic unrest has begun to emerge among some cultural groups, even those of European background.

Russian language and culture, Marxist-Leninist philosophy, and a ponderous bureaucracy are dominant forces in a country that seeks to create a new society from an assemblage of widely differing Asiatic and European inputs.

Australia-Oceania

The Australia-Oceania region includes Australia, New Zealand, the eastern half of New Guinea, and a host of islands scattered across the Pacific Basin. The relative isolation of most of this area led to the development and preservation of a unique wildlife and vegetation, late discovery, and belated colonization by Europeans. Although it is thoroughly Western in culture, its remoteness from Europe has led to its gradual drift into the Asiatic economic orbit. Its very emptiness is a strong regional trait; the sparse population reflects in part its low-utility environment.

East Asia

As stated earlier, the developmental differences exhibited across East Asia are astounding. Its smaller states—Japan, the Koreas, Taiwan, and the city of Hong Kong—have clearly entered the stage of full and advanced development. China, containing the bulk of the region's population and occupying by far the largest territory, has been in the process of developing for the last four decades. China was the dominant cultural force in the region over most of history, and it was also the dominant economic power. Japan's amazing growth and economic transformation have challenged China's dominance in the region on both grounds.

The fantastic density of settlement is a regional hallmark; the region contains one quarter of the world's total population. Here the landscape mirrors these densities in the intensity of land use.

Within China, the threat of overpopulation has been met with strong programs to limit family size. In this manner, China has been artificially hurried through the demographic transition. The immense population base, however, ensures continued growth despite falling birthrates. China's rapid industrialization is particularly evident in the huge and bustling urban centers, but the country remains, for the moment, a dominantly rural nation.

Even though the nations of the region harbor traditional animosities, all recognize the advantages of increased regional economic ties and trade. Culture, however divergent on the surface, binds the region together in various ways. All the nations share the Chinese system of written communication, the incredible regional work ethic, and the goals of a better living standard. They all have modified their culture to suit the chronic lack of space. Only portions of China's interior and Mongolia remain truly less developed, mainly because of distance from the core and difficult physical environments.

South Asia

South Asia can be convincingly justified as a region on a variety of grounds. It has an ancient historical continuity, because much of it was joined together in past empires. More recently, it was all a part of the

British Empire or, in the case of its fringe areas, strongly under its influence. Religion is a major cause of regional disunity in an area where Buddhism and Islam now coexist uneasily with Hinduism. Religion is often the basis for the existence of separate states.

Poverty remains an overriding concern since limited resources must be stretched to cover a host of problems. Particularly in India, the forces of culture and tradition have hampered development. Constant crisis and political turbidity are a nearly universal regional attribute. Several wars and political separation have not ended the Hindu-Moslem dispute. Disputed borders perennially complicate relationships. On careful analysis, the region demonstrates far more physical than cultural unity.

Southeast Asia

Southeast Asia is a physically dispersed area of peninsulas, island archipelagoes, and separate riverine lowlands. Three religions—Buddhism, Christianity, and Islam—dominate in three different areas of the region. There is great ethnic diversity, although many in the area have racial and cultural ties to China. Curiously, there is also strong anti-Chinese sentiment among many of its peoples. Residual Hindu populations in Bali, the ruins of Angkor Wat in Kampuchea and the architecture of Burma and Thailand, and the dominance of Buddhism on the mainland are all indicators of the strong Indian cultural influence on the region which preceded the advent of the Chinese. Indeed, it is this transitional nature that is the central unifying theme of the region. It is between two great cultures and the two largest population concentrations in the world.

Southeast Asia shares the diversity and unrest of South Asia and, increasingly, the industrial growth of East Asia. Rapid population growth now complicates the economics of what was not long ago an area of food surplus. The bitter Vietnam War is over, but combat and terrorist activity still rock the area. Guerrilla resistance movements are active in most of the region's states. Even at the national level, unity is superficial, though national identities are gradually building.

Africa South

The Africa South region includes a very large number of countries, and the number of individual cultures is immense. Yet it is not the cultural complexity that resulted in the creation of this great number of countries. They are instead the residue of 80 years of European colonial control. Claims of European powers to various areas were decided without regard to ethnic or linguistic patterns or even tribal jurisdictions. This maladjustment of national boundaries to cultural realities has been the central problem for many African states. Contending groups dispute leadership through politics and even civil war. This situation has greatly complicated the building of a national loyalty or popular consensus in the region's relatively new successor states.

Except for the southernmost part of the region, Africa South is preindustrial. It is the least developed of all the world's populated regions. The lack of development is perhaps the region's most outstanding characteristic.

Middle East-North Africa

The key to understanding the Middle East-North Africa region is its crossroads position at the juncture of three continents. This position has given the region strategic importance and a strong advantage in trade through the centuries.

The region is dominantly arid; dryness is its outstanding physical attribute. The population clusters in the better-watered fringe areas and along rivers that provide crop-supporting irrigation water. Oil is the outstanding resource; its peculiar pattern of distribution results in great economic differences among the region's states.

Islam is the overwhelmingly dominant religion, and the Arabic language is spoken in all but a few of its states. Despite this seemingly high degree of religious and linguistic unity, the region includes a host of individual countries. Portions of this region were the origin areas of Western civilization. The region also gave birth to three great monotheistic religions. Vast and powerful empires arose here and extended their influence far beyond the region's borders. The rich history and cultural heritage of the area have helped to create a series of individual national groups instead of a single political entity. The scattered nature of settlement compounds this separatist tendency.

Latin America

Latin America is a very large region, stretching from Mexico's border with the United States southward to the southern tip of South America. Mexico, the island states of the Caribbean, Central America, and all of the South American continent lie within this vast region. The region's characteristics include a predominance of Spanish or Portuguese (Iberian) culture and language. There are important admixtures of Amerindian peoples and cultures, and even some non-Iberian cultures and peoples (French, Italian, Dutch, English). With the important exceptions of the Caribbean islands and the three former European colonies in the Guianas, this region's countries achieved independence from colonial control early in the nineteenth century.

If there is a superficial cultural unity to the region, the diversity of economic circumstances among individual states tends to counter it sharply. Included within the region's borders are overpopulated, resource-poor islands with unidimensional economies and—in contrast—huge, resource-rich mainland countries racing toward the goal of development. Some of these larger states have immense foreign debts, because developmental projects were funded with borrowed capital. The region's economies, once dominated by a single mineral resource or crop, are rapidly diversifying. Individual countries, if not the entire region as of yet, appear to be nearing the ultimate goal of developed status, though not without a host of accompanying problems.

ENDNOTES

1. William D. Pattison, "The Four Traditions of Geography," *Journal of Geography* 63 (1964): 212.
2. Preston James, *All Possible Worlds* (New York: Bobbs-Merrill, 1972).
3. Preston James and Clarence F. Jones, *American Geography: Inventory and Prospect* (Syracuse, NY: Syracuse University Press, 1954).
4. Richard Hartshorne, *The Nature of Geography* (Lancaster, PA: Association of American Geographers, 1939).
5. William D. Pattison, "The Four Traditions of Geography," *Journal of Geography,* 63 (1964): 211–216.
6. Walt Rostow, *Stages of Economic Growth* (Cambridge: Cambridge University Press, 1960).

REVIEW QUESTIONS

1. Identify the unique perspective of geography in examining the changing world patterns of production, technology, innovation, consumption, and standards of living.
2. Why were the environmental determinists incomplete at best in their interpretation of physical-cultural interactions?
3. Why do some spatial correlations not necessarily imply a cause-and-effect relationship?
4. Briefly describe recent technological innovations that greatly increase the detail of information available to geographers and that facilitate the search for spatial correlations.

5. What are the fundamental differences between a regional approach and a systematic one?
6. Why is some system of regionalization a necessary tool in organizing a geographic study of the world?
7. How would you identify a neighborhood and determine its boundaries? Is there a similarity to identifying and bounding major world regions?
8. Why are national boundaries still an important consideration in regionalization, despite increasing international flows of raw materials, manufactured goods, technology, and people?
9. Identify and briefly describe physical landscape dynamics.
10. Identify and briefly describe cultural landscape dynamics.
11. Why is the definition of natural resources changing—usually expanding—over time?
12. Why does diffusion from a center of innovation virtually never occur at an even pace related to distance?
13. Briefly describe the demographic transition.
14. Why does the death rate of a society decline earlier than the birthrate?
15. Give historical and contemporary examples of nations within each of Rostow's stages of economic growth.
16. Why is it almost impossible for any state or region to become independent of international flows of materials and technologies?
17. Why do higher-technology societies become progressively more dependent on international raw material flows rather than more self-sufficient?
18. Choose any world region as defined here and speculate on revising its boundaries, providing logical geographic reasons for the revisions.

SUGGESTED READINGS

Bennett, Charles. *Man and Earth's Ecosystem*. New York: Wiley, 1975.

Berry, Brian; Conkling, Edgar; and Ray, Michael. *The Geography of Economic Systems*. Englewood Cliffs, NJ: Prentice-Hall, 1976.

Brunn, Stanley, and Williams, Jack. *Cities of the World: World Regional Urban Development*. New York: Harper & Row, 1983.

DeSouza, Anthony, and Faust, Brady. *World-Space Economy*. Columbus, OH: Merrill, 1979.

DeSouza, Anthony, and Porter, Philip. *The Underdevelopment and Modernization of the Third World*. Washington, DC: Association of American Geographers, 1974.

Gould, Peter, *The Geographer at Work*. Boston: Routledge & Kegan Paul, 1985.

Hall, Peter. *The World Cities*. New York: McGraw-Hill, 1977.

CHAPTER TWO

The United States and Canada

Statue of Liberty.

The United States and Canada are close neighbors in several important respects: they are each other's largest customers in international trade; each is the other's largest source of international tourists; and 5,525 miles of common borders are undefended against each other (Figure 2–1).

These good neighbors share many characteristics. Both are huge in area: Canada is the second largest country on earth, following the Soviet Union, and the United States ranks fourth in area, after third-ranking People's Republic of China. Together, the United States and Canada total nearly 7.5 million square miles of land. Those square miles contain, in very uneven distribution, over 262 million people, who are, on average, among the most productive and wealthiest people on the planet. Together, citizens of the United States and Canada number only about 5 percent of the world's population. Their rates of population increase, slightly under 1 percent per year for the United States, slightly over 1 percent a year for Canada, are both below the world average.

Both countries are democracies with an emphasis on individual rights and freedoms. With basically capitalist economies, both have strong government regulation of private enterprise, and both have created extensive social welfare systems. The United States and Canada are predominantly English-speaking, but Canada is officially bilingual (English and French), and Spanish is commonly a second language in government notices and publications in the United States.

The overall population growth rates in Canada and the United States have been about the same for a century. Canada has had 1 resident for every 10 residents in the United States, and that ratio continues. The United States and Canada produce similar surpluses, yet there is a **complementarity**—a situation in which two units produce different goods and develop an exchange of production in certain areas—which results in a brisk trade between the two.

There is a dynamism associated with the development of both countries. Droves of immigrants came from overseas to colonize the new farmlands. Generations of their descendants and newer immigrants continued the process until most of the arable land was cleared, cultivated, and integrated with the economy and political jurisdiction of the **core** (the densely settled, productive economic and political heart). This emphasis on expansion, growth, and development became a hallmark of the region; it carried over beyond agriculture and territorial conquest into the industrial societies that evolved. Progress, change, and material betterment are seemingly synonymous in the minds of the region's inhabitants, who have evolved societies not tightly bound by tradition, land, or even regional ties.

GROWING IMPORTANCE OF PACIFIC BASIN RELATIONSHIPS

Both Canada and the United States historically have had an Atlantic basin focus. Although Amerindians and Eskimo (Inuit) originated in Asia, the European discovery of the Americas resulted in strong, continuing trans-Atlantic migration. The dominant cultures, languages, and political philosophies of both countries are of European origin. Until the mid-twentieth century, migration to the region was overwhelmingly from Europe. The major exception was the involuntary migration of slaves, mostly from West Africa, and that too was a trans-Atlantic event.

The great port cities of the eastern United States and Canada were long dominant as financial and cultural centers; for centuries, their close trans-Atlantic political ties, trade relationships, and cultural interchanges far overshadowed trans-Pacific ties. The Pacific basin has always held some economic attractions, of course, but until after World War II clearly

FIGURE 2–1   →
Subregions and Cities of the United States and Canada. The United States and Canada region is divided into eleven subregions on the basis of internal cultural, economic, and physical similarities—seven within the United States and four within Canada.

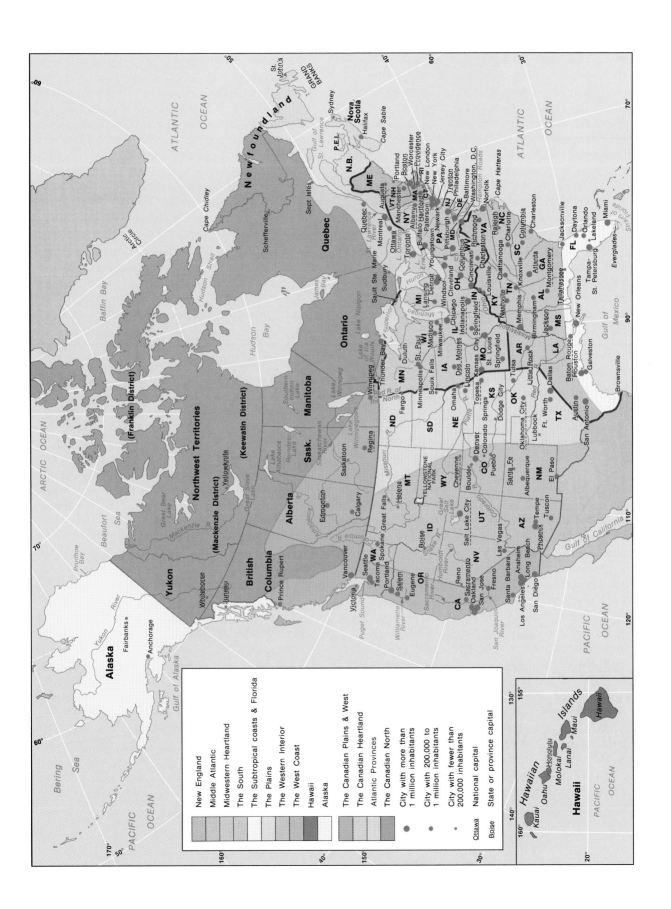

these were secondary to trans-Atlantic interests. These long-standing Pacific basin interests have expanded, gradually at first, and now at a quickening tempo. The trans-Atlantic focus has not been replaced but rather supplemented by the rising importance of trans-Pacific relationships.

America's last two wars, Korea and Vietnam, were fought partly to resist perceived threats to the security of American interests in the Asian rim of the Pacific. The second most important trading partner of the United States, after Canada, is Japan. The most dramatic shift in U.S. foreign policy in the last 30 years was the 1971 "China Initiative" seeking to restore normal diplomatic relations and to greatly expand trade. The rising significance of trans-Pacific relations is nicely illustrated by the challenge of West Coast cities to the commercial and cultural leadership of eastern cities.

MULTICULTURAL BLENDS OF IMMIGRANTS

As noted before, both Canada and the United States have expanded, historically, through strong flows of immigrants from Europe. In each country, the European share of continuing immigration has declined. Latin American and Asian immigration now far outpaces European migration to the United States. In Canada, European migrants now form about 40 percent of new Canadians, in contrast to 80 percent a quarter of a century ago; Asia is the source of many new Canadians.

Today, there are about 900,000 Amerindians and Inuit (Eskimo) in the United States, representing 0.4 percent of the total population, and 300,000 registered (officially recognized) Amerindians and 19,000 Inuit in Canada, representing about 0.9 percent of the Canadian population. Thus, 99 percent of the population of both countries is composed of relatively recent migrants and their descendants.

In each country, the volume and source of these migrant inflows have changed over time. The respective sizes of ethnic groups today are related to the length of time since the groups first arrived. For example, the original French migration to Canada, before the British victory at Quebec in 1759, is estimated to have involved no more than 15,000 people.

Despite the small number of French immigrants since 1759, the French in Canada are now 27 percent of the total, or about 6.3 million. Similarly, although it is believed that fewer than 500,000 Africans were imported as slaves into the United States, African-Americans now number some 12 percent of the U.S. population, or about 28.5 million.

In 1800, there were only about 5.5 million people living in the area now included in the United States and Canada. In the century between the end of the Napoleonic wars and World War I (1815 to 1914), 30 million immigrants came to the United States. At first, they came primarily from the British Isles, Germany, and Scandinavia; the most important source areas for immigrants shifted to southern and eastern Europe by 1900. The predominant source area for the most recent arrivals has shifted radically from Europe to Middle America (especially Mexico) and Asia (Table 2–1). As indicated in Table 2–1, Africa has provided the smallest proportion of immigrants of any major region after 1820.

Canada's major immigrations occurred between 1900 and World War I, in the 1920s, and after 1945. The first period was linked with the opening up of the prairies to settlement, an opportunity that attracted many Germans, Ukrainians, and Russians. Although a full third of Canada's total immigrants over time have come from Britain, the later waves of immigrants have been primarily from Italy, Germany, eastern Europe, and now Asia.

TRENDS AND PERCEPTIONS

Cultural geographer Wilbur Zelinsky observed that, "by definition, all places are unique; but the United States is unique in a manner that transcends the parochial pride or interests of its citizens."[1] But what are the cultural origins of this great society, this richly innovative, vigorous hybrid of peoples and cultures from around the world?

Though many cultures were present in the colonial history of the United States, three were of widespread significance at the onset—Amerindian, British, and African. A fourth, German, was primarily of regional importance, but in an area that became a major cultural hearth for all America—the Middle Atlantic region.

TABLE 2-1

Percentage of total immigration to the United States from important source regions.

	Overall Immigration, 1820–1984 (% of total)	Modern Sample, 1971–80 (% of total)
Northwest Europe	40.4	10.9
East and South Europe	30.1	6.9
Africa	0.4	1.8
Asia	8.4	35.3
Middle America	9.1	33.2
South America	2.7	7.1
Canada	8.1	3.8

SOURCE: U.S. Immigration and Naturalization Service.

The most valuable contributions of the original Americans to the "invaders" from Europe and Africa were cleared land and geographical information. "Indian oldfields," as the colonists called them, were agricultural clearings that spared the invaders some of the labor costs of transforming a forested land into farms. Amerindian knowledge of soil qualities, wild and domesticated food resources, and transportation routes was invaluable, as reflected in many Amerindian place names that survive. It would be difficult to imagine an America without "Indian corn"; other foods adapted by non-Indians include cranberries, sugar maple, squash, and pumpkin. Canoes and moccasins quickly crossed the cultural gap as well, but on the whole, the Euro-African impact on Amerindians seems to have been greater and longer-lasting than Amerindian contributions to the evolving American blend.

African contributions to American life were virtually ignored—rather deliberately, it appears—until fairly recently. The "myth of the negro past," so eloquently destroyed by Melville Herskovits, presumed that blacks arrived in the New World without memory or "cultural baggage."[2] It must be remembered, however, that most of the slave traffic originated in West Africa—the most advanced part of Sub-Saharan Africa except Ethiopia. It was a portion of the black African culture realm that had developed advanced metallurgical technologies and elaborate social-political organizations before the disruptive tragedies of the slave trade. The full input of African culture will not be appreciated until much more research effort has been made, but the obvious spheres of major impact are speech patterns, musical forms and instruments, dance, some types of worship, and burial customs and grave decoration.

Many other, earlier European cultural groups almost disappeared under the numerically larger, aggressive onslaught of the British. The major continuing exception, and one with obvious consequences to contemporary cultural and political life, is French Canadian Quebec. Other islands of cultural distinctiveness surviving from before the British-American cultural tidal wave are the Hispanic-Amerindian Southwest and Cajun Louisiana.

To these citadels of antecedent cultures still retaining their unique qualities must be added later regional accretions of non-British cultural groups. Mostly nineteenth century in origins, these local ethnic group concentrations which impart a distinctive flavor to the cultural landscape include Germans in the Upper Midwest; Scandinavians in Wisconsin, Minnesota, and the eastern Dakotas; and Slavic groups in the industrial cities of southern New England, the Middle Atlantic-Upper Midwest, and Appalachian coal towns.

The Cultural Heritage

In a book published in 1781, Hector St. John Crevecoeur, a Frenchman who had lived in America much of his life, posed an often quoted question: "What

THE U.S. AND CANADIAN ECONOMIES: COMPLEMENTARITY AND INTERDEPENDENCE

Canada purchases about 20 percent of the exports of the United States, and the United States buys almost two thirds of Canada's exports, which make up about 23 percent of the United States' total imports. The long common border, which cuts across major transport arteries and divides some relatively dense population clusters, is a very busy one.

Trade between the United States and Canada once was based on an exchange of Canadian raw materials for American manufactured goods, but this situation has changed. Although wheat exports earn almost $2 billion a year for Canada, autos and auto parts earn more than $10 billion. Industry is now the leading segment of the Canadian economy, producing a greater value of goods than agriculture, forestry, mines, and fisheries combined. In both countries, service industries are expanding employment faster than is manufacturing.

The automotive industry is an excellent example of U.S.-Canadian economic cooperation. U.S. automakers originally established assembly plants in Canada because high import taxes were placed on assembled cars, while low taxes were levied on parts. However, the technology of auto assembly and the character of the auto market combined to make totally separate car production inefficient. Because the Canadian car market was one tenth that of the United States, a typical Canadian plant repeatedly had to change over to making different sized cars in order to produce a full

then is the American, this new man?" Note that the question was *what,* not *who,* almost as though Americans were some sort of new species of humankind. Americans and foreign visitors alike long have recognized that Americans are different in new and important ways.

The land that Americans settled, developed, and shaped is uniquely American, reflecting the values, dreams, energies, and ambitions of this new culture and these new people. "New" is a powerful concept in the way Americans see themselves and interpret their interaction with their environment. As a society, we have repeatedly stressed that we are a new nation in a new world. *Novus Ordo Seclorum* ("The New Order of the Ages") reads the motto on the Great Seal of the United States, engraved on the back of

every dollar bill. This motto certainly refers to the new political philosophy and the new political system created here, but America also represents a new order of the ages in the popular conception of what life is meant to be, and how a free and prosperous life could be pursued on this magnificent and bountiful continent. Traditionally, Americans disdain convention and precedent; we are experimental and open in our approaches to solving problems. America's unique cultural traditions have shaped a bold new world whose cultural landscape shows the impact of new ways, new ideas, and the dreams of this "new man" (and woman)—the American.

One of American culture's most enduringly popular and evocative images is that of the road. The open road is symbolic of the personal freedom that range of automobiles just for the Canadian market. Then, in 1965, the United States and Canada agreed to free trade in autos and auto parts. The result was a huge increase in the two-way trade in automotive products. Now, for example, a decision may be made that all full-sized cars for U.S. and Canadian markets will be manufactured at the Canadian plant, and all other models for both markets made in the United States. Canada now produces more than its proportional market share of autos, though less than its market share of trucks. In a sense, this automotive industry agreement was a forerunner of the U.S.-Canadian "Common Market."

The close economic ties between the United States and Canada, together with the evident close cooperation and friendship of the two countries, should not mislead Americans into regarding Canada as a satellite. Canadians are highly conscious of the advantages and disadvantages of living next to the United States. Although there is no doubt that American markets and American capital investments have sped the remarkable growth of the Canadian economy, Canadians are concerned that many vital economic decisions affecting them are made in New York, Washington, DC, Detroit, Pittsburgh, or Chicago rather than Ottawa, Montreal, Toronto, or Vancouver. Former Canadian Prime Minister Pierre Trudeau once likened living next to the United States to sleeping next to an elephant; no matter how friendly and well intentioned the beast, one must remain sensitive to its every move!

is a cornerstone of our national life. The Declaration of Independence proclaimed not only that all people are created equal, but that their unalienable rights include life, liberty, and the pursuit of happiness—and that pursuit is often literally a mobile effort involving traversing great distances geographically and socially.

Political freedom to move about is usually matched with the technological and financial means to do so. One of the first things a foreign visitor, or an American returning after a long absence, notices is the immense scale of things in the country, and this is true of Canada as well. Great multilane, divided, nonstop highways sweep through cities, suburbs, small towns, farming districts, vast deserts, and sprawling mountain ranges—and the traffic seems to pound on, day and night. The world's greatest and busiest network of commercial air transportation links all sizable urban centers. Before major holidays, urban billboards advertise "Home for the Holidays" airline specials, to cities a continent away. For many Americans, the concept "home" entails memories of several places, perhaps in different neighborhoods and suburbs of a single great city or in different states on opposite coasts of the continent. The geographic consequences of this determined mobility and the search for the good life—the pursuit of happiness—have led to large-scale redistributions of population at the metropolitan, regional, and national scales.

Another theme affecting the geography of the United States and Canada is the frontier and the emphasis on expansion—a kind of optimistic national mobility. There are many positive aspects of the appeal of the frontier; many courageous Americans and Canadians endured great hardships in the titanic struggle to settle and develop their immense national territories. But new kinds of frontiers now challenge the imagination and energy of contemporary North Americans—frontiers of managing expanding populations, diminishing resources, decaying inner-city neighborhoods, and sprawling suburbs in the midst of a technological revolution and fast-changing patterns of world trade. The image of an expanding frontier with virtually limitless resources is hardly realistic, but its assumptions persist in many Americans' attitudes toward the environment, both natural and cultural. The "throw-away" mentality is only now giving way to the resurgent conservationist movement and associated recycling efforts.

Americans' careless optimism that more good land and abundant resources lie on the horizon long has been accompanied by a persistent faith in science and technology. Science, we expected, would save us from the consequences of our wasteful exploitation of land and resources, and of our shortsighted pollution of our own environment. Practical applications of science earn Americans' admiration; we like to think of ourselves as preeminently practical people. Our heroes feature creatively pragmatic men such as Ben Franklin, Thomas Edison, and Henry Ford. Another observant Frenchman, Alexis de Tocqueville, in his classic 1835 critique of America, noted that: "In America the purely practical part of science is admirably understood, and careful attention is paid to the theoretical portion which is immediately requisite to application. . . . Americans always display a clear, free, original and inventive power of mind."[3] We make mistakes, but as a people we are willing to experiment with a variety of solutions. This pragmatic, let's-see-if-it-works attitude has also been important in our reshaping of the environment.

Americans take justifiable pride in the flexible integrity of their Constitution, and this combines with an awareness that American society is an ongoing, pioneering, political and social experiment. This awareness, in turn, leads to the assumption that everyone else would and should benefit from adopting our ideas and systems of governance. The full title of the Statue of Liberty, *Liberty Enlightening the World,* says it well: Americans understand America as a beacon of democracy to those unfortunates living in the political darkness of dictatorship and oppression. Americans proselytize on behalf of the virtues of democracy; on occasion, this zeal can be annoying to its recipients. More important, this attitude can lead to unrealistic American expectations in foreign affairs.

Individualism is the salient trait of Americans; it marks the perceptions of society, government, and economic opportunity. Individualism in the pursuit of a home surrounded by attractive scenery and recreational amenities helps to explain interregional migration and regional growth. Individualistic desires for a single-family, detached house in a garden setting propel suburban growth ever outward from cities. Individualism remains a fundamental and essentially positive natural trait. In a remarkably prophetic essay, de Tocqueville summarized American greatness and Russian power:

> The Anglo-American relies upon personal interest to accomplish his ends and gives free scope to the unguided strength and common sense of the people; the Russian centers all the authority of society in a single arm. The principal instrument of the former is freedom; of the latter, servitude. Their starting point is different and their courses are not the same; yet each of them seems marked out by the will of Heaven to sway the destinies of half the globe.[4]

Major Trends

Two significant and interrelated trends will continue to have an important impact on the lives of Americans and Canadians alike: the ongoing redistribution of population in both nations and the technological revolution. The technological revolution, or "second industrial revolution," is a fundamental shift in the employment structure of the two countries. Two associated shifts in employment are notable: the overall increase in service occupations over manufacturing, and the rise of high-tech, research-oriented modern industry along with a decline in the traditional heavy industries in the face of intensive foreign competition.

Heavy industries, such as steel, must be highly sensitive to the locations of raw materials and markets, because both their bulky, low-value-per-unit-weight raw material inputs and relatively heavy products are expensive to transport. Access to cheap transportation is a must; most basic steel mills are located on navigable waterways and are served by railroads.

Location considerations for a plant that manufactures computer components would feature a quite different prioritization. The raw materials this time are not low-value bulk; instead, the silicon wafers that will become computer chips can arrive economically by air freight or truck. Access to navigable water and railroads has little significance. Instead, the ability of modern, high-tech industries to attract highly educated personnel might become the key location consideration.

Population shifts from 1970 to 1980 gave rise to the Rust Belt-Sun Belt scenario. This perception saw the general decline or stagnation of the Northeast and Midwest—the old Manufacturing Belt, or Rust Belt, of obsolescing, mostly heavy industry. Two northeastern states, New York and Rhode Island, were the only states of that period to have a net loss, along with the District of Columbia. The "Sun Belt" of both the Southwest and the Southeast, during the same period, showed large percent gains, topped by Nevada, Arizona, and Florida.

The map of 1980–87 population growth by states, however, shows some signs of a coasts-versus-interior shift (Figure 2–2). Census Bureau estimates show that, since the 1980 census, no seaboard state has lost population while five interior states have sustained net losses. The rates of growth of the Rocky Mountain and Western Interior states have slowed considerably. In Canada, the leading recipients of interprovincial migration were Ontario and British Columbia.

Although amenities can induce migration decisions in advance of finding a job, employment opportunities are the key factor for many people. Are the high-tech and R&D (research and development) industries leading or following interstate migration trends? Industrial geographers John Rees and Howard Stafford studied factors influencing the location of new high-technology plants. Labor skills and availability ranked first, as was true for other types of manufacturing. Quality of life, or amenities, are also an important factor, as are academic institutions, especially those emphasizing scientific and technical training and research.[5]

Quality of life has both physical and cultural aspects. Physical and cultural amenities are in the eye of the beholder. Many beholders agree, however, on seacoasts and mountains, climates supportive of outdoor recreation, and the stimulating intellectual "climates" of urban areas with a varied and sophisticated cultural life and high-quality educational and research facilities. The top 10 metropolitan areas in high-tech job growth, 1972–77, were San Jose (1), Anaheim (2), and San Diego (4), California; Houston (3) and Dallas (6), Texas; Boston (5) and Worcester (7), Massachusetts; Oklahoma City, Oklahoma (8); Lakeland, Florida (9), and Phoenix, Arizona (10).

There are some indications that regions with high growth rates in high-tech jobs are also likely to be leaders in all-industry growth rates. Between 1976 and 1980, for example, New England's high-tech employment grew by 27 percent; its overall industrial growth was 22 percent at a time when the national average was 9 percent.

Every city and region would like to attract high-tech, R&D activities. Three major successes are the "Silicon Valley" (Santa Clara County, south of San Francisco), the Route 128 Corridor around Boston, and North Carolina's "research triangle" formed by the closely spaced cities of Chapel Hill, Durham, and Raleigh. Success in each case is intimately associated with the presence of major research facilities, laboratories, and universities. Excellent transportation-communication facilities, especially access to busy metropolitan or regional airports, are another common factor. A dynamic, richly varied cultural life is an additional common asset, along with generally attractive recreational-leisure time opportunities. The Phoenix-Tempe area would like to be known as the "Silicon Desert;" Florida's southeast coast terms itself "Silicon Beach;" Dallas-Fort Worth counters with a claim to being the "Silicon Prairie"; and the Boulder-Denver-Colorado Springs complex aspires to "Silicon Mountain."

The geographic implications are considerable, as high-tech industries become more important in regional economic growth and population shifts. As stated earlier, low-cost transportation, for example, by waterway or superior rail service, is almost insignificant to high-tech; transport by air freight or truck

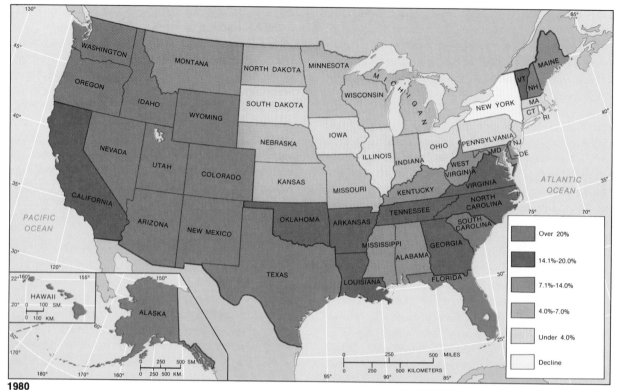

1980

1987

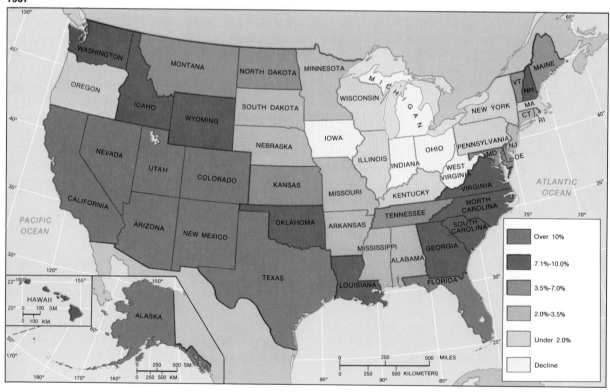

Over 20%
14.1%-20.0%
7.1%-14.0%
4.0%-7.0%
Under 4.0%
Decline

Over 10%
7.1%-10.0%
3.5%-7.0%
2.0%-3.5%
Under 2.0%
Decline

is economical for high-tech products. Ideas, the major product of R&D, are the least costly of all to transport! Scenic, recreation-oriented regions with pleasant climates and reasonable closeness to metropolitan centers with several prestigious universities and a busy airport will be the centers of high-tech, and also regions of net population growth.

THE NATURAL ENVIRONMENT

At the broadest level of generalization, the United States and Canada can be easily regionalized in terms of physical patterns. Mountain systems parallel the east and west coasts, with a large interior lowland in between and a broad coastal plain along the southern and southeastern edges. Climatically, three quarters of the region is well watered; only the southwestern portion of the United States is truly arid (Figure 2–3). In the more humid area, climatic zones run east-west, with progressively longer and colder winters from south to north. In the drier quarter, the climatic zonation is more strikingly oriented north-south, reflecting the effects on climate of mountain systems. Vegetation corresponds to climate in detail except for the extension of grasslands eastward into quite humid areas.

The United States and Canada contain the largest areas of highly productive soil in the world. Soil is one of this region's most important natural resources and one of the major keys to understanding its progress and development. Currently, it is the chief area of surplus food production of the world, and can continue as such with careful management of soils, minimizing erosion.

The soils of the wetter three quarters of the region are fairly heavily leached, meaning that many soluble minerals vital to plant growth have been dissolved and carried away by water. Soils of the drier quarter are characterized by the opposite problem—mineral accumulations near the surface to sometimes poisonous concentrations.

The southern two thirds of the region have many productive soils despite heavy leaching. Essentially, much of the northern one third of the continent has been virtually scraped bare by the continental glaciers. The Great Lakes form the approximate divide between the areas dominated by glacial erosion and glacial deposition. The northern Rockies and the Missouri and Ohio rivers generally separate the area of glacial deposition from nonglaciated areas.

Topography

Along the Atlantic and Gulf of Mexico coasts of the United States, a low, relatively flat coastal plain stretches from the Mexican border, narrowing to the offshore islands at New York's Long Island (Figure 2–4). The Outer Coastal Plain is a flat, sandy expanse that is often poorly drained; for much of its length, it is fringed by sandy beaches and offshore sandbars. The more erosion-resistant materials of the Inner Plain produce a gently rolling landscape and more fertile soils.

The Mississippi River has created a giant inland delta that reaches up to southern Illinois. This delta, called the Mississippi Embayment, is a landscape of meandering streams, temporary lakes, and backswamp that extends for over 100 miles in either direction from the streambed in a huge deltaic plain.

In the Florida arch, gently folded limestone has risen above the sea to form a long, peninsular extension of the plain. The glaciated coastal area extending from New York northward to Cape Cod has a highly irregular coastline, with islands and peninsulas along the continental margin.

The whole East Coast-Gulf Coast region of the United States is fringed by a broad **continental shelf**—shallow water over a surface that is, geologically, part of the continent—which contains some offshore deposits of oil and gas and rich fishing grounds. Whereas the plain widens southward, the shelf tends to widen northward. A line of small falls marks the inland boundary between the softer sediments of the Coastal Plain and the more resistant

← **FIGURE 2–2**
***Population Change for the United States,
1980–87.*** States of the United States experience various rates of population growth (or loss) as the economies evolve and shift, geographically, and as Americans change their perceptions of the relative desirability of regions.

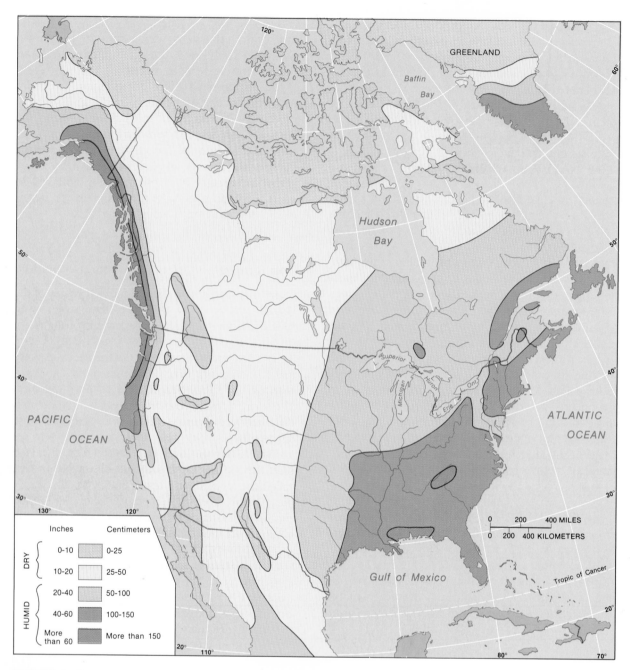

FIGURE 2–3
Annual Precipitation in the United States and Canada. Generally, the
eastern half of the region, and the northwest coastal area, is humid, with a major
moisture-deficient region in the western interior.

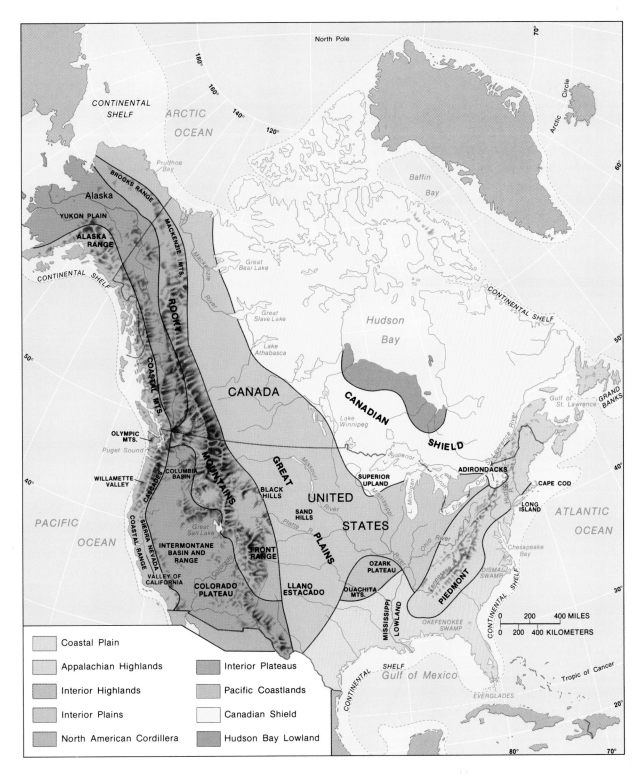

FIGURE 2–4
Land Surface Regions of the United States and Canada. The physiographic
diversity of the United States and Canada—together with the great variety of climate,
vegetation, and soils—provides a rich texture of different environments and natural
resources.

rocks of the Appalachian system. These falls or rapids offered water power potential and blocked upstream shipping; both factors supported urban growth at the fall line, such as at Philadelphia, Baltimore, Richmond, Virginia, and Columbia, South Carolina.

The Piedmont—the foothills of the Appalachians—is built of ancient rocks that have been eroded to a rolling countryside of low hills. The towering Blue Ridge marks the transition to the Ridge and Valley Country, where sedimentary rocks have been folded into a series of parallel ridges and valleys many miles long. The western edge of the folded Appalachians is marked by the Allegheny Front, the eastern edge of the Appalachian Plateau. This plateau has been eroded to form mountainous terrain and contains some rich coal deposits and gas and oil.

Though separated from the Appalachians, the Ozark and Ouachita mountains are of essentially the same age and structure. The Ozarks are a lower, much eroded plateau surface equivalent to the Appalachian Plateau, whereas the Ouachitas and other ranges are folded mountains or fault block mountains. This mountain complex provides an attractive forested recreation area for midwestern and southern residents; like the Appalachians, it has highly developed tourist functions.

The vast level-to-rolling stretches of interior plains and low plateaus are generally referred to as the interior lowland despite the fact that the Great Plains are considerably above sea level.

The Canadian Shield is composed of ancient crystalline rocks that have been heavily eroded by continental glaciation, exposing a bare rock surface. Dotted with lakes and bare rock hills, the shield is rich in metallic minerals and has enormous hydroelectric potential. A few areas of former lake plain have clay soils capable of supporting farming.

The Adirondacks of New York are a detached segment of the Canadian Shield, connected to it by an axis that underlies and forms the Thousand Islands of the St. Lawrence. The Adirondacks are flanked by lowland troughs that extend to Lake Ontario and the St. Lawrence at Montreal.

South of the shield, the interior lowland is covered with glacial deposits of varying thickness and texture. Gravelly hills of glacier-deposited debris mark varying stages of glacial retreat. Most of the surface, however, consists of broad plains formed under temporary meltwater lakes; poorly drained areas are quite common.

With a few exceptions, mainly in Kansas, the area south of the Ohio and Missouri rivers remains unglaciated; eroded by streams, it is rather more like hill country at its southern margins.

Under the High Plains, really a sedimentary plateau, which reach elevations of up to 5,000 to 6,000 feet, sediments arranged in horizontal sheets stretch eastward from the Rockies. Most of the high-lying surface remains intact as a broad, level plain; the heavily eroded Badlands of the Dakotas are among a few stark exceptions.

Younger by far than the Appalachians, the western mountains widen from north to south. There are essentially two main mountain chains, the Eastern and Western Cordilleras, separated by rough plateau country north of the Columbia River and a vast disrupted Great Basin to its south. The westernmost range is flanked by a series of depressions paralleling it, including Puget Sound and the Central Valley of California, which in turn are paralleled by low coastal hills. There is little coastal plain except in northern Alaska. The Great Basin is largely a desert.

The Rockies, of the Eastern Cordillera, developed when a tremendous uplift occurred. On the eastern edge of the Rockies, this disturbance resulted in two parallel ridges of fault block mountains running north-south: the Front Ranges. Their abrupt rise from the level Great Plains adds to their visual grandeur. The Rockies south of Yellowstone National Park consist of alternating plateaus and basins. Some of the plateaus have impressive, glacier-sharpened peaks, and the largest of these basins, the Wyoming Basin, is underlain with rich fossil fuel deposits. The southern Rockies are narrower and are composed of lava flows incised with deep valleys.

The Arctic portion of the intermontane ("between the mountains") plateaus in Alaska and the northern Yukon is a low-lying eroded surface that is hilly rather than mountainous. South of the U.S.-Canadian border, broad lava flows spread out over almost the entire Columbia basin.

The southern, largely desert portion of this intermontane basin is the Great Basin; here fault block mountains, oriented roughly north-south, alternate with sediment-filled valleys. Salt flats are the remains of old lake beds; large lakes once occupied much of Utah and Nevada during wetter climatic periods. The Great Salt Lake is the residue of a once much larger freshwater lake, Lake Bonneville.

In the southwest is the Colorado Plateau, composed of thousands of feet of horizontally bedded sedimentary rocks. The Colorado River has incised its Grand Canyon, exposing millions of years of geologic history while sculpting a fantastic and beautiful landscape.

In Alaska and British Columbia, the westernmost coastal ranges appear only as a series of peninsulas and offshore islands, while the valleys are submerged. South of Vancouver, three alluvium-filled lowlands separate the Western Cordillera from the coastal ranges.

In the Pacific Northwest, the Western Cordillera is known as the Cascades, where glacial erosion of volcanic cones has created snowcapped craters. The Sierra Nevada, mainly a series of large block mountains, is quite steep along its eastern side; the western side is known for its ice-carved valleys, falls, and canyons, such as Yosemite.

The Hawaiian Islands are a chain of volcanic peaks that have risen above the sea floor. The mammoth outpouring of lava here is associated with the sliding of the Pacific plate over a hot spot.

Climate

The U.S.-Canadian region has no truly tropical areas except for Hawaii and the very southern tip of Florida. Much of the region is humid, receiving adequate moisture, and continental, receiving limited moderating influences from the ocean (see Chapter 1, Figure 1–17). And most of the region has four well-defined seasons.

The southeastern one fourth of the United States has a subtropical climate with year-round distribution of rainfall and a 9- to 12-month frost-free season. Only a few miles inland from the sea, the moderating effects of the ocean are largely lost; large seasonal and daily temperature ranges reflect the **continentality** of an interior location.

The northeastern quarter of the United States and Canada's St. Lawrence lowland experience a more intense winter. Because of the prevailing westerly winds and continentality, the East Coast and the Canadian Maritimes experience some oceanic moderation, but the rigors of winter intensify rapidly inland. Summers are moderately long but grow rapidly shorter north of the edge of the Canadian Shield. The northern two thirds of Canada and most of Alaska experience long, cold winters and short, cool summers. The Arctic fringes of the entire continent have very short frost-free seasons, forming the large area of **tundra** (treeless, slow-growth vegetation).

A broad swath of **steppe**—semiarid grassland—fronts the eastern side of the Rocky Mountains south of the 55th parallel of latitude and west of the 100th meridian of west longitude. This grassland area experiences hot summers almost throughout its length, and intensely cold winters except south of Kansas.

West of the Rockies, the land becomes a desert south of the 40th parallel; however, occasional thunderstorms mean that there is not a total absence of rainfall. From northern California north, the west coast of the continent has cool, rainy weather all year round—a marine west coast climate. In the center and south, California has a Mediterranean climate with parched summers and moderate winter rains.

Almost three fourths of the United States has sufficient average rainfall for unirrigated agriculture, and virtually all of it has sufficient warmth for some agriculture. Exceptions occur in the highest mountain areas and in most of Alaska where winters are long and bitter. Canada is less favored; over two thirds has a very limited or almost no growing season. Overall, however, the region is one of great potential for agricultural productivity.

Drainage Patterns

The Mississippi-Missouri-Ohio system is one of the world's most important navigational systems (Figure 2–5). It drains the richest agricultural portion of the United States as well as the western half of the Manufacturing Belt. Although the system is not suited to oceangoing traffic beyond Baton Rouge, Louisiana, armadas of barges ply its waters carrying bulk materials. The western, drier, portion of the basin is important primarily for irrigation rather than for navi-

FIGURE 2–5

Major Inland Waterways and Rail Routes in the United States and Canada. The economies of both nations have benefitted from the extensive networks of natural and improved waterways and the relatively dense rail net.

gation. Some Appalachian tributaries, such as the Tennessee, have been developed as multipurpose water projects used for hydroelectric generation and recreation.

The St. Lawrence system, in contrast, is open to navigation by moderate-draft oceangoing vessels over its entire length. It serves the Canadian industrial heartland and the huge industrial cities strung along the Great Lakes; navigation is its most important function despite its being icebound for three to five months each year.

The Columbia-Snake basin generates proportionately the largest amount of hydroelectric energy of all the basins. Also, a vast area in eastern Washington and Oregon and large parts of southern Idaho are irrigated by the waters of this system. The Colorado is the master stream of the entire Southwest, the largest in drainage basin and in importance; its waters are heavily used for irrigation.

In general, the most important use of western rivers is irrigation, whereas eastern rivers are used primarily for industrial and navigational purposes. Despite such major power projects as Grand Coulee and Hoover dams in the West, the amount of electricity generated by rivers in the eastern half of the continent is much greater. This fact is particularly evident in Canada, where large, new hydroelectric projects in Quebec and Labrador generate an enormous power supply in the eastern portion of the Canadian Shield.

THE STRUCTURE OF SUBREGIONS

At the scale adopted in this book, the U.S.-Canadian region is conveniently divisible into 11 subregions, 7 in the United States and 4 in Canada (see Figure 2–1).

THE UNITED STATES

The American Core

The northeastern quadrant of the United States contains the core of the American population and economy and includes four distinctive geographical areas:

New England, the northeastern Megalopolis, northern Appalachia, and the midwestern heartland.

New England. New England's cultural heritage includes an unusual initial settlement pattern — the **town** rather than relatively isolated family farmsteads. Early New England settlement was carried out by groups rather than by individual pioneers. These groups, organized prior to settlement, planned a compact community so that farmers would live close to one another and together support a church and a school. Larger, outlying plots of farmland required farmers to commute daily for most farm activities. The town provided a **common** — a central area where farm animals could be conveniently pastured close to home, generally for a short time. Boston Common, for example, survives as a park.

New England's Puritan heritage can be seen in the cultural landscape. Old villages of clapboard houses of simple design, painted white, focus on a Congregational church. The church itself is an expression of Puritan theology; it is classically proportioned with its towering spire but plain rather than decorated. The Puritans were reacting to their perception of the established church as so caught up in worldly pomp as to have forgotten the central points of the gospel message, and so Puritan churches had no stained glass, elaborate vestments, or interior decoration. Another regional trait associated with the Puritan heritage is the alleged frugality of New Englanders; avoiding waste was pleasing to God as well as sensible in a physical environment marginal to the production of most grains and other basic foods. Contemporary New England, with its more recent inflows of Irish, Italians, and French Canadians, is more typically Roman Catholic than Puritan.

Natural Resources and Population Distribution. The nature and qualities of the core's resource base have played an important part in regional development, both by their presence and absence. New England has a very short list of natural resources yet was in the forefront of American industrialization. Although, after centuries of exploitation, three quarters of the land is again forested, the slow-growing trees are more important for pulp and paper than for lumber. New England has no fossil fuels (coal, petroleum, natural gas), though offshore exploration holds promise.

THE "EUROPEAN INVASION" AND THE LAND USE REVOLUTION

The European invasion of North America produced profound changes in the human ecology of all regions; New England was historically among the first to be densely settled by Europeans and thus serves as a good example. American Indians had made a relatively light impact on the land and its resources. Living at a Stone Age technology level, native Americans had achieved a harmonious balance with nature. The men specialized in hunting and fishing, and women tended the small, scattered plots of vegetables and corn. Small areas were cleared for agriculture by girdling trees or by controlled burning; corn, pumpkin, squash, and beans were all grown together. Land was not fenced but held by tribes and clans rather than by individuals, who owned only a few simple personal possessions (tools, jewelry, clothing). The relatively light Indian population densities seldom harvested wild game or fish to the point of permanently reducing their numbers.

All was to change with the arrival of the Europeans, because not only was their technology much further advanced, but they were part of a rapidly expanding global economy. The Europeans had an almost insatiable demand for trade items such as beaver pelts. Beavers were almost exterminated, and in the process former beaver ponds were drained as beavers no longer maintained their dams. The drained ponds made lush meadows for the newcomers.

Land clearance for farms reduced the habitat of the deer on which the Indians were highly dependent. Worse, the Europeans also hunted deer, but with guns. Still worse, from an Indian viewpoint, the Europeans introduced cattle and pigs, which they allowed to roam the woods to compete with deer for food; but the Indians were not permitted to kill cattle and hogs. In the bargain, the Europeans regarded Indian men as lazy playboys because they hunted and fished, which the Europeans regarded more as sport, while Indian women did the "serious work" of farming. Trampling of vegetation in the woods by cattle and hogs, combined with exposed, cultivated soil in the fields, led to serious erosion problems as the Europeans completed their "taking up of the land." The clearing and fencing of land on a large scale, with an insistence on private property rights, was alien to Amerindian concepts of ownership, and the native Americans must have been horrified by the ruthless exploitation of the land and its resources by the white settlers. The European invasion produced a real revolution in the human ecology of the United States and Canada.

The natural resource base of the Middle Atlantic coast and neighboring Piedmont of the Appalachians is only slightly more impressive. As in New England, the forest resource was more important in the past than it is in the present. Small iron ore and copper resources were largely worked out by the early twentieth century; natural gas and oil offshore have just been discovered and are not yet in full production. This area's major resource is its situation on the seaboard, encompassing the historic heartland and with ready access to the coal and other minerals of northern Appalachia. As in most other regions, the superior railroad system serving the densely settled portions of the United States and Canada greatly enhances the situational advantages (see Figure 2–5).

The steam-powered early maturity of the industrial age in America was fueled first by the coal resources of Pennsylvania; later, exploitation moved westward to West Virginia, Ohio, and Illinois, supplemented by coal from the Upland South (Kentucky, Tennessee) (Figure 2–6). The first oil well was drilled in western Pennsylvania in 1859, and that state remained the largest producer until 1895. Although Pennsylvania, New York, Ohio, Indiana, Illinois, and Michigan all produce some oil and/or gas, their production is minimal compared to their present consumption (Figure 2–7, p. 102). The Adirondacks of New York have long produced iron ore, as has eastern Pennsylvania, but the iron that supported the huge expansion of America's industrial output in the late nineteenth and early twentieth centuries was from the Upper Great Lakes (see Figure 2–6).

It is in population distribution that the Core is clearly dominant; the list of major cities is the longest of any American subregion and illustrates the industrial power concentrated here. Although densely built-up urban areas occupy a minute fraction of the total area, their low-density, sprawling suburbs tend to reach out toward their neighbors, creating the appearance of city strings or constellations. Most of the people, money, and productivity of the area are concentrated in these enormous urban-dominated regions.

The great cities of the Core were primary benefactors of the great age of European migration to the New World, giving rise to the strong ethnic flavors of their neighborhoods. Heavy immigration in the early twentieth century tended to flow into the great ports and industrial cities, because the agricultural frontier was then closing. The source areas of migrants now have shifted from Europe to the rural South, Appalachia, and the Caribbean. Once in place and adjusting to the urban-industrial society, the new arrivals begin to plan their "escape" to the less crowded outer rings of urban-suburban expansion. The move to the suburbs has been an American dream for generations, but only since World War II has it become possible for the great majority of the middle class to leave the old city behind for the newer type of low-density city we call suburbia.

The Northeastern Megalopolis. To anyone flying over the area on a clear night, the northeastern seaboard of the United States presents a dramatic spectacle. Great starbursts of light mark the cities. Bright webs of light outline the chains of smaller cities, towns, and suburbs that almost link the metropolitan centers from Boston to Washington, DC. Although this great urban-dominated region still contains a great deal of open space, the interconnection of cities is evident. A composite of nighttime satellite photography shows the northeastern Megalopolis with stunning clarity; it also shows that other city-chains or urban regions are evolving.

Megalopolis was the name applied by geographer Jean Gottmann in a landmark book published in 1961.[6] It described a new phenomenon—a giant urban-dominated region in which a series of great and small cities were growing toward one another along their dense webs of transport arteries. The northeastern Megalopolis, prototype of others to follow, was first identified as the region stretching from metropolitan Boston's New Hampshire suburbs southwestward to metropolitan Washington, DC's Virginia satellites (Figure 2–8, p. 103). Gottmann did *not* mean a continuous urban region. Because urban-suburban development tends to follow highways, many of which radiate outward from the urban centers, there is a large area of open space between the spokelike highways. Between New York and Hampton Roads (the mouth of Chesapeake Bay), the more

NEW ENGLAND AND THE "SECOND INDUSTRIAL REVOLUTION"

New England can be characterized as resilient in its continuing adaptation to a changing and more competitive industrial world. This area has numerous small rivers, which were easily dammed in the early days of the industrial revolution to provide direct water power. New England had the critical availability of investment capital thanks to its fleet of trading vessels, and it had a relatively large population, forming both a market and a potential labor supply. Textile machinery was fairly simple at first, and most women already had developed relevant skills at home in spinning, weaving, and design. When industry switched from water power to steam, the region imported coal. Eventual obsolescence affecting plants, locations, machines, and transport systems later combined with high taxes, unimaginative management, and unionized wages to favor a shift of the textile industry to the southern Piedmont. By the 1940s, the South led in low-value-added goods, and New England's remaining factories specialized in fancy and special-purpose goods.

The new and growing industries of the area are in electronics fields. Here, skill is important; access to research labs, libraries, and university consultants may be critical, and high inputs of metals and energy are unimportant in proportion to the weight or value of the final product. **Value added**, the difference in cost of raw materials and finished goods due to labor and skill expenditures, is a common measure of the importance or regional impact of manufacturing. High-value-added industry is based on human skills and engineering more than on energy, a chronic deficit requiring expensive imports for New England. New England's prestigious universities, especially in metropolitan Boston, have become a major factor in attracting high-tech industry.

Western New England early developed a metalworking specialization. Colt revolvers were a typical early product. Weapons, aircraft engines, scientific instruments, and tools are still important here. Machine tools—machines that make other machines—exemplify the products of an established industrial region with accumulated skills. Nuclear submarines, produced at New London, Connecticut, are both a holdover from shipbuilding days and an example of the application of fine skills in demanding, precision work.

concentrated development does not fringe the coast but parallels it 50 to 100 miles inland. Large "empty" or lightly populated spaces exist in southern New Jersey's pine barrens, the rural Delmarva (Delaware-Maryland-Virginia) peninsula, the Appalachians to the west and north, and the Adirondacks.

Most of Megalopolis's major cities are seaports; their historic growth impetus was to connect the complementary economies of the developing American Core with the rest of the world, especially with Europe. Although New York, Philadelphia, Baltimore, and Boston are still great ports, their industrial

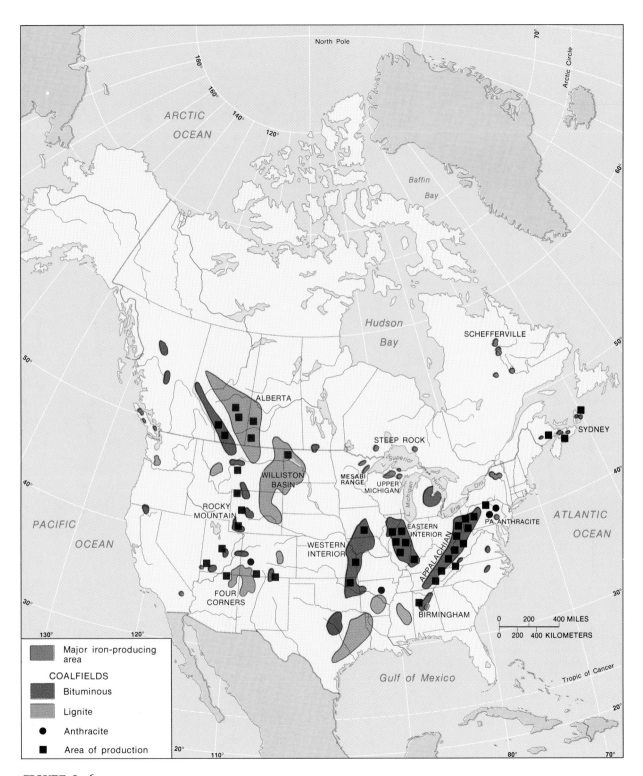

FIGURE 2–6
Major Coalfields and Iron Ore Areas in the United States and Canada.
Coal and iron ore are two critical minerals in the development of modern industrial states. Note the geographic relationships of this map and Figure 2–10.

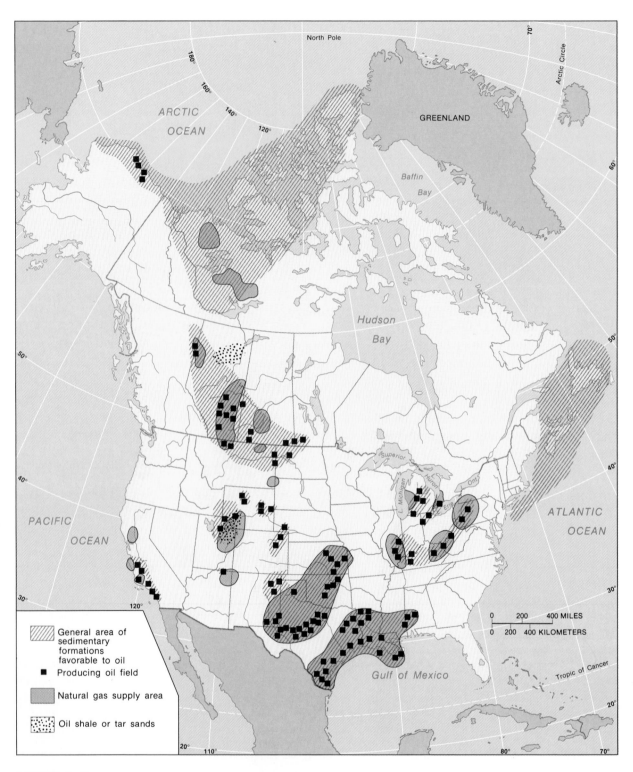

FIGURE 2–7

Important Oil- and Natural Gas-Producing Areas and Prospective Areas in the United States and Canada. Although the United States is an important producer of petroleum, it must import large volumes to meet its soaring demand; Canada is a major supplier of oil to the United States. Oil shales and tar sands may become exploited on a large scale in the future.

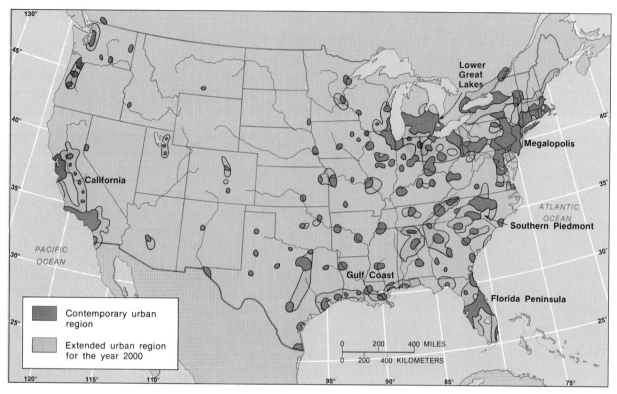

FIGURE 2–8
Urban Regions of the United States, 1980 and 2000. The continued outward
growth of metropolitan regions suggests further evolution of huge metropolitan
areas not only in the Northeast but in the Upper Midwest and West Coast as well.

functions are more important, and their service functions still more so.

Gottmann's epic study also identified a new category of economic activity to explain megalopolitan functions. Traditionally, there were three employment categories: **primary**—production of raw materials (agriculture, mining, forestry), **secondary**—processing raw materials (manufacturing), and **tertiary**—all services (e.g., distribution, retailing, entertainment, medical and other professional services, administration). The third, or services, category was much too broad. It included persons performing routine services that require relatively brief training and little intellectual creativity, such as retail clerks, waiters, and business machine operators. It also included the highly educated, specially trained professionals and executives who give advice, make

decisions, or educate, such as lawyers, accountants, medical personnel, teachers, and executives of all kinds. Clearly, these two groups are different in fundamental ways, and so Gottmann labeled the decision makers the **quaternary** (fourth) category, noting Megalopolis's significance in this employment category. Quaternary activities are increasingly important in the northeastern Megalopolis.

For half a century, light industries, warehousing, wholesaling-retailing, entertainment, and even office functions have been moving toward the outward-growing edges of the metropolis. Even earlier, space-consuming, polluting industries (oil refineries, chemical works) had sought peripheral locations.

A return flow of economic functions and affluent residents now is evident in the cores of many large cities. Urban revitalization is an encouraging fact,

MEGALOPOLITAN DYNAMICS

T he low-density sprawl of contemporary North American cities is intimately related to changes in transport technology and associated changes in lifestyles. For most of urban history, cities tended to be compact with high-density development, owing to the costs of commuting. Urban historian Sam Warner characterized the major growth stages of Boston: a "walking-scale city," very compact, with most people living near their work; the "streetcar city," with linear, compact strips of outward growth as streetcars reduced commuting costs; and, finally, "automobile cities," the filling in of low-density areas between transit lines, with a sprawl much further into the countryside.*

Megalopolitan dynamics emphasize the advantages of locations on transport axes between cities. Accessibility to market is one of the most powerful locational factors for modern industry. Metropolitan areas are more than markets, though. Their pools of skilled, semiskilled, and unskilled labor; the range of services available; the presence of related industries; and their commanding positions within transport webs are additional advantages to locations close to them.

*Sam Bass Warner, Jr., *Streetcar Suburbs: The Process of Growth in Boston, 1870–1900* (New York: Atheneum, 1971).

from Hoboken to Pasadena, as the prime asset of accessibility outweighs obsolescent buildings and onetime rundown surroundings.

Megalopolis's linear series of cities essentially doubled the incentives inherent in suburban location; a position between two big cities, astride the transport routes connecting them and accessible to both of their markets, along with other locational advantages, was nearly ideal for many economic activities.

This accelerated growth in the "connector zones" between cities tends to snowball; that is, development and growth attract more development and growth. The original megalopolis of the northeastern seaboard, sometimes called "Bos-Wash" (Boston to Washington, DC) is the prototype of more to come. A southern Great Lakes to Appalachia megalopolis, "Chi-Pitts" (Chicago-Detroit-Cleveland-Pittsburgh), is

developing (see Figure 2–8). Growth trends indicate the possibility of future supermegalopolises or city-chains connecting the Great Lakes and northeastern megalopolises via both the Ontario Peninsula of Canada and the southern Lake Erie shore and Mohawk Valley-Hudson Valley. Interestingly, the original megalopolis has a parallel, smaller version of its growth dynamics along its seashore fringes, particularly in New Jersey, where a series of resorts are growing toward one another in a "leisureopolis."

A major business of Megalopolis is decision making; the area has far more than its proportionate share of "brain power"—government and business executives, research scientists, and other highly skilled and highly educated people. It serves as the headquarters for a third of the 500 largest corporations in the United States, contains the national capital, and has half of the largest banks in America.

A serious problem for an urban area's "growing edge" is the set of pressures placed on farmers and farmland on the suburban fringe. For the great majority of cities, their original founding and much of their growth are partly attributable to the quality of agricultural resources in their vicinity. The physical spread of the city is almost always at the expense of good-quality farmland. Because residences, shopping malls, and industrial parks can pay far more for land per acre than can farmers, not only does suburban growth make expansion of individual farms prohibitively expensive, but farmers are pressured to sell out by rising prices and, especially, rising taxes.

Agriculture varies widely in its specializations, productivity, and prospects throughout the region. New England's peak farm population was reached 150 years ago; many rural areas steadily lost population for a century before the overspill of nearby towns or cities reversed the trend. Cultivated fields gave way to hay or pasture because New Englanders could not compete with midwestern grain production. Improving transportation systems, coupled with easily worked soils, gave economic advantage to Midwest farmers at the expense of New England farmers. The New England physical environment—summers short, cool, and damp, soils thin and stony resulting from glacial deposition—is unfavorable to grains. The farm economy has shifted to dairying and such specialty crops as potatoes in northern Maine, cranberries on Cape Cod, and poultry in New Hampshire and Maine. The Megalopolis nearby is a good market for fresh milk and fresh fruits and vegetables (Figure 2–9).

Before refrigeration and in the infancy of canning, truck farmers and producers of highly perishable fresh fruits and vegetables had to be close to markets to survive. Land use, whether in rural or urban areas, reflects land values and the relative ability of potential users to compete for the most attractive or accessible locations.

High-value crops are essential for farmers on the urban fringes, because they must frequently pay high real estate taxes reflecting the potential market value of their land rather than its agricultural productivity. Suburban areas constitute a large market for young trees and shrubs for transplanting, as well as for cut flowers, flower bulbs, and other horticultural specialties. This market occurs around any growing met-

Nighttime Satellite View. *This nighttime satellite photo of the eastern United States clearly shows the interconnected points of light that outline the sprawling metropolitan areas of the northeastern megalopolis.*

ropolitan area, not just Megalopolis. Poultry are important in the Delmarva peninsula, eastern Connecticut, and eastern Long Island.

Northern Appalachia. The farms of Appalachia have traditionally been rather small. Both northern and southern uplands have had cropland go out of production in the more general farming and livestock areas of the mountains. There are many specialty areas such as the grape, fruit, and vegetable area of New York's Finger Lakes and the apple orchards of the Great Valley of the Appalachians from Pennsylvania to Virginia. One of the most productive and picturesque areas of general farming, with its variety of crops and livestock, is in a three-county area of southeastern Pennsylvania known for the Pennsylvania Dutch—really "Deutsche" or German in ancestry—and particularly the Amish. Shunning

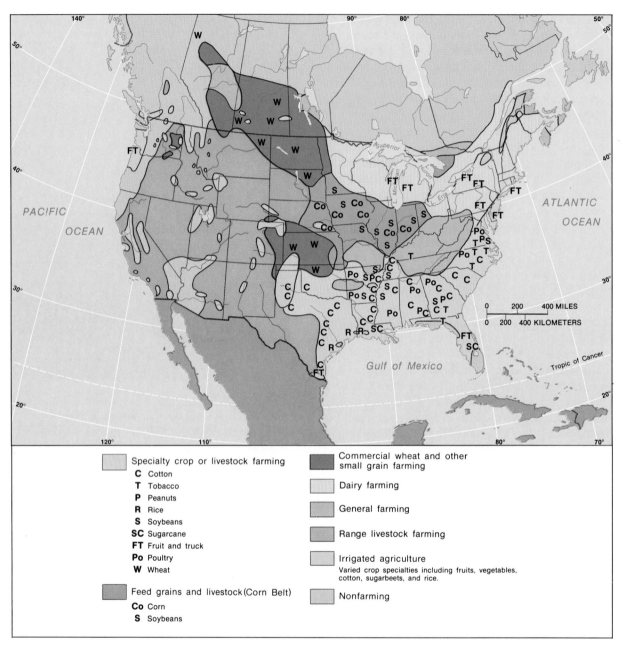

FIGURE 2–9
Agricultural Regions of the United States and Canada. The diversity of
natural environments, in combination with access to markets, has contributed to the
present generalized pattern of specializations.

Legend:

Specialty crop or livestock farming
C Cotton
T Tobacco
P Peanuts
R Rice
S Soybeans
SC Sugarcane
FT Fruit and truck
Po Poultry
W Wheat

Feed grains and livestock (Corn Belt)
Co Corn
S Soybeans

Commercial wheat and other small grain farming

Dairy farming

General farming

Range livestock farming

Irrigated agriculture
Varied crop specialties including fruits, vegetables, cotton, sugarbeets, and rice.

Nonfarming

many modern conveniences, these people have created a charming cultural landscape. Pressures of megalopolitan fringe expansion and large-scale tourism are countered by an extremely conservative, thrifty, and closely cooperative society intent on maintaining its traditions.

The Midwestern Heartland.

The Midwest is the premier agricultural area within the subregion. Without the prodigious outflow of corn, hogs, cattle, wheat, soybeans, and other farm products from this area, the United States would be far less well fed. There are other vital agricultural regions, of course, but the American heartland is the most productive in the world.

The changing nature of farming is nowhere better illustrated than in this classic agricultural region. The Midwest, particularly in its western portion, has a higher proportion of regional income from farming than any comparably sized area in the United States.

A survey of trends in American farming shows an industrialized agriculture with huge investments in machinery. Mechanization has increased productivity per acre but has also greatly increased the capital requirements of commercial farming. Industrialization has also affected the farmer by providing efficient chemical fertilizers, cheaper irrigation technologies, and pesticides that are generally efficient if controversial. Scientific breeding of plants and animals has produced wheat that is less susceptible to diseases, corn that grows faster, vegetables less vulnerable to damage by mechanical picking, cattle that gain weight faster, hogs with less fat, and turkeys with bigger breasts. All of this technology requires that modern farmers respond to information on plant and animal genetics, soil chemistry, soil-water balance management, weather forecasting, marketing, and financing. There is a growing list of federal regulations and programs that presents opportunities as well as frustrations. Mechanization makes larger farms possible, and using mechanization efficiently requires large land units.

Much of the current crisis in farming is the result of the interaction of some of the trends just noted. Many farmers, struggling to grow large enough to be economically viable, took on a heavy debt burden at a time when land values had peaked, in the 1970s.

Heavy mortgage payments, along with debt on machinery, proved almost impossible for many to meet.

Manufacturing.

The Core corresponds closely to the traditional American Manufacturing Belt (Figure 2–10). Certainly, there are other manufacturing regions, but the North American Manufacturing Belt, which also includes portions of Ontario and Quebec, still contains the majority of total U.S. factory production and over half of its 20 largest industrial centers (Figure 2–11). Manufacturing here is found in small cities and towns as well as in metropolitan areas; major centers such as Detroit or Chicago are simply large nodes in an industrial matrix.

The old industrial cities of the Manufacturing Belt have been stereotyped as the "Gritty Cities," and the whole region has been characterized as the "Rust Belt." As with other stereotypes, these negative views are not uniformly true, if they ever were. The northeastern Manufacturing Belt cities are, however, mostly those that emphasized industries presently under intense pressure from foreign competition—steel, tires, autos, railroad equipment, and heavy machinery.

The media have popularized a Rust Belt image of soot-stained old snow mantling an abandoned steelworks, contrasting it with a picture of burgeoning high-tech industries in palm-lined industrial parks in the Sun Belt. The heavy-industry focus of the older industrial centers meant that cheap water transport of bulky raw materials was a necessary locational consideration. The Gritty Cities are almost invariably ports—seaports, lake ports, or river ports. Philadelphia, Detroit, and Jersey City are good examples. Each of the industrial centers had, and has, its distinctive array of specialties related to its locational advantages. Whether these centers are in the Rust Belt or Sun Belt, their prosperity depends on two key factors: diversification and flexibility.

The New York-northeastern New Jersey area was based more on transportation and market than on local raw materials. This area also enjoyed excellent location relative to western European markets. The 1825 opening of the Erie Canal gave New York an important advantage in serving a developing Midwest hinterland, although the canal was more important in handling regional exports of agricultural pro-

FIGURE 2–10
Manufacturing Regions and Districts of the United States and Canada.
Industrial specialties reflect the influences of markets, resources, transport, labor, energy, and historical factors.

FIGURE 2-11

Shifts in Manufacturing Employment. An important feature of ongoing economic change has been the decline of the former concentration of industrial employment in the classic Manufacturing Belt as industry shifts south and west.

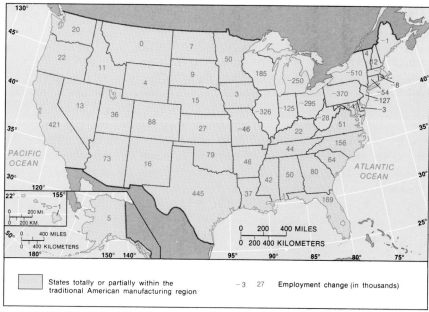

States totally or partially within the traditional American manufacturing region

−3   27   Employment change (in thousands)

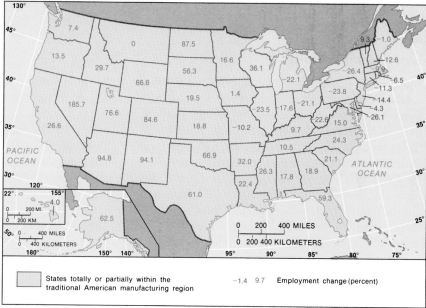

States totally or partially within the traditional American manufacturing region

−1.4   9.7   Employment change (percent)

duce than in handling imports. The region is quite diverse, with only primary metals absent, and the huge market and port functions combine to support a wide array of consumer goods and import processing. Paterson-Passaic, Jersey City-Hudson County, and Newark-Elizabeth were all early industrial centers in neighboring New Jersey, and New York continues to decentralize outward to both New Jersey and southwestern Connecticut.

Metropolitan Chicago funnels the exchange of imports and exports between much of the U.S. West and Southwest with the East and the rest of the world. Early industries were based on agricultural processing and were related to Chicago's transport center functions. Chicago's concentration of transport junctions is highly attractive to industry. The St. Lawrence Seaway has made seaports of Chicago and its lakeshore neighbors. Connected by a canal with the vast

Mississippi River system, Chicago is a major convergence point for waterway, rail, and highway traffic, and it may have the busiest airport in the United States (Atlanta is a close contender). Diverse products include iron and steel, appliances, farm machinery, transportation machinery, machine tools, and processed foods.

The Pittsburgh-Cleveland heavy-industry section extends from the Pittsburgh suburbs through Youngstown and Akron to Cleveland. It has added related specialties such as chemicals from coke and coal. Heavy equipment, such as mining machinery, and heavy trucks are manufactured to meet the demands of industrial customers. Machine tools have advanced, partly taking up the employment slack of declining, obsolescent steel mills and railroad equipment shops. Old, inefficient steel mills proved unable to compete with new plants at tidewater locations which were accessible to the increasingly significant imported ores.

The Dynamic South

Southerners tend to be highly conscious of being southerners, and non-southerners also are acutely aware of the special qualities of the environment, both physical and cultural, that identify and characterize this region. Non-southerners frequently are amazed at the persistence of southern memories of the Civil War. There is no doubt that the war was traumatic in its physical destruction. The transport infrastructure—railroads, bridges, port facilities—was largely destroyed, and entire cities were left heaps of smoking rubble. Of more lasting consequence, however, was the loss of national political power. Not until 1976, with the election of Georgian Jimmy Carter, could a southerner win the White House, in contrast to the pre–Civil War era, when 9 of the nation's first 12 presidents were southerners. A common southern viewpoint was that the North, or at least some northern capitalists and politicians, treated the South like a colonial territory, and there was some justification for that perception up until World War II. The self-consciousness of southerners can be observed in displays of the Confederate flag and in the pride taken in southern popular culture, such as gospel music, the old-timey melodies of the southern Appalachians, and regional food specialties.

If modern Yankees were to drive south, heading perhaps for the Florida resorts, what visual clues would indicate when they were truly in the South? Some travelers might accept state boundaries as regional boundaries—if we've crossed into Virginia, we're in the South. But at what point does the landscape look obviously "southern"? Many details of the physical environment, such as soil color and texture, are not always observable at 55 miles per hour (or more!) on the interstate. It is the cultural landscape that offers the best and most readily apparent evidence of such a distinctive region.

Traditional southern houses, for example, are more likely to have porches than are northern houses, reflecting the pre-air–conditioning need for a shady retreat from summer heat. Many rural southern houses were raised off the ground on bricks or blocks to allow air circulation underneath, a sensible building technique in a land of long, hot, humid summers. Unlike their New England counterparts, older southern houses usually have an exposed chimney, built outside the wall so that the brickwork or stonework is observable from ground level to chimney top. New England design enclosed the chimney so that only the part projecting above the roof could be seen from the outside. In colder New England, the purpose was to retain as much heat from the fireplace as possible; in the South, the object was to minimize house heating by summer cooking fires.

The religious geography of the South shows Baptists predominant, with ethnically related islands of Roman Catholicism in Louisiana reflecting the French heritage there, and in Southern Florida, with its concentration of Cuban-Americans. The Baptists spread quickly during the early settlement period because they were not hampered by a hierarchical structure with rigid training requirements. Any person who sincerely experienced a "call" to preach did so, and energetic and dedicated, if not seminary-trained, clergy spread the word effectively.

The Rise of the New South. The demands of World War II on the economy greatly accelerated the emergence of the New South. Industry, particularly

Chicago Board of Trade.
Cities develop complex interactions with their hinterlands, or trade-service areas. Chicago grew partly to serve the richly productive Midwest farm region. Agricultural commodities are traded at Chicago's Board of Trade.

textile manufacturing, had already begun the move south before 1917, attracted by milder climates with lower heating costs, cheaper labor, a lower cost of living, and a varied resource base. The southern pine forests became more valuable when a new sulphate process made it possible to produce better-quality paper from the trees by dissolving out the tarlike gums in their sap. The South has a longer growing season than the North, and genetically engineered trees reach pulp market size in as few as 10 or 12 years. The phosphate mines of Florida provide material for fertilizer and chemical use; reserves are adequate for many decades, and some is exported to Japan. A series of major oil and gas fields has been found in the western portion of the region, including the great East Texas Field, the largest ever found in the United States. Offshore wells tap a huge series of fields extending from the Mississippi Delta west and south to the vast new offshore fields of Mexico.

The population of the South long remained largely rural and small-town, reflecting the long regional dominance of agriculture and slower industrialization. A consequence was a prevalence of poverty and an outward flow of people, particularly from severely eroded and impoverished rural areas. These internal refugees from declining mining areas or mechanization and other agricultural revolutions sought better lives in the cities, as had earlier generations who left farms in other areas.

One of the greatest migrations in history had been sustained (with few temporary reversals) by African-Americans responding to a powerful set of push-pull factors. Among the "push" or negative factors were decreasing demand for agricultural labor, inadequate housing, deficient social welfare programs for the poor of rural areas, and generally bleak prospects for improving life. "Pull" or positive factors included access to industrial employment, superior-quality housing in cities (even in slums), less pervasive discrimination in jobs (particularly federal government jobs), more generous social welfare programs, and a generally optimistic outlook for one's children. It must be remembered that southern whites responded to the same set of attractions and motivations, minus most of the discriminatory patterns.

The South's metropolitan areas have grown impressively, partly because that region 20 years ago, with its slow start in industrialization, still had less than its share of truly large cities. Most big cities of the South were peripheral to the region; they were "hinge cities" connecting the subregional economy of the South with the Northeast (Baltimore, Washington, DC), the industrial heartland (Cincinnati, St. Louis), or the West (Kansas City, Dallas, Houston). New Orleans was a large city primarily because it functioned as a port for much of the highly productive Midwest, not simply as a southern port. Similarly, Miami was *in* the South but not really *of* it, its citizens and functions being quite atypical of the region. Atlanta is the best example of a truly southern city; it functions as a regional center for much of the Southeast, just as Dallas-Ft. Worth is a regional center for the western part of the Old South. In fact, Atlanta's atypical size is related to some southerners' attitude that it isn't really a "southern" city anymore.

Diversification, along with many changes in the southern economy, has long since reduced the once-continuous Cotton Belt to a few highly localized favorable areas and to nontraditional new cotton-producing areas in California, Arizona, and West Texas. Soil depletion and erosion after a century or so of cotton production particularly affected the Piedmont. The newer, highly mechanized and irrigated cotton production to the west was far more efficient. At the same time, federal programs, state universities, and county agents were advocating diversification, particularly the switch of worn-out cropland to pasture.

Coal is a major resource in much of the Upland or Appalachian South. While underground mines are heavy users of machines, surface mining is even more efficient in production per labor-hour. Unfortunately, the strip mining of coal has a high potential for environmental degradation.

The Tennessee Valley Authority (TVA) has demonstrated the positive contributions to the regional economy possible through coordinated conservation, energy development, flood control, and navigation improvement projects. Thanks to the TVA, Knoxville, Tennessee, is a river port central to a large and growing industrial cluster. Cheap electricity generated through the dams attracted industry, including the Atomic Energy Center at Oak Ridge, and brought prosperity to an impoverished part of southern Appalachia.

An impressive highway-building program, including Appalachian Access Roads, has accelerated development in this subregion. Land suited to recreational use is being bought up by urbanites looking for a peaceful mountain retreat. The tourism-recreation boom can pit well-financed outsiders against the mountaineers, outbidding them and raising taxes to provide services to the newcomers. Local tourism job opportunities are mostly low-paying, seasonal service trades.

Tennessee and Kentucky, in particular, seem to be making rapid progress in job creation, with both a Japanese automaker and General Motor's new division having chosen Upper South locations for assembly plants.

Caribbean minerals such as oil and bauxite, and agricultural products including bananas, sugar, and coffee are helping to stimulate major port improvements at Houston-Galveston, New Orleans, and Mobile, while Miami becomes a major center of Latin American culture and a prime attraction to Latin Americans visiting the United States. The subtropical coast sector of the South will continue to attract seasonal tourists and permanent (largely retired) residents. Some intraregional shifts in growth are evident. Fewer retirees can now afford Florida's rising real estate prices. The steep increase in the number of Cubans and Haitians immigrating to metropolitan Miami, with its attendant social costs, may alter some perceptions about the desirability of southern Florida, sending more retirees to the Gulf shores of Texas, Louisiana, Mississippi, and Alabama, and to the Florida panhandle (Figure 2–12). Miami and New Orleans are link cities between the United States and its Latin neighbors; their wholesale-retail and service functions and light industries reflect the growing importance of this Latin American orientation.

Water—fresh, clean, and plenty of it—has become a major asset of the subtropical coast region as chemical and pulp and paper industries grow and capitalize on the other material assets of the area. Conflicts concerning water management, from the endangered Everglades to the pollution-threatened Texas coast, will capture regional interest for some time, as for example the "developers versus preservationists" struggle over Padre Island, a long, sandy barrier beach off the South Texas coast.

The Changing Plains

Where does "The West" begin—that image of great open spaces, the migration target of the restless and adventurous, a series of golden opportunities amid

SHOTGUN HOUSES

The folk architecture of the South includes the shotgun house; shotguns are narrow houses, one room in width, with the front door and gable end facing the street or front of the lot. They are almost always one story, although a small second floor may be constructed at the rear of the house. Supposedly, the term *shotgun* comes from the fact that the interior and exterior doors are all aligned, so that, if all the doors were open, one could fire a shotgun through the house without hitting anything. Shotgun houses appear throughout the South, from East Texas to Louisville, even into southern New Jersey, but the oldest such styles are found in New Orleans. Clearly, the style spread outward from New Orleans, but did it originate there? Evidence uncovered by John Michael Vlach shows that Haitians emigrating to New Orleans after a revolution in Haiti introduced the style.* Nicely suited to narrow urban house lots, the shotgun is actually a West African house style, transported to Haiti by African slaves, and then brought to New Orleans by free blacks. Shotguns are quite common even today in the Yoruba tribal areas of Nigeria.

*John Michael Vlach, "The Shotgun House: An African Architectural Legacy," *Pioneer America* (1976):47–56.

FIGURE 2–12
Perceptions of the American Sun Belt. The Sun Belt is a perceptual region, not an official one. This map records geographers' perceptions of the Sun Belt, a region typified by high rates of in-migration, particularly by retired people.

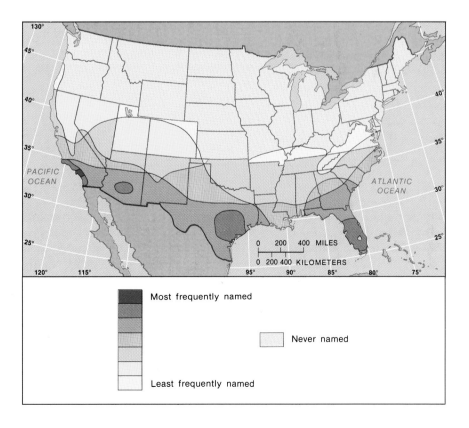

SELECTING THE SITE OF WALT DISNEY WORLD

Choosing the best, that is, most profitable, site for any economic activity is an exercise in the practical application of geography. The optimum location will minimize costs, reach the largest possible market most effectively, and produce the greatest income. A poor locational decision could spell low profits, or even failure. Choosing a site for Walt Disney World, the largest single investment in resort-recreation services, required an analysis of many geographic factors.

The first Disney theme park, in Anaheim, within the Los Angeles metropolitan area, was a great success. Of its visitors, 80 percent are Californians, most from the greater Los Angeles area. The Anaheim site is only 230 acres, so it is convenient for visitors to buy food, souvenirs, and gasoline from non-Disney businesses across the street. The company determined that its second effort should be much larger, but where should this second park be located? Clearly, if two Disney parks would together serve a national market, the second should not compete with the first. That consideration, plus the national population distribution, suggested an East Coast or central location.

Florida was a rapidly growing state even before Walt Disney World was built; a popular tourist destination, it was well served by highways, railroads, and air services. Although a location in New Jersey would be close to more people, Florida had the critical advantage of a year-round outdoor activity season. A capital investment on the Walt Disney World scale makes better sense if the park can be operated the entire year. Although early Florida resorts were mostly winter resorts, air conditioning was helping to expand summer season patronage.

The site selected, over 27,000 acres near Orlando, was in relatively inexpensive orchard and pasture land, but in the Interstate 4 "Growth Corridor" between Daytona and Tampa-St. Petersburg. The huge success of Walt Disney World testifies to the wisdom of this central Florida location.

---

frequently harsh, challenging environments? To early colonists along the Atlantic seaboard, the "rugged West" began at the Appalachians. The concept of "West" moved westward with the advance of relatively dense farm settlement. The advancing wave of thickly settled rural landscapes faltered at the Great Plains. If, psychologically, the West is perceived as relatively empty, then it begins at the edge of the vast brown midriff of the continent. The plains were first viewed negatively by early European explorers. For

people accustomed to relying on ample wood for houses, furniture, farm buildings, fences, and fuel, the almost treeless plains were perceived as a nonsupportive, even threatening, environment. The iron-tipped wooden plows of the early nineteenth century could not turn over the heavy sod of the plains, and lower precipitation seemed to imply inevitably lower yields.

A revolution in land use occurred just after the Civil War, when the "cattle kingdom" spread quickly up the plains from Texas. Railroads entered the edge of the plains, connecting the immense pastures with the burgeoning cities farther east. Great cattle drives brought cattle to the railheads of the time, such as Dodge City, Kansas.

The cattle kingdom flourished only a few decades before being partially displaced westward by the second great revolution in land use systems—the wheat-farming frontier. Steel plows, mechanical reapers, and barbed wire fencing (to keep out cattle) combined with railroads to transform the economic possibilities of the great grasslands. Homestead lands and inexpensive purchases from railroads, which had been given government land as subsidies to construction, made it possible to acquire farms that averaged larger than those back east. Having more land partly compensated for lower productivity per acre. Hybrid seed, better adapted to drought, also boosted production per acre.

Unfortunately, settlement by wheat cultivators pushed westward into semiarid lands that might better have been left in pasture. A feature of semiarid environments, discovered the hard way by pioneering farmers, is the unpredictability of precipitation from year to year. Long-term averages mean little, because erratic cycles of drier-than-average years alternate with wetter cycles, both phases being interrupted by atypical years and all being uncertain in length. Optimistic expansion of wheat tillage in wetter periods was followed by serious droughts. The "dust bowl" of the 1930s sent recent arrivals back east or to the West Coast as their farms literally blew away.

Wheat remains the major crop, although more land is actually in pasture than under cultivation; sugar beets, vegetables and, in the Texas panhandle, cotton, are alternatives where irrigation is available. Small grains and oil-seeds are the basis for diversification in nonirrigated areas. Relatively little wheat is irrigated, and the uncertainties of precipitation are minimized more by improved hybrid seed, cropping in alternate years, and crop insurance underwritten by the federal government.

In many parts of the Great Plains and the high, dry West, the surface water supply has long been inadequate to maintain the irrigated acreage. Subsurface water is being brought to the surface and used much faster than natural recharge can replace it in this semiarid environment. Much of the High Plains, for example, is underlain by the Oglala aquifer, a great underground reservoir containing up to 2 billion acre-feet of water. This water is being tapped at an increasing rate because, in a water-short area, water is money. Irrigated land produces about 70 percent more sorghum, for example, than does unirrigated land in this region. Unfortunately, the Oglala water is

Florida Retirement Community. *The Sun Belt's rapid growth has been due in part to large scale in-migration of retired people. These Gulf Coast seafront condos in Naples, Florida, are occupied primarily by retired sun-seekers.*

Pivot Irrigation on the Great Plains. *Great irrigation sprinklers repeatedly sweep around fields, pivoting on a central well or water source and creating huge circles of lush green within a drier matrix of semiarid plains. What will happen as the water resource approaches exhaustion?*

deeper wells boosted production during the boom of the 1970s. Natural gas is also produced in Wyoming, Kansas, and the panhandle area of Texas. Geologists long have known that enormous volumes of relatively low quality coal underlie the northern Great Plains. Because of the geographic location and the low quality, little attention was paid to this reserve until the 1970s (see Figure 2–6).

The geographic concept of **intervening opportunity** rests on the fact that it makes no sense to transport anything to market across or through a competing producing area closer to that market. The great urban-industrial centers of the southern Great Lakes and the Northeast had higher-quality coal closer in the Appalachian and Eastern Interior coal fields. Great Plains coal is not of metallurgical (coking) quality, and its heat energy content is low; although 70 percent of the tonnage of U.S. coal reserves lies west of the Mississippi, 55 percent of the coal measured in energy content, exclusive of Alaskan coal, lies to the east. Technological advances in surface-mining machinery such as more efficient power shovels and transmission of electricity now make it possible to use Great Plains coal to generate power near the mine and transmit electrical energy to distant markets.

Enormous new strip mines are being opened, and power-generating plants rise in Montana and Wyoming. A negative impact on the environment seems inevitable. Coal stripping in itself tends to lower water tables, and regeneration of vegetation in the semiarid stripped lands is not as readily achieved as in Appalachia.

The Western Interior

The Western Interior subregion includes the Rockies and intermontane basins. Lightly settled in comparison to the Core, the South, and the West Coast region, the Western Interior is nonetheless one of great change, vitality, and significance in the American economy (Figure 2–13). The oil and gas potential of the "overthrust belt" of the Rockies, and the oil shales of the Rocky Mountain basins, could prove essential in meeting the future energy needs of the United States.

The Rockies and intermontane basins are characterized by scattered, occasional energy booms, continuing summer tourism complemented by rising

"fossil water" which originated in much more humid conditions that prevailed a million years ago. It cannot be replaced, any more than a deposit of copper ore could be replaced; depending on the rate of withdrawal, one day the great pumps extracting water from the Oglala may run dry.

The enormous scale of agriculture is a striking characteristic of the plains landscape—huge grain elevators, cotton gins, storage bins, and machinery sheds, combined with the grand geometry of field boundaries and circles of pivot irrigation. There is a larger-than-life aura to the subregion.

Conventional oil wells have been producing in the subregion for many years, but new fields and

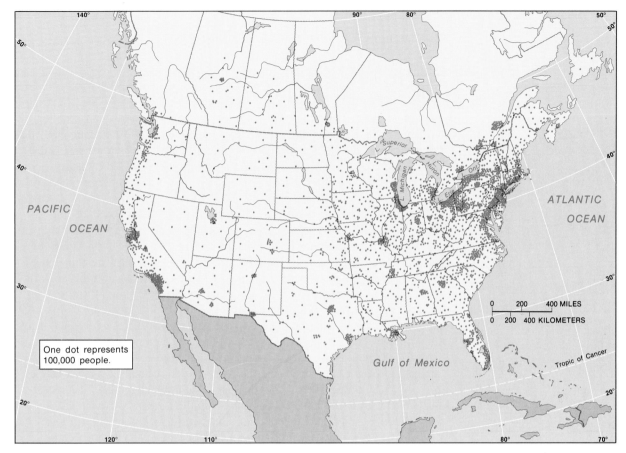

FIGURE 2–13
Population Distribution of the United States and Canada. Population density in the contiguous 48 states diminishes sharply from east to west at the transition from the Midwest to the Great Plains, reflecting the change in climate and, thus, agricultural systems. Historic settlement, resource distributions, and urban and industrial growth are other key factors in interpreting population density.

winter tourism, and grazing and forestry industries. The front ranges of the Colorado Rockies are particularly popular tourist areas. Some of the subregion's cities are booming as Americans come to appreciate the great natural beauty of the area and move there. Population, as a market and labor source, becomes an attractive factor for some industries, especially electronics and defense and space-related industries. These "footloose" industries can bear the high transport costs of location here between major population nodes. As happened earlier in California, in-migrants attracted by the amenities of the area spurred economic growth, rather than economic opportunity attracting people.

Dryland soils are characteristically rich in calcium but can contain such high concentrations of soluble minerals as to become harmful to most plants. In all dryland areas, the danger of **caliche** formation is present whenever irrigation is used. Caliche is a concentration of mineral salts in a tough, concreted hardpan at or near the surface, making the soil unusable.

In Nevada, Reno and Las Vegas may be the largest cities in the United States to be so dominated by resort functions. The availability of virtually every form of gambling enabled Nevada to capitalize on its accessibility to California cities and its position astride important east-west transit routes in luring

THE MORMONS AND WATER ENGINEERING

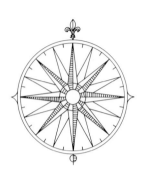

Two maps, one of the distribution of Mormons and one of the distribution of irrigated farmland in the intermontane West, would show approximately the same concentration patterns. This correspondence would not be mere coincidence. Understanding the dry West and the vast irrigation-based culture built there requires some understanding of the Mormons.

The Mormons sought out the isolation of the Great Salt Lake portion of the Great Basin to avoid religious persecution. The group's prophet, Joseph Smith, had received his revelations in upstate New York. The Mormons moved repeatedly to avoid conflict; Joseph Smith was murdered by a mob near the Mormon settlement of Nauvoo, Illinois, and the growing band of believers moved to northern Utah. By necessity, the Mormons became the first Anglo culture group to tackle the problems of water supply and distribution in the dry West. The cohesiveness and close cooperation of the Mormons, forged under persecution, helped them to succeed as a group where individual efforts would necessarily have failed—in creating and maintaining elaborate water engineering projects. A little more than half a century after the Mormons came to Utah's Wasatch Valley, they had spread over much of the drylands and brought 6 million acres under irrigation. Still, although America contains one of the world's most ambitious desert civilizations, there are severe limitations on available water and suitable lands—only 3 percent of Utah is cultivated. Will it be cost effective to attempt to bring more land under irrigation? The easiest, cheapest irrigation projects are usually the first attempted. Adding more irrigated land becomes progressively more expensive because water must be brought from greater distances or pumped up from deeper wells. Western historian Walter Webb warned that the West was "a semi-desert with a desert heart" and advised against emptying the federal treasury in an attempt to make it look like Illinois.

---

tourists. Retirement to the dry sunshine of the southern intermontane areas has also spurred growth. Shortages of fresh water will be the important depressant of growth rates in the future unless further large-scale water transfers are funded by the federal government.

The Southwest has a strong Amerindian-Hispanic flavor (Figure 2–14). This region has a distinctive culture created by the layering and interleaving of its three predominant cultures (Amerindian, Hispanic, Anglo mainstream) rather than a blending of them.

With the exception of Oklahoma, no other part of the contiguous 48 states has such a strong Amerindian imprint. While sizable groups of Hispanic-Americans are increasingly familiar in cities across the continent, the Southwest has a numerous and long-established rural Hispanic population as well. New Mexico is the only U.S. state that is officially bilingual (Spanish and English). Late-nineteenth century Anglos arrived to expand irrigation agriculture, and many more flocked in with the twentieth century tourist-retirement and high-tech industry booms.

The West Coast

The West Coast has for a long time had a special appeal to both Americans and Canadians. California is more than a state—it's a cultural ideal, a model of desirability in the American consciousness. Many Americans have acted on their glittering image of California and moved there. Likewise, Canadians have made "B.C." (British Columbia) an important domestic migration objective. Geographer Peirce Lewis suggests that the West is special to Americans because, in a sense, it is the most perfectly American. As Lewis interprets it, Americans have repeatedly left behind the parts of their "cultural freight" that they found irrelevant, inappropriate, or constraining. Vestiges of culture and character imported from Europe are removed or modified by passing through the filter of space. The selective nature of migration—whose participants would score high on self-reliance, determination, and independence of thought—has resulted in, perhaps westerners in general and West Coast residents in particular, being the least influenced by cultural traditions and assumptions transported across the Atlantic. Lewis proposes that the American West contains the final distillation of the American motifs identified by Wilbur Zelinsky:

1. an intense individualism
2. an emphasis on the desirability of mobility and change, sometimes for their own sakes
3. a mechanistic vision of the world, seeing it as a machine that can be fixed and improved

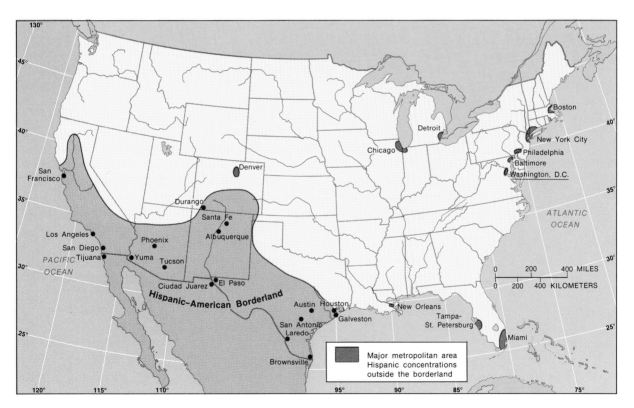

FIGURE 2–14
Generalized Distribution of Hispanic-Americans. Hispanic-Americans in the Southwest border area, itself once part of Mexico, include long-established residents and recent arrivals from Mexico. Eastern urban concentrations represent mostly post–World War II migration from Puerto Rico, Cuba, and Mexico.

WHAT IS A NATURAL RESOURCE?

Simply stated, a natural resource is any material or energy source occurring naturally in the environment that people have identified as economically useful and retrievable. The definition is thus economic and technological. Rising prices redefine the resource to include that which was too expensive to produce before but is now within the economic range. This change over time in the qualitative and quantitative definitions of economic resources is quite common. There is a tendency to exploit at first only the richest, easiest-to-mine, most-convenient-to-market parts of a total resource base. Later, as demand escalates, prices rise, and the technology of exploitation advances in efficiency, what was overlooked or discarded before is redefined as an economic source. Exploration is also accelerated.

Occasionally, the decision to exploit a particular mineral or fuel source will involve political considerations as well. A domestic mineral deposit, for example, might be utilized even if a lower production cost could be achieved in a less stable, foreign area.

The oil shales of the Rocky Mountain basins are a good illustration of the definition of a resource by a constellation of economic, technological, and political factors. The oil shales of Colorado, Wyoming, and Utah contain vast quantities of shale oil (kerogen), which can be refined into the familiar list of petroleum products. The total reserve is estimated at 2 trillion barrels—enough to last the United States (at present rates of consumption) about 3,300 years! There are a few catches to this bonanza, however. Shale is porous and can absorb and hold a liquid or a gas, but it is not permeable, that is, it will not allow a liquid to move through it readily. It is like a one-way sponge—what gets absorbed is trapped. Producing shale oil requires mining and crushing the shale into small chunks, then heating the pulverized rock to drive off the kerogen as a vapor. The process is expensive and leads to serious environmental concerns. A large-scale commitment to shale oil mining and processing would require a huge capital investment that is seen as simply too risky, at least until the next oil shortage.

4. a messianic perfectionism—the idea that the United States can and will realize its dream of self-perfection and share its message with others around the world.[7]

This subregion includes California, Oregon, and Washington. Agriculture in Oregon and Washington is divided between production of wheat and fodder in eastern dryland areas and humid area specialty crops and dairying. Lumbering, wood products, aluminum production, and aircraft industries dominate in an economy far less developed than that of highly varied California. All three states share in the markets of Alaska, Japan, and China. As California continues as a center of innovation and a mecca for migrants, Washington and Oregon will undoubtedly share in the decentralization of California's population, industry, and functions—a process already in progress.

The problems of perception—trying to sort out the various popularly held images of a region in relation to people's cultural background and values—have fascinated geographers for some time. California has a vividly positive image to the overwhelming majority of Americans. This image has changed in emphasis over the years but remains powerful. California is a special place perhaps because its physical and cultural landscapes appeal to, and reflect, popular American ideals about desirable landscapes.

It is said that the most enduringly popular picture postcard scenes of California show the edge of the Los Angeles Basin with orange groves in fruit in the foreground and snowcapped mountains behind, all under the bright blue sky typical of Mediterranean climates. That scene is now more nostalgic than representative. The air temperature inversions common over the basin make clear skies a rare treat, and many of the orange groves have been cut down to make way for freeways, industrial parks, shopping centers, and housing. Even the pines of the fringing mountains have suffered blight associated with high levels of atmospheric pollution. Clearly, the reality has changed faster than the cherished image. But the image, although altered and dimmed, is still a positive one. California, specifically the "Southland" (metropolitan Los Angeles-San Diego), is, to many people, an exciting, innovative place.

Plush California Lifestyle. *The California image of a lush lifestyle in a pleasant climate helped attract in-migration in search of the "good life"; amenities displaced jobs as the primary motive for migration. San Francisco continues to attract migrants to its sophisticated cultural life and magnificent setting.*

Urbanization in the Los Angeles Basin features many small centers rather than a single major core. This polynuclear structure reflects the formerly separate towns, most with a different agricultural specialty, now embedded in a matrix of automobile-fostered suburban sprawl.

Los Angeles's growth has occurred explosively in the past 80 years. Los Angeles early exhibited its trend-setting, lifestyle leadership in the auto age. By 1928, Los Angeles already had one car for every 2.25 people. Los Angeles opened its first freeway in 1940; the metro area now has more land in freeways than the total area of land in Miami.

In cultural leadership, Los Angeles has mounted a powerful challenge to both New York and Chicago. The production of films and television programs shown around the world makes the area an international factor in popular culture.

With a few exotic exceptions such as coffee in Hawaii, limes in the Florida Keys, and cane sugar, all agricultural products grown elsewhere in the United States are grown in California. The range of climate and soil types in the Golden State supports a full variety of temperate and subtropical plants. California is first in total value of farm products; it is the leading producer of 30 crops and the sole producer of some.

Colonial California's major agricultural export was cowhide. There was no practical way to deliver fruit and vegetables to distant markets. The successive developments of transcontinental railroads, refrigerated boxcars, commercial canneries, freezing plants, and refrigerated trucks traveling interstate highways transformed California's agriculture. The Mediterranean climate, with its relatively warm winters, supports a specialization in winter season fresh fruits and vegetables. California, like Florida and Texas, can ship fresh salad vegetables to northeastern and midwestern markets at times when northern fields are snow-covered.

California has a long tradition of large land units. Large land holdings, called "latifundia," were favored under Spanish and Mexican rule. Some of these huge land grants, confirmed in the treaty ending the war with Mexico in 1848, have continued intact into the twentieth century.

The reshaping of California by people has taken place on a massive scale. The California Water Plan is one of the greatest engineering works ever undertaken by a U.S. state; it supplements extensive water engineering systems built by the federal government. California has a massive imbalance in water availability and water demand. Northern California, north of Sacramento, has most of the precipitation but little of the population. Two thirds of the total flow of water in the Central Valley is in the northern half—the Sacramento system. Federal and state water systems transfer "surplus" (not needed for irrigation) Sacramento Valley water southward to irrigate semiarid San Joaquin Valley (Figure 2–15). Some of the nation's most valuable farmland now is in the San Joaquin Valley.

A second important irrigation project is in Imperial Valley-Coachella Valley, east of San Diego. This is the last major oasis supplied by the much-used Colorado River before it flows across the Mexican border. The All-American Canal helps feed this lush subtropical "winter garden," where three crops a year can be produced with irrigation.

Oregon and Washington are relatively less urbanized than California. These northwestern states are notable for their enthusiasm for environmental protection. This attitude of controlled growth may have been inspired by the spectacle of rapid growth accompanied by environmental deterioration in California, a process Oregonians call "Californication"!

Although industrial growth is evident in smaller urban communities as well, five major metropolitan areas still dominate: Los Angeles, San Francisco, and San Diego in California, Portland, Oregon, and Seattle-Tacoma, Washington. Important factors encouraging industry include a richly productive, highly varied agricultural base; a local, relatively cheap energy supply; increasing trade ties with the Orient; Alaskan development; the large local market provided by the subregion's burgeoning population; the attractive force of climatic and other amenities; the growing pool of technical and research personnel; and the growing economic and political influence of the subregion's inhabitants. As a center of innovation, it also generates inventions, fashions, and fads that lead to industrial production. California's

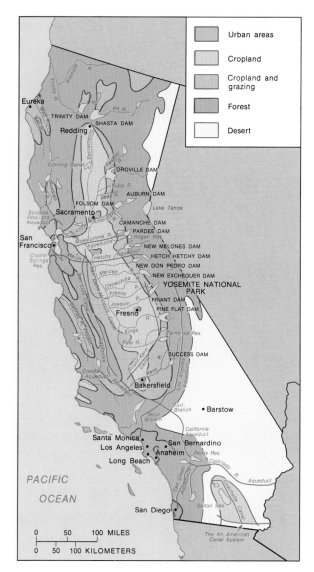

FIGURE 2–15
California Water Plan and Other California Water Engineering Projects. With a water surplus north and a densely populated and growing southern coast, California's water supply and water demand are out of balance, geographically. State and federal projects connect the supply with the need.

appeal to migrants has produced a polyethnic, poly-racial population (Figure 2–16).

By far the largest industrial center is metropolitan Los Angeles. Recently discovered new reserves supplement older oil fields, and Los Angeles remains a major center of oil refining and related petrochemical industries.

Nearly perfect flight weather, the year-round, made Los Angeles an ideal place to test aircraft. Over the years, Los Angeles has become a major production center for aircraft and aeronautical components. The advantages of the area soon became those of skill rather than climate. Bolstered further by space industry contracts, the region's aircraft industry has continued to grow.

Innovation in electrical components and systems for planes has given birth to a full-fledged electronics industry, one of the largest in the country. The large and affluent local market has attracted assembly industries of all kinds; automotive plants are representative of the market draw.

San Francisco overshadows Los Angeles as a port but is not nearly as important an industrial center. Food processing still forms an enormous segment of its total industry, and its position as the natural outlet for the Great Valley of California will likely result in a continuation of this function. The presence of famed institutions of higher education in the area has led to the establishment of research-oriented industries, most notably pharmaceuticals and electronics. Silicon Valley south of San Francisco, a pioneering computer technology center, continues as an important high-tech manufacturing and research center.

San Diego thrives on government military contracts from nearby armed service bases and assembly industries using unskilled labor supplied by recent migrants. The aircraft, chemical, and electronics industries also are fairly well represented. San Diego functions as a major **entrepôt** (collecting and distributing center) for raw materials and semimanufactures from Mexico as well as a production point for components shipped to Mexico for assembly.

Portland functions mainly as a manufacturer of lumber, paper, and a large array of wood products. Seattle, Tacoma, and their suburbs constitute a major and growing industrial complex. Cheap power from

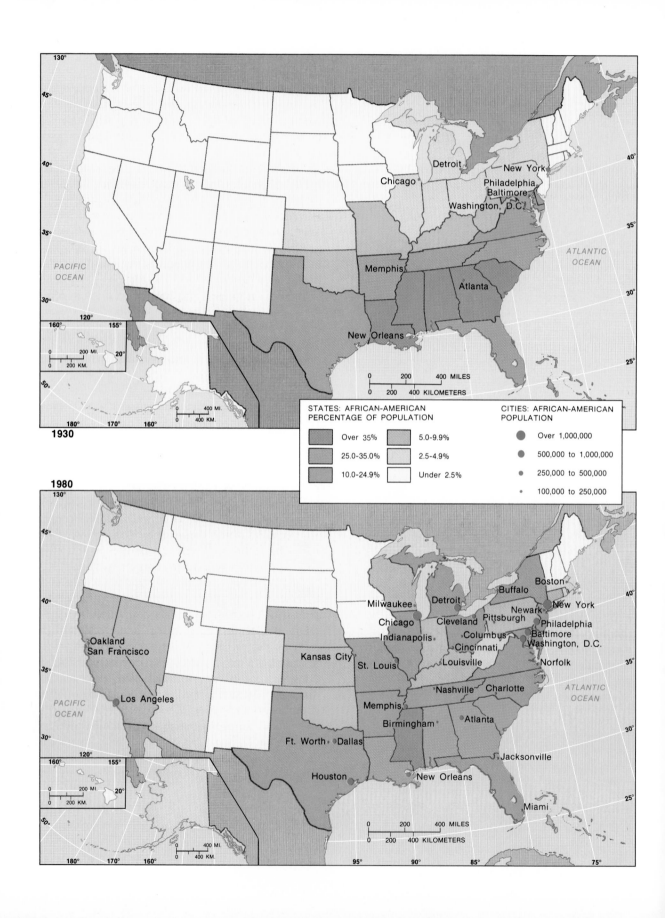

1930

1980

STATES: AFRICAN-AMERICAN
PERCENTAGE OF POPULATION

- Over 35%
- 25.0-35.0%
- 10.0-24.9%
- 5.0-9.9%
- 2.5-4.9%
- Under 2.5%

CITIES: AFRICAN-AMERICAN
POPULATION

- Over 1,000,000
- 500,000 to 1,000,000
- 250,000 to 500,000
- 100,000 to 250,000

1930 map labels:
PACIFIC OCEAN
ATLANTIC OCEAN
Detroit
Chicago
New York
Philadelphia
Baltimore
Washington, D.C.
Memphis
Atlanta
New Orleans

1980 map labels:
PACIFIC OCEAN
ATLANTIC OCEAN
Boston
Buffalo
Milwaukee
Detroit
Newark
New York
Chicago
Cleveland
Pittsburgh
Philadelphia
Indianapolis
Columbus
Baltimore
Oakland
San Francisco
Cincinnati
Washington, D.C.
Kansas City
St. Louis
Louisville
Norfolk
Los Angeles
Nashville
Charlotte
Memphis
Birmingham
Atlanta
Ft. Worth
Dallas
Jacksonville
Houston
New Orleans
Miami

Alaskan Oil Boom. *Over half of the world's petroleum is produced from wells drilled into the sea floor. This offshore oil rig is built on an artificial "ice island" in the shallow Beaufort Sea off Alaska's North Slope.*

local hydroelectric installations has attracted electroplating and metal refining activities. The area is a major contributor to U.S. aluminum production, an industry attracted by cheap power. Boeing, a giant in aircraft assembly, is itself a major market for aluminum.

Woodworking industries are still of great importance in this area of superb forest resources. Alaskan development, the growth of nearby markets in British Columbia, and proximity to (plus a tradition of interaction with) Japan favor continued industrial development.

Alaska

Once owned by Russia, Alaska was purchased by the United States in 1867, a highly controversial move. Alaska's fur resources, at least those accessible from

← **FIGURE 2–16**
Distribution of African-Americans. From an original concentration in the pre-Civil War Plantation South, African-Americans have migrated to urban areas in the Northeast, Midwest, and all the way to the West Coast, indicating that they are responding to the same set of economic/amenity attractions as others in this highly mobile society.

the coast, had been seriously overhunted. In the words of one American critic of the purchase, Alaska was a "sucked orange," drained of its value. This dreary image of a worthless, remote, frozen wasteland was dramatically shattered by the gold rush of the 1890s. The oil boom touched off by the Prudhoe Bay discoveries led to the most recent new appreciation of Alaska. Unlike the gold rush, however, oil production in association with ongoing expansion of logging, fisheries, and metallic ores seems to have achieved a major permanent increase in the state's economic growth.

Although Alaskan lands long have been open to homesteaders by both federal and state programs, the state has fewer than 400 farms and produces the smallest proportion of food relative to its population of any state (about 10 percent). Alaska's population is primarily urban, and it is likely that Alaskans will continue to flow into the state's largest city, Anchorage. No other major city seems likely to evolve, and the "last frontier" will be a quite different kind of frontier. Rather than the wavelike advance of farmers and ranchers into the plains, the Alaskan frontier will be characterized by widely scattered mineral exploitation sites embedded in wilderness. The harsh realities of the Alaskan frontier will result in only temporary excursions by long-range or seasonal commuters into truly wild country, while the few cities continue to function as support-supply bases for this extractive economy. Exceptions will be many small settlements along the coasts serving forestry and fishing activities. The sheer size of Alaska will continue to support subregional centers such as Ketchikan and Juneau, but Anchorage has such an unchallenged position of leadership that even an earthquake could not retard its growth for long.

Hawaii

America's 50th state lies more than 2,000 miles west of San Francisco and occupies a chain of great volcanic mountains built up from the sea floor. Hawaii is the only U.S. state in which Caucasians are not the largest racial group. Native Hawaiians are a tiny minority now, and interracial marriages are common, producing a society in which the largest census group (60 percent) is "mixed."

The tremendous expansion of sugar plantations in the late nineteenth century was associated with rising importation of labor—first Chinese, then Japanese and Filipinos. Plantation profits were invested in trading companies that once controlled the islands' shipping, banking, insurance, real estate development, hotels, and other services. Of serious concern to the state and its people is the fact that almost half of Hawaii's land is owned by fewer than a hundred private landowners. Small private owners hold less than 5 percent.

Metropolitan Honolulu has far outstripped the rest of the state in attracting new "mainland" immigrants as well as rural Hawaiians from the other islands. Considering Hawaii's positive image and popularity with tourists, relatively few mainlanders move there permanently; high living costs and relative isolation are possible explanations. Industries other than agricultural processing are almost entirely service industries based on tourism, retirement, and military expenditures. Hawaii, conscious of the overconcentration of tourism on Oahu, has begun to deflect new facilities to the other islands. Military spending is likely to remain high, guaranteeing an inflow of money even if the tourist boom should slacken.

CANADA

Although Canada has one tenth as many people as the United States, it has 110 percent as much space. With over 3 million square miles, Canada is the second largest country in the world, stretching 4,500 miles east-west and nearly 3,000 miles north-south. The 25 million Canadians, however, are concentrated near their southern border—in four distinct clusters (see Figure 2–13).

Canada is a country of vast scale. Its smallest province, Prince Edward Island, is nearly twice the size of the smallest U.S. state, Rhode Island. The province of Quebec stretches as far as the distance from New York City to Omaha, Nebraska; Ontario's maximum east-west distance would match the distance between Los Angeles and Dallas, Texas. The Yukon and Northwest Territories make up nearly 40 percent of Canada's area, yet they contain a mere 0.2 percent of the country's people and are ruled directly from Ottawa. Together, these two territories are larger than India, but their combined populations would not overcrowd the New Orleans Superdome.

The relative distributions, as well as relative size, of the populations of Canada and the United States have contributed to a heightened Canadian awareness of its southern neighbor. About 80 percent of Canadians live within 100 miles of the U.S. border, which many cross frequently to shop, vacation, visit friends and relatives, and even attend sports events. The broadcasting range of many U.S. radio and television stations extends well into the populous border region. On the other hand, only about 12 percent of Americans live within 100 miles of the Canadian border. Although U.S. citizens form most of Canada's international tourists, they seem less aware of the distinctiveness of their northern neighbors. As an old saying goes, "Americans are benevolently ignorant of Canada; Canadians are malevolently well informed about the U.S."

Canada is a country of vast distances, formidable physical barriers, and large tracts of nonagricultural land separating the major settlement regions. The mainly coastal populations of the Atlantic provinces are separated from the main concentration—the upper St. Lawrence Valley-Ontario peninsula—by densely forested, thinly settled lands, while to the west of the St. Lawrence-Ontario heartland lies an even more formidable barrier. The glacially eroded, rugged terrain of the Canadian Shield forms a gigantic crescent around the southern coast of Hudson Bay, dipping as far south as the northern peninsula of Michigan in the United States (see Figure 2–4). Its scraped, rocky surface supports agriculture only in isolated pockets of clay soils originally deposited in lake beds. Its rugged but not particularly high surface has been a major obstacle to east-west development. Until 1960, for example, when the trans-Canada highway was extended across the shield, traffic had to detour south across the U.S. border. The Canadian Rockies separate the western coastal population concentrations from the population of the Canadian Plains, which in turn lies beyond the western side of the shield, separated from the heartland.

Contrasts among Canadian Provinces

With only 10 provinces and two territories politically subdividing Canada's vast territory, the country's subregional geography can be portrayed only at a very general scale, given the nature of provincial data. Still, the contrasts shown by geographic data available on a provincial scale are both dramatic and representative of Canada's subregions. The populations and economic activities of territorially huge Ontario and Quebec are concentrated in the provinces' portions of the "Divided Heart" (see Figure 2–13). Similarly, the people and economic activities of the three "prairie provinces"—Alberta, Manitoba, and Saskatchewan—are mostly if not exclusively within the Canadian Plains and West Coast subregions, along with booming British Columbia. The two territories—Yukon and the Northwest Territories—represent the northlands and Arctic fringe. The Atlantic Provinces subregion includes the territories and peoples of New Brunswick, Newfoundland, Nova Scotia, and Prince Edward Island, except for sparsely settled Labrador, the mainland part of Newfoundland. Table 2–2 presents key economic and demographic data on a provincial basis.

The overwhelming dominance of Ontario is at once apparent. With over 35 percent of Canada's total population, Ontario produces over half of the na-

tion's manufacturing output. Ontario gains most from interprovincial migration. Neighboring Quebec, second in percentage of population, total economic output, and manufacturing, continues as a net interprovincial migration loser, but its net loss has declined from the 1977–78 figure. The relative poverty of the Atlantic provinces is apparent in average incomes, with Newfoundlanders enjoying only 62 percent of the income of an average Ontarian. Second richest British Columbia also has the second largest net in-migration from other provinces. Newfoundland's daunting 20 percent unemployment is related to its low income and moderate out-migration. Among the prairie provinces, Alberta's recent reversal of formerly high net in-migration reflects the sharp decrease in opportunities in energy resource employment there.

The Atlantic Provinces: A Bypassed Subregion

Much of Canada's dynamic economic growth has been concentrated in Ontario, Alberta, and British Columbia. The four Atlantic provinces—New Brunswick, Newfoundland, Nova Scotia, and Prince Edward Island—seem to be relatively stagnant backwaters. The Atlantic Provinces are at the bottom of the list of average provincial income (see Table 2–2).

There are two fundamental geographic reasons for the relative poverty and slow growth of the Atlantic Provinces. The location of the subregion relative to transport routes and the better-quality land of the interior has resulted in the provinces being "bypassed." Unlike the early colonies of America's seaboard, the Atlantic Provinces did not lay astride the routes westward. They did not control the all-important hinge function of connecting sea routes with interior routes and thus did not benefit from their headstart. The great St. Lawrence Valley was the obvious route to the interior, so the Atlantic Provinces lay aside the major European penetration westward. With the notable exception of Prince Edward Island, the Atlantic Provinces have little good-quality agricultural soil. The people of the Atlantic Provinces

TABLE 2–2

Economic status and demographic data by province.

	Per Capita Income, 1983 ($)	% Unemployed, 1984	Net Interprovincial Migration		% National Population, 1984	Gross Domestic Product, 1983 (billions $)	% Total GDP	GDP, Manufacturing, 1982 (billions $)	% Total GDP, Manufacturing
			1977–78	1983–84					
Ontario	14,784	9.1	+29,658	+42,078	35.71	151.6	37.7	30.4	51.75
Quebec	12,531	12.8	−45,042	−19,077	25.97	90.4	22.5	15.6	26.55
Atlantic Provinces									
New Brunswick	10,040	14.9	+1,537	+1,387	2.83	7.3	1.8	0.85	1.45
Newfoundland	9,179	20.5	−3,101	−2,444	2.28	5.4	1.3	0.48	0.82
Nova Scotia	10,889	13.1	+540	+4,668	3.47	9.4	2.3	1.0	1.7
Prince Edward Island	10,056	12.6	+948	+484	0.49	1.2	0.3	0.06	0.1
Prairie Provinces									
Alberta	14,652	11.2	+33,973	−42,784	9.25	56.5	14.1	3.1	5.3
Manitoba	12,603	8.3	−4,634	−708	4.21	16.3	4.1	0.64	1.0
Saskatchewan	12,686	8.0	+2,704	+4,202	4.02	15	3.5	1.6	2.7
British Columbia	14,339	14.7	+19,159	+13,125	11.41	47.3	11.8	5.0	8.5
Yukon and Northwest Territories	14,282	—	−384	−932	0.28	1.4	0.3	0.009	0.02

SOURCE: Statistics Canada, various years.

relied more on their development and use of sea resources, forests, and mines.

Although Halifax is Canada's largest ice-free, deep-water port, it does not dominate Canadian trade to the degree that would seem logical. Because sea transport is much cheaper than land transport, the eastward thrust of the Atlantic Provinces does not necessarily attract cargoes to and from Canada's major industrial and population centers. Although seriously handicapped by winter freezing, the St. Lawrence Seaway has advanced the head of navigation from Montreal to Thunder Bay, the western end of Lake Superior. Though winter port functions are significant, the volume of bypassing sea traffic increases for the spring opening of the St. Lawrence.

About three quarters of the region is forested. Paper, pulp, and lumber have been mainstays of the economies of all provinces except Prince Edward Island which, with 60 percent of its land in farms, provides potatoes and hay. Both agriculture and forest products are slow-growth industries, at best. A tree-destroying insect, the spruce budworm, is a major problem in New Brunswick and Nova Scotia. Many of the rich fisheries of the region, which first attracted and sustained European settlement, have experienced declines in catches owing to overfishing by foreigners. However, Canada has modernized fishing fleets with grants to fishing cooperatives, and fisheries are reviving after Canada expanded its jurisdiction over fishing grounds to a 200-mile limit. The size of catches should continue increasing, and their value will increase even faster.

In an attempt to compensate for climatic disadvantage, limited agricultural area, and distance from markets, the farms of the Atlantic Provinces are beginning to emphasize high-quality products, such as seed potatoes and pedigree dairy and beef cattle. Substantial new coal reserves have been discovered at Cape Breton. Although Nova Scotia's peak coal production was in 1917, a major new mine was opened in 1974. The steel plant in Sydney was taken over by the province in 1968, modernized, and may again be viable. Oil exploration off Nova Scotia and Newfoundland has produced encouraging finds. Oil was discovered in 1979 under the Grand Banks, 168 miles southeast of St. John's, Newfoundland.

Bleak, almost unpopulated Labrador, the mainland portion of Newfoundland, contributes income through the sale of hydroelectric power. The largest hydro-generating station in the world is located at Churchill Falls; almost twice as high as Niagara Falls, it generates 7 million horsepower. Energy is exported via the Quebec Hydro system. In addition, Labrador shares a major iron ore resource with Quebec at Schefferville (see Figure 2–6).

It must be remembered, though, that moderate expansion of fishing, forestry, and onshore mining of metals and coal will not lead to an economic growth rate matching that of Ontario or British Columbia. Even a major oil boom, if it occurs, would employ relatively few people permanently, once wells and pipelines were in place. Federal "equalization" benefits remain an important prop.

The Divided Heart

The heartland of the upper St. Lawrence Valley, southwestward from the city of Quebec through the Ontario peninsula to the U.S. border at Detroit and Port Huron, dominates Canada's cultural, economic, and political life. Canada's most serious problem lies in the cultural-political geography of its great heartland, because the heart is divided. Quebec's voters defeated a 1980 referendum on the continuation of a drive toward complete independence from Canada, but the province seems determined to assert a high degree of autonomy within the Canadian confederation.

Quebec, nearly twice the size of Alaska, is Canada's largest province. It contains one of Canada's two largest metropolitan areas, Montreal, and accounts for about 26 percent of the country's population. In manufacturing, Ontario dominates the rest of Canada with 80 percent of the national total value added by manufacturing (see Table 2–2). No single U.S. state approaches that concentration of economic power within the national economy. Not only are the two rivals for largest city, Toronto and Montreal, located within this heartland, but the national capital, Ottawa, whose metropolitan area is fourth largest (after third-ranked Vancouver), and Quebec City, the seventh largest metropolitan city, are here.

Quebec has had to cope with a slow-growth economy, large-scale abandonment of farms, and sizable out-migration of French Canadians to the other provinces and to the United States. The heavy involvement of U.S. corporations in the Canadian economy has strongly favored English as the language of commerce. The "new Canadians"—recent immigrants from the Ukraine, Russia, Poland, Italy, China, the Caribbean, and elsewhere—are most likely to immigrate to "English" provinces. They learn English first, French either later or never. The strong influence of American popular culture is expressed in radio and television, movies, and magazines. To the French Canadians, it seems clear that although they are pressured to be bilingual to get good jobs, most other Canadians do not bother to learn much French.

Legislation requiring the use of French in Quebec produced a flight of offices and service industries that serve national markets. Many of these jobs have moved to Toronto. Industries with heavy investments in purpose-built structures such as oil refineries are not, of course, as footloose as insurance companies and cannot respond nearly as quickly to a changing political climate. It has been service industries, however, that have been growing fastest in the Canadian economy, typical of a country whose economy is maturing.

Fully half of new Canadians initially choose Ontario, while 15 percent choose Quebec. This inmigration follows a cultural cleavage: immigrants from France and Italy favor Quebec, whereas Ontario attracts a more cosmopolitan array, including East Europeans, Southeast Asians, Caribbean blacks, and Chinese.

Provincial government-owned Quebec Hydro produces huge amounts of hydroelectric power from the rivers that flow toward the St. Lawrence from the shield and from the seaway project. A newer source is the network of rivers draining westward across northern Quebec to James Bay. One of the largest hydro power projects in the world is being constructed on La Grande Riviere; this hydroelectric energy is particularly significant to the Canadian heartland because the great St. Lawrence Valley has no oil or coal of its own. Hydro power is also important in Ontario, although neither province has fully developed its potential.

Quebec has large iron ore deposits, mainly near the Newfoundland (Labrador) border. Ontario's iron ore resources are not as great as those of Quebec, but they are being exploited. A variety of metal ores, including nickel, copper, lead, zinc, silver, and gold, are known to exist in isolated, rugged terrain. Exploitation in most cases awaits higher world market prices to support expensive transport development. Ontario has one of the world's largest reserves of nickel at Sudbury. This deposit also produces smaller quantities of copper, lead, zinc, gold, and silver as by-products.

Logging is a significant activity in both Quebec and Ontario. Pulp and paper are more important than lumber and veneers, and the vast areas in forest are partly counterbalanced by slow growth. Trees in northern Quebec and Ontario may well take a half-century to grow big enough to harvest for pulp, whereas southern pines in Georgia could be cut at 10 to 12 years' growth.

Quebec's generally cooler climate and often thin, rocky soils are a considerable handicap. The province has some minor advantage in the large urban markets nearby, and Quebec-produced butter and cheese are marketed throughout Canada. The agricultural frontiers of Quebec are in retreat. Marginal farms are being abandoned, although the farmhouse may remain occupied by rural nonfarm people who work in forestry, mining, or manufacturing, or it may become a seasonal home for urbanites.

Ontario has Canada's best agricultural land outside of the prairies (which contain almost four fifths of Canada's farmland). The Ontario peninsula, south of a line from Georgian Bay to the Thousand Islands, is similar geologically to the Central Lowland of the United States. It has been glaciated, but the effects of glacial deposition, as in much of the American Midwest to the south, have generally favored agriculture. The Great Lakes that flank the peninsula moderate its climate; the water acts as a reservoir of warmth in the early fall and helps to provide a longer growing season and a lower seasonal temperature range than in neighboring regions. The lakeshore and hilly edge of the Niagara escarpment (over which flows the Niagara River in the famous falls) are excellent orchard lands. The orchards, however, are disappearing under suburban and urban expansion in the "Golden

Quebec City. *The French Canadians have maintained their distinctive culture which gives Quebec City an attractive foreign flavor appreciated by tourists as well as by French Canadians.*

Horseshoe" from the Niagara frontier with the United States to metropolitan Toronto.

This Golden Horseshoe is a prospering industrial region. As in the U.S. Midwest, lower Ontario's relatively dense, prosperous farming population helped support agricultural machinery industries and farm product processors. Canada's auto assembly industry is concentrated in metropolitan Toronto and Windsor, with the most recent expansion taking place in Montreal. Highly accessible to the huge markets of the U.S. Midwest and northeastern megalopolitan regions, the Ontario peninsula is well served by rail and highway routes with international connections and by the St. Lawrence Seaway. Toronto has become an important port because the larger ships cannot proceed farther west due to the narrowness and shallowness of the Welland Canal bypassing Niagara Falls. Toronto has developed huge, interconnected underground malls to counter the long, cold winters.

The Canadian Plains and the West Coast

The great Canadian Shield, pitted, gouged, and scraped by glacial erosion, has considerable local relief here. The shield dips south of the international border into the upper peninsula of Michigan and into northern Wisconsin and Minnesota so that the easiest route between the already settled Ontario lowlands and the plains lies south of the Great Lakes, through U.S. territory. Not surprisingly, many Americans migrated northward into the Canadian Plains in the late nineteenth and early twentieth century to take advantage of Canadian homesteads after virtually all American homesteads had been claimed. A railroad from the U.S. plains reached Winnipeg in 1878, seven years before the Canadian Pacific Line from Ontario was completed. Canadian Plains farms followed the U.S. experience in the same expansion-contraction pattern of farming versus ranching occasioned by fluctuations in precipitation.

The mineral resources of the area are also considerable. Great deposits of coal and lignite underlie vast portions of the plains. Major oil and gas deposits are already exploited, and large new reserves have been found in the last few years (see Figure 2–7).

As would be expected, the Canadian West Coast bears a strong physical resemblance to the American Northwest, but it is also culturally similar to California in some ways. British Columbia is a relatively wealthy province, attractive to migrants from other parts of Canada, and it shares some of California's reputation as a culturally innovative, exciting place.

THE CULTURAL DISTINCTIVENESS OF FRENCH QUEBEC

A **national minority** is one that advocates some measure of political autonomy within a country, or even has ambitions of achieving complete independence. Quebec is North America's only instance of a national minority being in political control of a state or province. The reasons for Quebec's drive to assert its own identity and its potential, but now receding, threat to the integrity of Canada are cultural. French settlement in North America began early in the seventeenth century; Quebec City was founded in 1608. The climate of "New France," though, was considered quite harsh by immigrant Frenchmen. The French in Canada, who were later to develop a distinctive French Canadian culture, numbered only about 65,000 at the end of French rule; French immigration to Canada virtually stopped thereafter. Thus, most French Canadians today in Quebec and the other provinces (and the more than half a million who migrated to New England just in the nineteenth century) can trace their ancestry in the New World at least back to 1763. The French Canadians have, historically, maintained a high birthrate, although their rate of natural increase now is below the national average.

The cultural distinctiveness of Quebec can be seen in architectural styles and in settlement patterns. Land division in the St. Lawrence Valley of Quebec, for example, clearly shows the French colonial imprint in the "long lots." Rather than the British pattern of square or rectangular lots, the French assigned land in "rangs"—each farmstead was located on the river bank, lakeshore, or road built parallel to the river or lakeshore. The river, lake, or road frontage was narrow, but the long lot was deep. In this way, each farmer had access to transportation, while road construction and maintenance costs and taxes on waterfront land were minimized.

The French in Canada have resolutely maintained their culture and language despite the change of sovereignty, for two fundamental reasons. First, there is the inherent strength of ethnic pride, reinforced by a language different from that of the majority of Canadians and most of the neighboring Americans; second was the British recognition of French Canadian cultural integrity in the treaty ending the Franco-British struggle for the St. Lawrence. There was agreement that, within the structure of the British Empire, French Quebec could maintain its identity through controlling its educational system and was guaranteed freedom of religion and equality of language. The bitter resentment of the French Canadians is based on their allegation that the equality of language agreed to has not been maintained. Perhaps in retaliation for this perception of past discrimination, the French Canadians have made Quebec the only province that is *not* officially bilingual; French is the only language recognized officially.

Vancouver. *Vancouver is Canada's (and North America's) largest Pacific port. This sparkling city helps make British Columbia a net gainer in Canada's interprovincial migration.*

Vancouver, like Los Angeles, is a city whose growth occurred primarily in the auto age. Vancouver is Canada's second-ranking port, serving as the principal western terminus of transcontinental transport routes. It is also the nation's fourth largest industrial center.

Both developed and potential hydroelectric power are impressive in British Columbia; the abundant energy and scarcity of competition thus have attracted construction of a huge aluminum smelter at Kitimat, near Prince Rupert. Like its U.S. West Coast counterparts, British Columbia has outstanding timber resources; more than half of Canada's softwood reserves are here. Large reserves of coal (presently inaccessible), some oil and gas, and relatively minor deposits of metallic ores complete the resource list.

The Far North

Canadians perceive several successive zones of the northlands approximately paralleling the U.S. border to the east and trending more northwest-southeast in the west. The "near north," which is adjacent to the Canadian ecumene, is an established, aging pioneer zone whose settlements include agricultural activities. The "middle north" is a zone of even thinner settlement, almost all of whose pioneering population is in mining towns. The "extreme north" is almost uninhabited, with isolated groups of Inuit and Indians and the very small numbers of military personnel keeping watch over the potential transpolar trajectories of intercontinental missiles. This last, most remote, and economically least significant area

NORTH AMERICAN GRAIN SURPLUSES
AND INTERNATIONAL MARKETS

In the early 1970s, doom-and-gloom scenarios for the future disturbed many people. Among the alarms sounded was the prediction of famine in much of Asia before the end of the century. Canadian and American farmers were encouraged to invest heavily in expanding their production capabilities. However, world food output rose, partly due to the "green revolution," by 25 percent from 1971 to 1982. China alone increased its output by 40 percent in five years, and India and Pakistan became nearly self-sufficient in grain. Per capita food production rose 16 percent in South America. In the face of this sudden market saturation, grain exporters saw markets and prices stagnate by the late 1970s, and then fall in the early 1980s. The largest exporter, the United States served 41 percent of the international grain market in 1981. By 1987, this market share had declined to 25 percent, as former purchasers became self-sufficient, or even exporters. The present, if perhaps only temporary, world wheat glut is more serious for Canadian farmers than for farmers in the United States, because Canada exports 80 percent of its wheat, in contrast to the U.S. figure of 50 percent.

The basic costs of Canadian and American grain producers are among the lowest in the world thanks to natural advantages, a high degree of mechanization, and superior transportation systems. Canadian and American wheat farmers should enjoy lucrative markets in heavily industrialized Western Europe, whose farmlands are generally not as well suited to grain production. The EC (Common Market) countries, however, chose to subsidize their farmers heavily, and to protect them from import competition. As a result, Western Europe became self-sufficient in basic grains and went on to control 17 percent of world exports by the mid-1980s. This European achievement, however, does not reflect any comparative advantage; it is entirely the result of huge subsidies and unnecessarily high consumer prices. In 1986, Canadian Plains farmers received $130 (U.S.) per ton of wheat, while their EC counterparts got $450 a ton and were protected against cheaper imports.

Thus, the domestic political decisions of foreign countries, and their climate, terrain, and soils qualities, can influence international markets in agricultural commodities. This, in turn, can mean prosperity or recession to farm regions that are major suppliers of exports. Unfortunately, some of the people most in need of food imports cannot afford them, as in many African states. For this complex of reasons related to international trade, some American and Canadian farmers are being forced out of business because of declining prices, while many Third World residents go to bed hungry.

SOURCE: "The Grain Crisis," *Canada Today,* vol. 18, no. 2, 1987.

could be the most affected by radical change in areas of potential mineral exploitation, especially fossil fuels (see Figure 2–7).

The environmental handicaps of short, unpredictable growing seasons, permafrost, and lack of summer heat are compounded by summer droughts. Much of the Far North is characterized by low availability of moisture, which becomes critical during the short growing season. When these climatic problems are accompanied by thin, poorly developed soils over the shield, agriculture is practically eliminated as an economic possibility. The farming fringe advances from the north only in the extremes of the plains in the famous Peace River district.

In the "new frontier" of the middle north, mining towns are the major modern intrusion. About two thirds of the population of the Far North are Amerindian, Inuit, or Metis (half-Indians). The animal life of the Arctic, both land and sea, is relatively sparse and can be locally eradicated by overhunting or overfishing. The biological processes that bring decay and dispersal of waste operate slowly.

The extreme north has a high potential for oil and gas exploitation. Geological conditions similar to those in Alaska's Prudhoe Bay extend eastward along the Arctic coast of Canada to the Atlantic. Given the fragile nature of the environment and the extremely high costs of establishing temporary settlements, many in the work force of future oil and gas fields may be long-distance commuters. It could prove cheaper to fly workers to Arctic sites for an intensive three- or four-day work week and then return them to their homes and families by air rather than construct complete communities in the Arctic oil and gas fields. Such environmental hazards as sewage disposal and large freshwater demands thus could be minimized.

The question of maintaining a humane and realistic relationship with the Far North's original inhabitants is a serious one for both Canada and the United States. The contact of a high-technology society with one essentially in the Stone Age has resulted in a massive adoption by the less advanced culture; now armed with rifles instead of harpoons or spears, the Inuit began to kill more of the sparse supply of sea mammals. Amerindians, along with Inuit whose food supply was based on land animals, also overhunted those animals. Under these circumstances, it was perhaps inevitable that some of these people would slip into a dependent relationship and welfare.

Political-strategic occupation is another type of settlement on the Arctic frontier. Joint Canadian-U.S. defense forces man radar installations and other military bases.

The world's largest island, Greenland, is 85 percent covered by an ice cap averaging 1,000 feet in thickness. In 1979, Denmark accorded home rule (local control over domestic affairs) to its long-time colony. Now known as Kalaallit Nunaat, the new country depends largely on fishing to support its population of about 55,000.

U.S.-CANADIAN COOPERATION: A SUMMARY VIEW

Although Canadians emphasize that their international relations are neither dictated by, nor copies of, U.S. views, the differences must be regarded as relatively minor. Geopolitical reality binds the two as one in event of any aggression against North America. An independent, democratic Canada could not exist at the side of a hostile, conquered United States, nor could the United States long tolerate hostile dictatorship or external control in neighboring Canada. Just as the sources of foreign investment in Canada's growing economy shifted dramatically from Britain to the United States during the twentieth century (accelerated by the liquidation of British investments to help pay for two world wars), so too did Canada's ultimate security shift to a closer relationship with the United States. Any exchange of intercontinental ballistics missiles between a warring United States and USSR would almost certainly take place over the polar routes (the shortest routes between most major industrial centers of each country), which would also cross Canadian territory. This reality led to the joint U.S.-Canadian North American Air Defense Command. This factor combined with their close investment and trade relationships, culminating in a "Common Market," ensures that U.S.-Canadian relations will be based on mutual self-interest.

ENDNOTES

1. Wilbur Zelinsky, *The Cultural Geography of the United States* (Englewood Cliffs, NJ: Prentice-Hall, 1973), p. 4.
2. Melville Herskovits, *Cultural Dynamics* (New York: Knopf, 1964).
3. Alexis de Tocqueville, *Democracy in America* (first published in 1835), ed. and abridged Richard Heffner (New York: Mentor Books, 1956), p. 164.
4. Ibid., p. 20.
5. John Rees and Howard Stafford, "Theories of Regional Growth and Industrial Location: Their Relevance for Understanding High-Technology Complexes," in John Rees, ed., *Technology, Regions, and Policy* (Totowa, NJ: Rowman & Littlefield, 1986).
6. Jean Gottmann, *Megalopolis: The Urbanized Northeastern Seaboard of the United States* (New York: Twentieth Century Fund, 1961).
7. Zelinsky, *Cultural Geography*.

REVIEW QUESTIONS

1. Why has the trans-Atlantic cultural interaction been traditionally more important than the trans-Pacific interaction? How is this situation changing?
2. What are the cultural, economic, and political factors contributing to the "good neighbor" relationship of the United States and Canada?
3. Why has the character of the trade relationship between the United States and Canada changed in the past decades?
4. Contrast New England's industrial assets in the early industrial revolution with its present industrial assets.
5. How did the industrialization of the United States and Canada reshape the nature of agriculture in this region?
6. What are the economic problems of the American Core? What developmental strategies can revitalize the core?
7. Why is the Dynamic South developing at a relatively fast rate now?
8. Why does the long-term migration outflow from rural and small-town America seem to have at least partially reversed?
9. How did the interstate highway network revolutionize location requirements for industry wholesaling and retailing?
10. Briefly describe the sequential changes in Great Plains land use since the first exploration by non-Amerindians.
11. What physical and cultural geographic factors underlie California's special appeal to many Americans?
12. In what ways is California agriculture an innovative, pioneering fringe of U.S. agriculture in general?
13. Briefly describe the role of Ontario in the Canadian economy.
14. Contrast the resources, population, and developmental strategies of Alaska and Hawaii.
15. Why is Canada's heartland a "divided heart"? What are the implications of this split?
16. What geographic factors have contributed to the slowdown in Quebec's economy?
17. Is British Columbia the "Canadian California"? Contrast the economic development and migration rates of these areas.
18. Why were Canada's Atlantic Provinces unable to build on their early settlement to become hinges for the support of western expansion in Canada?

19. What are Canada's developmental strategies with regard to the near north and the middle north?

SUGGESTED READINGS

Berry, Brian. *The Changing Shape of Metropolitan America*. Cambridge, MA.: Ballinger, 1977.

Brunn, Stanley. *Geography and Politics in America*. New York: Harper & Row, 1974.

Clay, Grady. *Close-Up: How to Read the American City*. New York: Praeger, 1973.

Davis, George, and Donaldson, Fred. *Blacks in the United States: A Geographic Perspective*. Boston: Houghton Mifflin, 1975.

Garreau, Joel. *The Nine Nations of North America*. Boston: Houghton Mifflin, 1981.

Gottmann, Jean. *Megalopolis: The Urbanized Northeastern Seaboard of the United States*. New York: Twentieth Century Fund, 1961.

Hart, John Fraser, ed. *Regions of the United States*. New York: Harper & Row, 1972.

Hart, John Fraser. *The South*. New York: Van Nostrand, 1976.

Irving, Robert, ed. *Readings in Canadian Geography*. Toronto: Holt, Rinehart & Winston, 1972.

Jackson, John Brinckerhoff. *Discovering the Vernacular Landscape*. New Haven, CT: Yale University Press, 1984.

McCann, Lawrence, ed. *Heartland and Hinterland: A Geography of Canada*. Scarborough, Ontario: Prentice-Hall of Canada, 1982.

Meinig, Donald. *Southwest: Three Peoples in Geographical Change, 1600–1970*. New York: Oxford University Press, 1970.

Miller, E. Willard. *Manufacturing: A Study in Industrial Location*. University Park, PA: Pennsylvania State University Press, 1977.

Muller, Peter. *Contemporary Suburban America*. Englewood Cliffs, NJ: Prentice-Hall, 1981.

Oxford Regional Economic Atlas: The United States and Canada. New York: Oxford University Press, 1975.

Platt, Rutherford, and Macinko, George. *Beyond the Urban Fringe: Land Use Issues of Non-Metropolitan America*. Minneapolis: University of Minnesota Press, 1983.

Robinson, J. Lewis. *Concepts and Themes in the Regional Geography of Canada*. Vancouver, British Columbia: Talonbooks, 1983.

Rooney, John; Zelinsky, Wilbur; and Louder, Dean, eds. *This Remarkable Continent: An Atlas of United States and Canadian Society and Culture*. College Station, TX: Texas A & M Press, 1982.

Stansfield, Charles. *New Jersey: A Geography*. Boulder, CO: Westview, 1983.

Vogeler, Ingolf. *The Myth of the Family Farm: Agribusiness' Dominance of United States Agriculture*. Boulder, CO: Westview, 1981.

Warkentin, John, ed. *Canada: A Geographical Interpretation*. Toronto: Methuen, 1967.

Watson, J. Wreford, and O'Rioran, Timothy, eds. *The American Environment: Perceptions and Policies*. New York: Wiley, 1976.

Webb, Walter. *The Great Plains*. New York: Ginn, 1931.

White, C. Langdon; Foscue, Edwin; and McKnight, Tom. *Regional Geography of Anglo-America*. Englewood Cliffs, NJ: Prentice-Hall, 1985.

Yeates, Maurice, and Garner, Barry. *The North American City*. New York: Harper & Row, 1980.

Zelinsky, Wilbur. *The Cultural Geography of the United States*. Englewood Cliffs, NJ: Prentice-Hall, 1973.

CHAPTER THREE

Europe

French Riviera.

T o understand Europe, one must recall the time, less than a century past, when European powers held dominion over much of the world. A map of colonial empires in 1914 shows the zenith of Europe's political hegemony over large parts of the world (Figure 3–1). Great Britain was political leader of one fourth of the earth's surface; its navy ruled the world's oceans. France ruled an African-based empire with outliers from the Caribbean to Southeast Asia; the Netherlands governed Indonesia,

Germany and Belgium controlled African territories much larger than their home states, and the reduced empires of Spain and Portugal were still extensive.

Europe also led the world in industrial technologies and productivity. One or another European state was then first in the world's output of industrial products. Although the United States, Japan, and Russia were expanding industrial states not to be underestimated, economic and political geographers looking at the world in mid-1914 could hardly have been

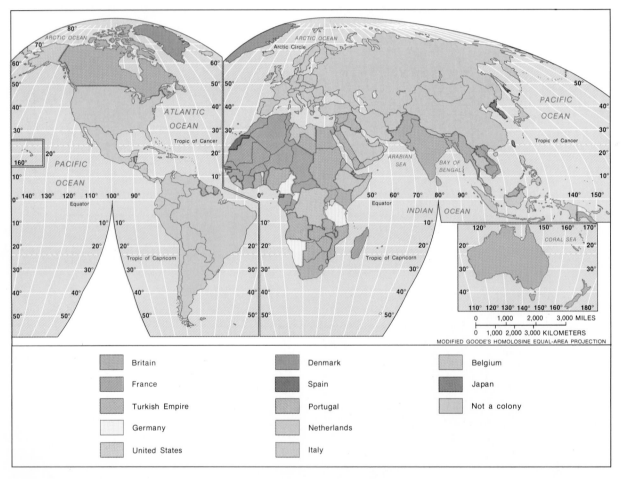

Britain

France

Turkish Empire

Germany

United States

Denmark

Spain

Portugal

Netherlands

Italy

Belgium

Japan

Not a colony

MODIFIED GOODE'S HOMOLOSINE EQUAL-AREA PROJECTION

FIGURE 3–1
Colonial Possessions, 1914. When World War I began, European powers controlled colonial territories throughout much of the world, particularly Africa, South Asia, and Southeast Asia. Britain alone ruled one quarter of the earth's surface.

expected to foresee the destructive effect of two world wars on Europe.

Within the memory of a single generation, except for minute remnants, colonial territories directly administered by the home country have advanced to independence. Seemingly, economic leadership and industrial power also have been eclipsed. It is not surprising that many non-Europeans view individual European states as shadows of their former glory. Each seems almost a pygmy in comparison to the United States or the Soviet Union. The end of World War II saw an exhausted, largely ruined Europe caught up in the new struggle for world leadership between the United States and the Soviet Union. Much of Europe's industrial and transportation infrastructure had been wrecked, including whole cities, such as Rotterdam, Warsaw, and Hamburg. Much accumulated wealth and overseas investment had been liquidated to finance wars. The enfeebled colonial powers clearly could not long hold their colonies as independence movements gathered momentum. Is today's Europe a collection of the once-important overshadowed by the new giants?

By long tradition, Europe is classed as a continent. However, many geographers speak of "Eurasia," the single, massive continent of which Europe is merely the western peninsula and associated islands. This approach to Europe's continental status symbolizes a relative demotion of Europe in non-Europeans' perceptions. A few generations ago, Europe was perceived as the source of civilization. Indeed, it was there that the industrial revolution was initiated, along with associated changes in living standards.

It would not be accurate, however, to categorize contemporary Europe as an unimpressive array of small states living in the shadow of the superpowers and incapable of meeting the economic and technological challenge of Japan and the rising new industrial states of the Third World. Europe can be understood as a boldly conceived mosaic. Viewed up close, each state is characterized by highly varied physical, economic, ethnic, religious, political, and cultural landscapes. These landscapes often stand in sharp contrast to those of neighboring states. It seems to be a culturally fragmented continent, split by different languages, historical conflicts and rivalries, and religious and other cultural contrasts. Yet, collectively, the states and peoples of Europe represent a major

economic force in the contemporary world. But it is a disunited giant. If Europe is viewed as a unit, even though it may never achieve complete economic or political unification, then a generalized picture of great economic power emerges from the pieces of this mosaic. Though they represent little more than 10 percent of the world's population, the 495 million Europeans account for an astonishing 30 percent of the world's total gross national product (GNP) (the money value of all goods and services produced in a year). In comparison, the United States, with less than 5 percent of the total world population, produces a little over 25 percent of world GNP, while the USSR, with 5.7 percent of the world's people, accounts for 13 percent of the planet's GNP.

Among the world's major regions, Europe's average GNP per capita ranks it third, behind the United States and Canada (ranked together) and Australia-Oceania but ahead of the USSR and far above the world average. With most European states in the concluding stage of the demographic transition, Europe is the only world region in which total population is currently projected to decline (slightly) after the year 2000. In the future, the proportion of Europeans among the world's people certainly will be smaller. Their economic and cultural contributions to the world, however, are less likely to decline.

A troubled past sometimes seems to weigh heavily on the collective consciousness of Europeans. Prideful memories of a rich heritage and immense cultural contributions to the world have a dark side: old grudges remain vivid in memory. Partisans of one side will annually celebrate a famous victory, while the descendants of the opposing side pledge to someday avenge their defeat; a street brawl may follow. It often astonishes outsiders to discover that the particular battle, seemingly so fresh in memory as to inspire passionate threats and counterthreats, took place during the Hundred Years' War (1338–1453). The anniversary of the Battle of the Boyne (1690) is still the cause of riots in Ireland. Patriotic Scots still grimly remember unpleasant details of the Battle of Culloden (1746), and Poles can recite a long list of grievances against the Germans going back to the invasions of the medieval Teutonic Knights. Such long-cherished animosities are common in Europe. Unfortunately, a long list of much more recent wars, invasions, occupations, territorial transfers, organized looting, official discrimination, and general friction sustains suspicions and resentments. It is remarkable, if not a miracle of sorts, that the governments and peoples of Europe have managed to achieve any cooperation among states, or even within states!

However, since 1945, there have been no major wars in Europe, a long and fairly calm period for this region. An amazing amount of cooperation *has* been achieved, and the future seems bright. Some of this international cooperation, it is true, was imposed on European states, the military alliances directed by the USSR in particular. Other multinational agreements may have come about partly in response to Western Europeans' concerns about possible Soviet invasion, but many forms of cooperation were positively, and internally, inspired.

The non-Communist European states frequently and vocally demonstrate their differences with the United States, exhibiting a full range of opinions and loyalties from faithful ally to neutralist. Even the Eastern European Communist regimes often manage to bend and adapt the usually more rigid ideological dogma of the USSR. Europeans are coping with their new geopolitical environment by reaching out in many directions, frequently pragmatically ignoring the East-West political split. As a unit, or more realistically two economic blocs with a few neutral abstentions here and there, Europe can still play an important role in the world. Europe should not be underestimated.

The industrial revolution, that complex association of significant and rapid changes in economic, political, social, and technological developments, was born in northwestern Europe. This great revolution in the way people used raw materials, energy, transportation, and technology to improve the average living standard is one of the unifying factors of this region. Not all the states of Europe have made the same level of progress through the industrialization-modernization process, but all have been affected by it.

Traditionally and historically, Europe has been considered one region. As in the contrasts of the EEC and CMEA, already noted, or in the sharp split between Warsaw Pact countries allied with the USSR and Western European states joined in NATO (North Atlantic Treaty Organization) with the United States, Europeans live with a sharp divide in their region. More than four decades of Communist governments have led the eight Eastern European states of East Germany, Poland, Czechoslovakia, Hungary, Yugoslavia, Bulgaria, Romania, and Albania on a different, sometimes divergent, path from the Western European capitalist and mixed (capitalist-socialist) economies. This East-West divide is more than one of political systems and alliances. Whereas Western European economies emphasize light-industry consumer goods, the centrally planned Communist states tend to concentrate on heavy industry for further industrialization. Average personal incomes in Western Europe far outpace those of Eastern Europe.

TRENDS AND PERCEPTIONS

If a single phrase can serve as a capsule description of Europe, it is "small but complex." Europe's physical geography, and, indeed, its economic and political geography, have been strongly influenced by the complicated intermeshing of land and sea. This is a peninsula subdivided into more peninsulas and islands (Figure 3–2). Europe's twisting, convoluted coastline gives it an unusually high ratio of coast to area. Almost all of Europe's cities are readily accessible to maritime trade. The deeply indented coastline separates peoples, yet also provides mutual access.

EUROPE WEST AND EUROPE EAST

Eastern Europe is the zone of Moslem-Catholic-Protestant-Orthodox contact. The linguistic meeting ground of Romance, Slavic, Germanic, and other languages, it was the traditional buffer area against attack by non-European elements (Figure 3–3, p. 146). Eastern Europe is thoroughly varied culturally because of its position astride the major routes of the Crusades and Near Eastern trade as well as important east-west routeways. Traditionally a land of food surplus, major parts of it were incorporated into the Roman, Byzantine, Holy Roman, and Ottoman empires for lengthy periods of time.

FIGURE 3–2
Countries, Principal Cities, and Subregions of Europe. Europe has been
divided into six subregions for ease of comprehension.

TOWARD ECONOMIC FEDERALISM

The many states of Europe (there are 27, not counting the microstates of Andorra, San Marino, Monaco, Liechtenstein, and Vatican City) have, under various pressures, moved toward several varieties of economic integration. As in so many other aspects of their economic and political life after World War II, the western and eastern states have gone their separate ways, although east-west frictions seem to be easing now.

The two superpowers—the United States and the Soviet Union—have demonstrated the economic power of huge units. These great states grew with an unfettered internal flow of raw materials, finished goods, capital, labor, and energy. All could be freely interchanged, unhampered by political borders. An economic union looked like the solution to the serious recovery problems of Europe after World War II. Prosperity through cooperative effort could help build stability and reduce old rivalries.

The United States, facing deteriorating relations with the Soviet Union, its wartime ally, sought to rebuild the industrial democracies of Western Europe as quickly as possible. This desire to help Europe recover led to the Marshall Plan of 1947. Europeans were to allot the aid and organize recovery; this challenge was met by the Organization for European Economic Cooperation (OEEC), a forerunner of the Common Market.

The European Economic Community (now the preferred term for the Common Market) is abbreviated either EEC or, more commonly in Europe, EC. As a unit, the EC outproduces the Soviet Union and ranks second only to the United States in overall economic-industrial power.

The EC, whose full members are the United Kingdom, West Germany, France, Belgium, the Netherlands, Luxembourg, Italy, Eire, Greece, Spain, Portugal, and Denmark, is not a federal state like the United States (Figure A). Although some forms of federalism exist, such as the Assembly or European Parliament, they function more as consultative bodies than legislative ones.

In 1949, the Soviet Union led the Communist states of Eastern Europe into an economic plan called the Council for Mutual Economic Assistance (CMEA). This communist version of the Common Market, known as Comecon, is intended to integrate effectively the Soviet and Communist East European economies. The USSR, Poland, Czechoslovakia, Hungary, Romania, Bulgaria, and the German Democratic Republic (East Germany) are full members. Yugoslavia is an associate member, while Albania refuses to be considered part of the Soviet bloc in any fashion. Whereas the EC is based on multilateral, cooperative agreements, Comecon is less an association of equals but rather is dominated by the Soviet Union.

Britain once took the lead in organizing the European Free Trade Association (EFTA), with Switzerland, Austria, Denmark, Iceland, Norway, and Sweden; Finland was an associate member. Both Britain and Denmark have since left to join the EC. Since 1973, there have been no industrial product customs barriers between the EFTA and the EEC.

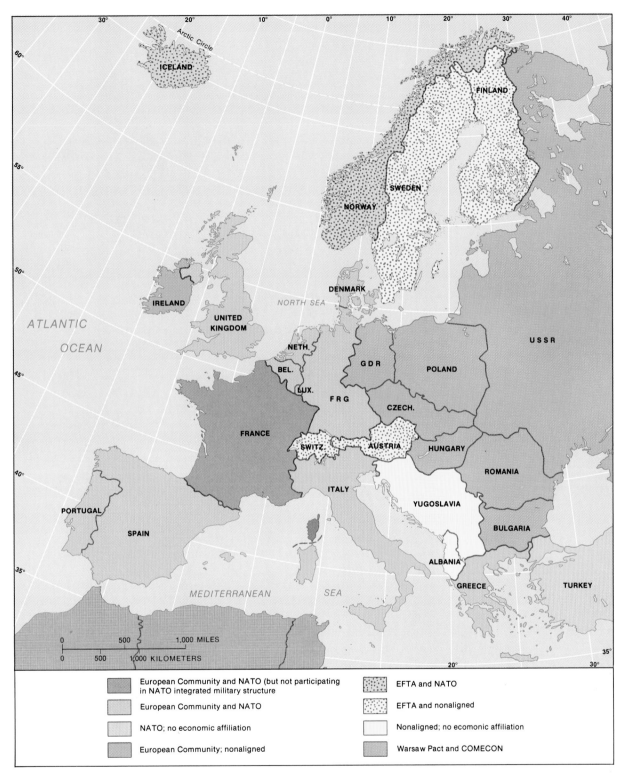

FIGURE A
European Economic and Political Affiliations. Many European states are
members of a transnational economic bloc or political-military alliance.

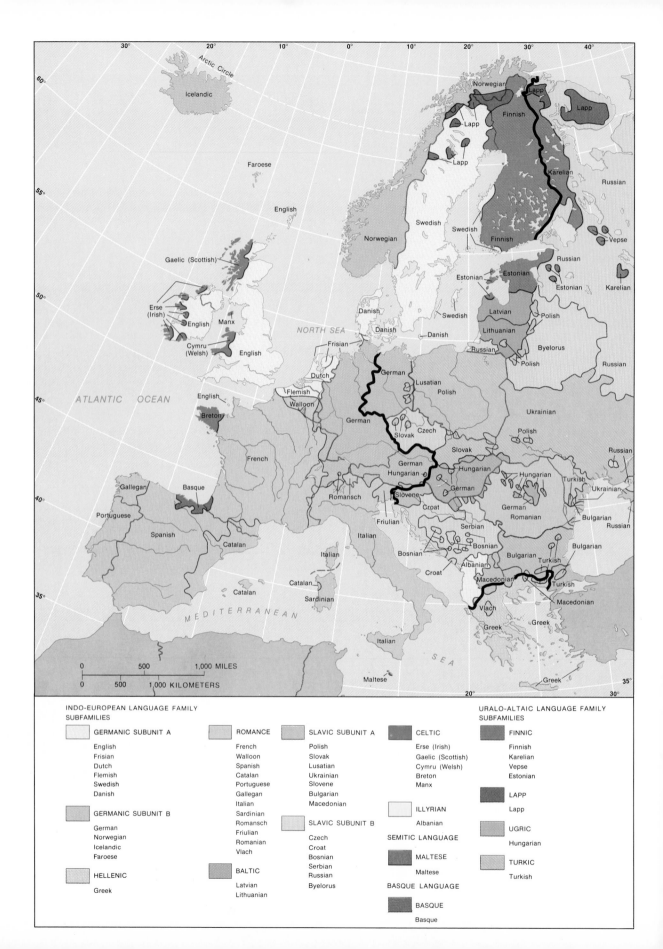

INDO-EUROPEAN LANGUAGE FAMILY
SUBFAMILIES

GERMANIC SUBUNIT A

English
Frisian
Dutch
Flemish
Swedish
Danish

GERMANIC SUBUNIT B

German
Norwegian
Icelandic
Faroese

HELLENIC

Greek

ROMANCE

French
Walloon
Spanish
Catalan
Portuguese
Gallegan
Italian
Sardinian
Romansch
Friulian
Romanian
Vlach

BALTIC

Latvian
Lithuanian

SLAVIC SUBUNIT A

Polish
Slovak
Lusatian
Ukrainian
Slovene
Bulgarian
Macedonian

SLAVIC SUBUNIT B

Czech
Croat
Bosnian
Serbian
Russian
Byelorus

CELTIC

Erse (Irish)
Gaelic (Scottish)
Cymru (Welsh)
Breton
Manx

ILLYRIAN

Albanian

SEMITIC LANGUAGE

MALTESE

Maltese

BASQUE LANGUAGE

BASQUE

Basque

URALO-ALTAIC LANGUAGE FAMILY
SUBFAMILIES

FINNIC

Finnish
Karelian
Vepse
Estonian

LAPP

Lapp

UGRIC

Hungarian

TURKIC

Turkish

Eastern Europe's obvious natural riches and strategic trade importance were such that Rome, Austria, Germany, Sweden, Greece, Persia, France, Italy, Turkey, and Russia each attempted to control all or part of it at various times. Repeated attack and nearly constant control from outside the region eclipsed native political control and resulted in poverty for many eastern Europeans.

In western Europe, Germanic and Romance languages prevail (see Figure 3–3). The major exception is the "Celtic fringe," including Erse, the official language of Eire (although most Irish speak English). Smaller groups using Celtic languages are found in Scotland, Wales, and Brittany. Basque, a language that does not share the ancient Indo-European roots of other European languages, is spoken by an ethnic minority in both Spain and France, and Catalonian is spoken in northeastern Spain. A Finno-Ugric language is spoken in Finland, though many Finns also speak Swedish, another Germanic language. In eastern Europe, Slavic languages are spoken in Poland, Czechoslovakia, Yugoslavia, and Bulgaria. The Hungarian language is related to Asian languages, whereas Romanian is in part a Romance language. Greek and Albanian round out the list of languages.

European states have shared the cultural heritage of Greece and Rome, blended with important contributions from Teutonic, Byzantine, and Saracenic cultures. Much of Europe was at one time part of the Roman Empire, and almost every portion of this region was influenced to some degree by Roman trade and culture.

Christianity, in its Roman Catholic, Orthodox Catholic, and Protestant forms, is almost the universal religion, at least nominally, of Europe (Figure 3–4). Judaism remains the most important minority religion despite the tragic holocaust and large-scale migrations to Palestine-Israel. Relatively small minorities of Moslems are found in Britain, France, and the Netherlands; like Britain's small Hindu minority, these people are primarily products of recent immigration from former colonies. The influence of Islam is much stronger in Albania, Bulgaria, and Yugoslavia, formerly dominated by Turkey. Secularism is common among Europeans, who, though nominal church members, may pay little attention to formal religion in everyday life.

Western Europe shares a strong maritime orientation. Excepting landlocked states, many European states have developed ports of international rank. Many are both major shipbuilders and operators of extensive merchant fleets.

One of western Europe's salient characteristics is its continuing aspirations for world leadership in cultural affairs. European art, ancient through contemporary, still carries prestige. Rome remains the administrative center of Roman Catholicism, with over 500 million adherents worldwide. Canterbury remains an important influence on worldwide Anglican church communities.

Eastern Europe's eight small political units each usually contain more than one major ethnic group. While virtually all these states contained minorities, two—Czechoslovakia and Yugoslavia— are actually mosaics of national groups. In general, ethnic complexity reached its greatest intensity in the Balkans.

Economic Characteristics: The East-West Contrast Intensifies

Although all of Western Europe has state-owned railroads, state-owned or state-subsidized airlines, and varying degrees of ownership of basic industries such as electric generation, steel, and coal mining, there remains some reliance on entrepreneurial capitalism. Privately controlled businesses and corporations may be subjected to comprehensive regulations, but only selected parts of the economy are directly owned and managed by governments. Private ownership of housing, farmland, and retail business remains. Democracy is the universal ideal.

In Communist Eastern Europe, agriculture has been collectivized in most cases, and the level of investment in that sector has been low. Yet, all Communist states have made compromises with the peas-

FIGURE 3–3
Languages of Europe. The areal distribution of languages, a major component and carrier of culture, frequently is reflected in national borders, territorial claims, and past wars.

FIGURE 3–4
Religion in Europe. The geography of religious identity in Europe features three major variants of Christianity, Judaism, and Islam.

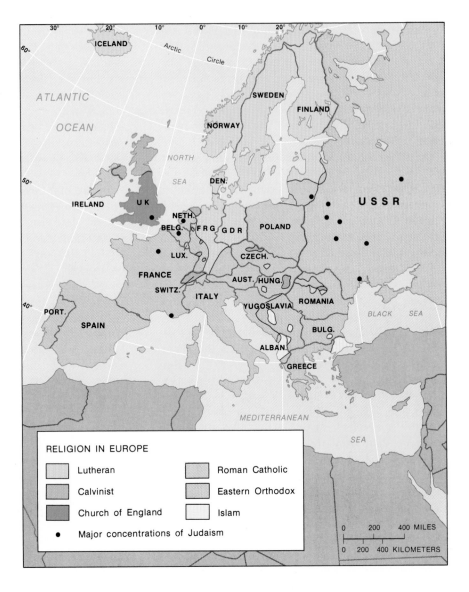

RELIGION IN EUROPE

- Lutheran
- Calvinist
- Church of England
- Roman Catholic
- Eastern Orthodox
- Islam
- Major concentrations of Judaism

antry in implementing their particular type of governmental ownership and direction in the farming component of the economy.

At the onset of Communist control, individual self-sufficiency and basic industry were emphasized. Each Communist European state attempted to develop a heavy industrial base as well as a full range of other industries to ensure self-sufficiency. Later programs emphasized specialization, and each Communist government attempted to develop a series of export-oriented specialties.

There is increasing trade with the capitalist nations of the West. Meanwhile, the Soviet Union, in a reversal of the traditional colonial trade pattern, has become the major supplier of raw materials to Eastern Europe's industries while purchasing finished goods and sophisticated manufactures from its so-called satellites. Soviet natural gas and oil move through pipelines, supplying not only Eastern European economies but Western countries as well. There is some concern in the West about some Western European countries becoming dependent on Soviet gas supplies, a potential strategic problem if East-West tensions mount.

THE NATURAL ENVIRONMENT

Europe, essentially a westward peninsula of the huge landmass of Eurasia, itself splits into peninsulas and fragments into fairly large islands (Figure 3–5). The east-west orientation of the continent is reflected in the similar orientation of its greatest continuous lowland, the North European Plain, and the directional bias of some of its mightiest mountains, the High Alps. Blocks of land alternate with bodies of water, adding to the complexity, as seas and bays extend far inland.

Landforms

Over one-third its extent, Europe is a level plain; even its mountain ranges rarely block movement to any great degree, although their "barrier" effects were important in the past. Much of the area is usable for cultivation or at least pasture. The regions are interconnected by land corridors, by sea, by river systems—all of which have acted more to enhance movement and interaction than to deter it, although ease of movement is far greater now than in the formative period of this mosaic of cultures.

The Alps occupy an arc stretching from the French and Italian Rivieras to the eastern borders of Austria. Sharpened into peaks by glaciation, this highest portion of the Alpine system is *the* Alps, the most rugged part of Europe. Lower, related chains spread out in several directions. The Apennines, forming the central spine of Italy, the Sierra Nevada of southern Spain, and the Pyrenees, which form the Franco-Hispanic border, are all parts of the same system; so are the Pindus of Greece and the Carpathians of eastern Europe.

In Spain, the lowlands of Andalusia and the Ebro Valley are shallow basins between the Alpine ranges and the old, rugged Iberian massif. Still larger is the basin of the Po River of northern Italy.

The Hungarian Plain lies between the Carpathian Mountains, to the north, and the Dinaric Alps, to the southwest. The Romanian Plain, cradled within the reverse S-curve formed by the Carpathians, the Transylvanian Alps, and the Balkan Mountains, is a rich and productive agricultural area.

Just to the north of the Alps lie the Central Highlands, a remnant of ancient mountains. Major highland areas form the bulk of the Spanish Meseta, the Massif Central of France, the Franco-German highlands, and the Bohemian massif in Germany and Czechoslovakia.

Stresses accompanying Alpine mountain building often fractured the crust into a series of fault blocks, some thrust up, others dropped down. Where zones of weakness allowed, volcanoes formed or lava flowed over the surface. These events resulted in the most complex geologic features of Europe. Up-tilted fault blocks form the Vosges of France; down-dropped blocks form large, linear valleys such as the Rhine Graben (Trench) and the valley of the Rhone. Much of central Europe is a country of mountain, basin, and low plateau alternating across the landscape. The Bohemian Block is a mass of ancient crystalline rock whose steep sides form the Sudeten, Ore, and Bohemian Forest mountains, all of which are rich in ores and minerals.

The European Plain (European Lowlands) stretches from the Pyrenees to the Urals, widening out at the Polish-Soviet border to encompass the width of the entire continent. North of the delta of the Rhine and the London Basin, the plains are composed of glacial deposits left by the continental ice sheets of Pleistocene time. Old glacial spillways, valleys formed by great rivers of meltwater at the close of the glacial period, form a series of east-west valleys stretching in a series of gentle arcs from the North Sea to western USSR. In places, they are interconnected by short north-south spillways oriented at right angles to the major valleys. Many have been excavated to form Europe's dense network of canals. Along the northern edges of the plain are found gravelly morainal hills which form the Danish islands, the eastern edge of that country's mainland peninsula, and the hills of the Baltic coast. Small outliers of the plain are found in southern Sweden and the Swedish islands.

The oldest mountains of Europe, the Northwestern Highlands, are really two sets of mountains. The older portion forms most of Sweden and Finland.

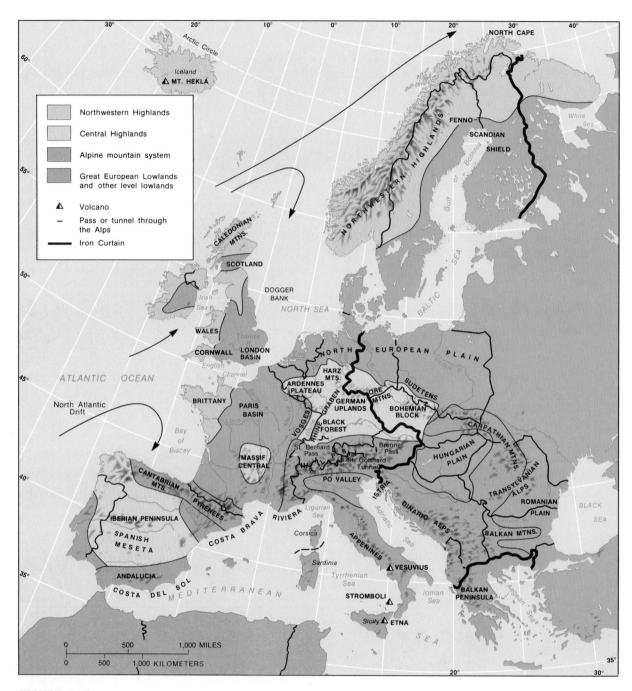

FIGURE 3–5
Selected Physical Features and Physiographic Regions of Europe. Europe,
itself a peninsula of the Eurasian landmass, is fragmented into many peninsulas and
islands, complicated by a complex pattern of highlands and lowlands.

Over a billion years old, repeatedly eroded by glaciers and constantly eroded by water, the mountains were reduced to a plateaulike surface. Called the Fenno-Scandian Shield, it is similar in age, origin, and landscape appearance to the Canadian Shield. Only along the Norwegian borders do the features rise to a significant height.

The Caledonian Mountains, eroded almost level at one time, were later uplifted into low mountains. Glaciers deepened the valleys. Uplifted blocks form the Scottish Highlands and the mountains of Norway. Iceland, a recent volcanic creation, dates to the time of great volcanic activity elsewhere in Europe. The extreme complexity of this landscape is reflected in the diversity of European land use, mineral wealth, and crop patterns.

Climate

Flanked to the north, west, and south by oceans and seas, Europe is unusually warm, considering its northerly latitude. The continent's western and northwestern coasts are bathed by warm currents. The North Atlantic Drift brings relatively warm water all the way north to Norway's North Cape, and even seeps into the USSR's White Sea. The westerly winds thus transport the relative warmth, as well as the moisture, of maritime air onto the continent (Figure 3–6).

Maritime influences are felt throughout the continent to some degree. Seasonal and daily temperature ranges increase with distance inland from the sea. Rainfall is greatest in mountain areas and along the west coast. Sheltered valleys and basins in the north receive less precipitation, as do the lands of southern Europe. Winter precipitation maxima occur in the Mediterranean lands, the mountains of the British Isles, and along the Norwegian coast. Summer maxima occur in the lands along the Baltic Sea and in the Fenno-Scandian Shield. Only the Mediterranean lands have a definite dry season, which occurs in summer. The rest of the continent receives precipitation throughout the year, some of it with monotonous regularity. Total rainfall generally decreases eastward.

Few would call the climate of northwestern Europe pleasant, but there is no doubt that it is **temperate**. Seasonal and daily temperature ranges are

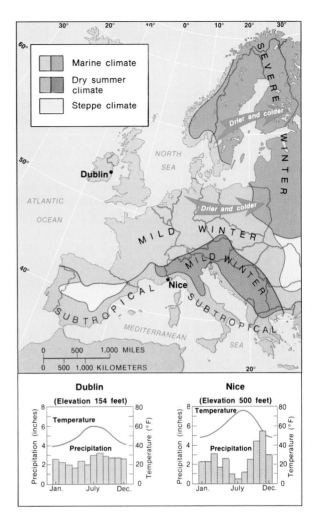

FIGURE 3–6
Climates of Europe and Average Monthly Temperatures and Precipitation for Selected Cities. Latitude, ocean currents, wind direction, mountain barriers, and distance from the sea are factors in Europe's complex climatic patterns.

small. Winters are rarely bitterly cold, but they are damp and uncomfortably raw. Summers are never hot; they are damply cool. Rains seem endless, if seldom intense. The skies are cloudy—virtually all day. Winter skies tend to be dull and overcast, and rain comes as a mist or drizzle. Fog, present at all seasons, is most frequent in winter. Summer skies have few sunny days. Total rainfall is surprisingly light, some 10 to 40 inches except in the mountains and along the northwestern coast. Nineteenth century environmental determinists, noting the spatial coincidence of this climate with their ethnocentrically defined center of civilization, called the climate "stimulating." Perhaps more would agree with Karl Marx's verdict: "tubercular"! Areas north of the Alps and south of the Baltic, the western British Isles, and the Norwegian coast have a maritime west coast climate—cool, cloudy, and wet virtually all year. The Scandinavian countries, the mountains, and sheltered interior portions of Germany are subject to a more continental climate with larger daily and seasonal temperature regimes, warmer summers, and sharply cold winters. There are no dramatic seasonal changes. Spring creeps imperceptibly forward. In fall, the leaves turn from green to dull yellow to brown and disappear in the drizzle and fog, and it is winter again.

A few areas fare better. The southern half of France, transitional to Mediterranean conditions, is much more sunny (and considerably warmer) in summer. Bright, sparkling weather does occur in the Alps in winter. Any southward-facing slope is a little warmer in summer. Any leeward area (facing away from prevailing winds) is somewhat drier and a little less cloudy.

There are considerable advantages to this climate, including a six- to nine-month growing season almost everywhere except in higher altitudes. Weather disasters are few. The Mediterranean climate, with its dry, sunny summers and mild winters, provides a complementary agricultural area and a tourist magnet during the long winters of the north. Most of southern Europe is adequately watered, and has mild winters and a long growing season. The Mediterranean-type climate is sunny and pleasant throughout the year. Most rainfall occurs between October and May, though even then the skies are sunny between brief showers. In summer, drought is the norm. Winter temperatures are mild and frosts are infrequent. Sheltered or coastal areas are generally warmer than interior areas. Even in winter, the nights are cool rather than cold, and snow is rare.

The climates of north-central Europe are somewhat more typical of this nearly subpolar latitude. Yet even here, the modifying influences of the ocean are not absent. The generally east-west trend of mountains enables the prevailing westerly winds to carry the ocean's moderating influence far eastward. The shallow Baltic, frozen for over half of its extent in the winter season, still has a tempering effect. The result is a climate that is more nearly continental, yet not as extreme as that of similar latitudinal areas in Canada or Siberia.

Further from the Atlantic, the winters of eastern Europe are much colder than those to the west, and the seasonal and daily temperature ranges are also greater. Nonetheless, the general maritime influence is felt, and the area has a much warmer climate than might be anticipated at that latitude.

East Germany, Poland, and most of Czechoslovakia experience relatively mild winters. Temperatures decline eastward, indicating the very real significance of the maritime effect. Summers are relatively cool, and the transitional seasons are very long.

Southeastern Europe, with the exception of Greece, has summers that are long and hot, with precipitation peaking in spring and early summer. Generally, the Balkan countries, astride the Mediterranean-continental climatic divide, experience the greatest fluctuations in weather conditions.

Soil and Vegetation

Each of Europe's more populous countries contains a core of superior soil, which frequently becomes the economic and political core of the country as well. Complex interrelationships of climate, vegetation, drainage, and parent rock material make soil. European soils, however, have been greatly modified by thousands of years of human use. Population pressures encouraged the clearing of forests, terracing of slopes, fertilization of poorer soils, drainage of swamps, and even reclamation of land from the sea floor. Perhaps only in East Asia have people modified the original soil to a comparable degree.

Faced with limited supplies of land in relation to population, Europeans have resorted to technical efforts to improve land quality, increase the arable area, and improve yields. The results are high food yields and the ability to support large populations with the produce of relatively small areas. The ability of most European states to hold down imports of food to a relative minimum is a tribute to European technology in particular and human ingenuity in general.

In few areas of the world have people so completely modified natural vegetation as they have in Europe. The distinctly Mediterranean areas produced forests of oak, cypress, and pine. Human use (or misuse) often reduced that forest to a cover of drought-adapted shrubs. The northern, or boreal, forest consisted of pine, spruce, and birch, with oak probably a late invader. Commercial plantings have favored spruce because of its rapid growth and value for pulp and paper. The zone between, often called temperate forest, is a matrix of broadleaf deciduous trees with islands of pine or mixed forest. Higher altitudes produce conifer forests.

A major environmental concern, particularly in central Europe, is *waldsterben,* or "forest death." Industrial pollution, including acid rain resulting from incomplete combustion of coal and oil, is blamed for widespread destruction of trees. While some trees succumb directly to pollution, many more are weakened by it, becoming easier prey to insects and plant diseases.

Heath and moor extend discontinuously across northern Europe in areas where glaciation occurred. Both are associations of low, scrubby vegetation. Heath is a complex of heather and other shrublike plants. Moor is generally found in rolling, broken countryside and contains bogs in lower spots. It is probable that the spread of these vegetational types over large areas is the result of land misuse.

The barren grasslands of the Yugoslav karst (limestone) country are caused by the absence of surface water. Sparsely wooded steppe was the normal vegetational cover in most of the Danube basin.

CULTURAL LANDSCAPES

Although, as we have seen, there are considerable variations among natural environments in Europe, the cultural landscapes present the greatest variety. Perhaps one reason that European states seem almost overwhelmed by tourists is the strong association of culture with distinctive, appealing landscapes.

The scale of the cultural landscape makes a striking impression on visiting North Americans; the scale of the countries themselves seems small. The cultural heritage of a people makes an impact on the landscape that they shape, build, and create. There is a fine texture of cultural landscapes in Europe, a legacy, in part, of times past when communications and transportation were slow and inefficient. Relatively small, specific environmental complexes, for example, may have presented few options in building materials. Slate may have been available from nearby quarries, and used almost universally for local roofing. Only a few score miles away, the combination of custom, taste, and previously high transport costs may have favored roofs of wheatstraw thatch, while in a neighboring district, baked clay tiles or wood shingles may have been the preferred material.

In architectural traditions established well before the time of central heating, air conditioning, and relatively inexpensive fuels, people sought architectural solutions to environmental challenges. Houses in the south of France, for example, were often built on southern slopes. Floor-to-ceiling windows allowed the winter sun to flood into the rooms, providing passive solar heating. In summer, louvered shutters were closed over open windows, permitting free circulation of air but keeping out the direct rays of the sun.

Architectural and decorative traditions can be spread by migration, conquest, or colonization. A house type common in North Africa and the Middle East, featuring a central, open patio, was brought to Spain by the Moors. The enclosed open space allows outdoor activities in privacy; inhabitants may even choose to sleep under the stars in complete security. Arcades around interior walls of the patio keep direct summer sun off the walls of rooms, and the lavish use of stone and tile flooring provides surfaces that are cool to the touch. A patio fountain, with its splashing, running water, helps cool the air during hot summers while soothing the soul.

Each national culture in Europe has evolved at least one, often several, regional cultural landscapes. The particular complex of building styles, urban planning, traditional materials, and decoration comes, in time, to symbolize the nation. Central Warsaw, for example, contained many fine old buildings that summarized Polish architectural traditions. The retreating Nazis, toward the end of World War II,

__Historic Area of Central Warsaw.__ The unique architecture of Warsaw's "old city"—a prized symbol of national culture—was meticulously restored after wartime destruction by the Nazis.

deliberately reduced the central area to heaps of smoking rubble. The Poles gave high priority to meticulous rebuilding using old photographs to make exact reproductions; the buildings were invaluable as an expression of cultural pride and identity.

THE STRUCTURE OF SUBREGIONS

Proximity, history, cultural similarity, and developmental levels have been used here in subdividing Europe into six subregions (see Figure 3–2).

The Continental Core

One of the greatest concentrations of cities and industrial centers of the world lies in France, Switzerland, Austria, northern Italy, West Germany, and the Benelux countries (Belgium, the Netherlands, and Luxembourg). To a significant degree, this complex is the geographical, economic, and cultural **core** of Europe. France has long taken a leadership role in European culture. West Germany, strongly resurgent after the wartime destruction and postwar division of Germany, is one of the world's foremost industrial

states. Indeed, the French and German nations have contended for continental leadership for centuries.

Together with the Benelux countries on their North Sea flank, France, Italy, and Germany were the original members of the European Community. The nucleus of the EC was formed by the European Coal and Steel Community, in which the two historic antagonists—France and (West) Germany—cooperated in economic recovery. Although major European industrial powers (Britain, Sweden, East Germany, Poland) and emerging industrial economies (Spain, Yugoslavia, Czechoslovakia) lie beyond the Core, this heartland is the greatest concentration of economic power in this important region.

The significance of international flows of raw materials throughout the subregion is underlined by the geographic distribution of the resource base (Figure 3–7). Although France has a fairly large array of mineral resources, it must import large quantities of coal from West Germany. Iron ores are plentiful; the largest fields (in Lorraine) are well situated for interchange with West Germany whose complementary abundance of good-quality coal is not far from the French border. This geographic juxtaposition suggested the formation of the European Coal and Steel

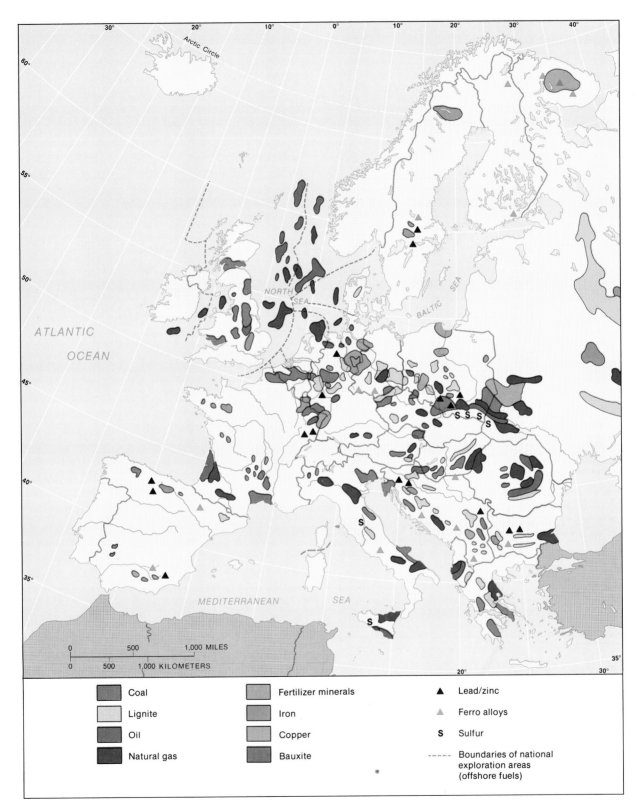

FIGURE 3–7
Mineral Wealth of Europe. The large, diverse stock of minerals has helped to
support industrialization throughout Europe.

Community of 1951, beneficial to both French and West German steel industries.

In addition to its coal production, West Germany ranks as one of the world's largest producers of lignite, or "brown coal" (low in thermal energy content but used to generate electricity). Neither France nor Germany has any immediate or foreseeable chance of following Britain and Norway into petroleum self-sufficiency. In natural gas production, Germany is third, after the Netherlands and Britain.

Belgium's coal deposits have been worked for so long that production has been in decline. The Netherlands' natural gas deposits (at Groningen) fuel its economy and earn foreign exchange through gas exports via pipeline to France, Belgium, West Germany, Switzerland, and even Austria.

Iron ore is Luxembourg's major resource, although production is down from the peak of the 1960s. The Netherlands has no important resources other than its fossil fuels, and Belgium is little better off, with only minor iron ore supplies.

Italy is a large producer of both hydroelectric energy and natural gas; there is little potential for further expansion of hydro power, and natural gas deposits are being tapped so rapidly that they are unlikely to last beyond the end of the century.

Switzerland has no significant mineral resources under current exploitation. Austria, on the other hand, has a considerable range of minerals, including oil, natural gas, lignite, and iron ore, but quantities are rather small. Swiss hydro potential is almost completely developed; climbing consumption will be met by adding new nuclear plants. In contrast, little more than half of Austria's hydro potential is used.

France, once regarded as the premier example of a nation that had reached the very low growth stage at the end of the demographic transition, has recently begun to grow again, but with a doubling time—the time needed to double the population at current growth rates—of nearly 200 years! Population experts warn that much of Europe is entering a **demographic winter**—a period of an aging, numerically declining population. The ratio of retired people to active workers is increasing, producing a strain on entire economies. Health care costs are rising sharply. Deaths outnumber births in several European states, a situation which, if not reversed, will cause national groups eventually to disappear. As with Britain, so with France: withdrawal from empire accelerated a return of French settlers and also an influx of ethnic groups from the former colonies.

West Germany's population expanded after 1945, largely through in-migration. Nearly 2 million fled East Germany before the infamous wall shut off escape. Employment prospects have attracted inflows of Italians, Spaniards, Greeks, Yugoslavs, and Turks. Rising social problems and economic slowdowns have led to legislation controlling this influx of foreign *gastarbeiters* ("guest workers"), but EC countries only reluctantly modify the right of open borders to fellow EC workers. West German birthrates are now so low that the government is expressing concern for the future of the German nation. The population has actually declined since 1975.

Belgium's population density is one of the highest in Europe. This country straddles the shatter zone of Romance and Germanic culture and language. Unlike Switzerland, whose multilingual population achieved harmony and identity before nation-states based on unity of language crystallized around it, Belgium has a potentially serious division. A relatively sharp east-west line separates the Flemish-speaking majority in the north from the 45 percent of the population (including bilinguals) who speak Walloon (a French dialect) or French. Brussels, the capital, is approximately astride the linguistic-cultural divide.

Luxembourg's third of a million residents have a very low rate of increase. Almost 20 percent of the work force is of foreign origin. The Netherlands, by contrast, has had one of Europe's highest average growth rates over the last 100 years. More than 14 million people are crowded into its borders.

Italy's population, approaching 56 million, is densely settled over most of the peninsula. Huge concentrations live in the great industrial cities of the Po Valley and in central Italy, where Rome has a population of 3 million. The natural increase rate is low by world standards. Current projections to the year 2020 show a slight decline. Emigration continues to relieve population pressures but takes place on a much smaller scale than the outflow earlier in this century. Internal migration flows remain high, the south being the sender and the north and Rome being recipients.

Harvest in a French Vineyard. *Highly varied, highly productive agriculture makes France a net food exporter. Typical of European emphasis on high-quality, specialized agriculture is France's world-ranked wine industry.*

Rome, by virtue of its immense historical importance, its continuing international role as headquarters of the Roman Catholic church, and its capital function, would be considered a city of international, not just national, importance—a true **world city**. Milan, headquarters of multinational corporations and a leading industrial design center, is also a world city.

Switzerland's population of some 6.5 million includes about 1 million foreign workers drawn by the booming Swiss economy. This imported work force is generally employed in lower-skill, lower-paying jobs such as construction, maintenance, and tourist service. The Swiss are often multilingual. Two thirds speak German as their native tongue, nearly one fifth speak French, and most of the rest, Italian.

Austria's slowly growing population, currently about 7.6 million, includes a much smaller contingent of foreign workers. Both Austrians and the Swiss are predominantly urban. In Austria, the largest city and capital, Vienna (Wien), declined in population following the loss of empire after World War I.

France's relative wealth and power over the centuries can be traced, at least in part, to its variety and quality of agricultural resources. French farms, though small by American standards, are larger on the whole than those of most of its neighbors. France is among the top 10 wheat producers in the world. The Mediterranean environment of the southeast is suitable for winter fruits and vegetables, olives, and rice, while the more maritime and continental climates of the northeast favor apples and dairy products. This variety and productivity places French agriculture in a strong position, at least within the EC.

West German agriculture is far more productive now than it was a few decades ago, with greater use of fertilizer and machinery. German farms emphasize grains, sugar beets, and animal fodder; livestock products bring most farm income.

Agriculture in the Core region occupies a great deal of land but supplies little of the total gross domestic product. More than half of the Netherlands is agricultural land, and more than half of this is pasture. The Netherlands' exports to the rest of the EC include beef, butter, cheese, and eggs. Horticulture (the commercial gardening of fruits and vegetables) is well advanced in the Netherlands, which has half of Europe's total greenhouse horticulture. The Dutch, with their neatly bordered fields, lush meadows, and fields of flowers have created an appealing and colorful cultural landscape in an otherwise boringly flat land. The Netherlands' agricultural specializations take full advantage of the country's location relative to the great urban-industrial centers of Europe.

In Italy, government policy is aimed at enlarging the many tiny, inefficient holdings but must then contend with an increased flow of displaced peasants into cities with already crowded slums and high unemployment rates. Government involvement in, and funding of, agricultural reform is impressive, and mechanization is increasing rapidly. The flow of poor farmers to the industrial cities has been so great that in some areas there is actually a labor shortage in agriculture. Italy is the largest exporter of wine, by volume, in Europe, and the Italian wine industry is now building a market for higher-priced quality and specialty wines.

Little of Switzerland can be farmed; much is too steep, too rocky, or in permanent ice and snow cover. Switzerland subsidizes its agriculture, directly and indirectly, especially by land tax regulation and agricultural product tariffs. There is both strategic logic in this (no nation wants to run short of food in wartime) and a desire to maintain a rural society and landscapes as part of the tourist-attracting milieu. Austria manages to supply nearly 85 percent of its own food needs from a very limited land base. The high percentage of pasture land makes Austria, like Switzerland, self-sufficient in dairy products and a net exporter.

Europe's more successful economies remain flexible and alert to continuing changes in the **global economy** that Europeans did much to initiate. Common European problems include fairly dense populations relative to farmland, minerals, and fuels. Aging, increasingly inefficient industrial plants can pose difficulties as can modernization of the infrastructure. Europeans must compete in a world of rapidly expanding East and Southeast Asian economies powered by disciplined, highly productive, and inexpensive labor. The two superpowers have, in comparison with Europe, vast and varied resource bases. The European solution is based, in part, on a carefully polished reputation. Salesmanship, reputation, quality control, and fine craftsmanship can go a long way in offsetting other regions' advantages in labor cost, raw materials, and more modern plants.

Consider wine, for example. French and German wines command premium prices wherever wine is consumed, even in the United States with its great variety of fine domestic wines. Some of the premium over non-European wines may reflect the accumulated skills and knowledge of European winemakers, but surely some of the price differential reflects glamour and mystique rather than substance.

Superb design and a carefully maintained reputation for high-quality engineering are the basis for much of Europe's industry, especially industries geared specifically at export markets. The risks of high R&D costs can be eliminated by buying licenses to use designs of others or minimized by joining in international consortiums. This does not mean that Europeans lack the ability to innovate and create; there are plenty of examples of such inventive skills. Marketing skills remain an important partner to engineering skills, however.

France's iron and steel industry is moving toward tidewater, as is Italy's. This location enables steelmakers to import raw materials at the lowest transportation costs and ship finished products to world markets. There is a general surplus in steel capacity throughout Europe, because each industrial nation has expanded capacity. In France, as in Norway, the aluminum industry is attracted to inexpensive hydroelectric power in mountain valleys where competition for the energy is minimal. The French aerospace industry, in addition to cooperating with the British on the Concorde, is involved in consortiums building the much more successful "airbus," an intermediate-range, high-capacity airliner. Military aircraft are a major export; France sometimes seems to specialize in supplying countries and regimes out of favor with the United States, Britain, or the Soviet Union. The French reputation for quality and innovative design is excellent.

Increasingly, in Europe and other high-tech societies, with their emphasis on high-value, high-technology goods and on services, airports and vicinities are becoming important locations for industrial and service activities. Air freight can be used efficiently for shipping the relatively lightweight, valuable products, parts, and components of high-tech industry. Busy international airports, such as those of London, Amsterdam, Frankfurt, Rome, and Paris, attract many new economic activities to their environs. The new Euro-Disneyland, for example, is located east of Paris in the French countryside near Charles de Gaulle Airport. Well served by highways, this resort is highly accessible to vacationers arriving by car or by air.

West Germany's industrial economy is strongly oriented toward export and heavy industry. West German steel plant expansion and new sites have not

THE HIGH-VALUE-ADDED INDUSTRIAL STRATEGY

Switzerland is a premier example of resourceful, hard-working, and well-educated people overcoming the handicap of a poor raw material and energy base. The Swiss strategy, also used by many other resource-poor countries, is to specialize in products with a high markup or **high value added** by skill. The Swiss took full advantage of their position astride major north-south routes and significant east-west routeways within Europe. The Rhine routeway links Basel with the world's greatest port at Rotterdam and with the extensive European canal network. This transport linkage is as vital to the Swiss economy as is the superb railway system linked with France, Italy, West Germany, and Austria. The Swiss are still building Alpine tunnels for both rails and highways.

The Swiss watchmaking industry once dominated world markets but now experiences strong competition from Asian-made digital watches. Watches are the best example of high-value-added goods, because the high price of a fine timepiece is due largely to skilled workmanship and design rather than to any intrinsic value of raw material used. The Swiss have overcome having to import raw materials and fossil fuels by specializing in such high-value-added products.

The Swiss reputation for quality of design and workmanship is reflected in the success of their machinery industries. Electrical machinery, instruments, surgical appliances, and electric locomotives are exported. Machine tools, that is, the machines that shape, press, drill, and cut metal and other material to make other machines, are a Swiss specialty.

The Swiss chemical industry, like its metal-fabricating industries, is oriented toward high-value-added products. Chemicals high in bulk but low in value added must locate near raw materials, but pharmaceuticals are very high in value, low in bulk, and can be transported economically over large distances. Research and quality control are keys to success in drugs, and the Swiss control a surprising fraction of the world's ethical drug markets.

High-value-added, fancy food products, based at least partly on imported materials, are also major factors in the Swiss economy. Switzerland has created huge, worldwide corporations in these fields. Swiss milk surpluses have been used in the manufacture of fine chocolate and cocoa for a century and a half, with cocoa beans and sugar from the tropics. Cheese manufacturing also consumes large quantities of milk in producing another expensive specialty food for export.

INVISIBLE EXPORTS

One way in which resource-poor industrial states can pay for large-scale imports of raw materials and food is to "sell" picturesque scenery, famous resorts, quaint old cities, and superb service in prestige hotels and restaurants. Most highly industrialized nations already have dense networks of modern, efficient transportation systems to serve industry; these systems become assets when serving visitors as well. When a country provides tourism services to foreigners, it earns foreign exchange but does not export anything more tangible than souvenir trinkets. International tourist services are **invisible exports**, as are financial services such as banking and insurance.

For example, countries such as Switzerland and Austria have converted agriculturally unproductive or marginally productive mountains into economic advantages as tourist attractions. The Swiss have developed tourism not only into a major industry but also into an art and science to be taught in hotel, restaurant, and tourist service schools to both Swiss and foreigners. Despite a very high value currency, which makes visits expensive for most foreigners, the Swiss tourist business continues to thrive. Austria is one of Western Europe's premier tourist areas, with an even larger volume of tourists than Switzerland, and France and Italy annually host millions of visitors. Switzerland's policy of armed neutrality, coupled with an impressively stable democratic government, has helped make the country a haven for bank accounts from the wealthy of less stable countries. There has been such a strong inflow of money into Switzerland that some Swiss banks began charging "negative interest" on foreign-owned accounts; a charge was levied on the account rather than interest paid. The Swiss are active in the international re-insurance industry; insurance companies purchase insurance on their higher risks to reduce their outlay in the event of unusual disasters.

---

emphasized coastal location, instead maintaining a strong orientation toward coalfields. The geographic advantages of coking-quality coal located on navigable waterways have led to one of the world's largest concentrations of heavy industry there. The internationalized Rhine River is a heavily used transport artery, supporting a string of industrial cities culminating at Europoort, the new port section of Rotterdam.

Both the Dutch and the Belgians have made maximum use of their prime locational assets, developing international ports in the delta of the Rhine River, a major routeway serving one of the world's greatest industrial regions. German chemical technology ranks among the world's first and best. German leadership in the development of coal tar derivatives, for example, introduced many new synthetic products from this one source—dyes, pharmaceuticals, and artificial fibers. The West German chemical industry remains one of the largest in Europe. Three huge multinational companies market and package their products throughout the world, with manufacturing facilities in many foreign countries.

The Volkswagen auto plant at Wolfsburg is the largest single auto plant in Europe. "VW" now is an

Alpine Pastoralism. *European farmers must contend with small farms in crowded lands. Every type of landscape must be used if possible. Alpine pastures, covered in deep snow in winter, are used in summer only.*

international producer with plants in Mexico, Brazil, and elsewhere. On the other hand, U.S. companies (General Motors and Ford) operating in West Germany hold about a third of the domestic West German market. Precision instruments and such optical products as cameras, microscopes, and telescopes are traditional German specialties now under severe competition in world markets.

Transportation system development has given West Germany one of the highest densities of transport routes in the world. The superhighways (*autobahns*), begun in the 1930s, are among the finest

anywhere but are also heavily traveled. The West Germans also have saturated the extensive canal network that links practically every major city. With highways and canals both unable to take much more freight load, the Germans have modernized their rail network for both freight and fast passenger service.

Belgian industry has suffered the destruction of two wars in this century but, paradoxically, this very destruction meant that there was no way to maintain obsolete facilities. Modernization has ensured high efficiency. Many multinational companies have chosen Belgium as a final assembly point, especially for cars, because of its central location in Western Europe. The Netherlands' economy is closely tied to its important port functions and transshipment activities. The Dutch occupy the Rhine Delta; that river is Europe's busiest waterway, serving the industrial heartland of West Germany and even supplying Switzerland with a port. The multinational hinterland of the largest port in the world, Rotterdam's Europoort, helps support the Dutch industrial base. Steel and chemicals benefit from tidewater locations in the Netherlands. The Dutch have maintained a viable aircraft industry by specializing in smaller aircraft. The huge Philips electrical appliance firm, headquartered in the Netherlands, makes everything from electric razors to x-ray machines.

In a sense, this subregion is **overdeveloped**. Much of the world must cope with problems of **underdevelopment**, in essence, the inadequacy of technological levels and industrial and transport infrastructure to develop the resource potential fully. The crowded, older industrial nations, however, must deal with a diminishing resource base and heavy dependence on imported raw materials. The Netherlands is perhaps the best example of this phenomenon in which human resources of skill and efficiency are developed to the maximum. The Holland Randstadt ("ring city") and the agglomeration of heavy industrial constellations in the Ruhr Valley are particularly significant in that they represent a multinuclear form of urbanization that seems to be growing rapidly in Europe just as in the megalopolises of North America (Figure 3–8).

Industry in Italy is so heavily concentrated in the Po Valley of the north that two thirds of all industrial workers are located there, while the region contains only 50 percent of the total population. The Italian government has tried to shift industrial expansion to the south. Although Italian industry includes such giants as Fiat, Dunlop-Pirelli, and Olivetti, there are also large numbers of small companies. The Italian government has taken over failing, inefficient firms and attempted, through consolidation and modernization, to make them competitive. This nationalization has not always boosted profits or productivity; the political goal of higher employment frequently overrides economic considerations. The government

bureaucracy itself is widely regarded as bloated and inefficient, a perception that does not foster confidence in its ability to manage industrial concerns successfully.

Clothing has long been a large Italian export industry, based on Italy's reputation for fashion design and craftsmanship. Italy ranks third, after the United States and Japan, in output of household appliances, dominating EC markets. Olivetti typewriters and office machines are sold around the world and are also assembled overseas. Fiat supplies about one fifth of the entire EC auto market; it is the most international of all auto manufacturers, having set up plants in Yugoslavia, Spain, Turkey, and Libya. Fiat has built entire auto plants for Poland and the Soviet Union, making Fiat designs the most familiar in the world.

In contrast to Switzerland's predominantly free-enterprise industries, Austria's basic industries are nationalized. Steelmaking is the largest industry in Austria, specializing in high-quality steels from an oxygen blast process invented in Austria.

The United Kingdom and Eire

More than eight centuries of English colonialism in Ireland have left a legacy of mistrust between the Irish and English, leading to the tragic civil war in Northern Ireland. However, the very fact of British domination for centuries and its effect on the development of contemporary Eire, plus the close ties of these two countries in trade, tourism, and migration, make this a logical subregion.

The United Kingdom (UK) includes England, Scotland, Wales, and Northern Ireland. Great Britain is the island on which are located the once-separate states of England, Wales, and Scotland. *Britain* and *British* are used interchangeably with the lands and residents of the United Kingdom including Northern Ireland.

The UK has had to adjust its economy, internationalist philosophy, military expenditures, and self-image to the rapid liquidation of empire. The largest empire the world had ever seen presented the British with the problems of huge expenditures for immense naval forces and colonial administrations and the potentials for enormous markets and diverse raw

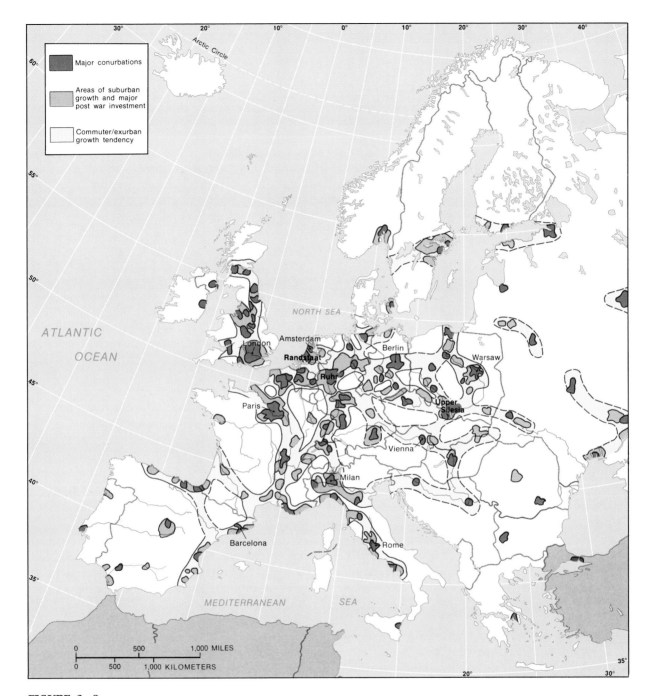

FIGURE 3–8
City-Chains in Europe. The location advantages that propel megalopolises in North America—development of cheaper land between cities astride interconnecting transport routes—operate in Europe as well, producing giant city-chains.

PRIMATE CITIES

Many French people tend to think of Paris as virtually the center of the universe. Most Parisians are convinced of it. Paris is one of many European capitals that are their country's **primate cities**. A primate city is always the largest in the country, and significantly larger than the second-place city. Primate cities almost always serve as the national capital. European examples include London, Warsaw, Athens, Stockholm, and Brussels; Mexico City, Tokyo, Cairo, and Buenos Aires meet primate city criteria in other regions.

The primate city is much more than just the largest city in the country. Primate cities *dominate* their respective country's economic, cultural, and political life. Representing their national identity to citizens and foreigners alike, they often give expression to much of the nation's culture, in architectural monuments, famous museums, and prestigious theaters, opera houses, and concert halls. Primate cities serve as splendid display cases and perpetuate culture through great universities.

Because they form huge local markets and contain the cream of human talent, attracted by urban glamour, primate cities are centers of industrial and service activities whose market may be the whole country. Necessarily, they are at the center of national and even international transport and communications networks. This accessibility to distant material and market sources enhances the concentration of yet more economic enterprises. Diseconomies of scale develop as congestion, high land values, high labor costs, and high taxes accompany high desirability. Commonly,

material supplies. As in its construction, the end of empire has presented both problems and potentials. Transition from colony to fully independent, equal, and voluntary member of the British Commonwealth has not always been smooth for either the ex-colony or the parent country.

Economically, the UK has been among the slowest growing Western European states. Its transition from imperial splendor to a rather small state by super-power standards has been caused by four factors: (1) a long insistence on retaining trade relationships within the Commonwealth; (2) the associated failure of the UK to join the Common Market at its outset (Australia and New Zealand pressed to retain their preferred status in British markets while the personal antagonism of General de Gaulle resulted in delay-

ing entrance into the Common Market); (3) the hardship of operating a largely obsolescent industrial plant; and (4) an economy besieged by a bitter, sometimes irrationally self-destructive class warfare. The psychological transition to the late twentieth century's realities has not yet been fully made. This fact must remain in the forefront of any geographical analysis of the UK. Similarly, Eire (the Republic of Ireland) has not yet recovered from its long colonial status, although its evolution of more suitable economic and political policies may outpace that of the UK.

The resource base of the island of Britain is not particularly impressive for a state that led the first industrial revolution. Coal has been exploited on an industrial scale for three centuries; obviously, the

London. *London is a good example of a primate city and a world city—a city of international rather than national or regional importance. No longer the seat of a vast empire, London remains important as an international banking, insurance, and cultural center.*

national governments try to dispense some functions to provincial cities as they attempt to limit continued centralized growth.

For example, Paris has so long dominated French political, economic, and cultural life that the French government has taken active steps to decentralize industry and service functions (including government) into areas outside the Paris region. Industrial building permits are deliberately restricted for Paris but easy to obtain outside that metropolitan area. Recent government policy allocates relocation-expansion aid almost in direct proportion to distance from Paris.

preceding centuries of mining have removed the most accessible coal, and that least expensive to exploit. The remaining coal lies deep and in narrow, broken seams. Mechanization is progressing, however, and Britain's coal production is at a cheaper overall cost than Germany's. Tin has been mined in Cornwall for several thousands of years; high prices recently have rejuvenated some mines. Iron ore was sufficient to supply Britain's early industrialization, but the remaining stock is of rather low iron content, with many impurities.

The brightest parts of the resource picture are natural gas and petroleum. Production of natural gas was so high by the early 1970s that the British government required households to shift from burning coal for home heating to natural gas. This shift not

only guaranteed a market to sustain heavy investments in gas exploitation, but also reduced air pollution levels in the cities. Oil discoveries have been most encouraging. The UK's section of the North Sea includes three large, proven fields (Figure 3–9). These hold at least 15 billion barrels of oil and make the UK the only Western industrial leader self-sufficient in oil *and* a major exporter. Extensive test drilling has been carried out in the south of England, the southwest, the Channel, and the shipping lanes leading from the open Atlantic to the Channel.

The North Sea oil boom has focused activity on Scotland's east coast, while the west coast sinks into relative decline. West coast cities such as Glasgow thrived in the days of imperial Britain's colonization of North America but are now burdened by derelict

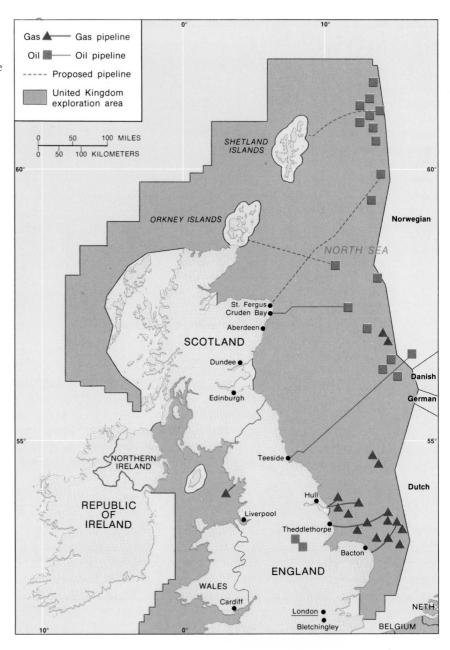

FIGURE 3–9
North Sea Oil and Gas.
Britain has become a major exporter of oil, producing huge surpluses beyond its own substantial needs.

shipyards and abandoned docks. East coast Aberdeen, an old fishing port, now bustles with the business of servicing oil and gas rigs.

Eire's natural resource base is less encouraging as that republic makes its long-delayed entrance into the modern industrial world. Some natural gas has

been discovered and exploited off the southwestern coast near Cork. Exploration is accelerating in the Irish Sea. Eire has little coal but uses large quantities of peat (turf), taken from bogs, to generate about one fourth of its electricity. Historically, Ireland mined some gold, but metallic minerals had not been exploited on an industrial scale for centuries before copper, lead, zinc, and silver were discovered in the 1960s.

The population of the UK is about 56 million. The UK's slow population growth rate and low birth- and death rates identify it clearly as being in the last stage of the demographic transition. Over 16 percent of the population has passed retirement age. Eighty percent live in towns or cities. Seven giant **conurbations** (continuous urban areas) that have grown toward one another, creating an urban chain or a large city expanding to include older, once-separate cores, house about 30 percent of the population. In common with other mature urban-industrial economies, the UK has had slow growth or decline in population in these central cities.

The qualities of European urban landscapes are well illustrated by London, the region's largest city and one of its oldest. Two of the oldest urban functions are trade and administration. Although London is no longer among Europe's busiest ports, it remains an internationally important center of trade, finance, and communications industries. The architectural texture of such old cities is a rich blend of the styles of many centuries. While seventeenth century founding dates earn American or Canadian cities the aura of antiquity, many European cities trace their origins back to the Roman Empire, or even earlier. A stroll through London will delight the visitor with a Victorian neighborhood, a modern high-rise complex, a seventeenth century church, a house in the Tudor style of the first Queen Elizabeth, and a shopping street planned when many of North America's great city sites were remote wilderness. With a little knowledge of building styles and materials, it is possible to observe the "growth rings" of the city's fabric just as it is easy to count growth rings in tree trunks. Its spurts of rapid growth are behind it now, but London remains a vital and fascinatingly complex city.

By 1961, Eire's population had declined to only 2.6 million after more than a century of heavy migration to Britain, North America, and Oceania, following the famine that resulted from a potato blight in the 1840s. Population began to grow again in the 1960s and is presently 3.6 million. About half of Eire's people live in towns, and one fifth of the nation lives in Dublin. Irish emigration has resumed as a result of high unemployment rates in the 1980s.

In addition to the steady inflow of Irish to the UK, there has been a large immigration from India, Pakistan, Jamaica, and Trinidad. Commonwealth passports allowed these citizens of former colonies unrestricted access to Britain.

Britain also has had to cope with "brain drain"—the international migration of highly educated, skilled people in search of better opportunities.

About 80 percent of Britain's land is in agriculture, and of that, less than one third is cultivated, with the balance in pasture and meadow. It is lowland Britain that is cultivated in barley, wheat, oats, potatoes, sugar beets, and hops (used for flavoring beer). Highland Britain, which is most of the Pennines, Wales, and Scotland, is mainly pasture for sheep and cattle. Britain produces about half of its total food requirements, an impressive figure considering its high population density. Productivity has been significantly increased since World War II.

Although, as indicated by a voter referendum, the British generally favor membership in the EC, the EC's agricultural policies are a main point of disagreement because of artificial barriers erected by other members. British farmers would be in an excellent competitive position if French and German farmers were less insulated from marketplace realities.

Eire uses about 70 percent of its land in agriculture; over three fourths of this is in pasture. Farming employs over 20 percent of the Irish, a much higher proportion than in the more industrial nations of Europe.

"British and Best" was once a proud boast of the quality of British manufactured goods. Allowing for some patriotic overstatement, it was largely true up until World War I. The UK was the first industrial nation, and the British take justifiable pride in this fact. There is a penalty attached to being first, however. Unless care is taken to maintain a sufficiently high rate of modernization, the first becomes the oldest in industrial plant, and the least efficient. In Rostow's scheme (discussed in Chapter 1), Britain had reached, by 1914, the final phases of industrialization, that is, full maturity. But it was too dependent on coal mining, steel manufacturing, textiles, and shipbuilding. In 1914, Britain launched over 60 percent of the world's total commercial tonnage of ships. It now supplies about 4 percent.

The textile industry, of course, is one of the first established by industrializing countries. British textile production, a mainstay of the economy, has slumped drastically because only the high-quality, specialized textiles retain a future in a highly competitive world industry.

Some of Britain's impoverished old industrial cities and urban regions, such as Glasgow, Newcastle, and the Midlands, have a markedly grim appearance; their obsolete economics require assistance from the national government (Figure 3–10). The sheer density of urban-industrial populations living at fairly high standards by world averages generates traffic congestion and deteriorating pressures on scarce recreational space. The quality of much of the workers' housing of the Victorian Era is quite low. The coal-based industrialization of a century and a half has left a grim, scarred landscape in many areas.

Adjustment to the EC, which Britain entered in 1973, has not been easy. Although membership opens up the entire market to British exports without tariff, it works both ways: inefficient manufacturers in Britain will close.

Ireland's industrialization began quite late, a fact partially related to the country's long subordinate status. Peripheral to industrial Europe, and without the coal that aided manufacturing in England, Scotland, and Wales, Ireland remained almost untouched by the industrial revolution. Its labor migrated to opportunities in Britain and overseas. Eire's per capita income is lower than the UK's and is comparable more to that of Mediterranean than northwestern Europe. Irish wage rates, though rising, are still lower than in most highly industrialized nations. This, along with Eire's policy of "tax holidays," lured many factories from EC corporations. Eire's highly attractive combination of tariff-free access to the whole EC and low wages boomed its industrial growth in the 1970s, but this growth has slowed. Food processing remains the largest employer, but probably not for long as electronic equipment, engineering, and textiles become prominent.

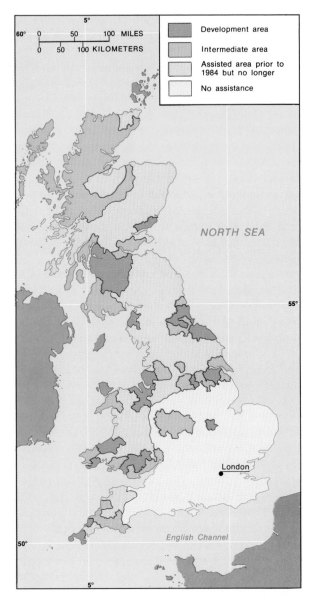

FIGURE 3–10
Industrial Development Areas in the United Kingdom. Areas of declining, obsolescent industries with high unemployment rates are assisted by policies aimed at encouraging new industrial locations there. Tax holidays, loans, grants, and other government aid in such cases are common throughout Western Europe.

More general changes in Britain's industrial structure can be exemplified by the experience of the auto industry. The British did not expand their auto industry as quickly as the Americans in the early twentieth century, and the size of the domestic market in the United States supported far more efficient mass production. The British auto industry has gradually shifted location by building new plants in Wales, Scotland, and Lancashire, deemphasizing the older locations of the Midlands, Oxford, and metropolitan London. Cars are the most valuable item of trade among EC countries, and competition with the large German, French, and Italian companies has been intense. Many EC car manufacturers, including branches of Ford and General Motors, produce standard models or "Eurocars." Whether cars are actually assembled in Britain, Germany, or Italy, parts flow freely among countries to assemble interchangeable autos. Even Britain's nationalized automaker assembles some cars in Belgium. Production continues at a high rate, but mainly for the domestic market.

Of Britain's traditional industries, coal, steel, and automobiles seem reasonably modernized and productive. Other healthy enterprises include chemicals, machinery, electronics, oil refining, pharmaceuticals, and clothing. Britain's economic future may lie in those high-technology industries in which the expensive research is available cheaply through licensing or through consortiums.

Prime Minister Margaret Thatcher revised labor union regulation and tax laws in a controversial attempt to modernize further Britain's industrial economy. Scotland is enjoying an economic expansion based on offshore petroleum and natural gas, though the oil glut of the mid-1980s has slowed the boom.

Northern Europe

The five northernmost nations of Europe, known collectively as **Norden**, have had an intertwined history and share some economic and cultural traits. The Swedes gained independence from Danish kings in 1523 and became a major European power in the next century. Sweden ruled Finland before 1809, when Finland became a grand duchy of the Russian Empire. Norway was linked with Denmark between 1381 and 1814, when it was united with Sweden; independence was finally achieved in 1905. Iceland was an independent republic until 1262, when it joined with Norway, and then was associated with Denmark from 1360 to 1918. Only in the twentieth century have all five states consolidated their independence; their cultures remain strongly influenced by their long association.

Norway, Sweden, Denmark, and Finland have quite different economies, and their resource bases vary enormously. Denmark, whose peninsulas and islands are mostly glacial outwash and debris from the great glaciers that eroded the Scandinavian peninsula to its north, has the best agricultural land, but otherwise the poorest resource base. The Danes' only domestic raw materials are food products processed there and a cement industry based on limestone deposits. Only 9 percent of the Danes are farmers, but they produce enough food, most of it very high quality, to supply triple Denmark's population. Denmark has specialized in high-value-added animal products such as butter, cheese, eggs, hams, pastries, and bacon. Excellent quality control has earned Danish products premium prices throughout northwestern Europe.

Denmark has had to develop highly specialized industries whose reputation for excellence of design and quality of craftsmanship support the high prices resulting from the Danes' poor raw material and fuel situation. Danish furniture, for example, emphasizes graceful designs that use slender, narrow supports and legs, stretching the wood resource, often imported tropical woods such as teak. Fine-quality porcelains, such as collector plates, turn glacial outwash clays into foreign exchange.

Machinery, such as diesel engines, and stereo components are among Denmark's most successful exports. The country is also a large beer exporter and a major exporter of specialty and high-quality foods. Copenhagen, the capital—literally "merchants' harbor"—is an important entrepôt for much of Scandinavia's trade.

Denmark is the most visited of the Scandinavian countries, although, as in many Western European countries, a heavy outflow of tourists comes close to eliminating the favorable balance of tourist expenditure. Denmark's population of a little over 5 million

THE CIVIL STRIFE IN NORTHERN IRELAND

The most serious threat to the political stability of both the UK and Eire is the civil war in Northern Ireland. Part of the problem is economic; an unemployment rate of 11 percent may be almost as important as the centuries-old cultural-religious split. Considered separately, the six counties that make up Northern Ireland have a 2:1 Protestant majority. The entire island has an overall Catholic majority. Roman Catholics in Northern Ireland complain of pervasive discrimination, politically, socially, and economically. Working-class Catholics tend to suffer higher unemployment in declining textile and shipbuilding industries. Northern Ireland, with its Scottish "connection," was the only part of Ireland to industrialize in the nineteenth century; it suffers from obsolete plants and no industrial raw materials other than flax (linen).

Protestant Irish are fearful that a political merger with the Irish Republic would submerge them as a cultural-religious minority and that they then would suffer discrimination. The British army has been in Northern Ireland for more than 15 years in an apparently fruitless attempt to eliminate terrorism.

This undeclared civil war, with little realistic hope of resolving the conflict, is hardly the environment to attract badly needed new industrial investment. Both sides need jobs, and the economic drain on the UK is serious in terms of welfare costs, subsidies of all kinds, and military expenditures. The Republic, although advocating the union of the divided island, disavows the terrorist activities that are bleeding the people and economy of the north. It is possible that terrorism could spill over on a larger scale to the Republic and even destroy the democratic government in Dublin. Clearly, the "Irish question" must be solved for both the UK and Eire if they are to advance at rates comparable to those of their continental partners in the EC. The 1985 agreement between the UK and Eire, which gives the Republic a voice in determining the future of Northern Ireland, is a political step in the right direction. A reasonable solution may be a federal structure that would merge the Irish Republic and Northern Ireland, but with local autonomy for the north, and appropriate guarantees of freedom of religion and civil rights for all.

is relatively homogeneous, with only a small minority of Germans in the south.

Norway is Europe's most thinly populated country; much of its industry is small-scale, and most of it is related to abundant hydroelectricity. Aluminum is refined using large amounts of electricity. Surplus electric power has led to the creation of a large-scale chemical industry in Norway; fertilizer and a variety of electrochemicals are the main products. Fishing is a major industry. Norway, Sweden, and Finland have developed a wood-based material culture, using lavish amounts of wood in building construction, interior paneling, and decorative arts. As in the similarly forest-based culture of medieval Germany, Scandinavian folklore is rich in tales of the woods. The dangers of the dense, mostly coniferous forests with their past threats of animal and human predators, remain in such everyday sayings as, "not out of the woods yet."

Norway has an excellent hydroelectric potential created by heavy rainfall along its mountain spine. Although it has one of the world's highest production rates of hydro power per capita, only half of its capacity has been harnessed. Iron is mined near the Soviet border in the far north, with a surplus exported after Norway's own steel industry is supplied.

Little of Norway is warm enough to produce grain, but output on this small area has been increasing because of better fertilizers and farming techniques. Norwegian agriculture is heavily subsidized to maintain some domestic production. Most Norwegians live near tidewater, with each **fiord** (long, narrow, glacially eroded valley now partially inundated by the sea) containing a pocket of population at its landward end.

Norway's sparse population, spectacular scenery, and largely unpolluted environment make it a continuing tourist lure. The limited economic possibilities of the environment have encouraged the Norwegians to develop the invisible exports of services.

Norway's merchant fleet is one of the world's largest, including tankers, container ships, liquified gas carriers, and highly specialized cargo carriers of all kinds. The Norwegians also operate a large fleet of cruise liners in the Caribbean and Mediterranean as well as along their own beautiful coastline. Their huge fleet makes Norway especially vulnerable to any sudden changes in the volume of international trade.

Norway has struck it rich in its large section of the North Sea oil and gas fields; Norway's per capita GNP is second in Europe (Switzerland is first). Norway was the first European country to export a surplus of oil. The energy glut of the mid-1980s was a particularly difficult time for Norway, which had depended heavily on its oil resource.

Sweden, whose population of about 8.5 million approximates that of New York City, has an amazing variety of industry. It is an important auto manufacturer and exporter, aircraft manufacturer (under license with foreign designs), and telephone equipment manufacturer (one of the world's largest exporters). Sweden has encouraged industrial consolidation to improve competitive advantage in world markets; it has also embarked on a program of nationalization of industry. Swedish steel manufacturing specializes in high-value specialty steels, such as stainless steel and steel for machine tools. Spe-

*A **Norwegian Fiord**. Fiords, glaciated valleys partly drowned by the sea, make Norway's coastline particularly scenic. Norway has almost no flat land for agricultural or other uses but possesses a great hydroelectric power potential because of heavy precipitation along its mountain spine.*

cialty goods include ball bearings, electrical appliances, and typewriters.

Sweden is relatively rich in metallic minerals. Iron ore is mined at several locations; the largest mines are at Kiruna, Gällivare, Svappavaara, and the Bergslagen district. Copper and lead are also mined. To date, oil prospecting in Sweden's offshore waters has been disappointing. Like Norway, Sweden has great hydro power potential, half of which is developed.

Unlike Sweden, which has a negative attitude toward nuclear power, Finland intends to develop this energy source. Two plants were built by the USSR and two by Sweden in a classic example of Finland's delicate maneuvering between the USSR and the West (Figure 3–11).

Finland's glaciated lowland, highly similar to the Canadian Shield, offers relatively little agricultural land. Less than 10 percent of Finland is used for food production, but the Finns have been successful in pushing northward the limit for crops such as wheat, barley, and sugar beets by scientific plant breeding.

The population of Finland, 5 million people, is less than that of metropolitan Chicago. Although Swedish is still an official language, the majority of the people speak Finnish, a language related to Hungarian and Estonian. Finland's forest products industry is its largest industry; two thirds of the country is in forest, mainly privately owned. The industry is handicapped by a short growing season and a relatively high production cost. Building on its own forest exploitation experience, Finland has specialized in the production of paper-making machinery, exporting whole factories. Finnish shipyards specialize in ferries and icebreakers.

Iceland is one of the most homogeneous nations in the world; Icelanders are descendants of Norwegian settlers who arrived in the ninth and tenth centuries and absorbed the earlier Celtic settlers from the British Isles. With less than a quarter-million people, Iceland boasts one of the world's most literate populations (99.9 percent); and infant mortality rates, a classic measure of the quality of life, are lower than those of the United States and most other European states.

The Icelanders have built a reasonably prosperous economy on a largely barren island with very few natural resources. Iceland is of recent volcanic origin. About three quarters of the island consists of glaciers, lakes, a lava desert, and other wasteland. Hot water and steam from hot springs and geysers have long been used to heat buildings and greenhouses. Iceland now generates some geothermal electricity from the interior heat of the earth to supplement its hydroelectric power. These energy sources are exceptionally valuable on a volcanic island with no fossil fuels.

Iceland's major resource is the rich fishing grounds around the island. Eighty-three percent of the value of Iceland's exports are supplied by fish; over 1.5 million metric tons of fish are caught per year. Icelanders consume large amounts of fish and rely on fish for export income. For this reason, they have been quite aggressive at zoning out foreign fishing boats. Only 4 percent of Iceland is cultivable, and climate limits crops to potatoes, turnips, and hay.

Mediterranean Europe

The Mediterranean Europe subregion includes **nation-states**—that happy coincidence, geographically, of the nation, or people unified by cultural, historical, linguistic, and other common traits, and the state, the political unit—states of long standing, such as Spain and Portugal, along with Greece and southern Italy. These last two, though of ancient roots, were unified as modern states relatively recently. Greece achieved independence in the modern era in 1830; modern Italy was born in 1870.

Contrasts between northern Europe and the Mediterranean basin can be seen in the great public festivals of each area. The public holidays of Mediterranean Europe, the old Roman Empire, tend to be religious (the word *holiday* is a version of *holy day*). The celebration itself may have become largely secular, but at heart, it is a religious feast day or otherwise part of the church calendar. It is said that there were so many holy days, during which workers were not working, that the Protestant Reformation, influenced by a more work-oriented ethic, prohibited many holidays to boost output. Street carnivals, parades, public dancing, and mass celebrations typify the joyous, more spontaneous holy days of Roman Catholic Europe, while such uproarious, zestful gai-

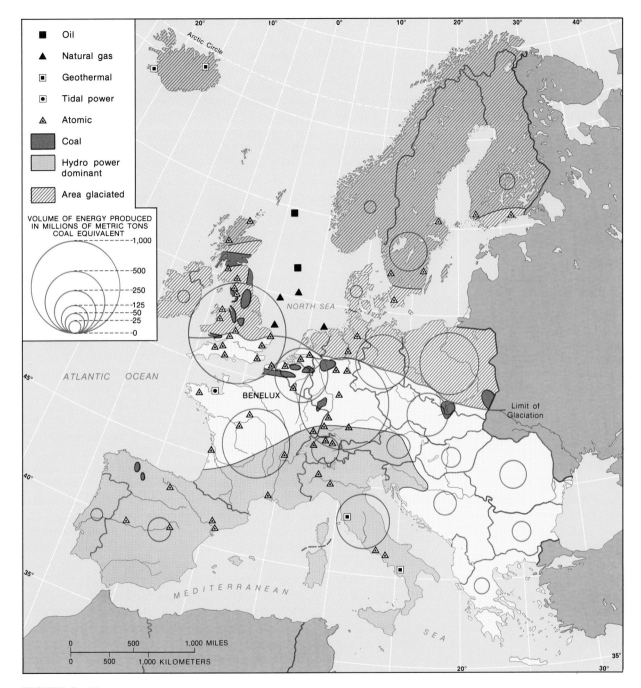

FIGURE 3–11
Energy Sources and Production in Europe. Glaciated highlands offer many
hydroelectric site potentials, while Europe's many coal basins fueled the early
industrial revolution. North Sea oil and gas have become important regional energy
resources in the last 20 years.

ety is much rarer in northern, more solidly Protestant Europe. There, public holidays are more secular in celebration, and often secular, or non-Christian, in origin. Germany's Oktoberfest, for example, is not overtly Christian, though it may have origins in pre-Christian religious beliefs. And even Oktoberfest is most famously and enthusiastically celebrated in Munich—in southern, more Catholic, Germany.

The Iberian Peninsula is shared by Spain, with nearly 39 million people, and smaller Portugal, with a little less than 11 million. Spain is a relatively poor nation by European standards but has a per capita income more than twice that of Portugal, Western Europe's poorest country. The two countries shared in the cultural heritage of Moorish occupation, the battle to expel the Moors, and the race for world trade and colonial empires.

Spain is an important producer of mercury, copper, lead, and silver, and a relatively minor source of coal, iron ore, zinc, and bauxite. Coal output has fallen because imports are cheaper, despite an assured market in state-owned steel and electric generating plants. Some oil has been found in the Ebro River basin and offshore, but this supply is expected to provide less than 10 percent of Spain's annual consumption. Spain does possess large supplies of uranium and has embarked on a nuclear power program.

Portugal's minerals, unfortunately, are not commercially exploitable in most cases. Tungsten ores are important, and large reserves of iron ore could support higher production. No oil or natural gas had been discovered by the mid-1980s, but exploration continues.

Spain's population is projected to grow to 42 million by the year 2000; it has one of Europe's higher rates of natural increase. Nearly 1 million Spaniards have emigrated, at least temporarily, in search of jobs. Internal migration is also high, flowing from less developed rural areas of the interior to the coastal industrial and tourist centers.

Spain has long had serious problems with strong centrifugal forces, cultural and political. The Catalans of the northeast, with their "capital" at Barcelona, consider themselves quite different from the people of northwestern Asturias or southern Andalusia. The choice of Madrid, a geographically central location, as capital (in the hope of overriding the conflicting claims of the great cultural and economic centers such as Barcelona, Seville, and Zaragoza) has not unified the contending provinces sufficiently to allow relaxation on the part of the central government. The most serious contemporary threat to Spain's stability and civil peace is the Basque independence movement. The Basques are a culturally distinctive group that speaks a language unrelated to Spanish (see Figure 3–3). Their ancient homeland straddles the border of France and Spain at the western end of the Pyrenees. The Basque desire for regional autonomy has been partly met by Madrid, but not fast or far enough for the Basques.

Portuguese workers outside Portugal once sent home more foreign exchange than Portugal earned through tourism. Faltering economic conditions elsewhere can send thousands of Portuguese back to the home labor market. The same is true for Spain, Europe's largest exporter of labor after Italy (Figure 3–12).

About 40 percent of Portugal is in cultivated land, with one third in forest and most of the remainder in meadow and pasture. Droughts are common in the south where precipitation is only one-fifth the northern average. The land tenure situation is varied. Small family farms dominate the north, and large estates are common in the sparsely populated south. Portugal is a major wine producer, exporting large quantities throughout Europe and North America.

Mediterranean Europe has a characteristic cuisine that faithfully reflects the combination of environment and culture that emphasizes particular foods. Wheat, a widespread basic crop since antiquity, is consumed as bread or pasta. Wine accompanies several meals a day, and fresh table grapes in season are complemented by raisins or currants (the latter, a term derived from the sun-dried grapes of Corinth). Onions, garlic, and tomatoes are used in salads and main dishes and as spices. Olives—fresh, preserved in oil, or pickled—are also pressed for the regions' most popular cooking and salad oil. Citrus crops are large and have long been major exports to cooler, northern Europe, along with quantities of wine. A traditional Spanish blend of wine and citrus juice has become popular in the United States.

Spain has as high a proportion of land under cultivation as does Portugal. There is more pasture and meadow, because the Spanish have not preserved

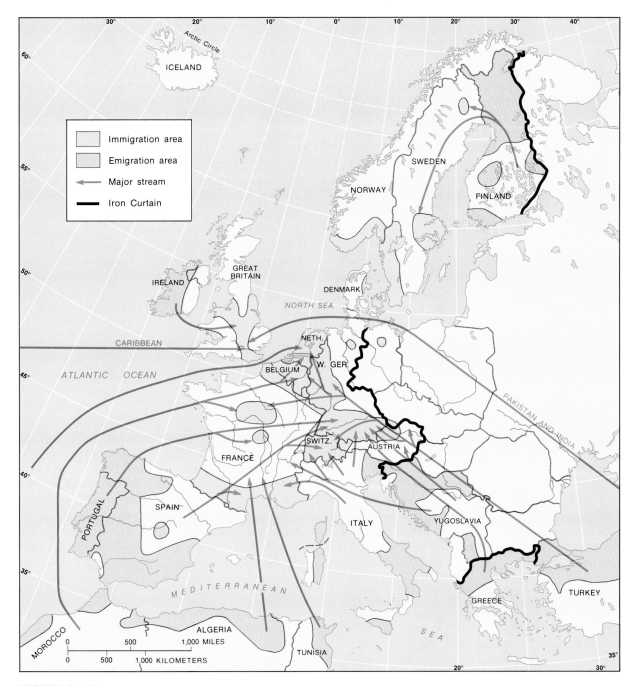

FIGURE 3–12
Guest Worker Flows in Europe. Intraregional differences in income and living standards, along with former colonial ties, influence labor migrations to and within Europe.

their forests as carefully. Low precipitation over the interior and south has led to large areas of extensive farming combined with smaller, irrigated areas of intensive use. Mechanization levels in all of Iberia are much lower than in northwestern Europe.

Spain is one of the most rapidly modernizing economies of Europe. Its application to the EC was implemented in 1985. As in other western European countries, the government has acquired control of, or significant interests in, a broad range of industrial enterprises. The government has at least a one-third interest in petroleum refining, coal mining, automotive production, and chemicals, and also owns almost all of the shipbuilding industry and controls the nation's airlines.

Steel production in Spain has risen rapidly, and Spain's chemical industry has been growing, associated with oil refineries and local raw materials. Shipbuilding and auto production are both healthy industries with up-to-date physical plants.

Portuguese industry is far more concentrated in the highly competitive areas of textiles and clothing. Strong competition from such industrializing nations as Brazil, India, and the Philippines is posing serious problems for Portugal. The Portuguese government has nationalized large companies and is investing heavily in building new industrial complexes to reduce unemployment rates. Portugal has attained full membership in the EC.

Tourism is Spain's most important industry and supplies more than half of its foreign exchange. The warm sunshine of southern Spain and Portugal is a powerful attraction for the residents of the highly industrialized northwest European countries. There, gloomy, short winter days are a burden that increasing numbers can afford to escape for a winter holiday. Even in summer, the marine west coast climates of Britain, Scandinavia, and other areas produce overcast, drizzly days, in sharp contrast to the almost cloudless skies of the Mediterranean.

Rising incomes in highly industrialized countries, progressively cheaper air fares, and the relatively low hotel and restaurant rates of lower-labor-cost Spain and Portugal have contributed to the boom. The Costa del Sol and the Costa Brava have become linear cities with wall-to-wall hotels and vacation home condominiums.

One custom evident to tourists is the siesta, common throughout Mediterranean Europe but most strongly associated with Spain. The siesta, a midday break from work, usually features a nap following a leisurely noontime meal. Often interpreted by outsiders as a sign of laziness, the siesta is actually a practical means of coping with the intense heat of early afternoon. People go to work early, while it is still fairly cool. As the heat built up in the pre-air conditioned towns, it was sensible to take a break, then return to work in the cooler early evening, working for another few hours with renewed vigor before the late supper so typical of the region.

Spain's participation in the EC, its heavy tourist inflow, and remittances from Spaniards working overseas have been instrumental in maintaining a viable economy. If a policy of government subsidies to evidently inefficient industries continues, and tourism does not sustain growth, Spain could come under financial stress.

The shock Portugal experienced in losing its huge African colonies has not yet been fully overcome. There is room for an expanded tourist industry to take up some of the slack employment, but Portugal must expand its industrial base to improve living standards.

Cultural leadership in times past and its continuing legacy in the political and economic thought of Europe characterize Greece. Ironically, this fountainhead of Western culture, which long served as a vital link between Eastern and Western civilizations, has achieved only moderate economic progress in the modern world.

Greece, the second poorest country of the subregion after Portugal, has a fairly favorable resource base. It has relatively large deposits of lignite; reserves of peat are even larger. More than half of Greece's electric energy is supplied by these domestic reserves. Although oil from Greek portions of the Aegean Sea supplies less than 10 percent of consumption, there is optimism about future exploration. Greece also has at least a billion tons of bauxite reserves. This ore is processed into aluminum for export as metal and metal products.

Greek population has expanded from less than three quarters of a million at independence to about 11 million currently. Over one fourth of the total live

Barcelona, Spain. *Barcelona, the "capital" of the Catalan region of Spain, is representative of cities embodying regional cultures within Spain. Madrid was constructed as a new national capital because it had no such regional, and therefore partisan, identity.*

in metropolitan Athens, while the mountainous areas have exceptionally low densities. As in Spain, there are many densely populated small nuclei separated by areas of low density. Greece is part of the "southern tier" of labor-exporting countries, along with Portugal, Spain, Italy, and Yugoslavia. The long-term rural-to-urban migration continues both internally and internationally. About 400,000 Greeks work in other EC countries, mostly in West Germany (see Figure 3–12).

Agriculture, though declining in relative importance and in work force, provides one fourth of Greece's exports by value. Less than one third of mountainous Greece is cultivable. Fortunately for

Greece, air masses rising against the highlands provide heavier precipitation than would be expected in a Mediterranean country, and the runoff is used in irrigation projects. Greece is self-sufficient in cereal crops; cotton and citrus production is rising, while exports of fruit, vegetables, wine, and tobacco increase.

Greece's industrial strategy emphasizes processing and packaging of agricultural products. Other main industries are textiles and chemicals. Government policy favors decentralization of industry from Athens into smaller cities and towns and domestic production of consumer goods to reduce imports. The two policies are related, resulting in small but modern factories scattered through the country. Large shipyards produce ships for Greece's large merchant fleet as well as specializing in repairs.

Greece enjoys a combination of low prices, abundant sunshine, and an array of cultural artifacts high on every tourist's list. As with Italy, the Greeks are attempting to spread tourism more evenly across both calendar and country.

Greece has had a turbulent political history and must cope with the fact that it is bordered by communist countries and its long-term traditional enemy, Turkey. Tensions with Turkey over their respective national populations in Cyprus and over division of potential oil fields in the Aegean are serious problems.

Rural-to-urban migration within Italy often features a south-to-north movement as well. The relatively poorer, more rural south has a higher rate of population increase than the urban, highly industrialized north. Personal incomes in southern Italy average two thirds of the national norm. Traditionally, the northern industrial cities such as Milan, Genoa, and Turin have drawn much of their entry-level work force from the south. The agriculture of southern Italy and Sicily is quite different from that of the north; traditional Mediterranean specialties such as citrus, olives, almonds, wheat, and grapes flourish in the south.

Long-term, widespread rural poverty in the south has led to a severe shortage of capital necessary for farming to modernize. Hillside erosion must be controlled, and more dams built for irrigation projects. Still, the picture is far from bleak. The Italian government is acutely aware of developmental and income disparities between north and south and directs new public and private investment to the south where feasible. Recent oil and natural gas strikes in Sicily and just offshore are helping to produce a level of prosperity unprecedented in modern times for Sicily.

Industrial East Europe

Each country in Industrial East Europe—East Germany, Czechoslovakia, and Poland—has different problems, potentials, and production specializations. Nonetheless, their economies are to a large degree complementary, and there has been a relatively high degree of integration of production. All three are tied to the Soviet economy yet trade heavily with the West. Poland is the largest of the Eastern European countries; with firm cultural ties to the West and large emigree colonies in North America, Poland has been able to act in a more independent fashion than most of the region's other states. Its most noteworthy deviations from the Communist norm are the open practice of formal religion and the dominance of a peasant, freehold farming system.

With the country positioned between northern German Protestantism and the Russian tradition of Orthodox Christianity, Polish nationalism and Roman Catholicism have become closely associated (see Figure 3–4). The ties of the people to the land and religion are deeply entwined with the national mystique, and the Soviets are acutely aware of the fact that the violation of Polish nationhood ultimately precipitated Western involvement in World War II. Located between the Germans and the Russians, the country has always suffered a precarious existence (Figure 3–13).

Over half of Poland is under cultivation, and in normal climatic years, Poland is self-sufficient, even producing a small, exportable surplus. In bad years, grain must be imported, while contracts to supply the USSR with meat, dairy products, and vegetables must still be fulfilled. Eighty percent of all farms are privately owned. Most farms are small, and rapid rural depopulation continues. In general terms, Polish agriculture has greatly increased its variety and productivity over the past three decades. Commercial dairying and meat production are increasingly important.

The government sees its problem as too much land used for industrial crops. The private sector sees low commodity prices and the high cost of fertilizer and machines as its greatest hindrances to production. Private farmers are allowed, even encouraged,

FIGURE 3–13 →
Border Changes in Central and Eastern Europe.
The state of Poland disappeared from the map for more than a century due to the expansion of the Russian, German, and Austrian empires. The eventual dissolution of imperial power led to the creation of many successor states in this culturally complex portion of Europe.

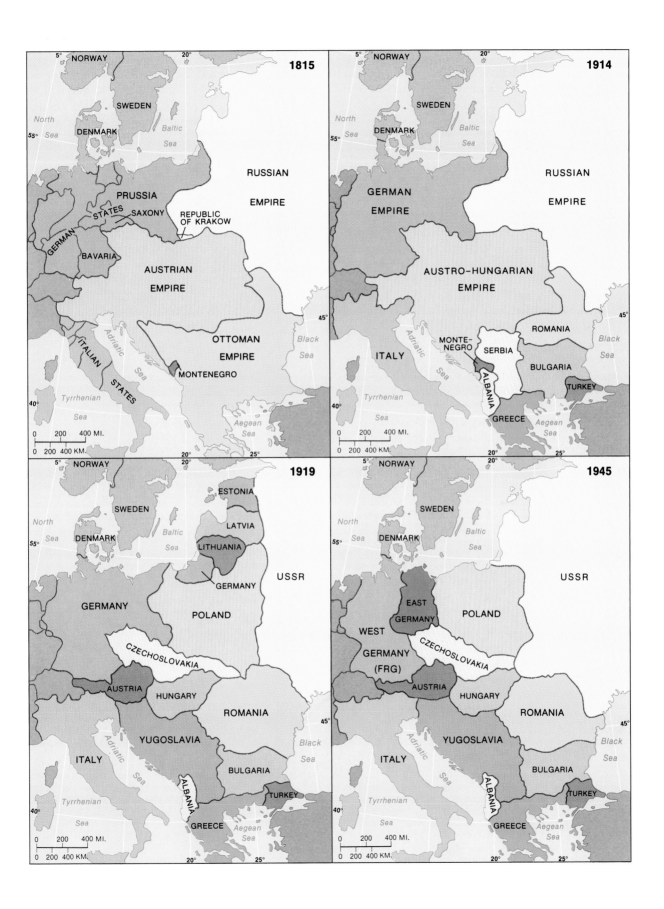

to increase the size of their holdings through purchase or rental of land.

One of the newer aspects of Polish agriculture is the commuting of workers who reside on the farm. Generally, one marriage partner commutes to an industrial job while the other farms. Rural space and food supply coupled with industrial wages and fringe benefits allow such commuters the best living standard.

Food preservation technologies and a tendency for even rural areas to import more foods from beyond the immediate region have meant that the seasonal rhythm of food consumption is less marked now than in the past. A century ago, most European diets had considerable seasonal variations depending on local availability of fresh food. Such seasonal variations in diet often were tied to the religious calendar. In northern Europe, including Poland, the late winter was generally a time of dwindling food supplies. Stored food from the previous fall harvest was almost used up; the winter fields were not productive, and farm animals had no fresh pasture but subsisted on shrinking amounts of stored hay and feed. The season of Lent, with its emphasis on a restricted diet, was a perfect match for the realities of food supply. Fresh meat could not be on the table until after the spring delivery of calves and lambs; farm animals, even chickens, were scrawny after their meager winter rations.

By Easter, the coming spring had produced increases in egg production, and hence the tradition of dyed and decorated eggs as gifts. The Polish and Ukrainian folk art of elaborately hand-decorated eggs may also reflect a temporary availability of extra time for such handicrafts, before the high labor demands of planting and cultivating began in earnest. At Easter, the last of the pickled and smoked meats preserved from the slaughtering season of fall would be used up, giving many Americans and Europeans the tradition of ham at Easter. Such customs as special holiday dishes thus may continue long after the relationship of land and livelihood becomes less intimate.

Polish coal production fuels eastern Europe's power plants and steel mills as well as earning foreign credits through export. The large Polish petrochemical industry is dependent, overwhelmingly, on Soviet-supplied petroleum. Poland, a major steel producer but modest iron ore producer, uses imported ores for most of its needs; Soviet ore is available but high-priced.

Upper Silesia has a Ruhr-like concentration of steel mills, metal refineries, and chemical works that create a polluted, if prosperous, industrial district of world rank.

Urbanization has taken place at a spectacular rate; metropolitan Warsaw, with 2.2 million people, remains the prime attraction for migrants, while most of the rest crowd into the port cities and the high-wage industrial areas of Upper Silesia. The cities of the Baltic combine scenery and climatic attraction with the best opportunities to buy foreign-produced luxuries and consumer goods—a kind of unofficial amenity in a society where many items are still scarce.

Solidarity, a labor union movement that began in a Gdansk shipyard, grew quickly as an expression of deep dissatisfaction with growing shortages in the consumer economy. The Polish government had borrowed heavily to expand industries of steel, artificial fibers, coal, and petrochemicals (based on Soviet oil), all of which ran into worldwide overproduction or severe competition from cheap-labor Third World countries. Expansion of auto production, again financed by foreign loans, was also poorly timed. The Polish government moved relatively cautiously to crush Solidarity as Lech Walesa, the movement's leader, became an international figure, winning a Nobel Prize in the process. That Solidarity was once crushed is typical of Communist regimes; that it arose at all is typical of discontented, individualistic Poland.In 1989, Solidarity candidates were elected to the Sejem, one house of the two-house Polish parliament, filling all available seats of a minority of overall membership. Although Solidarity does not hold power, it does have influence.

The German Democratic Republic has the highest standard of living of any Soviet bloc nation. East Germany was the more agricultural portion of Germany, whereas what became the Federal Republic of (West) Germany had most of the mineral wealth. East Germany does contain the largest area of superior soil found in prewar Germany, but it also has been thoroughly industrialized.

East Germany is the former Soviet zone of military occupation, a leftover of the division of Germany among the Allies after World War II. Some 4.5 million expelled Germans, mostly from Poland, entered East Germany between 1945 and 1950; yet the population still declined, as millions more migrated westward. This heavy emigration led to the construction of the Berlin Wall in August 1961. Currently, some 20,000 migrants (largely the aged and infirm) are allowed to emigrate legally to West Germany each year.

As in West Germany, deaths exceed births in a now-aging population. There were low birthrates in the immediate post–World War II period because of the heavy wartime losses of males. Over half the country is arable, and meadows and pastures are intensively fertilized. Collectivization takes an unusual form in East Germany: land may be collectively owned, but each collective is divided into lands farmed by individuals rather than cooperatively. The independence of decision allowed is rather broad. Yields are comparable to those of Western Europe, the only such case in a Communist economy. Livestock numbers (especially of pigs and beef cattle) have increased radically in the last 15 years in an effort to reduce the imports of meat. However, imported grain and livestock feed are necessary to maintain this high level of production, and the country remains a net importer of agricultural commodities.

East Germany's mineral base is not large, nor varied, but it is sufficient to meet most of the nation's fuel needs and some critical raw material needs for its exceptionally well-developed chemical industries. Despite the lack of a domestic supply of coking coal or iron ore, East Germany has a major steel industry, which serves the large market formed by metal fabrication industries.

Chemicals lead in terms of value of production. There is a huge chemical fertilizer industry, and myriad plants produce consumer and industrial chemicals for export and domestic markets. Cheap power generated with lignite is the most important domestic resource for industry. Soviet oil, brought by the Friendship pipeline, supplies the other main ingredients.

Every major type of industrial or consumer machinery is produced somewhere in East Germany. Carrying forward its prewar tradition, East Berlin remains the major center for the production of electronics and office equipment as well as appliances. Leipzig and its satellites produce lenses, scientific-medical laboratory equipment, and machine tools. The two major cities (apart from East Berlin), Dresden and Leipzig, have never recovered their prewar population sizes. Dresden's famous porcelain and china, with which its name is so closely associated, are still made. However, the city is now a major center for chemical and machine production. Leipzig still holds its annual trade fair, a holdover from medieval times. The Leipzig Fair today is a showcase for Communist bloc industrial technology and a place for the exchange of goods, machinery, technology, patents, and processes between the East and West.

A curious effect of the post-1945 border changes has been the development of Rostock. This ancient port, long stagnant, has become East Germany's major port and, through modernization, is seeking to attract Czech trade away from the Polish port of Szczecin (Stettin).

Czechoslovakia unites two Slavic language groups. The Czechs, long a part of the Austrian Empire, occupied a highly developed and heavily industrialized district. The Slovaks, governed by the Hungarians, were largely farmers or small-town artisans. Despite animosities between the groups, the state was one of the most successful creations of the Treaty of Versailles (1919) (see Figure 3–13).

However, Czechoslovakia contained large ethnic minorities; some 3.3 million Germans occupied the rimlands of Bohemia and Moravia, which Hitler seized in 1938. Together, the Sudeten Germans, the Hungarians, and some half-million Ukrainians accounted for about one third of the total population. Since 1945, Czechoslovakia has become more ethnically uniform. Nonetheless, it remains a coalition of two distinct nationalities.

There are still twice as many Czechs as Slovaks among the population of 15 million, despite the higher Slovak growth rates. Many Czechs are Protestants, whereas the overwhelming number of Slovaks are Roman Catholics. Most important, despite heavy government investment, the Slovak portion is still largely rural and definitely less developed than the Czech portion. Unity apparently is now accepted by both groups because other problems have taken precedence.

Virtually all agriculture is collectivized, though some rural districts remain dominated by private agriculture in the most mountainous areas of Slovakia.

POLISH BORDER CHANGES AND POPULATION RELOCATION

At the close of World War II, Poland's eastern and western borders both were shifted westward (see Figure 3–13). The Soviet Union expanded westward at the expense of Poland, and Poland was "compensated" for this lost territory by having its western border moved farther west at the expense of defeated Germany. The result of this territorial exchange was one of the largest politically motivated forced migrations of all time. It was as though Poland was picked up and transported westward. Virtually no other country has experienced such a transfer of population. Between 1945 and 1947, some 8 million Germans were expelled or migrated from the formerly German territories awarded to Poland after World War II. At the same time, some 3.5 million Poles were transferred from the western USSR after border changes in the east (Figure A). The gap between expellees and immigrants was filled by volunteer migrants from rural central and eastern Poland, so that by 1960, Poland could declare the former German "western territories" fully repopulated.

About 1 million of the ethnic Germans who remained in the recovered territories after the border changes of 1945 and their descendants have chosen to migrate to West Germany over the last 25 years. This migration has been arranged officially.

The changes in borders and consequent transfers of people have eliminated a major problem for Poland. In 1939, some 30 percent of Poland's citizenry belonged to a different ethnic group. Today the country is 99 percent Polish. There are some Poles still living outside the country, some 1.6 million in the USSR and a few tens of thousands in Romania and Czechoslovakia.

Although the borders of present-day Poland have been officially accepted by treaties with both East and West Germany, the territorial question conceivably could be reopened at a future date. Polish-German relations are officially cordial, but Polish public opinion is still decidedly influenced by World War II.

Despite mechanization, labor shortages continue as youth migrate to the city. Although the land is rich, with a history of careful use, the country still must import food. Large areas sown to industrial crops infringe on food production. Animal husbandry is well developed, though meat is often in short supply because of contracted exports to the USSR and East Germany. Wine and beer are both major export commodities.

Large deposits of low-grade iron ore in the Slovak Ore Mountains form the supply for the large new steel mill at Kosice. There are minor deposits of oil

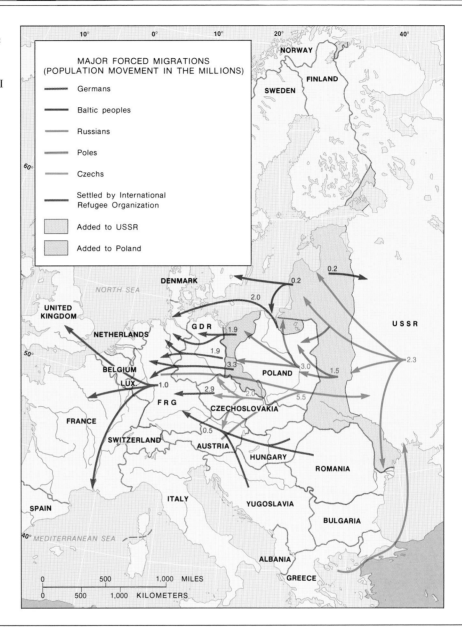

FIGURE A
Major Forced Migrations in Central and East Europe.
Millions were forced to migrate as a result of post–World War II territorial changes.

and gas, though the USSR supplies most of both. Fully one third of the land is covered with some of Europe's best-managed, highest yielding forests.

Czechoslovakia produces and consumes more steel per capita than any other country in the world, except Luxembourg. Transportation equipment and armaments are two of its most important lines of production. The world-famous Skoda Works at Pilzen and other cities produces many types of machines, though it is best known for its cars and light armaments. Prague (Praha) manufactures aircraft parts, buses, motorcycles, and machine tools; most of

THE QUESTION OF GERMAN REUNIFICATION

There is one serious question remaining for both Germanies—that of ultimate reunification. To achieve normalization of relations with East Germany, it was necessary for the Federal German Republic to recognize all post–World War II border changes and to enter into nonaggression treaties with the Soviet Union and the Communist countries of Eastern Europe. Further, West Germany extended large interest-free credits to East Germany and paid some outstanding indemnities to individuals and to the East German state, recognizing that entity as an independent, sovereign state. To avoid any disruption of the treaty provisions, the question of reunification was left open. Both states are members of the United Nations and have a full range of diplomatic relations with each other and most of the world's other states. West Germany is the most important trading partner of East Germany after the USSR and the other states of Eastern Europe.

Several factors either enhance a separate status or block reunification. It is a decided advantage for the USSR to maintain the current situation; a divided Germany is viewed as less of a military threat, and a sophisticated East German industrial plant continues to supply Soviet needs on highly favorable terms. The East German state enjoys a monopoly on certain items of trade within the Soviet bloc under CMEA agreements.

There is a considerable body of opinion in both Germanies that favors military noninvolvement. Again, one element of public opinion feels that West Germany has become too Americanized. There is also a tradition of separatism. Until the mid-nineteenth century, Germany was many states rather than one. Separation, then, is not new, and it conceivably could be thought of as normal. Currently, the prospects for reunification are not good. However, the economic prognosis for the separate Germanies continues to be excellent. Each, in its own way, is the economic power of its own respective group of associated states. There are powerful forces in both states who see the status quo as beneficial, and there are even larger numbers who fear that any change might have detrimental effects.

this industrial production enters the export market. A country known for quality, its high-priced lines of production are shipped to the Soviet Union and Western Europe, while low-priced items are sent to Third World markets. Long renowned for quality glass, Czechoslovakia still produces and exports crystal, Christmas tree ornaments, and costume jewelry. Continuing the glass tradition on a more Soviet-style line, Czechoslovakia also manufactures plate glass and insulated storm windows.

The Balkans

Hungary and Romania: The Non-Slavic Center.
The Romanian language is Latin based, a cultural residue of Roman control. The language of Hungary, Magyar, is an ancient Asiatic tongue not related to any Indo-European language (see Figure 3–3). Truly distinct Hungarian and Romanian cultures have emerged, and the national unity of both

states has in some ways been built on the perception of being different from surrounding Slavic groups.

Both states suffered long periods of dominance by other political units. While the Turkish occupation of Hungary lasted only a little over 100 years, the defeat of the Turks was followed by subjugation by the Austrian Empire. Romania spent almost 400 years under Turkish control. Unlike Hungary, it was isolated from the cultural trends and technology that emerged in western Europe over that time period.

Some 14,000 square miles of Romanian territory were seized by the Soviet Union in 1940. Much of it now constitutes the Moldavian People's Republic, incorporating approximately 2.5 million Romanian-speaking people within the Soviet Union.

The Hungarians are Roman Catholics or Protestants, as are the northern Slavs, whereas the Romanians are largely Orthodox Catholics like the traditional Russians, Greeks, and most South Slavs.

Both with originally agricultural economies, Hungary and Romania have sought to develop industry at a rapid pace in recent decades. Both have tended to create distinct and innovative versions of communism that suit national needs. Both perceive themselves as something of a bridge between East and West. They occupy the center between the Balkans and the North European Plain. They are at the contact zone of Islam and three variations of Christianity.

They control some of Europe's most important traditional trade routes and occupy some of the most fertile and productive of that continent's lands. Linked by the Danube, astride the steppes that brought Asian conquerors and Euro-Asiatic trade, these countries are the epitome of cultural crossroads.

These varied cultural contributions helped produce an exceptionally rich and colorful folk culture—the arts, crafts, and customs that evolve within a society over a long period, growing out of long tradition. It was formerly common for popular travelogues to emphasize displays of colorful folk dress of the Balkans peoples—"sturdy peasants in their quaint costumes." This was more than a Western criticism of less industrialized people; it was misleading because the "quaint costumes" involved were more ceremonial than everyday. The daily routine was carried out in much more practical clothing; the intricately embroidered and lace-trimmed apparel of special feast days bore the same relationship to everyday life that high school prom dresses and tuxedos bear to everyday life in contemporary America.

Long the food supply area for western Europe, Hungary and Romania produce large grain crops on rich loess or on black earth soils. Over 10 percent of all Romanians live in the capital, Bucharest, which

The Infamous Berlin Wall.
The divisions of Berlin, like the division of all of Germany, originally reflected zones of military occupation following World War II. Now the East-West split is personified by the Berlin Wall, shown here looking eastward toward East Germany.

contains one quarter of all Romanian industry and is a classic primate city. Budapest, the capital of Hungary, contains a quarter of all Hungarians and over half the country's industry. The capitals of both countries have long been **developmental enclaves**, areas of intense and advanced development far beyond that found in their respective hinterlands. Both have reopened Western trade contacts as a means of economic expansion, while unable to forego their economic ties to the Soviet Union; both have made enormous strides in industrializing.

Agriculture has been almost completely collectivized or put under state control; baronial estates once occupied most of the land, and a landless peasantry was the norm until the Soviet occupation. Collectives, now by far the dominant type of farm, own their own land. Each individual is guaranteed a private plot, which in Hungary can be unusually large; some 10 percent of agricultural land is in private plots. Horticulture has taken on a new importance in recent years, with large acreages converted to orchards, vineyards, and truck crops for export to CMEA nations.

Much power is imported from the USSR, which has linked Hungary to its power grid, and two pipelines feed Soviet oil into Hungarian industrial enterprises. Bauxite is the only mineral resource produced beyond local needs, and the aluminum industry has been developed as a joint Soviet-

Hungarian enterprise using Hungarian ore and Soviet power at locations in both countries.

On the losing side in World War I, Hungary was reduced in size and population, leaving a solidly Hungarian state with few minorities. On the other hand, large numbers of Hungarians were left outside the homeland, leading to potential border disputes. Again a German ally during World War II, Hungary came under complete Soviet domination in 1945. An abortive revolution came in 1956 when a rapidly liberalizing Hungarian regime was crushed by Soviet tanks. Since that time, change has occurred at a slower tempo.

Population change has become negative due to a very low birthrate and a relatively high death rate in an aging population. Migration from the countryside to the city has been proceeding at a rapid rate, so that some 55 percent of the population is now urban. Hungary is densely populated, reflecting its high fertility and exceptionally high degree of agricultural utility; almost three quarters of the land is in farms.

Consumer goods have become dominant; Hungary produces the highest quality television sets in the bloc. Hungary has reestablished a private fashion clothing industry and enjoys a good export market in high-quality clothing.

Hungary essentially scrapped centralized planning in 1960 in the face of an economic slowdown and renewed domestic unrest. State ownership is re-

Budapest, Hungary.
Budapest retains some of its imperial heritage through war and revolution into a Communist society. Hungary has managed to modify traditional Communist emphasis on heavy industry by specializing in quality consumer goods.

tained under this new plan, the so-called New Economic Mechanism (NEM), but factory management and production is left in the hands of production units. Factories are generally free to set prices, to buy parts from any source, to introduce technological changes, and even to increase wages so long as they show a profit. Under this system, personal income has risen, but it is still lower than Portugal's.

Of all the CMEA countries, Hungary is most dependent on foreign trade. Over half is still with Soviet bloc states; the Soviet Union is also its major supplier of raw materials. The real dynamics of Hungarian industry, however, are exhibited in its aggressive marketing of consumer goods, luxuries, and technical equipment of high-quality craftsmanship.

Romania's Roman conquerors, who called it Felix Dacia ("Happy Dacia"), left an indelible mark on the language. In succeeding centuries, invading Asians pillaged the lowlands, forcing the Roman culture peoples into the safety of the Transylvania Basin, an area surrounded by protective mountains and forests.

The impetus to nationalism came from the old Transylvanian core. There, a local noble, Vlad the Impaler, defended independence in an area never fully under Turkish control. The "Impaler" made a practice of displaying the bodies of Turkish troops and tax collectors along roads, each victim with a stake driven through the heart. Immortalized in a corrupted version of history as Count Dracula (his actual title), he was in reality a national hero who managed to stave off Turkish control.

Economic policies are relatively orthodox Marxian, and no overt disobedience is shown toward the USSR. Romania's chief goal is development; economic and political liberalization has not accompanied these policy changes. Theirs is a different kind of independence than that demonstrated in Hungary, Yugoslavia, or Poland.

Population increase has now slowed to 0.5 percent yearly; the government has toughened abortion laws to prevent depopulation, since the population had actually declined for a few years. Romania is less densely populated than most of Europe.

Private plots have been enlarged and peasant production for the free market encouraged; some 10 percent of all farmland is in individual hands. Increasing rapidly is the area devoted to orchards and vineyards. Growth in wine production has been rapid yet of relatively high quality, allowing Romanian inroads into Western European markets.

Romania is one of the world's oldest oil-producing states. The Ploesti fields, which played an important role in World War II, are all but exhausted. Because demand has increased, Romania must now import Soviet oil. Exploration, however, has uncovered one of the world's greatest gas fields in Transylvania. Gas has become an important raw material resource for the burgeoning chemical industry.

Romania must import most of the iron ore required for its steel industry. Convinced of the need to develop heavy industry, Romania now rivals Czechoslovakia in production. Prospecting has uncovered major deposits of bauxite, and an aluminum industry has been constructed.

After a long tradition of distrust and political animosity, Romania and Yugoslavia are developing a joint hydroelectric and navigation scheme at the Iron Gate gorge and rapids on the Danube, the most serious navigational impediment on that river. Trade between the two states has increased rapidly. Contacts and trade with both Germanies have been increased, and foreign investment is actively sought.

Yugoslavia, Bulgaria, and Albania. These three countries are governed by Communist regimes, although each has developed or accepted a different brand of communism. Two of the three countries are overwhelmingly Slavic in speech and culture.

Yugoslavia is a composite for which there was no historical precedent prior to 1919. Its name means "land of the South Slavs," and it does indeed include the major portion of South Slavs. Yugoslavs like to joke that their country has two alphabets, three religions, four languages, five nationalities, and six constituent republics, but even this jest underestimates Yugoslav diversity (see Figure A, p. 189).

Serbs (40 percent) and Croats (22 percent) are the two largest ethnic groups. They speak the same language but use Latin or Cyrillic alphabets in accordance with their past religious traditions. Five other major ethnic groups are recognized by separate autonomous republics or districts, and 17 other minorities are acknowledged.

The country is so culturally diverse that it has had to institute a form of ethnic federalism not unlike that

ETHNIC DIVERSITY AND DISUNITY IN THE BALKANS

The Balkans subregion contains a bewildering patchwork of ethnic, religious, and linguistic groups. Seldom do cultural boundaries among different groups coincide precisely with national political boundaries.

There are disputes over borders, irredentist claims, historical claims, and a general lack of coincidence between the boundaries of the states as now constituted and the distribution of each country's ethnic groups. For example, although Hungary is peopled 97 percent by ethnic Hungarians, there are a quarter-million Austrian Germans in residence. In this instance, the minority does not form the basis of any ethnic claim; unfortunately, this peaceful coexistence is not the norm. Numbers of Serbs, Slovaks, Croats, and Romanians live around Hungary's fringes but are counterbalanced by much larger numbers of Hungarians residing in Romania and Yugoslavia. Some 1.7 million Hungarians live in Transylvania and the western borderlands of Romania, while 500,000 live in Czechoslovakia, 400,000 in Yugoslavia, 150,000 in the USSR, and a like number in Austria.

Romania has both internal complexity and external ethnic claims to complicate its national unity. In 1939, about 30 percent of Romania's population was non-Romanian, but postwar border changes, reparations, voluntary migrations, and population exchanges have simplified the ethnic picture: some 88 percent of its residents today are Romanians. Hungarians, almost 8 percent of the total population, have been granted cultural autonomy. Yugoslavia and Romania have exchanged minorities on an amicable basis over the last five years, while the cession of territory to Bulgaria removed most of that minority. The other, smaller minorities are gradually becoming Romanianized, though the repatriation of the small Turkish minority has been the subject of negotiations. Ethnic diversity in Yugoslavia, Bulgaria, and Albania holds real potential for further disputes. Borders have stabilized to some degree, although many claims and counterclaims remain unsolved. Nationalism is a potent force in the subregion, but nationalities are not uniformly well developed. There are many cores, often mountain or valley basins, but large areas of disputed hinterland separate them, and one nationality may occupy all or a part of several cores.

Yugoslavia has the most complex intermingling of ethnic groups, and the most difficulty in balancing ethnic desires for autonomy with a need to maintain unity. Countering governmental attempts to conserve unity is a series of cultural and historical claims made by surrounding states.

FIGURE A
Ethnic Federalism in Yugoslavia. Like the Soviet Union, Yugoslavia is composed of many ethnic groups, whose aspirations to self-government have led to the establishment of a federal state whose member republics' boundaries reflect ethnic distributions. Kosovo and Vojvodina are autonomous provinces governed by the Republic of Serbia.

Internally, the most serious challenge comes from nationalistic Croatian elements. There is a disquieting tendency for economic development to conform to ethnic lines, with the Slovenes and Croats more highly educated and skilled than other groups. In an attempt to reconcile economic disparities, each republic within the federation has received some major government investment (Figure A). Despite this effort, initial industrialization of the two northern republics has tended to maintain a developmental gap among the republics.

of the USSR. There are six ethnic republics. It has been the explicit policy of the government to reduce tensions by allowing for maximum cultural autonomy.

In Yugoslavia, as in all areas south of the Carpathians and east of Vienna, corn is the most important crop, with wheat second. Yugoslavia has chosen to increase its specialization in temperate fruits and wine grapes, with heavy production of grapes, apples, pears, plums, and a series of wines and brandies, which increasingly enter the export market. Farming is not collectivized, though peasant cooperatives and state farms exist. There is an upper limit (25 acres) on the size of private holdings.

Yugoslavia's climate is mild, similar to that of Virginia in the interior but with a Mediterranean climate on the Adriatic coast. The growing season is long enough for two crops annually along the coast and in the south. Fresh fruits and vegetables are grown and marketed out of season in northwestern Europe's urban centers.

Yugoslavia possesses a wide variety of mineral resources, but fuel resources are present in only moderate supply. There is a large hydro power potential—the second largest in Europe after Norway—and about half the electricity generated is hydroelectric.

It is in metal ores, however, that Yugoslavia is richest. Some 2.5 million tons of bauxite are mined annually, almost exclusively along the Adriatic coast. Yugoslavia is a major producer of copper, and major investments are continuing to open up large reserves of iron ore.

In contrast to the situation in most Communist states, Yugoslavia's industry is profit motivated and consumer oriented. Industry is quite diverse, and virtually all major branches are represented.

In 1948, Yugoslavia's leader, Marshal Tito, refused to follow Soviet dictator Stalin's orders. Yugoslavia broke with the Soviet bloc and began to pursue a neutralist, nonaligned, self-interest-motivated brand of communism. Yugoslavia was the only European state in the Third World movement, which it helped organize, and it played a far more important role in world politics than its size and power would seem to imply.

The most interesting aspect of Yugoslavian industry is the decentralization of management into the hands of elected workers' councils. This structure is far closer to the orthodox interpretation of Marx than the Soviet industrial model, which establishes professional managers, technicians, and production planners.

Italian, American, German, and Japanese firms have invested in joint enterprises in Yugoslavia. Exports to Western Europe tend to be inexpensive clothing, footwear, and furniture; those to Eastern Europe tend to be more sophisticated—appliances, cars, electrical equipment, and high-value canned foods and wines. Many of Yugoslavia's manufactured goods enter Third World markets, where Yugoslavia has been an important source of aid, technology, and credit since the mid-1950s.

Yugoslavia's population growth rate is now lower than that of the United States. At the same time that population growth has slowed, the GNP and real income have been increasing at rates of 3 to 4 percent per annum, allowing for a steady increase in the standard of living. Urban dwellers are almost half of the total, and the declining percentage of labor employed in agriculture indicates the strides made in economic development.

The most obedient of the Soviet allies, Bulgaria is in many ways atypical of Eastern Europe. Like the other Balkan states, it has several national cores; yet, unlike Yugoslavia, it is relatively uniform in ethnic composition, with 86 percent of its people Bulgars.

The combination of near-level land, fertile soil, and mild climate results in a strong set of agricultural advantages. The country is of a size and fertility similar to the state of Ohio, but it is less level and also less populous, with only 9 million people.

Bulgaria underwent extensive land reform before World War II. Small farms were grouped into production cooperatives that produced in part for the export market, emphasizing high-quality horticultural products. This early tradition of cooperative agriculture may have made collectivization somewhat easier.

In an attempt to improve efficiency, state and collective farms are being integrated with industrial processing firms and consumer outlets in what are termed agricultural-industrial complexes. Agriculture now employs 25 percent of the work force.

Bulgaria, like Yugoslavia, relies on brown coal and hydroelectric power for most of its energy supply. It has constructed a large refining and chemical industry based on imported Soviet oil. The largest metallurgical works are located near Sofia, the capital. Bulgaria meets most of its steel needs; the integrated plants supply metal for Sofia's engineering and machine plants.

Bulgaria has doggedly resisted its assigned role of raw material and grain production within CMEA. Without straying from the ideological lead of Moscow, Bulgaria has quietly built up its industrial strength. Almost 25 percent of all industrial employees are now employed in machine building. A cigarette and tobacco industry has world rank, while wineries, canneries, and other processing plants still account for much of the industrial employment.

Yugoslavia, Romania, and Bulgaria rival one another in tourist trade; each combines sunny climate, excellent beaches, and low cost into a vacation package with broad appeal. Bulgaria has opened its doors to Western tourists on a large scale. Sandy beaches stretching along the Black Sea form the bloc's "riviera."

Last and least important of the Balkan states is Albania. Long the most backward corner of Europe, it is still backward after more than 40 years of Communist rule.

Albania has a population of 3 million in an area a little larger than Vermont; many Albanians live in neighboring Yugoslavia and northern Greece. The people have a somewhat uncrystallized nationality, and tribal groups and dialects distinguish the peoples of the north and the south. Religious differences between Moslems and Christians have split the country, but the current regime has diminished the influence of religion. In the last 35 years, a standardized Albanian language has evolved, and the literacy rate has risen, yet portions of the country remain outside the modernization programs despite government efforts.

It is difficult to obtain any data on the Albanian state. It does export some oil. Deposits of chrome, copper, and ferro-nickel ores are exploited, and concentrates of these ores (and some refined metal) are exported. There are large deposits of brown coal and a superior hydroelectric potential, yet per capita income is thought to be the lowest in Europe.

One hindrance to economic development is the rapid population growth rate (2.2 percent annually). Further, a disproportionate share of national income is channeled into defense spending in a country that perceives itself as surrounded by hostile states. In 1961, Albania broke with the Soviet Union. It is now, belatedly, attempting better trade relationships with its neighbors.

CONCLUSIONS

The generally sophisticated, high-technology industries of most European countries, combined with various European states' invisible exports of transportation, brokerage, insurance, and tourism services, and some high-value-added food specialties, mean that Europe, East and West, is a highly active region in international trade. Similarly, some regional or at least subregional shortages of raw materials contribute to heavy imports and large exports. Soviet gas and metal ores flow into both Western and Eastern Europe; Persian Gulf and North African oil go to the majority of European states that are net importers. Tropical agricultural commodities and metal ores come from Africa (south of the Sahara), Southeast and South Asia, and Latin America. Europe's roles as supplier, servicer, and customer have together forged strong links to all other world regions.

Europe remains a potent force, economically and culturally. The next century may be the "American century," or the "Japanese century," or the "Chinese century"; but the Europeans will be an active and prospering part of the world to come.

REVIEW QUESTIONS

1. Some Europeans feel that there has been a reduction of European power and influence in the world. Are they correct? Compare European living standards and production capacity with those of other world regions.

2. What are the cultural and political relationships of Europe (East and West) with the superpowers of today?

3. What factors, other than Communist forms of government, distinguish East Europe from West Europe?

4. It has been said that Europe is one of the world's most usable continents. What attributes of its physical environment bear out this statement? Why is Europe's climate so warm relative to its latitudinal location?

5. Compare and contrast the climates of Northern and Mediterranean Europe. How do maritime and continental influences differentiate Europe's climate?

6. Which six states were the original members of the European Community? What was the role of the geographic distribution of mineral resources in the development of this organization? What new members have since been added?

7. Using France as an example, review the stages of the demographic transition. What factors have contributed to continued economic prosperity in France?

8. Discuss the problems, prospects, and productivity of West European agriculture. What factors have contributed to this modern resurgence of agriculture?

9. What steps have the states of Europe taken to maintain their competitive position in the arena of world trade?

10. What are invisible exports? Give examples and explain their role in maintaining a favorable balance of trade?

11. What are the environmental and economic problems facing the small, densely populated and heavily industrialized Benelux states? What are their chief assets?

12. How does the European automotive industry illustrate the internationalization of production? Give examples.

13. Know the concept of *overdevelopment*. Give examples of this phenomenon as it exists in the European Core. Discuss the degree of concentration of both urban populations and industrial production within the core.

14. Discuss the role of the United Kingdom's resource base in the development of the industrial revolution. What is the significance of that country's resource base today?

15. What four factors have served to retard the economic growth of the United Kingdom since the loss of its empire?

16. Eire, Iceland, Italy, and France have fewer (and poorer) sources of conventional fuel than most of Europe's states. What innovations have these countries developed to counter their lack of traditional fossil fuels?

17. What are the demographic characteristics of the population of the UK and Eire? How have the currents of migration influenced the respective populations of these two countries?

18. Describe the effects of the industrial revolution on the distribution of the population within the UK. What are the consequences for the UK of having been an early leader in the industrial revolution? How is this early industrialization mirrored in the landscape?

19. What are the healthy and unhealthy segments of today's British economy?

20. What are the historic and cultural roots of the political civil strife in Northern Ireland?

21. What factors contribute to the sustained economic growth of West Germany? Contrast the cultures and economies of West Germany and Switzerland.

22. What are the five national states composing Norden (Scandinavia)? What are their common characteristics and major differences? How has their industrial development differed from that of other parts of Europe?

23. What are the root problems of Spain's internal disunity? What are this country's economic strengths and weaknesses? Why hasn't Portugal shared in the recent Iberian economic development?

24. What states compose the Northeastern Industrial Zone? Where is its productive core? Compare the energy bases of these three states.

25. Discuss the shift in borders that occurred in Poland after World War II. What ramifications have these changes had for the population, economy, and ethnic makeup of the country?

26. Which states of East Europe are amalgams of different ethnic groups? How have the individual governments dealt with this ethnic complexity? What major ethnic discord (and disputes) remain in East Europe?

27. Discuss the differences in the organizations of agriculture in East Europe. Where have farms been decollectivized? Analyze the relative success or failure of this decollectivization.

28. What are the prospects for German unification?

29. In what political respects do Albania and Yugoslavia differ from the rest of East Europe? In what important respects do they differ from one another?

30. East Europe is a zone of both cultural contact and cultural diversity. What are the religious, linguistic, and historic similarities and differences that characterize various parts and components of this area?

SUGGESTED READINGS

Beckinsale, Monica, and Beckinsale, Robert. *Southern Europe: A Systematic Geography*. London: University of London Press, 1975.

Clut, Hugh, ed. *Regional Development in Western Europe*. New York: Wiley, 1981.

Demko, George, ed. *Regional Development: Problems and Policies in Eastern and Western Europe*. New York: St. Martin's, 1984.

Diem, Aubrey. *Western Europe: A Geographical Analysis*. New York: Wiley, 1979.

East, W. Gordon. *An Historical Geography of Europe*. London: Methuen, 1966.

Gottmann, Jean. *A Geography of Europe*. New York: Holt, Rinehart & Winston, 1969.

Hoffman, George. *The Changing European Energy Challenge: East and West*. Durham, NC: Duke University Press, 1985.

Hoffman, George. ed. *A Geography of Europe*. New York: Wiley, 1983.

Hoffman, George. ed. *Eastern Europe: Essays in Geographical Problems*. New York: Praeger, 1971.

Houston, James. *The Western Mediterranean World*. New York: Praeger, 1964.

Jordan, Terry. *The European Culture Area*. New York: Harper & Row, 1973.

Malmstrom, Vincent. *Norden: Crossroads of Destiny*. Princeton, NJ: Van Nostrand, 1965.

Nystrom, J. Warren, and Hoffman, George. *The Common Market*. New York: Van Nostrand, 1976.

Pounds, Norman J.G. *An Historical Geography of Europe, 1500 to 1840*. New York: Cambridge University Press, 1979.

Pounds, Norman J.G. *Eastern Europe*. Chicago: Aldine, 1969.

Stamp, Dudley, and Beaver, S.H. *The British Isles*. New York: St. Martin's, 1971.

White, Paul. *The West European City: A Social Geography*. New York: Longman, 1984.

CHAPTER FOUR

The USSR

Red Square.

arger than the United States and Canada combined, the USSR (Union of Soviet Socialist Republics) is the world's largest country (Figure 4–1). With over 280 million people, it is also one of the world's population giants. Yet, because an immense area is involved, much of it of difficult environment, it is relatively lightly populated. As geologically varied as it is vast, it contains virtually every industrial raw material needed, making it potentially the most nearly self-sufficient country in the world. Vast areas of land are farmed, yet, ironically, it is unable to produce enough food to feed its own people. Although the political system, with its centrally planned economy, may be partially responsible for many Soviet problems, it is nevertheless true that the system has moved the USSR forward, within 70 years, from a comparatively underdeveloped country to the status of an economic and military world power.

Tremendous size and distance complicate the governance, development, and production of the USSR (see Figure 4–1). Ethnic diversity threatens its unity. Most of its territory is poorly suited for farming; its cold, northerly location adversely affects the productive prospects of much land, and the southern, warmer portion of the country has a generally insufficient moisture supply. Great distances separate farm, resource, factory, and market. Population-resource imbalances plague an economy in which 70 percent of the people live in the developed west and only 30 percent inhabit the resource-rich eastern lands which await further development (see Figure 4–1).

It is necessary to understand the full scope of all positive and negative geographic factors operating in order to fully comprehend the USSR. Both its society and its economy are as complex as its physical environment. No simple changes or answers can quickly solve Soviet economic dilemmas. The Soviet government, through its educational system, has tried to eradicate the baggage of prerevolutionary history. However, at least some of the cultural legacy, developed over a thousand years, is still present. No political system, nor any technology, can operate obliviously to the constraining realities of the country's harsh physical environment. Even under a democratic capitalism, the USSR would not, could not, be another United States. Its cultural and physical geography, not just its ideology, is vastly different. The questions that the Gorbachev reforms attempt to address include, Would a more profit-motivated system propel production upward?

Comparisons with the United States are frequent. Both nations control huge territories richly endowed with natural resources. Both are originally offshoots of European culture, though both have created distinctive new cultural amalgams incorporating non-European peoples and cultures.

There are as many strong contrasts as similarities, however. Most Americans would interpret the second phase of the Russian Revolution, the part that brought the Communists to power, as a failed revolution. Few could sympathize with the oppressive and corrupt tsarist regime that was overthrown. But the citizens of the Soviet Union have not achieved political or economic freedom since the Revolution, though they have, in general, enjoyed a rise in standard of living. There is no political alternative to the Communist party, and there can be no organized opposition to its policies. Dissidents are dealt with harshly. Religious freedom, guaranteed under the Soviet constitution, does not exist in reality. While the United States struggles to deal with an incoming flood of both legal and illegal immigrants, the Soviet Union rigidly restricts the out-migration of its own citizens.

The grandchildren of the Revolution (almost no one alive has direct experience of the 1918 Revolution) live in a world of continuing shortages of consumer goods for all but the elite. There is much evidence of the power and privileges of class in the

FIGURE 4–1
(a) *Place Names and Member Republics of the USSR.* The USSR, the world's largest country in area, is a federal state, composed of 15 member republics, called Soviet Socialist Republics (SSRs). (b) *Subregions of the USSR.* The USSR has been subdivided into six subregions for purposes of geographic analysis.

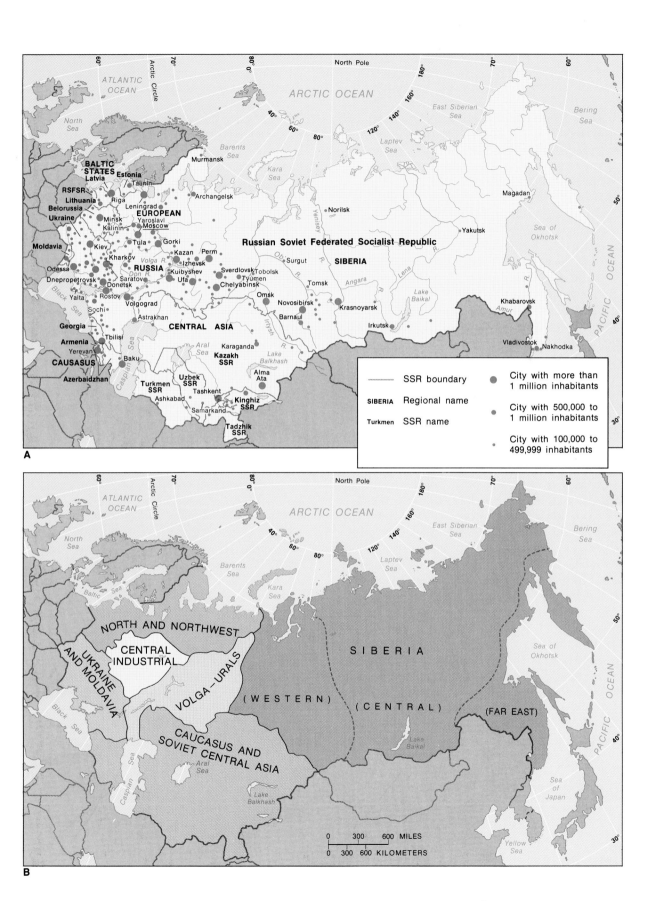

A

ATLANTIC
OCEAN

Arctic Circle

ARCTIC OCEAN

North Pole

Bering
Sea

North
Sea

Barents
Sea

Kara
Sea

Laptev
Sea

East Siberian
Sea

Murmansk

**BALTIC
STATES** Estonia
Latvia Talinn

Archangelsk

Magadan

RSFSR
Lithuania Riga
Belorussia Leningrad
Ukraine Minsk
Kalinin Yaroslavl
EUROPEAN Moscow

Norilsk

Yakutsk

Moldavia Kiev Tula Gorki
Kharkov Kazan Perm
Odessa **RUSSIA** Izhevsk
Dnepropetrovsk Saratov Sverdlovsk Tobolsk
Donetsk Kuibyshev Tyumen
Yalta Rostov Ufa Chelyabinsk
Sochi Volgograd Omsk
Astrakhan Novosibirsk
Barnaul

Surgut

Russian Soviet Federated Socialist Republic

SIBERIA

Tomsk

Angara

Lake
Baikal

Sea of
Okhotsk

Krasnoyarsk

Khabarovsk

Irkutsk

Georgia Tbilisi
Armenia
Yerevan
CAUSASUS
Baku
Azerbaidzhan

CENTRAL ASIA

Karaganda

Aral
Sea

Lake
Balkhash

**Kazakh
SSR**

Vladivostok Nakhodka

PACIFIC OCEAN

Alma
Ata

**Turkmen
SSR**
Ashkabad

**Uzbek
SSR**
Tashkent

Samarkand

**Kinghiz
SSR**

**Tadzhik
SSR**

⎯⎯⎯⎯ SSR boundary	● City with more than 1 million inhabitants
SIBERIA Regional name	● City with 500,000 to 1 million inhabitants
Turkmen SSR name	· City with 100,000 to 499,999 inhabitants

B

ATLANTIC
OCEAN

Arctic Circle

ARCTIC OCEAN

North Pole

Bering
Sea

North
Sea

Barents
Sea

Kara
Sea

Laptev
Sea

East Siberian
Sea

Baltic
Sea

NORTH AND NORTHWEST

Sea of
Okhotsk

**UKRAINE
AND
MOLDAVIA**

**CENTRAL
INDUSTRIAL**

VOLGA–URALS

S I B E R I A

Black
Sea

(WESTERN) **(CENTRAL)** **(FAR EAST)**

**CAUCASUS AND
SOVIET CENTRAL ASIA**

Caspian
Sea

Aral
Sea

Lake
Baikal

PACIFIC OCEAN

Lake
Balkhash

Sea
of
Japan

Yellow
Sea

0	300	600 MILES
0	300	600 KILOMETERS

supposedly classless society. The everyday experiences of average Soviet adults feature long waits in lines by those hoping to buy such scarce commodities as fresh oranges, toilet paper, meat, or toothbrushes. Access to good-quality goods and services is a privilege of the Communist elite, as are spacious apartments and trips abroad.

The USSR controls what is likely the world's last colonial empire; its troops support the stability of Communist governments in Eastern Europe, and its occupying army suffered casualties before withdrawing from the guerrilla war raging in Afghanistan. Economic subsidies to friendly regimes and military aid in Cuba, Kampuchea, Vietnam, Angola, Afghanistan, and Eastern Europe add up to $20 billion a year—a heavy burden for a sluggishly expanding economy. Many of the USSR's own people are increasingly cynical about their government's ability to deliver the promises of the Revolution.

The USSR, and Russia, its historic predecessor, have both been characterized as xenophobic—mistrustful of foreigners and foreign governments and reluctant to interact with areas beyond the home base. A political unit of its sheer size was almost always able to attain self-sufficiency in food, fuel, and raw materials. Russia's only shortcoming, traditionally, was a lack of specific technologies, obtainable only from its more developed neighbors. The first Russian state was founded on trade, exchanging its surplus raw materials and grains for the technology and consumer goods of more advanced eastern neighbors, including China, Persia, various Arab kingdoms, and the Byzantine (Eastern Roman) Empire. Modern Soviet trade also is often viewed as the exchange of raw materials for the technologies of better developed states of the West—a generalization that is only partially correct.

The Tartar invasions and the fall of Constantinople in 1453 saw ancient Russia sink into isolation. For almost four centuries afterward, Russia maintained itself separate from the currents of Europe, developing its own considerable resources for its own needs. Periodically throughout its history, Russia has vacillated between isolation and participating in world trade. An outward-looking Russia, seeking modernization and military equality with the great powers of Europe, actively engaged in trade during the eigh-

teenth and nineteenth centuries. Its capital was moved to the seacoast when a great port was founded at St. Petersburg (now Leningrad) to facilitate contact and trade.

In 1918, the Revolution brought trade to a halt as the new USSR returned, only partly through choice, to renewed isolation. The USSR was treated as an outcast by fearful Western powers, and the Soviets were ideologically convinced that capitalism was evil. Briefly, during World War II, wartime relief and military supplies entered through ports that functioned mainly as military bases. The immediate postwar years were characterized by minimal trade, even between the Soviet Union and its Communist client states. That limited trade moved overland, rather than through long-idle Soviet ports.

Now, slowly at first, backed by an enlarging merchant fleet, the USSR has returned to the position of a trading power. In 1960, over half of all Soviet trade was with Eastern Europe, and over three quarters was with Communist bloc states. Grain and raw materials were traded largely for East European manufactured goods. Trade with non-Communist states represented only about one fourth of the total, although, from the first, it was heavily weighted toward technical imports from Western Europe. The USSR, trading cheaper commodities for more sophisticated goods, had a negative trade balance at that time. By 1984, it had a positive trade balance, and now has increased its total volume of trade over 15 times the 1960 level.

Over the past five years, the Soviets have rolled up a $50 billion trade surplus. Oil and gas, both sold at a relatively high price until recently, are the major exports to Europe. The Soviets also ship surplus cotton, vegetable oil, and industrial fibers to Europe. These relatively high priced agricultural exports earn enough profit to more than counter the cost of importing cheaper North American grains. Fertilizers and chemical feedstocks, semimanufactures, and machinery are of increasing importance in the list of Soviet exports. Sturdy, inexpensive Soviet-made Belarus tractors have begun to appear at American agricultural equipment dealers. Raw materials still exceed manufactured goods as a percentage of total exports, but obviously Soviet exports are no longer limited to those items. Trading partners have shifted as well. The Communist bloc now accounts for just over half of all Soviet trade, while the capitalist nations of the developed world account for about one third of the Soviet imports and exports.

Soviet equipment and technology now move to Third World nations in increasing amounts. Return imports, invariably less valuable than Soviet exports, consist of tropical commodities such as coffee and bauxite. Raw materials shipments to Eastern Europe, although still strong, are gradually being supplanted by the outward flow of manufactured items. Even trade with the West is changing as Soviet-produced furs, plywood, luggage, and forklifts join the more traditional raw materials in a westward flow. Despite American opposition, Soviet oil and gas now flow, via pipelines, to Western European economies. Western technology still flows eastward in the form of West German steel furnaces, Italian automotive assembly machinery, French electrical equipment, Austrian arms, and Scandinavian oil drilling equipment.

Barter occurs on a large scale. The East European satellites have financed Soviet pipelines in return for gas, and West Germany finances Soviet steel mills in return for high-grade ores and oil. The Japanese, long excluded from Soviet markets, pay for Siberian rail construction and mineral development in return for coking coal, timber, and aluminum.

Increased involvement in trade appears, in some measure, to be a new route to Soviet internal development. Soviet investment capital, stretched thin by years of isolation, is being augmented with the profits of trade. Indirectly, Soviet exports defray (or delay) the cost of developing new, expensive technologies. Industries can grow on a base of both legally and illegally imported machinery and advanced Western manufacturing systems much more quickly and cheaply than if created wholly from domestic resources. Trade has brought increased prosperity; it also has the potential to enrich the prospects for world peace. It will not likely "convert" the USSR to capitalism, nor result in large-scale Soviet indebtedness or overdependence on foreign sources of supply. With domestic gold reserves sufficient to cover any short-term deficit, and an almost endless resource base, the Soviet Union now trades from a position of power. Soviet-Western trade, like any other rational trade, has benefits for all involved. The

USSR is becoming an integral part of the increasingly interdependent world economy.

TRENDS AND PERCEPTIONS

The Soviets long viewed the world through the prism of Marxism, expecting a worldwide revolt of the working classes. The failure of a revolt to materialize in the 1920s led to an isolationist attitude. Non-Communist governments were viewed as a threat. Isolation gave time for the country to develop its internal structure and to consolidate effective political control over the world's largest state. To be fair, it must be noted that the Western powers did their best to isolate the USSR as well. The United States, for example, did not officially recognize the Soviet government until 1933. Soviet paranoia was hardly calmed by the temporary incursions onto Soviet soil of U.S., British, French, and Japanese troops during an abortive campaign to aid the "White" (non-Communist) forces during the civil war following the collapse of their wartime ally, the tsarist regime, in 1917.

The Soviets viewed participation in World War II as a means of survival while waiting for the ultimate class conflict to arise; they would not have entered that war without being attacked. They saw it as a war among imperial powers in which they should take no part.

At the end of World War II, the Soviets sought to surround the USSR with a series of "friendly" states; in purely political terms, it sought buffers. Only when China came under a Communist regime did Soviet success result in new goals—the piecemeal change in the balance of power. The United States, for its part, saw containment—the creation of a series of military and economic alliances with states surrounding the USSR to prevent its further ideological expansion—as the only alternative short of all-out war.

Khrushchev (Premier, 1958–64) sought to alter the world balance and inadvertently recognized the Third World as a separate entity. Economic competition and skillful diplomacy replaced "brush wars" as the means of expansion; the Soviet Union had reached a level of development where it felt that it had much to lose in an armed conflict of major scale. It had reached equality in power.

The arms competition between the United States and the Soviet Union has moved to advanced technological levels and into space. Cuba was more a useful thorn in the side of the United States than a strategically located ally, until the establishment of a Marxist regime in Nicaragua. African nations have changed ideologies and allegiance swiftly, and there have been no lasting loyalties or commitments to either the USSR or the United States. What began in Yugoslavia in 1948 bore fruit in China in the mid-1950s as communist ideology became a matter of national interpretation rather than uniform dogma. Moscow's onetime unchallenged dominance over other Communist states received a setback.

Straightforward world conquest and full-scale armed conflict have lost any appeal they may have once had before the nuclear age. Generations of Soviets used to an improving living standard find ideological purity outdated and unrealistic. World power is less appealing if it will cost great sacrifices, and it could result in total destruction.

GEOGRAPHIC PROBLEMS AND POTENTIALS: AN OVERVIEW

Soviet central planning is being called upon to deal with regional inequalities in population growth and development. Siberia's severe physical environment, shortages of goods and housing, and lack of amenities pose strong disincentives to population growth. Since 1965, more people have left than have migrated to Siberia. Only very high non-Russian birthrates and higher than normal birthrates among the young, rural Russian settlers who remain prevent the area from actually losing population (Figure 4–2).

The Caucasus-Central Asia subregion presents a different picture. Enormous growth rates among non-Slavic ethnic groups, coupled with the Soviet "Sun Belt" migrational tendency, present the prospect of an almost unmanageable future growth (Figure 4–3).

Investment in European Russia is most productive; it is, relatively, the cheapest with the greatest immediate economic return. Water shortages in the Caucasus-Central Asia subregion are compounded by massive labor training needs and cultural resistance to industrialization, making that region's costs and potential benefits more expensive. Siberian physical conditions and the great distances to the highly developed, relatively densely populated Soviet core result in high construction, production, and marketing costs that reduce the benefits of investment in that region. Investment in Siberia would require the greatest cost and yield the least short-term return.

Extremely rapid population growth in the Caucasus-Central Asia subregion requires either massive investment or massive out-migration to induce real economic development within the region, as well as simply to provide sufficient future employment.

The contrasting pulls of Siberian resources, high living standards in European USSR, and attractive Soviet Sun Belt climates are not matched by any significant push in a society that guarantees employment and provides the bulk of social welfare needs wherever one resides. Siberian wage incentives are often eaten up in increased living costs.

Developing Siberia implies massive new construction, in particular of transportation infrastructure, to develop resources and industry. The expediency of maintaining economic growth for the whole USSR has favored development and expansion in the European core over the last 25 years; the core's last corners are undergoing development and redevelopment right now. The Soviets appear to be doing a little in all three areas at once, rather than emphasizing any one area. Before new major investments in the Siberian and Central Asian regions can be made, the Soviets apparently are experimenting on a smaller scale.

West Siberian resources already have been developed to the point where oil, gas, and timber are amortizing earlier investment costs, though fluctuating world oil prices may alter that profitability. A metallurgical base (the Kuzbas), an urban service and distribution network (anchored by Omsk and Novosibirsk), and a strong regional food supply are already in place. The most sorely needed transportation lines have been built. The already thoroughly developed Urals industrial region acts as a nearby

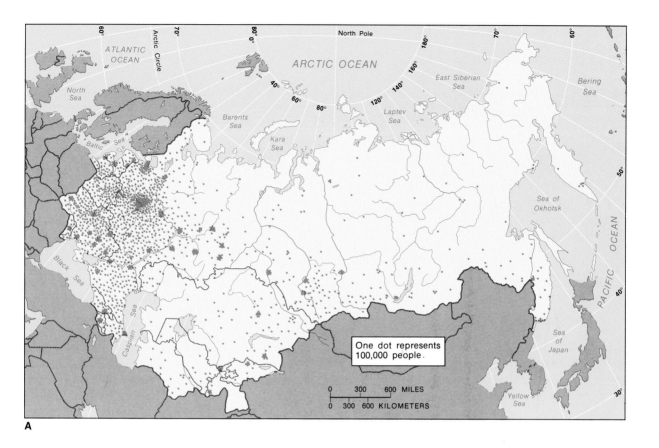

A

FIGURE 4–2

(a) ***Distribution of Soviet Population.*** The Soviet ecumene (densely settled, well-developed territory) takes the form of a great triangle, with one segment formed by the western political frontiers of the USSR, another by the sharply decreasing population density of northern European Russia and Siberia as a result of increasingly severe winters, and a third by the sharply diminished precipitation along the southeastern desert borders.

(b) ***USSR Population Pyramid.*** The Soviet Union, as this population pyramid shows, has a relatively youthful population, but one in which the birthrate has declined recently. The excess of women in older age groups partially reflects the higher death rate of men in wartime.

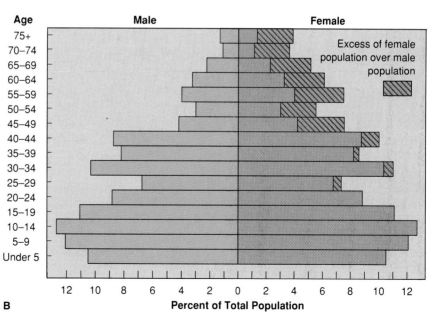

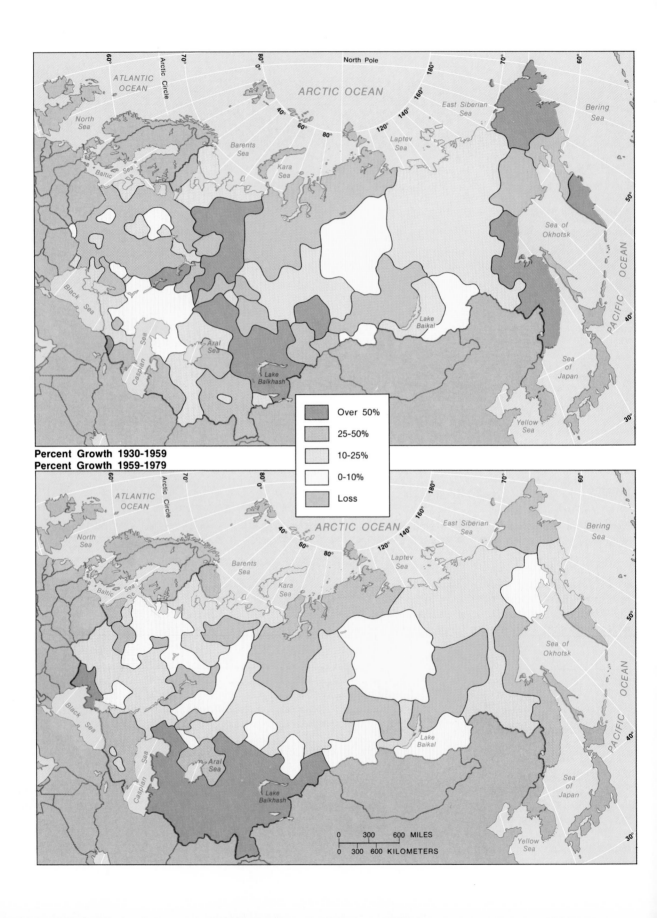

Percent Growth 1930-1959
Percent Growth 1959-1979

Over 50%

25-50%

10-25%

0-10%

Loss

0   300   600 MILES

0   300   600 KILOMETERS

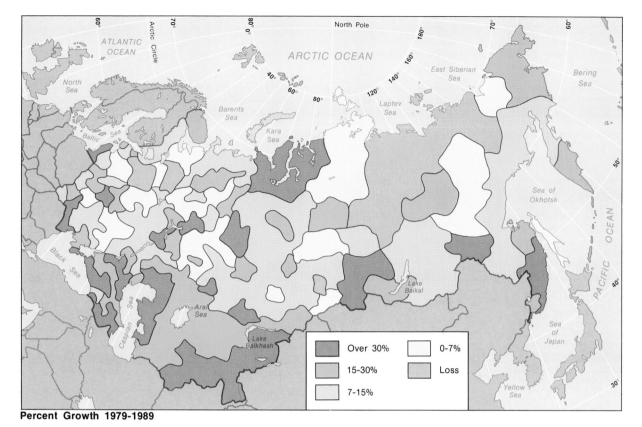

Percent Growth 1979-1989

Legend:
- Over 30%
- 15-30%
- 7-15%
- 0-7%
- Loss

FIGURE 4–3, pp. 202–203
Comparative Population Growth Rates. This time series of population growth
shows the rapid increase of most non-Russian ethnic groups compared to the "great
Russians" (ethnic Russians) and other western, Slavic groups. Note especially the
recent growth in the Caucasus and Central Asia.

supply base, so the costs of further moderate expansion of West Siberia's economy are significantly reduced. The region is gradually nearing the point at which it will be fully integrated with the European core.

Eastern Siberia will likely lag in receiving significant new investment. Environmental conditions there render development costs even higher than in Western Siberia. The Far East will likely continue limited expansion, relying on Japanese capital and Japanese markets wherever possible. Siberian growth, then, can continue at a moderate pace without major in-migration or massive investment. Lower world prices for timber and oil, alone, would seem to discourage implementation of any grand-scale schemes.

Limited out-migration of the native population, increased tourism, and moderate furtherance of the R&D, high-tech, and labor-intensive industries that already exist there should keep the Caucasus region's economy in balance. The Caucasus is not a major segment of the Soviet Union in area or population, nor is it a great distance from established Soviet markets, and integration with the core requires only improved transportation across the mountain barrier.

Soviet Central Asia, however, poses immediate needs on a large scale. Regional oil and gas resources along the eastern shore of the Caspian have already been developed and will likely be expanded. Production of cheap thermal power in the northern Kazakh areas is also expanding. Production of minerals and ores in the Kazakh plateau is being actively developed as a replacement for the already heavily exploited resources of the Urals. Massive past investment in the automotive, petrochemical, and machinery industries of the Volga Valley can be easily and cheaply expanded if additional water and power can be made available. The Volga Valley is the "natural" supply base for expansion in Central Asia.

Soviet planning often is of a grand scale, involving vast distances, huge areas, and massive investments. If these plans seem to be grandiose examples of **planetary engineering** (rearranging nature on a grand, perhaps arrogant, scale), it must be remembered that the scale of the country *is* huge. Planning the country's development requires bold concepts and visionary designing, at least in the eyes of Soviet bureaucrats.

Although many problems remain, and despite the fact that the European USSR may appear more shabby and less advanced than Europe—East *or* West—the grand development plan of the 1970s and 1980s is virtually completed. The remaining challenges are in the east, and, if they are ever to be addressed, grand projects appear to be necessary first steps in the process.

THE NATURAL ENVIRONMENT

There is little doubt as to the physical diversity of the USSR. The longitudinal extent of the country dwarfs its latitudinal spread; continentality maximizes east-west climatic differences. The long, cold winter experienced by over 90 percent of the country provides a degree of climatic similarity, but the sheer size of the USSR makes generalization difficult.

Most of the population is concentrated in a limited belt of land of relatively high utility. Composed of **steppe** (natural grassland) on the south and grading through wooded steppe and mixed forest to the **taiga** (dense, coniferous, boreal forest) on the north, this populous zone stretches across the Soviet Union from one end to the other, narrowing sharply eastward. Level to rolling throughout, it provides a similar narrow range of climatic and physiographic milieus. The lack of sharp breaks gives it a degree of uniformity. Only areas outside this belt, generally beyond intervening physical barriers of low utility, are markedly different.

Topography

There is less topographic variety than one might expect within the 8.5 million square miles of the Soviet Union. Three large plains cover more than half the territory (Figure 4–4). In the west, the North European Plain (redesignated as the East European Plain) widens to encompass virtually all of the European portions of the USSR, from the Arctic to the Black Sea. The northern half of this plain has been glaciated and is generally rolling, boggy country with occasional strings of morainal hills (ridgelike glacial deposits) and a few low mountains. The southern, unglaciated half is mainly a platform covered with rich soil, deeply eroded along stream valleys.

A large plain occupies the depression from east of the Caspian Sea to the mountain lands at the Chinese, Afghan, and Iranian borders. The West Siberian Plain occupies the areas between the Urals and the Yenisey River; it is a flat, sodden, poorly drained area, commonly flooded during the thaw season. These three plains are separated from each other by the narrow north-south, folded mountain chain of the Urals, the Caspian Sea, and the Kazakh Plateau, which stretches eastward at right angles to the Urals.

Farther eastward, the great rivers of Siberia have eroded the Central Siberian Plateau into a rough countryside that was once notoriously difficult to cross. The Lena River drains a large lowland basin at its eastern edge, while the Soviet Far East is composed of impressive rows of parallel mountain chains that reach to the Pacific.

Most of the southern border area is mountainous; the Caucasus and related mountain systems stretch from the Crimea (or Crimean Peninsula) through the Turkish borderlands to eastern Iran. The jumble of mountain ranges at the core of Asia, known as the Pamir Knot, extends into Soviet Central Asia, with strings of related mountains forming parts of the Soviet-Chinese border. An intensely folded area of mountains and high plateaus, quite similar to the difficult country of the Soviet Far East, extends from Mongolia to the headwaters of the Amur River, giving

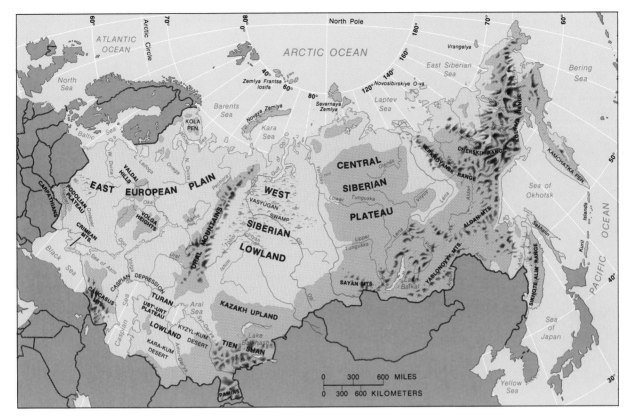

FIGURE 4–4
Landform Regions of the USSR. The open, rolling plains of most of the western border with European states is in marked contrast with the more rugged qualities of the USSR's Asian borders.

rise to the headwaters of the great Siberian rivers. Ringed with mountains on its Asian frontiers, the USSR has been most vulnerable to attack from the plains of the west. To some extent, this fact has conditioned Soviet military and economic thought.

Land-Sea Relationships

One of the most overplayed lines of Western European reasoning about Russian geopolitical motivation has been the "drive" for ice-free ports. Russia long has had ice-free ports, but they are in peripheral, awkward locations such as Murmansk, or on the Black Sea, exiting through narrow straits subject to foreign military closure. Openings to seas *other* than the Arctic or the Black Sea, in other directions, was far more the reason for Russian expansion. Only re-

cently, however, have the Russians become sea oriented in their thinking and military reasoning. Some of the seas surrounding Russia are singularly unattractive. The Arctic and its seas are navigable for only around 30 to 60 days east of Novaya Zemlya, the island extension of the Urals. The Caspian and Aral seas are giant salt lakes isolated from the world ocean.

At times, Germany and Sweden or Denmark so dominated Baltic trade there that Russia was effectively squeezed out. Leningrad (then St. Petersburg) was built to give access to the Baltic. Russian conquest of Finland and the Baltic states enlarged this foothold, but at a time when the importance of the Baltic was waning rapidly.

A major Russian foreign policy objective has always been the bypassing of the Turkish Straits (through either territorial acquisition in the Balkans

Soviet Icebreaker. *Among the Soviet Union's geographic problems is the fact that it has few ocean ports that are not closed seasonally by ice. The USSR's longest shoreline, along the Arctic Ocean, can be navigated for only a few weeks each season. This ice breaker is one of a large fleet operated by the Soviets.*

or Iran or the development of friendly allies in those areas). Since 1918, the straits have been open to trade in times of peace, though the Turks are still in control of the crucial waterway and less than friendly toward the Soviets.

The Pacific coastal waters are open to navigation for varying periods of time. In most years, Vladivostok has been kept open all year with icebreakers. Border revisions after World War II awarded southern Sakhalin and the Kuriles to the USSR along with clear access from Vladivostok to the open Pacific.

The Soviet Union now possesses the world's largest fishing fleet, a powerful navy including the world's biggest armada of submarines and a considerable merchant marine. Soviet ships traverse (and fish) all oceans. Increasing Soviet trade with Western

and Third World nations has taken Soviet interest in the sea far beyond any search for ports. Soviet military naval presence is felt in all oceans. Nonetheless, the USSR is still relatively self-sufficient economically, and domestic movements (largely by land) of goods overshadow its international marine traffic. The seas around the USSR have been far less influential in that country's development and outlook than they have been in the United States or Western Europe.

Climate

One of the primary facts about the Soviet Union is the limited utility of much of the country. Only two population concentrations of any size (the forelands of the Caucasus and the oases of Soviet Central Asia) exist outside the Leningrad-Odessa-Novosibirsk triangle of development. Nothing better illustrates the significance of the climate of the Soviet Union than the population distribution. North of the triangle, the land is generally too cold for commercial agriculture; the region south of it is generally too dry (see Chapter 1, Figure 1–18).

Most Russians live, in effect, at the warmer margins of the cold zone and the wetter margins of the dry zone, in neither case highly desirable but in both cases at least livable. Both these zones range eastward as continentality increases and moderating maritime influences decrease.

Much of the USSR is characterized by long, extremely cold winters and short summers. Continentality increases eastward, with greater seasonal and daily temperature ranges and greater extremes. Precipitation declines southward and eastward, grading into the semidesert that extends along the triangle's southern side.

The polar climatic type known as **tundra** occupies the Arctic fringe and many areas of higher elevation, and a subarctic climatic zone forms the transition between tundra and the triangle. Deserts occupy most of the lowland in Soviet Central Asia. Only a few sheltered areas along the Black Sea experience a warm, subtropical climate with a long growing season and plentiful rain.

Moscow is fairly typical of the more humid parts of European Russia. The average January temperature is 13.5° F, similar to that in Winnipeg, Canada. In July, the temperature warms to the mid-60s, a degree of warmth similar to that found in London. Only a little more than four months are frost free. Because of the Atlantic influence, temperatures are more moderate than average at that latitude; for the same reason, skies are frequently overcast (80 to 100 percent cloud cover on three quarters of the days of winter). Snowfall is relatively light, but it stays on the ground for five months.

Summer features bright, sunny days. Most rain falls in late summer or early fall, unfortunate in that it coincides with harvest time. The northerly location brings long hours of summer sunlight whereas winter daylight periods are extremely short.

Southward, winter is shorter and less intense. Kiev has winter and summer temperature averages similar to those of Chicago, 23° F and 72° F, respectively. Precipitation is light, 18 to 20 inches annually. Despite the obviously warmer average weather, seasonal extremes are almost as great as those experienced in Moscow.

The subarctic areas possess a climate even more extreme. In Irkutsk, a large Siberian city, the January temperature averages −12° F. Areas near Oymyakon have reached −90° F in winter. Summers are fairly hot; Irkutsk has the same July average as Moscow. Oymyakon has experienced a temperature range of 185° F from the hottest to the coldest day of the year. Midsummer frosts can occur, and the growing season varies from two to four months. Most of the area is dominated by the Siberian high-pressure system in winter. Winds move from the cold land outward to the continental margins. With no significant water source from which to draw moisture, it is quite dry, averaging less than two feet of snow per winter. Summer rains are brief, heavy thunderstorms in much of the area, but the Pacific coast has almost monsoon-like heavy and frequent summer rains.

In steppe areas, cold winters and hot summers are characteristic. Rainfall is light but comes in the best possible season for grain crops, early and midsummer. Climatically, the steppes of the USSR are similar to Montana and the Canadian Prairies.

In the midlatitude deserts, rainfall is everywhere less than 8 inches yearly. January temperatures average below freezing except in a few favored spots. Summer heat, however, is intense, and the more

southerly settlements average over 80° F in July. Temperatures of over 100° F can occur during summer afternoons. Everywhere, it is intensely dry. Only in a few locations in the foothills of the mountains is there sufficient rain to approach steppe conditions.

There are three small enclaves of subtropical climate: the Crimea southeast of the protecting coastal mountains, the eastern Black Sea coast, and the lowlands along the southern Caspian Sea. Use of the word *subtropical* here is relative. January temperature averages are only a few degrees above freezing, and frosts are common. Only by comparison, then, are winters mild. The Crimean coast has a Mediterranean regime with a dry summer. The other two areas have more evenly distributed rainfall.

These areas have developed extensive resort functions and specialize in subtropical crops, regardless of the frequency of frost hazards. In sum, the Soviets have had to develop their economy in an area that is subject to climatic extremes and is largely marginal.

Vegetation and Soils

The correlation between climate zones, vegetation, and soils is marked. As with climate, vegetational and soil zones exhibit an east-west latitudinal distinction.

The northernmost, or tundra, zone contains vegetation of little economic value—mosses, lichens, and a bush of dwarf willow and birch. Much of the area consists of barren rock. Plants must tolerate drought, poor drainage, extreme winter cold, midsummer frost, a long dormant season, and acid, shallow, poorly developed soils. This area seems better adapted to raising mosquitoes than vegetation; the former are a plague in summer (Figure 4–5).

Covering a vast extent in Siberia and the northern one fifth of European Russia, the taiga is a forest of coniferous needle leaf trees; because of the very cold climate and the **permafrost** that underlies the area, the forest grows slowly and contains few species. Bogs are common. The acid-loving conifers shed needles that decompose slowly, yielding little humus. Abundant groundwater percolates through the soil, leaching almost all nutrient value. Taiga soils are ashen white, highly acid, rich in iron and silica, low in humus, and devoid of calcium. Agriculturally, they yield poorly under most circumstances. Chemically,

THE SHORT, SWEET SUMMER AND THE LONG, COLD WINTER

The relatively short summer can be quite warm; almost every Soviet citizen, wherever they live, can be comfortable in short sleeves in July, and even in Siberia, sunbathers wear the briefest of bikinis on lakeshore beaches. In most of the country, summer temperatures surpass those of any place in Britain. After a long, bitterly cold winter, Soviets revel in the warm sunshine, spending as much time as possible out-of-doors. Suntans are pursued with as much enthusiasm as in Southern California, if with less prospect of maintaining the tan. After a winter spent enveloped in layers of heavy outdoor clothing, it is a great pleasure to forget about coats, sweaters, gloves, scarves, hats, and heavy boots, at least for a while.

The relish with which Soviets greet summer, however, does not mean that they sullenly endure winter. Rather, the prevalent attitude seems to be: because winter is unavoidably long and harsh, let's make the best of it. The great parks in and around the cities are filled with Soviets on winter weekends, and ice-skating seems to become the national sport. There is in evidence everywhere the slightly self-conscious, robust determination to enjoy bitterly cold weather that one can also observe in Minnesota, Wisconsin, and much of Canada.

In January, it would be possible, at least in terms of climate, to ice-skate from Leningrad to Vladivostok. Unlike the situation in the United States, bordering mountain systems to the south tend to prevent tropical air from reaching most parts of the Soviet Union, helping to produce amazingly uniform winter temperatures over vast expanses. In terms of cloud cover, however, Far Eastern Siberia has only about one-third the cloudy days experienced in Moscow in winter. Siberian residents have reason to be thankful that, due to winter's Siberian high-pressure cell, with its strong outflow of air, their skies are a clear, deep blue, often with enough solar radiation to melt snow from roofs and paved surfaces. In contrast, Western Europeans, although experiencing much milder winter temperatures, live under an almost continuous cloud cover with miserably damp, cold, and dreary days of sleet and snow.

For those lucky Soviets who can manage a winter holiday, the choices of winter resorts are distinctly limited. Virtually no Soviet territory can match the winter climates of southern California, southern Arizona, or Florida. The Soviets must make do with what they have. At such Crimean resorts as Yalta, popular since tsarist days, sun worshipers set up three-sided tents on the beach, with the open side facing the uncertain winter sun. Yalta's January temperature average is comparable to Seattle's, though Yalta does enjoy more sunshine in that month than Seattle. It is customary for winter visitors in the Crimea to wear bathing suits under heavy winter coats. When the sun is out, coats are opened or discarded; when the sun disappears in clouds, they are donned and closed until the next sunny interlude. It's not Palm Beach, but it's not Moscow either.

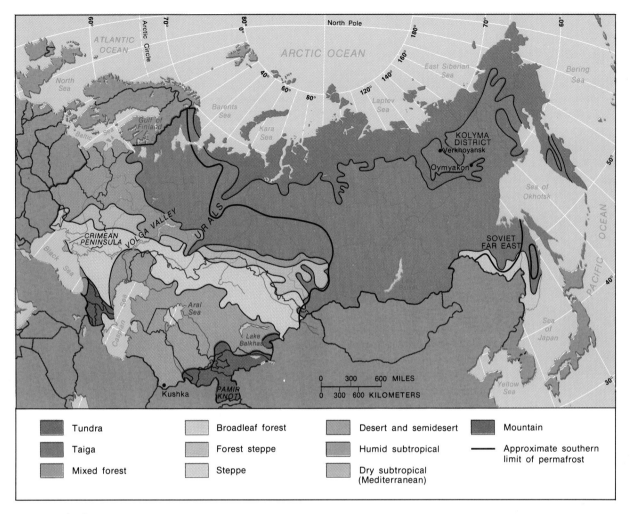

FIGURE 4–5
Vegetation Regions of the USSR. The USSR's great size and latitudinal extent
provide a wide range of physical environments. Note the correlation of this map
with the map of world climates (Chapter 1, Figure 1–18).

Legend:
- Tundra
- Taiga
- Mixed forest
- Broadleaf forest
- Forest steppe
- Steppe
- Desert and semidesert
- Humid subtropical
- Dry subtropical (Mediterranean)
- Mountain
- Approximate southern limit of permafrost

cultivation is akin to trying to grow garden plants in
ground glass, shredded steel, and aluminum cans.

Farther south is a zone of mixed forest. The soils
are less acid and the winters less severe. With no
permafrost, drainage is better, and trees are more
deeply rooted. Much of this area has been cleared for
agriculture; forest remains only in poorly drained
areas, on steep slopes, and in sandier soils. Leaching
is less intense, and soils are thicker and better
drained. The Soviet Union is fortunate in that much
of this area has been enriched by glacial action. Rock

particles transported from distant source areas
added desirable nutrients.

The steppe is a belt of almost endless grass. The
high humus content of the soil results from accumulation of the decayed roots and stalks of annual
grasses. The unleached humus creates a thick, dark,
waxy soil of incredible fertility—the famous black
earth. These soils are naturally rich in calcium, nitrogen, phosphorus, and potash—virtually all the minerals a plant needs. They are deep and generally well
drained, and have good groundwater retention. They

have the potential to yield enormously, but the absence of sufficient water, the very condition that keeps them fertile, is a detriment.

Sagebrush and xerophytic (drought-adapted) shrubs are the most common cover in the desert proper. Vast areas of the Central Asian desert have shifting dunes or such sandy or rocky soils that little vegetational growth is possible. After the thaw and the early spring rains, however, the desert flowers and turns green, leaving patches of nutritional grass that support herds of sheep.

Drainage Patterns and River Uses

The USSR possesses an enormous hydroelectric potential, great irrigation possibilities, and significant transportation advantages in its riverine system. Yet most of the rivers are icebound from 4 to 10 months of the year, limiting navigation. In spring, they flood vast areas, hindering plowing at a crucial time in a short growing season. In summer, many rivers are shallow because of the lack of rain. The strong seasonal fluctuations in flow reduce their energy-generating utility as well. Most even flow the "wrong way" (Figure 4–6b). The giant rivers of Siberia dead-end in the almost useless Arctic Ocean. They thaw first in the south (headwaters) and flow toward ice-jammed mouths in the north, creating a backup that floods incredible acreages. The Volga (which flows southward) empties into the Caspian Sea, which has no outlet. The Soviet Central Asian rivers, fed by glacial meltwaters derived from the border mountains, die out in the sands or the shallow, salty Aral Sea. The rivers in general flow due north or south in a country where the dominant traffic direction is east-west. The Soviet rail network functions to complement the river transport system with the vital east-west links (Figure 4–6a).

In the Communist era, the USSR has become a world leader in fishing, hydroelectric power generation, irrigation, maritime traffic, and naval might.

Dams along the Volga have created a series of great lakes that extend in an unbroken chain from Volgograd northward to Kalinin, northwest of Moscow. Canal systems link the Black Sea to both the Arctic and the Baltic (see Figure 4–6b). It is planned to link the Soviet river system to the pan-European canal and river system. Thus, the entire European continent, with all its markets and resources, would be directly tied by cheap water transport to the Soviet Union.

However, there are problems. The shrinkage of the Caspian Sea is in part attributed to the projects on the Volga. Scientists have opposed the northern canal links, fearing the unknown effects on the water table, groundwater supplies, and underground mining. Agricultural lands and entire towns have been inundated by reservoirs.

If the sheer scale of European river engineering is staggering, that of the Siberian River Reversal Scheme is even greater (Figure 4–7). To date, three major steps have been taken. The first was control of the headwaters of the Yenisey along its tributary the Angara. A dam now backs up the water of the Angara River to the city of Irkutsk. The second step in this water stairway is already completed. The Bratsk Dam on the Middle Angara has a capacity of over 4 million megawatts. A third dam of even larger power capacity has been built farther downstream, and another is under construction. Dams on the Yenisey proper have been constructed in the headwaters and at Krasnoyarsk. The entire system is to be tamed and utilized in this fashion. Surplus power is shipped westward over the national grid. Lake Baikal, the world's deepest lake, ensures a steady flow of Yenisey water, even during the intense cold of winter.

The second phase of the scheme is the irrigation of the deserts using the Amu Darya, Syr Darya, and other rivers of the region. With these streams fed by glacial meltwaters from some of the world's highest mountains, the hydroelectric potential is quite large, and huge dams add a new source of power to the region at the same time that they increase irrigation.

FIGURE 4–6

(a) ***Railroads of the USSR.*** Whereas the United States has emphasized its highway system, the USSR remains far more dependent on its rail network. Note the contrast between the dense rail net in western Russia and the few but heavily used lines in Siberia. (b) ***Navigable Waterways of the USSR.*** Soviet engineering has created a vast network of interlinked navigable waterways. Their utility is limited by two factors: a long winter during which they are unusable and the tendency for rivers to provide north-south movement but not east-west potential.

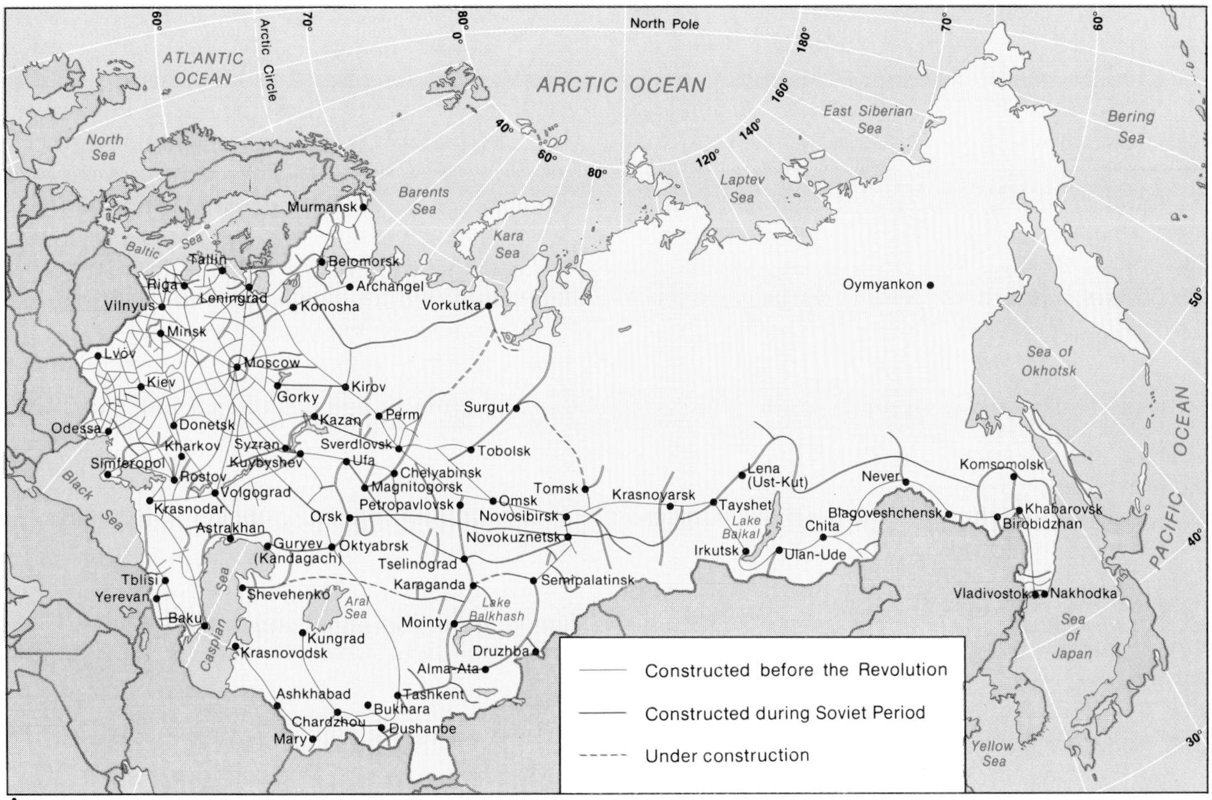
North Pole
ARCTIC OCEAN
ATLANTIC OCEAN
Arctic Circle
North Sea
Barents Sea
Kara Sea
Laptev Sea
East Siberian Sea
Bering Sea
Baltic Sea
Black Sea
Caspian Sea
Aral Sea
Lake Balkhash
Lake Baikal
Sea of Okhotsk
Sea of Japan
Yellow Sea
PACIFIC OCEAN
60°
70°
80°
0°
40°
60°
80°
120°
140°
160°
180°
70°
60°
50°
40°
30°
Murmansk
Belomorsk
Archangel
Tallin
Riga
Leningrad
Vilnyus
Konosha
Vorkutka
Minsk
Lvov
Moscow
Kiev
Gorky
Kirov
Kazan
Perm
Donetsk
Odessa
Kharkov
Syzran
Sverdlovsk
Kuybyshev
Ufa
Simferopol
Rostov
Chelyabinsk
Magnitogorsk
Volgograd
Krasnodar
Petropavlovsk
Omsk
Orsk
Astrakhan
Guryev
(Kandagach)
Oktyabrsk
Tbilisi
Yerevan
Shevchenko
Baku
Kungrad
Krasnovodsk
Ashkhabad
Chardzhou
Mary
Bukhara
Tashkent
Dushanbe
Alma-Ata
Mointy
Karaganda
Tselinograd
Semipalatinsk
Druzhba
Novokuznetsk
Novosibirsk
Tomsk
Surgut
Tobolsk
Krasnoyarsk
Tayshet
Lena
(Ust-Kut)
Irkutsk
Ulan-Ude
Chita
Blagoveshchensk
Never
Komsomolsk
Khabarovsk
Birobidzhan
Vladivostok
Nakhodka
Oymyankon
Constructed before the Revolution
Constructed during Soviet Period
Under construction

A

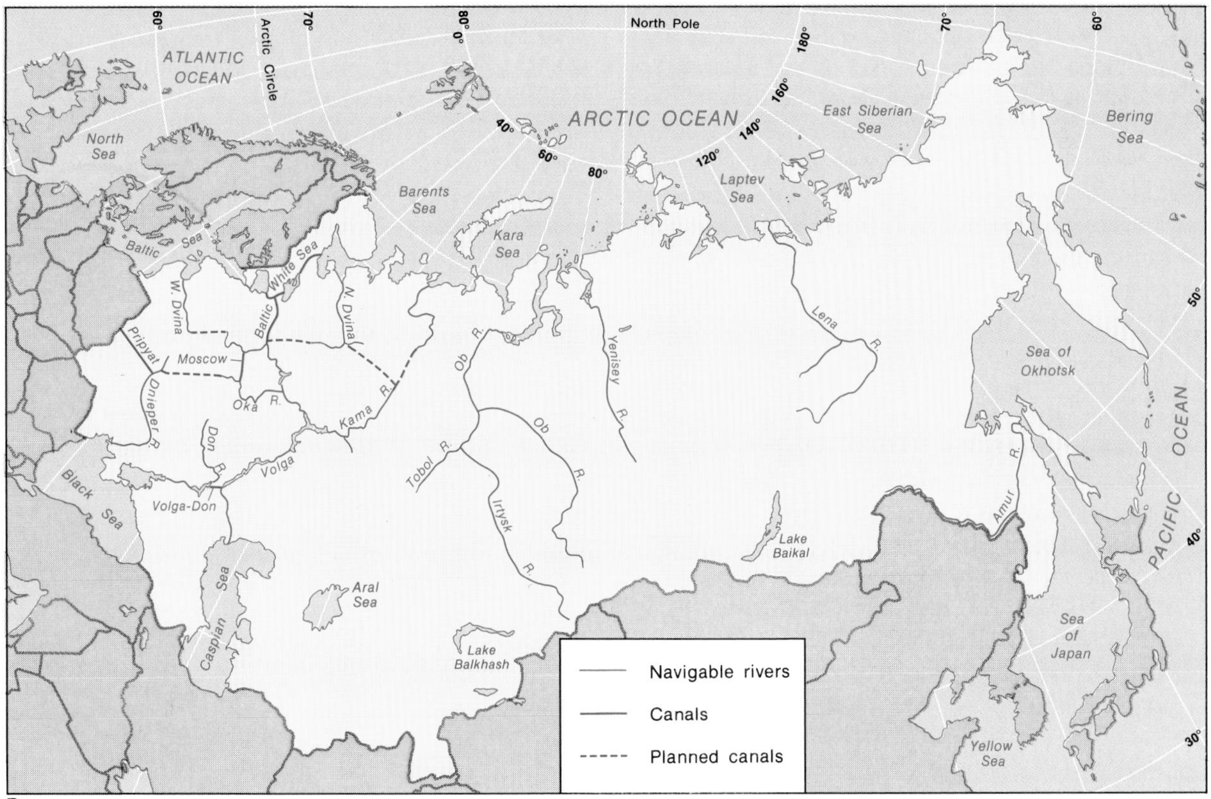
North Pole
ARCTIC OCEAN
ATLANTIC OCEAN
Arctic Circle
North Sea
Barents Sea
Kara Sea
Laptev Sea
East Siberian Sea
Bering Sea
Baltic Sea
White Sea
Black Sea
Caspian Sea
Aral Sea
Lake Balkhash
Lake Baikal
Sea of Okhotsk
Sea of Japan
Yellow Sea
PACIFIC OCEAN
60°
70°
80°
0°
40°
60°
80°
120°
140°
160°
180°
70°
60°
50°
40°
30°
W. Dvina
Baltic
N. Dvina
Pripyat
Moscow
Dnieper R.
Oka R.
Kama R.
Don R.
Volga
Volga-Don
Ob R.
Ob R.
Tobol R.
Irtysk R.
Yenisey R.
Lena R.
Amur R.
Navigable rivers
Canals
Planned canals

B

The third phase is the planned link of the Yenisei system to that of the Ob; surplus waters would be reversed through to Soviet Central Asia. Waters of all three systems would drain into the Caspian and link the western two thirds of the Asiatic USSR to the almost-completed network of the European USSR.

There are certainly problems. Valuable timberland will be inundated. There will be little water left to redress the shrinkage of the Caspian Sea. Anticipated costs will be very large. It is a calculated gamble that, if it pays off, could pave the way for future schemes of even larger scope.

The scheme is not impossible. Given the enormous costs involved, the real question is whether it is economical. Immediate problems of grain supply and consumer demands may cause the planners to shelve these grandiose schemes.

THE UNITY AND DISTINCTIVENESS OF THE USSR AS A REGION

The Soviet Union is the only country that is here given the status of a region in itself. Its population represents only about 6 percent of the world total, yet it is larger than four of the seven continents and occupies portions of two continents.

An overriding problem faced by the Soviet government has been the political and economic integration of all parts of such a huge and diverse territory. These very problems have created a distinct outlook and set of practices that are not operative in any other area governed by a similar political system.

Despite the fact that Soviet culture has clear roots in an older Russian culture, the postrevolutionary government has gone beyond inherited tradition Following an interpretation of Marx peculiarly suited to the scale of the country and the level of development inherited from the past, it has created a new cultural hybrid. Acculturation has proceeded apace. The collective, the party, the Five-Year Plan, even the monotonous architecture, have created a common Soviet experience. The government has attempted to refocus the loyalty of the people beyond the ethnic nations to Soviet national culture.

If the Soviet interpretation and administration of Marxist ideology has produced an often dull uniformity throughout the country, the diversity of the citizenry has produced a rich cultural amalgam. The culture of today's Soviet Union is different from the Russian culture of the past. The state is modern, urban, and technologically advanced. Controls on religion have restricted the contact of many Russians with one of their great sources of traditional culture. Much inherited culture has been discarded in favor of something new. Influences of the distant past have their place, but the society has become conditioned more to present and future.

The borders of the USSR are approximately the same as those of the Russian Empire of prerevolutionary times; with rare exception, the ethnic composition is virtually the same. The land and peoples that fell to Russian conquest in the sixteenth through nineteenth centuries are a part of the USSR. Russia was indeed an empire; it is legitimate to ask whether the successor government has changed this situation.

Many ethnic groups are quite small; many are scattered over large, noncontinuous areas. Many seem never to have had a state of their own, and only among a few groups was a national consciousness fully developed. Nonetheless, some groups now included in the USSR have had a clearly developed nationalism at one time or another (see Figures 4–1 and 4–8).

Lithuania, Latvia, and Estonia were small independent states in the modern era from 1918 to 1940. Among the older generations, there must remain some legacy of past independence. The Moldavians are Romanian in speech and culture, though long association with the tsarist empire left as a legacy the Cyrillic alphabet.

Briefly (1919–23), an independent Ukraine appeared in the aftermath of the Revolution and the civil strife in Russia. Similarly, each of the three Transcaucasian republics—Armenia, Georgia, and Azerbaidzhan—declared their independence at the time. A coup temporarily established two Turkic language entities in Soviet Central Asia during the 1920s.

Russians are the overwhelming majority in their own republic (the Russian Soviet Federated Socialist Republic, or RSFSR), the largest in both area and population within the federation. They also form the largest group in the Kazakh SSR, where they significantly outnumber the native group after whom the republic is named. They are the plurality, even the

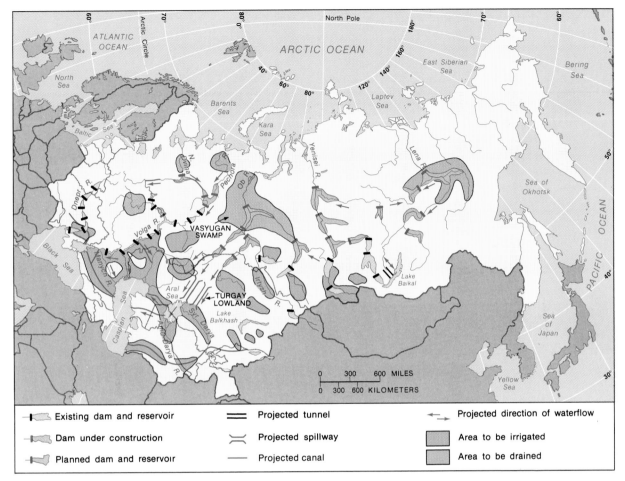

-◄━◢ Existing dam and reservoir	══ Projected tunnel	◄━ Projected direction of waterflow
-◄━◢ Dam under construction	⋈ Projected spillway	▨ Area to be irrigated
-◄━◢ Planned dam and reservoir	— Projected canal	▨ Area to be drained

FIGURE 4–7

Siberian River Reversal Scheme. Soviet engineers and regional development
planners have taken advantage of the fact that although the great Siberian rivers flow
northward, the need for irrigation water is to the southwest of these rivers'
headwaters. The directional flow of three great rivers has been partially reversed in
this mammoth engineering scheme.

majority, in many of the smaller autonomous republics and districts that are subunits within the larger
republics. They are a significant minority in virtually
all the republics; and, in the areas east of the Volga,
they are the majority in virtually all cities, even in the
non-Russian republics. Russians have been active col-
onizers.

It has not always been necessary to be a Russian to
advance to the top governmental or party ranks. Stalin was a Georgian, Lenin had some Tartar ancestors,
and most major national groups have had represen-
tatives at the most influential levels.

Russia in 1918 was faced with a dilemma. The
entire country was racked with civil strife. Separatist
ambitions on the part of non-Russians could have
destroyed the tenuous control of the Communists.
Stalin supplied the answer; the system that he created
was one of ethnic federalism with (ultimately) 15
republics based on ethnic groupings. Each republic
would have the theoretical right to secede. Other
national groups, those who did not meet the requirements of population size or a common border with
a foreign country, were given autonomous units of
lesser stature (see Figure 4–1).

GEOPOLITICS AND SOVIET EXPANSION

Geopolitical analysis attempts to relate international political power to the geographical setting. Geopolitical thought shows repeated change. A prominent scholar of geopolitics, Halford Mackinder, observed: "Each century has its own geographical perspective. To this day, our view of geographical realities is colored for practical purposes by our preconceptions from the past."*

Long before Lenin took power, the Russian state was strongly expansionist. Soviet expansionism is the twentieth-century continuation of long-established geopolitical objectives. The heart or core of the Russian, now Soviet, state always has been in Europe. Traditionally, the economic and population focus was around Moscow; for the last few centuries, territorial expansion has been less difficult to the south and east than westward.

Mackinder, a British geographer, proposed that the political control of the center of the great Eurasian landmass gives potential geopolitical advantages to whatever state is in possession. He modified his theory several times, but his most famous capsulization of his thinking is the dictum, "Who Rules East Europe Commands the Heartland (which occupies most of the Soviet Union, much of Iran and Pakistan, and western China); Who Rules the Heartland Commands the World-Island (Eurasia and Africa); Who Rules the World-Island Commands the World."

When Mackinder promulgated this concept, the heartland, with a frozen sea to its north and guarded by great mountain chains and vast deserts to the south, was almost invulnerable to attack. Now, with nuclear submarines patrolling beneath the pack ice of the Arctic, and intercontinental missiles capable of reaching any target, the heartland looks less immune from attack.

Implicit in this theory is the assumption that the only effective counter to the heartland's dominance is control over the periphery of the Eurasian landmass, and control over the "World Ocean"—the interconnected oceans and seas. Control of the seas would counterbalance the heartland's advantages in internal transport and communications.

Whether consciously accepting the heartland premise or not, U.S. policy since the end of World War II has focused on containing the USSR and its satellites with encircling alliances and military bases. U.S. emphasis on ocean and airborne power makes the rapid expansion of the Soviet navy an ominous and serious challenge. If nothing more, the heartland concept should stimulate continued reevaluation of the relationships between geography and power.

*Halford Mackinder, *Democratic Ideals and Reality* (New York: Henry Holt & Co., 1942), p. 29.

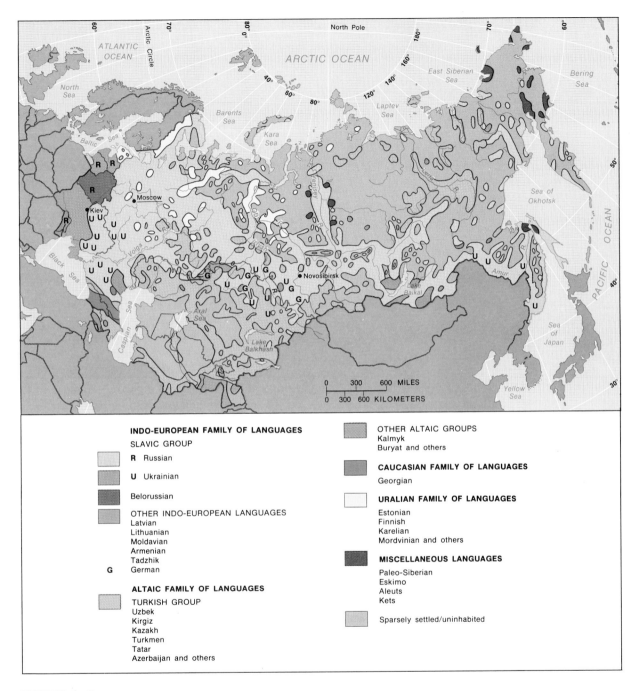

INDO-EUROPEAN FAMILY OF LANGUAGES

SLAVIC GROUP

R Russian

U Ukrainian

Belorussian

OTHER INDO-EUROPEAN LANGUAGES
Latvian
Lithuanian
Moldavian
Armenian
Tadzhik
G German

ALTAIC FAMILY OF LANGUAGES

TURKISH GROUP
Uzbek
Kirgiz
Kazakh
Turkmen
Tatar
Azerbaijan and others

OTHER ALTAIC GROUPS
Kalmyk
Buryat and others

CAUCASIAN FAMILY OF LANGUAGES

Georgian

URALIAN FAMILY OF LANGUAGES

Estonian
Finnish
Karelian
Mordvinian and others

MISCELLANEOUS LANGUAGES

Paleo-Siberian
Eskimo
Aleuts
Kets

Sparsely settled/uninhabited

FIGURE 4–8

Major Language Groups of the USSR. This map of language groups illustrates
the cultural diversity of the Soviet Union. It should be noted that some large
territorial extents of language groups in Siberia may, in fact, contain very few
people.

THE SOVIET ETHNIC DILEMMA

The Soviet Union contains over 100 ethnic groups, which speak eight different languages and write in five different alphabets. Despite government pressure and antireligious propaganda, Soviet citizens practice Judaism, Buddhism, Islam, and many sectarian varieties of Christianity. The internal passports that all Soviets carry specify ethnic background; although *Russia* is used interchangeably with *USSR,* the Russians are but one of many ethnic "nationalities."

Since January 1, 1983, the Russians have been officially recognized a minority, though still the largest ethnic group in the USSR. The highest growth rates now are found among the Turkic and Iranian language groups that inhabit Central Asia and the Caucasus Mountain region (see Figure 4–3). An era of rapid growth also has begun among Siberian native groups. Death rates among these peoples have undergone the dramatic decline typical of the early stages of the demographic transition. The predictable, but later, drop in their birthrates has not yet occurred, largely because of the essentially rural nature of the peoples in question. Growth rates of the Central Asian and Caucasian peoples have remained high for well over two decades. Unless there is a reversal of Soviet regional developmental attitudes, coupled with a reversal in minority growth rates, this rapid growth will likely continue for some time.

The situation is not uniform throughout the country, nor is rapid growth characteristic of all non-Russian ethnic groups. Ukrainians and Byelorussians have birth- and death rates similar to those of the Russians. The Baltic peoples exhibit even lower rates of growth. These slow-growing groups have been most likely to assimilate. Jews, largely assimilated to Russian nationality in the past, are exhibiting a renewed ethnic

Each group received schooling (through eight grades) and the right to trial in its own language. Native customs were virtually glorified. Every ethnic group had its own museums, and native literature was republished. In short, the very things often used to build nationalism were enhanced, not forbidden. The apparent idea was the removal of the sting of subjugation by soft-pedaling Russian nationalism while retaining (even encouraging) the trappings of nationalism among non-Russian groups.

European Tartars and most of the Finnic peoples have increased their assimilation rates. After generations of eking out a poor living as farmers in a difficult physical environment, the Finnic peoples of the Volga and Kama valleys have become caught up in the general industrialization of that area. As urban dwellers and industrial workers, they now exhibit the lower birthrates associated with urban living. Surrounded in the growing cities by Russian neighbors, they are assimilating rapidly to Russian language and culture. Tartars have tended to assimilate, as early as the fifteenth century, when they began to filter into the Russian military and bureaucracy. Their Islamic faith did little to deter assimilation, with their strongest cultural bond being the clan rather than their faith. Their military prowess is legendary, and, as such, they are prominent in the armed services. As a minority that was widely feared and openly hated,

Bukhara, Uzbek SSR. *Russian expansion in the past created a multiethnic empire that included many non-Western peoples. These Uzbeks are part of one of many ethnic groups that are growing faster than the ethnic Russians.*

consciousness. Large numbers are attempting to leave the Soviet Union for Israel or the West although there are many obstacles to this emigration. In either case, the result is to diminish the total potential numbers once classified as Russian.

they have always played down their own nationality. Ambitious, they have sought education and advancement.

Tartars are valued as a bridge between Russians and the Islamic nationalities. Their growing numbers in Soviet Central Asia attest to deliberate Soviet placement of Tartars in managerial positions in that region's farms and new factories. More sensitive to, and familiar with, Islamic cultural traditions, Tartars are more acceptable as bosses to Central Asians than are Russians.

Education and advancement are strong attractions to assimilation. Although all native groups are schooled in their own language through the first eight grades, they must also learn Russian. In high schools, the native language becomes simply one of several subjects, with other subjects taught in Russian. University studies are almost entirely in Russian, and higher-level jobs assume the use of the Russian language. Long-distance migration submerges ethnics in overwhelmingly Russian cultural surroundings; even the move from farm to city may do the same. Compulsory military service acts to familiarize all non-Russians with the language and culture of the dominant group.

Nonetheless, native ethnicity is retained among the overwhelming majority of non-Russians. This is particularly true among Central Asians, because few go beyond eighth grade. Mothers impart native, rather than Russian, culture and language to successive generations.

Urban birthrates among Central Asians are half those of rural dwellers but still twice or more those of Russians in the same cities. Similar tendencies occur among the native peoples of the Caucasus, even the fairly heavily urbanized Azerbaidzhan.

Among the culturally Christian Armenians, despite high levels of education and urbanization, abnormally high growth rates persist. Memories of the Turkish decimation of Armenians that followed on the heels of World War I combine with the continued fear that high birthrates among Islamic neighbors present a threat, both physical and cultural, to Armenian existence.

Siberian native ethnic groups, the last to benefit from improved dietary, health, and medical conditions because of their extreme isolation, have only recently begun to experience rapid growth rates. Their original numbers were so small, however, that they pose no great threat to Russian dominance, either regionally or nationally.

Rapid non-Russian growth has proven to be a multifaced dilemma. Will it lead to increased separatist tendencies? Can native peoples be persuaded to migrate—either to colonize new territories or to the urban centers of economic-industrial growth wherever they are located? Will continued Russian outmigration from Siberia, coupled with strong native growth, change that area's ethnic makeup? Seventy years after the Revolution, ethnic diversity still presents problems.

THE CULTURAL HERITAGE

Historical Unifiers

The Russian core area (essentially the European portion of the RSFSR without the Arctic fringe) has been centrally controlled by one government since roughly the time of Columbus. Most of Siberia has been under the same control for 300 years or more. It is only the peripheral areas (where all the ethnic republics are located) that came under Russian control later (Figure 4–9).

Historically, the route to unity under the tsars was cultural, involving the spread of Christian Orthodoxy and Russian language. Overt **russification** (the Russianization of culture) was official policy, and church and state worked together toward this goal. The Orthodox liturgy asked God's blessings on the rulers and the country as well as on the people. The emphasis on community, as opposed to the individual, found in the Orthodox liturgical service is striking. The designation of the country as Mother Russia and the tsar as the "little father" throughout Russian literature is a manifestation of this all-encompassing drive for unity. The advent of Communist control in 1918 was a change in the philosophy and leadership of government, not necessarily the interruption of a political continuity.

The determination to mold "Soviet man" (and woman), a selfless person dedicated to achieving collective rather than individual goals, results in an attempt to replace religious ceremonies with ones that reinforce the official Communist creed. Although Soviet couples may also arrange a church or synagogue ceremony, they must first wed in government reception rooms or "wedding palaces." Most brides wear a traditional white gown, and the happy couple are driven to a patriotic monument to pose for their wedding pictures. Many a bride and groom are pictured in front of a "Great Patriotic War" (World War II) monument such as a tank, fighter plane, or artillery piece. In Moscow, couples frequently visit Lenin's tomb, where the wedding gown guarantees admission at the head of the waiting line.

Economic Unifiers

The drive for economic self-sufficiency and economic development has been the central guiding theme at least since 1918, though the economic unification of the realm was also a tsarist goal. An early tsarist attempt at economic unity was the Trans-Siberian Railroad, yet it was the Communists who fully developed it as an economic unifier after 1918. The rivers, which crossed the railroad at virtually right angles, served as collectors and distributors. The railroad network of European USSR essentially follows the rivers and the trade routes established in ancient times, focusing on Moscow, the traditional

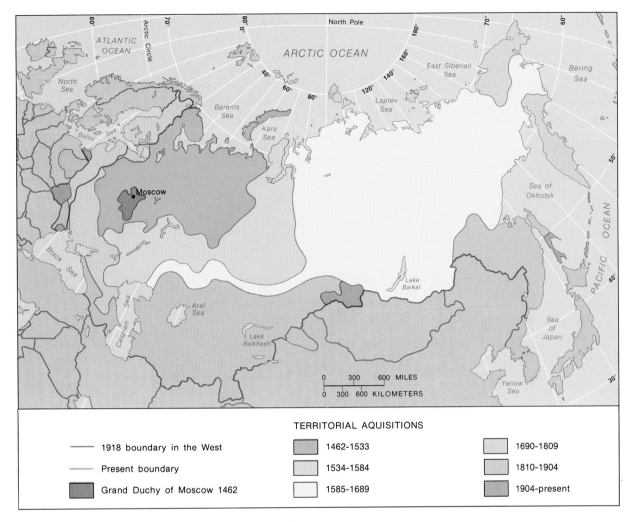

FIGURE 4–9
Growth of the Russian/Soviet State. This map portrays the territorial expansion
of the Russian state and its successor, the USSR. Note that expansion was more
difficult toward the west than to the east and southeast.

center of power from the fourteenth to the eighteenth centuries. It focuses only secondarily on Leningrad, the "late-comer" capital founded by Peter the Great as St. Petersburg in 1703. Even though the railroad net of the western part of the country was built when Moscow was not the capital, the focus shows the strong unifying effect on Russia that Moscow had come to possess through years of centralized rule.

The Trans-Siberian came to be the official mechanism for tying the lightly settled Asiatic domains to the Moscow economic core. The proof of Communist intent can be seen in the flurry of postrevolutionary railroad building. Several new lines across the Urals tied European Russia to Western Siberia. In turn, the building of the Turk-Sib (Turkestan to Siberia) Railway (1931) linked Soviet Central Asia (earlier linked to European Russia) to the Siberian lines. New lines linked all the republics of the European USSR with each other and with Moscow. Economic regions were created and new rail links were con-

GROWING UP SOVIET

Before the Revolution, Russian Orthodox homes featured a "red corner," a corner in which religious icons were displayed, usually on a red cloth backdrop, accompanied by votive candles or oil lamps. In the Russian language, the word for *red* has the same root as the word for *beautiful,* and in ancient usage, *red* and *beautiful* were the same word. The "red" or "beautiful" corner symbolized the family's religious beliefs.

In contemporary USSR, schools set aside a special room for meetings of the Young Pioneers and Komsomol groups, clubs that emphasize the new orthodoxy of political ideology. These meeting rooms always have a red corner—called that too—featuring a picture of Lenin. Books on Lenin and communist ideology are displayed in the red corner. Lenin and the cult of Leninism have literally displaced the holy icons as symbols of the beliefs to be instilled in youth. Lenin and his successors have expressed far more profound objectives than imposing a new political system: they aim to change society and human nature itself.

The earliest experiences of Soviet children help to prepare them for life in a collectivist society, one in which individual initiative is actively discouraged. Toddlers are given little freedom or scope for imagination. Parents, grandparents, and *detsky sad* ("children's garden" or nursery school) teachers intervene continually in children's play. Children are lovingly enveloped in a warm, but bossy attention from adults.

School children quickly learn that standing out in a crowd does not pay. An old Russian proverb, popular among Marxists, says, "In a field of wheat, only the stalk whose head is empty of grain stands above the rest." In art, children learn to make precise copies of pictures displayed for that purpose; neatness and accuracy are rewarded, not imagination. Soviet citizens are strongly discouraged from seeing things differently, literally and figuratively. Schools grade each day's efforts, and grades are posted for all to see. Political philosophy permeates all studies, and successful young scholars learn to sprinkle quotes from Marx, Lenin, and the current leadership in papers and essays on any topic.

One of the advantages of joining the Young Pioneers and Komsomol is the chance to go to summer camp. Those camps offer sports and craft programs that would be familiar to many Americans, but the daily regime is more regimented. Mandatory political education sessions occupy several hours a day. Soviet citizens are given large and frequent doses of political indoctrination throughout their lives. There is no dissension possible at these sessions, only a predictable rehash of the same viewpoints.

structed to ensure the flow of raw materials to manufacturing or marketing regions. Thus, the system of economic planning and the rail network combined to create a new economic unity.

The plan was to further unity through economic integration. So-called **kombinats** were developed for the exchange of complementary raw materials to maximize the utilization of freight cars. The system shipped iron ore from the central Ukraine to the Donbas coalfield and steel complex (Figure 4–10). The reverse haul carried coal to the then newly developed steel base of the central Ukraine. The **complementarity** of resources, then, was exploited to the fullest. After a time, kombinats were set up all over the country, regardless of the distance involved. However

wasteful this may have been of both time and energy, it served to integrate the country's economic regions.

Similarly, at a later date, unified power grids and oil and gas pipelines were created to shift fuel and power throughout the populated portions of the Soviet Union. The river network, however seasonal in utility, was (and is) gradually being developed to the same end.

Planned programs to reverse Siberian rivers and increase irrigated acreage in Soviet Central Asia could create a contiguous agricultural core that would span both sides of the Urals, encompassing almost equal areas. Of all the individual factors helping to unify the USSR, economic unification may be the most important.

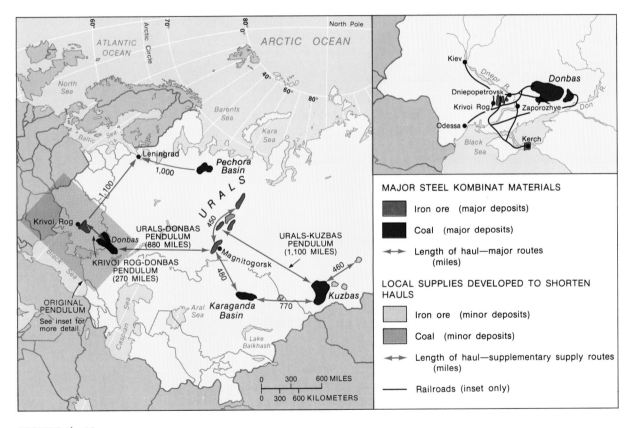

FIGURE 4–10
Manufacturing-Resource Kombinats. Vast distances can pose problems to assembling of bulky raw materials for heavy industry, such as steel. A Soviet solution is to locate steel mills at both iron ore deposits and coalfields, and use railroads to haul coal to iron ore-based mills and return with iron ore to coalfield-based mills.

SOVIET INDUSTRY: WINNING THE WRONG RACE?

The list of Soviet achievements in engineering, industrial production, and development of the USSR's huge natural resource base is long and impressive. The USSR ranks second only to the United States in overall industrial output and is first in steelmaking, traditionally a measure of basic economic power. When the Soviet system assigns high priority to a project, the results can be astounding. The list of Soviet firsts in space, for example, is breathtaking: first artificial earth satellite, first lunar probe to hit the moon, first man in orbital flight (also, first woman in orbital flight), first space walk, and first linkup in space of two manned spacecraft.

The Soviets have a love of mammoth projects; they cherish having constructed the largest, tallest, most powerful, fastest-completed, highest, broadest, and other superlatives in engineering applications. For example, they planned and constructed (albeit with technical advice and machinery from Europe and America) the world's largest self-contained heavy-truck factory 600 miles east of Moscow on the Kama River. This complex, which sprawls over 23 square miles, has its own foundries and forges; typically, the Soviets also built a new city from scratch to accommodate 90,000 people where there had been only a village. The Kama truck plant was built in only five years, at a cost of $5 billion (its original projected cost was $2.2 billion). The plant, with a capacity of 150,000 heavy trucks a year, was more than a year behind schedule in producing its first truck, however. Some equipment ordered from the United States arrived before buildings to house it had been completed; the machinery was left to rust in the fields. Bratsk High Dam, another example of frantic effort poorly coordinated, was a high-priority, massive hydroelectric project on Siberia's Angara River. Through heroic efforts, it was completed 10 years before its main power customer, an aluminum plant, was finished.

The Soviet, or more accurately Stalinist, plan to accelerate the timetable of industrialization was to emphasize heavy industry at the beginning rather than to build light (consumer-oriented) industry to stimulate heavy industry. When heavy-industry capacity was built up, then the economy could be shifted toward consumer production. This strategy can work well in the rapid industrialization of a developing country. However, it is an inflexible system, unresponsive to popular wants. Such a system can produce vast quantities of steel but too little in the way of consumer goods. For example, the USSR in 1980 had the same number of cars on the road as the United States had in 1920. By 1990, the USSR should match America's 1925 achievement. The USSR has built the largest armada of tanks in the world, the world's largest fleet of naval vessels, and vast quantities of all varieties of weapons, yet its citizens must line up to buy scarce consumer goods such as toilet paper, toothpaste, and razor blades. The USSR devotes about 1/8 of its total GNP to the military; the United States spends about 1/16. On the other hand, in a year when the United

Volga River at Volgograd.
The Soviets take pride in superlative engineering feats, setting records in largest, highest, and fastest-constructed. This view of Volgograd shows the scale of typical construction projects.

States spent $950 per person on health care, the Soviets spent $40 per person.

Although the Soviet Union has succeeded in doubling average calorie consumption since the 1940s, and provided three quarters of its citizens with refrigerators and two thirds with washing machines, the quality of consumer goods is often poor, and the quality of life may be in decline as well. Some analysts assert that average longevity has actually declined in recent years, so that Soviets now have shorter life expectancies than Cubans or Albanians.

The centrally planned or "command" economy (production responds to government decree rather than to market demand) experiences severe imbalances. The system emphasizes quantity rather than quality. The quota is sacred—it must be met at all cost, and so a jelly factory will add water to meet its quota, producing large quantities of useless, watery jelly instead of fewer jars of good-quality product.

Bankruptcy is the final solution to inefficiency in capitalist countries, but Soviet managers do not have to worry about profit, just meeting the quota. Goods produced surplus to actual demand simply pile up in a warehouse, while severe shortages are not necessarily translated into new factory orders.

Marshall Goldman, an American economist specializing in Soviet studies, observes that the Soviets have won the wrong race. They have become addicted to heavy industry, cranking out huge outputs of producer goods. But they have lost the race in terms of standard of living.*

*Marshall Goldman, *USSR in Crisis—The Failure of an Economic System* (New York: W.W. Norton, 1983).

Monument to Peter the Great, Leningrad. *Tsar Peter the Great, determined to modernize Russia and reorient it to the west, built St. Petersburg (now Leningrad) as his new capital on the Baltic. Leningrad is the USSR's "second city," with a proud tradition of sophisticated culture and a heritage of magnificent architecture.*

SOVIET AGRICULTURE: THE ACHILLES' HEEL?

Periodic droughts, hailstorms, floods, unseasonal frosts, and a dozen other natural hazards represent stumbling blocks to Soviet self-sufficiency in food production. Soviet agriculture, under any economic system, would be a relatively risky undertaking. Nonetheless, the full potential for food production has not been reached. Many Western observers, however, believe that privately held, profit-motivated farms could produce higher quantities and better-quality products. Generally, attempts to improve production involve two types of schemes: colonization of new (and therefore more marginal) areas and improvisation of new schemes in established areas through organizational changes or improvement of the land itself (Figure 4–11). Agricultural investment was assigned one of the lowest priorities until the last decade. Ideologically, Marx saw labor as the only determinant in production costs; philosophically, he envisioned society as able to overcome environmental problems and to exist essentially independent of the natural environment. Marx, an urbanite, knew and cared little about agriculture. Lenin, interpreting Marx to fit Russia's situation, was also urban by nature. As a result, Soviet agrarian policy evolved over time as a curious, often conflicting, set of principles—all compromises between philosophical aims and physical or cultural forces.

Collectivization of Agriculture

In tsarist times, much of the land was in large estates farmed by near-slave laborers called serfs. Feudalism of a sort, this ended in 1863, leaving many free but without land or work. Others received land but at high prices with high-interest mortgages. A communal system of landholding also existed in ancient Russia and, in a revised form, succeeded serfdom in some areas; this was ownership of the land and operation of the farm by the farm village community. These two factors, land-hungry peasants and the tradition of communal ownership, were eventually to shape Soviet land tenure policies.

The first "answer" to the problems of Soviet agriculture was found within the framework of the N.E.P. (New Economic Policy), the brief return to capitalism under Lenin (1921–28). Large farms were nationalized, and landless peasants received lands from this government land bank. Lenin resorted to this policy to gain the support of farmers. Many city workers were either without food or paying enormous food prices. The proletariat, supposedly the primary beneficiaries of the Revolution, were thoroughly an-

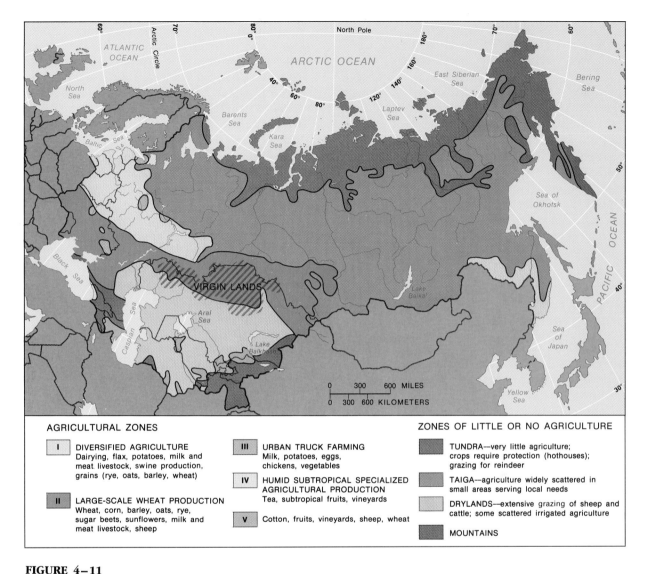

FIGURE 4–11

Agricultural Zones of the USSR. Much of the Soviet Union is too cold, too dry, or too rugged for crop production, at least without major environmental engineering. The Soviets have made massive investments in expanding agricultural lands, including the Virgin Lands Program and the Siberian River Reversal Scheme (see Figure 4–7).

noyed with the high prices and the scarcity of food in the cities. Stalin initiated the program of mechanization and collectivization. Tractors failed to increase yields; few could use or maintain them. Fuel was scarce, and so was general farm machinery.

Collectivization was an even worse failure, at least at its inception. The unannounced (but most impor-

tant) aim was the formation of capital for industrial investment from agricultural profits. The government bought crops at low prices and sold at high prices.

Theoretically voluntary, collectivization was actually forced on the peasantry. Resistance was rife as livestock were slaughtered, farmers refused to plant or harvest, and city dwellers starved. But Stalin, whose brutal policies resulted in the deaths of millions, succeeded in carrying out the plan.

Overall, state farms are larger than collectives, but the major differences are in wage structure. State farm workers are paid flat hourly rates. Collectives pay in share of profit; wages of a collective, therefore, fluctuate with farm productivity. Both sets of farms (in particular the state farms) are increasingly specialized in type of production.

Total food production has increased. The grain exports of tsarist times were drawn out of the public diet, not from any greater productivity of farms. Production is greater than ever. The Soviets produce more wheat per capita than do Americans; they simply consume far more bread. Total per capita production of all grains is less than in the United States, however, and meat production per person is about half that of the United States. Soviet production of fruits and vegetables is much lower.

Recently, Soviet imports of grain have set new world records for food imports; much of this imported tonnage originated in the Great Plains of the United States and Canada until an American embargo in the late 1970s; Argentine grain has partly replaced the American sources. The Soviets are not short of bread; they are trying to upgrade the national diet further to include more beef, dairy products, eggs, and chicken, all of which require a heavy input of grain and other animal feeds. The economic and political implications of this new reliance on imports from capitalist societies are intriguing.

Fertilizer usage is now about 75 percent that of U.S. farms. There are about half as many tractors per unit area of farmland in the Soviet Union as in the United States, but many Soviet tractors are larger and used more intensively. Productivity on Soviet farms is lower than that of U.S. farms, but climatic conditions are sometimes as responsible as the differing systems.

The greatest handicaps to Soviet agriculture remain capricious weather, a poor distribution system and, above all, the low priority assigned to investment in the agricultural sector. Planning reforms, increased autonomy in decision making, and improved prices for farm products have all led to greater production. Increased capital investment is the likely key to greater productivity.

A series of schemes to expand agriculture have proven only partially successful. The Virgin Lands Campaign that plowed up millions of acres of dry steppe for grain cultivation was successful to some degree. However, one third of the original area has had to be abandoned, and the remaining two thirds have highly variable harvests.

Current investment seems more rational as the Soviets stress seed and breeding stock improvements, increases in the areas of irrigation and drainage, and the intensive application of fertilizers. Conservation, at long last, is receiving major emphasis.

Despite all these positive changes, distribution remains a weak point. The infrastructure is extremely weak. There is a critical lack of grain storage facilities. The transportation system, particularly rural roads, remains grossly inadequate. Movement of agricultural commodities remains a low priority; industrial goods still receive preference on the railroads. Decision making is still bogged down in a bureaucratic morass. Parts procurement is still a problem. Theft is frequent, and prices paid to farmers are unrealistically low. Adverse environmental conditions are not fully recognized in prices paid or quotas assigned.

THE STRUCTURE OF SUBREGIONS

Official economic planning districts, designated by the government, have been combined into larger subregions that exhibit similar characteristics, problems, and prospects (see Figure 4–1).

The Central Industrial District

The Central Industrial District is the home of Russian culture and the heartland of what came to be the modern Russian state. The road and rail networks focus on Moscow, as does the subregion's economy (Figure 4–12). No other Soviet subregion focuses so completely on a single urban center.

Moscow, with 8.5 million people, is surrounded by a 30-mile-wide greenbelt of farmland, parks, and forests—an area in which development is restricted.

THE PARADOX OF PRIVATE PLOTS

Both state and collective farm workers have the use of private plots. All collective members and most state farm workers till a piece of ground for their own food needs, though each receives a share of grain, potatoes, dairy items, or other foods from the production of the collective or state farm itself. These private plots produce a disproportionate share of certain commodities, reflecting a great intensity of production on a small area—brought about by hand-intensive, high-labor-input techniques and selection of crops (those that bring high prices, yield well, and are not grown by the state). Much of this private plot production (originally intended for family subsistence) is sold at the free bazaars. Collectives emphasize grain; cattle and other livestock have remained largely in private hands, so that 40 percent of meat and milk, 80 percent of potatoes, 80 percent of eggs, and 40 percent of vegetables still come from private plots. The private plot has become a fringe benefit to generate extra cash income for low-wage farm workers; it also extracts extra production from them. As in Oriental agriculture, no mechanized farm (private, corporate, or public) could produce as much food per acre as hand labor horticulture under any economic system. Labor-intensive agricultural pursuits, then, have been left to the private plots. In actual fact, some grains produced on collectives are *assigned* as feed for livestock on private plots. Much feed grain for private operations is simply stolen, or bribed, from government stores. Thus, the high productivity of private plots is less of a miracle than it seems.

Still, an obvious question arises. If private plots are so phenomenally productive, why doesn't the Soviet leadership accelerate the extension of private management, if not private ownership, in the farm sector? Clearly, the profit incentive produces a great deal of food. Will ideology remain more important than productivity?

The greenbelt serves "day trip" recreation needs. Beyond it is a planned ring of satellite towns, generally of 50,000 to 100,000 or more people, that serve to partially process goods for finishing and final assembly in Moscow or as specialized scientific, technical, or educational suburbs.

Moscow produces almost 10 percent of the Soviet Union's industrial goods by value. It is also the largest educational center, the home of the government, a major port, the largest transportation node, and the dominant center of Russian and Soviet culture.

Its fortified center, the Kremlin, is its nucleus, although most of the buildings within the Kremlin's walls are now museums. The Kremlin is still the governmental nerve center, although most of the bureaucracy works outside its walls. Adjacent Red Square, with Lenin's tomb and GUM, the giant department store, is the focal point for parades, state occasions, and shopping.

In the city inherited by the Communists in 1918, the skyline was dominated by the Kremlin and the splendid onion-domed churches. The first step in

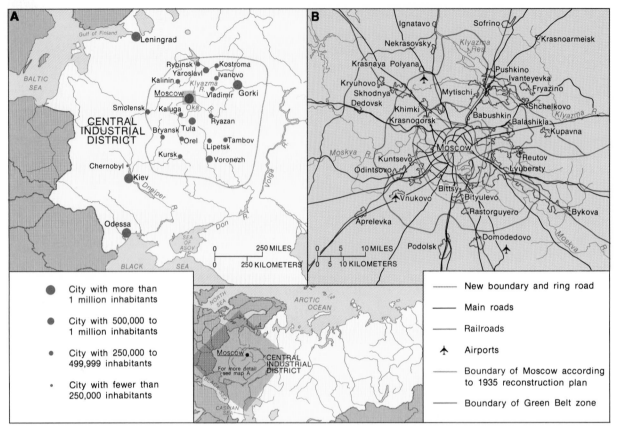

FIGURE 4–12
The Central Industrial District. This key subregion of the USSR contains the
original core of the Russian state, its capital, many important industrial centers, and a
dense population.

modifying it (1930–40) was the development of
broad boulevards radiating from the city center,
lined with new construction. Seven large towers (in
a style known as Stalin Gothic in the West) were
constructed to give a new skyline, symbolically lowering the importance of the church and raising that
of the state.

Beyond the city center, progressing slowly because of the huge demands for nonresidential construction, are row upon row of high-rise dwellings.
Each development is relatively self-sufficient, with
schools, workplaces, medical services, and shopping
for the entire neighborhood. The whole city is interconnected by a splendid, ornate subway system, itself

a tourist attraction. Moscow's metro has almost 200
stations and provides 15 million rides a day in this
society of few private cars and fewer parking lots.

East and southeast of the city center are concentrations of heavy industry. Lighter industry, including
consumer goods, is scattered throughout the city.
New industrial and research suburbs have been developed at the outer edges. Many emphasize chemicals, plastics, appliances, and precision engineering—the newer branches of Soviet industry.

Moscow is the place where goods are most readily
available (in part to serve the emerging bureaucratic
bourgeoisie, in part to impress foreign tourists and
dignitaries). These amenities draw migrants from all

over the USSR. Moscow has an unusually open, green appearance, with 40 percent of its total area devoted to parks and public open space.

Moscow has long been a closed city, with migration by permit only. Yet, despite its low birthrate, it continues to grow. Many illegal migrants come to the city on forged I.D. cards. Thousands live there on "temporary" permits that never seem to expire. No new industry is allowed to locate there, but expansions are not forbidden. The burgeoning bureaucratic establishment demands new labor constantly. The citizenry of Moscow demand better services and more consumer goods, swelling labor demands. As a result, Moscow grows, despite official no-growth policies.

In the 1980s, growth slowed, allowing Moscow to begin to meet the demand for housing. Soviet policy on controlling Moscow's growth initially was in response to a doubling of the city's population since 1939. Among the 20 largest cities in the Soviet Union, only Novosibirsk now grows more slowly proportionately.

Highly attractive **growth poles** (cities chosen for large-scale industrial investment to stimulate economic growth in surrounding areas) are being established at some distance from Moscow in satellite development. Special housing priorities and the assignment of Moscow-produced consumer goods to these poles have lessened the Moscow-bound craze. In the 1970s, 95 industrial plants were moved out of Moscow to reduce pollution and congestion.

The Central Industrial District includes many major industrial centers in addition to Moscow (Figure 4–13). Tula and its hinterland have developed an important chemical industry; this function enlarges continuously with the increased availability of Siberian gas supplied by pipeline. Now that a new regional source of iron ore has begun to be exploited (the Kursk magnetic anomaly), the iron and steel industries, as well as associated metal engineering plants, have begun to develop rapidly. Ryazan (population 500,000), a diversified manufacturing center, also has grown rapidly in recent years.

Located at the junction of the Volga and the Oka River, Gorki is the location of the oldest truck factory in the country, the Volga automobile factory, and dozens of engine and auto parts factories. New plants elsewhere, particularly in the Volga and Kama valleys, have superseded Gorki's plants as the leading Soviet automotive center. Gorki continues as an important industrial center, diversifying its production. It serves as the contact point between the Central Industrial District, the Volga Valley, and the Urals—three of the Soviet Union's largest agglomerations of industry.

In general, this oldest (and largest) Soviet manufacturing region is undergoing diversification and redevelopment. Though the region is short of resources, other than iron, the network of oil and gas pipelines that pass through the subregion makes expansion both profitable and possible. The southern half, with its giant new metallurgical base, will be the major area of growth.

The agricultural picture has traditionally been less bright. In the north, summers are short, cool, and damp. Drainage has always been a problem, and the poor agricultural prospects in juxtaposition to the growing industrial cities led to long-distance commuting and widespread rural depopulation.

This area is scheduled for agricultural revitalization. The plan is to intensify production through massive investment. Dairying, feedlot meat production, and a renewed emphasis on market produce are expected to revive local agriculture.

Commuting workers here are less anxious to live in city apartments because the rural food supply is much better. At harvest time, city workers help on the farms after working hours. As often as not, pay is received "under the table," and a kind of extralegal, symbiotic rural-urban relationship is more prominent here than in any other area of the USSR.

The North-Northwest

In the eighteenth and nineteenth centuries, the North-Northwest had profited by its position between the Russian core and western Europe. It received the innovations and technology of the outside world, diffusing them throughout Russia. As the capital and chief port, St. Petersburg (Leningrad) became the most technically advanced manufacturing center. In the Stalin era, an emphasis on internal development shifted new industrial investments south and east, toward more defensible positions.

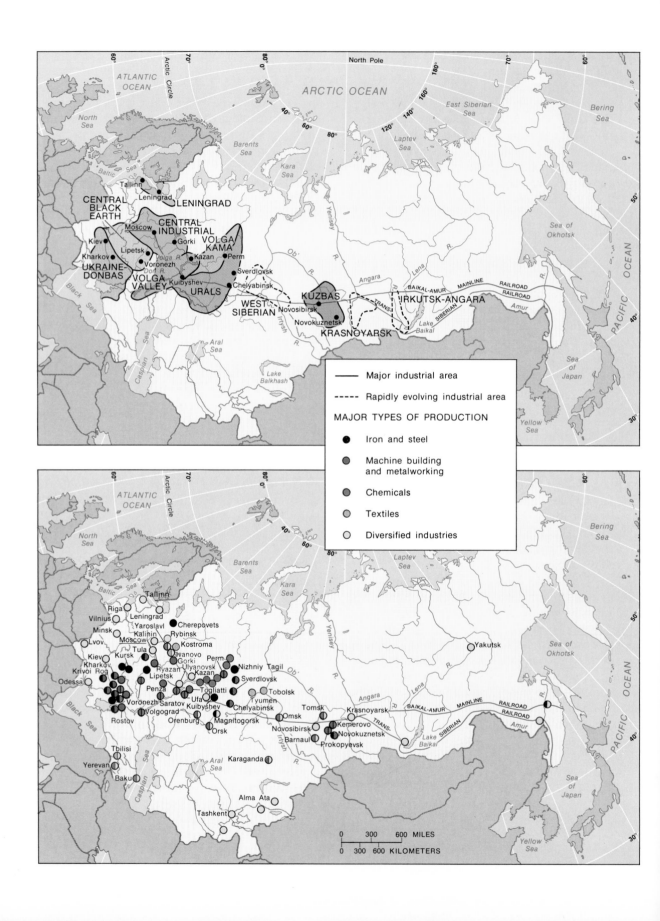

ATLANTIC OCEAN
Arctic Circle
North Pole
ARCTIC OCEAN
East Siberian Sea
Bering Sea

North Sea
Barents Sea
Kara Sea
Laptev Sea

Tallinn
Leningrad
LENINGRAD

CENTRAL BLACK EARTH

Moscow
CENTRAL INDUSTRIAL
Gorki
VOLGA KAMA

Kiev
Lipetsk
Kazan
Perm

Kharkov
Voronezh
Volga R.
Don R.

UKRAINE-DONBAS
VOLGA VALLEY
Kuibyshev
Sverdlovsk
Chelyabinsk
URALS

WEST SIBERIAN
Novosibirsk
KUZBAS
IRKUTSK-ANGARA

Novokuznetsk
Ob. R.
Yenisey R.
Angara R.
Lena R.
Amur R.

KRASNOYARSK
Irtysh R.
Lake Baikal

Sea of Okhotsk
PACIFIC OCEAN
Sea of Japan
Yellow Sea

Black Sea
Caspian Sea
Aral Sea
Lake Balkhash

BAIKAL-AMUR MAINLINE RAILROAD
TRANS-SIBERIAN RAILROAD

— Major industrial area
--- Rapidly evolving industrial area

MAJOR TYPES OF PRODUCTION
● Iron and steel
● Machine building and metalworking
● Chemicals
● Textiles
○ Diversified industries

Riga
Vilnius
Minsk
Lvov
Kiev
Kharkov
Krivoi Rog
Odessa

Tallinn
Leningrad
Cherepovets
Yaroslavl
Rybinsk
Kalinin
Moscow
Kostroma
Tula
Ivanovo
Gorki
Perm
Kursk
Ryazan
Ulyanovsk
Nizhniy Tagil
Lipetsk
Kazan
Sverdlovsk
Penza
Ufa
Tobolsk
Voronezh
Saratov
Kuibyshev
Tyumen
Chelyabinsk
Volgograd
Orenburg
Magnitogorsk
Rostov
Orsk

Tomsk
Omsk
Novosibirsk
Barnaul
Kemerovo
Novokuznetsk
Prokopyevsk

Krasnoyarsk
Yakutsk

Tbilisi
Yerevan
Baku

Karaganda

Alma Ata
Tashkent

0   300   600 MILES
0 300 600 KILOMETERS

BLAT AND THE UNDERGROUND ECONOMY

Although undoubtedly the Soviet economy has its success stories, its failures are most obvious in the consumer goods and services sections. The quota system under which all workers function emphasizes quantity, not quality. In free-market economies, customer demand in effect sets the quota. In the Soviet centrally planned economy, factory managers worry about meeting production goals, but not about marketing their product. There is no product competition, except in the form of relatively rare imports available to common people. In a planned economy, no one need be vitally concerned about customer satisfaction, or follow-up service.

An underground economy of undocumented, but obviously staggering, proportions has developed in the Soviet Union. The entrepreneurial spirit that is officially repressed has reappeared in a shadowy form. Virtually everyone, no matter what his or her job may be, can provide, or withhold, favors. Store clerks can buy up scarce merchandise before it is available to the public, then privately sell it at higher prices or trade it for other favors. A teacher can provide special tutoring (or higher grades) for the plumber's son in return for free plumbing repairs or a new bathroom built with stolen materials. Anyone who can provide favors, and thus demand favors in return, has *blat,* that is, clout. Truck drivers provide stolen state gasoline to clinic workers in return for scarce, imported antibiotics for a sick child. Just about everyone has some illegal sideline going to pay back favors and gain *blat.* Chauffeurs of official cars operate as gypsy cabs between official assignments; TV repairs are made on the spot for under-the-table tips, with the visit officially recorded as "estimate only" at a low fee. Such under-the-table activities are known to Soviets as *na levo,* literally "on the left"! The fundamental good humor of the Soviet people is illustrated by their ability to tolerate such an unwieldy system.

Once again, the area enjoys an intermediate position as the USSR has seemingly entered a new period of large-scale contact with the outside world. The relative economic position of the North-Northwest has shifted once again from the periphery to the zone of contact.

FIGURE 4–13
Industrial Districts of the USSR. Industrial resource location and transport development have been important considerations in the development of industrial districts.

Leningrad, with almost 5 million people, is often described as Russia's most Westernized city. It is indeed a great cultural center, second only to Moscow. The city occupies a series of low, marshy islands at the mouth of the Neva River. It is a pleasant city with an unpleasant climate, possessing a fine array of architecture from the eighteenth and nineteenth centuries. It houses one of the world's great art collections in the Hermitage museum. The citizenry possesses an elite mind-set; they still view Muscovites as less-than-cultured people, and even factory workers pride themselves on dress, educational development, and other status symbols. Townspeople have perhaps the highest level of knowledge about things Western.

The traditional industrial specializations of the city developed during the tsarist era with high-quality textiles, luxury goods, machine tools, shipbuilding, and engineering. These specialties remain the same, and Leningrad still accounts for some 5 percent of total Soviet industrial output. The combination of a highly skilled work force and an unimpressive raw materials base has resulted in Leningrad's continued production of the highest quality machine tools, scientific and medical equipment, measuring devices, and industrial controls. Even its textile and clothing industries specialize in the highest priced items and the latest in fashions. Increasing proportions of the subregion's industrial production enter export markets.

Archangel, a port on the Arctic, was the chief Russian port before the conquest of the Baltic coast. Closed by ice for five months of the year, it still ships a reasonable tonnage. Murmansk, the largest city in the north and a major Soviet naval base, is virtually ice-free because of the last vestiges of warmth from the North Atlantic Drift (the northernmost extension of the Gulf Stream). It handles much of the winter trade of icebound Leningrad and Archangel. Metal-processing industries provide employment based on the resources of the nearby metal-rich Kola Peninsula, which contains copper, nickel, and iron ores, and the large, new low-grade iron deposits being developed in Karelia.

The Baltic Coast presents an altogether more prosperous picture. Poor in resources, but with a strategic location and excellent ports, it is developing a well-balanced and technically advanced industrial sector. Different in culture and blessed with excellent beaches, it has become a Soviet tourist mecca.

The Baltic Coast's three ethnic Republics were independent from 1919 to 1940; their people possess a high degree of national consciousness. The entire coast is peopled by advanced, highly educated, and quite sophisticated groups whose attitude toward the Russians is a mixture of patronizing tolerance and uneasy acquiescence to the realities of power.

Peat and oil shale in Estonia are the only significant mineral resources, but oil and gas pipelines from the interior have solved the energy dilemma. It is not the resource base but the ports and position that have stimulated industry. The Soviets have sought to develop the area along the lines of industry in modern New England. High-value metal goods, electronics, and sophisticated consumer items are dominant.

Russians constitute 10 percent of the population of Lithuania, 75 percent of Latvia, and over 25 percent of the people of Estonia. Improved living standards are the draw, but the Soviet government also prefers to relocate Russians there in sufficient numbers to ensure regional loyalty.

Riga, Latvia, with almost a million people in its metropolitan area, is by far the largest city. It is a major port and contains a variety of industries. With a splendid medieval city at its center, it is one of the few Soviet cities with a significant night life.

The traditional Lithuanian capital, Vilnius, has over a half-million people and is an important university city and a major center of consumer goods production. Estonia boasts the highest income per capita of any Soviet republic. Tallinn, the capital and port of Estonia, functions almost as an industrial suburb of Leningrad. Access to large markets is booming the regional economy, and trade with the West is increasing demand even further.

Byelorussia was the home of the original Slavs. It has no history as an independent unit; what seems to make it different from Russia is its long historical association with Poland and Lithuania. The name, which means "white Russia," is an old designation. The ancient state of Kievan Rus was divided into "white," "red," and "black" portions, indicating the color of the soils of each part. This is white (Byelo) Rus, the land of ashen white, infertile soils. The area is a morass of bog and sluggish streams interspersed with gravel hills, hardly a fortunate environment for agriculture. Hay and pasture production are the dominant land uses. Poultry are fed on small grains and artificial feed in a U.S.-style broiler industry that is a recent innovation.

Several million acres have been drained and heavily fertilized over the last 15 years. Yields of most crops exceed Soviet averages, and the processed food products of the region are known for high quality. Thus, a poor environment has become highly productive, exhibiting the fruits of investment in Soviet agriculture.

Even more important than the area's agricultural progress is its industrial growth. Minsk, the capital, is

View of Riga, Latvia. *Western architecture and commercial traditions characterize Riga, the old capital of the Latvian SSR.*

the seventh largest Soviet city and has grown more rapidly than any other of the 10 largest. Unlike Moscow, its growth is based more on industrialization than on the growth of administration or services.

The Ukraine and Moldavia

The subregion comprising the Ukraine and Moldavia is the richest agricultural area of the Soviet Union and its second most important industrial area. The combined mineral assets of the area (including coal, oil, gas, iron ore, salt, mercury, manganese, potash, sulfur, and uranium) are staggering. Without a doubt, it is the most productive part of the USSR.

The term *Ukraine* means "land at the edge" (in the context of the Russian core). The area was the original Russian state (Kiev Rus), from the ninth through the thirteenth centuries. Out of diverse beginnings, a Ukrainian nationality has developed. The nation has its own language, literature, culture, and mythology. More people than ever before speak Ukrainian, and a strong cultural revival is apparently in progress.

The Soviet government reinstituted control over an old province of the tsarist empire with the annexation of Moldavia in 1945. Moldavia and the Ukraine are the old breadbasket of Russia; Moldavia is also the Russian "southland," an area of warm, temperate climate with an increasingly different type of agriculture from that found in the rest of the country.

Climatically, the area is warm by Russian standards; the growing season is generally six months or longer. Rainfall decreases from northwest to southeast, though the small area of Black Sea coast has heavy rainfall.

Agriculture is diverse. Traditionally, wheat has been the dominant crop, but corn covers an increasing acreage. After his visit to the United States in the mid-1950s, Nikita Khrushchev, impressed with the prolific Midwest, initiated the Corn-Livestock Program in the Soviet Union. Unfortunately, most of the Soviet Union is too cold for the ripening of grain corn, and many areas that are sufficiently warm are generally too dry for that crop. Of all the areas still planted to corn, this subregion is the most productive.

A series of canals leading from the giant reservoirs on the Dnieper are used to irrigate an increasing acreage in the southern Ukraine. This increased capacity will help to increase total and average yields and to stabilize annual production in this, the most drought-prone part of the Ukraine.

The level of mechanization is exceptionally high, even by Soviet standards. Crop and livestock patterns are regionally differentiated to match regional climatic and soil capabilities (see Figure 4–11). Northern areas engage in dairying and hog raising. In the central Ukraine, meat production is the most intensive in the USSR, with large herds of stall-fed cattle and swine. Sugar beet pulp is used as animal feed, and oil-seeds of various kinds are a major industrial and cash crop. The southern area emphasizes corn on wetter and irrigated lands, though wheat increases in acreage east and south. In Moldavia, the Crimea, and the Transcarpathian Ukraine, vineyards produce table grapes and wines while cooler and higher slopes are planted to orchards.

The oldest industrial development is associated with the Donbas coalfield; iron ore at Krivoi Rog 150 miles to the west, local limestone, and manganese ores within 125 miles meant that all raw materials necessary for the manufacture of steel were located in close proximity. Begun in the 1880s, this most important of all Soviet heavy industrial complexes has been growing ever since. Its relative importance has declined, but it continues to grow, if at a much slower rate. The coal basin itself became the location of heavy industrial plants that manufacture pig iron, steel, chemicals, and metal products, although the coal reserve is now quite small.

The Krivoi Rog iron beds, now largely mined out, have given rise to a second heavy industrial center in the central Ukraine. The so-called pendulum kombi-nat was designed to maximize the use of transportation by shipping coal on "back haul" or return trips to the iron district for use in a new steel base there. Electrical power available from giant dams on the Dnieper gave impetus to diversification. Dnepropetrovsk (population 1.2 million), Zaporozhye (population 850,000), Krivoi Rog (population 700,000), and a half-dozen other cities of over 100,000 now form a full-fledged industrial complex. Collectively, the industrial production of the Dnieper Bend is as important as that of the Donbas.

Kharkov and Rostov have excellent access to fuels, major national markets outside the region, and a cheap, nearby supply of metals. As such, they are major engineering centers of enormous scale. Both cities convert metals, chemicals, and semimanufactures into finished products.

Founded in the eighth century, Kiev was a defensive site where an important trans-European, east-west route met the north-south water route of the Dnieper. As the chief point of control on early trade routes, it grew relatively large at a time when London and Paris were villages.

When added to the Russian domains in the eighteenth century, Kiev became the center of the richest agricultural and, eventually, industrial domain of the country. It is now the third largest city in the Soviet Union (population 2.4 million). Kiev is a model of industrial diversity; it has a major tourist function and is a center of nationwide importance. Clean, pleasant, and sunny, it has an easygoing lifestyle and a culture-conscious citizenry; however, the worst nuclear reactor accident in history took place in April 1986, at Chernobyl, about 50 miles north of Kiev. The full dimensions of the disaster may not be known for years, because radiation-induced deaths will take place over decades.

In the southern portion of the district, intensive vineyard and vegetable production yields extra income for collectives engaged in general farming. As one travels south, the landscape takes on a Mediterranean appearance, with its white homes with tiled roofs, walled farmyards, and ubiquitous patches of vineyards and orchards. It is a pleasant land with less severity than the steppes of the eastern Ukraine.

Rounding out the subregion is a series of resort cities stretching along the southern foothills of Crimea and the Black Sea coast. Both areas are backed by sheltering mountains that reduce winter wind and cold. Yalta in the Crimea and Sochi on the Black Sea are the most famous and best developed. Busy all year, packed to overflowing from May to October, they are the premier internal tourist objectives in the USSR. Away from the centers of the tourist towns are the dachas (cabins) of the wealthier bureaucrats, officials, and factory managers—the so-called red bourgeoisie.

The Volga and the Urals

The Volga and the Urals—one a river valley and the other a mountain range—are linked by complementarity of both resources and functions. The Volga is icebound for much of the year; it flows the wrong way in terms of the delivery of raw materials to manufacturing and consuming centers and the return of empty barges. Overall, navigation was exceptionally difficult until greatly modified by dams and canals. Until the construction of the Volga-Don canal, it terminated in an interior sea located in largely desert country.

The Urals are a long, narrow range of folded mountains trending north-south. They are old, not very high, and only a slight barrier to transportation. The Urals are fuel poor yet rich in metal ores, and their economic development has been much more diverse (Figure 4–14, p. 238). Together with the Volga Valley, they exert an enormous influence on the Soviet economy. For historical Russia, Siberia began just across the Volga. Culturally, the meeting of European and Asian cultures is everywhere evident.

Agriculture and industry were late in developing. New cropland was added during the Virgin Lands Programs (1954–60), and existing agriculture has been intensified through irrigation. It is a land of large, mechanized grain farms except at its northern and western fringes where more mixed farming prevails.

The Urals, with rich, high-grade iron deposits, appeared to be a natural choice as a "second metallurgical base." It was necessary to bring in coal from Kuznetsk coal basin (Kuzbas) at the extreme eastern end of Western Siberia, a distance of over 1,000 miles (see Figure 4–10). Overcoming immense distances has long been a problem in the Russian, then Soviet,

economy. The Soviets must devote huge amounts of energy, literally, to overcoming the **friction of space**—the time, labor, and fuel costs of long-distance transport.

Four large integrated iron and steel complexes form the nucleus of the Ural's industry. The steel produced is now used locally or in the Volga Valley's aircraft plants, rolling stock factories, and a host of machine shops. The demands of a booming petroleum-petrochemical industry consume much steel in an important array of secondary manufactures.

The cities of the Volga Valley are metal engineering centers, and most have major petroleum refining and petrochemical complexes. Their industries use

Old Buildings, Kiev. *Cosmopolitan Kiev, an ancient trade center and now capital of Ukrain SSR, presents a lighter, more Western and more open atmosphere than Moscow.*

THE RADIOACTIVE AND POLITICAL FALLOUT
OF THE CHERNOBYL DISASTER

On April 26, 1986, the world's worst nuclear power plant accident occurred at Chernobyl. Thirty-one people died soon afterward as a direct result of the disaster, 1,000 were gravely injured by massive doses of radiation, and 135,000 were evacuated to safer areas. Two hundred fifty thousand Kiev school children were sent to summer camp early.

Chernobyl Four was the newest reactor of four at that site; it was ranked among the most efficient in the world, with a load factor (output of energy compared to full-time full-power potential) of 83 percent compared to an average lifetime load factor of 56 percent in the United States. The Chernobyl plants are the product of over 30 years of Soviet experience with nuclear engineering, and Chernobyl Four was only two years old. Many nuclear reactors around the world are much older and have greater potential for breakdown in safety and control systems.

The USSR's Chernobyl accident and the one at Three-Mile Island near Harrisburg, Pennsylvania, have a common cause—human error. In both cases, a chain of mistakes and poor judgment resulted in backup emergency systems being turned off, proper procedures ignored, and correct decisions critically delayed. At Three-Mile Island, however, the reactor's containment structure functioned as designed, keeping all but a tiny fraction of dangerous radioactivity inside the plant. At Chernobyl, the 1,000-ton concrete roof over the reactor was blasted aside as the damaged reactor core and graphite burned at 2,800° F. For the first time, the lethal radioactivity of a reactor erupted into the atmosphere, carried high by the intense heat from the fire. It took 12 days to extinguish the fire, finally achieved by pumping liquid nitrogen under the reactor to cool it down.

For nearly three days, Soviet officials denied that anything was wrong, despite the hard evidence accumulating in radiation-monitoring stations in Europe. This refusal to acknowledge the disaster was consistent with Soviet policy until that time. Routinely, airliner crashes, spacecraft accidents, even earthquakes had not been publicized or admitted. However, when the Soviet government did "go public," both internationally and within the USSR, it supplied technical details and self-criticism on a more open scale than resulted from the Three-Mile Island incident. Soviet leader Mikhail Gorbachev's policy of *glasnost,* ("openness") produced admissions of design flaws and gross human error.

The full cost of Chernobyl will be difficult to determine. U.S. experts predict at least 5,000 more deaths from radiation-induced cancer, though some of those deaths will occur decades afterward. Radiation levels in the vicinity remain 2,500 times normal, and the soil of the district will not be usable to produce food. Studies have shown that survivors of the Hiroshima and Nagasaki blasts are still developing new cancers; some scientists predict an eventual toll of as many as half a million cancer deaths attributable to Chernobyl. Replacing the plant, relocating evacuees, site cleanup, and loss of agricultural production will cost the Soviets at least $5 billion. Areas 1,000 miles away had enough radiation outfall to restrict use of crops and domestic animals. Swedish cattle were restricted to their barns and fed imported food all summer of 1986; direct Swedish losses in discarded food cost at least $140 million. English sheep had to be destroyed because they had eaten radiation-contaminated grass, and Polish farmers had to bury radioactive milk, vegetables, and fruit.

Around the world, enthusiasm for nuclear energy has declined sharply. No new contracts for nuclear power plants, not later canceled, have been signed in the United States since 1974. The Swedish government, once an advocate of nuclear energy, has reversed its position. Mexico, which once planned 20 plants, now plans 2, and 12 Pacific states, including Australia and New Zealand, have signed an antinuclear pact. A study released in 1987 predicts less nuclear generation by the year 2000, not more.*

Once in gear, the Soviet system functioned to evacuate efficiently and quickly 135,000 people from the most dangerous vicinity of Chernobyl. The authoritarian political system commanded speedy compliance with evacuation orders, and the fact that the Soviets are much less dependent on private cars meant that an efficient public transport system could move large numbers without traffic jams. There is one car for every 31 Soviets, but one car for every 2 Americans. There are much higher population densities around some nuclear plants in New Jersey, Pennsylvania, New York, and Connecticut than around Chernobyl; hectic traffic jams would soon follow announcement of a nuclear plant accident, critically slowing necessary evacuations and sending the radiation-related death toll much higher. What happened at Chernobyl could prove to be a valuable warning, if we heed it. Unfortunately, as the father of the atomic age, Albert Einstein, observed, "the unleashed power of the atom has changed everything save our modes of thinking."

*Christopher Flavin, *Reassessing Nuclear Power: The Fallout From Chernobyl* (Washington, DC: Worldwatch Institute, 1987), p. 7.

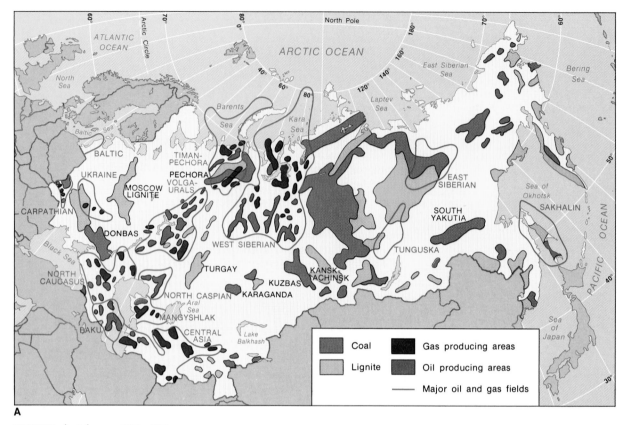

A

FIGURE 4–14, pp. 238–239
(a) *Fossil Fuel Resources of the USSR.* The Soviet Union contains some of the world's most important coal-, oil, and natural gas fields. (b) *Mineral Resources of the USSR.* The Soviet Union, already a rich storehouse of minerals, continues intensive exploration, especially in Siberia.

large amounts of electrical power supplied by the series of dams on the Volga. There are six metropolitan complexes of over 1 million inhabitants. Like most of the Urals' cities, they are largely the result of investment over the last 50 years.

The largest urban complex is located along the great bend of the Volga and centered on the city of Kuibyshev (population 1.3 million). Industry stretches in an almost unbroken line for 30 miles in either direction along the Volga. Kuibyshev is located where the Trans-Siberian Railroad crosses the Volga. Because it is the center of the Volga-Urals oil field, its regional importance is similar to that of the Dallas-Fort Worth area.

A recent addition has been the Soviet-Fiat auto-

mobile plant at Togliatti, located at the northwestern end of the bend, producing over 750,000 vehicles yearly. Even though the Volga Valley has been scheduled for reduced industrial investment in the future, Kuibyshev is likely to continue growing at an impressive rate. It serves as the chief link between the European and Asiatic portions of the USSR.

Volgograd (once known as Stalingrad) is most famous for the great battle fought there during World War II. It was the farthest eastward advance of German troops. The Volga-Urals region is now a surplus producer of food, agricultural raw materials, fuel, minerals, and machinery. What was a pioneering area 50 or 60 years ago has become a major industrial-agricultural complex fully integrated with the older

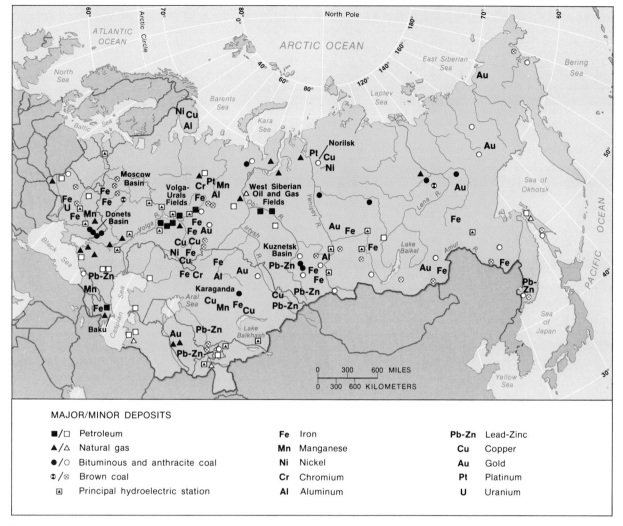

MAJOR/MINOR DEPOSITS

■/□	Petroleum	**Fe**	Iron	**Pb-Zn**	Lead-Zinc
▲/△	Natural gas	**Mn**	Manganese	**Cu**	Copper
●/○	Bituminous and anthracite coal	**Ni**	Nickel	**Au**	Gold
◐/⊗	Brown coal	**Cr**	Chromium	**Pt**	Platinum
▣	Principal hydroelectric station	**Al**	Aluminum	**U**	Uranium

B

centers of the European USSR. It also serves as the link between the older portions of the country and the rapidly developing realms of Transcaucasia, Western Siberia, Soviet Central Asia, and the Virgin Lands.

Soviet Central Asia and the Caucasus

The Transcaucasian valleys and some of the southernmost oases of Soviet Central Asia are subtropical, but most of the area has severe winters. A majority of its inhabitants are Moslems, but there are many important minority groups (see Figure 4–8). It is exceptionally diverse and complex in ethnicity and culture, yet large numbers of Russians and other Slavs dominate in many of its cities. Intensive, specialized agriculture and the resort and retirement functions are especially well developed.

The subregion has a population of about 60 million people and is growing at a rate almost double the national average through a combination of high

Heroic Monument to the "Motherland." *One decisive victory of World War II was won by the Soviets over Nazi forces at Volgograd (then Stalingrad). Destruction remains a prominent memory.*

birthrates and strong net immigration (see Figure 4–3). It has long been the aim of Soviet planners to open up Asiatic Russia with its great storehouse of mineral and forest wealth and its underutilized agricultural potential. This subregion is a part of that Soviet frontier. Of the two parts (Siberia being the other), the plan has succeeded best here. Despite inducements and regulation of migration, Siberia is a land of net out-migration. In Soviet Central Asia and the Caucasus, despite Slavic in-migration, local birthrates are sufficiently high to ensure the dominance of local ethnics.

The western portion of the subregion is one of the most culturally diverse areas in the world. The an-cient Christian peoples of Georgia and Armenia exist side-by-side with Buddhist Kalmucks, a variety of Turkic- and Iranian-speaking Moslems, Russians and other Slavs, and the remnants of Asiatic tribes. Occupying an isthmus between the Black and Caspian seas, the area's elevations run from below sea level to over 18,000 feet. The westward-facing valleys have a humid subtropical climate with heavy rainfall, while eastward-facing valleys are largely desert. The physical habitat is diverse, and sharp changes occur over short distances.

Separating the two portions is the Caspian Sea, the largest lake in the world. Because of the withdrawal of water for irrigation, the damming of the Volga, and perhaps even climatic change, the Caspian has been shrinking.

The eastern portion is a vast expanse of deserts surrounded on three sides by mountains or plateaus. The utility of the desert varies from range land suitable to sheep grazing to wasteland; a few areas are suitable to dry farming. Wherever streams fed by the glacial meltwaters from the high bordering mountains penetrate the desert, there is a dense settlement pattern, based on sedentary, irrigated agriculture rather than on grazing.

Cotton, fruits, early vegetables, and industrial crops are the most vital agricultural crops throughout the area. Most of the subregion's agriculture is dependent on irrigation. Geographically peripheral, the area is in some senses central to the economy. Most of it was a part of various Mongol, Persian, and Turkish empires in the past. It is the Soviet Union's piece of the ancient world, a colonial extension into the vast reaches of Asia.

The ethnic diversity is illustrated by the range of autonomous jurisdictions. There are three republics (Armenia, Georgia, Azerbaidzhan), eight autonomous socialist republics (a lower ranking), and several smaller autonomous oblasts. There are more than 20 million people in this portion of the subregion, and probably less than 15 percent of the total are Russians or other Slavs.

The highest slopes of the Caucasus Mountains are virtually uninhabited. Population is clustered in two parallel strips on either side of this empty core, with higher densities south of the mountains, where the climate is truly subtropical, at least in the lowlands.

Georgia specializes in tea, citrus, and other fruit. Large areas of vineyard and orchard occupy the eastern half of the republic along with silk culture and fine-quality tobacco. Subtropical crops are grown on the slopes, where the frost hazard is minimal. Rice and corn occupy the more frost-prone valleys, and livestock are pastured on the higher slopes during summer.

Drier Azerbaidzhan emphasizes irrigated cotton. Even here, though, sheltered areas are given to orchards and vineyards. Irrigation projects increase yields of the temperate crops grown in this republic.

Agriculture (and settlement) decreases in intensity beyond the Kama River. A thin strip of irrigated land follows the Volga from Volgograd to Astrakhan, but the rest of the countryside is poor quality grazing land.

Oil fields dot the northern Caucasus foreland, culminating in the great oil field at Baku. Long the greatest oil-producing region in the Soviet Union, Baku now accounts for only 10 percent of total production. Baku (population 1.8 million) is by far the largest manufacturing center. Drilling equipment, offshore drilling platforms, and pipe are its major specialties, reflecting its long-time association with the oil industry.

In contrast to the industrial slum of Baku, Tbilisi, the capital of Georgia, is picturesque. Situated along a river gorge and containing a wealth of historical sites and religious architecture, it is one of the USSR's great tourist attractions. Silks, quality textiles, appliances, food-processing equipment, and excellent wines are representative of its high-value production from small plants scattered throughout the city. This light, small-scale industry does not detract from the city's beauty and its tourism functions. Yerevan, the Armenian capital, specializes in aluminum and chemicals based on plentiful hydroelectric power. Textiles, wines, and fruit processing have become secondary, with electronics and computerized industrial systems new and growing specialties.

High demand and limited physical area have caused the Soviets to begin developing a string of resorts along the Caspian south of Baku. Astrakhan was once the capital of a powerful Khanate of the Golden Horde, a great trading center on the overland routes to the Far East, and bustling port on the

Caspian. Today it is on the dead end of the Volga, and its port is clogged with silt. Various canal schemes, rail, and highway links are planned to enhance its position as the pivotal exchange between Soviet Central Asia and European Russia.

The realm of Soviet Central Asia is as diverse and distinct as that of the Caucasus. Conquest by the Russian Empire came even later, covering the period 1824–95; this was virtually the last frontier, as Russia sought potential cotton-growing lands and a sphere of influence in Asia.

Most of the peoples speak a Turkic language, but a variety of non-Turkic ethnics also are found in the area. The Tadzhiks (and several other groups) speak Iranian dialects (see Figure 4–8). Russian and Soviet occupation have increased the diversity, and Slavs are a majority in the Kazakh SSR. Russians form 10 to 25 percent of the population in the other four republics, but their relative importance has declined in the last decade.

The physical core of this part of the subregion is the vast desert at its center. Population is concentrated in a crescent along the subregion's southern and eastern edge, sandwiched between mountain and desert (see Figure 4–2). There are two additional ribbons of dense population along the courses of the Amu Darya and Syr Darya that dead-end in the Aral Sea. Except in the extreme south and in sheltered valleys, the climate is severe. Far removed from the seas, the area receives little precipitation except in the mountains.

Agriculture is overwhelmingly dependent on irrigation: this region has over half of the country's irrigated acreage. Virtually every stream has been brought into use. Vast irrigated fields produce cotton, while areas near urban centers specialize in fruits, vegetables, and melons.

About 25 percent of all Soviet gas comes from the Soviet Central Asian republics; the chief fields are located near the city of Bukhara and in the Turkmen SSR. A series of small oil fields ring the Caspian Sea. Copper, lead, and zinc are found in the mountains of the Uzbek SSR and in eastern Kazakh SSR. Industry has just begun to develop in Soviet Central Asia. There is only one great industrial city, Tashkent, now the fourth largest city in the Soviet Union with 2.2 million people.

Street Market, Novosibirsk.
Open-air "free-markets" provide much of the typical Soviet citizen's fresh fruits, vegetables, eggs, and meat. Novosibirsk, the "USSR's Chicago," is a thriving modern city and the largest in Siberia.

Exotic cities, cloudless skies, and a superb backdrop of snowcapped mountains offer an array of amenities similar to those found in portions of the U.S. Sun Belt. Agriculture and construction are still the chief sources of employment for migrants, but tourism is rapidly increasing in importance. Official policy has long advocated decentralization of industry, and many government enterprises have placed expansion plants here to reap various benefits offered to managers and employees. Footloose industries with no great ties to raw material supply areas dominate; again, there is a strong similarity to the American Sun Belt. Thus, textile, light engineering, pharmaceutical, and consumer goods industries have expanded into this region.

Space program facilities, climatic and agricultural research stations, and geologic exploration facilities have established a research component in the local economy. This component serves to attract other research facilities that hope to share the brainpower as well as the research results.

The pull of **amenities** cannot be overestimated. Even in a controlled, centrally planned economy, people and agencies find ways to circumvent directives or to subvert intent. Outside the Soviet Far East, an area with a low original population base and intense investment in military and mining enterprises over the last 20 years, all areas of rapid growth are associated with Soviet Central Asia, the Caucasus, the Black Sea shores, and Moldavia—the Soviet equivalent of the Sun Belt.

Siberia

The name Siberia implies cold, forced labor, political exile, imprisonment, and a host of other distasteful images. Yet its original conquerors saw it as a land of economic opportunity, replete with gold and furs. The Soviets see it as containing the resources necessary to make the USSR the world's most powerful economic unit (see Figure 4–14).

Two schools of thought have evolved in the Soviet Union concerning the proper way to use Siberia's potential. One sees it as a colonial frontier awaiting colonization and settlement, with settlement essential to development and integration with the rest of the Soviet Union. The other envisions minimal settlement sufficient to exploit the resources, with shipment of Siberian resources to the developed parts of the country. Both schools of thought have periodically been dominant as Soviet investment in Siberia ebbs and flows.

The latest census shows that the Siberian growth rate over the preceding 10-year period barely exceeded the national average of 8.6 percent, despite the youth of the population and the much higher than average birthrates. Although the population grows, there is a net out-migration.

Only the Soviet Far East and the new oil and gas fields of Western Siberia showed truly high growth rates. The cherished Third Metallurgical Base (the Kuzbas) had a growth rate of a little over 1 percent, while the oldest and most favorable agricultural areas actually exhibited a small decline. The fundamental problems are distance and the rigorous climate, particularly in terms of the way in which they increase the costs of development, production, and transportation.

Siberia as defined here includes the Virgin Lands and all areas north of them and east of the Urals industrial complex. Though located in part in the Kazakh SSR, the Virgin Lands are included because of their Russian ethnicity and location adjacent to similar physical and agricultural area across the Kazakh border.

The resource base is best described in superlatives. Siberia contains 90 percent of Soviet coal, almost two thirds of the USSR's natural gas, and 80 percent of its hydroelectric potential. It contains the world's largest forest, largest coalfield (one of dozens), and the second largest gold and diamond fields. There are considerable iron and tin reserves, large deposits of lead and zinc, and many rare metals. Oil has been discovered in promising quantities, and aluminum ores are now exploited. All this is known on the basis of limited geologic prospecting; what is not yet known may be even greater.

Enticing as these facts may be, products must be shipped over the world's longest railroad or through the world's longest pipeline. Electricity generated there must travel the world's longest average distance over the world's longest power transmission grid. Siberia contains the world's coldest permanently inhabited city, the world's largest swamp, and its thickest permafrost outside Antarctica. It suffers some of the world's strongest winds and greatest daily and seasonal temperature variations. The resources are there, but getting them out is another matter.

The population is young, composed of fairly recent migrants and enormous numbers of temporary residents just out of school or college and repaying the cost of a free education by accepting duty in Siberia. Population concentrations follow the zones of usable agricultural land, essentially paralleling the Trans-Siberian Railroad. Each point at which the railroad crosses a major river has become the site of a large city. In Siberia, access is virtually as important as natural environment in influencing settlement patterns. The whole of northern Siberia is sparsely populated. A few large cities exist in unfavorable environmental areas solely to exploit mineral deposits. Primitive tribes pursue a nomadic lifestyle in the forested areas between the rivers.

Ethnic Russians dominate in all cities and commercial farming areas to the point of virtual exclusion of other groups. Just over a million aboriginal ethnics reside in Siberia out of a total of 34 million people. There is no well-developed nationalism among any of the native groups except perhaps the Kazakhs (called Cossacks in times past).

By 1700, the Russians had acquired most of Siberia, but it was treated more as a land for hunting and trapping than as an area for colonization. Missionaries followed the Cossack adventures and converted many native peoples to Orthodox Christianity. In the process, many of the indigenous population became assimilated to Russian language and nationality, adopting a sedentary lifestyle at the same time, thus becoming the first permanent agricultural settlers. Small villages grew at fur collection points and in conjunction with a string of defensive forts, much as in early North America. One can see many similarities in the process of settlement of the USSR and America, but the Russian frontier was more a series of nuclei strung out east-west rather than an advancing longitudinal line of settlement.

After the emancipation of the serfs in the 1860s, there was a movement to Siberia of landless peasantry seeking new farms. The Trans-Siberian Railroad, completed in the 1890s, allowed conversion of subsistence farms to commercial production and connected the scattered settlements. A second group of colonists was attracted to the remaining lands with access to the railroad. New urban settlements developed along the line farther east gave impetus to the development of agriculture in each urban hinterland.

World War II brought the evacuation of millions from the Ukraine, Byelorussia, and western Russia. Industry was evacuated to the east as well. While

LIFESTYLES OF THE PRIVILEGED AND NOT-SO-PRIVILEGED

A stated goal of communism is to eliminate class differences; the present reality in the USSR is not only a society with sharp class distinctions, but one that shows signs of developing hereditary classes as well. Sociology in the Soviet Union is a new and timid science, with findings that often run counter to official wisdom. Soviet data indicate, for example, that the children of the intelligentsia—the offspring of scientists, professors, high party officials, and military officers—have about 25 times the chance of admission to a university as the children of factory operatives. As elsewhere, much of this difference may be attributable to the level of parental support, expectations, and encouragement, but some doubtless reflects influence and contacts.

The party is a primary road to admission into the Soviet elite; another route is outstanding achievement contributing to Soviet power and prestige. Olympic champions, cosmonauts, prima ballerinas, prize-winning scientists, unusually successful economic managers, and the like can reach the elite class in living standards if not power. The *nomenklatura* (from *Nomenclature*), or secret roster of those holding sensitive, responsible positions, forms the central control system of who gets what. It is like a large, nationwide secret fraternity, which admits new members as it chooses, with no higher authority to make accounting to on its selections of privileges.

The privileged class lives in better housing, enjoys far better food, high-quality medical care in special clinics, educational opportunities in exclusive schools, and vacations in their own country cottages. Contrasts among classes in housing are representative of the widely differing Soviet lifestyles. In his perceptive book, *The Russians,* Hedrick Smith observes: "Housing, perhaps more than anything else, illustrates the peculiar combination of triumph and tragedy for Soviet consumers. Staggering achievements have been made, yet staggering shortcomings remain."* For decades, millions of housing units have been built every year. Astonishingly, many of these units are privately owned cooperative apartments or private *dachas,* country cottages built on land leased from the state. Members of the elite—groups of bureaucrats, professionals, and military officers—can get together and pay 40 percent down to have co-op apartment houses built to higher standards of space and quality than those accessible to the masses.

These co-op apartments can be sold or willed to one's children, as can country dachas.

For those without 40 percent downpayments or the contacts to get official permission, the housing picture can be discouraging. A 1985 study estimated that average Soviet living space is only one third of American standards and that 20 percent of urban dwellers live in communal apartments, occupying one room and sharing a kitchen and bath with two or three other families. To those crammed into communal flats and forced to schedule meal preparation and bath-taking around the plans of others, an apartment to oneself is a wonderful dream, even if it is small and poorly finished.

In their attempt to alleviate the housing shortage, the Soviets have standardized plans, bath fixtures, kitchen equipment, window units, door frames, and so on in order to mass-produce housing units. The result is a monotonous uniformity of style and construction. Ranks of 9- to 14-story apartments dominate the outskirts of Soviet cities from coast to coast. Soviet citizens could move from Leningrad to Novosibirsk and find themselves back in an identical apartment. The first residents of a new apartment often find that they must attend to details of finishing and decorating themselves. Whole blocks of apartments have been built in the suburbs in advance of neighborhood stores, cinemas, parks, or other facilities. On the other hand, Soviet rents have not increased since 1928! Housing costs in the USSR typically are 5 to 10 percent of family incomes, far less expensive than in Western countries.

Diet is another area in which rank has its privileges. The elite can shop in special stores, not open to everyone. There, prime meats, fresh fruits and vegetables (even out of season), caviar, and imported delicacies are available at subsidized, low prices. For most Soviets, the diet is heavy on bread, potatoes, and cabbage. Bread is very good and quite inexpensive—no one has forgotten that bread riots finally toppled the tsar—and potatoes and cabbages are readily available even in provincial cities, where fruit, vegetables, and meats tend to be more scarce than in Moscow and Leningrad.

*Hedrick Smith, *The Russians* (New York: New York Times Book Co., 1976), p. 75.

Wooden Houses, Siberia. *While modern, uniformly bland concrete and steel structures dominate new industrial centers such as Novosibirsk, rural and small-town Siberia retains the homely charm of wooden houses, which reinforce the perception that this was (and is) a frontier in a harsh environment.*

most of these refugees (and industries) moved to the Volga-Urals area, some 5 to 7 million people went to Siberian destinations. Despite the fact that most returned home, some 10 percent of the voluntary wartime refugees elected to remain.

The final large-scale migration came with the Virgin Lands Campaign of 1958–60. This program plowed and settled almost 100 million acres of steppes in Western Siberia and northern Kazakh SSR.

The new oil and gas fields and the mineral development areas are growing rapidly, but small numbers are involved. One Soviet study showed that over half the new migrants to the cities of eastern Siberia left after three to five years. The lack of housing, limited social life, poor services, and sheer boredom were the main reasons given. Most of Siberia remains an area of acute labor shortage. The recent trends toward capital-intensive projects in Siberia and shipment outward of raw materials have undoubtedly been influenced by this migrational pattern.

Agricultural patterns in Siberia have stabilized in recent years. Most of the steppes area west of the Kuzbas is planted to spring wheat, silage corn, and sunflowers. After a propitious start, crop failure became a recurring phenomenon in the Virgin Lands.

Erosion and depletion of soil fertility through constant cropping without fertilization took their toll. Farms there now use increased amounts of fertilizer, and land is fallowed in a short-cycle form of dry farming. Diversification with livestock is gradually taking place. Bad years are still expected, but drought rarely occurs in the Ukraine and the Virgin Lands at the same time.

Farming in Western Siberia is older, more general in nature, and usually less risky. Normally, about one third of the farmland is in grain and the other two thirds in hay and pasture. Dairying is of great importance, particularly along the northern margins. Following the European pattern, potatoes and hogs are important components of every farm's crop-livestock mix.

East of the Yenisey River, most of the area is more suitable to dairying and potatoes than to grain. Some of the steppes remain unfarmed, given over to grazing where native peoples have resisted sedentary agriculture.

The southern part of the Soviet Far East has great potential. A drainage program for the Amur Valley could almost triple agricultural acreage in this food-short region, but Chinese cooperation is necessary and unlikely. Experimental agriculture is found throughout this region. Fur farming is a lucrative sideline on many area collectives.

Industry has developed along highly specialized lines in each portion of Siberia. Coalfields and transportation centers concentrate most factory production. Primary production, in particular raw material processing, construction materials, and metallurgy, is best developed.

Western Siberia emphasizes oil refining, petrochemicals, wood industries, and synthetics. The new oil and gas fields of Western Siberia are now the chief raw material source.

The eastern district is also diverse but more oriented to primary production; the Kuzbas coalfield, the second largest producing Soviet field, is its fuel source. A dozen major industrial cities rim the edge of the coalfield. Despite the impressive array of industry, however, the Kuzbas is a stagnating economic area that never fully reached its planned scale of iron and steel production.

At the junction of two rail lines is Novosibirsk (New Siberia). It is Siberia's largest city (population 1.4 million) and the eighth largest in the USSR. Like Chicago, it is a large rail center adjacent to a navigable waterway and at the contact zone between a large coal and metallurgical base and a relatively rich agricultural area, compared to the rest of Siberia.

Eastern Siberia, with elementary industry and less well developed agriculture, is decades behind Western Siberia in its scope of development; distance from market centers is the obvious reason, because the area is not resource poor. Coal abounds; the small but productive Minusinsk basin is dwarfed by the almost unexploited Tunguska fields, the largest reserves in the USSR. Thus far, there has been little luck in the search for oil and gas, though rock origin, age, and structure are compatible. The Yenisey has a vast hydroelectric potential that is moderately well developed.

The function envisioned for this relatively remote region is to generate cheap electrical energy to feed into the national grid. What is not exported is to be used in a series of energy-consuming industries that employ little labor: paper and pulp plants, fertilizer plants, electrochemical works, alloy plants, and electroplating and aluminum reduction plants. Extensive forests supply sawmills and wood-working plants producing a third category of exports, all with the aid of great capital investment and little labor input. In this way, the chronic labor shortages will be less of a restraint on economic development.

Norilsk, a mining center in the north, boasts large mines and a smelting complex that obtains gold, silver, copper, nickel, and a dozen other metals from local ores. All goods and metals are shipped in and out over the Arctic Sea route, which is usable for only 30 to 60 days a year. Despite difficulty of access, it is a highly profitable operation because of the richness of the ores.

Fish processing, forestry, mining, oil refining, and paper are the chief industrial constituents in the Soviet Far East. Vladivostok, with over 600,000 people, is the chief port and industrial center; it is a major naval base as well. A new outport has been built at Nakhodka to serve the expanding Japanese trade.

Khabarovsk (population 580,000) is a river port on the Amur where the Trans-Siberian turns southward toward Vladivostok. A connecting railroad also reaches the new BAM line, the Baikal-Amur Mainline Railway. Magadan, a port that is icebound for seven months of the year, is important as the entrepôt for the rich Kolyma goldfields which rival those of South Africa's Rand.

A host of environmental problems limit Siberian development. The Soviets have shown themselves able to deal with most of them, developing special mining, pipeline, and construction technologies to deal with permafrost. High-pressure steam is used to thaw the permafrost long enough to sink 30-foot pilings. Concrete stilts lift buildings off the surface so that cold air can prevent the building's heat from thawing the permafrost and allowing foundations to sag. Storm windows are triple-paned for insulation.

Two problems of development remain distance and lack of local outlet for production. The Soviets are willing to make the largest investments to develop a plant, area, or region; yet it is not the initial cost, but the ongoing transportation costs, that often render endeavors unprofitable. The other great problem is labor supply: not many people are willing to work under such conditions at any price.

THE BAM (BAIKAL-AMUR MAINLINE) RAILWAY

A major new development effort recently completed is the long-planned BAM project (see Figure 4–13). An auxiliary route to the Trans-Siberian mainline, it will also be a developmental entry into the Lena River basin and a route for the export of surplus Siberian raw materials and semimanufactures to Japan. Constructed by the Young Communist League in a widely praised program, it was this generation's "great monument to Communism."

BAM was to alleviate the overburdened Trans-Siberian, which was jammed with long-distance cross-hauls of raw materials, finished goods, food, and supplies. The Soviets intend to use it as a land bridge for containerized freight moving between Japan and Europe. The BAM line is to be an outlet for growing petrochemical production shipped in tank cars; it will provide access to new nickel and copper deposits en route, and it has opened new coalfields whose total production will be exported. The new line has opened up timber reserves and attracted Japanese joint investment capital in coal and timber. The Soviets hope that, ultimately, it will develop the Soviet Far East as a raw material emporium for Japan; BAM also will likely help increase Russian settlement in an area disputed with the Chinese.

Problems of construction were enormous. The line is over 1,900 miles long, and the distance to the coast, with connecting lines, is over 2,500 miles. About a third of the route is constructed over permafrost, which buckles and heaves when vegetation is disturbed. The line must cross five major rivers (including two of the largest and widest in the USSR) and myriad streams, gullies, and gorges. It must also cross or tunnel through a dozen major mountain ranges; 40 tunnels and over 200 bridges and trestles were built. The area involved is earthquake prone and subject to landslides, mudslides, and winter avalanches. Temperature extremes with ranges of over 100° F will continue to affect workers, materials, and equipment.

The project was announced in 1974; the anticipated 1980 completion date was not met until 1983. The projected costs were scrapped, and real total costs may never surface. Communist centralized planning, however, is accustomed to long-range goals with even longer, large payoffs; the BAM exemplifies this kind of thinking.

CONCLUSIONS: WILL GLASNOST SUCCEED?

When Leonid Brezhnev died, late in 1982, he had ruled for 18 years. His next two successors, Andropov and Chernenko, had little time to affect major changes in the direction of the State, its economy, and its international relations. When Mikhail Gorbachev came to power in 1985, little momentum remained in a society and economy that had stagnated under Brezhnev and not been revitalized by Brezhnev's short-lived successors. Brezhnev had made only a few, cautious changes in policy. Substantial alterations in course were not his style.

A popular Soviet joke has Stalin, Khrushchev, and Brezhnev riding in a train. The train stops suddenly, for no apparent reason. The three leaders confer on how to get moving again. Stalin, the author of purges that killed as many as 18 million Soviet citizens, wants the engineer shot. Khrushchev, the remodeler and tinkerer, suggests that the engineer and conductor trade places. Brezhnev wants to reboard the train, pull down the blinds, and pretend that the train is moving. The joke is a capsule of Soviet economic policy direction for six decades: Stalin ruled by terror, Khrushchev attempted economic changes by relatively minor reshuffling of emphases in the economy, and Brezhnev implicitly accepted inefficiency, corruption, and waste. Initiative was discouraged; hard work seldom paid off to the individual worker. Work habits often were slovenly. In fact, the inefficiency of many farm and construction workers had contributed to the rise of the *shabashniks,* workers who made private deals at high prices to perform seasonal farm or construction work. Shabashniks became famous for their productivity—completing construction jobs in half the normal time and growing vegetable crops three or four times previous yields. Eventually, shabashnik ranks came to include skilled technical workers and engineers, who continued the disciplined work team tradition of the seasonal farmworkers who seem to have started the movement. So many of these unofficial work brigades came from the Caucasus republics that they are known jestingly as the "Armenian Construction Company."

The shabashniks worked hard and with good results. What disturbed the government was that these workers produced well without being directed by the bureaucratic apparatus of government ministries, trade unions, managers, and the Communist party. Shabashniks were not motivated by ideology, but by money; they turned in excellent results for excellent pay.

Glasnost—"openness"—has both domestic and international dimensions. Openness in the direction of the economy implies a willingness to explore revolutionary approaches to productivity problems. An early Gorbachev speech proclaimed: "Everyone, management, workers . . . all must . . . receive their pay according to the final product. If you do not get the final product, if profit is not created, then it means you cannot get paid."[1] In orthodox communist philosophy, it is heresy to be money conscious and profit motivated. The profit motive ends the power of centralized planning bureaucrats and stirs resentment within a work force accustomed to being accountable for quantity, but not for quality. The struggle apparent between Gorbachev and the entrenched, conservative bureaucracy will not be resolved easily.

ENDNOTE

1. Mark Frankland, *The Sixth Continent: Russia and the Making of Mikhail Gorbachev* (New York: Harper & Row, 1987), p. 219.

REVIEW QUESTIONS

1. What are the nature and scope of the USSR's "minority problem"?
2. Why is the Soviet-Chinese border tension a threat to the stability of some ethnic republics and autonomous areas?
3. How logical are the kombinats? What was their original intent?
4. In what sense was the Virgin Lands Campaign a calculated risk? Has it paid off for the Soviets?
5. What is the real nature of the supposed Russian drive for ice-free ports?
6. Why are the great Siberian rivers said to flow the "wrong" way? What are the implications of "wrong"?
7. What are the potential environmental problems associated with the Siberian River Reversal Scheme?

8. By what means do the Soviets attempt to control in-migration to Moscow and other attractive cities? Are they successful?
9. Criticize the view, popularized by the Soviets, that they took power in a backward, preindustrial society and then rapidly developed it.
10. Why was the creation of a new capital at St. Petersburg (Leningrad) a classic example of a political-cultural reorientation?
11. Contrast the present economic specializations of Leningrad and Moscow.
12. What qualities of the physical environment enhance the agricultural potentials of the Ukraine?
13. Briefly describe the Soviet version of the Sun Belt. Do its migrational and functional characteristics resemble those of the U.S. Sun Belt?
14. What engineering and environmental problems have been associated with the great dams on the Russian portion of the European Plain?
15. What is the nature of the glasnost policy in the USSR?
16. Contrast the Urals and the Appalachians in terms of mineral resources, effect on transportation, and nature of industrialization.
17. Why is Soviet Central Asia expected to develop faster than the alternative frontier area, Siberia?
18. While the Russian drive eastward to the Pacific can be compared to the American drive westward to the Pacific, what significant advantages did the Americans experience?
19. What are the industrial raw material assets of Siberia?
20. Critically evaluate the potential versus the cost of the Baikal-Amur Mainline Railway. Would such a project likely be undertaken in a capitalist or mixed economy?
21. How has the Soviet world view changed over time?

SUGGESTED READINGS

Cole, John. *Geography of the Soviet Union*. London: Butterworth, 1984.

Demko, George, and Fuchs, Roland, eds. *Geographical Studies of the Soviet Union*. Chicago: University of Chicago Department of Geography, 1984.

Demko, George, and Fuchs, Roland, eds. *Geographical Perspectives on the Soviet Union*. Columbus, OH: Ohio State University Press, 1974.

Dienes, Leslie, and Shabad, Theodore. *The Soviet Energy System: Resource Use and Policies*. New York: Wiley, 1979.

Harris, Chauncy. *Cities of the Soviet Union*. Washington, DC: Association of American Geographers, 1972.

Jackson, W.A. Douglas, ed. *Soviet Resource Management and the Environment*. Columbus, OH: American Association for the Advancement of Slavic Studies, 1978.

Jackson, W.A. Douglas. *The Russo-Chinese Borderlands*. Princeton, NJ: Van Nostrand, 1964.

Kelley, Donald; Stunkel, Kenneth; and Wescott, Richard. *The Economic Superpowers and the Environment: The United States, the Soviet Union, and Japan*. San Francisco: W.H. Freeman, 1976.

Kirby, Stuart. *The Soviet Far East*. London: Macmillan, 1971.

Lydolph, Paul. *Geography of the USSR*. New York: Wiley, 1977.

Mathieson, Raymond. *The Soviet Union: An Economic Geography*. New York: Barnes & Noble, 1975.

Mellor, Roy. *The Soviet Union and Its Geographical Problems*. New York: Macmillan, 1982.

Nove, Alec. *The Soviet Economic System*. London: Allen & Unwin, 1980.

Symons, Leslie, ed. *The Soviet Union: A Systematic Geography*. Totowa, NJ: Barnes & Noble, 1983.

Symons, Leslie. *Russian Agriculture: A Geographic Survey*. New York: Wiley, 1972.

CHAPTER FIVE

Australia-
Oceania

Sydney Harbor.

opular images shape our perceptions, or misperceptions, of the world's regions. A recent, immensely popular film had an Australian hero, and what a hero he was! Bold, self-reliant, highly independent, and individualistic, he was disdainful of pompous authority, and had a nice, wry, self-deprecating humor. A stereotype, to be sure, but like most friendly, nonracist stereotypes, an exaggeration and simplification of truth. But the hero's home environment in Australia was pictured as open, rather wild country with the smallest of small towns handy for a little rough-and-ready rest and recreation. This part of the stereotype was readily believed, but it is anything but typical. Eighty-seven percent of Australians are urban rather than rural; nearly three quarters of the continent's inhabitants live in cities of 100,000 or more, compared to a U.S. overall urban proportion of 74 percent, with 57 percent in cities of 100,000 or more. Australians thus are more likely to live in big cities than are Americans, yet the image of a raw frontier populated by macho, egalitarian ranchers, miners, and crocodile poachers persists. Perhaps the frontier image prevails because Australia and its New Zealand and Pacific island neighbors were "discovered" and brought into the global economy and international political system relatively recently. The Pacific Ocean, twice as big as its nearest rival, the Atlantic, and occupying more area than all the continents combined, was first seen by a European when Balboa crossed the Isthmus of Panama in 1513.

Not until the seventeenth century did Europeans have any direct knowledge of the "South Seas." The western coasts of Australia were so unimpressive to the first Dutch explorers venturing from Indonesia that, perceiving no trade value, the Dutch did not attempt settlement. Until that greatest of Pacific explorers, Captain James Cook, made his three voyages between 1769 and 1778, little was known of the Pacific basin as a whole. The first British colony in Australia dates only to 1788, and as late as the 1840s, most of the Pacific islands were unclaimed. The race for territory among the French, British, Germans,

and to some extent, the Americans, did not intensify until the 1870s. Thus, economic and political interest from outsiders in most of the islands of the Pacific is scarcely a century and a quarter old.

Oceania includes the "island continent" of Australia, the large archipelago of New Zealand, and the island realms of Melanesia, Polynesia, and Micronesia. Some islands, especially Melanesia and New Zealand, are "continental islands," rising from the relatively shallow continental shelf. Others, particularly in tropical Polynesia and Micronesia, are "ocean islands," rising directly from the ocean floor. "High" volcanic islands and "low" coral islands are two common types.

Sheer size and distance contribute to a perception of isolation. The Pacific Ocean is the greatest single geographical feature on the planet (Figure 5–1). At its greatest width, from Panama to the Malay Peninsula, it extends some 12,500 miles; from the Bering Strait between Siberia and Alaska to the Antarctic Circle is a distance of 9,300 miles.

These superlative distances and scales are related to two other measures of relative isolation: the biological uniqueness of the area and the recency of European discovery and colonial activities. Biological evidence indicates isolation over whole geologic eras. Australian native animal life is unlike that of any other continent. Australia has been isolated from Asia for about 25 million years, and New Zealand has been separated some 70 million years.

There is no native representative of the great cats, no indigenous antelope, deer, buffalo, or primates. On the other hand, there are animals unique to Australia. To the four dozen kangaroo species must be added such oddities as the world's only egg-laying mammals, the duck-billed platypus and the spiny anteater. Amphibians and freshwater fish are scarce, but bird species are numerous and varied. New Zealand is so remote from the biological mainstreams that it had no native land mammals save bats, and many of its birds were flightless, though shellfish and sea fish were abundant. Its streams had few edible fish spe-

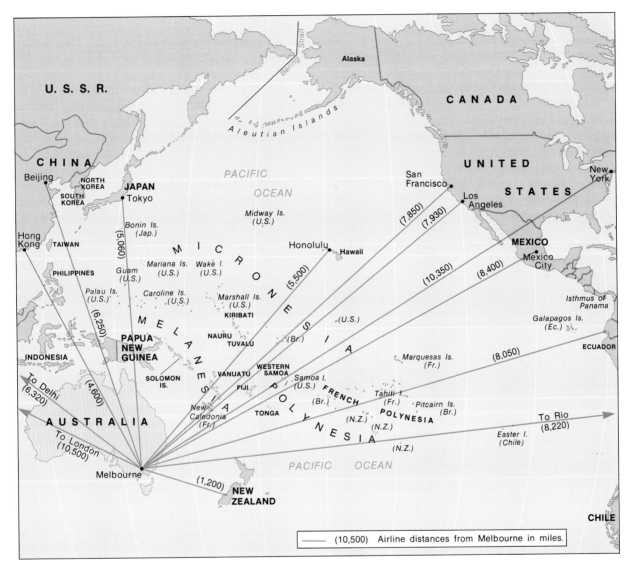

FIGURE 5–1
Australia, New Zealand, and Oceania in Their Pacific Basin Setting. Important to an understanding of the geographic problems and potentials of this region are concepts of situation—geographic location *relative* to other regions.

cies, except eels, which spend part of their life-cycle in the ocean and thus can migrate from one land area to another.

TRENDS AND PERCEPTIONS

Isolation, cultural distinctiveness, and relatively sparse population have been dominant features of this region's self-image. The region historically has had rather distant relationships with Asian neighbors in its "Near North" and close ties with distant Western, and especially British, "cousins." For most of their history, this region's two largest nations have had limited trade relationships with their neighbors. Immigration policy long favored Britons and Irish, with exclusion of Asians. But neither the so-called white Australia policy or New Zealand attitudes were ever totally exclusionary; about 5 percent of new Australians come from Asia and Africa.

The recent, small-scale relaxation of this white Australia policy is a gesture of friendly relations toward Asia more than a major shift in policy. Asian population pressure could not be significantly relieved by even a wide-open immigration policy. The image of an "empty" Australia, temptingly close to a teeming Asia, is not realistic, environmentally, if it is sometimes politically attractive within Asia. Australia has quite limited water resources. New Zealand's population, once ethnically almost homogeneous aside from the Maori, has been modified by an influx of Pacific islanders and Asians. Before 1945, Australia admitted relatively few immigrants who were not British; since then, it has been admitting, and even recruiting, new Australians, usually non-British Europeans.

Militarily, the primacy of Australia and New Zealand's links with Britain ended early in World War II. Japanese capture of Britain's main Southeast Asian naval base at Singapore and the Japanese invasion of New Guinea and the Solomon Islands made it clear that Britain could not defend its partners in the Pacific. Almost overnight, the United States became the strongest ally in the defense of Australia and New Zealand. The alliance of Australia, New Zealand, and the United States has been close until recently, when adoption of a policy of excluding nuclear weapons led to a U.S.-New Zealand dispute. The inability of a faltering British postwar economy to maintain high rates of investment in its Commonwealth partners, and growing investment from the United States and Japan, especially in Australia, helped seal the shift in orientation from Britain to the Pacific area, including the United States.

The relationship of this region with Japan is complex and controversial. Japan has developed a kind of economic colonial relationship with Australia. Australia's natural resources have been increasingly developed with the involvement of Japanese capital and technology. Many of these resources will go to Japanese markets. Manufactured goods from Japan, South Korea, and Hong Kong are common throughout the region. New Zealand's agricultural surpluses and Australia's mineral resources and agricultural surpluses appear to make them natural trading partners with Japan. Lingering memories of World War II have faded, and a prospering, democratic Japan is definitely in the interests of Australia-Oceania. The People's Republic of China is now the largest customer for Australian wheat. The United States supplies 22 percent of Australia's imports. Other East and Southeast Asian countries are also becoming more important trading partners, and the Middle East nations now purchase large quantities of mutton from Australia and New Zealand.

The balanced budget and low external debt of Papua New Guinea (PNG) is made possible largely by the generosity of Australia and New Zealand, which provide foreign aid to the sum of 40 percent of PNG's budget. Australia considers it essential that this nearest neighbor be a friendly and democratic country and, with New Zealand, supports the budgets of many island states.

INDIGENOUS POPULATIONS

The indigenous peoples and cultures of Oceania are commonly divided according to the four geographic regions of Australia, Melanesia, Micronesia, and Polynesia. Though geographic boundaries do not fit ethnic distributions exactly, Australia is the home of the dark-skinned Australoid ("southern") Aborigines; Melanesia, extending through New Guinea to Fiji, is the home of the black-skinned islanders. Some authorities now view these Australian and Melanesian peoples as closely related. The Micronesians and Polynesians have lighter, brown skins. The former inhabit the "tiny islands" north of Melanesia, and the latter occupy the "many islands" of the vast Polynesian Triangle extending from Hawaii to New Zealand and eastward to Easter Island.

Aborigines probably reached Australia at least 35,000 years ago, at a time when continental glaciers elsewhere had absorbed much water as ice and lowered sea levels, so that the Southeast Asian mainland extended to Borneo and Bali. Only narrow seas needed to be crossed to New Guinea and Australia, which were then joined.

About 300,000 Aborigines once had Australia's nearly 3 million square miles to themselves, an average population density of only 1 person to 10 square miles. Over 16 million people now occupy the continent, representing an overall population increase of 53 times the pre-European colonization figure. If Australia is a relatively empty continent now, it was historically "empty" also. Australia was a hard place in which to make a living for a nonagricultural people with an Old Stone Age technology. The Aborigines were hunters and gatherers, and the largely desert continent provided few edible plants to gather and a relatively sparse population of animals desirable for hunting. A nomadic life was necessary, with frequent moves to avoid overuse of the limited food resources of any one locale. With no draft animals available, the Aborigines' material culture was necessarily quite sparse—they had to be able to carry all their possessions. Wooden spears, the ingenious boomerang, stone clubs, wooden bowls, and string bags were the basic possessions. Clothing was seldom worn, and shelters were temporary lean-tos of brush and clay. Staying alive in a harsh environment required considerable skill and cooperation. An elaborate system of social obligations developed. What the Aborigines lacked in material possessions, they made up for in a fantastically rich and imaginative stock of folklore and ritual. They revered natural objects, such as rock pinnacles, which they regarded as personifications of godlike creatures that existed during the "dream-time," their creation myth. The Aborigines expressed themselves artistically in rock carvings, rock paintings, finely carved weapons and utensils, and painted shields.

As on other continents in other times, the technologically superior invaders, in this case the Europeans, quickly took the lands with the best agricultural potential. The Aborigines were displaced to the dri-est, most rugged and remote lands of the "center," where survival was exceptionally difficult, or ended up on the fringes, literally, of the white society. Most Aborigines have been detribalized and increasingly migrate to the urban centers of the south, although some reservations continue in the Northern Territory. They are entitled to government assistance and many have become dependent, some lapsing into alcoholism. Changing Australian policy toward the aborigines remains controversial.

At about the same time that Australia's earliest residents reached that continent, dark-skinned groups seem to have moved into New Guinea and the island constellations southeast of New Guinea. The Melanesian island chain was occupied next, with Micronesia and Polynesia being colonized later. Though there are uncertainties as to origins and undoubted mixtures, the Melanesians have both short-statured pygmy groups and taller peoples; their different languages and cultures form one of the most complicated mosaics on earth. The people of Micronesia also show strong variations, with Asian blood and cultural influences evident in the western islands, and Melanesian associations in the eastern islands. The tall, brown-skinned, wavy-haired Polynesians seem to be essentially mongoloids from Southeast Asia. Some authorities now believe that a distinctively Polynesian culture was developed only after Fiji, Tonga, and Samoa were occupied after 100 B.C. Between approximately A.D. 300 and A.D. 800, the vast Polynesian Triangle was occupied by a still-disputed blend of accidental and deliberate voyaging. Polynesians thus came to occupy the largest area measured in distances from one island to another. Considering Polynesia's far-flung island world, with some occupied island specks 7,500 miles distant from other outposts, Polynesian culture is surprisingly uniform. This uniformity is consistent with archaeologists' findings that the Polynesians probably completed their penetration of the Pacific basin relatively recently. This would establish Polynesia as by far the newest human habitat in the world.

To complicate this cultural and ethnic mosaic further, there is tantalizing, if inconclusive, evidence of pre-European-era contacts between Polynesians and South America. For example, the American sweet potato is widely distributed throughout Polynesia. Because the Polynesians managed to cross great ocean distances between sometimes minute islands, with-

AUSTRALIA: EUROPEAN ORIGINS AND
THE ASIAN-PACIFIC NEIGHBORHOOD

Australia, like America, is essentially a neo-European nation, historically and culturally linked with Europe. But in defense, it now looks more to the United States than to the United Kingdom as its most essential ally, and views its responsibilities as primarily in the Asia-Pacific area. New Zealand, also conscious of its Pacific setting, has sought to compensate for the relative decline in its British trade by diversifying its trade in the Pacific and with Japan and Australia. The island-states, now mostly independent in the political but not necessarily in the economic sense, are also seeking to diversify their economic and political relationships.

Australia exports about 14 percent of its gross domestic product (GDP), compared to 7 percent for the United States, 27 percent for Sweden and West Germany, and 24 percent for Canada. The traditional Australian export commodities were wool, grain, sugar, and meat, products characterized by steep competition. In the early 1950s, wool alone accounted for half of Australian exports by value, with other agricultural commodities forming another 30 percent. By the early 1980s, wool accounted for less than 10 percent, with other farm and ranch products totaling 42 percent. By 1981, minerals were 36 percent and manufactured goods 20 percent of total exports.

Not only did the product mix in export trade change, but there was a major shift in trading partners. The agricultural exports, which were traditionally absorbed by the British market, were being replaced by minerals and manufactures, which were mostly shipped elsewhere. In 1950,

out compasses and relying on the stars and observation of ocean wave patterns and water color and temperature characteristics, they rank among the world's greatest navigators.

As Westerners developed their economic interests in Australia-Oceania, they began to change ethnic and racial patterns in the islands. The newly established Australian and New Zealand colonists pressed the reluctant British government to limit the advance of other powers into their neighborhood, by annexing islands, if necessary. Economic development created demands for labor. Polynesians were recruited, often by kidnapping, to work in the mines of Peru and Chile, and Melanesians (Kanakas) were taken for Queensland's sugar fields. When the American Civil War reduced cotton exports to Britain, British settlers introduced hard-working Indian labor to their cotton and later sugar plantations in Fiji, as they also did in the South African province of Natal and in Guyana and Trinidad in the Caribbean region. French New Caledonia drew significant numbers of Vietnamese, Japanese, and Javanese to its mines and farms, and Chinese laborers were introduced to Tahiti. Australia, concerned about the proximity of densely populated lands and the troubles of plural societies elsewhere, sent home most of Queensland's Melanesians by 1900, and substantial numbers of introduced laborers were repatriated from New Caledonia, Tahiti, and Western Samoa. However, the demographic composition of the populations of Hawaii, New Cale-

36 percent of Australia's exports went to the UK, and another 22 percent to other Western European countries. Australia's imports were similarly dependent on the UK, which supplied 47 percent of all imports. This trade interaction with the UK and other European states later was handicapped by three factors: sheer distance, slow growth in the economy of the UK, and the formation of the European Economic Community (EC) and Britain's eventual membership in it.

Japan ranks as the major customer for Australia's exports, taking 27 percent of the total, while the UK accounts for under 5 percent. Japan ranks second as a supplier of Australian imports, following the United States. Other Pacific basin countries receive 16 percent of Australian exports and supply 12 percent of imports.*

As a major exporter of minerals (iron ore, bauxite, and coal are important), Australia is a natural partner to Japan, the closest major industrial nation and one with a notable reliance on imported industrial raw materials as well as imported food. American and Japanese capital has long since outweighed British investments among major foreign sources of development capital. American tourism to Australia has increased considerably, partly in response to a perception of terroristic activity in the Mediterranean region.

*This analysis is based on data published in Richard Caves and Lawrence Krause, eds., *The Australian Economy: A View from the North* (Washington, DC: Brookings Institution, 1984).

---

donia, and Fiji were substantially changed. In 1987, Fiji's government, elected mainly by the Indian-Fijians who form 49 percent of the population, was overthrown by a military coup backed by Melanesian-Fijians, who form 47 percent of the population.

THE DISTINCTIVENESS OF THE REGION

There were many reasons for colonization efforts in Australia-Oceania. Some areas were annexed to pre-empt take-overs by competing powers or, conversely, to bring law and order to disorderly fron-

tiers. Potential mineral wealth was the motivation in others, while some were sought to supply tropical commodities such as sugar, rubber, or palm oil. Still others were viewed as settlement colonies that could become **mirror images** of the home country—transplanted populations occupying similar landscapes and, in fact, reproducing the cultural landscapes of home. Commonly, colonies reflected some intermixture of these motives. New Zealand and the cooler, more humid portions of Australia seemed to offer preeminent possibilities for mirror image colonies, though the great distances involved tempered their attractiveness. The discovery of gold in the 1850s, however, added to an already growing migration from Britain in search of farmlands.

The small number of aboriginals (native Australians) and their relatively light impact on the landscape reinforced European perceptions of an empty land, a fresh canvas on which a British society could be reproduced following the end of the penal colony concept. The Polynesian people of New Zealand, the Maori, were cultivators, considerably further advanced than Australia's hunting and gathering aboriginals. Relative development levels and population densities made a substantial difference to British official policies toward native peoples. The Maori were more readily drawn into national participation than Australian Aborigines.

ECONOMIC CHARACTERISTICS

Among world regions, Australia-Oceania has the lowest ratio of population to area and resources. The continent of Australia has but 5 people per square mile; New Zealand, 31 per square mile; and Papua New Guinea, 20 per square mile (Figure 5–2). In contrast, Asia has an overall average of 150 per square mile, and Bangladesh has a density of 1,530 per square mile.

The combination of overall low density and great distance from important trading partners (Japan is 4,300 miles by sea from Sydney; California is over 6,000 miles, and the United Kingdom over 12,000 miles, away) has posed many problems for the region. Domestic markets are small and widely dispersed, and industrial production for export has faced two fundamental obstacles: distance from most markets and the much lower labor costs of the region's closest neighbors. New Zealand, for example, cannot hope to sell most low-technology or low-cost manufactured goods at competitive prices in Southeast Asia, though its sales in the Pacific islands seem promising. An alternative type of industrialization—**import replacement** for the local market—is more logical, but here the small population is a handicap. Import replacement attempts to create local manufactures to replace former reliance on imports. Expanded production could even generate exports. Thus, the two developed, technologically advanced states of the region attained and maintain their high standards of living mainly by exporting minerals and the products of technologi-

cally sophisticated, large-scale cultivation and especially pastoral farming.

Both Australia and New Zealand are characterized by development along their coastal fringes. The interior farmlands of Australia, indeed even some of the fringe regions, function as a traditional economic hinterland (see Figure 5–2). Ores, coal, wheat, wool, hides, meat, and refined metals are the major exports, reflecting the hinterland supply nature of the economies. Meat, dairy products, fruit, wool, timber, and wood pulp are exported from New Zealand in large quantities. In short, they are supply foods and raw and semiprocessed goods for more processing in populous and diverse economic heartlands elsewhere. Japan, Europe, the United States, the Middle East, and the Pacific basin countries are the major customers.

Tropical food crops, tobacco, and sugar have long been grown in Australia. Some of these commodities, until recently imported from Commonwealth partners in Southeast Asia, are becoming exports as Australian tropical agriculture diversifies. Petroleum and aluminum refining and chemical industries are expanding, again replacing imports with exports, not just developing import replacements. Ferrous and nonferrous metallurgy is following this trend, along with machinery and transportation equipment, areas where skill and mechanization may counterbalance relatively high labor costs. As they develop, the economies of Australia and New Zealand will likely become more integrated, and expanding links with other parts of the world are increasing the diversity of sea connections.

THE PHYSICAL FRAME

Australia, the smallest of the continents, is geologically one of the oldest landmasses, part of ancient Pangaea (see Chapter 1). It has a fairly compact form with a relatively regular coastline as compared to Europe. For millions of years, Australia has been geologically stable. Weathering and erosion have been at work without the counterbalancing force of uplift for so long that there are few truly spectacular landforms. Only on the east coast of Australia and at a few isolated interior locations have mountain-building activities occurred within the recent geological past

FIGURE 5–2
Selected Place Locations and Population Distribution in Australia-Oceania. Most of Australia's population is concentrated near the southern and southeastern coasts, leaving the interior almost empty of people. New Zealanders concentrate on North Island and the northern half of South Island.

(Figure 5–3). In contrast, New Zealand, New Guinea, and most Pacific islands are much more recent and rugged, if less complex, in character.

Topography

Most of Australia is a series of plateaus interspersed with lowlands. These lowlands include some low hill ranges but are generally sedimentary basins covered by recent debris deposited by water and wind. Extensive sandy or gravel deserts occupy much of this land. Dry lakes, occasionally filled with shallow water from thunderstorms, are the foci of internal drainage networks. The great Western Plateau occupies about half the continent; most of it is above 1,200 feet, with relatively small mountain ranges in the center and western portions. A few spectacular landform features, such as Ayers Rock, project as isolated rock formations above the general plateau level. In South Australia, a small area of faulted mountains with deep, narrow valleys and sea-filled gulfs creates an indented coastline. West of this area, along the Great Australian Bight, is an unusual desert landform, the Nullarbor ("no trees") Plain. Underground limestone drainage systems here ensure that practically none of the precipitation received stays anywhere near the surface. The eastern highlands, about one sixth of the total area, parallel the coast for nearly 2,500 miles. Relatively low in elevation—the highest peak, Mt. Kosciusko, reaches only 7,328 feet—the mountains reach their greatest heights in New South Wales and Victoria, sometimes plunging steeply toward the sea.

New Zealand is formed in part by a giant series of earth crust faults or fractures along which blocks of the earth's crust have been moved and thrust. The resulting landscape includes clusters of volcanoes on North Island, where ancient eruptions showered pumice ash on the surface. The immense ocean basin is bounded by the Pacific **Ring of Fire**, a titanic zone of frequent earthquakes, active mountain-building forces, and vulcanism. This area corresponds with the marginal zone where the Pacific plate meets other tectonic plates.

Along with New Zealand's North and South islands, the great islands of New Guinea, Fiji, the Solomons, New Caledonia, and Bougainville are continental islands—that is, they are geologically complex and related to the continental landmasses. Most continental islands are much larger than the oceanic islands—those not related to the continents but to the ocean floor in geologic origin.

Among the oceanic islands of the Pacific, the high-profile, steeply mountainous "high islands"—such as the Hawaiian Islands and Tahiti—are volcanic in origin. Other island types include uplifted coral platforms (e.g., Nauru and Tongatapu) and the "low islands," very low profile, commonly ringlike islands or atolls formed by coral reefs usually built upon some submerging volcanic base (Figure 5–4). The atolls are the most numerous but not usually the most populous or productive. Polyps that build the coral reefs through accumulation of billions of tiny lime exoskeletons are highly sensitive to temperature; living coral exists only in tropical seas. They do not tolerate dirty or cloudy water, and they cannot live more than 20 to 30 feet below the surface because they need ocean waves to bring them food. Wave abrasion of the coral builds up a sandy cover. The atolls usually have few species of plants or animals; only species capable of traveling as seeds in salt water, arriving by air (seabirds), floating on driftwood (insects, lizards, crustaceans), or that were transported by people have made the journey. The coconut palm, for example, is nearly ubiquitous along the atolls and beaches of the tropical Pacific because the coconut, a valued food, was carried by people. Coconuts may also have floated some distance and made accidental landings before salt water destroyed their viability.

Explorers and whalers from Europe and North America contributed their exotic plants and animals to the oceanic islands. European rats were an accidental contribution—a disastrous one for both native animals and plants in that rats are damaging coconut palms on many islands. Frequently, discovery by Westerners would mean the deliberate introduction of goats, sheep, or cattle. The animals usually were gifts to local people (who themselves already had introduced pigs, dogs, chickens, and edible plants), but even uninhabited islands were seeded with some domestic animals as ensurance against starvation should future crews be shipwrecked there.

Land-Sea Relationships

The parts of Australia most desirable to Europeans were (and remain) the better-watered and usually more temperate climatic areas—the southwestern and southeastern coasts. Due to geographic condi-

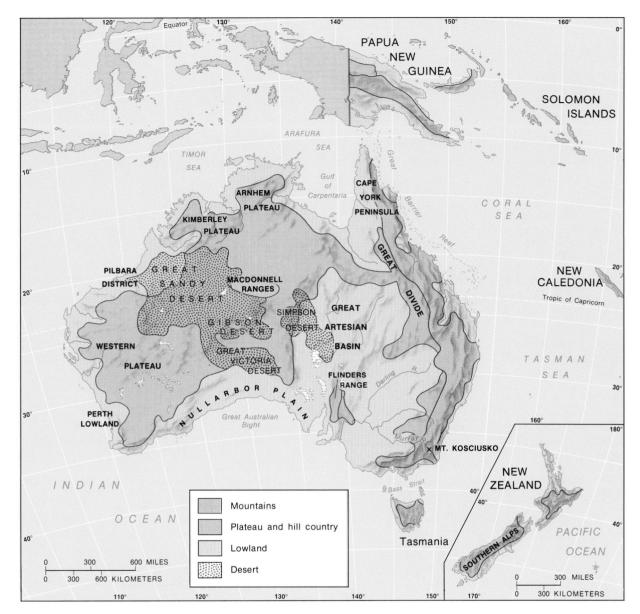

FIGURE 5–3
Topography of Australia-Oceania. This map depicts major landforms of Australia, New Zealand, and Oceania.

tions in Australia, initial settlement coincidentally took place at the greatest distance from northwestern Europe. The early sea routes back to Britain were of formidable length (12,000 miles from Melbourne to London via Cape Town before the Suez Canal, and over 11,000 via the canal—almost half the earth's circumference), so that seaborne trade back to home would be almost the greatest possible distance. (New Zealand is another 1,200 miles from southeastern Australia—not exactly a close neighbor.) Even by more direct air routes, London is 10,500 miles from Melbourne. Although many other regions of the world are physically closer to Australia and New Zealand, Britain remained economically, culturally, and politically closest, at least until World War II. Consequently, Australians and New Zealanders per-

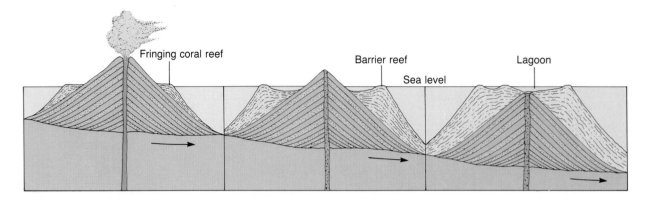

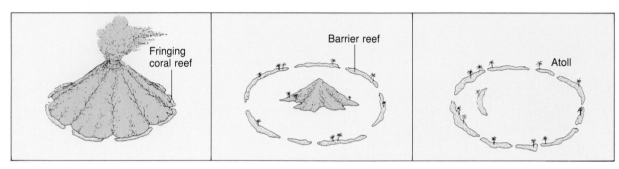

FIGURE 5–4
Formation of Coral Atolls. These cross-sectional views and oblique perspectives illustrate both the sequence of coral atoll formation and three types of Pacific islands. (From *Earth Science*, 5th ed., by Edward J. Tarbuck and Frederick K. Lutgens, Copyright © 1988 Merrill Publishing Co., Columbus, Ohio. Illustration by Dennis Tasa, Tasa Graphic Arts, Inc.)

ceived themselves as being at the end of the world's longest imperial lifeline. During some periods, the costly migration from Britain to Australia or New Zealand was heavily subsidized by the governments concerned.

The nations of Australia-Oceania are heavily dependent on overseas trade, and most of their major cities are ports, but they have not established merchant marines or navies that are large by world standards. In Australia and New Zealand, domestic coastal shipping is well established, although its role has been shaped by strong competition from (and integration with) efficient road, rail, and air services; bulky, heavy cargoes are characteristic in coastal shipping. However, the overseas role of locally based shipping lines, though long linking the region with Europe, remains modest. Exports and imports are handled largely by contract arrangements with inter-

national shipping companies. Australia and New Zealand subsidize shipping services to remote Pacific islands where cargoes are too small to pay. And though the islands are also served by schooners and small native craft which can navigate reef passes and enter lagoons, the island-states of the South Pacific Forum are jointly supporting the Forum lines to encourage regional enterprise and modernization.

Nevertheless, vast distances and especially travel times between scattered islands or from this region to other lands favor contact, trade, and travel by air rather than by sea. Only a few islands have international airports or seaports, limiting interisland links.

Climate

Except for the landmass of Australia and island areas of New Zealand and New Guinea, the influence of the sea is dominant, and both temperate and tropical marine climates are characteristic. Even islands such as the Bonins south of Japan, Midway Island northwest of Hawaii, and Easter Island, although technically poleward of the tropics, are tropical in climate because of warm Pacific waters and winds. Almost all the small islands of the Pacific basin, except those of the extreme north, such as the Aleutians, and cooler south, are clearly tropical in temperature, though some are dry or liable to drought.

The prevalence of warmth and moisture means that the high islands get heavy rain on their windward sides, and even the tiniest islands may be well watered by updrafts. Air is heated faster over the land than over water, expanding and cooling as it rises. Vertical development clouds are formed, and precipitation commonly follows. Even on small atolls that do not develop such strong updrafts, tropical storms and the pervasive moisture in the air may support a good vegetation cover. Thunderstorms are common, as are seasonal tropical depressions called typhoons in the Pacific (hurricanes in the Caribbean).

In the intertropical zone, strong updrafts are triggered by relatively high surface temperatures; air heated near the surface expands, rises, and cools, with resulting condensation and precipitation. This primarily vertical movement, lifting air off the surface, creates the intertropical convergence zone. The rising, cooling air circulating through the atmosphere forms a giant convection cell, sinking back toward the surface. The general zones of subsiding air sinking back to the surface give two more areas where vertical movement prevails, one in each hemisphere. All of these pressure and wind systems migrate seasonally, following, with a time lag of a month or so, the latitudinal shift in the location of the near-vertical rays of the sun. Areas of subsiding, warming air form cells rather than continuous belts of high pressure. Surface winds circulating from the subtropical high pressures toward the equatorial low are so strong and dependable that early sailors termed them the **trade winds**. The trade winds dominate the tropical ocean world of most Pacific islanders, though monsoons become important in Melanesia and Micronesia.

The climate of the Australian continent can best be described as arid, with an incomplete fringe of more humid climates (Figure 5–5). In Australia's winter (June through August), though, the southern fringe gets moisture from the westerlies. High pressure prevails over northern and central areas, and the resulting winds move primarily outward from the dry landmass toward the northern coasts. Descending, warming air moving over the dry land does not acquire much moisture for subsequent precipitation over that landmass. In summer (December through February), a low pressure replaces high pressure over northern and central Australia, drawing air onto the northern and northeastern coasts from the surrounding sea. This flow of moisture-laden air is of little benefit to the arid interior, however, because it penetrates only the coastal zone in the northwest, and in the northeast, the highlands concentrate precipitation. The interior falls in a rainshadow similar to that caused by the Sierra Nevada and Cascades for North America's intermontane region.

Australia's far northern coasts, nevertheless, experience the effect of the great monsoon atmospheric circulation. In summer, moist, northerly winds come in from the warm seas, providing a wet season. The winter in this region is much drier, because the monsoon circulation of air, with its seasonal reversal, is then predominantly offshore and draws in desert air.

Australia's east coast has adequate precipitation year-round, with only occasional droughts. The northern sector is influenced by monsoonal circulation and, hence, has a summer maximum; the southern east coast has a winter rainfall maximum which becomes more pronounced in the neighboring area of truly Mediterranean climate to the west. There are two areas of Mediterranean climate, in the regions centering on the cities of Perth and Adelaide. There, dry summers with bright sunshine are followed by mild, rainier winters. The seasonal shifts of subtropical high-pressure cells, subpolar low-pressure cells, and westerly winds account for this sequence.

New Zealand's climate is more influenced by the prevailing westerlies sweeping over ocean surfaces, which are relatively warm for these latitudes. This marine west coast climate, with its rainy, cool summers and mild winters, is shared by Tasmania. The rain-bearing oceanic air masses tend to influence the windward (west) coast more than the rainshadow areas of the east.

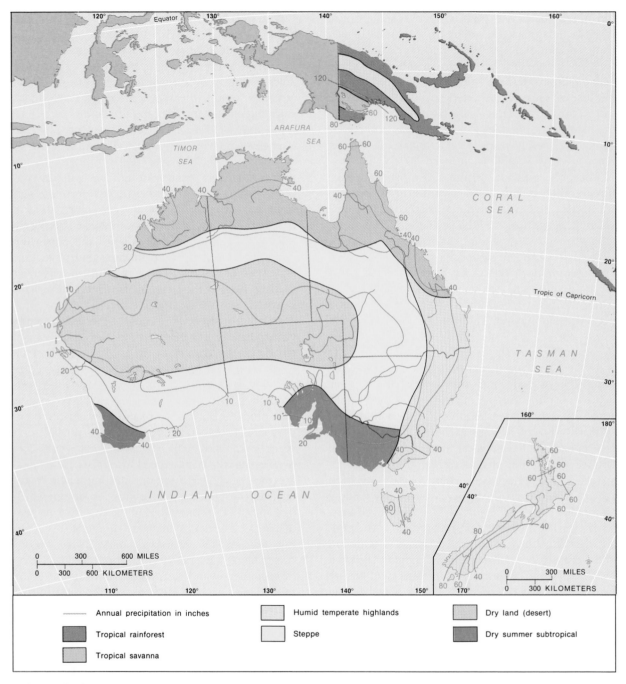

FIGURE 5–5
Vegetation Regions and Precipitation in Australia-Oceania. This map
shows the close relationship of natural vegetation and annual precipitation. Note
how much of Australia receives less than 20 inches of precipitation a year.

Drainage Patterns

There is only one major inland river system in Australia, the Murray-Darling complex of the southeast, which drains most of the states of New South Wales and Victoria as well as parts of Queensland and South Australia. The rest of the streams are either short, swift torrents running from the eastern highlands to the sea or seasonal streams of modest and erratic flow in the monsoonal and Mediterranean littorals.

Inland Australia is characterized by interior drainage. Intermittent streams dead-end in salt lakes or simply disappear into the sands. The combination of high mountain ranges and onshore, moisture-laden winds found in every other continent does not characterize Australia. Nearly three quarters of the surface lacks permanent streams.

Winding, mature, swamp-fringed, and sluggish streams dominate much of the lowlands in New Guinea but are rapid in the mountain core. Short, swift streams with steep gradients also prevail in New Zealand's mountains.

Australia's Underground Water Resources

In contrast to surface drainage, vast underground reservoirs of groundwater occur in Australia. The largest reservoir underlies the Great Artesian Basin in the northeastern interior. Seven other smaller, but important, artesian basins intermittently fringe the coast from western Victoria to northern Queensland. But reserves are limited and overutilized, and much of the western interior lacks both surface and subterranean water.

In some places, water is as deep as 7,000 feet below the surface and is quite hot when it first surfaces. Unfortunately, the bulk of this subsurface water supply is of very poor quality because of high levels of dissolved minerals.

Vegetation

Scattered areas of tropical rainforest are found in the coastal lowlands and adjacent slopes of Queensland. The northern fringes of Australia and parts of the Queensland interior are drier, with a mixture of savanna and straggly "monsoon" forest.

Eucalyptus and acacia varieties are most characteristic, however, prevailing in Australia's temperate and tropical Mediterranean zones. Many native broadleaf evergreens, with adaptations for seasonal drought, appear in Mediterranean climate areas. Low grasses ring the deserts of the interior. Where the natural vegetation is reduced to desert forms or semiarid grass which are found mostly in the transition between the desert and the northern, and southern woodlands, it is likely to be shrub.

New Zealand's native vegetation is very different, with lush, warm rainforests generally characteristic. The drier eastern side of South Island was covered with bunch grass and shrubby scrub-covered areas existed in both islands. Drier Pacific islands have dry, open, often evergreen forests, but wetter areas, including much of the large island of New Guinea, have a cover of luxuriant rainforests, though grassy savannas prevail in rainshadow areas.

Soils

Because many Pacific islands are volcanic basalt, they can have rich soil. Coral islands, however, are notoriously infertile and often lack fresh water as well. The northern extremes of Australia, parts of the islands of New Guinea, and the rest of Melanesia and elsewhere have the leached, largely infertile soils characteristic of much of the tropics.

Wetter subtropical areas of Australia and New Zealand develop soils of only moderate fertility unless alluvial deposition or volcanic basalt enriches them. They are similar in color, structure, and productivity to the soils of the southeastern United States. Some grassland areas of interior Australia are dominated by soils like those of the southern Great Plains. They yield well if supplied with sufficient water. The desert soils, often gravelly, are of varying fertility, with minerals sometimes present in salty concentrations.

THE STRUCTURE OF SUBREGIONS

The sheer size of Australia and the intrinsic climatic and economic differences between its developed, largely temperate fringe and its desert core necessitate recognition of two broad subregions. New Zealand and Papua New Guinea are sufficiently large (and different) to be treated as units. And, at this scale, their small size and similarity allow the thousands of Pacific islands, scattered across that ocean basin, to be grouped into one subregional unit.

Australia's Dry Interior

The driest of the inhabited continents, Australia has a sparsely populated arid center. Almost all of this vast interior region averages fewer than two persons per square mile; extensive areas are uninhabited. The aridity of the interior was especially forbidding to people whose economy depended on large domestic livestock, because forage and water were scarce.

Less than 15 inches of precipitation a year falls over most of the interior. High temperatures mean high potential evaporation. Average summer temperatures exceed 86° F everywhere, and even winter maxima are over 77° F. Where the climate is transitional from monsoon to desert, cattle raising has long been the dominant land use. As in other tropical areas, Brahman cattle have been increasingly cross-bred with European varieties. Beef is shipped to the United States and Japan as well as to domestic markets.

Where the climate is transitional from desert to subhumid, in the south and southeast, sheep for wool rather than for meat are the important agricultural product. Sales to Britain have stagnated, but rising incomes and increasing urbanization in the

Sheep in the "Outback". *Although sheep remain an important product of Australia, they no longer dominate exports in an economy now more dependent on mineral exploitation.*

THE HAZARDS OF THE PIONEER FRINGE: FARMING THE DESERT-HUMID TRANSITION ZONE

The pioneer fringe was a phenomenon of Australian settlement during the 1870s through the 1890s, just as it was in Canada and the United States. In each nation, the subhumid, semiarid transitions were being tested for wheat growing and other potential uses. Around the globe, similar dry grasslands, formerly regarded as viable only for extensive grazing economies, were being reevaluated as wheat fields. Relatively little was known about them, and, as noted in the discussion on the United States and Canada, the cyclic nature of precipitation in these semiarid lands was little understood. Much of the experimentation on these potential wheat lands was empirical—plow up the natural vegetation, plant wheat, and either harvest a successful crop or abandon the farm after a few unsuccessful attempts. South Australia has a Mediterranean-like climate of mild, wet winters and hot, dry summers. It quickly became apparent, though, that aridity increased sharply away from the coast in the vicinity of Adelaide. Colonization expanded rapidly in the 1870s. A cyclical advance-and-retreat movement of people occurred (even in this supposedly carefully planned and orderly colonization) due to bitter experience with variable rainfall patterns. Wheat acreages leveled off in the mid-1880s, and two decades of net out-migration followed one of heavy immigration. What was achieved, finally, outpaced the setbacks; the agricultural frontier had advanced 150 miles since 1870, and 2 million acres had come under cultivation, doubling the colony's settled area.* What was learned was just as valuable as what was produced; this advance of a farming frontier into the "margins of the good earth" must be done cautiously and with adequate experimentation with new technology, new techniques, or new seed. Today, that frontier is as much in the laboratory as in the field.

*Donald Meinig, *On the Margins of the Good Earth* (Chicago: Rand McNally, 1962), p. 203.

---

oil-producing states of the Middle East have created a large new export market for lambs and sheep from both Australia and New Zealand.

Much of the interior, however, remains unused, and probably unusable, for agriculture; the "empty heart" is likely to remain so (Figure 5–6).

Most of the Dry Interior of Australia is unlikely to experience revolutionary change in land use. The margins may be areas of minor expansion, but the great bulk will remain as almost empty land except for the spotty pattern of mining towns and isolated ranches. Large herds of sheep and cattle graze on vast acreages with a low carrying capacity.

Australia is rich in its variety and quantity of mineral resources (see Figure 5–6). Gold began the mineral rushes in the interior. Eastern fields opened up in 1851; western discoveries stimulated a second boom in the 1880s. In the decade following the discovery of gold, Australia's population almost tripled. Australia continues to rank fifth in world production;

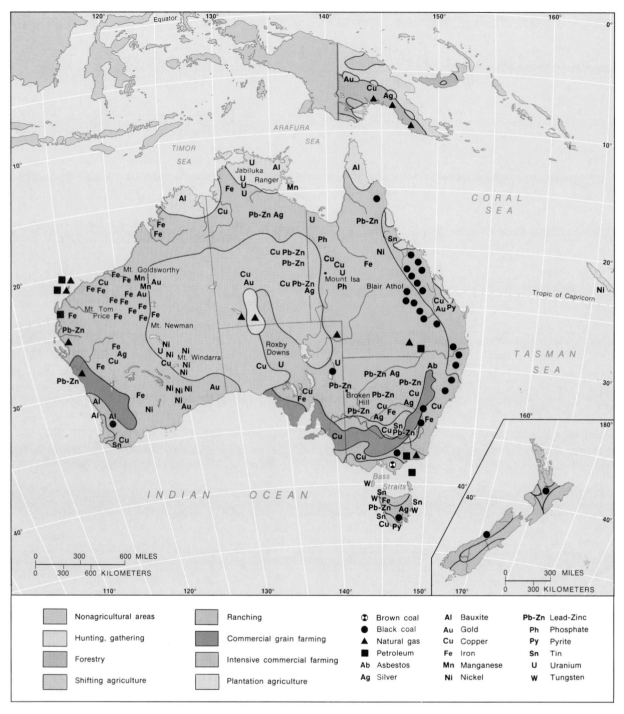

FIGURE 5–6
Agricultural Patterns and Mineral Resources of Australia-Oceania. This map portrays generalized agricultural regions and depicts mineral resources. Note the correlation of patterns of agriculture on this map with precipitation patterns, Figure 5–5.

most of the gold now is produced at the Kalgoorlie-Coolgardie area in Western Australia's interior. Broken Hill, in the interior of New South Wales, is a world-ranking mining center for lead, zinc, and silver, and Mount Isa in northern interior Queensland is a major mining center for copper, as well as for silver, lead, and zinc.

Iron ore has been mined for some time near the coast of Western Australia and in South Australia. One of the world's greatest concentrations of high-quality iron ore came into production in the 1960s, in the Pilbara region of Western Australia. Most of this ore is exported to Japan. Discoveries of these iron ores in large, rich deposits have given a new dimension to the importance of mining in the interior. Australia's iron ore production has soared since World War II; Australia is now the third largest iron ore producer in the world (after the USSR and Brazil) and the second most important iron ore exporter.

Australia discovered its first major oil field in southeastern Australia (Bass Strait), offshore, in 1964, at the same time that major new gas fields were found in central Australia. Oil is being pumped quickly from a modest reserve, and Australia will again be dependent on foreign oil if major discoveries are not made. Drilling and exploration have uncovered new, small fields as the search for oil quickens. New discoveries in the west show great promise, as do offshore sites along the northern continental shelf. Recently, a series of finds have been made in southeastern Queensland and around the edges of the Great Artesian Basin in the eastern interior.

Coal now surpasses all other minerals in export value and is used heavily in the domestic economy. To cap its extraordinarily good fortune in minerals, Australia has discovered huge deposits of bauxite at Weipa and elsewhere. The Cape York Peninsula bauxite deposits may be the largest in the world, and Australia has become the world's largest exporter. Northern Queensland also holds newly discovered deposits of phosphate, zinc, manganese, nickel, and copper.

Australia's Dynamic Fringe

Australia's population distribution shows a markedly peripheral pattern (see Figure 5–2). The Dynamic Fringe includes the highest, most scenic mountains of the continent and the better-watered parts of the areas with subtropical and temperate humid climate in the states of Western and South Australia, Queensland, New South Wales, and Victoria, as well as parts of Tasmania.

Australia's rapid population growth, a 100 percent increase since the end of World War II, testifies to the continuing importance of immigration. The government now limits immigration to persons sponsored by Australian relatives and to selected professional and occupational groups needed by Australia.

Australians are a heavily urbanized people, with fully 22 percent of the population living in Sydney, the largest metropolitan center. Aside from important mining settlements, huge, low-carrying-capacity ranches called "stations," and some evolving winter tourist resorts in the desert, the Dynamic Fringe is the demographic and economic core of Australia.

Sydney and Melbourne have maintained growth rates that place Melbourne a consistent, close second. Both are ports, and both are state capitals, located approximately in the center of their state's settled coastal areas. Both rank among the top 50 cities in the world. In 1970, Sydney (population 2.6 million) outranked Melbourne (population 2.3 million); Sydney contained 58 percent of its state's total population, while Melbourne held 66 percent of Victoria's total population. By 1986, the figures were 3.4 million and 3 million, with Sydney still leading. Between them, these two great cities dominate Australia's economic and cultural life. In a classic example of locational choice, Australians created their new capital between its two preexisting rivals for that role: the Australian federal capital at Canberra is between Sydney and Melbourne, but closer to the larger rival, Sydney.

Today, most Australian wheat farmers also graze some sheep, and many sheep ranchers grow some wheat. Market swings in prices for wheat and wool stimulate individual decisions to expand one at the expense of the other, so that Australian farmers continually expand and contract wheat acreages and sheep pastures. Wheat now occupies about half the total cultivated acreage and is grown on land with 10 to 20 inches of rainfall per year. The wetter margins of Western and South Australia are important producing areas, but somewhat better watered wheat lands occupy the corresponding zones of Victoria, New South Wales, and Queensland (see Figure 5–6).

THE PENAL COLONY ATTEMPTS AT SETTLEMENT

The earliest migration to Australia was in the form of transported convicts. The reasoning was that remote Australia was the perfect low-security prison. The convicts' offenses were generally minor, such as inability to repay debts, and the prisoners were to support themselves through agriculture, presumably reforming themselves in the process and certainly relieving Britain of the cost of maintaining them in prison back home. As with other schemes for reducing the costs of the criminal justice system, this one did not work out as planned. The penal colony at Sydney had to be supplied with food from Britain for a decade, partly because the poorer soils along the coast could not support the kind of concentrated settlement that the government favored. Free settlers were much more successful at farming, and so the government began to free prisoners ahead of schedule and encouraged free settler migration. Minimizing the penal colony image helped attract still more free settlers. The introduction of merino sheep by 1800 ultimately had the same continentwide impact on settlement and progress as the later gold discoveries of the 1850s. The first pioneer settlement was 1788, the largest transportations were in the 1830s, and even then the free settlers outnumbered the prisoners. The positive motives of opportunities, first in sheep farming, then gold, then wheat were, in the long run, far more important than penal transportation in populating Australia.

Sugar cane is a major crop in coastal Queensland. More cane could be grown, but Australia is cautious about expansion in this notoriously cyclical world market. At different levels, agencies of the federal and state governments set quotas and prices and handle marketing. Other subtropical and tropical crops, cotton, tobacco, pineapples, and bananas are also grown in Queensland.

Australia's climatic range enables it to produce not only midlatitude fruits such as apples, pears, peaches, and apricots, but also tropical fruits such as oranges. Vegetables are grown near most big cities, and Tasmania specializes in cool-climate crops such as potatoes and apples. A growing wine industry is based near Adelaide.

Australia can lead in world wool markets because, in addition to its role in mixed sheep and wheat farms, sheep are also grazed on the driest margins of pastoral land, land sometimes so dry with vegetation

so sparse that up to 30 acres are required to support one sheep.

Dairying is carried on near all big cities and in the wet coastal valleys southward from the central Queensland coast and especially in Victoria. The Tasmanian north coast is also a traditional area for dairy cattle, while coastal wetlands along the South Australia-Victoria border have been drained and improved for dairying in a seasonally dry area.

In addition to hydro power, Australia's large coal production relieves some of the pressure on its oil resources. Over 80 million metric tons of bituminous coal are produced each year, together with 36 million metric tons of brown coal (lignite).

Coal, the largest employer of miners, is mostly a product of the developed fringe, with heavy production in New South Wales and Queensland (see Figure 5–6). Fortunately for Australia, the most significant reserves in quantity and quality are relatively close to the great coastal cities. Coal had been exported as early as 1801, and exports to Japan recently have increased rapidly.

The limited oil production of the Bass Strait fields has intensified an extensive exploration program, on and offshore. Australia has a major deposit of oil shale on the east coast of Queensland. As with the U.S. Rocky Mountain oil shales, there are some serious environmental pollution potentials in large-scale exploitation. The total reserve is estimated at 2.3 billion barrels; it is projected to produce about one quarter of Australian consumption by 1990.

Despite Australia's small area of forest cover, Australians supply about 70 percent of their lumber requirement through careful management. Fishing resources have just begun to be exploited extensively. Historically, Australia has exported pearl shell from its tropical northern waters, and Japanese investment and expertise have encouraged a cultured pearl industry there. Crayfish tails and tuna have become major fisheries products, as has commercial oyster farming. The Australians expanded their territorial fishing waters from 3 to 12 miles as Japanese and Russian trawlers began to take large catches offshore.

Sixty percent of all Australians live in the eight largest cities, of which six are state capitals. The unusually high proportion of population in state capitals is related to economic development patterns. In the early colonial period, settlement progressed from separate nuclei, each a seaport linked directly to London, and thus each colonial capital was not just a state capital but also a transport node. Each separate political unit built a railroad system designed to channel the hinterland products of farms, mines, and forests to its economic and political capital, invariably a seaport or adjacent to one. It was not until much later that a unified national rail system became a goal. Achieving the goal has not been easy, because the early single-focus rail nets were often of different gauges, or widths, than those of neighboring states. The first entirely standard-gauge transcontinental railroad was completed only in 1969.

The urban economy, as well as the rural extractive sector, is characterized by a mixed economy philosophy of free enterprise combined with a strong regulatory role for, or even ownership by, the federal or state government. These governmental controls are especially significant in banking, minerals, and energy. Federal and state governments own railroads,

utilities, telephone communications, the international airline, and one domestic airline. The small domestic market (Australia has far fewer people than California) limits the opportunities for achieving economies of scale in production. Isolation and more especially the small size of the domestic market have restrained the development of Australian manufacturing. The nearest neighbors, geographically, are characterized either by dense populations and cheap wages, as in Southeast Asia, or by an even smaller market, as in New Zealand and the Pacific islands.

Australia's Developmental Strategies

The export of manufactured goods from Australia is expected to increase as Australian production of motor vehicles, heavy machinery, farm equipment, and machine tools increases. The shift toward the export of sophisticated manufactured goods to supplement raw and semiprocessed materials is uneven, however. It would appear that the high standards of living and wage rates can be supported in highly mechanized, export-oriented mineral production but not necessarily in all export-oriented manufacturing.

Interruption of normal trade relationships during World War II accelerated Australia's development of import replacement industries. Australia now assembles autos, builds machinery of all kinds from its own steel, and produces a wide range of industrial and consumer chemicals, clothing, processed foods, tobacco, and paper products. Wine and beer are produced for domestic and export sale. Australia and New Zealand are working toward forming a common market.

Australia's Problems and Potentials

The environment is a pressing issue because water is scarce, vegetation and animal life are vulnerable, and the economy is dependent on agricultural and mineral development. The new emphasis on mining and industrial growth is presenting new problems of pollution and new potential threats to both urban and rural environments.

The exploitation of resources, with little thought of conservation or the future, that characterized the nineteenth century United States has sometimes seemed prevalent in twentieth century Australia.

A Sing-Sing in New Guinea.
Each year several tribes of New Guinea gather on Mount Hagen to take part in a festival called a sing-sing. Here, members of a Western Highlands tribe participate in a dance, one type of exchange of tribal traditions.

Overgrazing is a serious problem, especially when it is remembered that 58 percent of Australia is in grazing land, with only 5 percent cultivated. Little land has been set aside as national parks or forest preserves. A growing conservation movement may reverse the exploitative tradition before serious, long-term environmental damage becomes more severe.

Australia has become an important donor of economic aid in its part of the world. Naturally, the Australians hope that this enlightened generosity will help achieve a stable, prospering regional setting for their prospering, stable country.

New Zealand

Quite similar to "home" in many ways and only lightly populated by an indigenous people, New Zealand considered itself the "Britain of the south." People of European heritage, primarily British, form 85 percent of the population. The remaining 15 percent are largely Polynesians, mostly Maori, who formed the pre-European population. About the size of Colorado, New Zealand has about 3.25 million people. The Maori are increasing at a faster rate than people of European heritage. Life expectancy and the literacy rate are high, and most of the population is urbanized, though there are still some differences between Maori and European-heritage New Zealanders.

New Zealand's resource base, other than its agricultural potentials, is quite narrow. Minor soft coal deposits in South Island, some lignite on both islands, some natural gas, some coastal ironsands, and a scattering of nonferrous minerals (including gold) on both islands constitute the resource base (see Figure 5–6). The lack of oil fields, the limited resources of natural gas, and the absence of major metallic ore sources other than a fair-sized deposit of ironsand make New Zealand an unlikely location for heavy industry. The hydroelectric potential is relatively high in proportion to population and is about one-third developed, though further expansion will be costly. New Zealand's position on the Pacific ring of fire means that earthquakes are common, although usually minor in scale. As noted before, volcanic activity has been a major force in shaping North Island. A few volcanoes remain active into the modern period, and there are numerous geysers, hot springs, and steam fumaroles; the country is among the few nations that have begun to harness their geothermal power potential.

About 70 percent of all New Zealanders, including almost all the Maori, live on the somewhat warmer, better-watered, and less Alpine North Island (see Figure 5–2). The Auckland peninsula of North Island is marginally subtropical, while the Southern Alps of South Island and the extreme south verge on the subantarctic. The British formally annexed the islands in 1840 and, that year, began systematic settlement. A series of land wars with Maori tribes, who resisted white settlement, ensued in the 1860s. Since 1870, natural increase has been larger than immigration and now accounts for three quarters of the population growth.

Agriculture remains the dominant sector of the export economy, although manufacturing is increasing. New Zealand is usually the world's largest exporter of lamb, mutton, and dairy products, and is the second-ranking exporter of wool; together, these products make up almost two thirds of the exports. New Zealand is heavily dependent on exports, which account for 23 percent of its GDP. This fact is even more remarkable considering the narrow range of produce and the vast distances between New Zealand and most of its trading partners.

The oceanic, high moderate climate of New Zealand makes it ideal for growing nutritious pasture grasses. About one third of the country is in cropland or sown pasture, one third in rough pasture and commercial forest, and one third in mountainous terrain, unproductive for agriculture but not for tourism. Apart from South Island's Canterbury Plains, there is limited flat land, a factor that also encourages grazing rather than cultivation; heavy precipitation and steep slopes carry a high risk of erosion.

Distance from other sources of industrial goods in the colonial era and reduction of imports in wartime helped to foster a variety of manufacturing enterprises aimed at the small local market. Large-scale food processing developed after 1882 as refrigerated ships made the export of meat and dairy products to distant markets feasible. Meat packing and freezing, butter and cheese manufacturing, and fruit and vegetable processing have become important industries based largely on export markets.

New Zealand's industries now include electronics, auto assembly, chemicals, paper, fertilizers, and printing. The economy faced a double economic threat in the 1970s. Britain's decision to join the EC in 1973, long delayed precisely because of its trade relationship with its Commonwealth partners, has drastically reduced that market for New Zealand's agricultural products. The loss of free access to British markets was serious. Governmental decisions to maintain almost full employment under these trying economic conditions worsened New Zealand's balance-of-payment problems. The economy became

stagnant, and inflation moved into double digits. Recent economic deregulation, removal of subsidies, and privatization of the public sector are related to significant structural changes as the nation seeks to stimulate growth. The results are not yet clear.

Auckland, with almost a million people, is the largest city. Both Auckland and Wellington, the capital (population 400,000), are on North Island. The largest city on South Island is Christchurch (population 335,000).

The economic problems of New Zealand, together with communication and transportation systems much superior to those available in the colonial period, have led to serious consideration of a political and, more especially, an economic union of Australia and New Zealand. Closer economic ties would probably benefit New Zealand somewhat more than Australia. Although not exactly geographically close, the two share a language and a highly similar cultural heritage. The countries are close in the context of their relative isolation from Europe and North America, and in their common sharp distinctions, racially, culturally, and economically, from Southeast Asia. However, New Zealand has refused entry into its ports to all naval vessels, U.S. ships included, that refuse to reveal whether any nuclear weapons are on board. This serious strain on the important alliance with the United States and the United Kingdom is worrisome to Australia, but antinuclear feeling is widespread in the region. The South Pacific Nuclear Free Zone Treaty seeks to exclude nuclear testing, storing, and dumping in the region. Australia, New Zealand, and most self-governing island-states have adopted this policy.

Papua New Guinea

Papua New Guinea (PNG), 100 miles to the north of Australia, shares the great island of New Guinea with Irian Jaya, the easternmost territory of Indonesia. PNG is included with Australia-Oceania rather than with Southeast Asia (which includes Indonesia) because of its proximity and close economic and political ties with Australia. It was administered by Australia until independence was achieved in 1975.

A single copper and gold mine on the island of Bougainville produces one fifth of PNG's gross domestic product and more than half its export value. Forty percent of the country is in commercially valu-

Making a Dugout Canoe, Papua New Guinea. *Papua New Guineans are a wood-based culture, as demonstrated in this view. Note the wooden houses in the background. Unfortunately, PNG is permitting large-scale commercial logging with little thought to conservation of tropical rainforests.*

able forest, but transport systems are not yet adequate to support large-scale exploitation.

This country of over 3 million is growing at an annual rate of 2.6 percent. Except for about 1 percent Australians (and even fewer Chinese), the population is entirely Melanesian, ethnically. Culturally, however, the country is extremely complex. Separated by steep, mountainous terrain covered with rainforest, highly localized cultures have developed, often in apparent ignorance of other groups a few miles distant. "Pidgin," a language evolved from English, German, and native words, was originally a makeshift language of convenience serving commercial relations and government administration; it has become the official language, though the small educated elite speaks English.

Commercial agriculture has expanded, with emphasis on coffee, copra (coconut), cacao, tea, and rubber. Cattle are becoming more important as a means of reducing meat imports. However, 80 percent of the population is still little touched by the commercial economy, remaining in subsistence farming of sweet potatoes, taro, and other root crops.

Some subsistence farmers do grow small amounts of coffee, cacao, and coconut to have some cash income. Typical of developing nations around the world is the rapid growth of PNG's urban centers. Port Moresby, the largest city (population 160,000), is growing at a rate of 10 percent a year.

PNG's developmental strategies must take into account both the cyclic nature of tropical commodity prices and the proximity of Southeast Asia, with its lower labor costs. The decision to legislate a high minimum wage for urban industries may reinforce the fact that PNG can scarcely compete with Southeast Asia in export of manufactured goods. Current government planning for industrialization emphasizes consumer products for import replacement and the further diversification of agriculture.

There is some concern that overly rapid expansion of slash-and-burn cultivation, commercial agriculture, and cattle grazing in the tropical rainforest areas could lead to serious soil erosion problems. Rainfall averages 80 to 100 inches a year, and some areas receive more that 200 inches. Once the deceptively luxuriant rainforest is removed, the often infertile soil may not long sustain cropping or even pastures, and severe environmental degradation already characterizes some areas.

PNG's transition to a fully independent state has been remarkably smooth. Undoubtedly, the large subsidies provided by Australia and New Zealand help. Its economy is expanding and diversifying, and its outlook is bright if a true nation can be melded from its multitribal, multicultural society. Friendly relations with Australia, its former mentor, are critical to PNG, whose border with Indonesia is not always peaceful.

The Pacific Islands

The Pacific islands vary greatly in size. Many cover a few acres, while Guam has 210 square miles and New Caledonia exceeds 8,500 square miles (see Figure 5–1). Western Samoa, Fiji, and Nauru are independent countries, as are Tonga, Tuvalu (the Ellice Islands), Kiribati (the Gilbert Islands), and the Solomon Islands. A number of islands have chosen "free association" (self-government but with continuing economic support) with New Zealand and the United States. Former Japanese Micronesia became the American-administered Trust Territory of the Pacific Islands, now divided into four political units. The Northern Marianas have accepted status as a U.S. territory, while Belau (Palaus), the Marshalls, and central Micronesia have become separate free-association states.

France retains a stronger measure of control in French Polynesia and New Caledonia. There is limited development on most of the islands. Coconut and copra remain the major commercial products, but agricultural diversification continues. Fiji exports sugar, bananas and cacao are important to Western Samoa, and the Cook Islands specialize in citrus. De-velopment of fisheries is viewed as a viable means of economic expansion, though claims to sovereignty over wide stretches of sea are often challenged by American and Japanese fishing boats.

Tourism is a major source of income in Tahiti and on some of the other islands. Military bases, such as those on Guam, Midway, and Tahiti, can be a major source of employment, and military personnel along with tourists encourage the growth of other services. Indirectly, military bases are a source of raw materials because used and abandoned goods, packing cases, clothing, and transport equipment are absorbed into the local economy or used in crafts.

However, population growth, though often relieved by emigration, is often rapid, and local resources, with the possible exception of the sea, form an insufficient basis for diversified economic growth. Outside of tourism, a few highly localized mineral deposits, and coconuts, there appears to be little future development possible on most islands. Emigration to, and remittances from, such countries as the United States and New Zealand have become part of island life.

CONCLUSIONS

Australians and New Zealanders have modified their global thinking, much as their trade relationships have shifted. Their futures would seem to lie in an expanding Asian-Pacific economy in which Japan and the United States will be influential partners along with Southeast Asia and the Pacific islands. The Pacific islands are entering a new era of increasing political and economic independence, new forms of regional economic cooperation, and new potential for tourism expansion as terrorism elsewhere focuses new interest on this region.

REVIEW QUESTIONS

1. Why was Australia originally viewed negatively for European colonization?
2. Why have Australia and New Zealand developed technologically sophisticated large-scale agriculture as export suppliers?
3. How has the significance of distance to Europe changed over time? distance to East Asia?
4. What is meant by the "empty heart" of Australia?
5. Briefly describe the mechanisms and effects of tropical marine climates.
6. In what way is the Australian choice of a new capital similar to the situations of Washington, DC and Ottawa?
7. Why are import replacements part of Australia and New Zealand's development strategy? Has this strategy worked?
8. Why does the international trade importance of Australia and New Zealand far outrank their relative population size?
9. How is the prevalent Australian attitude toward exploration of the environment related to population density and resource base?
10. Briefly describe the mineral resource bases of the major states of this region.
11. How did Britain's membership in the European Economic Community affect the trade patterns of Australia and New Zealand?
12. Why did Australia strictly control immigration for many years, strongly favoring Europeans?
13. Why did a close American alliance suddenly replace Australia and New Zealand's former reliance on a British military shield?
14. In what way is this region an economic hinterland in relation to the developed economies of Japan and the United States?

SUGGESTED READINGS

Brookfield, Harold, ed. *The Pacific in Transition: Geographical Perspectives on Adaptation and Change*. New York: St. Martin's, 1973.

Cumberland, Kenneth. *Southwest Pacific: A Geography of Australia, New Zealand, and Their Pacific Island Neighborhoods*. London: Methuen, 1958.

Cumberland, Kenneth, and Whitelaw, James. *New Zealand*. Chicago: Aldine, 1970.

Heathcote, Ronald. *Australia*. New York: Longman, 1973.

Howlett, Diana. *Papua New Guinea: Geography and Change*. Melbourne: Thomas Nelson, 1973.

Levinson, M; Ward, R.; and Webb, J. *The Settlement of Polynesia*. Minneapolis: University of Minnesota Press, 1977.

McKnight, Tom. *Australia's Corner of the World*. Englewood Cliffs, NJ: Prentice-Hall, 1970.

Meinig, Donald. *On the Margins of the Good Earth: The South Australian Wheat Frontier 1869–1884*. Chicago: Rand McNally, 1963.

Perkins, James. *Australia in the World Economy*. Melbourne: Sun, 1968.

Robinson, K.W. *Australia, New Zealand, and the Southwest Pacific*. London: University of London Press, 1974.

Spate, O.H.K. *Australia*. New York, Praeger, 1968.

Ward, R. Gerard, ed. *Man in the Pacific Islands: Essays on Geographical Change in the Pacific Islands*. Oxford, England: Clarendon, 1972.

CHAPTER SIX

East Asia

Tienanmen Square, Beijing.

The East Asia region serves as a critical transition in this survey of major world regions. All of the regions studied previously, if the chapter order has been followed, are formed by relatively technologically advanced industrial countries. East Asia is the bridge between the developed and the developing world because it contains representative states from both economic categories.

East Asia includes preindustrial Mongolia; the People's Republic of China (PRC), in the process of industrializing; and industrializing North Korea. One of the world's fastest growing economies is South Korea (Republic of Korea). Taiwan (Republic of China) is a major factor in world trade, along with the bustling city-state of the colony of Hong Kong (soon to belong to the PRC) and prosperous, intensely industrialized, high-technology Japan (Figure 6–1).

East Asia contains the largest group of humanity on the globe, reason enough to warrant the close interest of Americans. And the economic vitality of these 1.25 billion people is even more astounding than their sheer numbers. By itself, Japan has established, by volume and value, a trade with the United States that has created the greatest unilateral imbalance ever between two countries, some $40 billion annually in Japan's favor. Trade between the United States and Taiwan, South Korea, and the PRC is also expanding rapidly.

Probably no other world region will have such a profound impact on the Western industrial world within the next generation. Just as Japan wove its own distinctive blend of Western ideas and technology with characteristic Japanese traditions and perceptions, China is in the process of melding Marxist dogma with the profit motive of classic capitalism.

A salient characteristic of Japan, Taiwan, South Korea, Hong Kong, and the PRC is the ability to adapt and modify imported ideas and technologies. The national culture is not shattered by its adaptations of Western industrial models, nor does it become a copy of an outside culture. An even more important, and universal, quality of this region is the highly disciplined nature of the work force.

Japan's excellence in building and managing a high-technology society, matched with the PRC's huge population and impressive resource base, and augmented by rapidly industrializing South Korea and Taiwan, could create the most powerful regional economic bloc in the world. Serious, perhaps disabling, political questions remain. Japan's onetime

Guizhou Province, China.
Bicycles were a popular consumer goal during the Maoist regime. New goals include televisions, refrigerators, and washing machines. Economic reforms have led to renewed demand for political reforms as well.

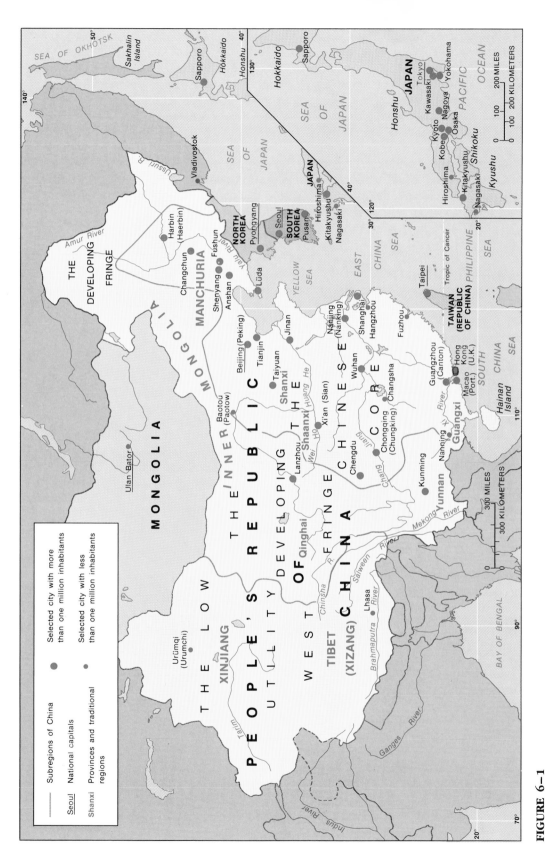

FIGURE 6–1

Countries, Subregions, and Cities of East Asia. This large, relatively densely populated region has been subdivided into subregions for greater ease of comprehension.

colonial control of Korea and Taiwan will not soon be forgotten, or entirely forgiven. Japan's history also includes a catastrophic invasion of China and long control over Manchuria; full reconciliation will require a considerable period of gradually increasing interaction with China. Some form of federation between the PRC and Nationalist China (Taiwan) may occur eventually, perhaps following the model of the British-Chinese agreements on the future status of Hong Kong. Even the bitter animosity of North and South Korea may be eased by the eventual death of North Korea's long-time dictator, Kim Il Sung.

THE GREAT DIVIDE: THE GAP BETWEEN DEVELOPED AND DEVELOPING COUNTRIES[1]

Customarily, the world's countries have been categorized as developed or developing. In **developed nations**, high per capita levels of production and consumption are accompanied by moderate to low population growth rates. These high-technology societies are characterized by low birthrates and low death rates; that is, they are in the final stage of the demographic transition. The traditional **developing countries** were typically preindustrial or at an early stage of industrialization. Their per capita levels of production and consumption were low, and their population growth rates generally higher than those of developing countries. They were in the second stage of the demographic transition, with continuing high birthrates, declining death rates, and consequent rapid growth. The prevalent assumption was that, given time, incomes would rise, birthrates would decline, and the developing country would achieve development, along with a lower population growth rate. This happy progress certainly has occurred repeatedly, and it continues to happen.

There is evidence, however, that some countries, in effect, are trapped in the second stage of the demographic transition. Their living standards are not showing continuing improvement, and their population growth rates are not dropping. The vision of developing nations lined up on an escalator labeled "industrialization," heading upward to developed status, is accurate for some, but not for all.

For example, in 1975, South Korea's per capita GNP was 7.8 percent of that of the United States; by 1986, it was 14.2 percent of the U.S. figure. For the same years, and by the same measure, Japan went from 62.5 percent of the U.S. figure to 71.6 percent. Hong Kong's comparable progress was 24.7 percent in 1975, 42.5 percent in 1986. Thus, heavily industrialized Japan further reduced the output (and income) gap with the United States, and South Korea and Hong Kong clearly made progress in comparable output and incomes. They, at least, are on the industrialization escalator to better living levels. The PRC, however, handicapped by uncertainty and repeated reversal in economic goals and basic philosophy of economic management (along with a recent poor harvest), declined in per capita GNP as a proportion of that of the United States, from 5.3 percent in 1975 to 2 percent in 1986.

Excepting China and Mongolia, East Asia turned in an economic performance superior to that of many of the non-oil-exporting developing countries. Mexico achieved only a slight increase, from 14.7 percent in 1975 to 15.8 percent in 1986. Several very poor states actually got poorer, and less productive, per capita as compared to the United States. Bangladesh fell from 1.2 percent of the U.S. per capita figure to 0.92, while Sudan declined from 3.7 percent to 2.8 percent, and Ethiopia fell from 1.4 percent to 0.99 percent. In other words, many of the least productive countries, those most in need of economic progress, made little or no such progress.

Perhaps the critical division in today's world is not between "rich" and "poor" but between countries in which per capita GNP is increasing and those in which it is not. The **older industrialized countries** (OICs) are relatively rich. With rare exceptions, they are progressively more productive, and thus richer, per capita, over time. The rich stay rich and, mostly, get richer still. The **newly industrializing countries** (NICs) are also for the most part increasing their productivity and living standards on a per capita basis; although they are still less rich than the OICs, they are less poor than they were. For both OICs and NICs, there are justifications for optimism about their future economies.

The OICs and many NICs share a significant, and most likely critical, demographic trait: they are slow-growth countries, with population increase rates at or below 1 percent per year. The world average is 1.7 percent per year. Countries that are not significantly increasing their per capita productivity and incomes almost always have rapid population growth rates. Is this a coincidence, or is there a cause-and-effect relationship?

With a few exceptions in each region, the slow-growth regions are Europe, the United States and Canada, the USSR, Australia-New Zealand, and East Asia. These regions' populations add up to about 2.3 billion people, growing at an annual average rate of 0.8 percent. The rapid-growth regions include South and Southeast Asia, the Middle East-North Africa and Africa South regions, and Latin America. The populations of these regions, some 2.7 billion, collectively are expanding at 2.4 percent, triple the average rate of the slow-growth regions.

East Asia is the pivotal region. It is the most recent addition to the list of predominantly slow-population-growth regions. The PRC is the pivotal country within East Asia, with four fifths of that region's total population.

The grinding poverty of most people in the pre-industrial, high-population-growth Third World can hardly be imagined by many in the comparatively wealthy OICs. In food available to the average citizen, expressed as daily calorie supply as a percentage of calorie requirements for active, healthy adults, the U.S. proportion is 138 percent. For the USSR, the figure is 132 percent; for most European countries, it is at least 120 percent; and for Japan, 122 percent. In these countries, abundance is the norm, and any malnourishment is usually the result of distribution and individual access problems, not of overall scarcity. The high-population-growth countries tend to have calorie availability ratios below 100 percent of daily requirements. For these people, hunger is the norm, and abundance the exception. For Ghana, the calorie availability is only 72 percent of requirements. In the South Pacific's Solomon Islands, it is but 77 percent; in Bangladesh, 83 percent.

The global economy is being reshaped by the rapid rise of the NICs. In 1963, the U.S.'s share of non-Communist world manufacturing output (Communist planned economy states do not publish comparable data) was 40 percent. The 10 most important NICs (Brazil, Spain, Mexico, Yugoslavia, South Korea, Taiwan, Portugal, Greece, Hong Kong, Singapore) to-

gether then produced 5 percent of global manufacturing output. By 1980, the United States (though still first, followed by Japan) produced 30 percent of the world total, while those 10 NICs had doubled their share of an expanding output to over 10 percent of the total. The fastest growing industrial economy of the 1970s and early 1980s was that of South Korea, where the share of total labor supply in manufacturing increased from 9 percent in 1960 to 29 percent in 1980.

In 1980, the top six non-Communist manufacturing countries (United States, Japan, West Germany, France, Italy, United Kingdom) produced 73 percent of the world's market economy output. Ninety percent was produced in the top 17 countries, only 4 of which (Brazil, Mexico, India, Argentina) were not among the traditional industrial nations of Europe, the United States and Canada, or Japan. With the exception of Japan, however, all of the countries with annual growth rates of 5 percent or higher in manufacturing output were NICs, led by South Korea, with an astounding 15.6 percent increase. Clearly, though the world's manufacturing is still heavily dominated by the OICs with both free-market and centrally planned economies, the dominance of these countries will now erode quickly. For some of the NICs, particularly in Latin America, a critical race is on between increasing output and increasing population, a race that must be won by output per capita if living standards are to rise.

It is in achieving the critical combination of lowering population growth rates and rapidly expanding manufacturing output that the East Asia region is leading the way as the developing world becomes a progressively more important part of the global economy. East Asia is the key to understanding the nature of the developed-developing world "great divide."

TRENDS AND PERCEPTIONS

East Asia has developed a major civilization which diffused knowledge and technology over a large area of Asia, creating a high degree of regional uniformity long before it came into contact with the Western world. The cultural interaction with the Western world was two-way, with both cultures exchanging ideas and technology.

A distinctively Chinese culture and a well-established state appeared about 1500 B.C. Chinese tradition and legend begin with the "Yellow Emperor" (about 2700 B.C.). The dynamic Chinese culture had a profound impact on China's neighbors in this region. China dominated the Korean peninsula culturally and politically for scores of centuries before Korea was first invaded by the Japanese in the sixteenth century. Korea clearly was a satellite of powerful China for much of its history until the late nineteenth century. Its alphabet (ideographs), clothing styles, religion, and art show strong Chinese influences, though Korean ethnic origins and language are clearly not Chinese.

Art, religion, technology, architecture, and other cultural elements flowed from the mainland to the islands of Japan, much as Britain and Ireland received cultural impacts from continental Europe. Buddhism came to both Korea and Japan via China, and Confucianism came to strongly influence both other nations.

Similarities among tools and household items used throughout East and Southeast Asia show the export of Chinese technology as well as of goods. The emphasis on paddy rice culture in all regional economies is another feature held in common, though not all parts of the major units grow rice. Each of the region's major states have had to deal with aggressive European colonial ambitions and trade initiatives. Each nation's modern industrialization has been powered to a degree by cheap, disciplined labor during its early stages of expansion.

East Asia is rather neatly defined by physical barriers of impressive scale. Mountain ranges, including the world's highest, and extensive deserts provide what earlier geographers would have categorized as natural boundaries (see Chapter 1, Figure 1–10). Rugged and lightly populated, these physical features have been used intelligently by regional political and military leaders.

The region's preeminent distinction is the unique blend of an ancient, non-Western culture and a dynamic adaptation of Western technology. In the past, Chinese culture spread to Korea, Japan, and Taiwan at the region's peripheries. Those areas in turn received the greatest impact of Western technology, selectively adopted portions of it, and modified it to fit Oriental society. Now this cultural amalgam spreads from the peripheral areas back to the ancient Chinese core. A resurgent modernization, begun in Japan, now reaches toward China through South Korea, Hong Kong, and Taiwan. The defeated Nationalist Chinese government fled to the island Taiwan (at one time called Formosa) and established there the Republic of China (see Figure 6–1). This state is known familiarly as Taiwan. China and North Korea also received Western influence via the Soviet Union. Thus, both Marxism and capitalism have had input and impact, yet the resulting regional interpretations of both approaches are decidedly different from their original forms.

In A.D. 2, the population of China was officially tabulated at 59 million, comparable to the population of the entire Roman Empire at the same time. Japan and Korea had proportionately dense populations early in their histories as nation-states. These relatively dense populations have tended to impress Westerners as overcrowded, a perception also frequently noted by East Asians themselves.

A result of the large numbers of people was the evolution of an agricultural system emphasizing the most productive use of land rather than the most efficient use of labor. Labor was plentiful; good, cultivable land was relatively scarce. Intensive agriculture is prodigal with human energy in order to maximize food production. Traditional oriental agriculture thus came to be more large-scale gardening than large-scale, commercial agriculture such as developed in North America in the nineteenth century (Figure 6–2). In North America, the emphasis was on maximizing the efficiency of the farmer through use of machines rather than on concentrating human labor on small patches of land for maximum production per areal unit.

East Asia's prodigious use of cheap labor influenced early patterns of industrial production. The adoption of labor-saving machinery now raises the potential for industrial production in the region to an astronomical level. Japan has essentially achieved this potential. Taiwan, Korea, and Hong Kong are well on their way. This modernization of industrial production is just beginning in the PRC.

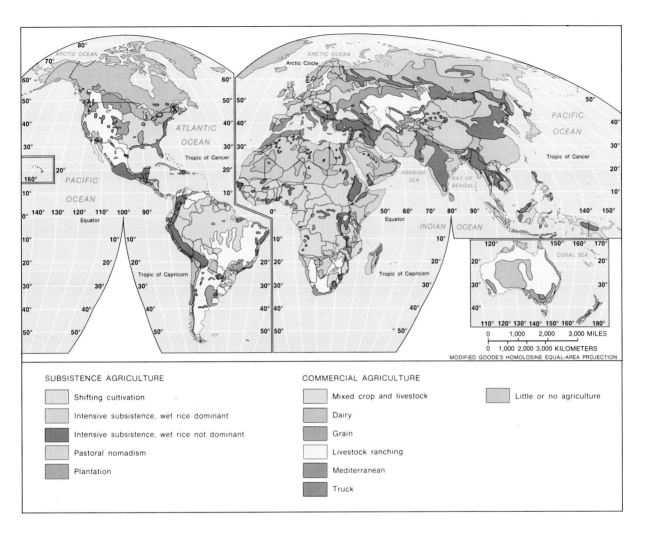

FIGURE 6–2

World Agricultural Regions. Perhaps the most globally significant division in terms of agricultural land use is that between *subsistence* (most production directly consumed by its producers; little surplus generated) and *commercial* (emphasizes production for market; generates large surplus beyond consumption needs of producers). Note the general association of developed countries with commercial agriculture, and the developing countries with subsistence agriculture.

THE PHYSICAL FRAME

The boundaries of the area of dense settlement have been defined by the ability of oriental society to control and shape nature. Dense settlement ends at the point where it becomes impractical to extend sedentary agriculture any further; yet within this densely settled core, the limitations of slope and water supply have been surmounted by human endeavor. Oriental society has expanded well beyond the most favorable natural environments, creating usable environments where none existed in nature.

Sichuan Province, China. *Wet-rice culture (paddy rice) in China still involves a great deal of human labor. This paddy is being plowed with the help of a water buffalo.*

Topography

About 80 percent of China is comprised of mountains and plateaus; this predominance of highlands also occurs in Japan, Korea, Taiwan, and even Hong Kong (Figure 6–3). China is rimmed by a series of mountain ranges, from the rugged mountains and high plateaus that lie to the south of the Siberian Plain, through the massive Altai and Tien Shan along its western borders, to the Himalayas and Xizang[2] (Tibetan) Plateau of the southwestern border reaches and the steep mountains and plateaus of Yunnan in south-central China.

The prevailing grain of China's topography is east-west, including much of the mountain rim frontier and the smaller, but significant Qinling range, which is an important climatic divide within eastern China. The Korean peninsula resembles a giant fault block, tilted up along its eastern Sea of Japan edge with a more gentle, though deeply eroded, back-slope down to the Yellow Sea to the west.

Japan and Taiwan are part of a series of great island arcs paralleling the mainland shores from the Kurils to Indonesia. Like the mainland of East Asia, these predominantly mountainous islands are geologically complex. The island arcs are geologically unstable; earthquakes and volcanic activity are common. The entire rim of the Pacific Ocean is characterized by frequent earthquakes and active volcanoes. Some magma thrusts up through lines of weakness in the crust in the unstable islands, emerging as lava from volcanoes. The island arcs are not

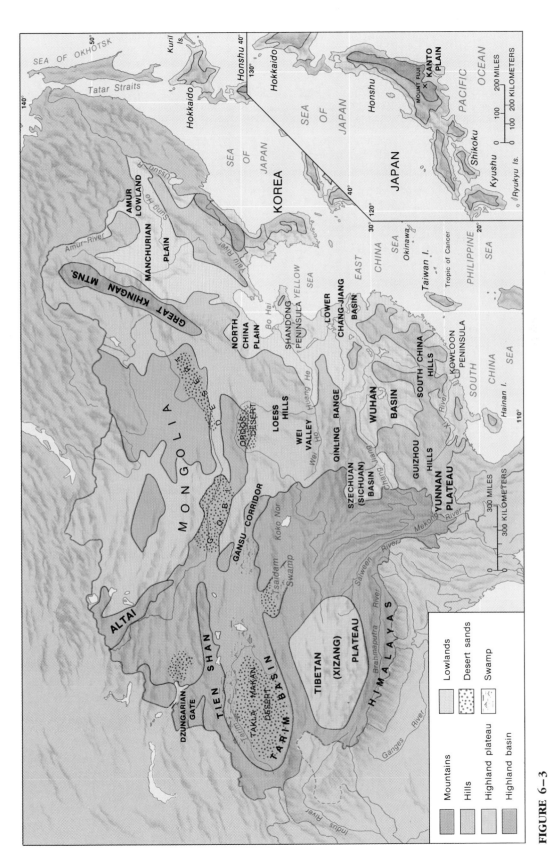

FIGURE 6–3

Major Landforms in East Asia. Note the combination of mountains and deserts that form most of China's land frontiers.

completely volcanic; some have older continental-type rocks and sedimentary strata as well. Their mineral resource base is correspondingly varied.

Whether island arc or continental mass, level land is at a premium in East Asia, and not all of that is usable for agriculture. Some 20 percent of China is level, but only 10 percent of that country is considered arable, given current levels of technology and investment. (It should be noted that millions of poor farmers in the world work land that others are convinced is not arable; it all depends on how desperate one is.) Population clusters into dense settlements along arable stretches, in particular in China with its relatively low degree of urbanization. The lowland of East China, composed largely of the delta lands of the Chang Jiang and Huang He, houses a high percentage of China's population; it is the primary core. Another large lowland forms the nucleus of Manchuria. Smaller, densely populated basins occupy the middle and upper portions of the Chang Jiang. Myriad smaller, level to rolling areas are scattered through hilly South China, where small rivers create usable delta and valley plains within a matrix of steeply sloping hills. Vast plateaus and basins in the arid west have little agricultural utility. Where possible, hills have been terraced, swamps drained, and lands irrigated and otherwise reclaimed under great population pressure.

Climate

Most of the East Asian region has a midlatitude climate, but China's extreme south coast and the offshore islands, Hainan and Taiwan, are subtropical. Northern Manchuria's climate resembles that of its neighbor, Siberia. The western deserts are seasonally cold, and the Himalayas and parts of the Xizang Plateau have a tundra climate or, in even higher areas, permanent snow and ice (see Chapter 1, Figure 1–18). The **monsoon**, a seasonal reversal of prevailing winds over most of southern and eastern Asia, is extremely important in understanding the agricultural problems and natural hazards of this densely populated part of the world. Monsoonal air circulation prevails from the mouth of the Red Sea to Japan; the term is of Arabic origin and refers to the seasonal wind direction in the Arabian Sea (see the map of Asian monsoons, Chapter 7, Figure 7–6).

The monsoon phenomenon results from the fact that land and water surfaces heat and cool at very different rates; surface temperatures tend to create atmospheric pressure differences (although not all pressure systems are the result of local surface temperatures). Atmospheric pressure differences in turn set up wind flows. In summer, landmasses heat up faster than the surrounding seas, where both vertical and horizontal ocean currents can distribute heat and store it in large quantities. The largest landmass, Eurasia, has the strongest seasonal pressure contrast with the ocean surfaces to its south and east. Warm air currents rise off the land in summer, generally producing a temperature-induced low pressure. The cooler sea surface in summer tends to produce temperature-induced high pressures. Air then flows from the ocean's high atmospheric pressure toward the continental low pressure.

In summer, oceanic air coming in over the coasts contains large amounts of water vapor. For most of island and coastal Asia, then, summers will be a time of heavy precipitation, cloudy skies, and warm, humid conditions. In the winter, however, the reverse is true, with the landmass cooling off much more quickly than the oceans (which can transfer heat energy within themselves, both horizontally and vertically). The winter sea surface is relatively warm, inducing lower atmospheric pressures than those over the colder land. Once again there is a flow of air from a relatively cool surface to a relatively warm one. This time, however, the origin and destination areas of wind flows are reversed. Winter skies for most of monsoon Asia are thus clear, with little chance of precipitation unless an outward flow of air from the continent crosses a sea surface before coming onshore again (western Japan and parts of Southeast Asia are examples).

The timing of the arrival of the summer monsoon is critical; it can vary greatly, and there is a relatively sudden wind reversal over South Asia. Over most of East Asia, the reversal is less dramatic. This concentration of precipitation over China in the summer has serious consequences in flooding and drought patterns.

Drainage Patterns

Two of the world's greatest rivers, the Huang He (Yellow River) and the Chang Jiang (Yangtze River), flow eastward across China (see Figure 6–1). The Yellow River is named for its muddy color, a product of heavy erosion in the general region of its great bend where it flows through an area of loess (wind-deposited, fine-grained sediment). As it flows out of its mountain gorges and onto the North China Plain, it moves from a steep, narrow valley onto a broad, gently sloping plain on its way to the sea. This river built much of the North China Plain by wandering back and forth across it and depositing silt. The river has changed course repeatedly throughout its history, sometimes abandoning one exit to the sea and creating another. Such shifts in course are accompanied by widespread flooding. In over 2,000 years of recorded history, there have been serious floods on the Huang He almost every other year. Diking the river behind artificially raised banks is not the perfect solution, however. Parts of the Huang He now flow on a raised ridge some 30 feet higher than the surrounding plain. When the river breaks through the walls of this ridge of mud, the resulting flooding is disastrous.

The Chang Jiang, nearly 3,500 miles long, is navigable over large stretches and has long been a major artery of trade and transportation for central China. The Chang Jiang is divided into three major basins—upper, middle, and lower—which together contain almost 40 percent of China's population. It is a major source of irrigation water for millions of acres of paddy rice and possesses an enormous hydro power potential that has barely begun to be developed. The Chang Jiang, unlike the Huang He, rarely overflows its banks.

Smaller rivers, such as the Xi, drain the extreme south of China and lower Manchuria, while the northern portion of the area is drained by the Amur and its tributaries. Much of arid western China and large parts of the Mongolian People's Republic have interior drainage, where rivers die out in sands or salt lakes, never reaching the sea. The Huang, Chang, Salween, and Mekong—major rivers of Southeast and East Asia—rise in the Xizang highlands and flow through spectacular gorges before emerging to form the rich lowlands of coastal Asia.

Vegetation and Soils

Over most of East Asia, it would be impossible to discuss purely natural vegetation because people have altered the vegetation so completely. They have destroyed entire forests, introduced new plants and grazing animals, and drained and irrigated soils. Overgrazing has seriously diminished grassland, extended deserts, and of course altered natural soil development. Most of China's core and developing fringe would have been forested, as was (and is) most of Japan, Korea, and Taiwan.

Soils have been altered as much as natural vegetation. Cultivation for upwards of 2,000 years has stirred the natural soil layers. Both draining swamps and irrigating dry soils have changed them. Oriental cultures have long fertilized soils by adding animal and human manure. Farmers also add "green manure," or plant refuse. The constant cultivating of some soils has changed their texture. Standing water in the rice paddies has also altered soil.

THE STRUCTURE OF SUBREGIONS

The combination of sheer size and diversity justifies a series of subregions, yet there are also the questions of separate nation-states, widely differing political systems, divergent historical evolution, and contrasting cultural elements requiring subregional treatment (see Figure 6–1). As the only non-Western, fully industrialized country, Japan serves as a transition between the developed and developing worlds; for this reason, it is discussed first.

Five political units constitute an East Asian **shatterbelt**: North Korea, South Korea, Taiwan, Hong Kong, and Mongolia. The Mongolian People's Republic is distinctive on both cultural and historical grounds. The people and their language are Mongolian, not Chinese. Even today, the Mongolian People's Republic is a client state of the USSR, not of China.

The two Koreas share a peninsula, isolated by water or rugged terrain from both China and Japan. At times in the sphere of influence of Russia, China, or Japan, Koreans are fiercely conscious of their cultural heritage. Split in two by cold war political fortunes, Korea is still a single cultural entity.

Chinese in language and culture, Hong Kong and Taiwan are politically separate from the PRC. The former is a colonial outpost, though it will be returned to China by treaty. The latter is the refuge of a government in exile. Both are rapidly industrializing, developing economies, and both are economically linked with the West, and with Japan. Each, in its own way and on a different timetable, will reach an accommodation with the PRC, one that will allow some local autonomy but move toward the goal of "one China."

The old core area, the developing fringe, and the sparsely settled periphery were chosen subregions within China. Han (ethnic) Chinese have long dominated the Chinese fringe area culturally and form the majority population in much of this zone. Nonetheless, the land outside the old 28 provinces is, or has been, a pioneer area of China, characterized by in-migration of Chinese and rapid new development.

Japan: A Paradox

Japan is very much an island culture; a strong insular theme runs through its long history. Korea, its nearest mainland Asia neighbor, is 100 miles away. In contrast, at its narrowest, the English Channel places but 20 miles of sea between Britain and France. Japan's culture shows strong influences from China. Characteristically, however, the Japanese have always adapted and modified their cultural imports, making them distinctively Japanese. The uniquely Japanese approach to industrialization has made this nation a giant in international trade. In industrial productivity, Japan ranks with the two superpowers, yet it has 122 million people, less than half the Soviet population, and a little more than half the American number. Japan's list of domestic industrial raw materials is almost laughably short compared with the variety, quantity, and quality possessed by both superpowers.

Japan enjoys many distinctions; it is first on many lists of economic achievements and qualifies for more than its share of superlatives. Its salient quality is the fact that it was the first non-Western society to undergo that complex, rapid modernization and change in industry, transport, agriculture, and society itself that we call the industrial revolution. It remains the only non-Western culture to achieve a thoroughly industrialized, high-technology society at the climax stage of both the demographic transition and the Rostow developmental model.

Japan is an ancient culture, proud of a recorded history stretching back almost 1,500 years. The Japanese believe that they are descended from a people they call the Yamato, who extended control over all of Japan in the third and fourth centuries A.D. A few centuries later, the Japanese had made contacts with Korea. Through these contacts, they received many technologies and arts from highly developed China. Thus, Japanese culture incorporates and adapts many ideas and technologies brought in from China and Korea. For example, the written form of Chinese, **ideographs**, was adopted. Each ideograph represents an idea or an object. (In contrast, Western alphabets communicate the sounds of a spoken language.) The Japanese, speaking a very different language than spoken Chinese, could study Chinese books on philosophy, sciences, astronomy, and medicine. The Japanese patterned their government on the Chinese model, again adapting it into a Japanese form.

In the late ninth century, Japan's contacts with China were interrupted, and Japan began a long period of assimilation of imported culture into one uniquely and characteristically Japanese. Power passed from the emperors to a series of feudal military rulers, called "shoguns," for some seven centuries. Portuguese traders had landed by 1543, followed shortly by Jesuit missionaries led by St. Francis Xavier. Soon, Spanish, Dutch, and English traders had joined the Portuguese. The Japanese government became alarmed that Christianity seemed to be spreading quickly, especially in southern Japan. The shogunate, fearful of the rapid spread of foreign ideas, decided to "close" Japan from such contacts in 1639.

Christianity was forbidden, and foreigners banned, with one exception. Small numbers of Dutch and Chinese traders were confined to a small island, Dejima, in Nagasaki Bay. For 250 years, this island was the only point of contact with the rest of the world. The foreign merchants, unable to leave their little island (under punishment of death), could not learn much of Japan; but the Japanese learned much from them. The Japanese ordered illustrated encyclopedias and models of machinery from the Dutch and, through them, from other European nations and America. It was an eminently practical arrangement, one through which the Japanese could continue to learn Western science and technology.

Thus, when Commodore Perry arrived in Edo (now Tokyo) Bay in 1853 to "open up" Japan to international contacts, the Japanese could be confident

GROWING UP JAPANESE

Traditionally, Japanese children are enveloped in the warm embrace of their parents, particularly their mothers. They are nursed for a relatively long time and are carried about by their mothers during household chores rather than placed in a playpen. In contrast to American practice, where infants are given their own rooms if space permits, Japanese children sleep among the family until quite old. American children might be given rules and warnings, then also relative freedom to explore their world and perhaps break the rules. Punishments and admonitions follow. But Japanese children learn acceptable behavior more by the patient example of parents who maintain intimate contact with them.

Japanese children are babied, not treated as miniature, independent adults. As a result, Japanese children are highly dependent on their parents, again especially their mothers. Growing up in Japan produces a child, and then an adult, who is accustomed to lavish approval from parents, and from the group. The power of the group, and the threat of group disapproval, are repeatedly emphasized as the child grows up. The warm indulgence of mother is replaced by the approval of the group—first one's schoolmates, then workmates.

Japanese parents chide poor behavior with the warning, "People will laugh at you." Japanese thus learn to discriminate between acceptable and unacceptable behavior in the eyes of the group. This OK–not OK division is more flexible, in different contexts, than the Western "hard line" between right and wrong. It has been said that guilt is the major deterrent to, and result of, bad behavior in Western societies; guilt follows disobeying the Commandments, thus disappointing the Ruler of the Universe. In Japan, shame before the group is the deterrent to, or result of, poor behavior. A Japanese wishes to avoid the disapproval of the group, and the group is willing to forgive and forget if the offender demonstrates shame and repentance. Mitigating circumstances carry considerable weight in group, and even court, judgments.

that they had a basic understanding of Western technology. They were ready to begin a period of rapid assimilation of this scientific-industrial basis for modernization. This habit of borrowing ideas and improving on them has long characterized the Japanese economy.

The shogunate collapsed in 1867, and full sovereignty was restored to the Emperor Meiji in 1868, who made Edo, the shogunate capital, his imperial capital, renaming it Tokyo. The Meiji era (1868–1912) was one of the most remarkable periods in the history of Japan, and of the world. In less than half a century, Japan, building on its Dejima Island knowledge, achieved a thorough modernization that had taken other modern nations centuries to achieve.

By 1894, the Japanese were confident enough to go to war with a disintegrating Chinese Empire; they won. Japanese power was further confirmed by their swift defeat of Russia in 1904–05.

The speed with which Japan ingested foreign influence was matched only by the speed with which they digested these influences, incorporating them into a definitively Japanese culture. This ability to switch over to new, unaccustomed ways of doing things has been demonstrated repeatedly, as in the democratization of Japan following World War II.

To Westerners, it sometimes seems as though Japanese social manners stand in lieu of an ethical code. There are few prohibitions in Japanese life, if moderation and suitable consideration of the appropriate time and place are present. Sexual activities outside marriage, including homosexuality, are not considered wrong as such. Sex is regarded as a normal function, one that is pleasurable, like dining—an activity to be enjoyed in the proper setting.

In business meetings and negotiations, the Japanese proceed cautiously until a sense of group decision is evident. Definite statements and strong opinions are avoided. The Japanese have developed such a strong sensitivity to the feelings and opinions of others in the group that nonverbal communication is highly efficient. Tiny variations in posture, facial expression, or gestures can accurately convey reactions, so that the group consensus can be finally voiced by someone without much discussion or a formal vote. To avoid an embarrassing outright rejection of a business proposal, the Japanese approach the subject elliptically; the American practice of initiating a discussion with an explicit price or set of requirements is as disconcerting to the Japanese as the seemingly mysterious arrival at a Japanese group decision is to Americans.

Work colleagues are expected to spend a great deal of time with one another after work, socializing in restaurants and bars; this expense account socializing among the group is as important as associated entertaining of clients and potential customers. All of this intimate contact enhances nonverbal interaction and focuses loyalties on the group. Status is defined by one's associates and the group's (or corporation's) prestige. It is significant that the Japanese words for "sweet" and "to look to others for affection" are closely related.

Japan is the only economic superpower that is not presently a major military power, in keeping with constitutional restrictions on military power dating to the post–World War II American occupation. The defense of Japan, in effect, is a responsibility shared by the Americans and Japanese, so that the Japanese economy is not burdened with the full cost of the military power defending it.

If anyone believed that there is a direct cause-and-effect relationship between the physical resource base and level of industrialization, Japan would be the best counterargument. Japan accounts for nearly 10 percent of total world trade, but it must import most of its coking coal, almost all of its raw cotton and wool, and virtually all of its petroleum.

If population density is expressed as **physiological density** (persons per unit of cultivable land) rather than as persons per total land area, then Japan has the world's highest population density, excepting only the city-states of Hong Kong and Singapore (Figure 6–4). Population distribution in Japan was once closely related to the limited availability of reasonably level, and hence easily cultivated, land. Because level land is a consideration for industry as well as for agriculture, the correlation of population with level land still has some validity. Much of Japan is made up of rugged forested mountains, with narrow ribbons of agriculture and population in the valleys. Japan's largest plain, the Kanto (Kwanto) Plain around Tokyo, is only roughly the size of the Los Angeles Basin, about 120 miles in length. Much of Japan's population is concentrated in the southern two thirds of Honshu, the largest island, and on the two southern islands of Kyushu and Shikoku (Figure 6–5).

The Land of Japan. Japan consists of four major islands, by far the largest of which is Honshu ("mainland"). Hokkaido, the northernmost main island, is second in area, followed by Kyushu and Shikoku. There are about a thousand, mostly very small, other islands, including Okinawa, which was returned to Japan by the United States in 1972, along with the other Ryukyu Islands held since World War II.

The Japanese islands are geologically young and are characterized by frequent earthquakes and no fewer than 196 volcanoes; Japan is about 80 percent mountainous. The highest volcanic peak (12,388 feet) is Mount Fuji, which last erupted in 1707.

The mountainous nature of Japan means that there are many short, swiftly flowing streams, and relatively little flat land. The broadest and most use-

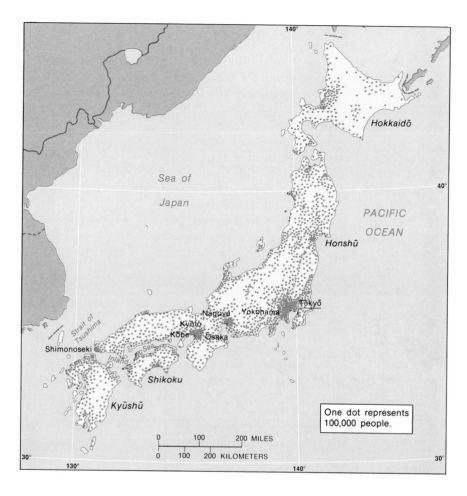

FIGURE 6–4
Population Distribution in Japan. Coastal lowlands concentrate most of Japan's people. Population density is extremely high if expressed as physiological density.

ful plains front on the eastern, Pacific coast. The Kanto Plain is only 12,475 square miles yet contains over one quarter of Japan's population.

Japan's coastline, 16,654 miles of it, is deeply indented. The Japanese have developed numerous good ports, although many on the edge of the larger coastal plains pockets must be dredged because of the alluvial fan nature of these lowlands, which extends recent sedimentary materials out under a shallow sea. The Setonaikai ("Seto inland sea") is used intensively as a water highway between southern Honshu, northern Shikoku, and northeastern Kyushu.

Japan's seaward orientation, compared to land-preoccupied China, is related to its heavy dependence on seaborne trade for imports of raw materials and exports of manufactured goods. For example, Japan operates nearly 500 tankers; the regular pro-

cession of oil tankers from the Persian Gulf and from Indonesia to Japan has been termed the **oil bridge**. Although Japan has developed most of its considerable hydro power, this source contributes only 5 percent of total energy demand. Japan also is third in number of nuclear power plants in operation (after the United States and the United Kingdom). Nevertheless, Japan must still rely on the oil bridge. Thus, Japan's traditional dependence on sea lanes has been strongly reinforced by its current economic position. Its limited military power, especially its naval and air forces, may soon be increased to protect these sea routes.

Japan often is perceived as a small country, but this perception depends on the comparison. In area, Japan is about 10 percent smaller than California, with which it is often compared. Although small compared to the superpowers (a favorite Japanese comparison is that Japan is only 1/25 the size of the United States), Japan is larger than most European countries.

Climatically, Japan experiences a seasonal temperature and precipitation regime similar to that of the eastern United States from New England to Georgia. The East Asian winter monsoon winds move outward from the mainland across the Sea of Japan before reaching those islands, and are thus warmed and humidified from their originally cold and dry state. Pacific maritime air masses dominate Japan in summer, bringing in warm, humid air from the southeast and south. Heavy rains accompany the occasional typhoons (hurricanes) in late summer and early fall. Precipitation ranges from 40 to 100 inches, with heavy snow in northwestern Honshu and Hokkaido in winter.

Natural Resources. Japan's natural resource base is one of the most meager among major industrial nations. Coal is the most important resource, but Japan has only 0.4 percent of world coal reserves. Most coking-quality coal is imported, principally from Australia and the United States. There are many small metallic ore deposits scattered across Japan; commercially produced minerals include lead, zinc, copper, sulphur, gypsum, limestone, and dolomite. Japan has only 0.01 percent of the world's known oil reserves; imports supply more than 99 percent of its consumption.

Forests cover about two thirds of Japan's total area; the national culture has always recognized the importance of forests, from prominence in the arts of all eras to the careful preservation of national forests (Figure 6–6). Varieties range from camphor, palm,

FIGURE 6–5
Japan's Core and Selected Cities. Virtually all of Japan's cities and industries are located on the seacoast, facilitating international trade via inexpensive ocean transport.

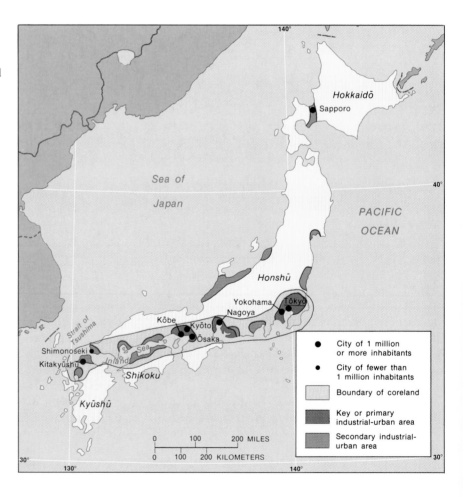

FIGURE 6–6

Land Use in Japan. Scarce flat land is used intensively for grain because three quarters of the land is in forest-covered steep slopes.

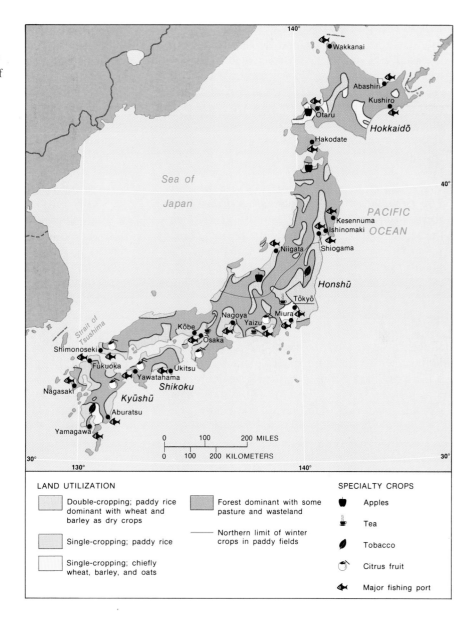

LAND UTILIZATION

- Double-cropping; paddy rice dominant with wheat and barley as dry crops
- Single-cropping; paddy rice
- Single-cropping; chiefly wheat, barley, and oats
- Forest dominant with some pasture and wasteland
- Northern limit of winter crops in paddy fields

SPECIALTY CROPS

- Apples
- Tea
- Tobacco
- Citrus fruit
- Major fishing port

and bamboo in sheltered areas of the south to maple, cherry, oak, and chestnut in the central area to conifers in the north. Replanted species are usually fast-growing pines.

Japan is rich in ocean fisheries, thanks to the mixing of warm and cool currents offshore. With so little farmland per capita, the Japanese have emphasized ocean sources of animal protein rather than beef, the Western equivalent. Hundreds of small ports send out fishing fleets to supply the fresh-seafood markets.

Unfortunately, the intensive industrialization of Japan, with almost all industry located on or near the coast (frequently built on fill), has polluted many coastal waters. For more than half a century, the combination of heavily fished, often polluted local waters and rising demand from a larger population has sent Japanese fishing fleets far from home, while finfish, shrimp, and octopus are "farmed" in local shallow waters. The wide-ranging, ubiquitous Japanese fishing fleets are running into increasing restrictions as coastal states extend territorial waters to protect their fisheries from overfishing and foreign exploitation.

Pearls and oysters are harvested from farms in protected bays. Eels, considered a great delicacy, are also farmed. Japanese excellence in scientific marine biology goes back to the 1880s, when pearls were first cultured in oysters. Freshwater fish farming is also common, in ponds, lakes, and even flooded rice paddies.

The Growth and Characteristics of the Japanese Population.

Japan's population was fairly stable at about 30 million until the period of rapid industrialization in the Meiji era. At that time, continuing high birthrates and falling death rates led to rapid expansion. Population tripled from 34 million in 1872 to 103 million by 1970, with most of that growth before 1955.

Following World War II, Japan's population showed a sharp surge as 3.5 million returned to the home islands from the former colonies, and there was a brief baby boom as returning veterans began long-delayed families. The 1945–50 census interval showed the highest gain ever recorded for Japan. The newly enacted Eugenic Protection Act in theory legalized abortions in instances of seriously deformed fetuses; in practice, it legalized abortions on demand. In modern Japan, the annual growth rate has further declined to 0.6 percent annually. Still, with a population base of 122 million, this very moderate rate adds over 1 million a year.

With birth- and death rates both low, Japan's is a rapidly aging population. In 1980, 9 percent of the population was over 65. By the year 2000, 16 percent of the Japanese will be in the 65-and-over category, and by 2020, a stunning 21 percent could be in that age bracket. In contrast, 11 percent of the U.S. population will be 65 or older in the year 2000, 15 percent by 2020.

The traditional Japanese respect for persons of advanced age may be eroded by the considerable burden of so high a proportion of the population in retirement. Those active in the work force must support, directly or indirectly, those in retirement plus pre-work force youth. As is typical in advanced, industrial societies, Japan's youth are increasingly likely to receive advanced training or higher education, stretching out their period of economic dependency.

The Japanese population is almost homogeneous ethnically. There has been little in-migration since A.D. 750. The Ainu, an ancient caucasoid group once dominant through the Japanese islands, are on the verge of cultural extinction. The only sizable group of non-Japanese residents are 800,000 Koreans, the remainder of a large group imported as labor during World War II.

A curious group called the *eta* or *burakumin,* are physically indistinguishable from other Japanese but are regarded as outcasts. This social prejudice, continuing despite legal equality, may date from discrimination against a clan of butchers and tanners, much as India's untouchables incurred discrimination when they necessarily violated religious prohibitions on taking life and handling carcasses. Also, Japanese families carefully investigate the lineage, educational achievements, and reputation of the family of prospective mates for their children. Even today, parents prefer to contact friends with marriageable children to arrange to introduce a couple whose families will have, by then, preapproved the possible match. If the two young people like each other, they will begin courtship, secure in the knowledge that each family finds the other to their liking.

Agriculture.

Japanese land units are minute by Western standards (the average Japanese farm is under 3 acres; 40 percent are less than 1.5 acres), but the productivity per acre is probably the highest in the world. Yields of rice per land unit are at least 50 percent greater than in the PRC, and twice those of India. The American occupation government, led by General Douglas MacArthur, pushed through a massive and thorough land reform that forbade ownership of more than 7.5 acres (except in Hokkaido's New England-like climate). This reduction in land-lordism, coupled with small-scale mechanization, has resulted in high rice productivity. With production increasing and per capita consumption falling, Japan often must cope with a rice surplus. This surplus, as in many industrialized European nations, depends on strict import controls and heavy subsidies. American rice could be delivered in Japan at costs less than those of raising it there. Japan, again like European industrial states, accepts the higher food costs that result from subsidies and import restrictions in order to maintain a stable rural society and to avoid a potentially disastrous reliance on imported food.

Japanese agricultural imports emphasize high-quality and luxury foods, especially proteins. Only about 2.5 percent of Japan's land is in pasture; beef cattle are stall-fed. The country imports about 20 percent of its food; however, if imported animal feed were also considered, the figure would reach 50 percent. Imported corn is used as animal feed to increase chicken, hog, and cattle production. Dairy products and meats are becoming more important in Japanese diets. Except for rice, Japan is becoming less self-sufficient than ever before.

At the same time, there is a drive to export more high-quality, high-value-added processed foods. Viticulture—the growing of wine grapes—is spreading rapidly as the Japanese determine to match European and American standards in wines. Japanese whiskey, fruit-based liqueurs, and beer already have achieved international reputations and markets.

Problems in Japanese agriculture include inherently inferior soils and severe urbanization pressures on farmland. Elderly farmers whose lands lie in the path of urban expansion (and in Japan it is difficult for fairly flat land not to be near a city) sell out to speculators as farm children show less inclination to take up agriculture and more desire to participate in urban social life.

Japan's Industrial Economy. Among Japanese strengths are the prevalent group spirit of cooperation, a long-term beneficial interaction of corporate and government planning, and the internal flexibility of giant conglomerate corporations.

As noted, Japan was never entirely cut off from the rapidly expanding technology of western Europe during its reputed isolation. The Japanese exported silk fabrics and porcelain to pay for European technology, and the government underwrote pilot projects in applying this technology. Government-fostered enterprises were sold off to reliable families for further development, starting a tradition of close government–business cooperation, which has persisted. The giant conglomerates, or **zaibatsu** that dominate Japanese business originated in this pe-

riod. Corporate size and concentrated power are encouraged, because they represent an advantage in world markets.

To praise Japanese labor as highly disciplined and motivated is almost an understatement. Japanese productivity per worker-hour would delight European, American, or Soviet managers. And production efficiency keeps increasing. For the 30 percent of the Japanese work force who work for the highly efficient, high-tech companies whose logos are recognized around the world, the corporation is the focus of lifetime loyalty. The corporation, in turn, provides fringe benefits including low-interest housing and car loans, low-cost vacations at company-owned re-

Corporate Logos, Tokyo. Japan's postwar "economic miracle" was led by giant corporations whose products and logos are recognized worldwide.

THE JAPANESE AND RELIGION

Western religious leaders observe, with dismay, the "secularization" of their societies—the demotion of religion from a central position to one on the periphery. In Japan, this secularization is far more advanced, and is centuries old.

Buddhism came to Japan from India via China and Korea around A.D. 538, transmitting a whole higher culture. Art and architecture were dominated by this new religion, and monasteries became rich landowners as well as strong influences on intellectual life. Buddhism remains in Japan, but only in a background role for many. Most funerals are conducted by Buddhist priests, and the ashes of cremated Japanese are buried in a family plot on the grounds of a Buddhist temple, so that most Japanese are nominally affiliated with one or another Buddhist sect. Between family ceremonies, however, the Buddhist temples are seldom visited. The post–World War II land reform also stripped the monasteries of their lands, and thus of incomes and prestige.

The viewpoint that Westerners call Confucianism, and East Asians term "the teaching of scholars," is the main reason for early secularization forces in Japan. Confucianism did not become important in Japan until the seventeenth century. It is more an ethical system than a religion; one thinks and lives correctly, following strict ethical rules. Confucianists believe in a rational natural order of which people are a harmonious part.

Most Japanese do not think of themselves as Confucianists. However, their society is so permeated with Confucian ethical values that, in effect, they are Confucianist in their belief in the importance of interpersonal loyalties, learning, and hard work. Japanese behave as good Confucianists while denying any formal affiliation, and are nominally Buddhist but ignore it most of their lives; their relationship with Shinto ("the way of the gods") is as casual as with Buddhism. Shinto's roots go back to a native Japanese worship of natural forces and fertility. There was no theology or even ethical system, but a variety of deities were worshipped, along with

sorts, medical benefits, and preferential hiring of one's offspring.

To the other 70 percent of the Japanese labor force, employment with smaller, less efficient, less well capitalized companies brings lower wages and few fringe benefits, if any. This second tier of Japanese enterprise is especially exploitative of female labor in assembly jobs. Many Japanese move from job to job in this tier (whereas workers in the first tier seldom move). Seventeen thousand business bankruptcies a year characterize this unstable industrial sector which is under increasing pressure from low-labor-cost economies such as South Korea, Taiwan, the PRC, and the Philippines. The duality of Japanese industry, with the modern industrial structure largely under the control of the zaibatsu and the traditional

Buddhist Temple, Japan.
Buddhism came to Japan in the sixth century of the modern era, brought by Chinese and Korean traders. It has had a major impact on Japanese culture, as evidenced by this temple, the oldest wooden building of this size in the world.

ancestors. Shinto was unconcerned about an afterlife and thus coexisted comfortably with Buddhism, which was willing to accommodate local beliefs that did not directly conflict with its tenets.

The leaders of the Meiji Restoration were anti-Buddhist and attempted to create a strongly nationalistic state Shinto, with enforced worship at shrines. Reverential treatment of pictures of the emperor was required, and the new worship of the state was far removed from original Shinto. State Shinto was abolished during the American occupation.

Today, brief visits to Shinto or Buddhist shrines are made to attend funerals, weddings, or birth celebrations, but it cannot be said that either religion is the focus of many people's lives. Visits are often motivated more by an interest in traditional art, folklore, and the patriotic association of the shrines than by their religious significance.

light industries of the second tier (textiles, toys, ceramic products, parts supplied to the zaibatsu) has existed for a long time.

Japan's dependence on a highly skilled and innovative labor force requires a superb, if rather rigid (by American standards), educational system. Entrance into the best schools is highly competitive; the reputation of one's school helps achieve a position with the most desirable companies. Japan is generally agreed to have the highest functional literacy rate of any country. A minor curiosity is that the Japanese score higher than Americans on American IQ tests (in translation).

Girls participate fully in the educational system through secondary school and form the majority of students in junior colleges. Many junior colleges, however, are regarded as polite finishing schools, preparing women for a good marriage. In the most prestigious four-year colleges, women are a distinct minority. Japanese women have progressed from their subservient roles in feudal Japan. However, although they have full equality under the law, there remains much economic disparity. Japanese tend to marry later (24 average age for women, 28 for men) than Americans. Most women spend four to eight years in the labor market between finishing their education and marriage. A later start in child rearing, and a much prolonged motherly supervision of children, means that women reenter the labor force much later than American women. Excluded for the most part from the privileges of lifetime jobs and the seniority system, women make up 40 percent of Japan's work force, but mostly in low-wage jobs.

A Changing Industrial Structure.

An outsider visiting Japan just following the end of World War II could be forgiven for not predicting that Japan would, by the 1980s, outproduce every other industrial country except the superpowers, and even outproduce them on selected products. Shipyards and steel mills lay in ruins. Transportation systems were destroyed. Entire cities were smoking heaps of rubble. The Japanese were at first slow to rebuild, apparently fearful that the victorious Americans would demand war reparations in the form of industrial products, plants, and machinery. Only after a few years of postwar apathy did rebuilding and modernization become commonplace.

Even Japan's mountainous island geography contributes to efficiency of plant operation. Shallow water offshore from crowded, narrow coastal plains encourages new industrial expansion through the dredging of deep ship channels close inshore and the filling in of adjacent tidal lands, providing industry with both new land and adjacent ocean shipping facilities. Coastal location using imported raw materials and supplying distant markets by sea is an extremely efficient transportation system (Figure 6–7).

Iron and steel were among the first industries to recover and modernize; imported ore and coking coal are unloaded from ships at tidewater plants, then carried to the furnaces by conveyor. Finished steel moves directly onto barges or ships, to be taken to other waterside plants for fabricating. Japan is the world's largest steel exporter, shipping directly from waterside plants. Shipbuilding boomed in the late 1950s and 1960s but has since shifted toward countries with cheaper labor. Total labor costs per hour, including fringe benefits and employment-related taxes, for Japan average 65 percent of U.S. costs. For South Korea, the cost is 10 percent of the U.S. average.

The repeated oil shocks of the 1970s contributed to an enthusiastic reception in the United States for the small, fuel-efficient Japanese cars. Cars and light trucks are now by far the leading Japanese exports to the United States. Other major items in the flow of Japanese manufactured products to the United States (and around the world) are office machines and consumer electronics.

The first-ranking U.S. export to Japan is corn, followed by chemicals, soybeans, computers, aircraft, coal, and lumber. The emphasis in U.S. exports to Japan on raw or semiprocessed materials is one reason for the huge trade imbalance in Japan's favor. The United States imports mostly high-valued manufactured products from Japan but finds it difficult to ship similar high-technology products into Japan, at least if the Japanese directly compete with those products. Though the Japanese have begun to dismantle the maze of tariffs and quotas used to protect their domestic manufacturers in their home market, a cumbersome bureaucracy still remains to frustrate foreign would-be suppliers. Detailed product standards seem designed to favor Japanese products, and foreign suppliers may be asked to submit applications for import qualifications that run to many thousands of pages.

Problems of Environmental Management.

At this time, Japan probably has the most severe and widespread environmental pollution problems of any major industrial power. The extreme crowding of Japan's industry, population, and agriculture onto a discontinuous series of alluvial fans backed by narrow mountain valleys is an important factor in pollution. Japan cannot spatially segregate polluting industries from its population centers, best farmlands, and offshore fisheries; they are all, necessarily, side-by-side. Also, the Japanese determination to remain competitive in world markets has retarded the retrofitting of existing plants with expensive, energy-

FIGURE 6–7
Urban and Industrial Patterns of Japan. Note the concentration of industry in the Tokaido megalopolis, comprised of Tokyo-Yokohama, Nagoya, and Osaka-Kobe-Kyoto.

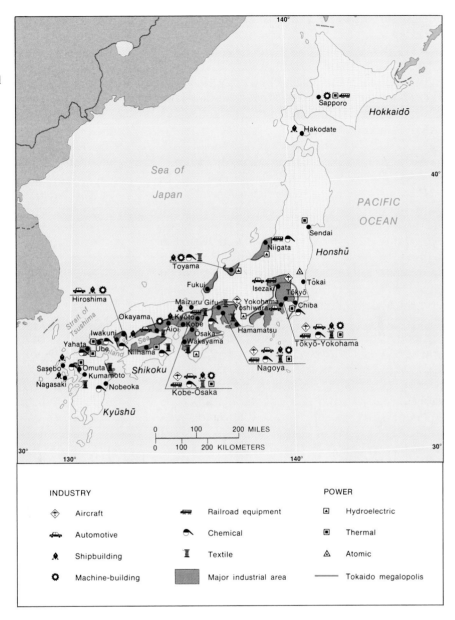

INDUSTRY

⊕ Aircraft

🚗 Automotive

⚓ Shipbuilding

⚙ Machine-building

🚃 Railroad equipment

🛢 Chemical

𝕀 Textile

▨ Major industrial area

POWER

▣ Hydroelectric

▢ Thermal

△ Atomic

— Tokaido megalopolis

consuming pollution controls. Further, public awareness of the extent and seriousness of environmental pollution lagged far behind the problem, as it did elsewhere in the industrial world.

Because so much of Japan's urban-industrial life is built on soft alluvial sediments near river mouths, and so much urban land has been reclaimed from shallow water by fill, subsidence has become a serious environmental problem. Buildings, even whole industrial complexes and city neighborhoods, are slowly sinking as groundwater is withdrawn at high rates.

Urban Japan. Officially, Japan's population is 76 percent urban, a little lower than in California or New Jersey, but far above the world average. However, some "urban" areas are rural-urban neighborhoods of apartment blocks, industrial and commercial land use, and the tiny fields and paddies that characterize Japanese agriculture. Only one third of Japanese farmers earn their living exclusively from agriculture; the other two thirds are part-time farmers with at least one adult householder commuting to urban-industrial work.

In Japan, as in other developed societies, cities, towns, and suburbs are merging into great metropolitan regions along connecting transport routes. Especially in the last two decades, the mammoth conurbation around Tokyo has become part of a "Tokaido megalopolis" reaching to Kobe on the Seto Inland Sea (Figure 6–7). Tokaido refers to both the megalopolis and the transport corridor that is its backbone. The "New Tokaido Line," for example, the 130-mph "Bullet Train," serves Japan's megalopolitan corridor as the Metroliner serves the U.S. East Coast's megalopolis.

About 75 percent of the Japanese live in the narrow corridor from greater Tokyo westward to Kobe, the Tokaido megalopolis, and then south and west toward Hiroshima. The cities are fantastically crowded and expensive in land and housing costs. Residential lots 30 miles from downtown Tokyo are offered at more than $150 per square foot. Typical two-room condominium apartments in Tokyo, averaging 600 square feet, can cost over $150,000. Many Japanese families live in apartments of 400 square feet or less (essentially one room with bathroom and kitchen cubicles, and built-in, curtained-off bunks for sleeping).

The Japanese have adjusted to living in cramped spaces by developing a "less is more" approach to art, including the design of gardens and interiors. A Japanese garden is designed skillfully to give the impression of spaciousness within a small area. The idea is to portray natural landscapes on a miniaturized scale. Raked sand or fine gravel might represent the sea, and large, picturesque rocks, islands. The mountains and the sea are common subjects for paintings and drawings, often suggested by a few brush strokes rather than portrayed in detail. Trees are relentlessly pruned and shaped, so that one decades old can grow in a pot or a windowsill—the famed "bonsai." Even in flower arranging, the ideal is a few flowers placed in elegant and delicate balance, rather than masses of flowers.

Strategies for the Future. Japan is a major factor in international trade, but it is not unusually dependent on that trade. The sheer size of its economy, and the fact that the United States is its customer, means that Japanese products are highly visible in North American markets. Japan exports 13 percent of its total GNP, a much higher proportion than for the United States (7 percent). However, Canada exports 29 percent, West Germany, 27 percent, and the United Kingdom, 20 percent. Still, Japan's export trade is important to Japan, if hardly unique to industrialized nations.

Several factors help explain Japan's success in export markets. In their home markets, Japanese producers must be intensely competitive. Japanese consumers are reputed to be the world's toughest customers, carefully inspecting all details of assembly, finish, and trim. Manufacturers know that, if they bring an innovative product to market, it will be only a short time before rivals market a close cousin, a little cheaper and quite possibly already improved over the pioneering model. In Japan's classic capitalism, the road to success lies in selling a better product at a better price. Although the huge zaibatsu have prospered since before the Meiji Restoration, and have immense capital resources with which to dominate markets, there are examples of new companies, founded on a vision and a few yen, that became giants through new products or cheaper, improved products. Examples are Sony and Honda, which serve as models for the thousands of small, hustling companies trying to carve out a place in Japanese, and then world, marketplaces. Survivors of competition at home are well equipped to compete successfully in international trade. The prosperous Japanese are their own best customers for many products, due, in many cases, to protectionist trade policies. The high savings rate of the Japanese means that Japanese banks have ample investment capital available to fuel the rapid expansion of successful innovators.

Government agencies share information on foreign economies with exporters and encourage the kinds of export price agreements, market allocation, and selling below cost to establish a presence in foreign markets that would put American corporations on trial for antitrust and anticartel violations.

The Japanese must maintain their reputation for high-quality, sophisticated products. Japanese cars regularly win top recommendations from American consumer products testing labs, and the Japanese seem to have reinvented the camera, so successful have been their new technologies in photography. The Japanese continue to purchase foreign technology rights and licenses (a tradition going back to the Tokugawa era), but they also innovate. They seem particularly adept at product improvement engineering. They take a proven technology (their own or an import) and refine the idea, making it more efficient, more durable, more appealing to a mass market, and then produce it in its improved version more cheaply than ever before available.

While improved engineering of existing products and technologies continues to be a profitable tactic, the Japanese are determined to innovate new computer technology. First-generation computers used vacuum tubes; the second generation used transistors. Third-generation computers use integrated circuits, and the fourth generation, just now coming into production, relies on very large scale integrated circuits (VLSIs), with chips that must be designed by another computer. These four generations were American innovations. The fifth generation, which the Japanese hope will be born in Japan, will use large numbers of VLSIs in parallel to achieve another breakthrough in power and speed. Fifth-generation computers will be able to run, for example, a voice-operated typewriter or scan and translate text into different languages.

Japan is a global economic power, but it is a society and government that often thinks in the old, insular patterns. The United States has long been Japan's most sympathetic, and largest, trading partner. American resentment at the Japanese insistence on unrestricted access to foreign markets while doggedly protecting and restricting access to Japanese markets is mounting. Japan may be near the pinnacle of its remarkably successful postwar "economic miracle." It may be time for a major shift in the Japanese economy to cope with potential social stresses at home; continually evolving relationships with the United States, China, and the USSR; and the rapid industrialization of the East Asian mainland periphery (South Korea, Taiwan, Hong Kong). The Japanese, however, have repeatedly demonstrated their collective ability to change their course sharply, quickly adapt to new ideas and new systems, and absorb and transform cultural impacts from abroad.

Traditional Japanese society did not develop concepts of inalienable rights of citizens or of representative government, or any of the philosophical foundations of democracy. Japanese feudalism committed the followers to obedience and loyalty and granted the leaders almost unquestioned authority. This structure was tempered by a long tradition of group rather than personalized leadership. It was made difficult, deliberately, for one person to exercise authority; group discussion would lead to a consensus of leaders, which seldom was at variance with what the population as a whole would accept. Group decision making was prevalent and remains characteristic of both the economy and the government. The 1947 Constitution, largely the effort of General MacArthur, functions quite well, and Japan is a true democracy, although many Western-imposed concepts of government have been turned to strictly Japanese applications and functions. There are occasional antigovernment demonstrations and some small, troublesome dissident groups. However, there are no internal cleavages of note, and neither government nor social instability is a problem so long as Japan grows and prospers.

It is fitting to end the discussion of the world's major industrialized societies with Japan. Japan already occupies a position, and copes with a set of problems, that lie in the future for many other industrial nations. Intensely urban and industrial, reliant on long-distance imports of raw materials, involved in an intricate network of international trade relationships, Japan may pioneer strategies of management that other countries, in turn, can profitably adapt.

The East Asian Shatterbelt

The term *shatterbelt* describes an area containing many small political units. The East Asia Shatterbelt consists of North and South Korea, Taiwan, Hong Kong, and Mongolia.

The Koreas. In common with Germany, Korea is a cultural, national unit split by cold war politics. As in Germany, the split grew out of what was intended to be a temporary division into military occupation zones. The Soviet Union entered the war in Asia only shortly before the surrender of Japan, which had occupied Korea since 1895. Both the Communist state

TOKYO AS A WORLD CITY

The aggregation of Tokyo-Yokohama-Kawasaki is the quintessential world city.* This great area contains between 25 and 30 million people (in bounding most metropolitan areas, determining how far out into the surrounding sprawl of suburbs and satellites one carries the official area is a complex problem) (Figure A). The first permanent nucleus of future Tokyo was established in 1456; the site is on a well-protected bay, backed by the largest plain in the islands. Its situation is on the largest island, Honshu, approximately central to the four main Japanese islands. Less than a century and a half after founding, it was the de facto seat of power in Japan under the Tokugawa regime, becoming the official capital with the restoration of the Meiji emperor in 1868. By 1920, Tokyo City had 3.3 million.

Tokyo's growth in the modern era has been powered by heavy in-migration as employment shifted from agriculture toward manufacturing and services. The service sector accounts for almost two thirds of employment in central Tokyo. Tokyo's notorious congestion and overcrowding of public transport has led to a large subway expansion program.

Ginza, Downtown Tokyo. *Tokyo may be the world's largest metropolitan area; rankings depend on data sources and definitions that are not exactly comparable. Tokyo has the world's most expensive urban real estate in this exceptionally crowded city.*

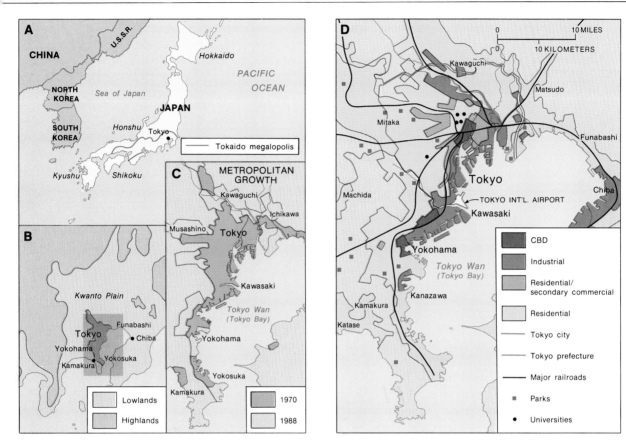

FIGURE A

Tokyo, A World City. This series of maps show Tokyo and its metropolitan area at four different scales to illustrate Tokyo's *situation* within Japan (a) and within the Kwanto Plain (b), its *site* on Tokyo Bay (c), and its generalized land use within the metropolitan area (d).

Housing problems reach nightmarish proportions; some companies resort to building apartment houses for exclusive rent to their employees. The explosive growth of the metropolis has severely strained sewage, water supply, refuse disposal, and traffic control. "Pushers" are hired to help pack in as many passengers as possible on subway and commuter trains. The Japanese have taken to wearing slick, nylon outercoats to facilitate sliding in and out of such dense crowds. The pragmatic Japanese are decentralizing some 80,000 government jobs to the far suburbs. Filling in large portions of shallow Tokyo Bay and building up, the city is able to continue its growth despite the scarcity of land. A recent example of building on filled, former bay is the Disneyland complex to the east of Tokyo.

*Peter Hall, *The World Cities* (New York: McGraw-Hill, World University Library, 1966).

(with its capital at Pyongyang) and the Republic of Korea (which uses the traditional capital at Seoul) claim jurisdiction over the entire peninsula (Figure 6–8).

Korea is not particularly rich in natural resources. There is a minor deposit of low-grade iron ore in North Korea (Democratic People's Republic of Korea), along with small, scattered deposits of iron alloys and anthracite coal. The considerable hydroelectric potential of the Yalu River (the boundary with China) was largely developed by the Japanese during their imperial control.

South Korea's resource base is even poorer than that of North Korea. A small anthracite coalfield and tiny deposits of tungsten, gold, silver, and manganese exist, but South Korea's remarkable industrial growth has occurred despite, rather because of, its resource base. Japanese investment had concentrated in what became North Korea, so the South started with little industrial infrastructure.

Changing Korean customs reflect the onetime high infant mortality rates of the first stage of the demographic transition. Households with a new infant once were quarantined for three weeks, to minimize exposure to diseases. When the baby was 100 days old (many did not survive till then), a lavish dinner was held in celebration, and the child's first birthday was the occasion of a large party. Now, most born alive do reach their first birthdays, but the custom of delayed celebration of a birth remains.

Despite their geographic position between two much larger, occasionally aggressively expansionist nations, the Koreans are a distinctive people with their own language and rich cultural tradition. Koreans still resent the wholesale destruction of many of their cultural artifacts in recent wars, and zealously preserve what is left of an ancient heritage.

Rice is the predominant crop in both Koreas, forming more than half the total tonnage of farm products, but North Korea has had a serious problem in feeding its population. Despite minor concessions including private gardens, collectivization into "cooperatives" has not produced the surge in productivity precipitated by South Korea's incomplete land reforms. South Korea's agriculture is becoming commercial and shifting toward a much greater emphasis on livestock raising.

South Korea's urbanization and industrialization can only be described as explosive. Seoul has a population of over 10 million, which is expanding rapidly. The industrialization of South Korea has set new world records in terms of pace of development. Whereas Japan already had an industrial base and a large skilled industrial labor force with which to build its postwar economic miracle, South Korea has surged from a largely preindustrial stage in 1945 to become an economic power by the 1980s. South Korea's progress from 1965 to 1980 equals a half-century of development in older industrial nations. In the 1990s, South Korea will probably become the third largest shipbuilder, and its steel industry will produce double Britain's. Korean automobiles entered U.S. markets in 1986. The total economy could match that of West Germany well before the close of this century. The stability provided by the U.S. defense forces, together with the strongly pro-growth attitude of both the government and the people, have attracted heavy investment from Japan, the United States, and overseas Chinese interests. South Korea's burgeoning tourist trade rests largely on Japanese tourists. The South Korean economy has recently shown some signs of the strain of being the fastest expanding in the world. Inflation and rising wages have resulted in the loss of some markets for cheap consumer products to Taiwan and Hong Kong.

North Korea's rate of economic growth has been less spectacular. True to the Marxist format, the North Korean government has emphasized the production of heavy industrial goods. A rapidly expanding chemical industry produces large amounts of fertilizer, some of which is exported to China. Intensive use of fertilizer from this new industrial branch has increased rice yields, and the general agricultural picture has brightened in recent years. Though North Korea has fewer people to feed from a larger area than does South Korea, much of its area is mountainous and ill-suited to agriculture, and the climate is colder.

North Korea is in the unenviable position of being the neighbor of both the Chinese and the Russians at a time of poor relations between those two countries. It has moved deftly, maintaining trade and an amicable relationship with both.

The Republic of China (Taiwan). Taiwan, an island 100 miles off the coast of mainland China, is about one-third the area of South Korea. A mountainous spine dominates the east coast, while the

FIGURE 6–8
Cities, Railroads, and Cultivated Area in North and South Korea. When Korea was divided into North (Communist) and South (pro-Western) following World War II, the North was most developed, industrially, while the South contained most of the cultivated land. South Korea now is one of the fastest expanding industrial economies in the world.

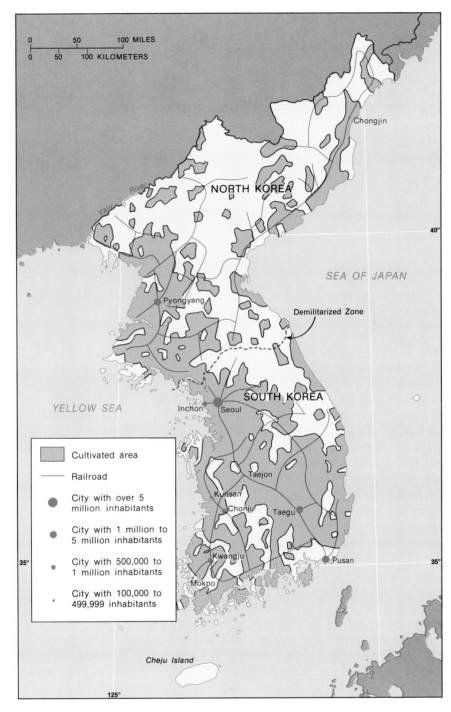

Seoul, South Korea. *South Korea has undergone one of the swiftest transitions into a modern industrial state ever witnessed. Seoul, the traditional capital, has rocketed to an estimated 10 million people.*

west coast has some coastal plain well suited to agricultural production.

The population of some 19 million includes about 2 million mainland-born Chinese who moved the capital of Nationalist China to Taipei after the Communist victory on the mainland. Following the loss of control over the mainland, the Chinese Nationalist government introduced a successful land reform in Taiwan. Tenant farmers could buy land at 2.5 times the value of annual yields, payable over 10 years. Landlords were compensated for the full amount received from tenants; many landlords whose wealth was then liquid became entrepreneur industrialists. Agricultural prosperity supported industrialization by reducing food imports and supplying raw materials for a growing food-processing industry. Subtropical and tropical Taiwan harvests two rice crops a year and is now more than self-sufficient in rice. Sugar, canned asparagus, canned mushrooms, canned pineapple, bananas, and tea are exported. Intensive agriculture is necessary in very high density Taiwan, whose land supports more people than all of Australia, which is 200 times its size.

In addition to food processing, leading industries are textiles, electronics, electrical appliances and machinery, motorcycles, and autos. Taiwan's shipbuild-

ing industry is capable of building tankers of 500,000 tons.

Taiwan's industrialization was spurred by the inflow of capital and skills with the arrival of the Nationalist Chinese. Political stability attracted investment from Japan, the United States, and Southeast Asia's sizable overseas Chinese business community.

Hong Kong. The island of Hong Kong and a small tip of the Kowloon peninsula were ceded to Britain by China following the Opium War, 1839–42; Britain fought the war to open China's markets to trade, including opium produced in British India. The opium helped pay for Britain's imports from China. The original cession of some 33 square miles was later augmented by the leasing of the adjacent 365-square-mile "New Territories"; this lease expires in 1997 (Figure 6–9a).

Hong Kong's original function, like those of the many treaty ports in which Europeans and Americans enjoyed special privileges, was as an entrepot, or transshipment, warehousing, and wholesaling point. In the face of a largely incompetent and corrupt civil administration in China under the last dynasty, the Western powers sought the security and trade advantages of their own small spheres of control. Trade

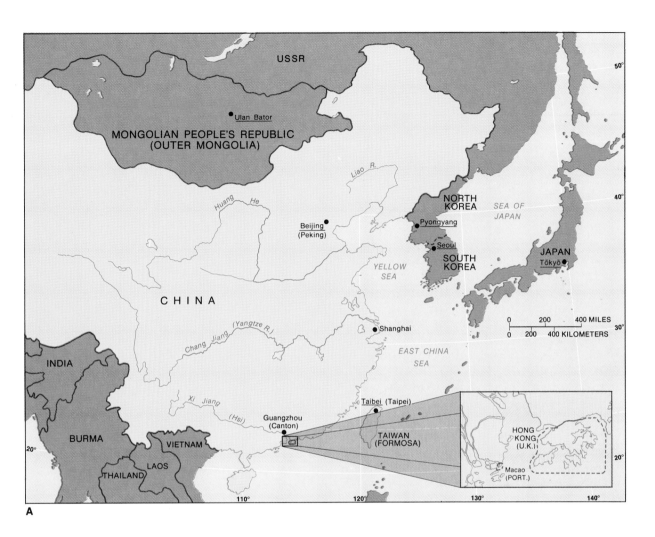

A

FIGURE 6–9
Site and Situation of Hong Kong. This pair of maps illustrates Hong Kong's situation—its position relative to China and East Asia (a)—and its site—the immediate, more local sense of location (b). Highly urbanized Hong Kong is a city-state awaiting an uncertain future with the PRC.

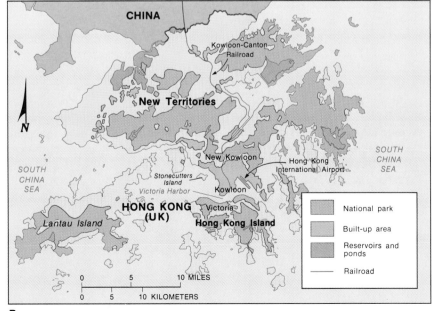

B

could be negotiated under Western laws and military protection; the treaty ports were run by Westerners for the benefit of Westerners.

The post-1945 civil war in China had several far-reaching effects on Hong Kong. Chinese fleeing the civil war and the repressive new regime boosted Hong Kong's population from under 1 million to 4 million. Many of these new arrivals were Chinese businessmen and entrepreneurs whose lives were in jeopardy under Communist rule. Sometimes these immigrants brought capital along with managerial talents and industrial skills.

Within a few decades, Hong Kong was transformed from a predominantly trading economy to an industrial one. The long nonrecognition of China by the United States made Hong Kong even more important as a neutral trading point. Chinese materials and semifinished goods were shipped to Hong Kong, finished and packaged, and sent on as "made in British Hong Kong" and thus acceptable in U.S. markets. Similarly, American technology could be exported to China via Britain and Hong Kong.

Hong Kong has no industrial raw materials within its borders. Most of the fresh food, and most of the water supply, flows across the border from China. Just by shutting off the water supply, Beijing could bring the colony to submission in a few hours (Figure 6–9b). There are only about 20,000 British in Hong Kong; it is a Chinese city whose unique political status offers more advantages to the Chinese than even to the British.

Hong Kong's industrialization is based on abundant, cheap, and hard-working labor, and on capital inflows from refugees and overseas Chinese. Textiles and clothing are major products. Plastic consumer goods, especially those requiring hand assembly (e.g., plastic flowers) are a Hong Kong specialty, along with electrical goods and transistor items.

The United Kingdom and the PRC have signed an agreement (December 19, 1984) on the future status of Hong Kong. China had already announced plans to resume full sovereignty when the lease expires in 1997. An advanced statement of China's policy toward Hong Kong was necessary to avoid chaotic economic conditions if fear of an unknown future led to large-scale flight of capital from the colony. China wishes to take over a thriving industrial and port city with a great concentration not only of capital but also of entrepreneurial and managerial ability. The agreement provides for a "one country-two systems" approach. Hong Kong will be a part of China, but can keep its capitalist system for 50 years after reunification with China. If the plan works as envisioned, the PRC will gain a valuable, if capitalist, asset, not an impoverished shell.

Mongolia. Mongolia, wedged between the Soviet Union and the PRC, has been alternately dominated by each of its far more powerful neighbors for the last several centuries. It was a Chinese province from the late seventeenth century to 1911 when, during the turmoil of the overthrow of the imperial Chinese state, it became an autonomous state under Russian protection. A 1946 plebiscite overwhelmingly reaffirmed Mongolia's independence of China. Since then, Soviet political and economic influence have intensified, with Soviet troops stationed in Mongolia since the 1960s. Ironically, Mongolia at one time was a world power that had conquered most of present-day USSR and China. The Mongols once ruled from the coasts of China to Poland and Hungary.

The only usable lands of Mongolia are in the north near the Siberian border. This inhabited part of Mongolia is separated from Chinese settlements in Inner Mongolia by several hundred miles of desert. Economic ties, as a result, are much stronger with the USSR than with China.

Mongolia's nearly 2 million people are predominantly Mongol ethnics whose population growth rate is 2.7 percent, one of the higher rates of increase in the world. The USSR and various Eastern European states have provided the technical assistance and financial aid to fuel Mongolia's rapid modernization.

Industry, mostly food processing and textiles, accounts for a larger share of the GNP than agriculture, which still employs half the labor force. Exports, however, are primarily agricultural—cattle, horses, wool, meat, hides, and grain.

The Soviet military presence and virtual control of Mongolia's economy have established a great salient in the border with China. The PRC regards Mongolia as an irredenta; which may prove to be a major point of contention between the USSR and the PRC.

Hong Kong. *The explosive growth of Hong Kong in the past 40 years, as the city's commercial economy added industrial functions, has led to severe housing shortages. Many Hong Kong residents live on boats like these.*

China

Western Perceptions of China. After the American diplomatic initiative to China in the early 1970s, the onetime enemy of Korean War days suddenly began to be looked upon as a potential friend. Long thought of as an alien, unfriendly society, the People's Republic of China suddenly became the focus of U.S. interest. A door long closed was again open.

Western cultures have long been fascinated by the Orient. Roman Europe had a small-scale but significant trade with ancient China and, ever since, much European economic and political energy has been directed toward developing trade with the Orient. Chinese technology (through the seventeenth cen-

tury) equaled or surpassed that of Europe. In ceramics and metallurgy, Chinese craftsmanship was vastly superior.

By the nineteenth century, European military technology forged ahead of Chinese military technology. Various European nations sought territory or trade concessions from the then-weakening Chinese Empire. The British controlled sea routes and acquired Hong Kong; the Russians developed overland routes; and the Trans-Siberian Railroad attempted, in a sense, to outflank British naval power and to secure some of the lucrative China trade to the Russians.

The emphasis on the China trade has shifted from China's once-superior export consumer wares and luxury items to its potential as a purchaser of Western manufactured goods and a possible supplier of oil and other raw materials. There was great excitement in the United States after the Nixon visit to Beijing. The sudden opening of the PRC can be compared to the famous opening of Japan in the 1850s. The prospect for significant shifts in world trade patterns, alliances, and the balance of power are just as high now as they were in the 1850s.

China as a Modernizing Country; China as a Civilization.

Only the Soviet Union and Canada are larger than China, and China surpasses all other states in the size of its human population (Figure 6–10). The high rate of agricultural employment, relatively recent industrialization, and comparatively low standards of living would indicate that it is a developing nation. Clearly, though, it is not an ordinary developing nation. The PRC has reestablished firm control over all portions of its territory; this unity had not existed for most of a century. Although there is still a pressing need to improve the quantity and quality of the average diet, for the first time in more than a century there is no widespread or serious famine. The PRC had detonated nuclear devices and has launched space satellites. There can be little doubt that, if current trends continue, China will become a superpower in every respect during the lifetime of Americans of average college age.

Although hardly unique to them, the Chinese have been notably ethnocentric. The two traditional names for their country, the "Middle Kingdom" and the "Central Flower Country," emphasize the Chinese view of China as the center of the universe. This ethnocentric view must be placed in a historical context. It is understandable that early Chinese civilization would have seen itself as uniquely advanced compared with the "barbarian" groups that surrounded it (and repeatedly threatened it).

The self-confidence of this great, culturally rich, and expansionist civilization must be understood in terms of a long history of cycles of expansion and contraction of centralized authority over the vast and varied territories of the Chinese state. The lands now ruled by Beijing are at least as large a unit as ever directly controlled by any Chinese central government. While a large, unified China has existed periodically over the last 2,000 years, there have been periods of internal disorganization, provincial independence, civil war, and temporary invasion by for-

eign powers. However, Westerners should not overemphasize the disunity and technological inferiority of the last 200 or so years. Placed within the context of over two millennia of Chinese history, this comparative weakness and vulnerability can be seen as a brief interlude. In terms of its significance to world culture, technology, arts, and economy, China has ranked among the major cultures for most of its existence.

Population.

China's population has always been considered burdensomely large, leading to various programs to control growth. The present Chinese annual growth rate of 1.1 percent is lower than the world average. The large original base, coupled with the youth of the population, however, makes even this moderate growth rate a matter of concern. Birthrates (19 per 1,000 in 1986) still are higher than in the United States (16 per 1,000), but they have fallen rapidly in recent decades and continue to decline.

Much of China is not usable for agriculture; a scant 10 percent of the total is currently considered cultivable. The Chinese work force is employed predominantly in agriculture (68 percent), so rural densities of up to 2,500 per square mile are found in the fertile lower and middle Chang Jiang basins. Densities of 1,200 per square mile are average in most of the arable country (see Figure 6–10). Some 21 percent of all Chinese now dwell in cities. There are 15 cities of over 1 million people, the largest of which, Shanghai, has over 12 million people.

Of an estimated world population of 5,011,000,000 in 1987, China's population of 1,067,000,000 represents 21.3 percent. Roughly one in five people alive on the planet today is a citizen of the People's Republic.

China's population was 500 million at the time of the Communist victory over the Nationalists in 1949. After three decades of Communist rule, China's population was closing in on 1 billion. Mao Zedong encouraged a baby boom; orthodox Marxism views the

FIGURE 6–10
Population Distribution of East Asia outside Japan. This map illustrates the extremely uneven distribution of the peoples of East Asia, excluding Japan. Note the divide between well-watered, densely populated eastern China and arid-semiarid western China with its much lower population density.

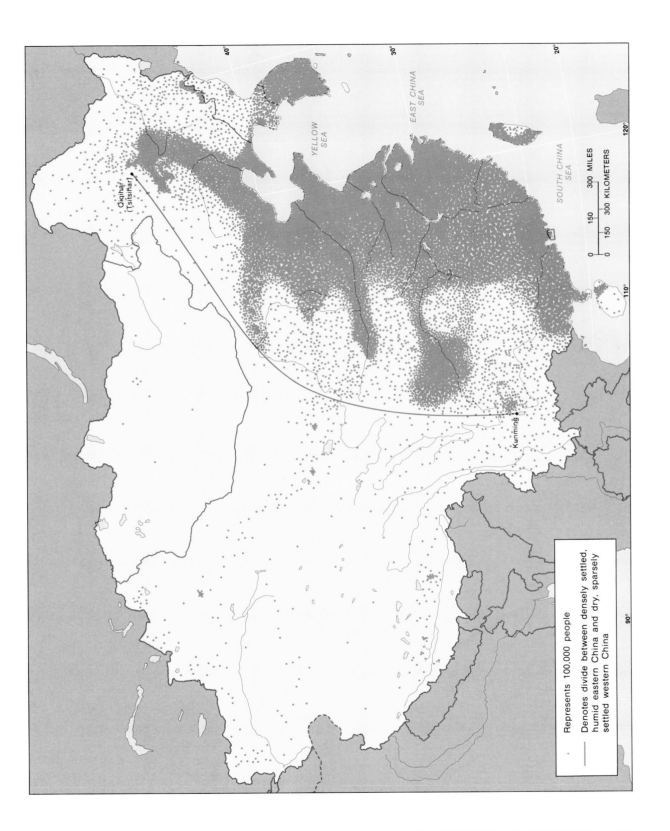

Diqihar (Tsitsihar)

Kunming

YELLOW SEA

EAST CHINA SEA

SOUTH CHINA SEA

300 MILES
300 KILOMETERS
150
150
0
0

• Represents 100,000 people

— Denotes divide between densely settled, humid eastern China and dry, sparsely settled western China

CHINA'S SECOND REVOLUTION

Chinese leader Deng Xiaoping's administration attempted to blend the seemingly opposite political-economic philosophies of Marxism and free enterprise. Sometimes termed the "second revolution," the new management techniques represent a sharp break with Communist ideology. The pragmatic Chinese see the problem in very practical terms. Why, they ask, despite immense natural resources, a large and industrious population, and a tremendous heavy-industry base, has the Soviet Union made relatively limited progress in improving living standards for its citizens? Seventy years after their revolution, Soviet citizens must still wait in long lines for laundry detergent or fresh fruit. Housing is still crowded and of low quality; private automobile ownership lags far behind that of the Western capitalist states. In China, why had dietary standards not improved noticeably in 27 years of Mao's ideological experimentation? Was something basically wrong with the Communist system? Why was free-enterprise Taiwan, with about 3 percent of the PRC's population at the time, outproducing the PRC on many consumer items? Why did the USSR have to import huge quantities of grain from the capitalist economies and purchase or steal computer and laser technology from the West?

Old Chinese folk wisdom says that it does not matter whether a cat is black or white so long as it catches mice. If private enterprise produced more rice, more televisions, and more good-quality housing, then perhaps some blend of private enterprise and socialism would work better for China. For example, collectivization of agriculture in China had been extreme under Mao. At one point, pay was based on a system of work points which emphasized quantity and ignored both quality and results. For example, work points were paid in relation to how many rice seedlings were planted, not how much rice was harvested. Each province was ordered to become more self-sufficient in grain, so forests and pastures were plowed in Ningsia province, although common sense had long ago concluded that the climate-terrain-soil conditions there were best suited to extensive animal husbandry. The result was a net decrease in food production as meat production fell and grain production was a failure. When the government decided to reward communes for improving crop yields per land unit, local managers simply underreported agricultural area to boost apparent yields. In the late 1970s, satellite photography supplied to the Chinese government by the United States showed that the PRC actually had about 20 percent more land under cultivation than official statistics showed!

The communes have been abolished. The state still owns all land, but it leases land to individual families or small production teams. Rent is paid by delivery of a specified quantity of crops to the state at an established (and low) price. Anything surplus to that rent may be sold in free markets at prices determined by the market, not by the state. Leases run as long as 15

Grand Canal, China. The Grand Canal was built many centuries ago to connect the lower valleys of China's two greatest rivers, the Huang He and the Chang Jiang. The exceptionally high population density here in eastern China reflects the high food productivity per acre.

years on cropland, 30 years on grazing land, encouraging leaseholders to care for the land in terms of fertilization, erosion control, and general improvement in quality. Unexpired leases can be passed on to one's children through inheritance.

Industrial reforms grew out of agricultural reforms, almost by accident. Farmers were permitted to build their own houses on state land. They could build small roadside stands to market produce, and they could buy a truck to haul crops to market. Soon, private entrepreneurs and village collectives were branching out into service industries (e.g., restaurants, inns, beauty parlors, tailoring) and small-scale manufacturing. Foreign companies are permitted to set up and operate factories, fast food franchises, amusement parks, soft drink bottling plants, and hotels.

Recent reforms have created a dual-sector economy in which factories, mines, and other state enterprises operate alongside private enterprises. In industry, production quotas for state delivery are known as "plan production." As with agricultural production surplus to state contract quotas, "Above-plan production" can be sold at higher, market-determined prices. Thus, the case of the auto plant that sat idle for three months, with its workers on full pay, because it had already fulfilled its quota, will not recur. Above-plan sales profits may be used for plant expansion, diversification, new tools, incentive bonuses, and such employee benefits as factory-built housing or retirement plans.

concept of **overpopulation** as a feature of capitalist societies. Overpopulation was equated with unemployment which, in Marxian interpretation, is a deliberately induced means of keeping labor costs low. The official line was, the more people, the more production.

The "more is better" Maoist approach to population policy may have been the most tragic of the many serious blunders of Mao's regime, including the ill-fated, ill-planned **Great Leap Forward** and the chaotic **Cultural Revolution**. Some Western scholars believe that food consumption per capita did not change appreciably in more than 50 years, until the mid-1980s. Average daily calorie consumption in China is estimated at between 2,000 and 2,100 per person. Americans average 3,240 and would regard 2000 calories as a rather spartan weight-loss diet. Not only is the quantity of food intake low, but the quality tends to be poor in China. More than three quarters of those Chinese diet calories are derived from rice, wheat, or corn, with little from vegetables, fruit, meat, or eggs. The government's goal is a minimum of 2,600 calories a day.

But official statistics show that the area of farmland, per person, in 1982 was only about half of what it was in 1957. Although some farmland had been added by reclamation projects, the population had increased dramatically. About 1/10 of total farmland had been subtracted, some converted to urban-industrial uses, and some lost to erosion. Recent changes in agricultural policy are credited with increasing food production per acre since 1978.

In many vital aspects of government policy—on the goals and organization of industry, on the various systems of agricultural land and labor management, and, especially, on population policy—the people of China seem to have been led in abrupt, repeated reversals. The relaxed, incredibly optimistic "the more the better" interpretation of Marxism has been replaced by a rigid population limitation policy. The goal is one child per couple in urban areas, two in rural areas. Ethnic minorities are not subjected to such a stringent quota. Intense pressure is placed on nonconformists from other members of their *danwei,* ("brigade" or "unit"), the group of co-workers and associates to which one belongs. The danwei is extremely important in Chinese life because it decides to issue (or not to issue) ration cards, building permits, occupancy permits, and so on. By law, women may not marry until age 26. Depending on where they live, the legal marriage age for men is 27 or 28. Couples agreeing to have only one child get a framed certificate attesting to their patriotism and cash rewards from their unit. Families with one child are supposed to get more housing space, and that child is promised better educational and job opportunities.

A long-standing preference for sons grew out of traditional family customs. Sons always stayed part of the family to which they were born. They would take over the family farm or business and look after aged parents. If families lacked a son, they tried to adopt one. Daughters, on the other hand, were temporary members of the family to which they were born. At marriage, each would go to live with her husband's family, under the legal protection of her husband and under the daily guidance of her mother-in-law, and subservient to her. Female infanticide was common in traditional China. Although against the law, female infanticide is again practiced under the pressure of the one-child policy. Some districts report that 6 of every 10 surviving live births are male, leading to a government warning that severe social strains will result when many men in the future cannot find brides due to this sex imbalance.

In China, all forms of birth control are available at no charge. Whereas in many countries additional children result in more deductions from taxable income, in China, a couple having a third child must pay 10 percent of their wages to the community. The official policy is that contraception is promoted, but abortion is permitted when contraception fails. The Chinese government strenuously denies that abortion is, in effect, nearly compulsory through danwei pressure or threats.

Whatever the means of persuasion, there can be no doubt of the government's determination to control the birthrate at a low level. Surgical sterilization is becoming a popular way to win praise, better accommodations, and better jobs. In their zealous and effective population campaign, however, the Chinese are sowing the seeds of a future social and economic problem. In a few decades, the age structure of the population will be extremely atypical, with many in the older cohorts and comparatively few in the younger.

MAO'S "IRON RICE BOWL" AND INDUSTRIAL INEFFICIENCY

The Chinese liken a steady job, or a prosperous farm or business, to a rice bowl. A good job is a good rice bowl; to lose a job is to "break your rice bowl." Mao had asserted that industrial laborers had an "iron rice bowl"—unbreakable, with lifetime job security. Despite China's break with the Soviets in the 1950s, Mao's emphasis on heavy industry featured a high rate of hidden, forced savings with which to finance rapid industrialization. The enforced "savings" came from low wages and high costs for consumer goods, with high factory "profits" a result. China is now the world's leading producer of cotton textiles, 2nd in radio output, 3rd in coal production, 5th in steel, and 10th in oil. China produces its own nuclear weapons and jet fighters. It manufactures computers and electron microscopes. And yet, expressed on a per capita basis, output is quite low and, for many products, of uneven quality. An enormous number of plants were built, but many produced little because of inefficiency, waste, and the unavailability of materials or components. There are many stories of factories "under construction" for 20 years, with local managers continually remodeling or expanding to disguise the fact that the factory had not been able to produce any workable products.

Western visitors and consultants often reported that plants had too much unused equipment and machinery, and were considerably overstaffed. The iron rice bowl philosophy kept workers on the payroll whether they were needed or not. *Dingti* (replacement policy) meant that the children of retiring workers inherited the parents' iron rice bowl, whether they had the needed skill or not.

Mao's legacy of ideological goals replacing common sense includes factories in the wrong places for the wrong reasons. In an industrial extension of his self-sufficiency directive to agriculture, for example, each province was to have its own auto and truck plants. China now has the largest number of auto and truck plants (130) of any country in the world. But total output is a quarter of a million vehicles a year, ranking China about 20th among world producers, behind Sweden and Australia. The provincial self-sufficiency concept led factories to stockpile machinery, tools, even labor, so that they could make their own spare parts or components, self-reliance being defined as independence from suppliers.

The Core. The core consists of most of the provinces of traditional China, plus the important industrial region once called Manchuria (Figure 6–11). Essentially, the core is agricultural China as opposed to the outer ring dominated by economies based on hunting, herding, or other pursuits besides sedentary agriculture. It is the homeland of the Han Chinese, a people who began their civilization in the valley of the Wei River, a tributary of the Huang He.

The two great river valleys and their associated lowland and fairly level areas concentrate most of the population. Most Chinese live on the "wet" side of the 20-inch rainfall line. Within the core, the mountains of southern China and the drier western margins of the core have considerably lower densities. The Chang Jiang corridor with its three basins and large, compound delta is the largest concentration.

Within the core, the dominant trend of movement is rural to urban. The Northeast (Manchuria), long a target of planned colonization, is now effectively settled. Youth are encouraged to migrate to new agricultural colonies and industrial developments in the west and southwest. The overseas Chinese, emigrants to other lands, total some 30 million, most in Southeast Asia. Recent restrictions (and sometimes out-and-out discrimination) in Southeast Asian countries discourage any further migration.

The large size of China, along with patterns of language distribution, might lead one to think of China as ethnically diverse. In fact, about 94 percent of the people are Han Chinese. The 6 percent who constitute ethnic minorities dwell in the fringe areas, where they often constitute a local majority. Tibetans and Mongols, the Uigur, and other Turkic groups dominate in the western half of the state. Chinese colonization has engulfed the Manchus and Koreans in Manchuria. The various groups related to the Thai are in southern China, in particular the areas bordering on the countries of Southeast Asia.

Agricultural Patterns. Chinese agriculture is perhaps the most intensive in the world. Virtually all available land is used, including seasonally exposed riverbeds, roadside verges, highway medial strips, and urban windowboxes. In Oriental intensive agriculture, the tools are primitive, but results are impressive. Individual per acre yields of rice or barley may be low compared to those achieved by Western mechanized agriculture. However, because barley, sweet potatoes, rice, and half a dozen other crops are produced on the same acre, the overall food yield in oriental agriculture is higher. Machines save time and labor, but they do not improve yields per se.

Oriental agriculture is extremely complex. Its heavy yield potential is based on an intricate system of rotations coupled with numerous techniques, including greenhousing, seedling selection, irrigation, drainage, selective cultivation, weeding, interculture, multicropping, intensive fertilization, and selection of complementary crops. Though all of these techniques are used to some degree in Western agriculture, it is the peculiar use of all of them in combination, and the way in which they are used, that makes oriental agriculture unique. There is almost no substitute for massive inputs of hand labor. Chinese production, despite heavy increases in mechanized agriculture in the production of industrial crops, in the dryland areas of non-paddy culture, and in the pioneer areas, still relies on the traditional system using hand labor.

Oriental intensive agriculture has evolved over thousands of years and has enabled countries such as China to be self-sufficient (or nearly so) in food supply, despite extremely high population densities. Grain imports from the United States, Canada, and Australia are used mostly to fatten animals and to enable farmers to grow industrial crops rather than grain. Processed foods and rice are among China's chief exports.

Within the country, there is a regional differentiation of crops (Figure 6–12, p. 324). The extreme south of China can grow two crops of rice a year. Tea is the specialty of southern hill lands too steep for most other crops. Most of the Chang Jiang Valley can grow two crops annually, rice and a winter grain, generally wheat. North China, colder and drier, supports only one grain harvest (generally winter wheat or some variety of millet), supplemented by vegeta-

FIGURE 6–11
Provinces, Railroads, and Cities in the People's Republic of China. Note the correlation of the relatively dense rail net with the population distribution, Figure 6–10.

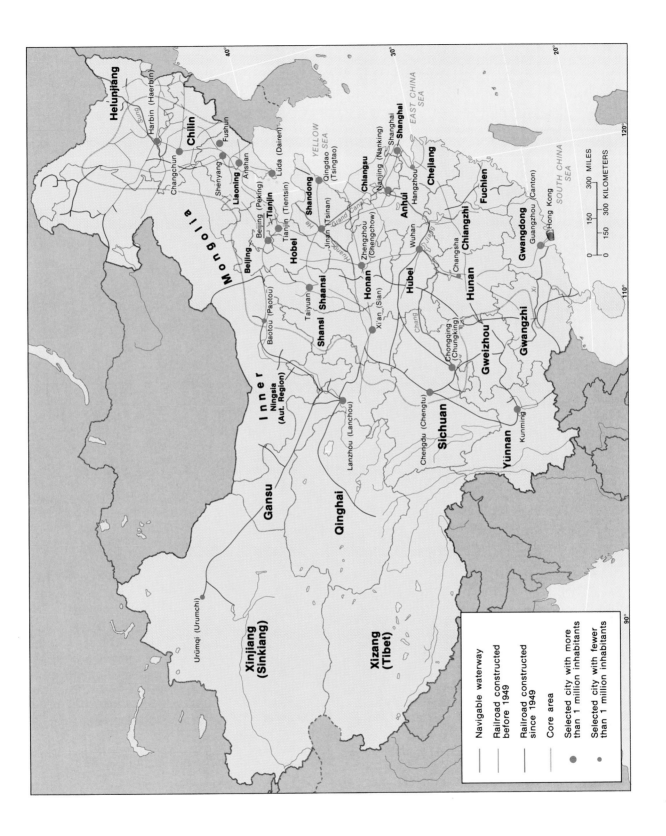

Helunjiang

Harbin (Haerbin)

Sung

Changchun Chilin

Shenyang Fushun

Liaoning Anshan

Lüda (Dairen)

Beijing (Peking) Tianjin

Beijing Tianjin (Tientsin)

Hobel

Qingdao (Tsingtao)

Shandong

Jinan (Tsinan)

Chiangsu

Zhengzhou (Chengchow) Nanjing (Nanking)

Grand Canal Shanghai

Shanghai

Anhui Hangzhou

Chejiang

Wuhan Hubei

Honan

Xi'an (Sian)

Shaansi

Taiyuan

Shansi

Baotou (Paotou)

Chiangzhi

Fuchien

Changsha

Hunan Gwangdong

Guangzhou (Canton)

Hong Kong

Gwangzhi

Chongqing (Chungking)

Gweizhou

Sichuan

Chengdu (Chengtu)

Lanzhou (Lanchou)

Ningsia (Aut. Region)

Inner

Mongolia

Yünnan

Kunming

Qinghai

Gansu

Urümqi (Urumchi)

Xinjiang (Sinkiang)

Xizang (Tibet)

YELLOW SEA

EAST CHINA SEA

SOUTH CHINA SEA

Chang

Huanghe

Xi

40°

30°

20°

120°

110°

90°

	Navigable waterway
	Railroad constructed before 1949
	Railroad constructed since 1949
	Core area
●	Selected city with more than 1 million inhabitants
•	Selected city with fewer than 1 million inhabitants

300 MILES

300 KILOMETERS

150

150

0

HEAVEN AND EARTH: CHINESE BELIEF AND PRAGMATISM

The Chinese live in a long-secularized society. They may honor several different religious traditions on special occasions, but rarely give religion an important role in daily life.

A family whose central beliefs follow the teachings of the scholars (Confucianism) might invite both Buddhist priests and Taoist monks to officiate at a funeral. This multireligious approach puzzles Christians, to whom one belief excludes others. The Chinese, while likely not wholeheartedly involved in any religion, pragmatically assume that if one religion may offer reconciliation with the gods, two or three are even better. Perhaps it is a form of hedging one's bets on the afterlife.

Jesuit missionaries arrived at the Court of the Chinese Emperors in 1582. They reasoned that if the ruler converted, most people would follow. The Jesuit effort was long and determined, and the emperors welcomed the Jesuits' technical advice on Western science and mathematics. The Chinese saw no necessary conflict between Christianity and their own religious and philosophical traditions, but did not share Christianity's central concerns. The Jesuits were interested in God; the Chinese focused on people. The Jesuits preached about salvation and the rewards of heaven; the Chinese were concerned about earthly prosperity and peace. The Jesuits were concerned about sin; the Chinese, who did not have the concept of personal sin, were worried about crimes among people. Successive emperors renewed the Jesuits' welcome at court, listening politely to theological lessons but always steering the conversation toward what the well-educated Jesuits could contribute on more worldly knowledge.

Confucius, as the founder of a school of philosophy, has had the most profound influence on Chinese thought of any single person. Confucius

---

ble crops and potatoes. The steppes of the northern fringe, similar in climate to Montana, grow spring wheat rotated with pasture. Hot summers and rich prairie soil encourage Manchuria to emphasize corn and soybeans. Increasingly, the Chinese seek to diversify their crops. Citrus, bananas, and sugar cane are rapidly becoming more important in southern China, and cotton acreage has increased enormously.

Problems of Land Tenure. Communist confiscation of landlords' holdings and redistribution to peasant farmers was quite popular. This initial land reform boosted farmer morale and farm output, but

not for long. Soon after the Chinese farmers at last enjoyed controlling their own plots of land, they were persuaded to join "production teams." By 1957, these small cooperatives of neighbors were being consolidated into collective farms. Later, it was determined to "collectivize the collectives," much as had been done in the USSR. The impact on the society was great, and the impact on the farm landscape, even greater. The complex, fine-grained rural hodgepodge—tiny fields separated by dikes, property boundaries marked by trees and shrubs, even the many small shady graveyards—was gone. Diseconomies of scale in the huge collectives and the lived about 500 years before Christ; he was an itinerant scholar, traveling about trying to find a job as an advisor to rulers. He thought that rulers should hire scholars and sponsor study of the classics. He did not convince anyone important in his lifetime, but he founded a tradition that succeeded brilliantly. About 350 years after his death, Confucianism became an unofficial state religion in China, and remained so until 1912. Rulers tended to be pleased by Confucian emphasis on submission, loyalty, and reciprocity in relationships between the people and their rulers. Only one of Confucius' "five relationships" is between equals—that of friend to friend. The others are hierarchical—between ruler and ruled, father and son, husband and wife, elder brother and younger brother. Confucian teachings stressed that only the qualified should rule; this teaching led to a system of rigid exams on classical studies to qualify one for the government bureaucracy.

Opposed to Confucianism was Taoism ("the way"), attributed to the legendary sage Lao-tse, said to be a contemporary of Confucius. Taoism teaches that people must accept what life offers and do nothing out of harmony with the flow of nature. (Taoists would agree with the contemporary advice to "go with the flow.")

Buddhism was introduced later to China; it was highly compatible with Taoism, with both sharing many values, such as compassion for others and reverence for the simple life. Elements of Buddhism, Taoism, and Confucianism became a religious amalgam that was generally, if casually, accepted by most Chinese.

resentment of peasants contributed to some dismal harvests in the 1960s and 1970s. Political commissars decreed growing crops where they had never been grown before—for good reasons apparently, because these crops failed. The disastrous emphasis on communist ideology apparently has now been shelved in favor of a more pragmatic approach. "Does it produce results?" seems to have replaced, "Is it ideologically pure?"

State farms (with salaried workers) occupy approximately 8 percent of the arable acreage. State farms are often huge and are frequently located in marginal areas. In addition to emphasizing industrial crops and breeding stock, they function as experiment stations, educational centers, and seed farms. They are heavily mechanized and still receive priority allocations of fuel, machinery, and fertilizer.

The latest phase in rural reorganization has sought to prepare the rural population for the transition to an urban-industrial society. The seasonal nature of labor demand in farming led government officials to view farm labor as an underused resource. Much as was done in Japan in the early twentieth century, cottage labor production is being marshalled into team efforts. Knit goods, clothing, canning, shoe assembly, wickerware, jewelry, farm

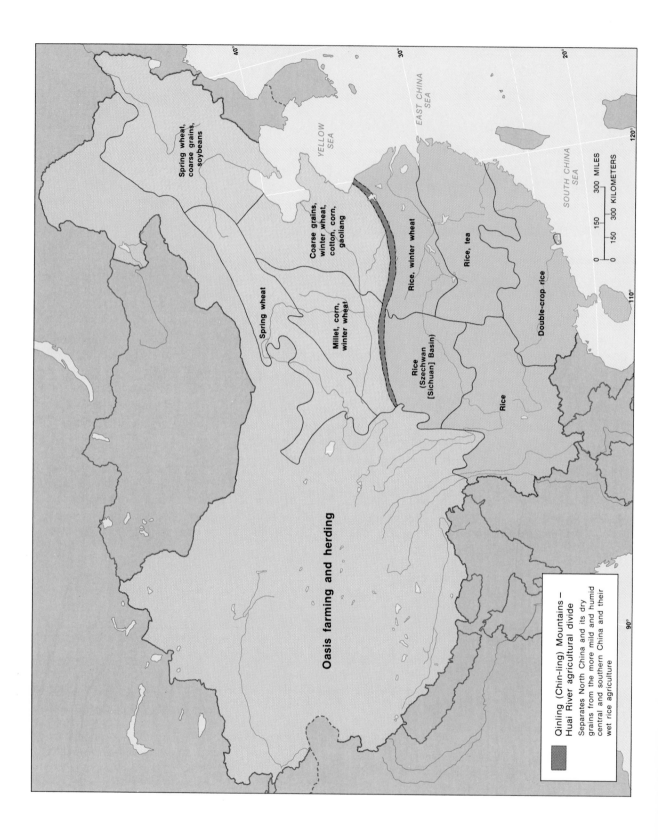

Spring wheat, coarse grains, soybeans

Coarse grains, winter wheat, cotton, corn, gaoliang

Spring wheat

Millet, corn, winter wheat

Rice, winter wheat

Rice, tea

Rice (Szechwan [Sichuan] Basin)

Rice

Double-crop rice

Oasis farming and herding

YELLOW SEA

EAST CHINA SEA

SOUTH CHINA SEA

40°

30°

20°

120°

110°

90°

150   300 MILES

150   300 KILOMETERS

Qinling (Chin-ling) Mountains – Huai River agricultural divide

Separates North China and its dry grains from the more mild and humid central and southern China and their wet rice agriculture tools, and electronics components are now being produced and assembled in villages. Peasants are learning new skills as well as increasing their earnings.

Privatization of the rural economy has its long-range perils. One farm family alone cannot maintain the elaborate system of dikes, canals, and dams necessary to water management and irrigation in much of lowland China. Maintenance can be deferred for short periods but, in the long run, disaster follows inattention. Some Chinese fear that the lusty pursuit of individual profit may undermine the necessarily collective approach to water engineering.

Likewise, the high prices for frogs in free urban markets have led to a serious frog depopulation of the countryside as the animals are energetically pursued by profit seekers. Without so many frogs, though, there are population explosions among harmful insects, which can lead to reductions in harvests. Formerly lush grasslands in the Northeast and Inner Mongolia are seriously overgrazed because pastoralists expanded herds beyond carrying capacity in an effort to cash in on rising meat prices.

The Resource Base. Chinese energy resources are impressive (Figure 6–13). Known coal reserves are 25 percent of the world's total. The largest reserves are in Shanxi and Shaanxi provinces and in the Shandong peninsula. Smaller, but strategically located deposits fuel the heavy industrial complex of Manchuria. Small deposits have been opened throughout the country as local fuel supplies to relieve pressure on China's forests, once heavily cut for charcoal. Usable reserves are estimated at over a trillion tons, and yearly production rivals U.S. output and surpasses that of the Soviet Union.

Gas production is small, but the potential for commercial finds is excellent. The search for oil has been more rewarding. Traditional source areas were in the far west. New fields were discovered in Manchu-

FIGURE 6–12
Generalized Agricultural Regions of China. While sparse precipitation results in livestock herding with oasis agriculture in the west and extreme northeast, the north-south contrasts in crop specialization are more the result of growing season and summer heat contrasts.

ria in the 1970s. More than self-sufficient, China now exports oil to Japan. Japanese technology produced oil from oil shale in Manchuria in the 1930s; production has continued and expanded. Reserves of oil shale are immense, though recovery is expensive. Currently, drilling is taking place offshore in both the Yellow and South China seas.

China's huge hydroelectric potential is a greatly underexploited resource. Immense rivers with steep gradients in narrow canyons located in southwestern China provide virtually ideal hydroelectric sites.

Iron ore reserves are abundant, but most deposits are small and quality is relatively poor. The principal advantages are their widespread location and the happy coincidence that deposits are frequently near large coal deposits. Some iron is exploited along with coal in the very same mines in Manchuria. The iron deposits near Wuhan (middle Chang Jiang basin) are extensive and of high quality.

Tin is produced in southwestern China in significant quantities, and China is still the world's largest producer of tungsten, a valuable alloy frequently used in electric light filaments. Bauxite, copper, lead, and zinc are produced in large amounts. There are new discoveries of alloys such as nickel, chrome, manganese, and antimony. China is proving to be a major mineral storehouse.

Wood has traditionally been a principal deficit resource. Land pressures and demand for domestic fuel resulted in the stripping of once-forested hillsides. Massive reforestation campaigns are necessary to reduce erosion and flooding and to provide a new resource for future generations.

Industrial Development Strategies. China's early industrial base focused primarily on light industry. Food processing, textiles, clothing, silk, and tobacco products were the most important elements. A few cities, Shanghai in particular, dominated industrial production. Japan developed a significant steel industry in Manchuria during its occupation of that area (1933–45), and the Wuhan district was an important producer of steel. Most of the rest of industrial production was best classified as crafts (jewelry, lacquerware, porcelain, ceramics, art objects), though they formed an important part of Chinese exports.

THE RATIONAL TRIUMPH OF CHINESE COOKING

China, like France and Italy, has developed one of the world's great cuisines; Chinese dishes and cooking styles can be enjoyed in most metropolitan centers in many of the world's nations. As would be expected in such a large, geographically varied country, there are several regional traditions within the Chinese cuisine. Southern China has an unusually rich variety of foods, perhaps the result of an unusual variety of ethnic groups in combination with a richly productive land and sea. Northern Chinese like to say that Cantonese "eat anything with four legs except tables, and anything that flies except airplanes." Southern Chinese food markets indeed are likely to include dogs, cats, and snakes, along with more familiar raw foods from land, sky, fresh water, and salt water. Western China has long had the reputation for preparing spicy-hot food, with plenty of red pepper and pungent spices. Northern Chinese food is regarded as unimaginative by other Chinese; because this is "wheat" rather than "rice" China, staple dishes feature noodles, pancakes, dumplings, and gruel, made from millet or wheat. The north is the home of sweet-and-sour dishes, spiced heavily with garlic, onions, and ginger. The Manchu contribution to northern cooking includes barbecued meats and mutton.

China has been an intensely crowded country for so long that prime farmland must be devoted to grain or vegetable production rather than to pasture. Traditional Chinese cooking uses meat or seafood as flavoring more than as main dishes. There is a strong tradition of vegetarianism, with roots in Buddhist admonitions not to take life. When meat is used, it is far more likely to be chicken or pork than beef, because chickens and pigs do not need extensive pasturage and may be treated as scavengers, finding much of their own food and being fed table scraps. Meats and vegetables are shredded or diced before cooking both to minimize fuel needs in cooking, and to make the use of chopsticks possible.

Firewood has been in short supply for a long time, so the Chinese developed a flared cooking pot with a much more efficient heat flow design than Western saucepans, whose vertical sides tend to burn the dish at the bottom but less fully cook food at the top. The Chinese narrow-bottom, flared-top pot dates to the New Stone Age and is known to Americans as the "wok."

Chinese cooking, with its emphasis on grains, plenty of vegetables, small amounts of meat, and little fat is quite healthy. That combination of foods, if not necessarily that cooking style, is often recommended to Westerners who have suffered heart attacks.

In sum, the Chinese habits of using every available food source, coaxing maximum food production from every square inch of land and water, and using fuel efficiently have contributed to a classic cuisine for the world to enjoy. More than that, Chinese dietary habits may be adopted of necessity by a world trying to feed more people more efficiently.

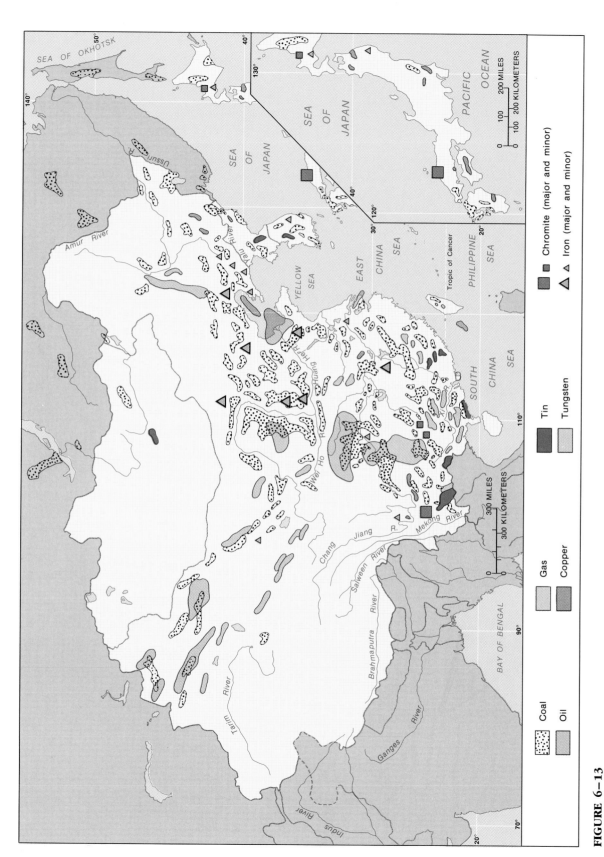

FIGURE 6–13
Natural Resources of East Asia. China contains an impressive quantity and variety of mineral and energy resources.

THE RHYTHM OF THE SEASONS AND CHINESE INTENSIVE FARMING PRODUCTIVITY

A typical unit of Chinese agriculture would be hard to identify. A theoretical example from eastern central China would probably follow a regime much like this one:

January/February—Seeds are planted in greenhouses, cold frames, or small, sheltered fields covered with plastic tenting. The seedling becomes the unit of planting at a later date; it allows for preselection of good, healthy stock. Because the rice varieties used here require 210 days for growth, the technique saves time, using a part of the season with frost to begin farming.

March/April—Rice seedlings are planted in a series of flooded ditches. Any late frost can be overcome by flooding the entire paddy. Flooding gives rice a guaranteed water supply, provides nutrients in solution and, most important, allows the growth of varieties with heavy grain yields (big heads) and a low ratio of straw (stalk). The rice will remain in the field (paddy) until harvest in early August. The hills between the trenches will be used to grow another crop, barley or oats—quick-ripening grains that like cool weather.

May—Barley or oats are harvested, and the entire paddy is flooded to hold up the now half-grown rice. Water chestnuts float among the rice. Fish, freshwater shrimp, or frogs are grown; they feed on weeds and plankton and can be eaten by the farmer, by larger fish, or by the ducks. Duck manure is a particularly rich fertilizer.

The first phase of industrial modernization in China began almost immediately after the Revolution. Following the Soviet lead, China embarked on a series of expensive, capital-intensive projects geared to establishing a heavy industrial base.

Disenchanted with the Russian-style plan's results by the mid-1950s, Mao initiated his own plan for development, attempting to capitalize on China's vast armies of labor. The program came to be known as the Great Leap Forward. It was designed to transform China instantly into a self-sufficient industrial economy, and its hallmark was the backyard steel furnace.

Outdoor bread ovens became miniature blast furnaces; much of the iron smelted was of such low quality as to be unusable. Agricultural production declined because the all-important ingredient in oriental agriculture—labor—was occupied with other tasks. Food supplies grew short, and piles of low-quality iron and steel began to mount. The Chinese economy came very near total collapse.

After this calamity, Chinese industrialization resumed a slow rate of growth. Agriculture was reemphasized, and fertilizer and farm tool industries received high priority. Private enterprise is now

June/July/August—The paddy is drained to aid in ripening the rice; the old trench is hoed up against the rice stalk to aid drainage and support the grain. The intervening ditch will yield lettuce or Chinese cabbage. After the rice is harvested, the area is given to seedling cucumber plants, beans, or sweet potatoes. Lettuce and cabbage will be harvested in early September, with some pickled for winter use. Cucumbers will be harvested through October. Vines from the sweet potatoes will feed the hogs.

Late fall—As the cucumbers and early sweet potatoes are harvested, a winter crop of radish or spinach is grown. Field peas are cultivated; if they do not ripen, they are still nutritious as a fodder crop. Often, varieties of winter onion and hardy cabbage are also grown. Cows feed on the pea vines and some of the oats or barley. The pigs, on a potato and table scrap diet, will also "fowl off" (search for and eat undigested matter in manure) the cow, and chickens (which act as an important insect control) will fowl off the pig, so that the least particle of food is not wasted through a lack of digestion. Pigs yield meat, leather, bone for buttons, fat (lard), and bristles for paint brushes, a viable export. Not only is the system a heavy yielder of food, but it provides for many other needs of the farmer. No piece of land is wasted; no growing day is not taken advantage of.

encouraged in the consumer goods sector; industry and industrial planning are deliberately decentralized. Both traditional and modern industrial sectors are given equal rank, as China continues its policy of "walking on two legs."

Despite the emphasis on decentralization of investment, distinct zones of industrial development have begun to emerge. Shanghai, together with Nanjing and Hangzhou, dominate China's largest industrial district in the Chang Jiang Delta lands. Traditionally a center of consumer goods, especially textiles, it has become China's most diverse industrial center.

The largest automotive center in China, it produces trucks, busses, and motorcycles. Shanghai leads in production of household appliances and electronics goods. Nanjing has become a major chemical production center while retaining its strong development in the manufacture of textiles and clothing. Hangzhou specializes in agricultural implements, food processing, and the manufacture of food-processing equipment.

Shanghai is China's largest industrial city at 12 million people; it may also be the world's most crowded metropolis. Many families occupy one-room apartments of 250 to 300 square feet and have no private toilet or kitchen facilities. Long the commercial and industrial hub of China, Shanghai has responded more slowly than Shenghen to the new dual industrial economy. The new industrial city of Shenghen, built a dozen years ago in former rice paddies in the "special economic zone" bordering Hong Kong, has many more joint ventures than Shanghai.

The second most important industrial complex is that of Manchuria. The name Manchuria is not used by the Chinese, who refer to that region simply as the Northeast. The original Russian and Japanese investment in iron and steel has been enhanced by later Chinese endeavors; the Northeast produces over a third of all Chinese iron and steel. Large new oil fields in the area have given rise to major refineries and extensive chemical complexes. As the oldest and largest center of steelmaking, it has become the major machine-building center of China. Anchored by the Taching oil field and the steelmaking center of Harbin on the north, it culminates in the dual port cities complex of Lüda, a great refining, petrochemical, and shipbuilding center. Intermediate centers include Changchun (automobiles and tractors), Jilin (fertilizers and synthetic substances), Shenyang and Anshan (steel and machinery), and a host of smaller centers arising as satellites to these major urban centers.

The much cooler average temperatures and the associated lower agricultural productivity per acre meant much lower population densities for the Northeast than were found in the North China Plain. Most Chinese did not consider the Northeast a desirable place, and the last dynasty, the Q'ing of the Manchus, was not eager to settle its traditional homeland and power base with larger numbers of Chinese either. The relatively sudden inflow of Chinese into the Northeast in the last days of the dynasty must rank among the great migrations of modern history. There is still room for expansion of agriculture in the drier and colder north.

Russian interests in the Northeast early in this century conflicted with a growing Japanese interest, helping to precipitate the Russo-Japanese War. The Russians had built a railroad across the Northeast as a shortcut from central Siberia to their Pacific coast port and naval base, Vladivostok (see Figure 6–11). Only the Japanese victory in their war with the Russians, 1904–05, removed the threat of Russian dom-ination. The Russian threat was followed by the reality of a Japanese take-over.

The discovery here of the largest oil reserves yet found in China has resulted in the rapid enlargement of chemical production. Recent discoveries of bauxite, molybdenum, lead, zinc, and manganese complement the huge coal reserves of the Northeast.

The third largest Chinese industrial area is evolving in North China. Tianjin, the outport for Beijing, is the largest center of the Chinese chemical industry. It has a huge iron and steel complex and a major emphasis on heavy machinery. The construction of a new port in 1949 has enhanced both the commercial and industrial functions of the city, which now has a population of over 5 million.

Beijing, the capital, has become a major industrial center in the last 35 years. There is a major integrated steel mill, enormous cotton textile and clothing mills, and myriad small plants producing electrical equipment, machinery, and a host of consumer goods. Still, administrative and cultural functions dominate over industry. The industrialization under the Communists has added a new dimension to the city, which now has over 9 million people.

As small satellite industries supplant agriculture as the chief employer, Tianjin and Beijing are gradually becoming one aggregation. The discovery of oil in the Huang He Delta has enriched the area's raw material base.

Thus, despite government plans to industrialize the interior, the coastal cities of old still dominate. The hinterlands have been enlarged, and major industrial districts have developed, but the old treaty ports and early investment areas are still of major importance.

Much of the rest of China's industries are in scattered nuclei in the interior and the south. Premier examples of large investments in the interior designed to develop the countryside and to integrate the country are Baotou, a major center of iron and steel production on the great bend of the Huang He, and Lanzhou, the gateway city to the dry west. The site of a major oil field, Lanzhou is a center for refining, petrochemicals, and the smelting of nonferrous metals. Both cities are on extensions of the railway network built since 1949 (see Figure 6–11).

Wuhan, in the middle of the Chang Jiang basin, is really a cluster of cities at one of the great crossroads of China—a major rail, water, and highway junction.

Major engineering industries, including large-scale factory machinery, use the steels produced in this urban complex of over 6 million people.

The Sichuan Basin contains two large cities, Chengdu and Chongqing. Supplied amply with coal, gas, and oil found throughout the basin, these cities are growing centers of heavy industry. Chongqing is considered the head of navigation on the Chang, and Chengdu is at an important junction on the north-south interior route. The Sichuan and middle Chang Jiang basins are viewed as the logical inland expansion of the lower Chang Delta complex.

Guangzhou (Canton), a major treaty port in the nineteenth century, is the regional capital of all of south China. Food processing is becoming important; the highly productive, subtropical and tropical agriculture of southern China forms the major raw material base. With more than 5 million people, Guangzhou is a major center supplying parts and semiprocessed raw materials to the burgeoning industry of Hong Kong. As labor costs rise in Hong Kong, Guangzhou may take over the assembly functions of the toy, clothing, and small machine industries.

Cultural and Economic Capitals. The Chinese core contains two great world cities, Beijing and Shanghai. Beijing is the greatest center of Chinese culture and education. Its name means "northern capital," and its status as the center of the Huang He lowland, as well as its position between China's two largest industrial districts, enhances its domestic importance.

At Beijing's center is the Forbidden City, once open only to the emperor and his court, now a museum of international fame. Filled with splendid architecture and displays, carefully landscaped and maintained, home of China's finest academic and technical institutes, it is the center of modern Chinese culture. Resplendent with history and tradition, Beijing is a memorial to China's past. Capital of the most populous state in the world and the center of an important modification of Marxist thought, it is a city with worldwide impact.

Shanghai is the largest city in China but was never a political capital. It grew rapidly after 1843 when it was first opened as a treaty port to foreign ships, merchants, and consulates. European influence was once so strong that many hotels, restaurants, and clubs, although they had Chinese employees, forbade Chinese guests. It is not surprising that many Chinese regarded it as a reminder of Western, exploitative imperialism. The Communist regime at first tried to restrict the rebuilding and continued growth of Shanghai, finally relenting when it became obvious that the imperialist image (and fact) had disappeared long before. Shanghai is a major industrial and commercial center, with a generally superior infrastructure of transport, communications, electrical energy supply, and support facilities installed during the period of Western dominance. In many ways, Shanghai was China's most Westernized or modernized metropolis at mid-twentieth century, and it continues to benefit from its superb situation.

Street scenes in contemporary China are much livelier than under Mao, whose rage against capitalism prohibited street vendors or private businesses of any kind. Recent liberalizations have led to a resurgence of street life as the Chinese, traditionally among the most energetic of entrepreneurial peoples, return to bargaining and the profit motive. Curbside restaurants, itinerant peddlers, knife sharpeners, barbers, and sidewalk booths selling T-shirts (often with American pop culture themes), folk remedies, audio tapes, and household gadgets now attract milling crowds of browsers.

The Developing Fringe

The fringe is essentially the drylands-wetlands contact zone—the land beyond the Great Wall, or the rugged areas of the southwest, peopled by non-Chinese minorities and only lightly used for agriculture until recently. The development of the fringe is confined largely to agricultural settlement and the search for minerals.

Inner Mongolia refers to the wetter grasslands between the traditional Mongolian semidesert environment and the wetter areas of grain culture. It is short-grass steppe, extending from Amur lowland to the foothills of the Tien Shan. Its southern and eastern limits are the Great Wall and the Great Khingan Mountains. Its northern border is with the (Outer) Mongolian People's Republic.

Two provinces, Gansu and Qinghai, can also be properly thought of as fringe China. The southwestern provinces of Yunnan, Guizhou, and Guangxi, though long conquered by the Chinese, also have

SELECTED ELEMENTS OF THE CHINESE CULTURAL LANDSCAPE

China historically was so rich compared to its neighbors that it was a tempting target for invasions and raids. One Chinese response was the Great Wall; another was the long tradition of the walled city. Indeed, the Chinese used the same word, *cheng,* for both "city" and "wall." Chinese city walls were thick at the base, tapering to a narrower top; they were usually of packed earth faced with cut stone or brick. Although the Chinese built more than 4,400 walled cities over time, few survive intact today. Beijing's elaborate wall survived only to be largely demolished by Mao's government during a fit of disdain for such cultural artifacts.

The preferred urban form was a square, aligned precisely with the cardinal directions—north, south, east, and west. Each compass direction had special significance to the Chinese. The city was always considered to be facing south, toward the sun, the symbolic life-giving force. City walls had four gates, one in each direction. The south gate was the "summer gate," facing the warm noonday sun and the rain-bearing summer wind out of the south. The west gate, or "autumn gate," had sacred associations, for the spirits of ancestors traveled west to join the gods residing in the western mountains. The east gate, facing the sunrise, was the "spring gate." The north gate, facing the bitter winter winds out of Siberia, was the "winter gate." Because fierce invaders had come out of the north, it was a "trouble" direction and the least lucky.

The symbolism of orientation in city walls and gates is extended to choosing sites for buildings of all kinds, gravesites, and even railroad lines. Eastern or southern orientations are preferred, because these directions represent brightness and vitality. Northern orientations are unlucky for houses, because they are favored for graves. The Chinese call this orientation science *fengshui* ("wind and water") and hire fengshui practitioners to advise on site selection, facade orientation, and even the best arrangement of rooms in the future structure. To most Chinese, fengshui is not superstition but the expression of proper concern that builders seek to place their structures in harmony with the natural order and with neighbors.

Fengshui principles also influence the architecture of houses and farmsteads. A single-story, rectangular room will serve all needs of a beginning family. As the family grows in numbers and in wealth, wings will be built at right angles to the original nucleus of the structure, forming an I or a U. These side wings are "protector dragons" and can be seen in both humble farmsteads and imperial palaces. As the household's wealth grows, a thatched roof may be tiled. Roof styles were so indicative of status that the upswept corner peaks of roofs strongly associated with Chinese architecture, known as "swallow's tails," could be built only on temples or on the houses of persons who had passed their Confucian-style civil exams and thus held the equivalent of higher academic degrees.

something of the frontier aspect and are more properly considered a part of this subregion.

The enlarged Inner Mongolia described earlier can be subdivided into the inner western slopes of the Khingan range, the developing center, and the western desert and semidesert. The western slopes of the Khingans, an undulating plateau surface covered with grass, shows little promise as a solution to Chinese food supply problems. Near the disputed frontier with the USSR and still peopled largely by Mongolians, it is primarily a military frontier zone.

Some 6 million people live in the entire region, but only 600,000 to 900,000 are Mongols; the rest are Han Chinese. The western, almost unpopulated third is largely desert. With little water available for irrigation, it derives its importance from recent oil discoveries.

Gansu province represents an early Chinese advance to the west. Hemmed in by mountain and desert, it is surprisingly well watered and has a milder climate than would be anticipated given its interior location. The soils are rich, wind-blown loess. Rice, cotton, and tobacco are grown under irrigation, and the area is known for its apricots, peaches, and melons. Irrigated areas are being expanded by tapping underground waters.

The province is rich in oil and gas and has developed a large chemical industry to use that resource. It is the development of industry, rather than increased cultivable area, that will most likely be responsible for the region's anticipated growth.

The southwest is a land that has been Chinese for centuries, yet it is not typically or thoroughly Chinese. This fertile but steep hill country is often relatively inaccessible and was not thoroughly incorporated into China until this century, though Han Chinese migrants long ago came to be the dominant ethnic group. Once famous for its opium (one of the few crops of sufficiently high value to bear the difficult journey and consequent high transportation costs to market), it grows two crops of rice a year and produces sugar as well as a wide variety of fruits and vegetables. High-grade deposits of tin, tungsten, antimony, bauxite, copper, lead, and zinc make it a mineral storehouse. Huge water power potential promises a bright future in the development of metallurgy.

The Guizhou Hills, east of Yunnan, have little level land, and virtually all slopes have been cleared and terraced. The Miao, Yao, and Thai minorities constitute a large percentage of the population; forced into the hills by Chinese in-migration, they pursue an extensive form of crop agriculture.

The Guangxi region and the island of Hainan are tropical China. Inhabited largely by Thai-related peoples in rural areas (if Han Chinese in the city), the area is among the most ethnically diverse in China. Intensive Chinese rice culture in the area yields three crops yearly. Sugar, bananas, citrus, and a wide variety of tropical fruits and industrial crops provide high-quality, hard currency-earning exports.

Collectively, then, the Chinese fringe represents an expansion of the core area into the colder, drier, and hotter margins, respectively, of the country. Intensification of agriculture and new irrigation schemes have enabled all the areas involved to absorb migrants from the overcrowded core area. At the same time, new employment has been created by mineral exploitation and industrialization.

Frontier China: The Sparsely Settled Periphery

China has a "Wild West" comparable in some respects with that of the nineteenth century United States. In one important respect, though, they are not comparable. The Americans could cross a wide belt of rugged mountains and plateaus of arid and semiarid climate with the goal of better-watered, good-quality farmland on the Pacific coast. For the Chinese, their western territories of desert and semiarid lands, mountain chains, and plateaus end in international boundaries across these rugged drylands, not a potentially productive and very attractive ocean shoreline.

Mineral prospecting is not yet complete, but there are promising finds of oil and metallic minerals. There is some opportunity for irrigation projects to add badly needed agricultural lands. Until recently, the Chinese have considered these largely barren lands, lightly populated by non-Chinese ethnics, as tribute territories of the Chinese state. Little attempt was made to assimilate these groups.

The high plateaus and mountains of Xizang (Tibet) and Qinghai are the highest land on the earth's surface. Most of Xizang has a tundra climate or is covered by permanent ice and snowfields except in a few sheltered valleys and lower basins. Much of the surface is bare rock or has only a dusting of infertile soil. The agricultural prospects are anything but promising. Much of Xinjiang is a desert basin, one of the driest in the world. Spots favorable to agriculture have long been settled, and expansion of arable area is a risky and costly process.

Because neither area is close to core China and both have scant transportation networks, these territories remain remote, beyond the fast-paced development associated with the rest of China. Now firmly controlled and integrated politically and administratively with the rest of the state, they remain peripheral to the economy.

Xizang's population of 1.8 million is mostly Tibetans. Only about 7 percent are Chinese, largely administrators and military. Some 100,000 Tibetans live in exile in India and Nepal, and an additional 2 million live in southwestern China and parts of Xinjiang. The distinctness of ethnicity results in its government as an autonomous region. The traditional religion is the Lamaist variant of Buddhism, characterized, in part, by mysticism and demon worship. The monasteries controlled much of the wealth and the limited agricultural land; nationalization of both took place almost immediately.

The government has made a number of massive investments in irrigation and built a skeletal, but modern system of roads. The area is now self-sufficient in grain (largely barley), and dairying has been introduced. Much of China's hydroelectric potential is in Xizang.

In Xinjiang, the largest group, the Uigur, a Turkic people, form about 45 percent of the total 13 million. The Chinese now constitute the second largest group (35 percent), whereas before 1949 their numbers were negligible. Most of the Uigur are Moslems. Most of these border peoples have tribal, linguistic, and cultural affinities across the Soviet border, a potentially dangerous situation for China (and for the USSR).

The area is encircled by mountains. The alluvial fans at the bottoms of slopes are irrigated and intensely farmed. These oases were important watering places for the caravans of ancient and medieval times, which brought silk and tea westward.

Large oil fields and moderate gas deposits are the most important mineral production; a pipeline flows oil eastward. Despite evidence of progress, the area is still a frontier region. The dead-end nature of the location (due to the Sino-Soviet dispute), the remoteness from markets, and the limited usable land will likely limit further development for some time.

THE FUTURE OF EAST ASIA

China's poverty is well illustrated by the fact that, in the East Asia region, China's per capita GNP of $290 is one of the lowest figures, outpacing only Mongolia. Hong Kong's is $6,000; Japan's, $10,100; South Korea's, $2010. The goal of the Beijing government is a per capita GNP of $800 by the year 2000; if achieved, the figure would still place the PRC on a much poorer level than contemporary Turkey, Guatemala, or Peru. China's present GNP per capita is less than 3 percent of Japan's, and less than 2 percent of the U.S.'s.

Still, there has been obvious progress for the Chinese. Average life expectancy of peasants is estimated at 64, above the world average. Infant mortality in 1949 stood at about 200 deaths out of 1,000 live births before the child's first birthday. It is now 38 deaths among 1,000 live births, one of the greatest accomplishments of any developing country. Economic progress associated with the second revolution has meant that the "three bigs" (in consumer ambitions) of Maoist days—a bicycle, a wristwatch, and a sewing machine—have commonly been acquired. The new three bigs are a TV set, a refrigerator, and a washing machine, and many people are achieving these desires as well. Although per capita incomes in less productive rural areas may be $100 or less, industrial wages in clothing, plastics, and electronic product assembly in the new city of Shenghen average over $900 per year, and prosperous families visit the local Sea World.

Until the government's violent repression of student demonstrations in 1989, it appeared that China might soon have reached an agreement with Tai-

Herding on the Xizang (Tibetan) Plateau. *Here on the cold, relatively dry "Roof of the World," land use is highly extensive—large areas of slow-growth, sparse pasture are needed for each grazing animal. Note the high Himalayas in the background.*

wan's Nationalist Chinese government to establish a confederation allowing Taiwan local autonomy and continued free enterprise. Certainly the Hong Kong settlement suggested a flexible attitude on Beijing's part, although Hong Kong residents, alarmed by the severity of the crackdown on student protestors, now fear unification with the PRC. The Portuguese colony at Macau, across the bay from Hong Kong, is the subject of negotiations aimed at the eventual return of Macau to Chinese control. Recent events in China suggest that political liberalism will be more difficult to achieve than economic liberalism. Moreover, armed suppression of Chinese dissidents will have a chilling effect on China's relations with its neighbors. East Asia, it would appear, has the potential to be the world's fastest changing region in terms of its economic and political significance to all other states and peoples.

ENDNOTES

1. This analysis is based on concepts and data contained in the following, highly recommended, sources: Lester R. Brown et al., *State of the World, 1987* (New York: A Worldwatch Institute Report published by W. W. Norton, 1987); Peter Dicken, *Global Shift: Industrial Change in a Turbulent World* (London: Harper & Row, 1986); *World Development Report 1986* (New York: Oxford University Press, 1986); *World Resources 1986* (New York: World Resources Institute and International Institute for Environment and Development/Basic Books, 1986).

2. Many Chinese place names, once familiar to Westerners and contained in books, atlases, and wall maps published as late as the late 1970s, have been revised according to the Pin-Yin system. This system more accurately reflects the preferred, northern dialect official Chinese pronunciation. A selected list of important place names in both "old style," and Pin-Yin is presented for convenience in using references and maps with the traditional names:

Traditional	Pin-Yin
Provinces	
Sinkiang	Xinjiang
Tibet	Xizang
Kiangsi	Jiangxi
Szechwan	Sichuan
Kansu	Gansu
Yenan	Yan'an
Shensi	Shaanxi
Hopeh	Hebei
Rivers	
Yangtze Kiang	Chang Jiang
Si	Xi
Hwang Ho	Huang He
Cities	
Peking	Beijing
Canton	Guangzhou
Nanking	Nanjing
Tientsin	Tianjin
Chungking	Chongqing

REVIEW QUESTIONS

1. How does East Asia function as a transitional zone between developed and less developed areas? What are its advantages over other less developed world regions in attaining full development? What is the "critical combination" that has led to this improved developmental level?
2. Where and when was Chinese civilization established? What non-Chinese areas did it come to dominate? What evidence exists in other cultures of the region to suggest this Chinese cultural dominance?
3. How usable is the East Asian physical milieu? How is the natural physical environment reflected in the pattern of settlement? Compare the degree of utility for agriculture of China and Japan.
4. How does the Chinese alphabet basically differ from that used in Western societies? What has been the major benefit of this type of alphabet to the people of China and the other peoples of the region? What is Pin Yin, who was it designed to benefit, and how does it differ from the traditional Chinese alphabet?
5. How do the practices of child rearing in Japan differ from those of Western societies? How do they affect the group behavior of Japanese adults? How is this manifested in the concept of group loyalty?
6. Analyze the Japanese resource base. What were Japan's early energy resources? Its current dominant energy source? What are its plans and projections for future energy resources? What countries are its chief sources of imported raw materials?
7. What are the demographic characteristics of Japan's population? How does Japan's growth rate compare to that of the rest of the developed world? What ethnic or social minorities are present?

8. How does Japanese religious practice and thought differ from that of the West? Describe and compare its major theological and philosophical inputs.

9. What programs have been developed to increase Japanese agricultural productivity? How much food is imported? What basic problems hinder increased Japanese food production?

10. Describe the two-tier organization of the Japanese industrial production system. What is the relationship of government to industry? What is the significance of the zaibatsu? What is the role of women in the Japanese work force?

11. Compare the flow of trade goods between the United States and Japan.

12. Analyze the effects of industrialization in Japan on that country's natural environment.

13. Where is Japanese population concentrated? What is its degree of urbanization? How is this crowding of people reflected in Japanese lifestyles? How has the city and region of Tokyo adopted to the scarcity of physical space?

14. What countries compose the East Asian shatterbelt? Analyze the relationships of each with neighboring countries.

15. Describe the influence of Japan on Korean development, past and present. Contrast the pace of development of South Korea and Japan. Compare the resource bases of North and South Korea and the relative degree of economic development of the two countries.

16. How did Taiwan evolve as a separate country? What are its leading industries?

17. Describe Hong Kong's evolution as a major industrial city. What has been the basis of this industrialization? What problems are foreseen for Hong Kong as the 1997 union with the PRC approaches?

18. How did the "colonial" experience of China differ from that of other less developed nineteenth century areas?

19. How important are ethnic minorities within China? Where are they located? Describe the traditional Chinese view of the relationship of the Chinese Core to surrounding, less populous areas.

20. What is the concept of the "Second Revolution?" How has it affected Chinese agricultural landholding and management systems? How have these reforms been extended to Chinese industry and other economic endeavors?

21. Compare Chinese population growth rates to those of other less developed countries and to those of nations of the developed world. How was this "premature" demographic transition accomplished?

22. Describe the policies of Mao Zedong's "iron rice bowl." What were its positive and negative effects on Chinese industrial production?

23. Describe the inputs into the religious amalgam that characterize traditional Chinese beliefs.

24. What are the major characteristics of oriental intensive agriculture? What are the regional agricultural specializations within China?

25. What were the Chinese communes? How have they been modified and improved? What is their role in developing the transition to an urban-industrial society?

26. Analyze the Chinese mineral resource base. What are its greatest strengths? List important recent mineral discoveries.

27. Discuss the programs of the Great Leap Forward. What, if any, were their positive results? What was the role of Chinese labor in these programs?

28. What role does the Northeast play in China's economy? Describe the impact of Russian and Chinese political control on this area. Why did the area remain undeveloped for so long?

29. List the major industrial areas of the Chinese Core. Relate them, where possible, to China's two major rivers.
30. How does China's "Wild West" compare with that of the nineteenth century United States? What has the government of China done to integrate this area into the Chinese economy? Analyze the productive potential of the Developing Fringe subregion.

SUGGESTED READINGS

Buchanan, Keith; Fitzgerald, Charles P.; and Ronan, Colin A. *China: The Land and People*. New York: Crown, 1981.

Ginsberg, Norton, and Lalor, Bernard, eds. *China: The 80's Era*. Boulder, CO: Westview, 1984.

Hall, Robert B., Jr. *Japan: Industrial Power of Asia,* 2d ed. New York: Van Nostrand, 1976.

Knapp, Ronald, ed. *China's Island Frontier*. Honolulu: University of Hawaii Press, 1980.

Kornhauser, David. *Japan: Geographic Background to Urban/Industrial Development,* 2d ed. New York: Longman, 1982.

Ma, Laurence J.C., and Hanten, Edward W., eds. *Urban Development in Modern China*. Boulder, CO: Westview, 1981.

Myrdal, Gunnar. *Asian Drama: An Inquiry into the Poverty of Nations*. New York: Twentieth Century Fund, 1968.

Pannell, Clifton W., ed. *East Asia: Geographical and Historical Approaches to Foreign Area Studies*. Dubuque, IA: Kendall-Hunt, 1983.

Pannell, Clifton W., and Ma, Laurence J.C. *China: The Geography of Development and Modernization*. New York: Halstead, 1983.

Purcell, Victor. *The Chinese in Southeast Asia*. London: Oxford University Press, 1966.

Samuels, Marwyn S. *Contest for the South China Sea*. New York: Methuen, 1982.

Smil, Vaclau. *The Bad Earth: Environmental Degradation in China*. Armonk, NY: M.E. Sharpe, 1984.

CHAPTER SEVEN

South Asia

Crowds bathing in the Ganges at Varanasi.

S outh Asia has acquired an image that suggests it is the poorest, most overpopulated part of the world. There is little doubt that the sheer size of population, compounded by rapid growth, is the central problem of this region. Clearly, the area has not completed the demographic transition. The classic phase of projected rapid population growth in South Asia began only 30 years ago, and the population is still rural, rather than urban. However, a survey of vital rates indicates that the birthrates of India, the dominant state in the region, and Sri Lanka, the most advanced, have begun to level off (Table 7–1). Government policies and education are being used to stem the tide of growth, hurrying the demographic trends ahead of the normal changes that would accompany economic development.

Although South Asia's growth rates are high when compared to those of the developed world, they are lower than those of the Middle East, Africa, or Central America. Significantly, though, the region's growth rates far exceed those of China, where the stages of development and demographic transition are comparable. India, with far less centralized control, has not been able to move as far, as fast, as China.

Despite efforts to curtail growth, the region's population is projected to double within 30 years (assuming stable death and birthrates). For comparison, the doubling time for China is 55 years, for the United States, 100 years, and for Japan, 120 years.

Vital rates alone do not illustrate the magnitude and nature of the region's population problems. India's population under 15, the potential childbearing group of the future, accounts for 40 percent of the total (Figure 7–1); in contrast, that of the United States is only 22 percent. A very young population (one with a large proportion of individuals under 15 years old) will have correspondingly high expenditures on education. However, although persons in their early teens are almost invariably dependents in urban-industrial societies, in traditional South Asia they would be active contributors to the family's income and food supply. This fact illustrates one aspect of the dilemma of developing nations: overcoming the traditional, agriculture-based, conservative cultural pressures to have many children. Children are viewed as an economic asset to agriculture. Quite young children can tend livestock, collect dung for fuel, weed, destroy insects, and generally make themselves useful while consuming little.

The alternative to a large family would be either a lessened capacity for working the land or the necessity of hiring outside labor, or both. Children also provide "social security" in less developed societies by helping to support and care for aging parents.

TABLE 7–1
Vital rates for South Asia.

Country	Population (millions)	Births per 1,000	Deaths per 1,000	% Natural Increase	% Urban
Afghanistan	15.4	48	23	2.5	16
Bangladesh	108.4	44	17	2.7	15
Bhutan	1.5	38	18	2.0	5
India	805.8	33	12	2.1	25
Nepal	17.8	42	17	2.4	7
Pakistan	104.6	43	15	2.7	29
Sri Lanka	17.0	25	7	1.8	25

SOURCE: *World Population Statistics,* Population Reference Bureau, 1988.

To these practical considerations must be added a historically high death rate. **Infant mortality rates** in South Asia (live births who do not survive to their first birthday) are still 12 times those of the United States. Parents tend to compensate for a low survival rate, consciously or otherwise, by having more babies than they can realistically expect to see reach maturity.

India, with a 2.1 percent rate of natural increase, must cope with almost 50,000 net additional Indians *per day*. This huge increase reflects the large population base to which the relatively modest rate of increase is applied, but it is typical of the region's problems. Pakistan and Bangladesh, with even higher rates of increase, but smaller bases, have net increases of 8,000 people daily.

Given the agricultural-economic pressures against small families, India is only likely to experience a slowly dropping birthrate in association with progressive industrialization and urbanization. It must be remembered, also, that in India, with a per capita GNP of only about $260, it is not likely that much money can be directed toward elaborate birth control programs. Yet, some progress is evident there. Sri Lanka, with equally pressing population problems, has made even greater progress in controlling population growth. It has the lowest birthrate in the region (25 per 1,000), and a growth rate that approaches the world average.

India's population density (655 per square mile) is similar to that of West Germany (635 per square mile) and much less than that of Japan (850 per square mile), nations with a much higher degree of prosperity (Table 7–2, column 1). India has an unusually high degree of agricultural utility, with a large percentage of its total (52 percent) in cropland (column 2). A more sophisticated measure, the **physiological density** (i.e., how many people there are in relation to a square mile of land sown to crops), shows an even brighter picture (column 3). India is better off in this respect than any of its immediate neighbors, many prominent Western European countries, and both China and Japan.

Physiological densities for the other states of the region are not hopeless, either (Table 7–2). Bangladesh's does not compare unfavorably to China's, and the figure for Sri Lanka is not unlike that of West Germany. Mountainous Nepal exhibits less population pressure than topographically analogous Switzerland, while India appears more fortunate than Italy, a country that exports a food surplus. None of the region's nations experience the productive land scarcity of Japan. Although development level helps explain the high densities supportable in Europe and Japan, there is no doubt that, given similar developmental inputs, most countries of South Asia are potentially capable of feeding their populations, and some may even come to generate a marketable agricultural surplus.

The situations are not completely analogous, however. European countries benefitted from massive out-migrations at times of population stress. No such migrational outlets are available to South Asian states. Technology is as available to South Asia as to any other area of the world, but little can be affordably purchased given regional incomes. Nor can the region's underdeveloped economies quickly or easily develop their own technologies.

The situation is far from hopeless, though. Most of the region's economies are self-sufficient in food. The resource bases of the region's three most populous states are better than that of Japan, and introduction of modern agricultural techniques and investments in irrigation have been made a priority in each South Asian economy. There *is* progress.

As in China, an immediate shift to mechanized agriculture is impossible. Cost alone makes such a change unlikely, and the resulting rural unemployment would be disastrous. What little investment capital is available must be shared by industry, agriculture, education, and services, and in order to develop, there must be progress on all these fronts.

THE COSMOPOLITAN NATURE OF SOUTH ASIA

Security of trade routes to an India perceived as fabulously wealthy was the motivation for the entire Age of Exploration. The Portuguese, in their quest for a route to India, circumnavigated Africa and claimed various African territories as convenient way stations. Similarly, the Dutch and then the British established themselves at the Cape of Good Hope largely to have

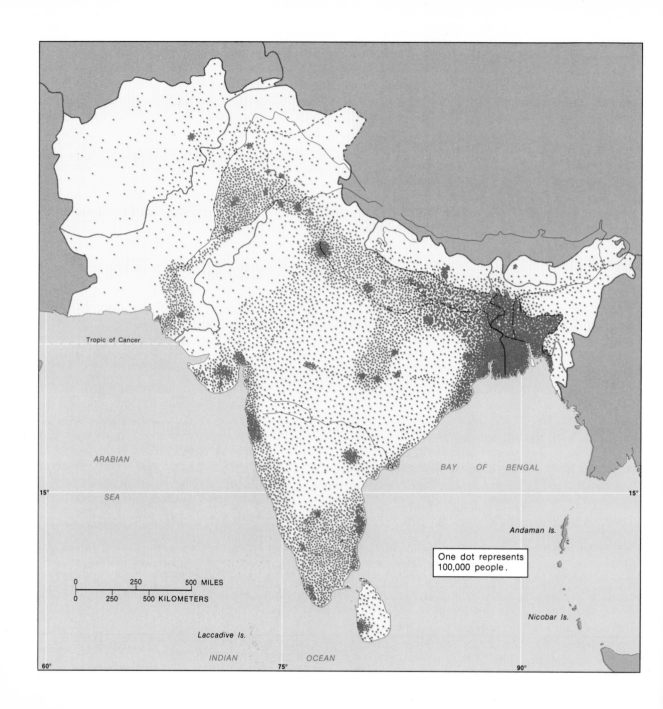

Tropic of Cancer

ARABIAN

SEA

15°   BAY   OF   BENGAL   15°

Andaman Is.

One dot represents
100,000 people.

0   250   500 MILES
0   250   500 KILOMETERS

Nicobar Is.

Laccadive Is.

INDIAN   OCEAN

60°   75°   90°

a service port for shipping in the India trade. Even the European discovery of the Americas was an accident of the race to India.

India, with its near-monopoly on diamonds and sapphires, seemed a dazzling place to Europeans of three or four centuries ago. A variety of spices, rare and beautiful woods, gold, and ivory were the things Europeans sought. India's fine steel weapons and cutlery and its gleaming brassware were world famous. Indian cottons and silks were the highest quality then available. The princes of India were conspicuous in their lavish display of luxury, though this image of wealth did not reflect the realities of life for the vast majority of Indians.

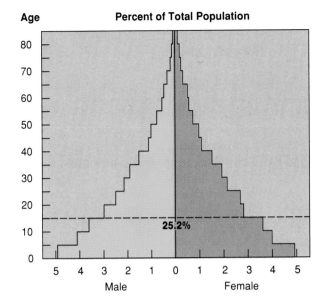

Age	Percent of Total Population		

Male — Female

FIGURE 7–1, pp. 344–345
Distribution of Population in South Asia. South Asia concentrates one fifth of the world's population. The map illustrates two striking facts about the population distribution within this region: the high overall density characteristic of the subcontinent south and east of the mountainous periphery, and the intense concentrations in the major river valleys and along the wetter coasts. The extremely high density in the delta region and the lower Ganges Valley implies both the exceptional agricultural utility of that area and the difficulties of supporting such a massive population concentration. The population pyramid indicates the seriousness of the overall situation in India. The wide base, illustrating the predominance of young people, suggests the enormous potential for continued growth in the immediate future.

TABLE 7–2
Population densities in South Asia and selected other countries.

Country	Overall Population Density (thousands per sq. mile) (1)	% Land Area Sown to Crops (2)	Physiological Density (thousands per cultivated sq. mile) (3)
South Asia			
Afghanistan	62	12	513
Bangladesh	1,878	65	2,822
Bhutan	83	2	4,096
India	627	52	1,185
Nepal	327	18	1,829
Pakistan	337	27	1,249
Sri Lanka	676	33	2,028
For Comparison			
China	291	11	2,607
Japan	850	13	6,480
West Germany	635	30	2,118
Italy	493	41	1,203
Switzerland	410	10	4,090
Netherlands	1,070	23	4,350

SOURCE: *World Population Statistics*, Population Reference Bureau, 1988.

Until the advent of European colonialism, India was not a unified state, though conquerors periodically had controlled large parts of South Asia. The Indus Valley was the location of one of the world's oldest civilizations, and other great centers of innovation arose in the Ganges Valley almost 3,500 years ago, subjugating still-older civilizations that occupied this rich agricultural area. This melding of earlier and Aryan cultures developed the classical Indian culture and civilization, including the Hindu religion with its reverence for life that is the hallmark of India throughout time. It is the base over which all else has been veneered.

In the third century B.C., the great Gautama, the founder of Buddhism, taught in India. His teachings were soon established as the official religion. Yet, ancient Hinduism never lost its foothold as Buddhism found greater acceptance in East and Southeast Asia than in India itself. Bhutan and the island state of Sri Lanka have Buddhist majorities, but Buddhism has only a small following within contemporary India.

Conquest and invasion from the west brought other cultures to the region. Today's Pakistan was a part of various Persian empires and that of the Greek Alexander. Later invasions paved the way for the introduction of Islam in A.D. 711 under an Arab conquest. Eight hundred years afterward, the Moguls (a Mongol dynasty) extended their rule to most of India, spreading Islam throughout parts of the Ganges Valley and some areas of the south. It was this religious dichotomy that resulted in the partition of the Indian Empire into separate republics of India and Pakistan in 1947. Bangladesh (formerly East Pakistan) in the delta of the Ganges and the area of the Indus Valley were regions of clear Moslem dominance (Figure 7–2). Scattered Moslem minorities remain in postpartition India, yet the overwhelming Hindu dominance in India attests to the strength of Hindu beliefs. Obviously, Hinduism withstood both the military might and the prolonged political control of Islam.

Christianity, brought by traders and missionaries from the Middle East, took root on the southwestern coast as early as A.D. 500 but never spread further inland. A later reintroduction by European conquerors made relatively little headway except in the enclaves of Portuguese control. Modern missionary efforts seem to have succeeded mainly among animist (nature-worshipping) tribes isolated in the hill districts of the center and extreme northeast of India.

The regional core remains attached to Hindu beliefs, whereas the periphery has adopted other religions (see Figure 7–2). Even within the core, however, there is some diversity. Parsis (descendants of Persian Zoroastrians), Tibetan refugees (Buddhists), native animists, and other minorities retain their beliefs. Hinduism, in common with other religions, has given rise to reformist and splinter groups; the Jains

and Sikhs are those most widely known in the West. Regionally, 60 percent are Hindu, while 40 percent are of other faiths. Within India itself, over 80 percent of the population is Hindu.

The religious heritage is only a part of the total diversity. Indians speak dozens of individual languages and hundreds of dialects (Figure 7–3). Thus, a common Hinduism binds together what different languages might divide. Yet, even among Hindus speaking the same tongue, the stringent social realities of the caste system compartmentalize society rather than unite it.

One must add to this linguistic, religious, and cultural diversity the sharp differences between urban and rural, the gap between rich and poor, and the phenomenon of widespread illiteracy coexisting with a highly educated elite. Rather like an Indian curry, the region is sufficiently homogeneous to be called by one name, yet the flavor of South Asia is composed of different ingredients, some sharp, some subtle, that collectively compose a unique whole.

Political Relationships and Regional Tensions

India's dominant position within the South Asia region means that relationships within the region, as well as with other regions, revolve around India's friendships and animosities.

Jawaharlal Nehru, India's first prime minister and one of the leaders in the struggle for independence, was chief architect of **nonalignment**. Mere neutrality was passive; India was to be an active force for a peaceful and progressive world. India has become a major voice for Third World countries and has played an active role in UN peacekeeping operations.

To many in the United States, India's nonalignment policy appears to take a highly moralistic, critical approach to American policy while saying little about Soviet policy. India's claim to moral leadership is flawed, in many Westerners' eyes, by its determination to retain Kashmir despite the Moslem majority of that state's population and by Indian action in Sikkim in the mid-1970s. India's seizure of Sikkim is an example of its determination to dominate the Himalayan frontier states in competition with China. In 1962, Chinese troops drove deeply into Indian territory before unilaterally proclaiming a cease-fire and

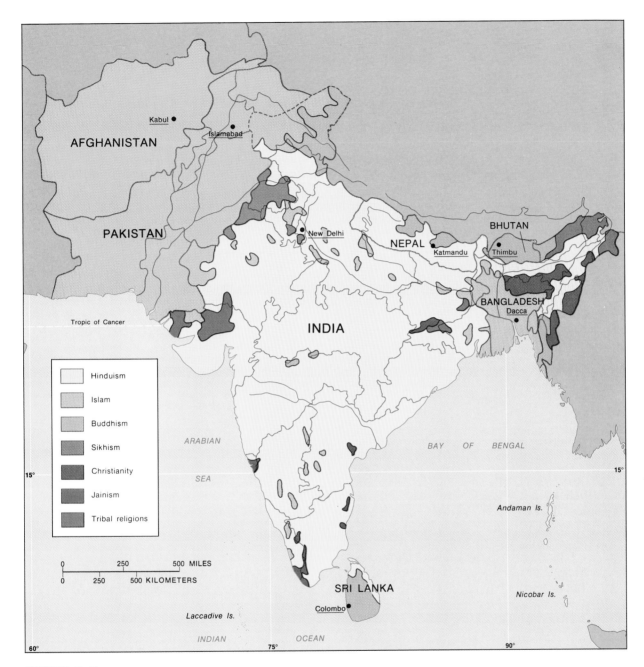

FIGURE 7–2
Major Religions in South Asia. Relatively uniform population distribution is not matched by ethnic or religious uniformity. Historically, the distribution of religions was even more complex. Independence brought civil conflict between the two largest groups, Moslems and Hindus, and resulted in the division of a once united Indian Empire into several smaller political units. Massive exchanges and migrations have simplified religious patterns. Political borders now coincide, generally, with the distribution of major religious elements. Nonetheless, disputed borders continue to complicate relations in the region, and remaining religious minorities in India and Sri Lanka have proven to be a source of internal conflict within those countries.

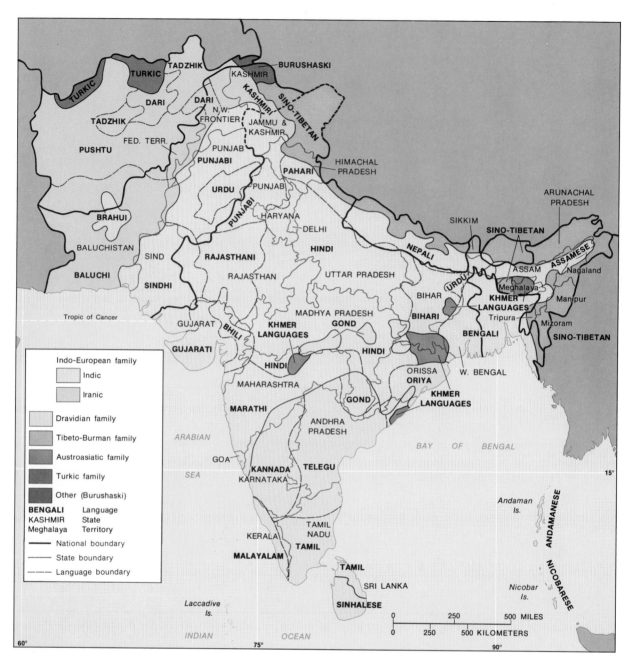

FIGURE 7–3

Languages and Administrative Boundaries in South Asia. The general language patterns of South Asia reflect a series of invasions in the historical past. The Dravidian languages, now relegated mainly to the south of India, were probably the original languages of the subcontinent. Later invasions by Aryans from the north and west spread their languages and culture to their area of control, the north and center. Within each major group, there are many individual languages. Most countries of the region have declared one language official, but all other languages persist. Strong internal differences in language are reflected in India's federal system of individual states.

withdrawing to previous positions. There is still a large area north of Kashmir that is claimed by India but controlled by the PRC (Figure 7–4).

In this clash of giants, the other nations of South Asia have tended to establish friendly relationships with either India or China. Opposition to India has helped foster a Pakistani relationship with the PRC. Soviet and PRC foreign aid, technical assistance, trade treaties, and military aid through South Asia tend to reflect the pro- or anti-India alignments that have developed.

Afghanistan and Iran seem to be replacing India as Pakistan's major concern. Afghanistan's civil war sent millions to Pakistan as refugees, and Soviet intervention poisoned Soviet-Pakistani relations. Withdrawal of Soviet troops did not remove all problems. Moslem fundamentalists in both Iran and Afghanistan will continue to be a concern to a more moderate, progressive Pakistan.

The Legacy of British Colonialism

After brief attempts by the Dutch, Portuguese, and French to control India and its trade, Britain came to be the dominating force during the eighteenth century. Beginning in 1609, the British East India Company, a private trading firm, secured commercial rights in parts of the Mogul Empire. Expanding British power relegated the French and Portuguese to a few footholds and enclaves by 1760.

The British East India Company prospered, sometimes returning a 200 percent dividend in one year. Such profits led to a strong desire to remain in control of trade which, in turn, necessitated putting down internal rebellion and safeguarding hostile frontiers. The British government began regulating the political activities of the Company in 1773, assuming full authority after the Indian Mutiny of 1857. The princely states, still under native rulers, were

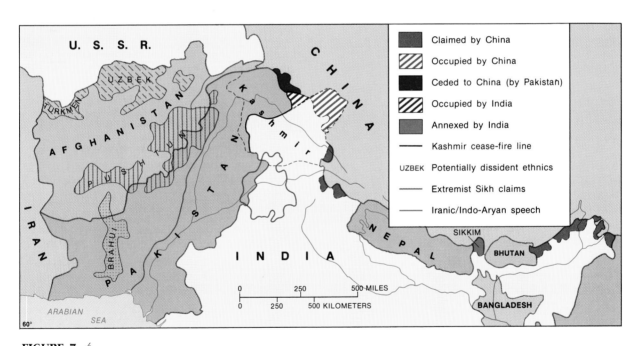

FIGURE 7–4
Border Disputes in Northern South Asia. The northern portion of the region, in particular, is affected by the claims and counterclaims of various countries. China has aggressively pursued its claims to various parts of India's Himalayan borderland. Kashmir remains the center of a prolonged dispute between India and Pakistan. Restive Sikh extremists promote independence for their religious group. An ethnically diverse Afghanistan presents a potential future problem for the region. No single linguistic group dominates in that country, while the largest language group, the Pushtun, is strongly represented in northern Pakistan. Iranian claims to territory, based on language and cultural affiliation, could further complicate matters.

also incorporated into the empire. Self-rule was permitted within strict limits of cooperation with the British.

In the 90 years of empire, Britain achieved a remarkable degree of economic development in India. To be sure, India itself literally paid for this progress, through taxes, exploitation of its resources, and the high profits made in trade between industrialized Britain and developing India. Nonetheless, the progress was real and unprecedented.

The Pax Britannica. India was an accidental empire, as British control was gradually extended to pacify the trade hinterlands of coastal trading cities. Prior to British control, warrior castes conquered a village, took over its land, rented it back to the villagers, and thus became relatively rich. This militaristic quest for land was one of the leading causes of the continued flare-ups and war that bled village India for centuries. Military strength produced wealth; unless control was ruthlessly maintained, the land was soon transferred to other military victors. The **Pax Britannica** (the period of peaceful stability enforced by British military power) put an end to this internal chaos and reordered the hierarchy of Indian society.

Independence and Partition

Prior to independence, the subcontinent was not composed of a few massive blocks of Moslem land flanking a huge Hindu territory. The pattern was more one of an incredibly complex mosaic of tiny, interwoven Moslem and Hindu parcels of land, buildings, and businesses. In many villages, both Moslems and Hindus, sometimes in company with a few other religious sects, lived out their lives literally side-by-side in comparative peace. It was only after the announcement of the approaching independence that animosity between the two groups became common; economics was a basic reason. Precolonial India had been controlled by Moslem rulers in the north and center. Under the British, both Hindus and Moslems had access to education and economic opportunities, but their respective responses to these opportunities were uneven. British control and the introduction of Western technology presented relatively greater opportunities to Hindus than it did to most Moslems. As independence ap-

proached, British India's Moslems feared submergence in a Hindu-dominated state. Although Gandhi suggested that the Moslems be reassured of their status within a united India by letting them run the new government, the Hindu majority, understandably, was adamant about one person, one vote.

Mohammed Ali Jinnah emerged as the leader of a movement for an independent Pakistan, a Moslem state. Pakistan was a created name, formed from the names of the Moslem-dominated provinces of Punjab, Kashmir, and Sind. Violence swept the subcontinent when independence was decreed, perhaps precipitated by an incredibly short timetable for British withdrawal. (The decision was made to grant independence to India and Pakistan quickly, in the hope of averting civil war.)

Boundary lines were drawn, under the pressure of a deadline, by British bureaucrats who knew little about the cultural and economic complexities of India. Had the boundary makers been more knowledgeable, though, the boundary would still have been an impossible job in terms of equity to Hindus and Moslems. Millions of people fled across borders that did not exist a few days before; this tragedy was compounded by the dispute over Kashmir, a strategically situated state where China, India, and Pakistan meet.

Creating Unity from Diversity. Unity is a basic ingredient in nationhood. Without it, there can be no nation, only a country struggling for survival, unable to turn its attention to other problems and opportunities. Although India is an independent state, it is not yet a nation-state in the sense of Sweden or the United States. Some would compare India with Europe rather than with a single country. Pakistan and Sri Lanka have also experienced problems of unifying their different ethnic groups, but their problems hardly approach those of India in scale or divisiveness.

Culturally, the region is extremely diverse, with India (the largest regional unit) being the least uniform. India has been forced to develop a federal state in response to its own internal diversity.

Cultural Fragmentation. Many Indians speak more than one language, and some of India's 782 languages and dialects are spoken by only a few thousand people. The largest single, and first official,

GANDHI AND THE DRIVE FOR INDIA'S INDEPENDENCE

Mohandas K. Gandhi ("Mahatma" is an honorary title meaning "great soul") was one of the world's least likely revolutionaries. A pacifist whose ultimate weapon was the threat to starve himself to death, he led one fifth of the human race to freedom. Gandhi was sent to England to study law in preparation for inheriting a bureaucratic position. While in London, he had a brief fling with cultural assimilation: he wore Western clothing, tried to copy a British upper-class accent, and even took dancing lessons. Unable to find work as a lawyer in India, he migrated to South Africa. There he was introduced to racial segregation of public facilities and developed his ideas and techniques for nonviolent protest.

In 1920, Gandhi began his long struggle to achieve an independent India. To Westerners, he appeared to underestimate the depth and rigidity of the fundamental religious split within India and to overestimate the average human's capacity for saintly behavior. Gandhi's now-famous tactics of passive resistance, featuring nationwide "strikes" in which everyone was to stay home or visit a temple for prayer, probably would have failed in almost any circumstances other than in British India. The indigenous population was so immense, the Empire so exhausted by World War II, and the British public so sensitive to world opinion, that independence did come with little blood shed between ruler and ruled. The tragedy of independence was the warfare between Hindu and Moslem residents, not between those seeking independence and the British.

Gandhi was assassinated in January 1948, only five months after India became independent. His assassin was a Hindu extremist, apparently dissatisfied with Gandhi's continuing counsel against violence during the huge transfers of Hindu and Moslem populations that accompanied the division into Hindu India and Moslem Pakistan.

language, Hindi, is spoken by fewer than one third of all Indians. The Indo-European languages that predominate are all descended from Sanskrit; they are Hindi, Bengali, Punjabi, Rajasthani, Marathi, Gujarati, Pahari, Kashmiri, and Oriya (see Figure 7–3). Each is at least as different from the other as are the Latin-based Spanish, Romanian, and Italian. There are four "major" languages in the south—the Dravidian languages of Telegu, Tamil, Kannada, and Malayalam. This basic Sanskrit-Dravidian divide follows the borders of various empires of the past. The communications problems brought about by multiple languages is illustrated by the continuing use of English,

the language of the departed colonial ruler. Although only about 2 percent of the population are literate in English, they tend to be the educated elite. Thus, India, which hardly needs another language, continues to use English as an "associate" language. Higher education and most government business are generally conducted in English.

Eighty-three percent of Indians are followers of Hinduism; a religion with at least as many different sects and denominations as Christianity. About 11 percent of Indians are Moslems, 3 percent are Christians, and nearly 2 percent Sikhs. Buddhists (along the Himalayan frontier), Jains, and Parsis are other significant, small minorities. All are growing at slightly faster rates than Hinduism.

The Hindu religion is considered native to India. There is little in the way of specific philosophy or dogma; Hinduism is tolerant of a wide range of individual beliefs and practices. It may have grown and prospered in this culturally fragmented land because it was able to absorb many different cultural traditions (see Figure 7–2).

Islam is philosophically much closer to Judaism and Christianity than it is to Hinduism. Originating in the Middle East, Islam sees itself as a continuum of Judeo-Christian traditions, enlightened through later revelations. Moslems actively seek converts to their faith; Hindus do not. Islam is strongly monotheistic. Modern interpretations of Hinduism see God as all-encompassing, expressed in many guises, exhibiting many personalities. To a Moslem, this appears to be the worship of many gods (and goddesses)—an unthinkable sacrilege. Islam is often a deeply private religion; its ceremonies are almost austere. Much emphasis is placed on a personal relationship with the Deity. Hinduism is a more public religion, with festivals, processions, and such ceremonies as mass bathing in the Ganges. Islam, like Judaism and Christianity, sees only one chance for salvation—life ends, and the individual moves to a spiritual realm beyond, in an afterlife. Hinduism sees life as cyclical—an individual returns, reincarnated as another life, often repeatedly, until perfection is reached and the soul is liberated forever.

Islam is essentially egalitarian; Hinduism, with its social castes, would have been viewed as an alien perspective. In Hinduism, caste is related to occupation, ensuring the distribution of all services in a complex society, and perpetuating all elements through the bans on marriage between castes. No doubt, the conversion of many to Islam was accomplished through the religion's appeal to those of lower caste. Islamic political control over a Hindu majority through the centuries produced resentment for past favoritism shown toward Moslems by Moslem rulers.

Extensive fractionalization of India's culture—its deep divisions of religion, its large number of languages (none numerically dominant), and its other cultural fragmentations—has presented many obstacles to unification. The primary unifying forces have been Islamic conquest, British imperialism, and the struggle for independence, each with a greater impact on unification than its predecessor. The fact is that India was never a unified state until 1947; even the split of Hindu India and Moslem Pakistan (and, later, Bangladesh) created fewer major states than had ever occupied the subcontinent until then. Certainly not free of restive minorities and cultural diversity, the other states of the subcontinent have experienced similar problems, if less frequently or severely.

Iconography and Circulation: Factors in Political Stability.

Geographer Jean Gottmann categorizes iconography and circulation as state-building factors. **Iconography** is the set of ideas, historical interpretations, heroes and heroines, and concepts of the state's role and goals that ties people together into a nation. A strong iconography shapes a strong, unified state. In Eastern Orthodox Christianity, icons are religious images, symbols of a shared faith. In the political-geographic sense, icons are symbols of shared beliefs that focus on a state and loyalty to that state. In terms of nation building, **circulation** refers to the physical communications and transportation systems and structures that literally tie the country together. In a strong, well-developed circulation system, internal communications and transportation are better than those with any surrounding state (across an international border).

A state with more than a dozen major languages, a sharp philosophical division between the two largest of many religions, a variety of cultural-racial heritages, and a history of political fragmentation has serious problems in establishing a strong, unifying iconography. The political unification problems of India illustrate the reasons for the lack of prior unity. Even India's circulation system, inherited from the colonial past, was a regional system rather than nationally oriented. Although considerably less diverse, the other states of the region seem to have had similar problems developing an iconography and a solid national identity. Smaller in size, with a less complex settlement pattern, each still has its isolated areas, and serious gaps remain in their respective transportation nets.

THE PHYSICAL FRAME

It is not an accident that early South Asian empires were more successful in dominating the northern and central parts of the subcontinent than in dominating the south. Two great alluvial plains stretch off from the Himalayas of northern Pakistan in opposite directions. The larger plain extends for 1,200 miles in the direction of the Bay of Bengal. The second, half its size, spreads southwestward to the Arabian Sea. These great lowlands form the core of the subcontinent and are the home of hundreds of millions. Three great rivers, the Indus, Ganges, and Brahmaputra, water the fertile soils. Successive empires spread their control over these vast plains and the peripheral hill lands. Extending their dominance southward across the Vindhya Hills to the great plateau of southern India proved to be a much more difficult task.

Physically, the subcontinent consists of a southern plateau, these great lowlands, and the mountain arc beyond. The plateau area is the triangle that gives India its characteristic shape. Sri Lanka is a detached piece of the plateau. A plateau in structure, the Deccan has the actual appearance of hills. It is an ancient block of crystalline rock material, a remnant of Gondwanaland, the supercontinent. Circulating currents of magma deep in the earth's interior moved this block up against the Eurasian landmass, and the collision has produced the Himalayas and the rest of the mountain arc (Figure 7–5, p. 356). This tectonic activity created two other features: an outpouring of lava that covered much of the northwestern portion of the block with thick layers of basalt material and a large geologic trough along the northern margins of the block.

The trough has been filled in with alluvial sediment in recent times and is occupied by the fertile and populous valleys of the Indus and Ganges rivers, the enormous northern plains referred to earlier. North of the plains is a relatively young series of complex folded and faulted mountains. Beginning as a series of hills, they rise in rows, higher and higher, culminating in the mighty Himalayas (see Figure 7–5). The Himalayas are only one of several ranges that resulted from tectonic forces in the area. The major mountain ranges of all of Asia fan out from the Pamir Knot, at the common borders of China, India, the USSR, Pakistan, and Afghanistan. Almost as high as the Himalayas, the chain that forms the spine of Afghanistan is called the Hindu Kush. Over 25 peaks in the Himalayas exceed 25,000 feet in elevation. Rivers have cut deep gorges in their flanks, and glaciers have sharpened their peaks. There are few passes, and those are high and difficult. Despite their rugged nature, the mountains have not been impenetrable cultural or economic barriers. They have, however, demarcated the Indian subcontinent from other cultural and political entities, allowing it to develop a different and special character. They also function as a major climatic divide between the wetlands of South Asia and the deserts of the continental interior.

The giant, level, compound delta of the Ganges and Brahmaputra rivers, flood prone and subjected to inundation by storm-generated waves, is the core of Bangladesh. It is both bounty and tragedy to that country— its rich soil supports dense settlements, yet its natural hazards bode disaster.

Fed by summer monsoon runoff, but more by snowmelt, the Indus never runs dry, supplying irrigation water even in the driest years. The Ganges, even more directly affected by monsoon rains, is subject to frequent floods. The Thar Desert of northwestern India and Pakistan is a true desert, whose level surface is covered with salt flats, playas (dry salt lakes), shifting sands, and sparse vegetation. The Cutch area at the southern extreme of the Indo-Gangetic Plain is a silted-in arm of the sea with semidesert conditions.

Weathering and erosion have created the rolling to hilly surface over the Deccan. Because this block that forms peninsular India tilts eastward, most streams flow in that direction. At the western, higher edges of this shield or block are mountains called the Western Ghats. With elevations of 3,000 to 8,000 feet, they are the crest of a scarp (a sharp ridge) with a steeply sloping western side and a gently sloping eastern (landward) side. The effects of altitude increase rainfall on their western slopes, resulting in a huge potential for hydroelectric power generation. The coastal plain bordering the Western Ghats is narrow, only 15 to 70 miles wide. The Eastern Ghats, on the other hand, are just a line of heavily dissected discontinuous hills. With a less steep slope and most

THE INDIAN RAIL SYSTEM

India has the fourth greatest total railroad track mileage in the world. Next to Japan, it has the best rail system in Asia. However, the system, inherited from colonial times, has several serious handicaps. Because the railroads were designed to expand the major port hinterlands, the lines radiated out from Bombay, Calcutta, Madras, and Karachi, all major ports (Figure A). They were built to facilitate transoceanic trade with Britain rather than to foster internal development. A secondary role of the railroads was to move troops quickly within India, especially to its borders. Because direct British internal rule never extended over all of India, service is incomplete. Rail lines were constructed mainly in parts of India that were ruled directly rather than in the princely states. The British made sure that the main lines connecting major cities and administrative-garrison centers were standard gauge. However, several narrower gauges were constructed in the princely states and in areas of difficult terrain. Although cheaper to build, in the long run the narrow-gauge lines are expensive for the country because they form entirely separate systems (see Figure A). Freight must be reloaded onto a different rail car when goods move from one gauge to another. Passengers must change trains in the same situation. About half of India's track mileage of 35,000 miles is standard gauge; the rest is meter gauge or narrower.

The Republic of India has nationalized all railroads and modernized them enough to have quadrupled the 1947 freight volume. India's passenger rail figures are also impressive because few Indians can afford to use domestic airlines and very few have private cars. Despite problems, the railroad system acts as an economic unifier, tying together (if with difficulty) all parts of the subcontinent, including the now-separate rail nets of Pakistan and Bangladesh.

FIGURE A

Railroads of India and South Asia. One of the positive legacies of British colonial control was the extensive rail system developed to aid commerce and troop deployment. The British also developed major ports at Calcutta, Bombay, Madras, and Karachi, where only villages had once existed. Rail lines interconnected these ports with the administrative capital at Delhi. Originally, these lines were designed to serve an export economy; the region's governments have since built new lines to adapt this colonial network to the internal needs of the region's countries. Postindependence political borders have resulted in a reorientation of traffic flows on the network.

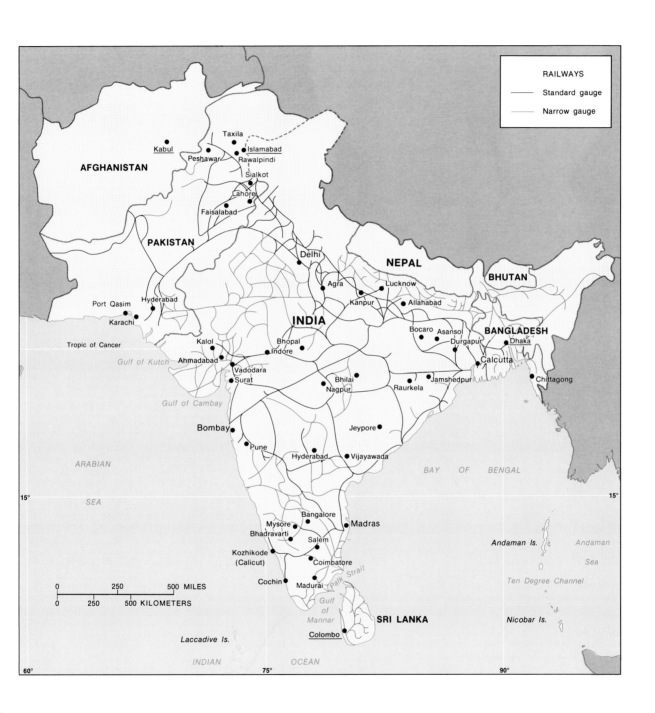

RAILWAYS

Standard gauge

Narrow gauge

AFGHANISTAN

Kabul

Taxila

Islamabad
Peshawar
Rawalpindi
Sialkot

Lahore

Faisalabad

PAKISTAN

Delhi

NEPAL

BHUTAN

Agra
Lucknow
Kanpur
Allahabad

Port Qasim
Hyderabad

Karachi

INDIA

Bocaro
Asansol
BANGLADESH

Tropic of Cancer

Kalol
Bhopal
Indore
Durgapur
Dhaka

Gulf of Kutch
Ahmadabad
Calcutta

Vadodara
Surat

Bhilai
Nagpur
Raurkela
Jamshedpur
Chittagong

Gulf of Cambay

Bombay

Jeypore

ARABIAN

Pune
Hyderabad
Vijayawada

BAY   OF   BENGAL

15°

SEA

15°

Andaman Is.
Andaman

Bangalore

Sea

Mysore
Madras
Bhadravarti
Salem
Ten Degree Channel

Kozhikode
(Calicut)
Coimbatore

Cochin
Madurai
Palk Strait

0   250   500 MILES

0   250   500 KILOMETERS

Gulf
of
Mannar

SRI LANKA

Nicobar Is.

Laccadive Is.

Colombo

INDIAN

OCEAN

60°

75°

90°

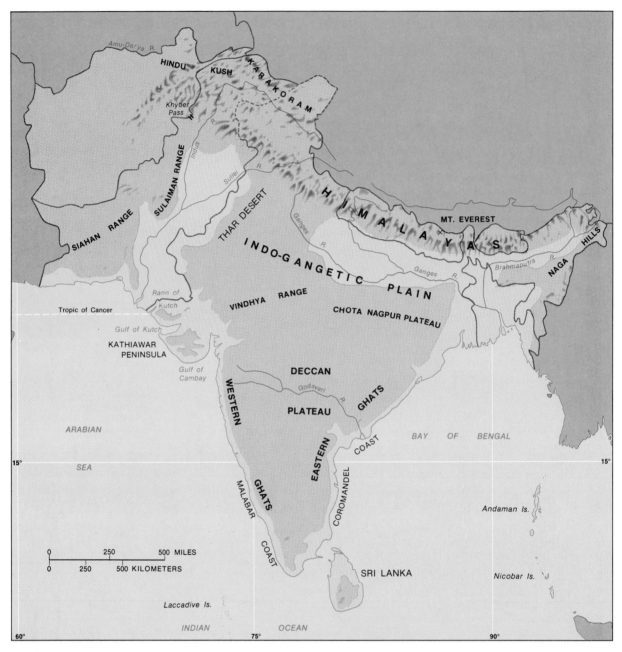

FIGURE 7–5
Physical Regions and Features of South Asia. Three major physical regions
encompass the bulk of South Asia: the Himalayas, the Deccan Plateau, and the broad
Indo-Gangetic Plain. Narrow, fringing coastal lowlands are backed by steep hills,
complicating access to the interior in many places. The central hill lands effectively
demarcate culturally dissimilar north and south India.

of the drainage, a wider (up to 150 miles) bordering coastal plain has developed along the eastern side of the Deccan.

On the island of Sri Lanka, the high mountains of the south-central area are fringed by dissected hill lands which are, in turn, surrounded by a broad coastal plain. The chain of islands and peninsulas that almost interconnect India and Sri Lanka has the fanciful name Adam's Bridge.

The Monsoon: The Season of Life

Most of the region is tropical or subtropical in terms of temperature. Altitude, not latitude, accounts for the winter cold of Afghanistan and the fringes of India's Heartland. Even in Nepal and Bhutan, the valleys have a pleasant climate. What is most important about the region's climate is the **seasonality** and total amount of rainfall. The drylands of north and west Afghanistan receive their snows and rains in winter. There the mountain heights squeeze the last drops of moisture from cyclonic storms that originated back over the Atlantic. The rest of the region receives its rain in the summer, from storms originating over the warm tropical waters of the Indian Ocean. The west of the subcontinent is desert; the east is lush and green in summer. In both cases, the mechanism is the same. It is the **monsoon** (Figure 7–6).

The term *monsoon* is derived from the Arabic word *mausim,* which means "season." To the Arab traders of the East African coast and the Arabian Sea, it was the season of the shift in winds and a consequent shift in the pattern of trade for their wind-powered vessels. It has come to mean the season of the rains for the people of South Asia, and understanding the monsoon is one of the keys to understanding that region.

A monsoonal tendency is noted by climatologists in each of the world's major continental masses. However, it is most pronounced in Asia, the largest of the continents. Monsoon winds are the major carriers of moisture from the ocean to large parts of East and Southeast Asia, but it is in South Asia where monsoon winds literally set the pace of rural life. The monsoon is the theme of much of India's literature and art. It brings relief from drought and searing heat. For many farmers, the monsoons are the literal difference between plenty and famine, between life and death.

In winter, the huge Eurasian landmass becomes intensely cold in its northern and central reaches. Semipermanent cells of high atmospheric pressure are reinforced by intense Siberian cold. Masses of bitter air, driven south by the jet stream, pour out of the Arctic reaches, flooding into interior Asia and western and northern China. Surface winds move from this cold, dry interior toward the relatively low pressure over the warmer oceans to the east and south.

In the winter pattern, winds circle clockwise out of the center of Siberia, across cool waters, and head toward South Asia's lands from the colder, drier northeast. The dominant flow is from land to sea. There is little chance of rain except in a few favored spots. Only where the path of the winds crosses warmer waters is there a possibility for the lower layers of the air to pick up enough moisture and instability for rain. Parts of western Japan and Korea, the equatorial portions of Southeast Asia, the island of Sri Lanka, and the Eastern Ghats receive rainfall in the winter season (see Figure 7–6a).

In summer, the pattern is reversed. The hot Asian interiors, baking in the strong rays of the late spring

FIGURE 7–6, pp. 358–359 →

(a) ***Winter Monsoon.*** Perhaps nothing in the physical environment is so influential in the daily life of South Asians as the seasonal changes in pressure and rainfall that accompany the monsoons. In winter, high pressures dominate over the continental interior of Asia, and winds move from land to sea. Protected by the Himalayas, South Asia is shielded from cold temperatures, but the winds bring no rainfall in their journey toward southern seas. Where winds are moistened by contact with warmer ocean waters on their way to fringing islands and coasts, there is a brief winter rainy season, but most of the Asian continent experiences drought during this part of the year. (b) ***Summer Monsoon***. In summer, intense continental heating creates a low-pressure situation over the interior of the Asian continent. Winter patterns are reversed, and moisture-laden winds proceed from sea to land. The intertropical convergence (ITC) is drawn northward over the land, creating a massive heat pump that draws in moist, unstable air masses, bringing heavy rains to the land. Where altitude causes a lifting of these air masses, precipitation is particularly intense. Agriculture, and with it relative prosperity or poverty, depends on the arrival of the summer monsoon in South Asia.

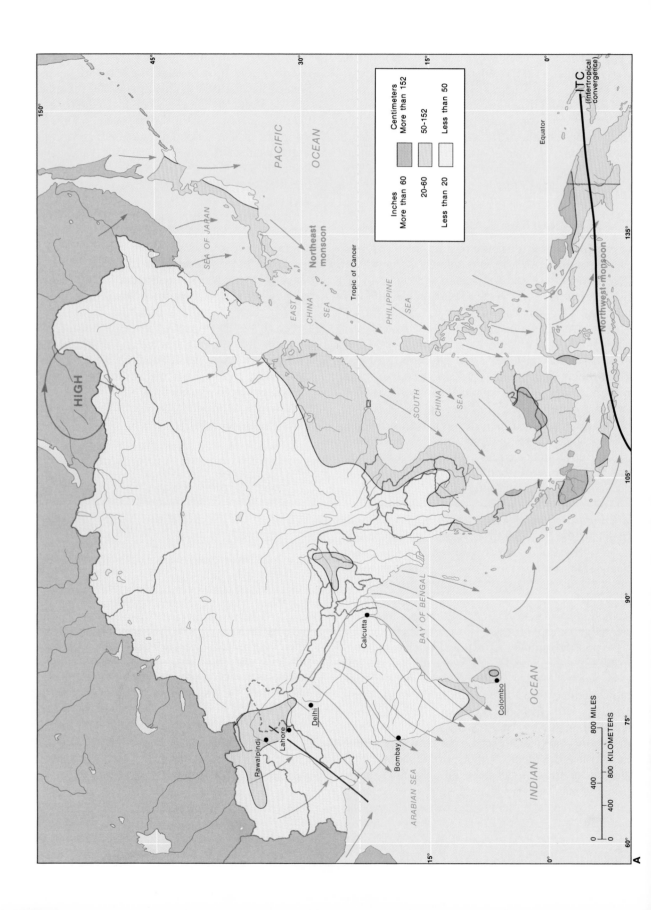

Inches	Centimeters
More than 60	More than 152
20-60	50-152
Less than 20	Less than 50

ITC
(Intertropical
convergence)

Equator

Tropic of Cancer

Northwest monsoon

Northeast monsoon

HIGH

PACIFIC

OCEAN

SEA OF JAPAN

EAST
CHINA
SEA

PHILIPPINE
SEA

SOUTH
CHINA
SEA

BAY OF BENGAL

ARABIAN SEA

INDIAN
OCEAN

Rawalpindi
Lahore
Delhi
Bombay
Calcutta
Colombo

150°
45°
30°
15°
0°
135°
105°
90°
75°
60°
0°
15°

0   400   800 MILES
0   400   800 KILOMETERS

A

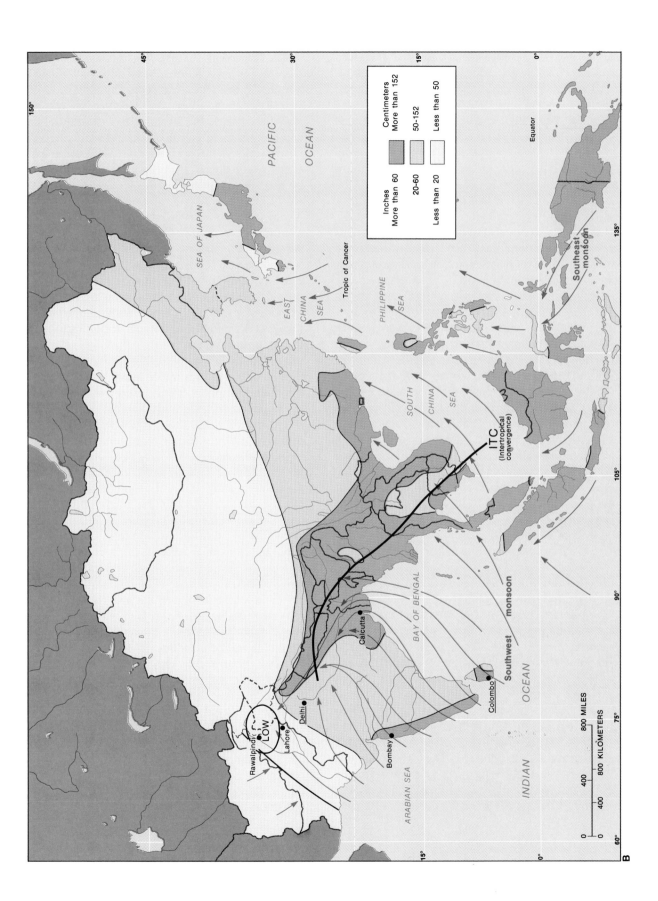

Inches
More than 60
20-60
Less than 20

Centimeters
More than 152
50-152
Less than 50

PACIFIC

OCEAN

SEA OF JAPAN

Tropic of Cancer

EAST

CHINA

SEA

PHILIPPINE

SEA

SOUTH

CHINA

SEA

Southeast
monsoon

ITC
(Intertropical
convergence)

Equator

BAY OF BENGAL

Calcutta

Southwest
monsoon

Colombo

INDIAN

OCEAN

Delhi

ARABIAN SEA

Bombay

Rawalpindi
LOW

Lahore

0   400   800 MILES

0   400   800 KILOMETERS

60°   75°   90°   105°   135°   150°

0°   15°   30°   45°   15°   0°

B

sun, lie beneath a trough of low pressure. The **inter-tropical convergence zone** (ITC) is drawn northward with the poleward migration of the sun's most intense radiation. The ITC is the tropical meeting place for prevailing winds of the Northeast and Southeast Trades; it is characterized by rising, moist air, often producing rain. At the spring and fall equinoxes, the ITC is near the equator. In the Northern Hemisphere's winter, it lies far to the south of the equator. In summer, it sits squarely over South Asia, reinforced by the heat-induced continental low.

As the ITC crosses the equator in its northward path, the direction of the winds changes. Southwesterly and southerly winds pump in moisture evaporated over the tropical Indian Ocean on their way toward South Asia. A low-pressure cell that builds over the southern Punjab in the upper Indus Valley effectively draws the winds from the warm, moist ocean toward the parched land. In an attempt to rise over the world's highest mountains, the winds are defeated. Force-fed updrafts of air drop their moisture over the subcontinent in the process. Air cooled in the process of rising over the Western Ghats drops tremendous quantities of moisture over western peninsular India. The ITC, in concert with the low pressure prevailing over land, the prevailing moisture-laden winds, and the barrier of the Himalayas, generates a massive weather machine that pumps needed moisture from the sea over the land (see Figure 7–6b).

Though the monsoons bring the moisture-laden air, it is the orographic lifting that actually causes the clouds and releases the precipitation. **Orographic precipitation** is simply mountain (or altitude)-induced precipitation. South Asia has three zones of heavy rainfall where moist air meets mountains. The first is along the Western Ghats, where heavy rains occur from north of Bombay to Cape Cormorin, the end of the peninsula. The second is along the flanks of the Himalayas, where streams that feed the Ganges have their sources. The third is a large area in northeast India, including the delta area of the Bay of Bengal and the hills of Assam. Over 400 inches in one year, mostly within four months, have been recorded on the lower slopes of the Himalayas. The long-term averages are less spectacular than the unusual year or season, but even they tell the story of abundantly wet summers followed by dry winters.

The Indian city of Darjeeling, just south of Sikkim, has a July average alone of 32 inches of precipitation, as much as many well-watered countries receive in a year. December, January, and February each have less than 1 inch per month. Darjeeling's 122-inch annual precipitation average includes a fantastically wet summer as the onshore winds from the tropical ocean sweep up the slope of the Himalayas, followed by a desertlike winter as the winds flow from the land toward the sea. Cherrapunji, on the slopes of Assam in the border area with Burma, records an even higher average—457 inches a year; July alone receives 107 inches on average.

Paradoxically, Indian farmers, whether in the almost unbelievably high precipitation areas mentioned or in more normal monsoon areas, are most bothered by drought. It is the timing of the onset of the summer monsoon, not the total precipitation, that is most important.

For the great majority of South Asians, the summer monsoon is the time of heavy, dark clouds and frequent downpours. Winter features clear skies. Before the monsoon, the subcontinent is a land of hard, baked brown earth. Crops must be planted before the arrival of the summer monsoon. The seed must have germinated and the tiny plant anchored by its root system before a possible downpour that could wash seeds away. If the monsoon does not arrive when expected, the soil will not have enough moisture to support the growing plant; the plant will simply wither and die. Because the reversal of wind systems and the rains depend on a combination of events—the weakening of the Siberian high pressure, the development of a Southwest Asian low pressure, and relative temperatures over the surrounding sea—the timing is predictable only in a general sense. A late arrival of the rains means crop disaster for millions of farmers. No wonder South Asians look forward eagerly to gray skies and torrential rains—the summer monsoon is their life-support system. Whereas Western cartoons portray persons with bad luck as having a dark cloud over their heads, South Asians understandably regard dark clouds as very good luck indeed!

Hot, humid weather prevails on the monsoon coasts of western India and the delta area around the Bay of Bengal. Sri Lanka and the Eastern Ghats, with year-round rainfall, are truly equatorial climates.

Most of southern India, away from the coasts, has a typical, tropical wet-and-dry climate. The northern area, although cooler in winter, can be unpleasantly hot in the fall as the monsoon wanes. Because the area is far from the moderating effects of the sea, there are greater seasonal, and daily, temperature ranges than in the south.

Subhumid climates form a belt of steppe in western India, transitional to desert at the farthest landward reaches of the monsoon influence. The Western Ghats induce heavy moisture from monsoon rains, but their eastern, more gentle slopes experience drought and generally dry conditions. Similarly, the upper reaches of the Ganges, at the tail end of monsoon influences, is a moisture-deficit area. Westward, into Pakistan, the country is desert. Only with irrigation can crops survive. Afghanistan, where mountains shield the valleys from rain that might arrive from any direction, is also desert country. Summer heat is intense, despite its altitude. The entire northwest depends on streams for irrigation, fed by snowmelt and meltwater from the glaciers of high mountains that form the northern boundary of the region.

Drainage and Natural Vegetation

All the largest rivers—the Indus, Ganges, and Brahmaputra—have their sources in the Himalayas. Thousands of short tributaries enter them from the Himalayas, keeping water levels perennially high. Portions of northwestern India, the west of Pakistan, and Afghanistan form an area of **interior drainage** (streams do not reach the sea).

The large volume and relatively steady flow of the major rivers provide an excellent and reliable source of irrigation water. Most rivers of peninsular India suffer erratic seasonal flows with commensurately limited opportunities for irrigation. In the areas of peninsular India covered by lava flows, the water table is very deep; consequently, storage tanks, ponds, and small reservoirs dot the landscape, providing some water for use during the dry season. New deep wells, drilled by the government, are used increasingly to tap deep groundwater pools and improve irrigation.

A variety of tropical wet-and-dry forest originally occupied all but the wettest areas of the subconti-

nent. Monsoon species, able to anchor themselves in the wet season, yet with natural adaptations against drought and fire, once predominated. Such species as teak, banyon, and sandalwood had notable economic value. Constant pressure on the resource has reduced vast areas to scrub and bamboo. In the northeast (Assam), there are stands of truly tropical rainforest that are being cleared rapidly for agriculture as pressure on the land increases.

Vast areas of alluvial soil fill the valleys of the three major rivers. New alluvium arrives with each flood, but the use of plant stalks and animal dung as fuel and building material rather than as fertilizer, coupled with constant cropping, has measurably reduced the original fertility in many areas of the Indo-Gangetic Plain.

The soils of peninsular India and Sri Lanka fall basically into two categories: tropical "black earth" (northwestern portion of Deccan) and red-yellow latosols. The famous black soil inland from Bombay is one of the world's great areas of naturally rich soil. Derived from deeply weathered volcanic rock, it maintains good yields despite constant cropping of cotton, a notorious depleter of soil fertility. The average soil of peninsular India is an impoverished clay. Its overall utility is quite low except in river valleys and a few other favored areas. Painstaking efforts of peasants are rewarded with only meager yields.

Northwestward, soils take on dryland characteristics, including a lower humus content. These soils vary from chestnut earths in the upper Ganges to light gray desert soils with overconcentrations of calcium and salt as one approaches the deserts. Unlike the almost totally man-made soils of the Chinese river valleys, South Asia's soils have been more depreciated than enhanced by long use.

THE STRUCTURE OF SUBREGIONS

A country as large and diverse (physically, culturally, and economically) as India must be divided into several subregions:

1. the Indian Heartland, including the Ganges Plain, the city of Calcutta, the Chota Nagpur

Hills (a heavy industrial district adjacent to Calcutta), and the lowlands of Orissa

2. the Indian Peninsular South, including the dynamic coastal ports and the more traditional (but rapidly changing) lands of the peninsular interior

3. the Indian Periphery, including the mountainous border districts, the drylands of the western frontier, and the wetlands of Assam, all essentially underutilized, potential pioneering areas of the subcontinent (Figure 7–7)

Islamic and predominantly desert, Pakistan and Afghanistan form a logical unit, though their recent history and level of development are quite different. The mountain buffer states of Nepal and Bhutan are also treated as a single subregion. Quite different in detail from the rest of the region, Bangladesh and Sri Lanka are given separate treatment.

The Indian Heartland

Identifying characteristics of the Indian Heartland include a very high population density, agriculture based on canal irrigation, a high proportion of land under cultivation, a relatively dense transport network, the earliest heavy industrial center, and a physical unity contributed most largely by the Ganges Plain. India's largest metropolis, Calcutta (population 9.5 million), and its third largest city and capital, Delhi (population 5.6 million), are located in this subregion. The Heartland contains 55 percent of India's population, and densities over 1,000 per square mile are not unusual.

The lower, easternmost part of the Ganges Plain is used for rice, whereas the upper, drier section (toward Delhi and beyond) is primarily sown to wheat (Figure 7–8). Even in the wheat area, rice is the second-ranking crop, and wheat is a common dry season crop in the rice-producing region, along with corn and, close to the Bangladesh boundary, the fiber crop jute. India is a major grower of sugar cane, with most production centered in the Heartland.

The westward transition to wheat is related to diminishing rainfall, since the monsoon's intensity and duration diminish inland from the warm sea. India's primary means of increasing agricultural production is irrigation. At independence, India had about 50 million acres (one sixth of the then-total 300 million

cultivated acres) under irrigation. By 1986, Indians cultivated 422 million acres, and of these, 120 million acres (28 percent) were irrigated.

In the past, irrigation schemes in the Heartland have been of a larger scale and more efficient than in the Deccan because of the dry environment and difficulty of obtaining groundwater in the plateau. Only low dams (barrages) were necessary to impound large quantities of water in the Ganges Plain. The best sites—the easiest to use and most efficient in terms of a benefits-to-cost ratio—were the first ones used. Further expansion means developing progressively less desirable sites. An educational campaign trains farmers to drain off irrigation water to avoid waterlogging and to minimize **salination** ("salting" of irrigated soils). A crust of soluble minerals can poison soil for plant growth. Irrigation, without careful long-range soil management, can be a temporary boost to agriculture, but a long-run disaster.

Village Life in the Indian Heartland. Nearly three fourths of the people of India depend directly on agriculture, and agricultural products provide about 40 percent of the GNP. Almost 80 percent of all Indians live in the more than one-half million villages. India is likely to remain an agrarian-based, village-dominated society for some time. While population grows at an annual rate of about 2.2 percent, the average annual increase in agricultural production has been running at 2.6 percent. Thus, India *is* making progress in one of its most critical areas—the race between increasing population and increasing food supply.

Even with the rising productivity of Indian agriculture in recent decades taken into account, average productivity per land unit is much lower than in developed countries or even in China. In a land where population pressure is immense, an estimated 100 million acres of uncultivated land is officially classified as land with the potential for cultivation. There are various reasons for this fallowed land not being used; deficiencies in traction power, irrigation water, or fertilizer are prominent. Some land is temporarily fallow because the landowners have not yet made satisfactory rental agreements with farmers. Traditionally, landlords, farmers, and moneylenders are of different castes. It would "break" caste to actually cultivate the land if one were an upper-caste landlord. Similarly, money lending would not be a suit-

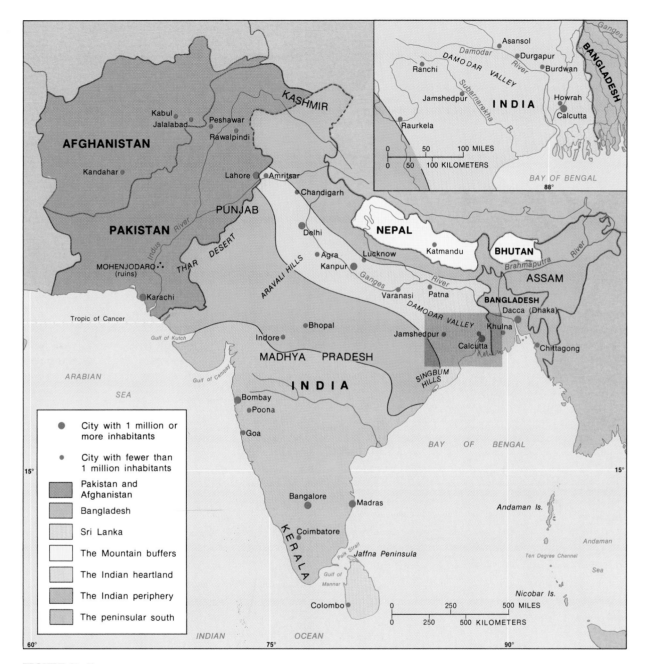

FIGURE 7–7
Subregions of South Asia. The physical unity of the region is not reflected in the
region's political diversity. Five successor states (Sri Lanka, Bangladesh, India,
Pakistan, Burma) replaced the old imperium of British India. A series of buffer states
(Afghanistan, Nepal, Bhutan) persist in the contact zone between South Asia and its
northern neighbors, China and the USSR. The sheer diversity and size of India
require that it be divided into subregions for accuracy of analysis. Though
dominantly rural, the region contains many large cities, including Bombay and
Calcutta, two of the world's largest. The inset displays the Damodar Valley, India's
largest industrial area, in a greater degree of detail.

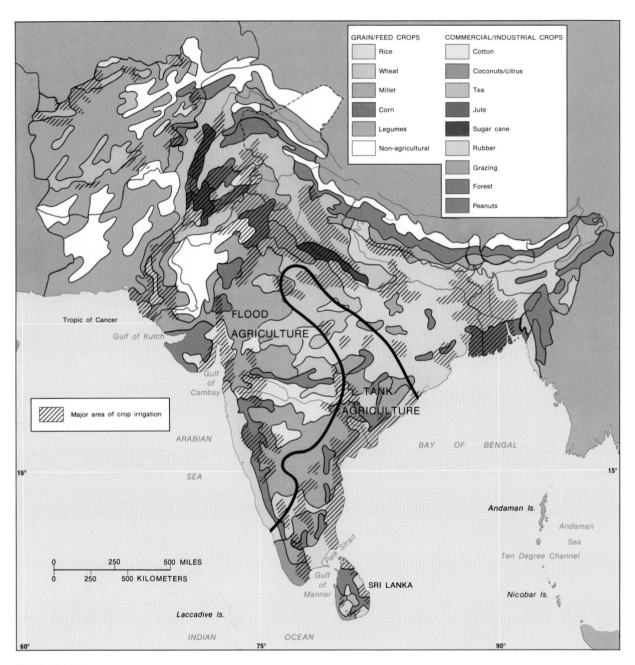

FIGURE 7–8

South Asian Agriculture and Rural Land Use. The map shows the enormous importance of irrigation to the region's agriculture. Without irrigation, production would often be confined to one annual crop, and yields would be considerably lower; irrigating extends the growing season beyond that provided by the rains of the monsoon. Cotton, corn, wheat, and millet are associated with drier areas. Rice and jute dominate in wetter regions, where grains such as wheat, barley, millet, and various sorghums are grown as a second crop during the dry season. In the river valleys of the north, water supplies are nearly constant. In the south, however, monsoonal rainwater must be stored in reservoirs or cisterns for use during drier periods. The drier and higher margins of the subcontinent are suitable only to grazing; crops are limited to oases or to small, favored patches of land that provide supplementary subsistence.

able occupation when one owned land and rented it to farmers. The cyclical nature of almost all categories of farming means that the cash return is not continuous through the year, but some expenses will be ongoing. For farmers with little or no savings, this means borrowing money to live until the crop is harvested.

A recent government survey showed that more than 6 percent of the rural households had no land, while almost half those sampled occupied, but did not own, "farms" of 2.5 acres or less. These tiny farms represented only about 6 percent of the total cultivated land. On the other hand, over a third of all cropland is occupied by only 4 percent of the households. Some of the smallest farms are worked by people who have an urban or industrial job as well. Both Hindu and Moslem laws of inheritance allow the subdivision of property among all the sons, which means that even a large holding can be reduced to mere patches of land within a few generations. Farms are often not contiguous. In one village, for example, 1,800 farms were divided into some 63,000 separate parcels. Landowners frequently have 20 or 30 separate plots of land.

Obviously, a farmer trying to use two dozen separate plots of land will spend considerable time "commuting." He will find it almost impossible to purchase fencing for all these small parcels, and wandering cattle will "harvest" his crops for him unless kept out by constant vigilance.

Most solutions suggested by Westerners are not acceptable to Indians. Mechanization does not necessarily increase productivity per acre; it is a labor-saving device. Using labor more efficiently in the Indian Heartland would make millions of farmers "surplus" labor. Urban labor markets of Heartland cities hardly need millions of additional unskilled laborers; unemployment and **underemployment** (holding only a part-time job or sharing a full-time job with several others) are already serious problems. Mechanization also would mean increasing imports of oil, already a large drain on foreign exchange.

Indian Land Reform. The cultural complexity of India is matched by the complexity of land tenure systems. In general terms, there are three large categories of land tenure: landlord-owned, tenant-farmed estates; small peasant proprietary farms; and village communal holdings. This system is complicated by the often separate ownership of water rights, draft animals, and equipment.

By law, tenants cannot now be charged more than 20 to 25 percent of the crop value. Tenants have land rights now (in the form of guaranteed, long-term contracts), and they are given the legal right to purchase the lands they farm under long-term mortgage arrangements. Farm sizes are subjected to strong controls, and a 50-acre ownership ceiling applies to farmlands that can produce two crops a year and have irrigation facilities in place.

Many miniholdings have been consolidated into large community-held units. Since 1965, inheritance laws have been changed to reduce the further fragmentation of holdings. Cooperative village management of lands is encouraged, and both legal restraints and incentives are in place to ensure the equitable distribution of water, land, income, and risks among all.

Of the 70 million rural households in India, one third still own less than one-half acre of land. Reforms have increased the average size of Indian farms, however, to almost seven acres. In areas that produce two or three crops yearly, such an average farm exceeds subsistence levels and produces a marketable surplus. Continued urban-industrial development will likely reduce further the number of mini-units, and India's increasing ability to feed itself has been aided in large measure by this progressive program of land reform. It is not perhaps ideal, but it is in harmony with Indian economic realities.

Agricultural Conservatism in the Indian Heartland. A common frustration to would-be leaders and innovators is the intense conservatism of most Indian farmers. Various efforts to introduce new crops, new plant varieties, chemical fertilizers, cultivation techniques, or other departures from the traditional seldom succeed quickly. Fertilization is one of the greatest deficiencies of Indian agriculture. After a millennium or more of constant crop production, Indian soils need heavy fertilization, deeper

THE SACRED COW: POINT AND COUNTERPOINT

The intertwined nature of culture, custom, and economy are nicely illustrated by the "sacred cow" in India. The role of the cow in the Indian agricultural economy is complex and easily misinterpreted by non-Indians. Westerners touring India see apparently homeless cattle wandering about, foraging for food. These cattle seem to be useless, and the Hindu religious ban on their slaughter appears born of ignorant superstition. Closer examination of the role of cows in the Indian economy, however, shows that the animals *are* useful. Almost all these wandering cattle have owners and will return home at night. They have been turned loose to scavenge food for themselves. Indian farmers cannot afford to feed their cattle grains, as is customary in Western farming; the animals subsist on a diet of mostly field stubble, weeds, and vegetable refuse that people would not eat anyway. Indian cattle are not needed full-time on the farm. Their main job is providing the seasonally necessary traction power for the poor farmer to plow his fields. The cow also provides fertilizer. Much of the manure is dried and used as cooking fuel, providing a clean-burning, slow heat.

Replacement of the all-purpose cow with a tractor would have several economic drawbacks. Tractors are expensive, need costly spare parts, and use diesel or gasoline fuel. They could not be used efficiently on the small separate plots of Indian farms. They do not contribute fertilizer or cooking fuel. A cow calves draft animals for the future; a machine just wears out. Contrary to Western belief, the cow eventually does become beef. **Untouchables** (the lowest caste) are untouchable largely because they deal in death; they haul away the carcass of an animal dead of natural causes, tan the hide, process the bones for bonemeal or charcoal, and eat the flesh. India is a major exporter of hides. Even the milk production, low by Western standards, is an important addition to diets normally low in fats and proteins. Considering that cows are necessary for plowing the fields and produce important by-products, the Hindu prohibition on slaughtering them can be seen as a way of making maximum use of the animal.

There are between 200 and 300 million cattle in India, the largest aggregate number of cattle found in any country in the world. All life in the Hindu religious philosophy exists on a continuum, and cows rank above other animals and many humans in the status accorded by caste. The words "sacred cow," and even the mild expletive "Holy Cow," entered the English language as a result of these cultural observations made in colonial India by Westerners.

Retained in its rural environs, the Indian cow is indeed of service to farmers. The "urban cow," however, presents increased problems, knocking over produce stalls, clogging arteries of both foot and motor traffic, halting trains, drawing flies, and littering streets with manure. With little forage available, urban cattle do consume food that is otherwise consumed by humans.

Cattle and Agriculture in South Asia. *Though a sacred animal in Hindu beliefs, the cow provides a variety of commodities and functions for Indian farmers. Its dung is used more often as a fuel or a building material than as fertilizer. It yields some protein for local diets as milk, but the flesh is not eaten by most Indians. Traction power is perhaps its greatest contribution. This cow is obviously less than a prime animal; nonetheless, it is functional. It feeds on rice stubble, the inedible remains left after grain harvest. It is pulling a simple seeding device that reduces the time and labor needed for sowing the second crop, in this case millet.*

The Indian government has recognized these problems. During the 1950s, it developed farms for apparently ownerless cattle and prepared legislation to remove cattle from urban surroundings. The purposes of the program were multifold: to replace indiscriminate breeding and thereby selectively upgrade the stock, to reduce the aforementioned hazards, to control the spread of human and animal disease, to channel manure into agriculture for fertilization, to provide export income from the sale of hides and quality Brahman bulls for breeding stock, and to provide a concentrated supply of raw materials (hides, bones, etc.) for selected Indian industries. Conservative Hindu elements found this program shocking and blasphemous, and it was abandoned under political pressure.

Viewed in this light, the Indian cow is still a net consumer rather than a societal producer. The objections of religious elements must be considered, however. Vocal, and often with the additional clout of high caste, this fundamentalist element could disrupt the government and perhaps lead to internal political unrest of major proportions.

THE CASTE SYSTEM

Caste is as old as Indian civilization itself. Though it may appear irrational to Western minds, its origins have an irrefutable logic. Its purpose was to ensure self-sufficiency in a largely rural society at a time when drought and plagues threatened survival, and when migration was difficult. Caste is based on occupation, and each necessary occupation was represented by a caste. Western society is characterized by both personal and job mobility; these are characteristics that emerged, however, with the development of an industrialized society. In an era comparable to that of historic India, such Western institutions as feudalism, slavery, the guilds, ranking clergy, and the nobility were similar occupational-status groupings held in place by the force of law and sanctioned by religion. They have all but disappeared or have been severely modified with the passage of time and the need for change in the West. The need for such changes in India is comparatively recent, and the persistence of caste, tied as it is to religious concepts of ultimate salvation, should not be surprising.

Caste is hereditary, and marriage takes place among members of the same caste. Two prominent ways to break caste are having sexual contact with, or coming into contact with the human wastes of, someone of a lower caste. Marriage outside one's caste assumes automatic downward movement in the caste hierarchy. Breaking caste also means returning as a person of lower caste, or even a lower, nonhuman form of life, in the next incarnation, lengthening the stay on earth and the time needed to reach nirvana, the eternal salvation. Breaking caste has the same impact as mortal sin or a serious offense in Western thought—punishment or retribution. Westerners expiate wrongdoing through penance, Hindus through returning to a more difficult life. When understood in these terms, caste is more readily appreciated as both an organized framework for society and a set of guidelines for personal conduct.

An outstanding example of Westerners' inability to understand the caste system was displayed in the development of the Indian rail system. Separate bathroom facilities, segregated by sex, were constructed at Indian rail stations (also, separated by Indian and British, for a total of four). Indians persisted in going to the bathroom outside; the British

plowing, and crop rotation. Indians have not, for reasons related to caste, used human waste as a fertilizer as fully as in traditional East Asia.

While such conditions are typical of much of India, they are particularly characteristic of the Heartland. With an average of a little more than half an acre of arable land per farm family adult and a very low cash income, Heartland farmers have little room for experimentation. Whereas Western farmers might spend $150 an acre on fertilizer, a small sum to them, similar expenditures for Indian farmers would represent their entire income. The Green Revolution is real, but it has its costs. And its initial costs favor relatively prosperous landowners, increasing gaps in misunderstood this as evidence of a low level of civilization. In an India then unaccustomed to public conveyances and facilities, one for each caste (not each sex) would have been the logical arrangement. The same would have applied to seating accommodations; first, second, and third class were simply not enough distinctions.

Gandhi was well aware of the problems that the caste system involved, and he was equally aware of the need for change if India were to develop. The lowest castes were to be given a special status in the new India. Sympathetic to their plight, Gandhi called them *Harijans*—"the beloved ones"—and saw to the passage of a series of amendments, or "schedules," that would assure them rights and equality. Given public intransigence and resistance to change, the untouchables—the lowest caste—inadvertently became enlarged to include all people covered by the schedules. "Scheduled castes," as they were called, became corrupted to "Shedulkhas," and that term (as well as "Harijan") took on the status of an epithet as it became synonymous with the old term "untouchable."

Who are the lowest castes? Those who deal with death (other than the military caste), the products of death, and wastes. Thus, butchers, tanners, undertakers, and persons who collect human wastes from urban dwellings would be in the lowest castes, as would hospital workers, given the need to change dressings and bedding and to dispose of hospital wastes.

The complications of caste for modern Indian society are legion. In harmony with the caste system, certain industrial plants hired all members of one caste. Villages went without blacksmiths (and scarce farm tools went without repair as a result) as members of that caste flocked to higher-paid employment in the country's early steel mills and foundries. Educational opportunities are guaranteed for all, but students of lower castes frequently sit in segregated classrooms, even at institutions of higher learning. Well-educated Harijans are forced to migrate to the anonymity of other cultures for lack of suitable employment or lack of clients due to caste differences. Governments can legislate, but enforcement is difficult; changing popular opinion may not be possible for generations.

Fundamentalist Hindus complain of new permissiveness and a lack of respect for traditions, but there is a sense of departure from strict interpretations of caste among India's young, the highly educated, and middle-class urbanites. Industrialization and commercial contact may be leading toward the evolution of a less complex, less rigid caste system as Indian society modernizes.

income and levels of living between the relatively few rich and the masses of poor farmers.

Industrial Development in the Heartland.
The Indian government invests in industry on a large scale because a **planned economy** was felt to be essential in a country where wealth is concentrated in the hands of a small, tradition-bound upper class. In an attempt to avoid regional rivalry and to alleviate rural unemployment, the government has attempted to disperse industry. Yet, rational economic planning and the realities of transportation, raw material distribution, and assembly cost have favored certain areas.

The Heartland reflects some of these divergent pulls. It has a large surplus of labor, yet most of it is unskilled. Rural areas here are intensely orthodox in their respect for Hinduism's religious and cultural practices. Industrial employment in some ways would violate caste, in terms of both occupation and the intermixture of castes under one factory roof. It is a seeming paradox that the Heartland receives less industrial emphasis than the peripheral area of divergent culture and, formerly, lower economic and political influence.

Yet there are sound reasons for this lack of industry. The Peninsular South has greater mineral resources except for fuel. The dense regional railroad net, most of it broad gauge, would appear to be a locational advantage. Yet, it is overburdened with passenger traffic and the cross-hauling of agricultural commodities. The dense population of the Heartland would appear to form the largest, most compact market in the country, yet many in this market area are poor and engaged in subsistence farming, almost wholly outside the commercial economy. The area is central to the Indian subcontinent but peripheral to its ports. Because industrial development hinges on needed imports and the need to raise currency through exports, the rimland, not the Heartland, is where the developmental action is concentrated. As in any developing economy, much of the impetus to the generation of economic activity is external, so the coastal regions of India are developing first.

Despite the fact that it generally lags behind the average for the nation, the Heartland contains the largest industrial region of the country at its eastern extreme (Calcutta-Damodar) (Figure 7–9). There is also a new and growing industrial complex centering on Delhi, the national capital.

Calcutta is a name that evokes strong, largely negative, reactions. The city, founded by the British in 1690, is situated on the Hoogly River, the best available harbor on India's east coast. It was sited to provide a deep-water anchorage safe from typhoons. Its explosively rapid growth has been fueled by a series of crop failures and famines in distant provinces; as a result, its population comprises many different ethnic and linguistic groups.

In addition to serving as the commercial, financial, and manufacturing center for most of eastern India, Calcutta is also the port for Nepal and Bhutan. Up until the 1960s, Calcutta accounted for almost half of India's imports, nearly a quarter of its exports, and one third of its financial transactions. Although this dominance has been reduced, Calcutta is still important in both India's domestic economy and international trade.

A century and a half of rapid, uncontrolled growth in the urban region has produced a planner's nightmare; almost half of Calcutta's population lacks clean drinking water. Crowded and congested, the city has come to epitomize the plight of urban poverty in India.

Calcutta and its suburbs, in particular Howrah, have long been established centers of industry. Jute milling was the major endeavor under British rule. Competition from Bangladesh (and the manufacture of plastic bags) has seriously cut into the market for jute. The 1947 division of India left most of the factories in India and most of the jute acreage in what is now Bangladesh, disorienting the jute industry from the inception of independence. The manufacture of cotton and synthetic textiles has replaced a considerable segment of the jute industry in Indian plants.

Recently, there has been much diversification of industry; light manufactures and consumer goods have become particularly important, largely as a byproduct of nearby heavy industrial development. Iron and steel, heavy chemicals, alloys, and energy-generating facilities in the nearby Damodar Valley supply the raw material needs for railway equipment, rubber goods, light machinery, paper, plastics, and pharmaceuticals produced in metropolitan Calcutta. Rapid expansion has pushed the limits of the industrial district far beyond the city; even small towns and villages now have assembly and packaging plants. Despite impressive economic growth, the city is still typified by massive unemployment and incredible poverty. It remains a city that only Mother Theresa could love.

In the Damodar Valley, a mixture of private and government capital has created India's major heavy industrial base, 150 miles west of Calcutta. Four major coalfields are found in the vicinity of the steel center of Asansol (see Figure 7–9). South of Jamshedpur, another steel center, are the great iron ore deposits of the Singbhum Hills. The fortunate coincidence of coking-quality coal and high-quality iron ore gave an early boost to India's steel industry. India's iron ore reserves are among the largest and richest in the world. Other iron ore fields, including

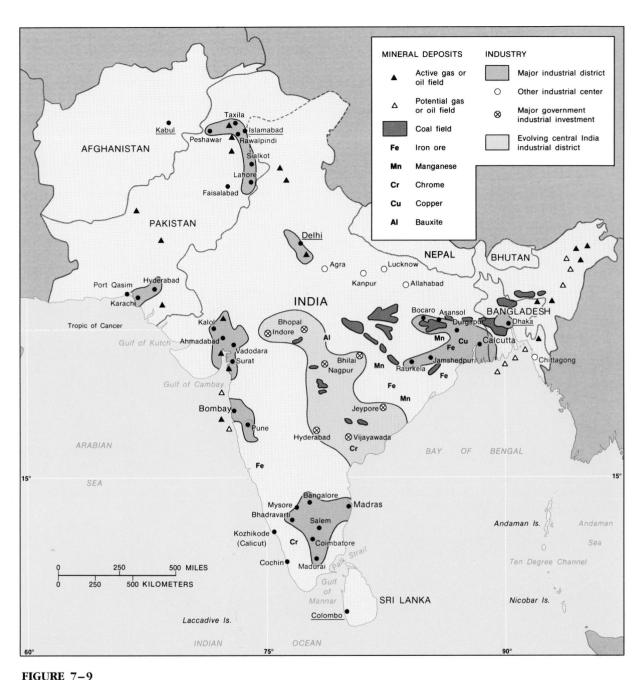

FIGURE 7–9
Mineral and Industrial Regions of South Asia. The coal of eastern India
provides a carbon fuel base for the country's industrial economy, particularly
important in a country that has discovered only meager petroleum and natural gas
reserves to date. Iron, bauxite, and alloys are present in plentiful supply. The
original dominance of Calcutta and the Damodar Valley in India's industrial structure
is being rapidly outpaced by the growth of manufacturing in the Peninsular South.
The densely populated Ganges Valley has little industry. Central India, with large
groups of Moslems, Christians, and animists in their local populations, develops
industry more as a result of the absence of caste considerations than as a response
to the presence of minerals. Industry is much better developed in India than in the
other countries of the region.

some beyond the Heartland, are being developed to support an expanded Indian steel industry and to supply exports to Japan. Discoveries of new iron ore fields have placed Indian reserves so high (relative to consumption) that India has literally thousands of years worth of ores in reserve. India is also self-sufficient in coal and limestone (see Figure 7–9). The original mill at Jamshedpur is still the largest steel-producing center. New centers have developed at Ranchi, Asansol, Durgapur, and Rourkela.

Impressive, if environmentally degraded, the Damodar Valley is frequently thought of as modern India. Yet even here, an enormous amount of hand labor is used in industrial production. Armies of mi-

grants still do the work machines could do in what is, inevitably, inefficient production. A positive social aspect of the steel industry is its willingness to employ large numbers of people of lower caste.

Delhi, like most capitals, is a city of bureaucrats. The local market is relatively affluent because government employment pays reasonably well by Indian standards. Clothing, pharmaceuticals, prepackaged food, furniture, and household appliances are manufactured there (see Figure 7–9).

Unlike Bombay and Calcutta, Delhi was an important center long before European colonial ambitions created coastal trading cities. India's capital is generally known as New Delhi because a new urban core

Downtown Delhi. *Delhi has functioned as India's capital since colonial times. It includes both ancient and modern districts and is a microcosm of all of India. The variety of alphabets and languages on the business signs and displays, as well as the diversity of dress, suggests the diversity of both class and culture that is typical of India. Delhi is one of India's most modern and prosperous cities.*

was created by the British to serve as the capital of British India. New Delhi is the eighth capital of India to be constructed in that vicinity.

Delhi occupies an important position within the subcontinent because it stands at the convergence of routes between the Himalayas, the Thar Desert, and the Aravalli Hills to the south. It is also near the watershed of the two great river systems, the Indus and the Ganges, and at the crucial junction of India's wheat- and rice-producing regions.

The Indian Peninsular South

The people of the Indian peninsula are generally shorter, darker, and of a somewhat different culture than those of the Heartland. They have experienced a different set of cultural inputs and are more completely a product of ancient, pre-Aryan India. The Peninsular South is the stronghold of Dravidian languages and perhaps the area of origin of the Hindu religion. The result has been a different landscape, with a settlement pattern, distinct architecture, and cultural flavor unlike those of the Heartland.

The Peninsular South has developed two major cultural centers—Bombay and Madras—at its extremes. Between is an area of tropical climate with no cool season. The South is a land of hills rather than plains. Yet it also possesses some areas of rich soils, rivaling or even surpassing the fertility of the Indo-Gangetic alluvium in some areas such as the black earth plateau country inland from Bombay.

Peninsular India is fuel poor, in contrast to the energy-rich Heartland. The greatest available local energy source is the hydropower generated in the rainy Western Ghats. The South's dominantly volcanic origins preclude deposits of fossil fuels except in minor basins and offshore. Still, it is not an area devoid of resources. One of the world's most imposing reserves of high-quality iron ore is found near Salem, inland from Madras. High-grade iron deposits are currently mined in what was the Portuguese treaty port of Goa, virtually at tidewater. It is the quantity, quality, and distribution of fuel, not iron, that retards an even more rapid expansion of local iron and steel industries (see Figure 7–9). Peninsular India contains some of the largest and highest grade manganese deposits outside the Soviet Union. Recently, large deposits of bauxite have been uncovered in the regions of Cutch and the Gulf of Cambay.

Among the major thrusts of Indian investment in recent years has been the development of increased electrical energy output to expand aluminum production and the making of chemical fertilizer. Both these initiatives are concentrated in the South.

Overall, South India is better developed and more commercially oriented in both its agriculture and industry. If tradition dominates in the Heartland, modernization and development increasingly dominate in the Peninsular South.

The Commercialization of Agriculture in the Indian South. The natural agricultural resource base of peninsular India is more varied and increasingly better developed than that of the Heartland. India now ranks fourth in world cotton production after the United States, China, and the Soviet Union. Although cotton is grown throughout India, the black earths of the Deccan are the overwhelmingly dominant producing region, and the district around Madras is also important. Here in the South, in an attempt to reduce imports, India has developed plantations for both tropical crops and temperate crops in high demand. Bananas, rubber, and coffee are grown in the peninsular southwest (see Figure 7–8). Tobacco is a specialty crop of the southeast coast, where cigar varieties similar to those of Brazil and Cuba are grown.

Renowned since the days of the spice trade, South India (in particular the west coast) grows ginger, mace, pepper, cardamon, cinnamon, nutmeg, and other spices for domestic use in highly seasoned Indian cooking and for export. Along with Sri Lanka and the northern districts of Darjeeling and Assam, the hills of South India near Madras are renowned for quality teas.

Despite this obvious and widespread emphasis on industrial and export crops, considerable area is given to more traditional farming pursuits. Rice is a major crop but only in the riverine delta areas of the South and some smaller, better-watered districts. Nowhere does rice production approach the dominance characteristic of the Heartland, and considerable southern production is in non-paddy, dry upland varieties of rice which are lower yielding. Often planted with rice, cotton, or other crops, millet can withstand drought to a remarkable extent and ensure some food production in even the worst years. Recently, peanut acreage has increased greatly.

Peanuts and sesame and cotton seed are the major sources of fats in an essentially vegetarian diet. The Indian Heartland is dominated by rice, barley, wheat, and sugar cane, whereas South India is dominantly a region of cotton, peanuts, millet, and tropical crops.

Proportionately more farmers in the South own their own land, and average farm size is larger than in the Heartland. Population densities are much lower in South India; high densities are reached only in the delta plains of the major rivers and in the coastal plains of Kerala state. Southern India contains a slightly higher proportion of urban dwellers. Of India's 11 cities of over 1 million population, 7 are found in the subregion including Bombay (population 8.2 million), India's second largest city.

The Booming Industrial South. Three of the four well-developed industrial districts of India are located in the South. Two, Bombay-Poona and the district of Madras-Bangalore-Madurai, are centers for the manufacture of cotton textiles and the processing of tropical goods (see Figure 7–9). In some ways, they are a legacy of British colonial investment policy.

Bombay was the most important colonial port. Seven small islands of lava, joined by natural and artificial accumulations of silt and sand, are separated from the mainland by a belt of poorly drained swamplands and salt marshes that provided some security from surprise attack by land. The harbor is one of the finest in the world, and Bombay-Poona is the second most important (and most rapidly growing) manufacturing region in India.

A cosmopolitan city in architecture and society, Bombay leads in motion picture production (India produces more films than any other country) and is the most diverse manufacturing city in all of India. Based on the black earth cotton district, it has long been the major cotton textile-producing center of India. Bombay is India's largest oil port and refining center, and recent offshore oil finds enhance the area's attractiveness as an industrial location. Petrochemicals, synthetic fibers, plastics, and chemical fertilizers are some newer industries that have developed around oil refining. With a food surplus in its hinterland, the region has developed an important food-processing industry. Diversification since the 1950s has added automotive, machine tool, electron-

ics, aircraft, and engineering industries to this most modern of all Indian industrial centers. The government plans to develop the old Portuguese port at Goa. The harbor and Goa's high-quality iron ores add to the attractiveness of that location. Goa is envisioned as the third pole in a triangular industrial district including Bombay and Poona.

The second of the old traditional manufacturing districts is Madras-Bangalore-Madurai. Madras is India's fourth largest city and its third most important port. Cotton textiles formed its first industrial base. The region has since diversified into consumer goods, aircraft, armaments, food processing, leather industries, and light chemicals. High-tech industries, such as computer components, transistors, and software, are produced in and around the interior city of Bangalore. Precision engineering and electronics are centered in Coimbatore and Madurai, making this interior part of the Peninsular South the Indian equivalent of the Silicon Valley. Madras, the regional port, has added automotive and engineering industries; along with Bombay, it has developed shipping facilities for containerized freight.

A third district of the South is centered around the Gulf of Cambay. Ahmadabad and Baroda are the chief cities of the area. It now functions as the entrepot district of Delhi and the upper Ganges Valley. Cotton textiles, the original base, have continued to expand, although at a slower rate than the manufacture of chemicals and related products.

South India: The Prospects. The French-British-Portuguese struggle for control of the India trade resulted in the development of at least a dozen ports, as well as rudimentary manufactures. After independence, a major road network was built in peninsular India, further enhancing regional accessibility. These attributes make South India the most economical and logical location for industrial investment.

Lignite and small bituminous coal deposits are being developed in the peninsula, and more hydroelectric sites in the Western Ghats have recently come on line.

New reservoirs in South India, improved tank irrigation, and thousands of deep wells have helped stabilize regional food supplies. Recent good national harvests are in part due to this regional stabilization investment.

Modern Bombay. *India's second largest city and greatest port, Bombay reflects both progress and modernity. Rapid industrial growth in the nearby district is the basis of its commercial success. The skyline reflects postindependence India's commitment to development. Boats in the foreground tell of the continued existence of more traditional India in its midst, but the buildings in the background attest to prosperity and strongly symbolize the break with the past.*

Japanese imports have greatly expanded the markets for South India's raw materials, citrus fruit, and tropical products. Japanese industrial investment in the area is also increasing. On the other hand, Japanese exports are slowing the growth of such industries as shoes and clothing.

The Indian Periphery

It is difficult to think of teeming India as having regions suitable to expansion, yet there are frontiers. Rain-soaked Assam, the mountain valleys and slopes of the Himalayas, the rugged Central Hills, and the Thar Desert district form the third subregion of India (see Figure 7–7). Markedly dissimilar in climate and topography, they have the common factors of low population density, unexploited resources, and nontraditional attitudes.

Cut off from the rest of India except for a narrow corridor, Assam is the classic borderland in that it contains intermingled but distinctive ethnic groups. Many Assamese are closely related to the Chinese and to the Shan of Burma. Assam is also atypical for India because it has large tracts of river valley land that could be cultivated, although currently they are

THE SOUTH ASIAN CITY

Cities were known in South Asia 5,000 years ago when a great culture already flourished in the Indus Valley—attested to by the ruins at Mohenjodaro in the middle courses of the river. Ruins at sites from Delhi to Patna attest to the existence of cities throughout the length of the Ganges Plain that predate the city-states of Greece. In the most ancient times, there was already a network of market towns and administrative centers throughout the subcontinent.

Many of the larger urban centers of today's India and Pakistan are descended from the earlier fortresses of the ruling Moguls, who had a practiced military eye for the selection of critical defense sites. Still others, such as Varanasi (Benares), were places of great religious significance, holy cities visited by pilgrims since the founding of the Hindu faith.

Some, such as Karachi and Calcutta, were once fishing villages whose advantageous sites were developed into great ports by the East India Company as contact points in the early Indian trade. Bombay's spectacular harbor was unused until the British set themselves up as competitors of the Portuguese.

Although these largest port cities of South Asia were truly colonial creations, they contain many of the same urban elements found in the older cities that occupy sites in the landlocked Indo-Gangetic Plain. In virtually all large Indian cities, there was a modern center of commerce established by the colonial ruler, full of nineteenth century European-styled buildings and offices. Adjacent to the busy wharves and harbors in the ports, in an inland setting, this business district developed next to the warehouses, which in turn occupied the land between the river and the main railroad terminal. The commercial center thus had access to both the major forms of transportation, old and new, for cargoes large and small, of both domestic and foreign origin. The commercial district always contained (or ended with) a suitably landscaped park that marked the transition to the governmental buildings, which underlined the prestige of the ruler. In the north, for example in Delhi, there is often an old, walled city, the citadel of

not. The Brahmaputra Valley, the easternmost extension of the Ganges Plain, is bounded on the north and east by rugged hill country. Rice and jute are produced in the lowlands and tea is grown on the hill slopes. Assam's forest resource, hydroelectric potential, and presently uncultivated lands give it a growth potential unusual for India. Rapid in-

migration is meeting resistance from native groups, and India has had to institute border patrols to keep out Bangladeshi migrants.

The northern bulge of India, Kashmir, has been disputed since the partition of British India in 1947; it was a major cause of war between India and Pakistan in 1947–48 and again in 1965. Kashmir became conquerors past. There is a (former) European sector of spacious homes, flanked with lawns and gardens on tree-lined streets that echo England. Beyond is the city of the people, unplanned and crowded, irregular in its pattern of streets, its absence of zoning, and its mixture of architecture. It may be segregated by caste. Because caste is linked with occupation, many crafts and functions may be scattered through the city, grouped by neighborhood instead of carefully lined along commercial streets. Where both Moslem and Hindu groups remain, the city often will still be segregated into religious neighborhoods.

At the edges of the old, the colonial, and the "native" cities, will be the newest construction—rows of functional modular apartments built by the government for people of modest means; neat, single-family homes for the rising middle classes; and, at some distance from the others, the hovels of the Harijans and other recent migrants to the city.

The whole is congested, bustling with traffic and teeming with people. Bicycles, pedicabs, busses, occasional trucks, and foot traffic intermingle in the crowded thoroughfares. Beasts of burden and loaded farm wagons have been legally relegated to the edges of the city, but cows still roam freely.

The European quarter now houses important government officials and a successful business elite. The business district has lost its colonial flavor. Sidewalks and storefronts are bedecked with impromptu stalls selling merchandise different from that offered by the more prosperous merchants within. Bombay is a little more staid, Calcutta a lot dirtier and markedly industrial, Delhi more definitively Indian, Madras more flavored with the culture of the South and Karachi with the signatures of Islam, but the basic elements of the subcontinent's cities are everywhere the same.

Planners in offices build models of subways and draft plans of new towns for construction at some undetermined future date, but the region's cities grow faster than it is humanly possible to plan for. Demand far outstrips limited national budgets.

a secondary topic of contention in the Bangladesh (East Pakistan) crisis and brief war in 1971–72. With the independence of Bangladesh, Kashmir's unresolved status has resumed its role as the major issue complicating relations between India and Pakistan. China also controls territory to the northeast of Kashmir proper that India claims (see Figure 7–4).

In 1947, the princely states of imperial India were pressured to join voluntarily Hindu India or Moslem Pakistan. The choice was usually obvious and based on the majority religion or culture of each state. The Hindu ruler of Kashmir opted to join India, despite a predominantly Moslem population. India has refused to consider a United Nations-administered

plebiscite (a poll of the citizenry on possible change in affiliation). The cease-fire line through Kashmir leaves most of the disputed territory within India.

Kashmir is not likely to be forgotten by Pakistan, and China is unlikely to relinquish its occupancy of land claimed by India. The northern frontier is thus likely to provide problems for decades. India has managed to normalize relations with both China and Pakistan despite these unresolved, conflicting territorial claims.

The entire Himalayan foreland has great potential for hydroelectric energy development. Mid-altitude slopes are clothed with a valuable forest resource. The upper slopes are barren, while lower slopes and valleys have long since been cleared for subsistence agriculture. There is little room for agricultural expansion because of steep slopes, but the surplus waters of the district could be used to extend irrigation in the Heartland. Outside the densely populated Vale of Kashmir, the low densities reflect the rigors and realities of agricultural development in mountain areas. Pastoral activities dominate now; only very limited intensification of use can take place in valleys. Pressures on the forest resource are already in evidence as poor peasants in the mountain valleys harvest wood for resale in the village and in Heartland cities. As slopes are pillaged in the search for wood and vain attempts to extend slash-and-burn agriculture, the land erodes and streams overflow, bringing disaster to local villages and the lands beyond.

The Central Hills, neither as isolated nor as lightly populated as the other peripheral areas, do not fit comfortably into the framework of South India. The area is served almost exclusively by narrow-gauge railroads, and its hilly nature makes both access and utilization difficult. The overgrazed Vindhya and Satpura ranges merge with the higher reaches of the Chota Nagpur Plateau; the rest is an interfluve of higher, drier land between the Deccan and the plain of the Ganges.

Almost half the land is suited to agriculture, but only 10 percent is irrigated. Yields are among the poorest in India. Large areas are forested, but the overall quality is poor. Still, the best-quality Indian teak comes from these hills.

The key word here is *potential*. There are extensive deposits of low-grade coal and a variety of minerals, mainly as yet unexploited. Diamonds, still mined, were the area's historic source of wealth. Today, the major sources of riches are coal and bauxite.

Six of India's 140 cities of over 100,000 population are in the Central Hills. Bhopal and Indore are the best known. Bhopal is the engineering center that produces most of India's generators and electrical transmission equipment. It is better known in the West for the tragedy that occurred there in 1984, when deadly gas killed over 2,500 people in the world's greatest industrial tragedy to date.

The dryland west comprises the states of Rajasthan and Punjab, areas of steppe and desert climate. At its center is the Thar Desert, an area of migrating sand dunes and wind-blasted, bare rock surfaces. There is little rain at any season. Around the edges of this desert, the steppe regions can support agriculture without irrigation, though irrigation is used wherever possible.

Hard (bread) wheats are the most common crop of the region (see Figure 7–8). Mechanization was introduced here early, as were techniques of dry farming. Population densities are quite low except in the intensely irrigated oasis of the Punjab. Less than half the arable land is cropped in any given year except where irrigation exists.

Petroleum has been found, but in disappointingly small deposits (see Figure 7–9). There is a small, but promising copper resource. Any future development of mineral or agricultural resources will require improvement of transportation.

This district is a problem area for India. There are large numbers of Moslems in the portion of the region that borders on Pakistan. The Sikhs, a people whose religion is a strongly modified offshoot of Hinduism, are the dominant group in the Punjab. The Jains, people of a strictly orthodox, ascetic offshoot of Hinduism, are also concentrated in this area. Thus, the dryland west is atypical in both religion and culture, and is in some ways a potential threat to Indian unity (see Figure 7–2). The possibilities for expanding irrigated agriculture are good, yet they run into opposition from local groups who fear an influx of culturally different outsiders.

The frontiers of the Indian Periphery offer both problems and promise. In all cases, they are ethnically and culturally mixed. Agricultural expansion is generally inhibited by difficult physical environments. Peripheral to the main concentrations of Indian population, these areas are often not fully integrated with the Indian economy.

Pakistan and Afghanistan: New Nations at Ancient Crossroads

The states of Pakistan and Afghanistan occupy the dry northwestern margins of the region. Uneasy neighbors and brethren in Islam, one is forward looking and integrated with the world economy, while the other has been best known for its isolation and adherence to tradition. The whole is a transition zone between the cultural realms of the Middle East, India, and the Soviet Union; it occupies one of the most strategic locations of the Asian landmass. In the colonial era, it was the frontier zone between the political spheres of Russian and British ambitions.

Populous Pakistan, the larger of the two states, is the modern heir to the ancient kingdoms of the Indus Valley and a residual portion of what had been British India. In keeping with its tumultuous history and cosmopolitan function, it is diverse in language and culture if uniform in religion. The people of its sparsely settled southwest, as well as those of its densely settled piedmont oases in the northwestern frontiers, speak an Iranian dialect. The overwhelming majority in the rest of the country speak Urdu, Sindi, Punjabi, or other languages related to Hindi. The seeming public unity in Islam masks an often bitter contest between the forces for progress and tradition. Pakistan has a desert climate and a dependence on irrigated agriculture, like the countries to its west. Like neighboring India, it is characterized by dense rural populations, and its agriculture is dominated by wheat and rice. An ancient cultural hearth itself, it has been invaded repeatedly by alien cultures from all sides; each invasion has left its mark on the landscape and lifestyle.

The separation of Pakistan from India in 1947 originally resulted in a split national territory whose two cores were almost a thousand miles apart. Unified only by religion, this Pakistan was perhaps too ethnically diverse to survive. The creation of an independent Bangladesh from former East Pakistan has resulted in a more homogeneous state, but the new Pakistan's national identity is still building.

Pakistan's mineral resources have not been fully explored. There are large gas reserves that currently fuel expanding basic industry and generate most of the country's power. Oil finds to date have been small, and there are some small deposits of low-grade coal. Hydro power, already moderately developed and with significant potential, is the current target of investment. Energy production is increasing rapidly because a crash program of investment in irrigation facilities includes power generation as a by-product.

Small amounts of iron ore support a small steel industry. Salt, gypsum, and sulfur are the domestic resources for Pakistan's expanding chemical industry. Chromite is the mineral export that generates cash. Geologic conditions indicate the potential for a much greater variety and quantity of mineral production, but prospecting is still in its infancy.

With over 100 million people, Pakistan is one of the 10 largest countries in the world. The size of its population gives it influence and prestige. About two thirds of its people speak Punjabi, though Urdu (spoken by only 9 percent) is the official language. Fewer than 30 percent of the people can be classed as literate in any language. English is still widely used among the educated and, as in India, is an associate language.

At 2.7 percent, Pakistan's annual growth rate is greater than that of India, if somewhat lower than in many other Islamic nations (see Table 7–1). Densities of 337 per square mile are really much higher when viewed in relation to **arable land**. Almost 30 percent of the population is urban; the two large cities of Karachi and Lahore account for 10 percent of the national total. Migration has been critical in Pakistan's development. Millions of Moslems left India for Pakistan in the chaotic period following partition. A much smaller migration took place after the Bangladesh split in 1971, with some 2 million Bengalis being repatriated from (West) Pakistan to Bangladesh, and a much smaller number of Pakistani repatriated in the opposite direction. The capital, located at Karachi in 1947, was later shifted to Rawalpindi and then to the newly created city of Islamabad, drawing functionaries and their families with each move.

The population concentrates in the triangular core of the Punjab and strings out from there along the Indus River. Much of Pakistan is too dry for cultivation. Paradoxically, much of Pakistan's farmland is

THE SIKHS

The Sikhs, 16 million strong, are one of India's more restive and articulate minorities. The formation of this group dates from 1519, in the days of the Mughal (Mogul) Empire of India. The Mughals were one of three great Islamic empires that flourished in the aftermath of the great Mongol conquest of much of the civilized world. The Sikh faith, which combines elements of Islam with Hindu beliefs, arose amid the persecution of Hindus and the forced conversions to Islam sporadically perpetrated by the conquering Mughals. Loyal to the Mughal rulers and noted as fierce fighters, the Sikhs were given tolerance, land, and even a privileged status within the empire. Some were great generals under the fabled dynastic leaders Akbar and Shah Jahan (the builder of the Taj Mahal). Renowned as architects as well as military leaders, they built great forts and monuments of their own. They took an active part in the great cultural renaissance that fused the art, culture, and technology of the Arab, Persian, and Hindu worlds into a characteristic form associated with the north of the Indian subcontinent.

Because of their admirable fighting skills (in a nation of pacifists) and their willingness to cooperate, Sikhs were highly esteemed by the British and awarded a status similar to that which they had enjoyed under the Mughals. From the earliest days of British India, they were employed as mercenaries, and were a potent force in the British conquest of the Punjab in 1849.

The Sikhs became concentrated in the Punjab after the development of that district's great irrigation scheme; many received land as a reward for military service to the Empire. The Sikh leadership, who were strongly opposed to inclusion in postpartition Pakistan, led a bloody revolt in the district in 1947. Aligning themselves with the Hindus, they fought to force the partition of the Punjab between India and Pakistan. The eastern Punjab went to India, which in turn granted it the status of a federated state. Some 2 million Sikhs fled to the new state from Pakistan as an even larger number of Moslems set out for that country. In the aftermath of the population transfer, Sikhs formed a large minority, but were not a majority.

In 1966, Sikh pressures forced the creation of a new Hindu state, Haryana, from the eastern, Hindu-speaking parts of the Indian state of Punjab. In the new, smaller Punjab, the Sikhs formed 60 percent of the population. The state contained Amritsar, the Sikh holy city, as well as several small districts that had long been under Sikh rulership. Sikhs, who have always shown an aptitude for business and banking, have developed their Punjab state at a faster pace than the rest of India. Punjab boasts India's highest level of literacy, longest life expectancy, a well-developed concentration of industry, and a per capita income that greatly exceeds the Indian average.

The Golden Temple at Amritsar. *The Sikhs of India follow a religion that combines elements of Hinduism and Islam yet differs from both. They are prominent in banking, industry, and commerce throughout India. Their homeland is in the northwest, near to the holy city of Amritsar; and the Golden Temple is their most sacred shrine.*

Sikhs have consistently pressured the central government for more autonomy. The government, fearing that similar pressures from Assam, Tamil Nadu, and other Indian states might ultimately lead to the dissolution of India, has uniformly refused to make concessions. Protest followed, and riots erupted in Amritsar and throughout the province. The Indian army was sent in to restore order and, in attempting to rout dissidents from their refuge in the Golden Temple in Amritsar, desecrated this holiest shrine of the Sikhs by killing those inside. In retaliation, Sikh extremists assassinated Prime Minister Indira Gandhi.

Sikhs continue to raise demands for autonomy, and extremists support an independent Sikh state. Sikhs are often in influential positions. There is little likelihood of a quick resolution to the Sikh question.

too wet, waterlogged after a half-century or more of continuous irrigation (see Figure 7–8). Inadequate drainage in fields has raised the water table too close to the surface for many crops; salination, combined with waterlogging, has taken 16 million acres of formerly good farmland out of production at the same time that expensive new irrigation projects race to add new acreage in the Indus Valley. Managing older and new irrigation areas is made more difficult by the high evaporation rates prevalent in the torrid summers. It is estimated that less than half the water entering irrigation canals reaches fields. Yet, irrigation is virtually the only route to a high-yielding agriculture.

Although Pakistan may become an industrial economy in the future, agriculture is still the economic mainstay. Between 25 and 30 percent of the total area is in farmland, despite the limitations of climate. Pakistan has built the world's largest irrigated food production system on foundations laid by the British in the 1870s. It has continued to develop the Indus and its many tributaries for large-scale irrigation. Land reclamation schemes, desalination projects, and deep well irrigation have greatly increased the available arable area since independence, but rapid population growth tempers economic benefits; Pakistan is still a net food importer.

The Green Revolution has succeeded here. Rice and cotton crops have doubled in the last decade, and over 10 million tons of wheat are produced yearly. Rice dominates the lower Indus Valley, while northern Pakistan is an extension of the wheat belt of northwestern India. Cotton dominates the zone between the two grain crops, but increasing acreages are being sown throughout the country as Pakistan pursues a policy of export production (see Figure 7–8). Oil-seeds, sugar, and tobacco are grown as industrial and cash crops, while dairying and horticulture are expanding near cities and in the mountain piedmonts to help feed urbanites and diversify the national diet.

Unlike in Bangladesh, the food supply picture in Pakistan is reasonably bright. An active program of land reform and resettlement has reduced rural poverty. Yet it is industrialization and the growth of cities that contributed most to improving living standards and reducing social inequities. Some 18 percent of

labor is now employed in factories, and 25 percent of the GNP is contributed by manufacturing.

Early industries were geared almost exclusively to the processing of agricultural materials. Since 1952, the government has made large-scale industrial investments. Karachi (population 6 million)—the seaport, largest city, and former capital—was selected as the heavy industrial development area. Chemicals, cement, and metallurgy are the basic industries developed to serve the needs of the entire nation. Giant fertilizer plants, built to fill crucial local needs, now export a surplus to the Middle East. Though Karachi is still the largest industrial center, recent investments are extending industry up the valley in the direction of Hyderabad, an old food-processing center, in what promises to become Pakistan's first full-fledged industrial zone.

Other manufacturing centers are found in the north. Electric power produced in conjunction with irrigation projects and local wool and cotton supply small plants there with their basic raw materials. Much of the investment is private. The northwestern frontier district can be characterized as an emerging industrial region as a dozen major and moderate-sized hydroelectric installations fuel a growing industrial economy.

Peshawar, Islamabad, and Rawalpindi are the regional centers that concentrate production. Here the real resource is refugee labor, with perhaps as many as 2 million Afghan refugees crowding into cities daily from the district's seemingly endless rows of refugee camps. Ambitious refugees weave oriental rugs in the courtyards of their tents and huts, swelling the production of an area specialty that enjoys a high demand in international markets. Among Pakistan's nearly 5 million Afghan refugees, talk is of war and the return to home. But many must work for a living in the meantime, because rations are scarce and the distribution of food relief is unreliable. Some will undoubtedly remain, even when the war in Afghanistan ceases.

The problems in Afghanistan have created serious tensions for Pakistan. American and other international aid has been generous, but much of the food and services for Afghan refugees is provided directly by the Pakistani government from its already overburdened resources. Local residents are sometimes embittered because labor costs are depressed by the sheer number of available workers. On the positive side, a brisk trade with Afghanistan, however illegal and unofficial, promises to continue in the advent of a more peaceful future.

There is a question of loyalties. The Pushtun peoples of the northwestern frontier, speakers of an Iranian dialect, are ethnic and linguistic kin of Afghanistan's largest cultural group. Will there be border troubles in the future? Temporarily, the bond of Islam transcends philosophical viewpoints, but a progressive, nontraditional Pakistani regime is housing and feeding a refugee population dominated by extremely conservative views. Will this conservatism find adherents and support within Pakistan? Certainly, relationships with the Soviets are complicated by the Afghan situation.

A series of coups have repeatedly overturned past governments; the military is doubtless the most potent force in the country. It has interfered in the electoral process in the past and now waits to pass judgment on the current head of state, Prime Minister Benazir Bhutto. Daughter of a past president who was overthrown in a military coup and executed in 1977, Bhutto is the first female head of state in an Islamic nation. A skillful, highly educated politician, she has wide public support.

Perhaps Afghanistan is the best example of the shifts in relative importance and power that can be brought about by changes in time and technology. In the days of the caravan trade, Afghanistan was the crossroads of the world. It was one of the richest and most coveted portions of the known world. Here, archaeologists have found evidence of once great cities and civilizations where now only nomads graze sheep. It is possible that the domestication of many grains and fruits first took place here and then spread outward. Afghanistan was once a leader and an innovator, a fount of knowledge and a seat of riches. The key to its success was its strategic location, lying as it still does at the contact zone between the realms of the Middle East, China, South Asia, and Russian Europe. Throughout its early history, it was a region of exchange for goods and ideas between cultures.

As the overland caravan routes were supplanted by sea trade, landlocked Afghanistan moved from center to periphery of the world's commercial inter-

ests. Its trade routes never stopped functioning; they simply became increasingly local in nature. The crops that originated there were surpassed by superior varieties developed elsewhere. The railroads, the means of landward extension of the lines of commerce, were never built into its territory. Traditional goods of trade were supplanted by new and different products. Relegated to a position of diminishing importance, Afghanistan stood still while areas around it moved forward.

For most of its existence, Afghanistan was a part of someone else's empire—Persian, Greek, Mongol, Indo-Aryan, Turkish, or Arab. Yet this subjugation must have contributed to its early greatness, with each empire imparting elements of its culture to the Afghans, and the area itself functioning as a place of exchange and transfer between the various parts of each empire. Two events seemed to have sealed Afghanistan's fate. In the eighteenth century, a native ruling dynasty arose and effectively imposed isolation by severing contact with the surrounding areas. The other event was the rise of Russian and British power in the areas surrounding Afghanistan. Fear of the infidel (non-Moslem) European may have been the reason behind this self-imposed isolation.

In the nineteenth century, Afghan forces repelled two British invasions, and two from Russia. In 1886, the country's boundaries were agreed to by London and St. Petersburg, and Afghanistan was recognized as a buffer zone between Russian interests in Central Asia and British interests in India.

Afghanistan's diverse culture is a reflection of its tumultuous history. Barely half its people speak Pakhtun (Pushtun), which is the official language. Pakhtun is distantly related to Iranian Persian and is something of a bridge between the languages of Iran and Pakistan. As many people in Pakistan as in Afghanistan speak the Pakhtun language, and the loyalties of this minority in Pakistan pose a serious question to that government. Another 20 percent are Tadzhiks, speaking a language even more closely related to Persian. One of the Soviet Central Asian republics is based on the Tadzhik national group, an uncomfortable situation for both the Russian and Afghanistani governments. Uzbeks and Turkomens, whose main linguistic groupings are also in the USSR, form another 10 percent of the total. There is

Struggle for Survival in Bangladesh. *By any measure, Bangladesh is consistently ranked as one of the 10 poorest countries of the world. Dense populations crowd the compound delta of the Ganges and Brahmaputra rivers that forms the bulk of the country. Most land is used to produce rice or jute. New crops, such as bananas, have been introduced to increase food production and diversify diets.*

no domestic linguistic or cultural unity. Worse, Afghanistan's neighbors can put forth claims to parts of Afghan territory on ethnic grounds.

Afghanistan is at the bottom of the development ladder, though it is not lacking in resources. It has common borders with Pakistan, Iran, China, India, and the USSR, yet it has had minimal contact with any of those states until recently. Its isolation is not determined by physical factors. It is in some measure the result of a desire to stay independent, since involvement with neighbors has historically meant conquest.

The rise of Russian influence in Afghanistan began in the 1960s, a product of cold war competition and the desire of Afghanistanis for development. The landlocked nature of the country was a hindrance to development. Worse, Afghanistan's internal population was scattered among several dozen densely populated oases, separated from one another by zones of lightly peopled grazing lands. Grazing regions in turn were grouped around a nearly empty mountain core which made travel between north and south extremely difficult. Central control of the country was hampered by this dispersed settlement pattern and the absence of a decent transportation system. The Afghan government reasoned that neither internal unity nor development could be obtained without a good road network. Afghanistan attempted to gain the maximum amount of foreign aid by playing off the cold war antagonists against each other. Both the Soviet Union and the United States launched impressive highway-building schemes. The United States interconnected the southern part of the coun-

try with Iran and Pakistan, and the Soviet program linked the northern half of the country to its territory. Four interconnections linked these two halves of the network. U.S. influence and interest waxed and waned along with commitments to Iran and Pakistan. Soviet influence grew consistently, especially among the military. A coup in 1978 established a firmly pro-Soviet leadership. Over the next 10 years, Soviet levels of support for the government rose steadily in the face of guerrilla resistance. It was only in 1988 that the Soviets agreed to pull out, a fact accomplished in 1989. U.S. support of the guerrilla movement, accomplished through refugee groups in Pakistan, played no small part in the Soviet decision.

The benefits of almost 20 years of competitive foreign investment have not been totally wiped out by 10 years of tragic civil unrest. There is a comprehensive road network. The area of irrigated land has doubled. Literacy rates have increased to about 20 percent, and a class of technicians has been developed by the schooling of Afghanis abroad. There are now a dozen airports. Still, the cost in lives and property destruction may long outweigh the benefits.

If peace follows Soviet withdrawal, there are some real possibilities for progress. The country's varied landscape and climatic conditions allow for increased quantity and variety of agricultural production. About 15 percent of Afghanistan is usable for crop agriculture. Most of this land is either in narrow valleys that follow the country's permanent streams or on alluvial fans that spread out at the foot of the mountain core. Much cultivable land is still uncultivated, and land pressure is not as serious as it is in

Pakistan and India. Citrus, sugar, and other tropical crops can be produced in lower elevations at Jalalabad and Kandahar oases (see Figure 7–7). The traditional crops—oil-seeds, nuts, and fruits—are in high demand in neighboring states. There are almost twice as many sheep as people, and there is a demand for both the wool and skins (karakul furs) in world markets.

There is no lack of resources. Only a small amount of oil is produced, but prospecting has indicated a much larger untapped reserve. Natural gas is plentiful and exported to the USSR by pipeline. There are good grades of chrome ore, and large iron ore deposits remain unexploited. There are known, commercial-scale deposits of coal, copper, and a great variety of other minerals.

There has been some industrialization in Afghanistan. The textile industry has expanded to meet some needs of the domestic market. Kabul, the capital, has increased its population sevenfold in the last 25 years; it constitutes a concentrated market of more than 2 million people. There has been a steady development of hydroelectric plants.

The framework for development is in place. The unanswered questions are crucial. Can internal peace be attained? Will the succeeding government be able to heal the scars of war and to develop a united Afghan nation?

Bangladesh

The name, and the country, are both relatively new, though the area has been settled for thousands of years. The great density of people in this small state attests to the fact that it is one of the most fertile, well-watered portions of the earth. Yet images of poverty and natural disasters are usually what first come to mind at the mention of the name.

Bangladesh is one of the most crowded parts of the world. With 112 million people, it is the eighth largest country, but its land area is equivalent only to the state of Wisconsin.

Until December 1971, this area was called East Pakistan and was a physically separated part of a then-larger Pakistani republic. In its 24-year association with West Pakistan, East Pakistan had always felt exploited. With the larger of the two populations, it felt unrepresented in the politics of power. It received a much lower proportion of total national investment.

The Awami League, based in East Pakistan, won control of the national assembly in 1971 by convincing local voters to support only their candidates, while West Pakistani votes were divided among several parties. The government repeatedly postponed sessions of the newly elected national assembly, and rioting broke out in discontented East Pakistan. East Pakistanis appealed to India for aid and, as a result, yet another war broke out between India and Pakistan. Pakistan withdrew in 1971, and an embittered Bangladesh declared itself an independent nation.

In the partition of British India, the predominantly Moslem areas near the eastern and western ends of the great riverine plain of northern India were designated Pakistan, although the two parts were separated by more than a thousand miles of dominantly Hindu Indian territory. Although united in their adherence to Islam, the citizens of East and West Pakistan held little else in common. The people of Bangladesh are Bengalis, related to the dominant ethnic groups of India in culture and language, if not religion. India saw an opportunity to weaken its enemy, Pakistan, by supporting the Bengali nationalists. After independence, India was a major source of aid for reconstruction.

Agriculture is the key to the Bangladesh economy, since 80 percent of the people live in rural areas and till the land. Although densely populated, Bangladesh has such fertile soil and ample water for agriculture that it can produce large volumes of food. Indeed, its very productivity explains its dense population. Food production has risen sharply since independence, with the application of Green Revolution technology. The major crop, rice, can be harvested three times a year. Better flood control and irrigation projects could result in even greater food production. India and Bangladesh have agreed to develop jointly the water resources of the delta area. A major problem in boosting food production further is the cost of chemical fertilizers needed for increasing yields. Bangladesh has developed some natural gas resources which presently are being used to manufacture fertilizers.

Dacca, the capital, has almost 4 million people and, as in India and most other developing countries, is growing faster than the employment base in industry or services. The industrial sector, mostly nationalized, is not considered very efficient. The main industry, processing jute into burlap and carpet backing, is Bangladesh's major earner of foreign exchange but is not highly profitable (Figure 7–10). Government-owned jute, sugar, cement, and textile mills have had economic problems. Government-encouraged private investment in pharmaceuticals, however, has produced a thriving industry, and this success has attracted considerable investment in pharmaceutical plants by large international firms.

There is a strong probability of oil offshore of Bangladesh (also a strong probability of a boundary dispute with India over the division of these prospective offshore oil areas). Hydroelectric power is fairly well developed, though the flat nature of most of the country means that the potential is limited.

Despite the serious problems of coping with a dense population and a high growth rate, the future need not be disastrous. Maintenance of good relations with India (especially concerning the equitable division of water), the good chance of striking oil, and the increasing food yields made possible by its gas-based fertilizer factories could draw Bangladesh back from the brink of disaster by starvation. If the population growth can be lowered, rising industrialization may be able to absorb enough of the surplus labor force to relieve pressure on the crowded farmlands. Yet, overcrowded city slums, jammed with urban unemployed, indicate that the pace of industrialization is still far too slow.

The rivers renew Bangladesh's soil each year with alluvial deposits spread across the land in seasonal floods. Bangladeshis are concentrated in the east-central part of their country, in the areas east of Dacca and Khulna to the Chittagong Hills (see Figure 7–10). The steep hills of the northeast and southeast remain in forests. The western delta, abandoned by the river distributaries, has degraded to infertile soil under severe natural leaching.

It is this delta location that both enhances fertility and places the country squarely in the path of disaster. Bangladesh fronts on the Bay of Bengal, directly in the path of yearly monsoonal winds. It is low-lying and therefore subject to easy flooding. Nearly all of the delta lies at elevations of less than 50 feet above sea level. Onshore winds create huge tides that back up the river waters, spreading them over the land. The result is tremendous damage to crops, dams, and ditches, and often a great toll in human life.

There are some answers, but the costs are prohibitive. Irrigating the infertile lands of the southwestern delta is a possible solution. The delta provides an almost ideal habitat for fish, and shallow offshore waters are equally rich. Fish production has reached a million tons yearly, but the harvest could be infinitely greater without damage to the resource. Fish farming in this environment could be extremely lucrative.

Perhaps the gloomiest aspect of the picture remains the high rate of natural increase. With a 2.6 percent annual growth and a continuing high death rate, the country's progress is almost immediately nullified by population increases. There is perhaps no greater example of a nation in need of a speedy demographic transition.

Sri Lanka

Sri Lanka was once one of the most highly sought after prizes in the race for colonies. The Portuguese, Dutch, and British controlled the island in succession, each for a period of 150 or so years, before independence in 1948.

The island is about half the size of Florida, and much of it lies unused. The physiological density is about one third that of Bangladesh, and slightly higher than India's. The northern part of the island and the southern coastal areas are lowland plains; a hilly and mountainous core occupies the south-central portion. About 40 percent of the island is under cultivation, but the arable area can still be increased.

Population density is greatest in the southwestern part of the country, where the heaviest rainfall is produced by the summer monsoons rising against the highlands. Favored by both monsoons, only the northern plain has a dry season of any length. Colombo, the capital, has reached the million mark, and there are a few other urban centers. The population growth rate has dropped to 1.8 percent, and the government has increased the availability of birth control services in an effort to reduce growth further. Sri Lanka has shown the most promising signs of growth control in the entire region.

About 72 percent of the people are Sinhalese, with about 20 percent Tamils, and 8 percent Moors (Moslems descended from sailors and traders originating in the Persian Gulf). The Sinhalese majority is Buddhist, while the Tamils are Hindu, and the Moors,

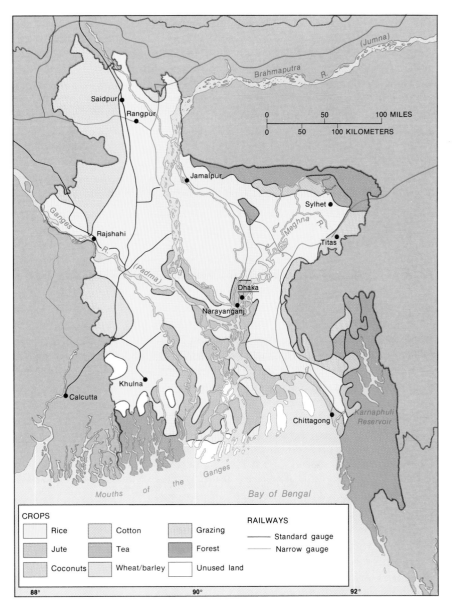

FIGURE 7–10
Land Use in Bangladesh.
Most of Bangladesh is a large and fertile deltaic plain. Its extremely high population densities are in part a reflection of the high degree of fertility (and utility) of the land. Rice occupies the greatest acreage; supplies are insufficient to meet domestic food needs. The natural forces that contribute to Bangladesh's fertility also bring disastrous floods that take a tragic toll in human life.

Moslem. This cultural divergence has been a cause of great ethnic strife, and some militant Tamils have pressed claims for a separate Tamil state or annexation to the mainland. About half the Tamils are Indian Tamils, whose ancestors arrived during the last century as plantation labor. These "recent" arrivals, in contrast to the Ceylon Tamils, who had been established there earlier and who are full citizens with votes, were not offered citizenship after Ceylonese independence in 1948. India also refused repatriation of these Tamils, claiming they were not Indian citizens either. India and Sri Lanka have gradually divided these stateless, hopeless people into repatriated Indians and Sri Lankan citizens, but much bitterness remains among Tamils in both nations. Sinhalese is no longer the sole official language; Tamil was granted official status in 1978. As in India, English is widely spoken among the educated. About 85 percent are literate, far above the average for South Asia.

The economy of Sri Lanka is dominated by plantation and cash crop agriculture. Tea earns 40 percent of export income, and rubber and coconuts account for another 30 percent. Traditionally, Sri Lanka has relied on foreign exchange earned by exports to purchase food, since normally half its food requirements were imported. For years, this exchange of high-value tea and rubber for lower-value rice worked well, and living standards were (and are) far superior to the regional average. The government began extensive land reform in the 1970s, nationalizing both foreign and locally owned plantations. Productivity has not risen as fast as expected, and an expanding population has outpaced increased food production. Food imports still use half of the country's export earnings as prices received for exports stagnate and food prices rise. Recently, real progress has been made in reducing food imports through increasing rice yields in older farming areas and colonizing new lands in the interior (Figure 7–11). Despite problems, Sri Lanka's per capita GNP is the second highest in the region (after Pakistan's), and its personal income is now the region's highest.

There has long been a historical link to India. The few miles of shallow sea that separate Sri Lanka from the mainland proved no barrier to movement even in ancient times. The Sinhalese are descendants of Indo-Aryans who came to dominate the island's few native inhabitants 2,500 years ago. Buddhism was transmitted from India and remains dominant, since migrations ceased before Hinduism returned to dominance in India. A great civilization arose under these Sinhalese conquerors, who established farms and cities on the northern plains. Ruins of cities and waterworks in that area indicate its once dense settlement and high degree of development. Many of the remains have been lost in the jungles for centuries because for reasons unknown, the plain was abandoned and the Sinhalese moved progressively south into the hill lands of the interior. Until recent government relocation schemes reopened a portion of these lands to settlement, only slash-and-burn farming took place.

Arab traders knew it as the land of Serendip, an appropriate name for a lush island that must have enchanted a drylands dweller with its waterfalls and blanket of green. European traders valued it for cinnamon, in high demand in the spice trade. The British seized the island in 1796, putting it to other commercial uses once the market for cinnamon was depressed by competition from other producers.

The Sinhalese population is growing as a proportion of the total, largely because the Tamils were expelled. Tamil riots in 1983 and 1988 demanded the creation of an autonomous Tamil homeland in the north of the island. Almost a million of the island's 3 million Tamils are crowded into the Jaffna Peninsula of the north. Densities are incredible, and farms there average only a quarter-acre each, the size of an average suburban lot.

Under such intense land and population pressures, the Tamils have taken to squatting on lands in the northern plains. Irate Sinhalese view this area as the national land reserve for future settlement; clashes have taken place between rival groups of settlers, and many lives have been lost. Open revolt of the Tamils began in 1987, and several thousand Sinhalese were killed. Tea production has declined because Tamils, fearing for their lives, have stayed away from fields and plantations in the southeast. Military expenses strain the budget and, as a result, Sri Lanka's plans for further development go unfulfilled. A lucrative tourism industry (mainly Japanese and Australians) that had shown much promise languishes as the threat of civil strife keeps tourists away. There are serious troubles in Sri Lanka, as the region's most prosperous economy flounders on the civil strife so prevalent in South Asia.

Nepal and Bhutan: The Mountain Buffers

The central Himalayas contain the two smaller states of Nepal and Bhutan, effectively client states of India. Both are mountain kingdoms that were never fully controlled by the British during the empire days. However influential India has become in the postindependence era, Nepal and Bhutan remain essentially self-governing. In an area of disputed borders, they constitute a buffer zone between India and China.

Only 20 percent of Nepal is cultivable, yet fully 90 percent of the Nepalese are farmers. One of the least developed countries of Asia, Nepal has an economy still dominated by subsistence agriculture. With both countries backed by the high Himalayas, the hydroelectric potential is exceptional. Both India and China have provided economic and technical assistance in power plant construction in recent years.

Although Nepal is not heavily populated by Asian standards, the ecological damage resulting from

FIGURE 7–11

Land Use in Sri Lanka. The diverse agriculture of Sri Lanka has helped to make it one of the region's most prosperous economies. Much of the interior is not developed for agriculture, though it is in the process of being colonized. The Tamils of the north, a Dravidian people who migrated from India in the distant past, are engaged in guerrilla action against the Sinhalese colonists and the Sri Lankan government. Internal prosperity and the tourist trade have been adversely affected by this ongoing dispute.

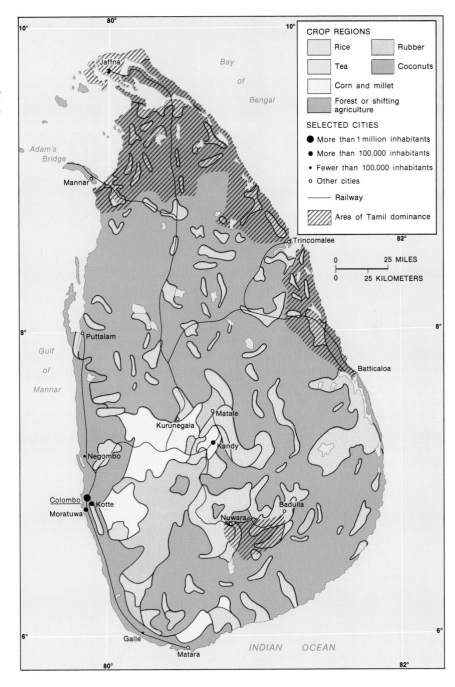

The Himalayas. This majestic mountain range includes the world's highest peaks. Their lower slopes exhibit some of the worst examples of soil erosion as peasants decimate forests in search of wood for household fuel. Intermediate slopes support meager herds of cattle, sheep, and goats on a sparse, rapidly disappearing cover of grass. Only in a few favored basins and valleys is there any significant potential for crop agriculture. Without proper conservation, erosion will worsen until only barren rock remains. The Himalayan foreland offers little possibility as a frontier for agricultural expansion from the crowded valleys of South Asia.

mounting population pressure is serious. Deforestation of steep hillsides has led to severe soil erosion and could pose problems for hydro dams through heavy silt accumulation in the reservoirs. India has made it clear that it will not tolerate any Chinese territorial aggression in Nepal, and both economic ties and cultural relations between Hindu Nepal and India are closer than those with the Chinese. However, the Nepalese are determined to maintain excellent relations with both giant neighbors. There is no other choice.

There are three zones of farming in Nepal: the foothills, the central basins and valleys, and the lower slopes of the Himalayas. The lowest zone is called the **terai**, the scrubby foothills that rise from the plains of the Ganges. Like the plain in India, it is devoted to rice, wheat, and jute. Malaria is still a problem on the swampy alluvial fans of the terai, so there are considerable areas in forest. Wild game, including tigers, find a natural habitat here, and Nepal has designated some of this area as national parks. A battle is going on over the proper use of the remaining terai lands between woodcutters, advocates of clearing the land for farms, and those who see its use as tourism.

The most densely settled part of the country is the zone of the central valleys and basins. The valleys have an irregular surface, where steep, deeply gul-

lied slopes isolate multiple levels of farmland—on, above, and below the average surface. Terraces create level land where nature does not provide it.

The growth of the country's three major cities, in particular Katmandu, the capital, has put additional pressure on the already crowded land of the central valleys. Here fields planted to grains, potatoes, and vegetables utilize every square inch of land to feed the urban areas. At least two crops are grown on every parcel of land each year.

The highest zone is found on the slopes of the Himalayas. The slopes are devoted mainly to the grazing of sheep and yaks. The winters there are so cold that barley and potatoes replace wheat and corn in fields that are only lightly tended. Livestock, not crops, constitute the basic food supply. Slopes are often barren as a result of uncontrolled woodcutting and overgrazing.

The people of Nepal are far from homogeneous; 12 languages are spoken. Ninety percent of the people are Hindus, but the religion as practiced here contains strong elements of Buddhism and earlier, animist beliefs. There is an air of tolerance and a frank acceptance of the culture's diversity.

Tourism is the budget-balancing item in the national economy. Indian pilgrims come to sacred Hindu shrines. Westerners pay for safaris in the cane breaks of the terai and, above all, there is the attraction of the Himalayas. Mountain climbers, campers, and more passive tourists flock to Nepal to scale, or at least photograph, their imposing heights.

Bhutan is far smaller, less developed, and much more mountainous than Nepal. By some measures, it barely qualifies as an independent state. Since 1910, foreign policy has been officially "guided," first by Britain, then (since 1947) by India. In return, Bhutan receives a subsidy which helps to balance its shaky budget.

The Bhutanese are definitely a part of the Sino-Tibetan world. The language is a variant of Tibetan, and the religion is dominantly Buddhist. Yet, economically, Bhutan is virtually a part of India, since India accounts for 99 percent of its foreign trade. Only 2 percent of the land can support crops. Not long ago, barter was more prevalent than cash. As in Nepal, tourism is a growing source of national income.

SOUTH ASIA IN RETROSPECT

India's self-confidence has been increased by the bumper harvests of the late 1970s and 1980s. It has reduced food imports, and lower oil prices have helped with its balance of payments. In 1974, India joined the "nuclear club." Domestic unrest remains a problem, and the new government has attempted to return India to a middle course in international relations, maintaining friendship with (and receiving aid from) both the United States and the USSR.

Pakistan's world view is geared to survival in what it sees as a hostile environment. There is an official policy of step-by-step normalization of relations with India. In a traditional "outflanking" gesture, it has sought allies outside the immediate region. Relations with China have grown much closer. The Afghan situation has resulted in increased American military and economic aid to Pakistan and closer ties between the two. The refugee problem will persist for quite some time.

Ever mindful of its Islamic heritage, Pakistan continues to seek closer ties with the Arab world. It pursues a rather uncomfortable course of being traditionally Islamic and progressively secular at the same time. In the face of Indian nuclear power, Pakistan feels that it must build an equal deterrent, yet it has recently entered into talks with India to discuss stopping nuclear proliferation in the area.

The entire region remains troubled as ethnic disputes rage in Sri Lanka and India, political instability characterizes Afghanistan, and poverty is the region's premier problem. Population pressures continue to mount in what will become the most populous region of the earth by 2025.

REVIEW QUESTIONS

1. Briefly describe the reasons behind Indian cultural diversity.
2. What cultural and political conditions unique to British India favored Gandhi's nonviolent protest movement?
3. List some major philosophical and cultural dissimilarities between Islam and Hinduism.
4. Why is India considered to have a severe population problem although its growth rates are lower than those in many other developing countries?
5. Which of South Asia's economies are subject to the most severe population pressure?
6. What are the economic incentives for many poor people of this region to have many children rather than few?
7. Why is the arrival date of the summer monsoon essentially unpredictable? What consequences does this unpredictability have for agriculture in much of this region?
8. What cultural factors act to retard modernization of Indian agriculture? Why would rapid modernization induce economic dislocations?

9. What is the rational basis for the Hindu proscription on slaughtering cows? In what ways is the cow problem detrimental?

10. In what manner does the Green Revolution tend to favor large landowners rather than the poorest farmers?

11. Why does the Indian Heartland seem to have economic growth below that of the Indian South?

12. How did the partition of British India affect Calcutta and the jute industry of Bangladesh? The internal ethnic makeup of India and Pakistan?

13. On the basis of raw materials, labor costs, and international markets, which of India's industries have the best export potential?

14. How does India's multicultural, multilingual population pose special problems for democratic government?

15. Why is irrigation in desert areas not always a permanent improvement in terms of soil productivity?

16. Assess the relative chances of this region's states for future success in adequately feeding their people.

17. What cultural and political tensions led to the separation of Pakistan into two states?

18. Of the region's states, which is most likely, in terms of present cultural, religious, linguistic, and racial tensions, to experience pressures for regional autonomy or independence?

19. How does the Indian-Chinese rivalry affect the foreign policies of the region's other states?

20. In what ways was British colonialism a positive factor in economic development within the region? Were there negative impacts of colonialism as well?

21. Assess the development strategies of each of the region's states, discussing their practicality as well as their problems.

SUGGESTED READINGS

Berry, B. J. L. et al. *Essays on Commodity Flows and Spatial Structure of the Indian Economy*. Research Paper No. 111, Department of Geography, University of Chicago, 1966.

Bhardwaj, S., ed. *Hindu Places of Pilgrimage in India: A Study in Cultural Geography*. Berkeley, CA: University of California Press, 1983.

Brush, John. "Spatial Patterns of Population in Indian Cities." In *Cities in the Third World,* edited by D. Dwyer, 105–30. London: MacMillan, 1974.

Chakravati, A. "Green Revolution in India." *Annals of the Association of American Geographers* 63 (1973): 319–30.

Chang, Jen Hu. "The Agricultural Potential of the Humid Tropics." *Geographical Review* 58 (1968): 333–61.

Dumont, L. *Homo Hierarchus: The Caste System and Its Implications*. Chicago: University of Chicago Press, 1970.

Faaland, J., and Parkinson, J. R. *Bangladesh: Test Case for Development*. Boulder, CO: Westview, 1975.

Farmer, B. *An Introduction to South Asia*. New York: Methuen, 1984.

Hall, A. *The Emergency of Modern India*. New York: Columbia University Press, 1981.

Harris, Marvin. *Cows, Pigs, Wars, and Witches: The Riddles of Culture*. New York: Random House, 1974.

Hutton, J. *Caste in India: Its Nature, Function, and Origins,* 4th rev. ed. New York: Oxford University Press, 1963.

Jahan, R. *Pakistan: Failure in National Integration.* New York: Columbia University Press, 1972.

Johnson, B. *Bangladesh,* 2d rev. ed. Totowa, NJ: Barnes & Noble, 1982.

Johnson, B. *India: Resources and Development.* Totowa, NJ: Barnes & Noble, 1979.

Neale, Walter, and Adams, John. *India: The Search for Unity, Democracy, and Progress.* Princeton, NJ: Van Nostrand, 1965.

Rose, Leo. *Nepal: Strategy for Survival.* Berkeley, CA: University of California Press, 1971.

Sopher, D., ed. *An Exploration of India: Geographical Perspectives on Society and Culture.* Ithaca, NY: Cornell University Press, 1980.

Spencer, Joseph, and Thomas, William. *Asia, East by South.* New York: Wiley, 1971.

CHAPTER EIGHT

Southeast Asia

Burmese Stupas.

S outheast Asia includes Burma, Thailand, Malaysia, Singapore, Laos, Vietnam, Kampuchea (Cambodia), Indonesia, Brunei, and the Philippines (Figure 8–1). Two traditional images have characterized Southeast Asia: the shatterbelt nature of the region, with its ethnic, religious, and cultural complexity, and the image of surplus as Asia's "rice bowl." Both these generalizations appear to be changing as industrial development, rapid population growth, national unity policies, wars, and forced migrations alter the old realities of the region. The region is undergoing a rapid transition on many fronts, albeit at different rates in different countries.

The basic physical geography of the region is a fragmented collection of islands and peninsulas surrounded by tropical seas. Southeast Asia's peninsulas

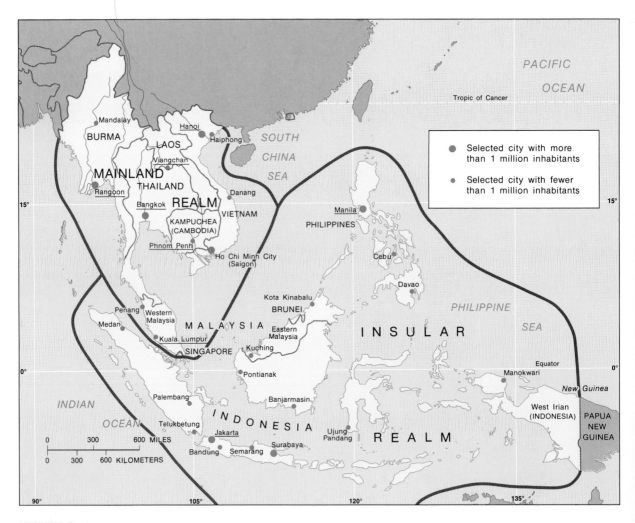

FIGURE 8–1
Subregions and Major Cities in Southeast Asia. Cultural variety matches the physical complexity of this region of mainland river valley cores and island archipelagoes.

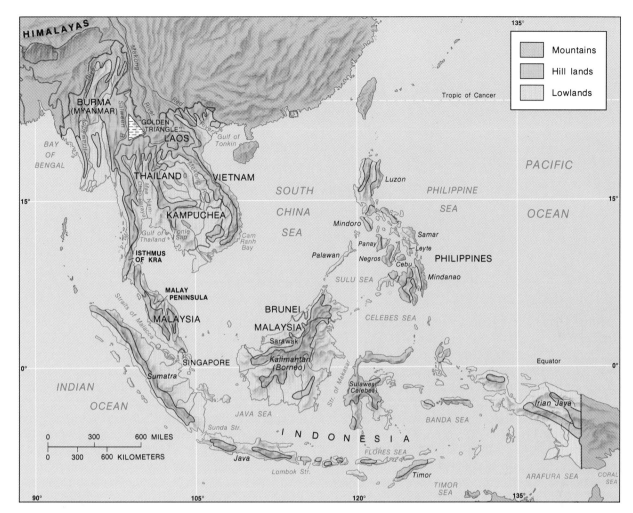

FIGURE 8–2
Land and Water Features in Southeast Asia. Note the intricate intermeshing
of land and sea in Southeast Asia—a kind of "Asian Mediterranean."

and archipelagoes may be regarded as an "Asian
Mediterranean." The region consists of nodes, separated by mountainous interfluves or surrounding
seas. It is tropical, well watered, and often more fertile than comparable tropical environments elsewhere.

This interpenetration of land and sea, in combination with a difficult topography, was often used to
explain the region's cultural diversity and political
fragmentation. From another viewpoint, this topographic compartmentalization has given rise to a series of discrete, rather uniform culture cores. In an
era of rapid development, the intermingling of land

and sea can be a definite advantage; it was the interpenetration of land and sea that became Europe's
chief developmental advantage, and a similar advantage is possible in Southeast Asia (Figure 8–2).

Interregional movement is relatively difficult over
land. Mainland populations remain clustered in fertile deltas elongated upstream in alluvium-enriched
valleys (see Figure 8–2). Within these cores of settlement, the rivers provide for trade and contact.
Each major mainland core is surrounded by a zone
of less dense population inhabiting rugged highlands, a less environmentally usable milieu. In turn,
these zones of greater and lesser concentration are

Country	Dominant Ethnic Group (%)	Dominant Religious Group (%)
Brunei	Malay (62)	Islam (64)
Burma	Burman (68)	Buddhism (85)
Kampuchea	Khmer (85)	Buddhism (86)
East Timor	Malay (90)	Buddhism (70)
Indonesia	Malay (90)	Islam (95)
Laos	Laotian (90)	Buddhism (70)
Malaysia	Malay (45)	Islam (58)
Philippines	Filipino (92)	Christian (84)
Singapore	Chinese (73)	(not available)
Thailand	Thai (82)	Buddhism (96)
Vietnam	Vietnamese (85)	Buddhism (70)

SOURCE: *Asia Yearbook* (Hong Kong: Far Eastern Economic Review, various years).

separated from one another by even greater physical barriers.

The island-states of the region are even more reliant on waterborne links. Zones of lighter population are fewer and often of lesser importance in the islands than on the mainland. Establishing economic links among the islands remains the major problem, but it is gradually being overcome. War, population transfers, migrations, national assimilation programs, and a growing sense of national identity have contributed to greater uniformity in most of the region's states. Excepting Malaysia and Burma, the region's emerging national units are clearly, often overwhelmingly, dominated by one ethnic group. To a remarkable degree, one form of religion clearly predominates in all the region's units but Malaysia (Table 8−1). It is true that dialect differences remain, but education is gradually producing a national language uniformity. Tribal and regional affiliations are still strong in parts of the region, but national programs to deemphasize such bonds are under way. Local ties remain strongest in Indonesia, though even there an emerging sense of national unity is evident.

Mainland Southeast Asia is frequently described as a cultural shatterbelt (Figure 8−3). Negrito native peoples may have occupied much of the area. Hindu conquerors mixed freely with the natives and formed a state known as Funan. The conquest of this state by neighboring peoples brought forth the great Khmer Empire, which included today's Kampuchea, much of Thailand, southern Vietnam, and parts of Burma. Beginning with a core in the Red River delta, the Annamese spread their empire along the Vietnamese coast and eventually into Laos. A series of wars were fought with the Khmer, a traditional enemy. The power of both was eventually eclipsed by China.

Little is known of early Thai history, but a state is thought to have existed since the twelfth century. The Laotians are culturally and linguistically related to the Thai; the areal extent of Thai linguistic groups (from southern China to the Malayan coast and from eastern Burma to the Red River valley) may give some hint as to the maximum extent of these peoples.

The Burmese are culturally tied to Tibet. Never spreading beyond its original core area, Burma has been ruled (in part) by Thai and Khmer empires as well as by the Chinese (thirteenth to sixteenth centuries).

This mosaic of empires, cultures, and conflicts is composed of units that are, consequently, ethnically mixed. An element of uniformity is provided by the

FIGURE 8−3

Mosaic of Cultures in Southeast Asia. Southeast Asia is a "cultural shatterbelt," that is, a fragmented region of intermixed cultural inputs. Five colonial empires once controlled territory here; four major religions intermix with local, native beliefs; many different languages are spoken here; and this region hosts the largest Chinese population outside China.

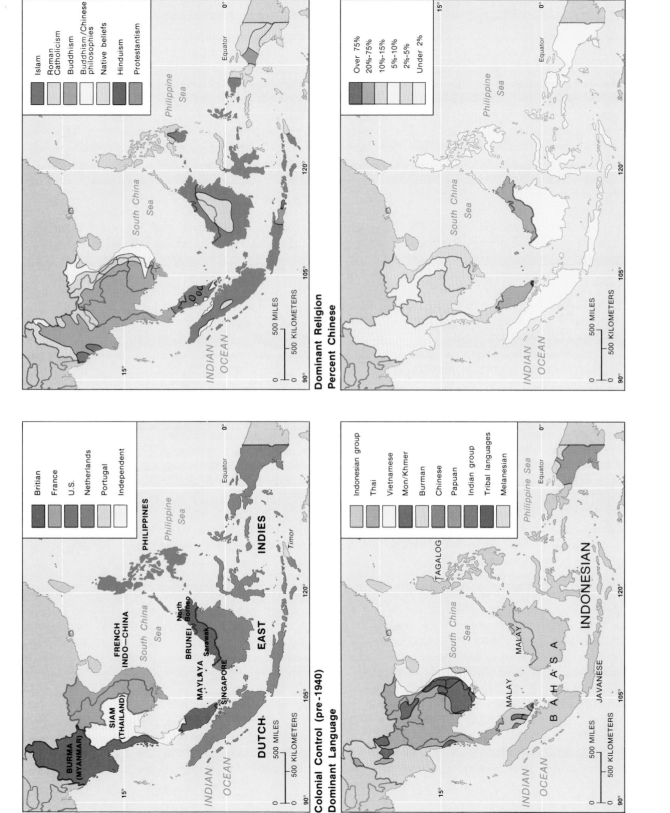

Colonial Control (pre-1940)
Dominant Language

Britian
France
U.S.
Netherlands
Portugal
Independent

BURMA
(MYANMAR)

SIAM
(THAILAND)

FRENCH
INDO–CHINA

PHILIPPINES

MAYLAYA
BRUNEI
North
Borneo
SINGAPORE
Sarawak

DUTCH   EAST   INDIES

Timor

South China
Sea

Philippine
Sea

INDIAN
OCEAN

Equator

15°

0°

90°
105°
120°

0   500 MILES
0   500 KILOMETERS

Indonesian group
Thai
Vietnamese
Mon/Khmer
Burman
Chinese
Papuan
Indian group
Tribal languages
Melanesian

TAGALOG

MALAY

MALAY

MALAY

B A H A S A   INDONESIAN

JAVANESE

South China
Sea

Philippine
Sea

INDIAN
OCEAN

Equator

15°

0°

90°
105°
120°

0   500 MILES
0   500 KILOMETERS

Dominant Religion
Percent Chinese

Islam
Roman
Catholicism
Buddhism
Buddhism/Chinese
philosophies
Native beliefs
Hinduism
Protestantism

South China
Sea

Philippine
Sea

INDIAN
OCEAN

Equator

15°

0°

90°
105°
120°

0   500 MILES
0   500 KILOMETERS

Over 75%
20%–75%
10%–15%
5%–10%
2%–5%
Under 2%

South China
Sea

Philippine
Sea

INDIAN
OCEAN

Equator

15°

0°

90°
105°
120°

0   500 MILES
0   500 KILOMETERS

once more widespread Buddhist religion. Even there, beyond the acceptance of Buddhism's simple doctrines, there is a regional difference. The primacy of monks and the narrow road to salvation for only a chosen few characterizes the Buddhism of the western four states; in contrast, the Mahayana Buddhists of Vietnam are a more liberal, Chinese-influenced sect with a wider variety of deities and an easier route to salvation.

The overseas Chinese constitute the largest minority in many of the region's states (see Figure 8–3). They are widespread throughout Southeast Asia, and their entrepreneurial skills often have led them to positions of relative wealth and power. Large numbers of Chinese were killed or expelled during internal conflicts in Indonesia and Vietnam, where they generally have been viewed as alien minorities. As a result, their regional importance (and numbers) have declined. The Chinese do not always live aloof from the ethnic majority. They have assimilated to a degree in Thailand and have made efforts to "lower their profile" in the Philippines.

Comparisons often are made between Southeast Asia and the Eastern Europe shatterbelt, but the similarities are superficial at best. Post-1918 eastern Europe included defeated states with bitter irredentist claims, and serious minority problems within the new states created by the winning powers. By way of contrast, the comparative ethnic and religious uniformity of most Southeast Asian units is quite remarkable. In most units, the dominant ethnic or religious groups account for 80 percent or more of the total (see Table 8–1). There are no composite states (grouping two or more related, but distinct, national groups) as is the case in Czechoslovakia or Yugoslavia. The potential for conflicting claims and future border rearrangements exists, but an ethnic sense of nationalism seldom is well developed among minorities, with the major exception of the Muslims of western Mindanao, Philippines. Vietnam appeared to pursue irredentist claims in Kampuchea, though other problems were involved in that military action. Insurrections present difficulties, particularly in Burma and the Philippines. Guerrilla groups in Southeast Asia, however, are often ideologically motivated rather than clearly identified with nationalist minority movements.

Southeast Asia, including a land area of over 1.7 million square miles, is almost as large as Europe (excluding the European USSR). Most of its political units are relatively large, comparing favorably to Europe's larger political units.

In Europe, there are 27 countries and a half-dozen micro (or city-) states; in Southeast Asia, there are only 9 sizable political units and 2 micro states. Indonesia, Southeast Asia's largest unit, equals in size the entire Eastern European shatterbelt's 11 units. A dozen well-developed Indonesian Malay dialects and 1 distinctly different language account for that country's linguistic diversity, while the Eastern European shatterbelt encompasses 16 distinct languages, an even larger number of ethnic groups, and dozens of dialects.

The onetime rice bowl, Southeast Asia is no longer a major exporter of rice. Wars and internal upheavals have lessened agricultural production in a part of the region, and the population growth of recent decades has resulted in a diminished ability to export. Rice is still produced in large quantities; it is now simply consumed more largely at home.

Physiological densities, though variable within the region, are often more comparable to those of South Asia than to those of the world's major food exporters (Table 8–2). It would appear that the physiological densities, even allowing for the longer tropical growing season, preclude large-scale future exports. Despite possible surpluses within the lower-density central core (Thailand and Kampuchea), this region would appear to be potentially self-sufficient rather than a major world food exporter.

The region is still not under the severe population pressures experienced in South and East Asia. Southeast Asia, now a world-scale population concentration (8.5 percent of the world total), is an area of rapid population growth, but also one of apparently accelerating economic development.

Diverse, but not hopelessly fragmented, there is an increasing melding of differences into national entities. The neighboring giants of South and East Asia, the source areas for the region's culture, constitute large potential markets for Southeast Asian resources and production. Although internal upheavals have been common, they have been successfully resolved in many of the region's states. Outside powers play an important role, but this role is not entirely negative. The region has achieved the successful end of colonialism and of the Vietnam War. If population growth rates are reduced, the area's prospects are reasonably bright. Industrial investment, both for-

TABLE 8–2
Physiological densities.

Country	Population (millions)	% Natural Increase	Geographic Density (persons per sq. mile)	Physiological Density (persons per cultivated sq. mile)
Major Food Exporters				
Canada	25.8	0.8	6.7	144
United States	242.3	0.7	67.0	335
Australia	16.1	0.9	5.4	90
Argentina	31.6	1.6	29.6	222
France	55.4	0.4	262.3	749
Southeast Asia*	422.1	2.2	243.1	1,430
Brunei	0.2	2.4	90.9	4,545
Burma	39.5	2.2	151.2	1,123
Kampuchea	6.5	2.1	92.9	547
East Timor	0.7	2.5	122.8	2,456
Indonesia	175.8	2.2	233.3	2,390
Laos	4.0	2.3	43.8	1,094
Malaysia	16.4	2.2	128.8	991
Philippines	59.6	2.5	514.7	1,319
Thailand	54.8	1.9	276.1	746
Vietnam	63.6	2.5	500.2	2,632
Other World Regions				
Europe	495.1	0.3	263.2	907
East Asia	1,279.7	1.1	281.1	2,740
South Asia	1,065.1	2.3	536.9	1,554
Latin America	424.9	2.3	53.5	595
Sub-Saharan Africa	446.0	2.9	53.1	781

*The city-state of Singapore is not included.
SOURCE: *World Development Report* (London: Oxford University Press, various years).

eign and domestic, is in the process of transforming some of the region's economies. Change, rather than any past geographic generalizations, seems to be the dominant characteristic of much of Southeast Asia.

TRENDS AND PERCEPTIONS: BETWEEN GIANTS

Southeast Asia has been very much a "land in between"; for much of its history, its culture has been shaped by a position between the repeatedly expansionist cultures and peoples of India and China. It is a region of sharp cultural differences and sharply contrasting economies, one in which topography plays an unusually important, if passive, role. Lowland-upland contrasts throughout Southeast Asia are typically acute.

A series of great rivers, many with headwaters in the eastern and southeastern flanks of the Himalayas, are channeled southward by steeply folded topography. They form wide valleys and alluvial plains as they approach the sea. These fertile alluvial soils support heavy concentrations of people, and several have become the cores of states (see Figure 8–2). Thus, the middle and lower valleys of the Irrawaddy formed the cores of Burma; the Chao Phraya, the Thai core; and the Red River, the original Vietnamese core territory. The central and lower Mekong (one of the largest river systems in all of Asia) flows through the densely populated sectors of Laos and Kampuchea, and the southern section and delta contain the second core of Vietnam. Many of the most productive soils of the island archipelagoes tend to be in the coastal areas and lower river valleys; thus, this part of the wet tropics, in contrast to those of much of South America and Africa, has always been highly accessible.

The fabled Angkor Wat ruins testify to the power of a great Hindu-influenced civilization that existed in the lower and middle Mekong Valley from the ninth to the fourteenth centuries. Hindu culture also reached to the central Vietnam area. The distribution of plow types shows the intertwined nature of the Chinese and Indian cultural penetrations. Indian plow types were common in Java, Sumatra, Kampuchea, and Laos, whereas Chinese plows predominated in Vietnam, the Philippines, and northern Borneo.

There are at least four different "cultural worlds" throughout the region: the isolated, primitive, shifting agricultural realm of the uplands with their distinctive ethnic groups (frequently not integrated economically or politically with the larger society); the crowded, intensively cultivated wet-rice lowland; the development enclaves of the cities, especially the capitals; and the less densely populated, gradually integrating zones between the cores and the mountains.

The rise of the great trading cities (Singapore, Manila, Bangkok, and Hanoi) is largely a product of European imperialism; western Europe began to replace Indian and Chinese trade and cultural influences by the eighteenth century (even earlier in the Philippines). The British in India, ever trying to stabilize and pacify the Indian frontiers, pushed into Burma; Burmese rice surpluses, so near to Indian food deficits, were an added attraction. The British and Dutch colonial empires split what was left of the Malayan culture area (the Spanish had occupied the Philippines in the sixteenth century) to acquire coastal trading stations for the lucrative spice trade (see Figure 8–3). Coastal trade centers expanded control into the interior to stabilize the hinterlands

supplying the spices. The successful transplant of natural rubber trees from Brazil led to further expansion of political control inland. The French expanded into Vietnam from an initial settlement (1858) at present-day Danang. By the end of the nineteenth century, they had moved into Laos, which had been loosely controlled by Thailand, and Cambodia. Thailand managed to remain free by a combination of able kings and its function as a buffer between British expansion from Burma on the west and Malaya on the south and French expansion westward from Vietnam.

Rubber to feed Western industrialization stimulated trade and migration. The demand for plantation labor led to imports of workers from British India and China. The commercial boom generated by rubber (and tin) helped establish great trading cities whose entrepreneurial middle class was almost invariably Chinese. Thus, some postcolonial states inherited a situation of increased ethnic diversity. New rubber plantations in Southeast Asia enabled the industrial states of Britain, France, and Holland to supply their own markets and other customers at prices that they then controlled. Spices, sugar, tea, palm oil, and other plantation crops were added, diversifying the exports but ensuring continued dependence on agriculture within most economies. Plantations and cities exist side-by-side with traditional farms in a dual economy. Thus, each of the successor states contained differing economic portions, some isolated and self-contained, some integrated with distant, world economies. Much of the energy of postcolonial governments has had to be devoted to economic integration and to blending the peoples into one nation. The region's international views range from the intensely neutralist, self-centered stance of Burma to the generally pro-U.S. views of Thailand and the Philippines to the virtual absence of a world view in devastated Kampuchea and poorly organized Laos. Vietnam is clearly an ally of the USSR and currently dominates Kampuchea and Laos; political logic would urge the PRC to establish friendships in Thailand, Malaysia, and Indonesia, although racial hostilities toward the overseas Chinese may complicate this effort. Japan has established many strong economic links throughout the region as market for minerals and forest products, source of manufactured goods, and entrepreneur in helping establish

Angkor Wat. *Angkor Wat is a huge, sprawling complex of temples and monuments built by the ancient Khmer culture. The Hindu influence is evident here in this "land between" India and China.*

light industries integrated with the Japanese economy. Memories of wartime occupation within Southeast Asia remain vivid, but the Japanese were not necessarily considered the enemy by the people of the region's then colonial units. Japanese money is welcome, and individual Japanese are treated politely.

Indonesia, with 40 percent of the total regional product, 78 percent of the region's oil production, and a position astride the sea lanes from the Indian Ocean to the Pacific (sharing the Strait of Malacca with Malaysia and controlling the Sunda and Lombok straits), is clearly the key to the region's harmony with East Asia and Australia-Oceania. The United States is Indonesia's second most important trading partner, after Japan, and is attempting to counterbalance the Soviet Union's Vietnamese alliance with a Thailand-Malaysia-Indonesia-Philippines combination of active allies and friendly neutrals.

U.S. relationships with the Philippines have remained close since Philippine independence. U.S. leases on an air base and naval base end in 1990, when the Filipino electorate will decide to renew them or not. Japan has recently displaced the United

States as most important trading partner, and a new Philippine government provides some uncertainty about future relations with the United States.

THE DISTINCTIVENESS OF SOUTHEAST ASIA

Southeast Asia is a region of contradictions. It is a probable ancient hearth of agriculture, with many domesticated plants and animals originating there. It is also a recent agricultural frontier with colonizing farmers pushing into lightly settled river valleys, deltas, and coastal lowlands within the last few centuries. It is both commercial and subsistence, partially developed and underdeveloped. Some borders overlap ethnic distributions; claims and counter-claims add political uncertainty. The area's states share the same basic problems; they are not functionally well integrated.

Historical-Cultural Characteristics

Southeast Asia is the meeting ground, or more accurately the intermingling and historical interweaving ground, of many religions. Hinduism spread through much of the area almost 2,000 years ago; the Hindu heritage remains apparent in the court rituals of Thailand and Malaysia. Hindu monuments and temples survive throughout mainland Southeast Asia and in Indonesia. Buddhism spread from India in the seventh century A.D. and remains important in much of mainland South Asia. Islam was brought to the Malay Peninsula and to Sumatra by 1250, spreading as far as Mindanao in the southern Philippines by the seventeenth century. Filipinos are 85 percent Roman Catholic, since Spain was the only European colonial power to attempt, and achieve, wholesale religious conversion (see Figure 8–3). Indonesia and Vietnam acquired Christian minorities through the less ambitious activities of missionaries under Dutch and French colonial rule.

Indonesia is the world's largest Islamic nation. Most mainland states are dominantly Buddhist. Islamic Malaysia has Christian, Hindu, and animist minorities. As noted with Thailand, colonialism was not universal, although it was once a dominant force in all other Southeast Asian states. Colonialism is still a major factor in the region in that many contemporary national policies (and problems) are reactions to its recent presence. Plantation economies were usually highly localized. For example, plantations were important in mainland Malaysia but almost nonexistent in the island portions. Commercial crop production for export revolutionized Java and Vietnam but hardly touched Cambodia and most outer islands of Indonesia.

Population Growth

All of Southeast Asia is characterized by rapid population growth, although the regional increase rate is lower than that of most of Africa or most of Latin America. High regional birthrates have not always been matched by a lowering of death rates. The great surge in natural increase characteristic of the second phase of the demographic transition is in process. Birthrates have begun to decline only in Thailand, Singapore, and Malaysia. No other area of the world has such an enormous variation in birth-, death, and growth rates among individual units. This variation may reflect most accurately the vast intraregional disparities in development (Table 8–3).

Political Instability

Political instability has been another distinctive feature of Southeast Asia. Ongoing guerrilla rebellions have occurred (and repeatedly recur) in many countries. Ruling regimes have changed with amazing rapidity.

The Vietnamese War is a case in point. France's colonial administration of Vietnam was glaringly exploitive, even within the frame of reference of European colonialism. The economy of all Indochina was reorganized to provide France with exports of rice, rubber, silk, spices, and minerals. Administration was in the hands of French nationals and a small elite of Vietnamese collaborators, mostly Christian converts, who earned a reputation for widespread corruption and insensitivity to the masses of poor, Buddhist peasants. Despite the French boast that theirs was a civilizing mission, literacy actually decreased under French rule. Public health received such low priority that French Vietnam had one doctor for every 38,000 people, while the U.S.-administered Philippines had one for every 3,000.

TABLE 8–3
Vital statistics for Southeast Asia.

Country	Birthrate (per 1,000)	Death Rate (per 1,000)	Per Capita GNP ($)
Brunei	28	4	21,140
Burma	37	15	180
Indonesia	34	12	560
Kampuchea	32	11	(not available)
Laos	41	18	(not available)
Malaysia	29	7	1,870
Philippines	32	7	760
Singapore	16	5	6,620
Thailand	25	6	810
Vietnam	34	9	(not available)

SOURCE: *United Nations Statistical Yearbook*, 1988.

President Franklin Roosevelt was firmly opposed to any return of French rule in Indochina following the end of World War II: "The case is perfectly clear. France has had [Vietnam] for nearly a hundred years and the people are worse off than they were at the beginning. [They] are entitled to something better than that."[1] Unfortunately, solid U.S. opposition to a return of French colonialism in Indochina died with President Roosevelt. French leader Charles de Gaulle insisted that Vietnam, Laos, and Cambodia, freed from Japanese occupation at the end of the war, be restored to France, threatening to withdraw France from the U.S.-led Western alliance against the Soviets if this was not done. Thus, geopolitical considerations (France was considered vital to any viable Western European-American alliance against further Soviet moves in Europe) led to American complicity in reimposing French control.

A fateful combination of American domestic political considerations and a general lack of appreciation of Southeast Asian political, physical, economic, and cultural geography culminated in America's tragic miscalculations in Vietnam. Immediately following the Japanese collapse in 1945, a Viet Minh Congress in Hanoi declared an independent republic, quoting extensively from America's 1776 Declaration of Independence. Ho Chi Minh hoped that the United States, pledged to grant freedom to the Philippines, would also oppose the reassertion of colonial control elsewhere in the region, but American troop ships and air transports delivered French troops to Vietnam. Successive U.S. presidents, concerned that they must not appear "soft" on communism, supported French and then Vietnamese anti-Communist elements against the Communist regime headed by Ho Chi Minh. The American evaluation of the political situation in Southeast Asia assumed that Communist Vietnamese were tools of Communist China, despite Vietnam's long history of resistance to Chinese control. The U.S. government seriously underestimated the determination of the Vietnamese, who had repeatedly risen in rebellion against the Chinese emperors (who regarded Vietnam as a tributary state), French colonialists, and corrupt Vietnamese rulers.

The lengthy Vietnam War did little to change the situation. Most Westerners assumed peace would follow U.S. withdrawal. Instead, there was an almost immediate resumption of war, this time against Kampuchea. Attacks have also occurred in Laos, and fighting at times spilled across the Thai border, where refugees from Laos and Kampuchea have sought asylum.

THE PHYSICAL FRAME

A decidedly usable portion of the tropics, Southeast Asia consists of five major river basins separated by long, parallel-folded mountain ridges, a major peninsula, and thousands of offshore islands that vary from tiny to areas larger than Great Britain. Most of the mountainous country, mainland or island, is re-

THE WAY OF BUDDHA

Buddhism has been a powerful force within Asian cultures for 2,500 years. Its influence on philosophy and the arts in Southeast Asia, South Asia, and East Asia has been long-standing and profound.

Siddhartha Gautama, born about 563 B.C. in Nepal, was raised in luxury as a Hindu of the warrior caste. At the age of 29, he decided to contemplate the riddle of life. After six years of meditation culminating in a transcendent revelation, he embarked on a life of itinerant preaching that lasted until his death at 80. When his meditation had led to a tranquil wisdom at age 35, he received the title "Buddha," or "Enlightened One."

Buddhism was comparable to a protestant revolt against orthodox Hinduism, itself already ancient when Gautama was born. Gautama disapproved of Hindu caste distinctions and preached that one could escape the endless cycle of death and rebirth into "nirvana," a state of blissful escape into infinity. Universal suffering results from selfish craving of earthly pleasures. When good works and inner peace result from self-discipline, enlightenment leads to nirvana. Legend holds that the newborn Gautama proclaimed that this was his last existence.

Gautama, who denied that he was divine, directed his followers to spread his philosophy as far as they could. Buddhism became the state religion of India under Emperor Asoka (third century B.C.), temporarily triumphing over Hinduism. By A.D. 800, Buddhism had spread to Korea and Japan to the northeast, southeast to Indonesia, south to Sri Lanka, and northward to Tibet. Islam displaced Buddhism in India in the twelfth century, but Buddhism remains important today in Southeast Asia and East Asia (Figure A).

Buddhism teaches tolerance, nonviolence, respect for the individual, and a love of all nature. About two centuries after the Buddha's death, some of his followers developed an interpretation of Buddhism that became dominant in Japan, Korea, and China. This new doctrine is known as Mahayana ("greater vehicle"), and its followers refer to orthodox southern Theravada Buddhism, prevalent in Sri Lanka, Burma, Thailand, Laos, Kampuchea, and part of Vietnam, as Hinayana ("lesser vehicle"). Hinayana Buddhism holds that followers of Buddha can achieve enlightenment only for themselves; it emphasizes the monastic tradition. Hinayana Buddhist males are expected to become monks for at least part of their lives, living very simply in a monastery and begging for their food daily. Theravada Buddhist males who have not yet served as monks are referred to as "unripe" and are likely to be scorned as potential husbands. Mahayana Buddhists, in contrast, teach that the monastic life is too rigorous for most, but that ordinary people can still achieve nirvana through faith and devotion to Buddha as their redeemer.

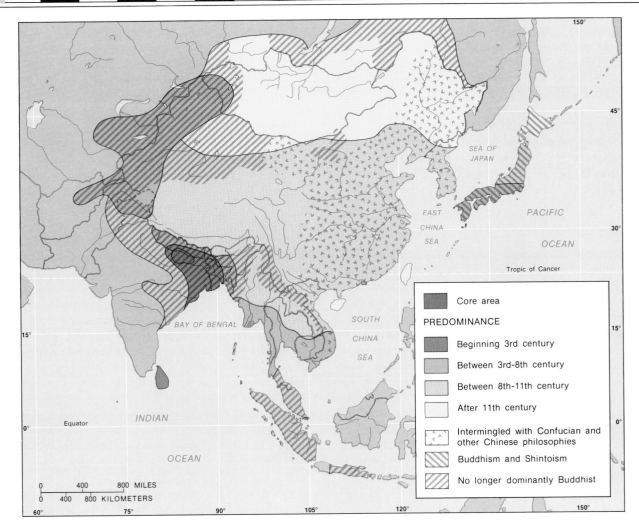

FIGURE A
Territorial Extent of Buddhism. Buddhism spread far to the north and east of the Buddha's home area, labeled "core area" on this map. Note the northern-northwestern and southeastern frontiers of Buddhism, where other religions later displaced the way of Buddha.

The word that best describes the aura of many statues of Buddha throughout Asia is *serenity*. The classic statues, some of gigantic size, almost always feature a Buddha seated on crossed legs, an enigmatic smile on the lips, and with the right hand resting on the knee, fingers pointed down. The fingers point to earth as a reminder that, throughout the onslaughts of demonic forces during meditation, the Buddha sat unmoved and tranquil, his right hand calling on the earth itself to witness his triumph over temptation. The way of Buddha is peace and acceptance.

lated to (and part of) a giant chain of folded structures that extends from the Atlas of North Africa through the Alps and Himalayas to the Malay Peninsula and continuing beyond as a series of island arcs off the Asian mainland.

Heat and high humidity persist year-round. Seasons tend to be marked by fluctuations in rainfall rather than in temperature. The mainland portion and most of the Philippines have a distinct winter dry season, but the seasons are better described as wet and wetter in the lower Malay Peninsula, the offshore islands, and parts of the mainland coast.

Topography

The riverine basin of Burma is Y-shaped, with two limbs separated by difficult hill country. The entire lowland associated with it is enclosed by a mountainous horseshoe-shaped rim that culminates only a few miles from the ocean. The Salween River creates a canyonlike valley that widens only into a narrow lowland, shared in part with Thailand. The Irrawaddy drains a long, somewhat wider valley. The Chao Phraya lowland of Thailand has been created by local rivers not of Himalayan origin, while the Mekong basin is shared by several countries. The short Red River originates in southwestern China and creates a fertile delta lowland as it enters the Tonkin Gulf. Over most of the mainland, except for the Isthmus of Kra and the central Vietnamese coast, there is reasonably good development of coastal lowland.

In Indonesia, the situation is more complex. The western limb of the Burmese horseshoe continues out at sea as the Andaman Islands, forks, and then becomes the mountain spine along the southern edge of the Indonesian island chain. In Sumatra, this spine has a generous coastal lowland eastward. In Java, alluvial fill, volcanic flows, and erosion have created a plateaulike surface, while eastward only the tops of the chain remain above water as a series of islands terminating at Timor.

The other limb of the Burmese horseshoe, after being split in two by the Salween, continues to form the Malay Peninsula before disappearing under the sea and reappearing as the mountain spine of Borneo (Kalimantan). Here the range splits into several other chains, making a series of alluvial basins separated by ridges—a miniature of the mainland. Celebes (Sulawesi) has the high mountain core with five prongs, but the intervening basins are largely drowned.

A single mountain spine with elevations over 15,000 feet runs northwest to southeast astride the center of the island of New Guinea. To its south is broad, swampy coastal lowland. To its north, a much lower ridge traps sediment in intermontane basins before it reaches the sea.

The complexity of the topography results from the contact of large, rigid crustal plates. Some plates are ocean bottom, whereas others carry continents above them. Where these plates collide, there is much deformation of the crust through folding and faulting. Earthquakes and volcanic eruptions occur as well, with one plate tending to slide under the other.

Southeast Asia is at the contact zone of at least four tectonic plates: the Pacific, Australian, Philippine, and Eurasian. Movement has generated mountain-building activity throughout this complex zone of mountain, basin, and island. Wherever else multiple plates collide (the Caribbean, the Mediterranean), the landscape is also one of peninsulas and islands. Even in topography, there is diversity, instability, and the mixture of positives and negatives so characteristic of the region.

The results of collision in Southeast Asia are not only spectacular, but also increase this tropical area's utility via volcanic material and alluvium. Tropical soils that would normally be leached of their fertility by the chemical reactions of laterization (aided by intense heat and large amounts of water) are periodically recharged with fertility by the overflow of streams or the addition of fresh, mineral-rich volcanic material.

Land-Sea Relationships

The sea has provided the most important linkages throughout the area, past and present. The Malay peoples, dominant on the island and the peninsula, are known as a great seafaring nation; they are thought to be an amalgam of negroid, mongoloid, and perhaps paleo-Asiatic, australoid, and some caucasoid racial elements. Related elements span much of the Pacific basin and reach Madagascar on the other side of the Indian Ocean off Africa.

The mainland cultures were not maritime in orientation. They seem to have spread down the mountains, rivers, ridges, and valleys from points of origin in western and southern China.

Most of the non-Malay cultural influence that came by sea was brought by traders. Over 2,000 years ago, pre-Aryan cultures from Ceylon and southern India had a great impact on the area. Hinduism was brought to Indonesia, Cambodia, and the Mekong Delta by traders and the migrants who followed. Architectural evidence of this is found throughout these portions of Southeast Asia. Later and continued contact with India is represented in the dominance of Buddhism on the mainland. The Chinese brought their culture by both land and sea beginning in the third century B.C. Again, the mechanism was traders followed by migrants. Arab traders were well established before the thirteenth century. Finally, Europeans began maritime contact in the fifteenth century, ending in a wave of colonialism. To this day, most of the region's economies are export oriented and rely on the sea for transportation.

The sea in this region is extremely shallow in the area between western Indonesia and the mainland, and shallow seas are again encountered between Australia and New Guinea. The Malay Peninsula and the outer islands virtually enclose the South China Sea and a series of smaller seas in the eastern portion of Indonesia. It is an area of intermixed land and water surfaces that separate the Pacific and Indian oceans (see Figure 8–1).

A dominant power could easily block a series of straits that control entry. By far the most crucial (and narrowest) entryways are those from the west: the Sunda Strait between Java and Sumatra and the Strait of Malacca between the Malay Peninsula and Sumatra. The British, to protect their routes to China, fortified Singapore and laid claim to the lower end of the Malay Peninsula (see Figure 8–3).

The Japanese, between 1941 and 1945, intended to protect their occupied lands in Southeast Asia by controlling the straits. A series of major battles were fought between the United States and Japanese fleets to open routes to these inner seas.

Today the strategic importance is just as obvious. Japan's oil lifeline extends through this sea. Indonesia, concerned for its fish resource, controls the size and routes of tankers through the straits. The United States clings tenaciously to strategic naval and air bases in the Philippines.

Three states in the region cannot reach all their domestic territory without sea or air routes. Another, Singapore, must receive all supplies, even most food, from overseas sources.

Climates

Most of the area is exceptionally well watered, but the Philippines has a prolonged, December-to-April drought. The circulation pattern set up by the monsoon (discussed in detail in Chapter 7) is such that parts of the mainland are characterized by a double maximum, receiving rains from both the winter and summer monsoons. The normally cold, dry air associated with the winter monsoon is modified by the time it reaches Southeast Asia. In passing over the warm waters of the North Pacific and the South China Sea, the air becomes warm, moist, and unstable. Winter rainfall is particularly heavy on windward slopes in the Philippines, and significant moisture reaches the coastal areas of Vietnam, Kampuchea, the Isthmus of Kra, and the upper Malay Peninsula.

Interior locations on the mainland are typically tropical wet-and-dry areas. In large areas of interior Burma, Thailand, and Kampuchea, there are two or three months without rain. The downpours of June through September constitute a truly rainy season. Monthly temperature averages are 10° F lower during the dry winter season, and seasonal humidity averages vary greatly. The dry season becomes more pronounced and longer from east to west. Indeed, the central Irrawaddy Valley of Burma is almost like a tropical steppe because of its combination of distance from a source of moisture and the strong rainshadow effect caused by the horseshoe of near-encircling mountains.

South of the zone of double maximum is a region of more typically tropical weather patterns. The island of Mindanao, the Republic of Indonesia, and most of Malaysia lie within the area of the **intertrop-**

ical convergence for most of the year. Much of this zone experiences a truly tropical rainforest climate of continuous high humidity and temperature with no marked dry season. Singapore, in many ways typical, has an average humidity of 80 percent and an average temperature of 80° F in virtually every month. Rainfall averages 6 to 10 inches per month and occurs on about one-half the days of every month. This monotonous, seasonless climate encompasses much of the area.

In eastern Indonesia, from Java to Irian Jaya (western New Guinea), there is a pronounced rainfall maximum in summer (this is in the Southern Hemisphere) from November to April, and a much lower winter rainfall, though no month is totally without rain in most years. The average monthly temperature cools a few degrees during a brief midwinter period, humidity levels drop 10 to 15 percent, and rain occurs on only two or three days per month.

In the Philippines, the western edge of the islands receives a summer maximum, while the eastern edge has a winter maximum. The central belt between has a generally drier climate because it is in the rainshadow of both sets of monsoonal winds. North of Mindanao, the climate tends to be monsoonal, with such heavy wet-season precipitation as to negate the dry season.

Northern Indochina has a climate rather like that of southern China or the Ganges Valley. It is sufficiently cool in the winter (except along the coast) to be classified as subtropical. Although frosts do not occur, average temperatures fall below the 64° F necessary to maintain some types of tropical vegetation. Winter days are mild, but nights can get cool.

Highland areas are everywhere subject to altitudinal moderation of temperatures. For example, in Indonesia there are rather large areas of temperate, high-altitude climate, though with virtually no seasonal variation in temperature or rainfall.

The Philippines and the east coast of Vietnam are particularly subject to Pacific hurricanes (typhoons). Virtually all lowlands are subject to seasonal summer flooding. The region, however, generally has less of the disastrous drought or perennially heavy flooding so characteristic of South and East Asia.

Rice and Southeast Asia's predominant climates are virtually perfectly matched. The prevalent heat and generally abundant rain are well suited to rice, and some variant on rice cultivation occupies most of Southeast Asia's farmers who, in turn, are the majority of the region's people. Most Southeast Asians eat rice every day; indeed, most eat little besides rice, with a few vegetables, some chilies for flavor, a bit of vegetable oil used in cooking, and, if they are lucky, a small fish for protein.

Drainage

Southeast Asia is one of the great reserves of undeveloped water power in the world. Burma alone has over 3 percent of the world's developable hydroelectric potential, much associated with the Salween Valley.

In Thailand and the three successor states of Indochina, the Mekong River forms the major potential. Any development here, unlike the hydro power dams in Burma, would require the agreement and cooperation of several countries.

Indonesia is a land of short, often swift rivers that end in swampy deltas and coastlands. Comprehensive hydroelectric development could serve the dual purpose of power generation and extension of arable land through drainage and flood control. As elsewhere in the region, little has been developed to date.

Regional deltas here are larger than that of the Nile, and three of them—the Irrawaddy, Mekong, and Chao Phraya—come close to the scale of the huge Ganges Delta (see Figure 8–2). The first two are rapidly extending the land surface seaward, and the last is virtually an inland delta. Yearly floods renew the soil, creating an excellent environment for wet-rice farming. Current delta population densities are one-half those of similar areas in China, India, and Bangladesh, implying that further population growth can be sustained. The region's rivers, with the exception of the Red River (northern Vietnam) and the rivers of the Luzon plain in the Philippines, have not been fully exploited as sources of irrigation water.

The Tonle Sap, a freshwater lake in Kampuchea, fluctuates greatly with the wet and dry seasons. It harbors a major fish resource in an area that traditionally has been short of protein. During the wet season, the Mekong floodwaters flow into the lake, while during the dry season, the flow is reversed.

Crops are grown on the exposed lake bottom during the dry season. Artificial control of this unusual water resource could improve Kampuchean agriculture and aid in industrial development.

Vegetation and Soils: The Critical Interaction

Virtually all the islands and the wetter portions of the mainland were originally covered with a dense growth of tropical rainforest or its monsoon variation. These magnificent forests existed even on areas of generally poor soil, because of deep root systems able to penetrate to nutrient-containing lower soil horizons.

The very lushness of the great variety of plant life in the tropical rainforest encouraged many serious misjudgments on the part of colonial-era Westerners in this, and other, tropical regions. To people accustomed to midlatitude ecological relationships, a dense forest cover usually was a reliable indicator of rich soils for agriculture. In much of the world's tropical rainforests, the vigor of the original forest cover does not necessarily mean that, once cleared, the soils will prove fertile for crop production.

The ability of many tropical soils to sustain a luxurious forest is based on a precarious balance of life and death, of growth and decay. There is constant recycling of nutrients vital to tree growth. Once the cycle is destroyed or significantly disturbed, the fertility of the underlying soils may prove to be an illusion.

Heat is a catalyst in many chemical processes, and heat energy is abundant in the tropical rainforest. Even more abundant is water—often 60 or 70 inches of precipitation per year, and sometimes much more. The result is that soluble minerals such as calcium, magnesium, and potassium rapidly go into solution. If not used quickly by vegetation, these soluble minerals are carried away by the huge volumes of water draining down through the upper soil levels. Over time, and unless renewed by fresh additions of stream-deposited alluvial materials or new outpourings of basaltic lava (basic rather than acidic), or located atop disintegrating limestone, tropical soils come to be dominated by insoluble compounds of iron, aluminum, and manganese in upper layers. Only the deeply penetrating roots of mature trees can tap any significant reserves of soluble minerals, and bring them back into the rapidly recycling exchange of nutrients that is the rainforest.

An almost continuous rain of dead leaves, flowers, and branches falls to the forest floor. This decomposing organic matter is vital to plant growth but will disappear if not continuously resupplied by organic debris falling from the forest above.

Once the great trees are gone, and large areas cleared for cultivation, the all-important nutrient-recycling mechanism is broken. Once-fertile soils lose their heavy inputs of organic material, which included vital minerals collected by deep root systems. The canopy of the branches and leaves that once broke the force of falling rain is no more; soil erosion becomes a problem. The now unscreened, nearly overhead rays of the sun bake top layers of soil into a bricklike consistency. Rapid soil deterioration can lead to abandonment of the new farm fields within a few seasons.

Most of the region's forests are broadleaf evergreen and multistoried, with rapidly growing, diverse species. Typically, there are three layers. Scattered trees between 100 and 180 feet high rise above an almost continuous roof formed by many species in the 50- to 75-foot range, with an understory of younger trees, shrubs, and smaller plants closer to the forest floor.

Climbing plants called lianas interlace the network of trees, using the last available sunlight. Mangrove swamps ring the coastal lowlands in many areas. Bananas, citrus, plantains, and coconuts are among the domestic species, but the rubber tree and the oil palms were imported from other areas.

Human intrusions into the environment have reduced the area of original rainforest; however, most of New Guinea, the interior of Borneo, the northern half of Celebes, and the east coast of Sumatra still contain large reserves of tropical hardwoods.

Teak is a widespread and valuable species harvested in Burma, Thailand, eastern Java, and the islands east of Java. Some eucalyptus have cash value, and sandalwood, a common species, commands a high price.

Indonesia's government owns almost all the forest land and has begun an ambitious program of investment and development. Over half the forest area has been granted as concessions to foreign (primarily

THE DEVASTATION OF TROPICAL FORESTS

About 6 percent of the earth's surface is covered by tropical rainforest. The continued existence of these forests—largely in Southeast Asia, tropical Africa, and Latin America—is threatened, and that threat has vital implications for everyone on the planet. The continuing ravage of tropical forests will lead to accelerated extinction of plant and animal species. It almost certainly will produce changes in rainfall volume throughout the tropics and their fringing (and seemingly expanding) deserts. The despoliation of this ecological treasure will also lead directly to diminished food production, increased poverty, and higher levels of desperation among people who already rank as some of the globe's poorest citizens.

Forty percent of the original rainforests have been cleared in the past 30 years, and the pace is approaching 100 acres a minute. In about 15 years, if the current rate of destruction is not slowed, all of the world's rainforests will have disappeared.

The tropical forests and neighboring tropical savannas are the habitat of a vast majority of the estimated 8 million plant and animal species that call earth their home. Allowing widespread habitat destruction and consequent extinction of species and their genetic characteristics is comparable to permitting random burning of the last remaining copies of many books left in the world's libraries. It cripples the ability of future scientists to find vital uses for presently unstudied or even now-unknown plants and animals, or to utilize their genetic traits in developing new varieties of plants or animals. It is possible that 25 percent of tropical plant species and 90 percent of animal species (mostly insects) have not been scientifically categorized and studied.

Japanese) concerns. Sawn timber and plywood are the chief wood industry products; conservation is largely lacking at present except in Java and Sumatra.

The forests of the Philippines have been largely cutover to supply the Japanese market. Reforestation efforts have been moderately successful, at best. Construction timber, paneling, plywood, lumber, and furniture are made from Philippine mahogany (luan).

Most of the mainland forest has been cleared for permanent agriculture, converted into tree plantations, or degraded by repeated slash-and-burn clearing for agriculture. Defoliation agents used during the Vietnam War destroyed much of the forest in the hills of the Indochinese peninsula. Their long-term effects on the vegetational cover are as yet unknown.

The natural grasslands of Southeast Asia were savanna or shrubby parkland associated with rain-shadow areas. They are frequently burned for shifting agricultural use and at times have been converted into bamboo plantations.

Much of the land of the Philippines and the smaller islands east of Java, and on parts of Sumatra, supports a sparse grassland cover on badly eroded land that has been cleared too often for shifting agriculture. Similar alteration of vegetational cover

Only about 2 percent of the world's rainforests are currently protected in national preserves. As botanist and conservationist Hugh Iltis observed, "the countries most desperately in need of money to create and maintain the gigantic national parks necessary to preserve their unique biological riches—both for themselves and for the whole world—are the very ones that are usually too poor to afford them."[*]

Rainforests directly influence local climates; indirectly, they impact global weather patterns. In one deforested part of Central America, average annual precipitation has dropped by 17 inches after tree clearance. The Sahelian droughts that may signal expansion of the Sahara could be related to rainforest destruction in West and Central Africa.[**]

Catastrophic soil erosion, siltation of rivers, dropping groundwater tables, even climatic changes—all results of unwise forest destruction to provide more firewood, expand plantations producing export crops, or create low-carrying-capacity pasture to produce cheap hamburger for the industrial world's fast food—have one long-term effect on human populations. Malnutrition and misery expand as tropical forest acreage contracts. Yet the greatest ecological tragedy in the world continues almost unabated. Southeast Asia currently supplies 70 percent of the world's tropical forest products, so these forests are under the most pressure.

[*]Hugh Iltis, "Tropical Forests: What Will Be Their Fate?" *Environment* 25, as reprinted in *Geography 86/87,* Dushkin Publishing Group Annual Editions, 1986, p. 78.
[**]"Ravage in the Rain Forests," *US News & World Report,* March 31, 1986, as reprinted in *Geography 88/89,* Dushkin Publishing Group Annual Editions, 1988, p. 69.

could occur anywhere that unwise forestry is practiced.

THE STRUCTURE OF SUBREGIONS

Traditionally, Southeast Asia is divided into mainland and insular portions. While the blending of pre-Aryan Hindu and Chinese cultural elements is apparent throughout the region, the dominance of the Chinese influence on the mainland is most obvious. A Chinese-related language predominates in only one state, Burma, but Chinese trade and culture, if not imperial control, have influenced all. The Chinese influence on agricultural techniques, clothing, and some cultural behavior patterns is striking. It is culture, therefore, on which subregions are based.

Malaysia, Indonesia, Brunei, and Singapore are a part of the Malay cultural-linguistic milieu, yet Singapore is a Chinese island in the midst of a much larger Malay realm. Most Malays are of the Islamic faith, differing thereby from the dominantly Buddhist mainland.

The Philippines, although Malay in language and some elements of culture, has been heavily Westernized by 50 years of American colonial rule. Ruled by

Spain for several centuries before 1898, the Filipinos are largely Christians.

Vietnam, Laos, and Kampuchea

Once united in French Indochina, Vietnam, Laos, and Kampuchea (Cambodia) are as dissimilar as any other states in the region. Poverty and the destruction of war are shared characteristics. Whereas the cultural influences of India are apparent in Kampuchea and Laos, Vietnam was most strongly influenced by China. Chinese expansion down the Vietnamese coast was much easier than in the canyonlike river country to the west. The original core of the Vietnamese (the Red River plains) was brought under Chinese control by the second century B.C. Even the name Vietnam means the "southern country," and obviously it is in relation to China that Vietnam is southern.

Vietnam has been likened to two rice baskets connected by a slender carrying pole (see Figure 8–1).[2] This image reflects the two fertile river plains and delta complexes, the Red River to the north and the Mekong Delta in the south. The Vietnamese traditionally have avoided the uplands, so Vietnamese ethnics occupy the two cores at either end of the country, plus pockets of valley lands facing seaward where the mountain spine meets the sea. The Vietnamese, to whom village life and its social relationships were highly important, could not move to the mountains and find enough continuous area for conversion to rice paddies to sustain an entire village. Their culture remained, then, a lowland one. The Vietnamese are the region's best hydraulic engineers, having built an elaborate system of dikes and canals to tame the flood-prone Red River of their primary core. The Mekong Delta has always been a more lightly settled frontier. By the mid-twentieth century, the Red River lowlands were so densely populated that importation of rice became necessary; the southern, secondary core of the settlement was a rice exporter. This natural complementarity was enhanced by the fact that northern Vietnam had the raw materials—anthracite coal, iron ore, tin, and chrome—to support industrialization.

It was relatively densely populated, culturally advanced Vietnam that was the star of French Indochina. Rubber plantations were added to its traditional exports of rice (Figure 8–4). Nationalism

among the 85 percent Vietnamese ethics had been well advanced by the long struggle against the Chinese and reinforced by language and an eclectic brand of Buddhism. Nationalist activity began against the French within 20 years of their conquest of Vietnam. Japanese occupation, not entirely unwelcome at first, demonstrated the colonial power's weaknesses and fueled determination to achieve freedom at the end of World War II. An anti-Japanese coalition led by Communists was quickly redirected to an anti-French movement, the Viet Minh.

The European colonial period introduced manufactured goods to the detriment of village crafts. This change undermined an important rural source of income, because rice culture is highly seasonal in labor demands. Off-season, chronic temporary unemployment resulted from the substitution of European manufactured goods for handicrafts. The decay of this communal craft industry contributed, as did the colonial regime's taxes, to an increasing indebtedness among the farmers. Land was typically in the hands of a few, with large numbers of peasants without land.

In the provinces south and west of Ho Chi Minh City (formerly, Saigon), two thirds of the farmers were landless, and 50 percent of the land was held by 1/50 of the people. Land rents were 40 to 50 percent of the crop. During the successful struggle to oust the French, the Viet Minh revolutionaries carried out a drastic land reform, confiscating land from landlords and giving it to peasants. South of the demarcation line drawn to facilitate French withdrawal (the partition was supposed to be temporary, but a free election to choose leaders of a unified nation, agreed to in the Geneva Agreements of 1954, never took place), the Diem regime instituted its own reform. Landlord influence was strong within the South Vietnamese government; understandably, the peasants of the South were not impressed by a reform that sold them land the Viet Minh promised them free.

This "temporary" partition became one of the sharpest boundaries in the world, a portion of the line of confrontation between Communism and the West. The South received lavish economic and military aid from the United States, but central government control was limited to the cities and interurban routes (at least in daylight), while the many-celled, self-contained villages were tied into neither the national economy nor political life. In 1964, the Hanoi

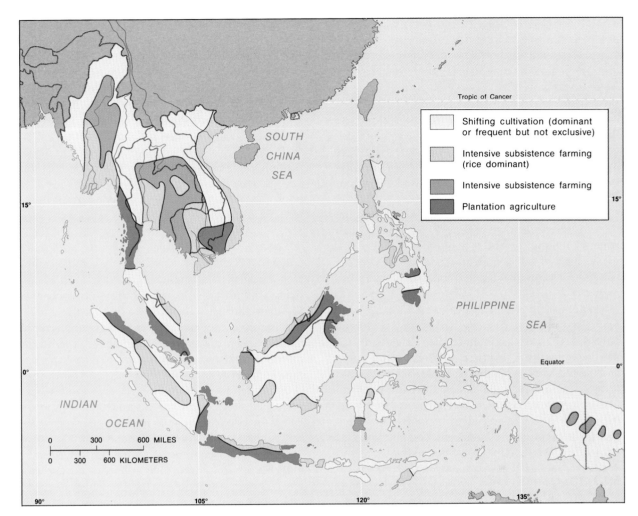

FIGURE 8–4
Major Agricultural Systems in Southeast Asia. Rice feeds most Southeast
Asians; plantation agriculture was introduced by Western colonial powers to supply
exports of rubber, coffee, spices, and tropical vegetable oils. While shifting
cultivation occupies large areas, it is intensive farming that supports dense
populations.

government decided to speed the work of the Viet
Cong Communist guerrillas in the South, committing
regular North Vietnamese army units to the South.
This invasion triggered an escalation in American
forces from "advisors" to more than half a million
personnel at peak. U.S. troop withdrawal began in
1972, leaving naval and air support forces to aid the
South Vietnamese until a peace agreement was concluded in 1973. Infiltration of more North Vietnamese troops into the South defeated the Saigon government in 1975.

Historian Barbara Tuchman argues that America's
folly lay in its official beliefs in the "domino theory"
(that if Vietnam "fell" to communism, so in turn
would each other Southeast Asian country and, ultimately, Australia and New Zealand) and that "nation
building" could be imposed from outside (it must,
obviously, be a force and commitment from within
the nation). Despite his penchant for quoting Thomas Jefferson, Ho Chi Minh never permitted a free
election of any official or on any issue. "The new
political order in Vietnam was approximately what it

would have been if America had never intervened, except in being far more vengeful and cruel. Perhaps the greatest folly was Hanoi's—to fight so steadfastly for thirty years for a cause that became brutal tyranny when it was won."[3]

Vietnamese communization of the south denied the large Chinese minority their traditional trading functions, now nationalized, and long-term animosities led to a mass flight of the Vietnamese-Chinese from the country, precipitating a brief war with the PRC in 1979. The Vietnamese also invaded Kampuchea.

Today's Vietnam has not fully recovered from the war or protracted civil strife. The population, now over 60 million, grows at a rapid rate. There are many, mostly small, groups of minorities in the country, and the degree of national cohesion remains unknown.

Contemporary Vietnam is desperate for U.S. dollars, due to a Western trade boycott and a general failure of Vietnam's economy under a rigidly doctrinaire communism. Tourism from Eastern Europe and the USSR flourishes, but attempts to lure Western travelers have had little result. Russians have replaced Americans as the white faces on the streets, though the average Vietnamese shows as little affection for Russians as for Americans, terming Russians "Americans without money." Although the United States has agreed to accept them, and Hanoi has officially agreed to let them go, in practice, the Amerasian children of U.S. service personnel are still trapped by bureaucratic sloth and deception. These highly visible reminders of the war generally are despised by the rest of the population and lead a semi-illegal existence since they are not given work permits or allowed to register for social welfare services.

Over 90 percent of the population is still employed in agriculture. There is an attempt to enlarge peasant cooperatives through consolidation, a plan that runs counter to old village loyalties.

The north continues mining coal and a few other minerals. Textile and food-processing industries have resumed normal production. All private firms have been nationalized, including holdings of U.S. firms. Only some French interests have been allowed to continue to function as foreign investments. Electronics and machine industries in the south that arose during the period of U.S. involvement continue

as government-owned operations. Japan, rather than the Soviet ally, is rapidly becoming the principal trading partner.

Vietnam is important in the region because it has by far the largest army in Southeast Asia, has occupied Kampuchea for so long, implicitly threatening Thailand, and has allowed the Russians to use the huge U.S.-built naval base at Cam Ranh Bay.

There are probably few states in the world whose nationhood is as shaky as that of Laos, perhaps the poorest state in the region (nobody knows just how poor Kampuchea might be). Laos consists mainly of wild hill country flanked to the west and south by the Mekong Valley. Two core areas have developed, in the northern Mekong Valley around the capital, Vientiane (Viangchan), and in the south-central Mekong near the Kampuchea border. Most people living within Laos's boundaries have no concept of loyalty to the state. The country has but the tiny beginning of a transport and communications infrastructure. Few towns have electricity yet; there is no railway, and only a very low ratio of paved roads to area. Industry is limited to the small-scale production of consumer goods (soft drinks, beer, cigarettes, rubber sandals, etc.) Laos was an unwilling tool in the East-West confrontation; its development prospects are as uncertain as its independence.

Kampuchea (Cambodia), unlike Laos, has had a long history as a distinctive culture and state (the great Khmer civilizations). The French were not as interested in either Cambodia or Laos as in Vietnam, with both positive and negative results. These states escaped some of the sociological problems that Vietnam experienced. A low level of investment in their economic development was the rule. Once independent, the Cambodians tried hard to be nonaligned, a virtual impossibility in this area. Up to 90 percent of the people are (or were) ethnic Khmers, while Chinese and hill tribe minorities were small. Population density is low by regional standards, 93 persons per square mile (547 per cultivated square mile). A complicated civil war, which finally ended as Vietnamese forces defeated the Pol Pot regime in 1979, has taken a severe toll, as have Vietnamese occupation and rampant refugee flight.

Premier Pol Pot's Khmer Rouge had instituted an unbelievably harsh regime in 1975. The country was cut off from all normal contact with the outside. Cit-

THE OVERSEAS CHINESE: A BUSINESS CLASS

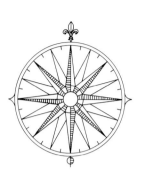

The overseas Chinese in Southeast Asia, some 12 million strong, are a proportionately small group within the overall population but are economically powerful in most countries of the region (see Figures 8–3c and d). Their economic power and strong tendency to retain a Chinese subculture amid some host cultures make them, at times, a persecuted minority. Although western European states acquired direct political control in most of Southeast Asia during the colonial period, the region was more a Chinese colony in terms of actual population migration. The Chinese who migrated to the Nanyang ("southern sea") constitute the largest group of Chinese living outside of China or Taiwan. These Chinese formed an urban middle class between the mass of indigenous peoples and the native elite and European rulers. Often imported by Europeans to work in tin mines and rubber plantations, the Chinese worked and saved to build small retail businesses. These businesses in turn supported a second generation of professionals and proprietors of larger, more prosperous businesses. Some Chinese also became moneylenders. The Chinese of today, then, are typically urban shopkeepers, craftsmen, market gardeners, bankers, professionals, and government bureaucrats.

Although few contemporary overseas Chinese were born in China, the Chinese language and culture are studied in community-run schools. Most often, Chinese are multilingual. Although some intermarriage and acculturation has occurred, the Chinese frequently are regarded as aliens. Their virtual control of the rice trade, obviously a sensitive issue throughout the region, contributes to local resentments as both farmers and consumers perceive the Chinese as profiting heavily from this everyday necessity.

The overseas Chinese often dominate trade between Southeast Asian nations. The hostility of their hosts interacted with Chinese determination to maintain their cultural identity and international contacts as opportunities in case of banishment from their "temporary" homes. It was an obvious trade advantage to an overseas Chinese merchant in Malaya, for example, to deal with another Chinese merchant in the Philippines. A shared language, a shared culture, and a knowledge of the other's reputation (and credit rating) within the close, intermarried communities of Chinese facilitate international deals.

It is not uncommon for overseas Chinese to hold several passports—their official citizenship in their host country and either (maybe both) Nationalist China (Taiwan) and People's Republic passports. This multiple-passport situation is often interpreted by the hosts as evidence of disloyalty. Resentments have resulted in expulsions (and even murders) in Indonesia, Kampuchea, Vietnam, and in the case of the "boat people." It is tempting for politicians in the region to focus discontent on the highly visible "alien" and prosperous Chinese.

ies and towns were forcibly evacuated and their residents chased into the wild countryside to fend for themselves with no tools or food. Money was officially abolished. Hundreds of thousands of people fled to Thailand; hundreds of thousands more died in the general upheaval and destruction of the economy. Kampuchea is the best example of near national suicide known in the modern world. The infrastructure of the nation—transport, public buildings, warehouses, hospitals, post offices—was senselessly destroyed along with unknown numbers of people.

Burma and Thailand

Burma and Thailand, the first a British colony conquered in three wars between 1824 to 1886 and freed in 1948, the second the only Southeast Asian country to retain its independence throughout the colonial period, share several characteristics and problems. Each has a sizable majority of its own national ethnics (the Thai are 82 percent of Thailand's population; the Burmans make up 68 percent of Burma's people). Each has minority peoples in strategic borderlands, but the minority problem is far more critical in Burma. Minorities are hill country tribespeople practicing more extensive agriculture, often of the shifting variety (see Figure 8–3). Much less concentrated than the people of the cores, they occupy far more territory than is reflected in their proportion of the national population. The Burman and (indirectly) Thai states were shaped by a European imperialism that determined their boundaries. Previous boundaries of the two had been influenced by Chinese expansion but were fluid and included tributary states of varying degrees of organization and loyalty. When Burma was attached to British India, boundaries were established with a view toward defensibility, not ethnicity. Thailand's boundaries reflect treaties with Britain and France based on imperial policies and objectives and not on who lived where.

Burma's densely settled upper core (centered on Mandalay) and lower core (the more recently settled delta area of the Irrawaddy, centered on Rangoon) (Figure 8–5) are occupied predominantly by Burmese-speaking, Buddhist Burman ethnics. The Arakan coast strip has cultural links with Bangladesh, including a large Moslem minority. The Shan states to the north and east enjoy considerable autonomy; they spread across boundaries from southwest China to Laos and Thailand as well as occupying part of Burma. There is little evidence that the Shan are effectively incorporated in the economic and political life of any state. Chinese influence may be strongest and could eventually cause problems if a Chinese-protégé "Greater Shan State" movement is fostered. The Shan are ethnic-linguistic relatives of the Thai.

The much older, more advanced cultures of the cores tend to disparage upland groups. For example, the Burmese label *Karen,* a designation for a group of tribes in the south of Burma, may be loosely translated as "slave-barbarians." These Karen have been in more or less continuous conflict with the central government. The Kachin of the far north are loyal, if not assimilated. In 1989, the name of Burma was changed to Myanma to include non-Burman ethnics in the proper name of the State.

Thailand faces similar, though less serious, problems. The Thai are a mixture of assimilated and unassimilated peoples, including the Shan (see Figure 8–3). The remote highlands, occupied by ethnic groups loosely controlled by or hostile to the central government, are not integrated into the national economy. Highlanders are notorious for smuggling and other illicit activities. The infamous "golden triangle" of rugged hills in the area where Burma, Thailand, and Laos converge has been a source of opium and heroin for decades; it is representative of the blurred boundary zones of uncertain loyalty between most mainland states. Malays and Vietnamese in southern and eastern Thailand complete the minority mosaic. This twilight zone of largely autonomous people is a simmering threat to both states.

Burma is still fairly dependent on the export of rice, making it literally a one-crop economy, highly vulnerable to fluctuations in world market prices. It has the potential for diversification; cotton, timber, rubber, and important (but underdeveloped) lead, zinc, and tungsten resources could make it one of the more prosperous developing nations, though its oil fields are near exhaustion (Figure 8–6). A growing population has reduced exports of rice. Diversification is necessary to even maintain the level of national income, much less raise it. Offshore drilling is underway, with tracts leased to foreign oil companies, a major departure from a former policy aimed against foreign investment. Burma is eager to maintain its tradition of neutrality and nonalignment,

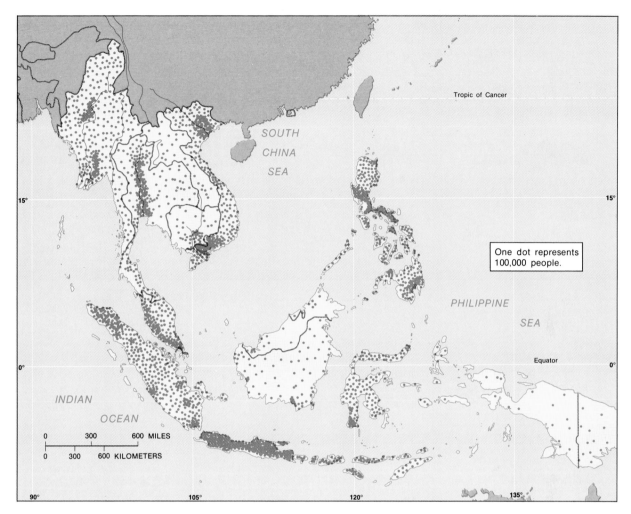

FIGURE 8–5
Population Distribution in Southeast Asia. Southeast Asian population
concentrations tend to reflect agricultural productivity. Note the extremely uneven
dispersal of people in this region.

though official policies may change as a result of the
upheavals of 1988–89.

Thailand, in contrast, is pro-Western and relies
more on private enterprise. Over one third of the
country is cultivated, and agricultural exports (rice,
sugar, rubber, corn) are foremost. Mineral production is expected to become much more important in
the near future. Several trillion cubic feet of natural
gas have been discovered under the Gulf of Siam,
replacing imported oil as fuel. Tin dredging is increasing, and forest product production has risen
sharply. Per capita GNP is almost $1,200 per year,

reflecting a high level of trade and consumption. An
impressive transport development program (both
highways and airports) has helped increase tourism.
Bangkok has become a major tourist destination, and
over 5 million tourists enter Thailand each year.

The Malay Culture Subregion

This subregion contains Southeast Asia's three
wealthiest states in per capita income: Brunei, Singapore, and Malaysia, along with much poorer Indonesia (see Table 8–3). The natural resource base is

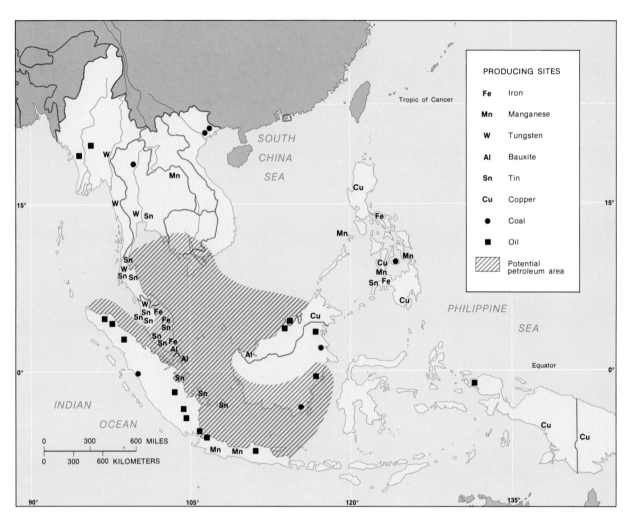

FIGURE 8–6
Mineral Production in Southeast Asia. Petroleum and tin are Southeast Asia's
major minerals; Indonesia leads in regional oil production and is second to Malaysia
in tin exports.

impressive. Indonesia's relatively low per capita income of $560 represents income from oil, timber, rubber, tin, and agricultural exports, diluted by a huge population (see Figures 8–5 and 8–6). With a population of over 170 million, Indonesia gives the impression of a densely populated, resource-rich, and potentially wealthy portion of the world. Malayan people, most of them Muslims, form the majority of the population in all of these states except the predominantly Chinese city-state of Singapore. The British and Dutch colonial empires effectively divided this Malay cultural-ethnic area among themselves

(excepting once-Portuguese East Timor and the Philippines). As a result, today's independent states essentially have borders conforming to colonial administrative limits. Nationalism, then, follows patterns established during colonialism rather more than strict ethnic and linguistic uniformity. The subregion consists of two small enclaves (Singapore and Brunei) at one extreme and two larger, but fragmented, political units (Malaysia and Indonesia).

Originally, Malaysia consisted of many separate colonial units. Malaya was a federation of small states that encompassed the strategic Malay Peninsula and included the fortified base of Singapore. Brunei, Sarawak, and North Borneo were three British colonial units on the northern edge of the otherwise Dutch-controlled island of Borneo.

The Chinese's economic success and relatively large numbers have made them a major concern to the Malays, who do not allow the Chinese to vote. Counterbalancing the large Chinese population has been a major goal in the portions of the subregion that were once British controlled.

British Malaya was an exceptionally prosperous colony. The mining of tin and development of extensive rubber plantations used large numbers of imported labor from China and India. As a result, the mainland portion of the onetime colony of Malaya is almost equally divided between Malay and non-Malay elements. When independence seemed imminent, the question of the Chinese was in the forefront. Singapore was in the original Malaysian Federation from 1963 to 1965, when it withdrew under pressure from the Malays. In a continuing effort to counterbalance resident Chinese, Malaysia incorporated former British colonies on the north shore of Borneo. Malaysia has over 16 million people. Some 36 percent are Chinese, and 10 percent are Indians and Pakistani. A variety of other groups are present, so that Malaysians still constitute a bare 50 percent of the total.

Initially, Indonesia was hostile to the new country on the grounds that is was a relic of imperialism that should have been merged with Indonesia into a single, Malay culture state. Such a merger would have given Indonesia the foreign-exchange earning power of Malaysia and added some potentially cultivable land. Indonesia's government pursued a policy of confrontation. It is possible that Indonesia will continue to covet East Malaysia (northeast Borneo) and Brunei as population pressures mount.

Malaysia's mainland is still dominated by the mining and plantation agriculture sectors. More diverse than formerly, peninsular Malaya produces iron ore, large amounts of bauxite and rare metals, gold, and tin (see Figure 8–6). The production of rubber, a traditional export now in decline, is being augmented by the increased production of palm oil, coconut oil, and tea.

There was little industry other than first-stage processing of raw materials until recently. During the

Bangkok Skyline. *Exotic yet modern Bangkok has become one of the great international tourist destinations, contributing to (and reflecting) Thailand's impressive economic growth in recent years.*

past decade, Japanese and U.S. firms invested heavily in electronics assembly plants, clothing and shoe factories, and other labor-intensive industries. Kuala Lumpur has an auto assembly plant that plans to market cars in North America.

Manufactured products now account for over one quarter of Malaysia's export earnings as it moves into the "upper middle-class" rank of national economies according to the World Bank. Malaysia's economy is booming to the point that this is one of the few developing nations that does *not* encourage population limitation schemes; the view here is that more people mean a larger labor force, which means accelerated economic development.

The territories of Sabah (once North Borneo) and Sarawak add little to the fortunes of Malaysia other than their function as dilutors of the Chinese minority. Most of their exports duplicate those of peninsular Malaya. Oil from Brunei is refined in Sarawak and exported as petroleum products.

Brunei's outstanding prosperity reflects a small population (only about 220,000) in relation to a large oil reserve, a situation similar to that of some Persian Gulf sheikhdoms. Brunei opted not to join Malaysia in 1963 to protect its high standard of living. Brunei would seem highly vulnerable to pressure from Indonesia.

The city-state of Singapore must import most of its food; a large-scale fishing program has been instituted to help increase its food supply. Over 75 percent of the population is Chinese. Almost 2,000 factories employ nearly half a million people in manufacturing; businesses, services, and transportation employ an additional 200,000 each in an economy estimated to be growing at almost 10 percent per year. Singapore refines and exports petroleum products from Indonesian oil; produces a host of electronics goods (mainly in Japanese-owned subsidiaries); exports crude, Malaysian-produced rubber; and builds and repairs merchant and military ships. Its machinery and transportation equipment are distributed throughout Asia. Singapore is the world's second busiest port (after Rotterdam); an important **entrepôt**, it acts as a collector and distributor of in-

ternational trade for Malaysia's scattered ports (Figure 8–7). Because of its political stability, Singapore has become an important international banking and finance center—an "Asian Switzerland."

Hosting more visitors each year than its number of citizens, Singapore is a major world tourism center thanks to its busy airport and duty-free shops. A classic example of a small, crowded island whose only resources are relative location and the enterprise of its hard-working people, it has achieved a much higher standard of living than oil-rich Indonesia, 3,000 times its size. By most criteria, Singapore is Southeast Asia's only developed country.

With an area of almost three fourths of a million square miles, Indonesia is by far the region's largest political unit. It comprises all or part of five major islands and thousands of smaller ones. The archipelago stretches 3,500 miles from east to west and almost 1,200 miles from north to south. Population distribution varies immensely. With over 100 million people in an area the size of New York State, the island of Java is one of the most densely populated

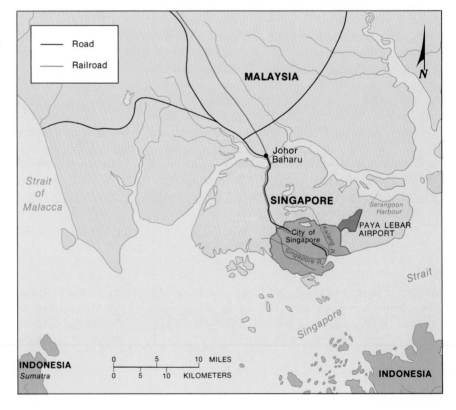

FIGURE 8–7
Singapore: Southeast Asia's City-State. Singapore thrives as a commercial city—an entrepôt—which has become a major industrial and banking center. Singapore's small area and rapid development have made it a carefully planned and regulated urban area.

Singapore Skyline.
Singapore, a world-ranking producer of consumer electronics, computer software, and information services, has prospered through a unique blend of capitalism and government planning.

areas in the world (see Figure 8–5). The so-called outer islands are much less densely populated, while Irian Jaya (western New Guinea) is virtually empty.

Known as the Indies or the Spice Islands, Indonesia was one of the most coveted pieces of international trade territory. The Dutch replaced the Portuguese as the colonial ruler in the seventeenth century, gradually extending their control to the outer islands from their original base in Java.

The world's sixth largest country in area, Indonesia ranks fifth in population. Its rather strong sense of nationhood, given its internal diversity, rests on a common language, the common experience of Dutch colonialism (except in formerly Portuguese East Timor), and Islam, shared by the vast majority of its citizens. The official language, Bahasa Indonesia, is based on "Market Malay," the lingua franca of traders throughout the archipelago, rather than on the language of Java or any other particular island. Because the language is not associated with any one region or group, it is acceptable to all. All official business and instruction in secondary schools is in Bahasa Indonesia, so the language is widely understood even among people who also speak a regional dialect.

Each of the religious "invaders" of Indonesia—Hinduism, Buddhism, Islam, and later, Christianity—has been modified and blended into ancient customs and beliefs. The Indonesians have a popular saying that, "Religion comes from the sea, but *adat* [customary behavior] comes down from the mountains." While some Indonesian Muslims are fundamentalist and militant, about half are nominal believers who pray only on religious holidays and who participate in traditional celebrations with strong Hindu and Buddhist overtones. The island of Bali remains predominantly Hindu, though without the rigid caste system of traditional Indian Hinduism. Although Islam is strongly opposed to idolatry in any form, the national emblem of Indonesia is the mythical bird Garuda, the mount of the Hindu god Vishnu. On the nation's coat of arms, Garuda holds in its talons a ribbon bearing the motto Unity in Diversity, a goal as much as a fact in this sprawling, immense nation.

Most soils of the outer islands are not particularly fertile. Java's soils (as those of southern Sumatra) are enriched by periodic volcanic eruptions. Erosion is a serious problem on slopes exposed to heavy, seasonal rains. Volcanic soils are only part of the explanation for Java's huge population. While the British set up plantations in Malaya by importing a labor force, the Dutch coerced the Indonesians into producing desired commercial crops. A large Dutch bureaucracy oversaw compulsory commercial crop production; Indonesians, especially on Java, were ordered to plant and deliver quotas of crops. Javanese peasants saw large families as necessary labor force

SOUTHEAST ASIAN CONTRASTS IN DEVELOPMENTAL PACE

Singapore is one of Asia's rapidly developing "four smaller dragons" (compared to Japan, the "great dragon," or great industrial economy and international trader), and the only one in Southeast Asia; the others are South Korea, Taiwan, and Hong Kong, in East Asia. Economic contrasts among the states of Southeast Asia are as great, or greater, than those among the states of other world regions.

In his landmark book, *Global Shift: Industrial Change in a Turbulent World,* Peter Dicken notes:

> The complexity of global industrial change has rendered simplistic notions of "core" and "periphery" rather less useful or capable of capturing the nature of today's global economy. The world is more a mosaic of unevenness in a continual state of flux than a simple dichotomous structure of core and periphery. . . .There are peripheries within the core, as in Western Europe, for example, and there are cores within the periphery, as in Southeast Asia.*

To the categories of OICs (older industrialized countries—most of Europe, the United States, Japan, the USSR), and NICs (newly industrializing countries—South Korea, Brazil, Mexico, etc.) must be added LICs (**less industrialized countries**—Burma, Bolivia, Zaire, Kampuchea, etc.). For many OICs, the primary economic problem is maintaining high employment rates as continuing mechanization and automation permit expanding production with fewer workers, and labor-intensive industrial production shifts toward Third World NICs. For the LICs, problems of coping with rapid population growth and extreme poverty command priority. In NICs such as Singapore, the essential problem is maintaining momentum in economic growth. With no natural resource other than its location, Singapore must emphasize a diligent, disciplined work force and attract a continuing inflow of foreign capital. No taxes or restrictions are imposed by Singapore on "repatriation" of profits on foreign investments; this policy is evidently successful. Singapore's ongoing success requires the continuance of an almost free trade privilege with all the states of South, East, and Southeast Asia.

*Peter Dicken, *Global Shift: Industrial Change in a Turbulent World* (London: Harper & Row, 1986), pp. 390–91.

for meeting the commercial crop quota and also growing food. Rural areas were thus forced into the commercial economy. This system was never as strong on the outer islands as it was in Java.

As the first in the region to achieve independence, Indonesia began its independence with considerable prestige among the nations of the developing world.

Three days after the Japanese surrender, an independent Indonesia was proclaimed. The Dutch, exhausted following a brutal and destructive Nazi occupation, could not reimpose control.

An early focus of the new Indonesian nationalism was western New Guinea, which, though ethnically and culturally different, had been part of the Dutch

East Indies. It was finally relinquished by the Dutch and added to Indonesia in 1962.

There is certainly sectionalism within the framework of Indonesia's 13,000 islands. The lesser-populated, export-earning islands of Sumatra, Borneo, and Celebes resent the fact that they export tin, oil, rubber, timber, natural gas, bauxite, and copper, in effect supporting densely populated Java (see Figure 8–6). Java's large population plus the control of the location of the capital (Jakarta) add up to control of the entire republic, including economic development policies and government spending. Resentments in the other islands weaken Indonesian unity. The government has galvanized nationalistic drive by repeatedly focusing national energies on land acquisition in Irian Jaya or on Portuguese Timor (annexed in 1976), union with Malaysia (unsuccessful), and past threats to "liberate" the Australian-protégé state of Papua New Guinea.

Despite the overpopulation problems of Java and the difficulties of communication among the scattered islands and peoples who speak diverse dialects, the state is inherently rich and a national sense of identity is fairly well developed. Indonesia is a major producer and exporter of oil and natural gas. Reserves are huge, and new discoveries have recently added to the total. Production exceeds 2 million barrels per day. Most oil is produced in central Sumatra, but newer fields are producing in New Guinea, western and eastern Java, eastern Borneo, and offshore in the Java Sea. There are large producing gas fields in northern Sumatra and southeastern Borneo (see Figure 8–6). Much promising territory, both on- and offshore, remains unexplored.

Indonesia has large nickel deposits on Celebes, and new copper mines have opened in Irian Jaya. Bauxite is mined on a moderate scale and is now domestically smelted in northern Sumatra. Over a half-million tons of coal are mined from deposits that could be greatly expanded. Ironsands are exported now, and a domestic steel industry is planned. By far the most important metal mined is tin on the islands of Bangka and Billiton off Sumatra and on the islands just south of Singapore. The huge lumber resource was largely unexploited before 1967.

Manufacturing is beginning to develop rapidly. The government has planned a dual approach: large plants, processing raw materials, are to be located in the outer islands, and consumer industries will be located in labor-surplus Java. A third, unplanned industrial grouping is developing rapidly in the villages that outlie the dozen large urban centers (eight of which are located on Java). These are small, private enterprises of a cottage industry nature, producing consumer items for urban populations.

Major foreign investors produce motor vehicles, tires, chemicals, and pharmaceuticals. Diversity of ownership, scale, and production are the hallmarks

Voting Day, Jakarta.
Indonesia has moved successfully toward establishing a strong national identity in a huge archipelago with many divergent regional interests. This is an election day scene in the capital, Jakarta.

of Indonesian industry. Almost 5 million people are employed in industry, but most of those would be regarded as craftsman rather than industrial operatives in a developed society. Nonetheless, industrial progress is both real and impressive.

Agriculture has come to be a problem area. Certain of the old Dutch plantations have been retained as government-owned estates that specialize in high-quality export and industrial crops. Many others have been divided and used to resettle landless peasants or to augment small, highly productive peasant farms. Irrigation and large inputs of commercial fertilizer are increasing yields, but there is a domestic food shortage because peasant agriculture cannot cope with the rapid increase in population. Until recently, commercial agriculture has been in a state of steady decline. Subsistence agriculture is increasing on the outer islands.

The government approach to agriculture is also dual: diversification of export crops and resettlement of population to new areas in the outer islands. The first program has been rather successful as coffee, tea, citrus, sugar, and palm oil acreages increase. The latter has not been quite as successful. It is estimated that over 1.5 million migrants have been resettled in Sumatra, but less than one half of these have been agricultural resettlers. The rest have sought employment in mining, services, and industry. Almost 2 million former peasants have packed into the capital district. Only Borneo of all the outer islands seems to have had any significant net agricultural in-migration (see Figure 8–5).

Overpopulation of Java exists alongside labor shortages in the outer islands. Short food supplies in Java's countryside are met with imported food, while grain and fruits rot in the fields and on the docks of Sumatra and Borneo. Solutions may appear simple, but problems, like the country itself, are immense, diverse, and complex.

The Philippines

The nearly 60 million Filipinos sometimes boast that they are the third largest English-speaking state in the world. All secondary and college-level education is in English, one of two national languages (Filipino is the other). All educated Filipinos speak English, and over half the population understands and uses some English. Filipino, based on the most widely spoken native language, Tagalog, is taught in all schools, yet English remains important because of the close trade and cultural ties with the United States.

The Filipinos are mainly of Malay racial stock, though almost 15 percent intermarried with Chinese. At least a thousand years ago, Filipinos were in contact with the Chinese, and later, with Indian and Arab peoples. European "discovery" was in 1521, followed by Spanish conquest in 1565. The Spanish introduced Christianity, which remains the religion of more than 80 percent of the people. The Spanish also introduced their emphasis on large holdings; land was handed out to loyal conformists as a reward. These large estates were a part of the motivation behind the Huk Rebellion of the past Communist guerrilla movements, but, because most land reform was accomplished in the 1950s and 1960s, other issues now prevail.

The United States acquired the Philippines in its war with Spain (1898). Under American occupation with promised eventual independence, the Catholic church was disestablished, and church lands were redistributed. A Filipino-staffed civil service was created, and self-governing commonwealth status was granted in 1935. Greatly improved health care, educational, and political systems were also implemented, contributing to strong loyalty to the United States. The Philippines distinguished itself as the only colonial territory in Asia that remained loyal against the Japanese invasion, and complete independence followed in 1946.

American reforms distributed much land, but in 1946, almost half the farms, occupying about 40 percent of the total acreage, still were operated by tenants or part-tenants. The typical split of farm income was half to the tenant, half to the landowner; tenants were deeply in debt. Crushing burdens of debt, coupled with a strong sense of helplessness and hopelessness, have been receptive ground for rebel movements. The first rebellion was put down as in 1953, but resistance groups again grew in strength under the oppressive Marcos regime's 20-year tenure. Rapid expansion of the population (2.5 percent yearly) cripples the ability of the government to provide land and livelihood for all.

The Philippine economy expanded rapidly just after World War II. A free-enterprise-oriented economy attracted American investment, particularly in forest products and minerals (iron ore, copper, chromite). A "birthday gift" to the newly independent state from the United States was guaranteed preference in American markets for sugar. Thus, political independence has not meant immediate or complete severance from the American economy.

Economic development strategy rests on four objectives: self-sufficiency in basic foodstuffs (rice, corn), accelerated exports of tropical specialty crops such as sugar, pineapples, coconut products, palm oil, forest products (the Philippines is one of the world's largest exporters of logs, lumber, plywood, and veneers), increased exploitation and export of minerals, and rapid industrialization. The Philippines has attempted to reduce imports with domestic consumer products, and to diversify into cement, steel, fertilizer, chemicals, and refined petroleum products.

The Marcos regime began auspiciously enough as a reform movement. During its early years, government policies attracted considerable U.S., Japanese, and European investment. Shortly, however, the military leadership and influential members of Marcos's political party began to accumulate land and wealth, developing into an economic elite. This elite invested in personal conspicuous consumption and land acquisition.

In 1986, the widow of assassinated opposition leader Benigno Aquino became the first woman executive of the Philippines, despite attempts of Marcos's political forces to falsify election returns. This somewhat unanticipated, relatively peaceful victory for the opposition reflected the deep disenchantment felt by the public toward the corrupt Marcos regime.

The new government is proceeding cautiously, emphasizing elimination of corruption while courting more U.S. aid. There are no new programs for development, though land reform and continued industrialization are significant, and constantly restated, goals. Rapid industrialization in concert with continued agricultural improvement will be necessary to cope with the growing population. Resettlement of Filipino nationals to the underutilized island of Mindanao, a project ongoing for 50 years, is likely on a relatively small scale, but not without concessions to that island's Moslem minorities (see Figure 8–3).

The population of the Philippines is highly educated by regional standards, and there is a relatively skilled labor force that has not been fully utilized. Rural overpopulation in the central islands is critical. The known resource base, although diverse, is not large, and fuels constitute a critical shortage. Food supplies are inadequate, but physiological densities are not high. Despite shortcomings, the economy shows promise for future development.

SUMMARY

Within half a century, Southeast Asia has gone from a region dominated by European and American colonialism to an arena of competition among the United States, the Soviet Union, and the PRC. The region is now fragmented into neutralist, pro-Western, pro-Soviet, and pro-Chinese states. These varying orientations and allegiances have been known to change drastically with changes in Southeast Asian governments.

China has long had active political interests in Southeast Asia, along with the strong cultural ties maintained by the overseas Chinese. Vietnam is China's major opponent in the region, having fought a brief war with the PRC and established close ties with the USSR, which has the use of the large U.S.-built naval base at Cam Ranh Bay. This naval base is approximately opposite the major U.S. base at Manila Bay across the South China Sea; it is a potential threat to both the United States and China.

At least while the United States and the PRC share an objective of counterbalancing Soviet power, the extraregional political relationships and tensions seem to have decreased in number from three to two. It is unlikely, though, that PRC and U.S. interests in mainland Southeast Asia can remain congruent for long.

ENDNOTES

1. As quoted in Barbara Tuchman, *The March of Folly: From Troy to Vietnam* (New York: Knopf, 1984), p. 235.
2. Robbins Burling, *Hill Farms and Padi Fields: Life in Mainland Southeast Asia* (Englewood Cliffs, NJ: Prentice-Hall, 1965), p. 107.
3. Tuchman, *The March of Folly,* p. 374.

REVIEW QUESTIONS

1. Why will forest destruction accelerate the problems of managing tropical soils?
2. How have the overseas Chinese contributed to economic development in the region? to internal discord?
3. Briefly describe how some mainland states have valley cores surrounded by peoples "outside" the national economy.
4. Where are the important straits into the "Asian Mediterranean"? Why is their control still so vital to industrial powers outside this region?
5. Why is Southeast Asia characterized as a cultural shatterbelt?
6. Why were the northern and southern cores of Vietnam complementary?
7. What were the geographic reasons for the greater colonial impact on Vietnam than on Kampuchea or Laos?
8. Why is the golden triangle border area of Thailand, Burma, and Laos essentially outside national control? What are the implications for these states?
9. What political considerations, based on cultural geographic factors, led to the independence of Singapore as a city-state?
10. Briefly review the geopolitical implications of Vietnamese historic (and continuing) animosity toward China.
11. How does Indonesia's extremely uneven population distribution affect its developmental strategy and viability as a state?
12. How does the intense social and economic interaction of some Southeast Asian villagers retard pioneering in more mountainous areas?
13. How did conflicting colonial ambitions enable Thailand to retain independence as a buffer state?
14. Why would a reputation for political stability be essential to Singapore's development strategy?
15. Which of the region's states is most likely, in terms of population, resource base, and level of development, to assert extraregional influence and power?
16. How can heavily populated, developing states such as Indonesia effectively use their populations as development assets?
17. Why were the Philippines the only Southeast Asian colony to remain loyal to the colonial power during Japanese occupation in World War II?

SUGGESTED READINGS

Dutt, Ashok, ed. *Southeast Asia: Realm of Contrasts*. Boulder, CO: Westview, 1985.

Fisher, Charles A. *South-East Asia: A Social, Economic, and Political Geography*. New York: Methuen, 1966.

Fryer, Donald W. *Emerging South-East Asia: A Study in Growth and Stagnation*. New York: Wiley, 1979.

Fryer, Donald W., and Jackson, James C. *Indonesia*. Boulder, CO: Westview, 1977.

Lee, David. *The Sinking Ark: Environmental Problems in Malaysia and Southeast Asia*. New York: Heinemann, 1980.

Spencer, Joseph E., and Thomas, William. *Asia, East by South: A Cultural Geography*, 2d ed. New York: Wiley, 1971.

Wernstedt, Frederick L., and Spencer, J.E. *The Philippine Island World: A Physical, Cultural, and Regional Geography*. Berkeley, CA: University of California Press, 1967.

Wolters, Oliver W. *History, Culture, and Region in Southeast Asian Perspectives*. Singapore: Institute of Southeast Asian Studies, 1982.

CHAPTER NINE

Africa South

Unique wildlife and splendid scenery are an important natural resource in Africa South.

There is no strong agreement among geographers on an appropriate name for this region. Black Africa, Sub-Saharan Africa, Intertropical Africa, and Africa South of the Sahara have all been used as titles to underline the separation of this cultural region from the rest of the continent. Race is not the critical factor. The designation Black Africa is applicable only in the sense that the region's population is composed primarily of darker-pigmented individuals. This pigmentation, thought to have evolved in response to the intensity of local solar radiation, has no meaning in the development of culture or economy. Further, black Africans are strongly represented in three other world regions, and other darkly pigmented indigenous peoples reside in Australia and parts of Asia. The origin of the term *Black Africa* lies in the traditional distinction made between two parts of Africa by Arabs: the "Beidan," the land of whites, and the "Sudan," the land of blacks. The color of the inhabitants was simply a convenient shorthand. Important and implied in the terms, however, was an enormous difference in the way in which the two parts of Africa made a living. Beidan was the land of the desert, irrigated agriculture, and herding of sheep and goats. Sudan was the land of rains, unirrigated agriculture, and the herding of cattle. This basic difference persists, separating Africa South from the drylands region to the north.

To define a characteristic, near-uniform set of cultural traits that apply regionwide is a virtual impossibility. One can easily identify cultural elements that are distinctly African, but they are as varied as Africa South itself. Western ideas about African culture often stem from contact with African cultural forms as expressed in the New World. These elements of art, architecture, music, food, custom, and religious practice are of undeniably African origin. However, they generally represent the particular culture area in which they originated and from which they were transferred; they are not often characteristic of the entire region. The austere lifestyle of the Bushmen hunters of Namibia contrasts sharply with that of the gregarious, commercially oriented Senegalese.

What of standard cultural elements such as language and religion? Many Africans have been converted to one of several major world religions. Even where traditional religions are maintained, there is enormous regional variation in observance and practice. Many shared basic elements of traditional African religions exhibit strong similarities to indigenous religious practices outside Africa; coming-of-age initiations, fertility rites, and cleansing rituals are common to most native religions. In much more stylized forms, they are also a part of religious observances among practitioners of major world religions everywhere. There is nothing so unusual about traditional African religious observances that would make them a basis for regional differentiation.

With languages, there is a fairly high degree of superficial uniformity over much of the region. The great Niger-Congo family of languages dominates (Figure 9–1). Individual languages, however, number in the hundreds and, as with the great language families of other parts of the world, individual members of the same family are not generally mutually intelligible. Khoisan, the continent's most ancient tongue, and the Sudanic family, spoken among the cattle-rearing cultures of parts of the region, are clearly African, though totally unrelated to Niger-Congo tongues, or to each other. The Hamitic tongues, once widely spoken, are now represented only by remnants, replaced by their distant cousins, the Semitic languages, including Arabic. Introduced through conquest, trade, and religion, Arabic is widely spoken in parts of the region as a language of trade. Migrations brought yet other Semitic tongues to the African Horn. European tongues, the languages of the colonial rulers, are either official or co-official in most of the region's states.

If the conventional cultural elements used to distinguish other regions are inadequate in this case, yet Africa South is more than sufficiently unique to constitute a separate region, what are its common characteristics? Certainly, there is a degree of uniformity in the physical environment. It includes a strong seasonality to rainfall over most of the region. In an area dominated by agriculture and where irrigation is not widely practiced, this pattern is of great significance. Moreover, rainfall is erratic over the bulk of Africa South, creating an air of uncertainty and almost guar-

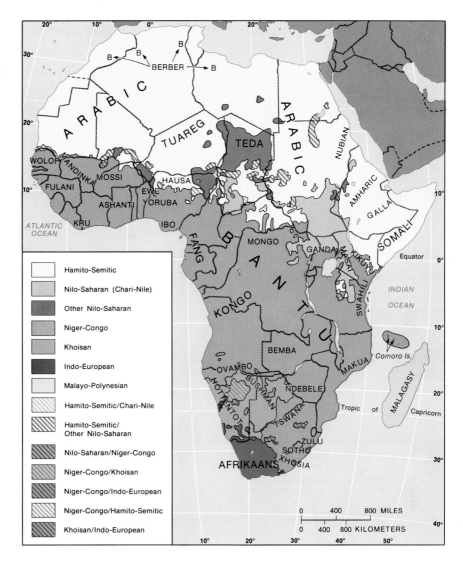

FIGURE 9–1
Languages of Africa. There are over a thousand individual languages spoken in Africa, but fewer than 50 are spoken by over a million people. The Niger-Congo language family dominates over most of Africa South, and one subfamily, Bantu, is spoken over almost half the region. Patterns of language, religion, and tribal territory coincide with political boundaries.

anteeing periodic droughts. Large areas of the continent have soils of low fertility, and the topography is most generally that of a plateau (Figure 9–2). Each of these physical elements has had some influence on the region's culture and economy. Each also occurs elsewhere; even their combination is not unknown.

Important is the sense of Africa South as a zone of transition. The region contains both deserts and rainforests, but most of it is transitional between the two. In response to this transitional and unreliable climate, Africa South appears to be in a state of constant economic flux, alternating between periods of starvation and plenty. The grazing economy, better suited to dry areas, competes with the crop agriculture of wetter areas almost everywhere. Herders and farmers battle each other, and nature, in an attempt to extend their respective domains. There are multiple, scattered, small cores with high population densities in a matrix of otherwise lightly settled terrain. Concentrations of dense rural populations occur where the twin advantages of fertile soil and consistent adequate rainfall coincide. The region is also characterized by large population voids in environments unsuitable to farming or intensive grazing. Yet, most of the region's area is transitional between these two population extremes. The region even seems to be in a state of economic transition, with

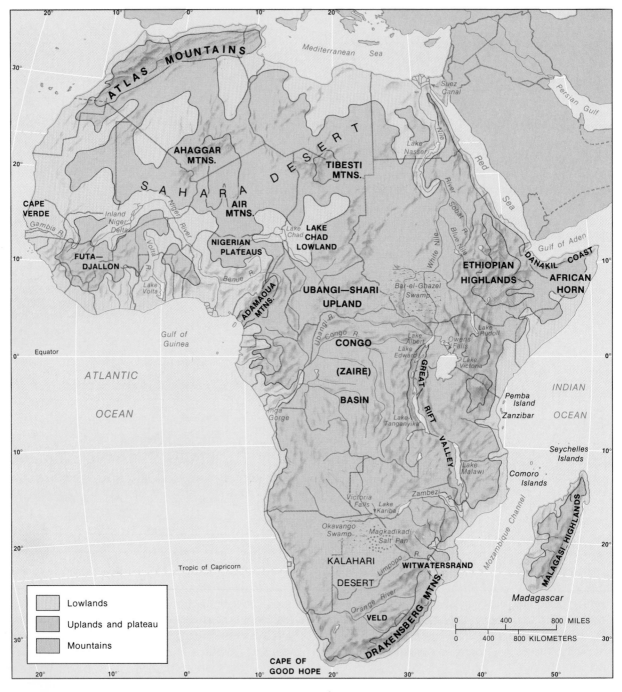

FIGURE 9–2

Topography and Drainage in Africa. Most of Africa's interior is a plateau
surface with elevations of 2,000 to 5,000 feet above sea level. Narrow fringing coastal
plains offer few good harbors, and penetration from coastal locations into the
interior is often difficult. Mountains are largely peripheral. The most prominent
ranges follow along the Great Rift Valley from the Red Sea to the Cape of Good
Hope. Four rivers—the Nile, Zambezi, Zaire, and Niger—drain most of the
continent.

the primitive and the ultramodern, the traditional and the technically advanced, all visible in the landscape. The great hope of Africa South is that it is in the process of transition from the underdeveloped to the developed.

An outstanding influence on the region's cultural separateness is its almost literal physical separation from all other regions by either vast ocean stretches or inhospitable Saharan wastes. Yet, isolation was never complete, even in the earliest of human times. Evidence discovered to date strongly indicates that East Africa was the original home of the human race; all populations of the earth diffused from this area.

The role of physical barriers in the diffusion process can be easily overemphasized if not analyzed with care. Goods, ideas, and people have crossed the Sahara since ancient times. But the movement was infrequent, and there was a selectivity to what was carried. Only the most costly and desirable goods made the difficult journey. Religion, however, without weight or bulk, easily diffused southward from the Arab-Islamic world. Saving souls obviously outweighed more worldly concerns at the time. Earlier, Christianity ascended the Nile from Egypt, converting the people of parts of East Africa. The transfer of these "outside" religions took place approximately 1,000 years ago in the case of Islam and about 1,500 years ago in the case of Christianity. Their spread seems to have been halted only by the rainforest, a physical barrier that took longer to conquer.

The crossing of the ocean barrier long presented a serious problem. As early as 500 B.C., Phoenicians may have reached the coasts of tropical Africa, but apparently made no landings. The first Europeans to circumnavigate the continent saw it as something to get around rather than as a place to establish contacts. Sustained contact with other cultures awaited the era of maritime commerce. When this era came, economics had outdistanced religion in the minds of commercial agents. Slaves were the primary commodity of interest to a class of traders bent on supplying labor to New World plantations. Cultural exchanges were minimal; African products were of no concern. Europe became interested in these products only after the end of slavery and the loss of its New World colonies. By the mid-nineteenth century, Africa was perceived as a potential replacement source for tropical raw materials, and full, constant, and consistent cultural contact was achieved. The two

culture realms, however, were not equal in European eyes. Africa was to *receive* European culture; Europe was not interested in receiving African culture.

The experience of colonialism has contributed an element of regional cultural commonality. In Africa South, colonialism lasted only a century, a far shorter time than in the New World. It did not involve wide-scale settlement of Europeans. Europeans were convinced of their cultural superiority; in terms of technology, that was indeed the case. It is why the stamp of colonialism is still so evident in the region's landscape. Europe "invested" in Africa, reorganizing the landscape to suit European aims and needs. It also reorganized Africa politically, for its own administrative convenience. The legacy of colonialism thus contributed to both the woes and prospects of the independent states that emerged in the aftermath of colonial decline. The colonial experience tended to magnify the differences between Africa and the rest of the world. Worse, it tended to breed a sense of separateness among cultures within Africa South.

There are multiple bases for classifying Africa South as a region. Certain elements of unity are convincing, including the region's sheer diversity.

Africa South is more than just south of the Sahara, or south of the Islamic world; it is south of the developed world—a crucial fact. Its internal conditions, however, are what is most important in distinguishing the region.

A REGION IN CRISIS

The average resident of this part of Africa is less well off than was the case 15 years ago. The World Bank estimates that the regional per capita GDP has declined by almost 5 percent since 1972. The U.S. State Department estimates that at least 20 million black Africans are "at risk"—in serious danger of starvation; another 20 percent of the region's people are malnourished. Africa South is the only major world region in which per capita food production has declined over the past two decades. Why? In its simplest form, there has been a drop in productivity accompanied by a rapid increase in population. The region experiences the world's highest rate of population gain, at over 3 percent per year; at the same time, its food supplies have decreased. Part of the region has been ravaged by prolonged drought. Grain imports have been increasing at a rate of 9 percent per year, not counting emergency food. Grain importation, however, is not confined to the states suffering drought.

Three quarters of the people in this region are directly employed in agriculture. It is the most important, yet most neglected, part of the region's economy. The present crisis has cultural roots that go back many years. Low, government-mandated prices for agricultural products in many parts of Africa South have discouraged expansion of food production, which the region so obviously needs. Drought and other weather disasters have driven millions from farms and grazing lands to the region's cities in search of employment and food aid. Urban populations have swollen to proportions far in excess of what might be expected. It is to placate the urban poor that many governments keep food prices artificially low, substituting cheaper imported grain, purchased on credits, for domestically produced food crops. Any rise in food prices would likely result in revolt. Devaluation of domestic currencies, a move that would reduce inequalities in prices paid for foreign versus domestic foodstuffs, raises the specter of defaulting on huge foreign debts.

Answers are not always simple. Return to a market economy in Malawi and Somalia has resulted in dramatic increases in production of food crops in those countries. Their success is not guaranteed in other nations. Foreign debt in Malawi was never excessive; there is no drought. A regime that does not tolerate opposition is firmly in control. Return to a market economy there could be accomplished without great disruption or dire consequences. Somalia receives huge amounts of foreign aid in its role as counterbalance to neighboring Marxist Ethiopia. Crop increases there were more the result of foreign aid and technical help than a return to a market economy.

Political crises and civil wars also are major causes of food shortages in Africa South. Crops in Uganda go unplanted as farmers' fears of sniper fire and undetonated land mines keep them away from their fields. The situation is repeated in many sections of the region.

Although the United States has pledged to provide half of all the region's food assistance, in the long run, Africa South will have to provide more of its own food. Africa desperately needs more agricultural research. World efforts to date have focused mainly on Asian conditions and environments.

Desertification (the "expansion" of deserts into the formerly more productive, semiarid fringes) is another growing problem. There are unpredictable dry and wet cycles of precipitation within the desert-savanna transition zones on both sides of the equator (Figure 9–3). Technology cannot remove this hazard. Desertification appears to be intensified by poor farming practices. A lack of conservation magnifies the effect of these natural dry cycles. Land better left in grazing has been plowed, exposing it to drying winds. Fragile balances of sparse vegetation and barely adequate moisture have been destroyed by overgrazing. Too many animals on the land have destroyed plant cover, increasing erosion. When vegetation is removed, sands migrate and engulf formerly productive land. Africans must develop better land management techniques to halt the desert's advance.

Governments have turned to borrowing in attempts to maintain living standards and modernize economies. Debt in the region increased at an average annual rate of 25 percent over the last decade. Many countries must devote over half their budgets to debt repayment. This heavy debt burden means that badly needed social services and necessary maintenance of roads and dams must be reduced. Banks project that debt service (paying interest and gradually paying off the principal) will use up 90 percent of some of the region's national budgets by the end of the 1990s. Clearly, that is the route to economic catastrophe.

It is a bitter irony that the most modern and most productive economy in the region—that of South Africa—is in political crisis. The state with the fewest monetary problems, and the greatest potential for developing new technologies that could benefit the region, is bogged down in civil disorder. Its rigid racial policies are being confronted by black South Africans with protests and economic boycotts. Even the most positive production conditions will be disrupted by unrest.

As strong and justified as the present "crisis" image of Africa South is, it should be seen in a larger view and in a longer historical context. The apparent end of the drought in large parts of Africa challenges the hasty conclusions of some environmentalists about "permanent modification of climates." The "darkest Africa" myth still persists in the minds of those who feel Western enlightenment is all that is needed to deal with the region's problems. Greater knowledge of Africa South is an obvious need. New technologies and new forms of economic organization may be called for to reverse ongoing regional crises.

For a time, it appeared that Africa South would become the latest cold war battleground—not an actual military confrontation, but a contest concerning which system was best in terms of developing still-undeveloped countries. The contest ended in a standoff. Apparently, neither system has had much success. The African milieu simply does not respond readily to Euro-American technology, in either its Western or Eastern variants. Political ideologies do not provide answers to environmental dilemmas.

There is nothing basically wrong with Africa South. Its problem is that there is no *one* answer to its problems; nor is there any definite beginning point. The correct answer is "all of the above," and the time framework is "all at once." There must be a concerted effort by the region's states and the world's developed nations to work toward positive solutions.

A REALISTIC OVERVIEW

A frank examination of the realities of Africa South is necessary. Africa is physically quite different from most other regions. It is tropical, but more dry than wet. A dominantly wet Latin America, or a dominantly dry, but more temperate Australia, do not provide climatic parallels. Africa's environment is unique. Weather disasters are a normal part of that environment, and soils vary greatly in their productive potential. Cultural differences also must be considered. Cattle are highly valued in most of the region's native cultures; substitution of other livestock for cattle could prove difficult. Currently, there are areas devoted to livestock that could support crops, but crop farming is an unacceptable lifestyle to some of the region's cultures.

The production of export crops is identical over large areas; new acreages will not help. Africa is not suited to many crops grown elsewhere, because varieties from other areas do not always adapt well to African conditions.

Birthrates are very high, and continued high growth rates can be anticipated for decades to come. Food assistance and health care programs reduce death rates but, inadvertently, also contribute to increased growth and population pressures. Governments in the region are in the unenviable position of having to decide between investment in social programs and investment in developmental ones.

Africa South is victimized by endemic diseases: yellow fever, malaria, sleeping sickness, bilharzia, hookworm, and now, AIDS (Figure 9–4, p. 440). A massive program for the eradication of these diseases is necessary before there can be orderly progress. Pest eradication will in some cases make increased food supplies more possible, by increasing worker productivity and by opening new lands. Such measures would also lead to increased growth, but to improved public health as well. If these measures are coupled with an improved domestic agriculture, the land can absorb projected growth.

Africa's political boundaries, drawn largely by European colonial powers in 1870, do not match the realities of ethnic, linguistic, tribal, religious, or economic distributions. Distaste for colonial powers was the unifying force in each country's nationalism. When colonialism disappeared, nationalism foundered on the rocks of internal diversity. A redrawing of borders, formation of economic federations, or educational programs designed to intensify national loyalties within existing countries are necessary steps.

Existing infrastructure is inadequate. The natural plateau structure of the continent actually creates difficulties. There are few good natural harbors; gorges and steep-sided valleys incised into the plateau complicate road and rail engineering and raise construction costs. There are few transcontinental links; those that exist cross multiple political boundaries along their routes. Lines of transportation often are open to sabotage by dissidents. An improved, integrated transportation system is necessary to increase domestic flows of food and goods and to reduce shipping costs.

The region is rich in metallic ores, but many of its countries suffer a shortage of fuel. Africa South must

FIGURE 9–3

Desertification. Rapid
population growth, increasingly
large livestock herds,
deforestation, and bad
agricultural practices have
contributed to an expansion of
Africa's deserts. Many scientists
suggest that these secondary
causal factors are less important
than ongoing climatic change
on a global scale. Prolonged
and recurring droughts in Africa
over the last 30 years have been
used as evidence to justify both
schools of thought. Because of
Africa's latitudinal location and
physiography, large parts have a
desert climate independent of
either climatic change or the
consequences of human activity.

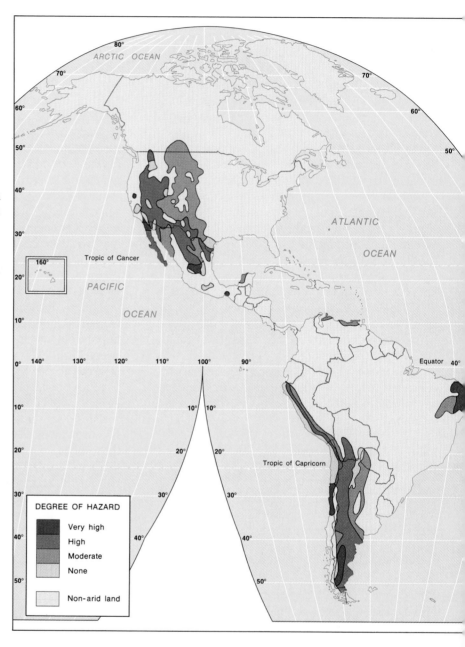

develop alternative power sources: hydroelectric resources, or even nuclear energy. Without cheap fuel
and energy, the transition to development will be
immensely difficult. Overall, the continent has been
only selectively prospected. There may be a lot more.

The labor force is large, but rates of literacy are
generally low. Skilled labor is in short supply. Given
its inadequate transportation, Africa South can com-

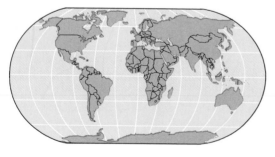

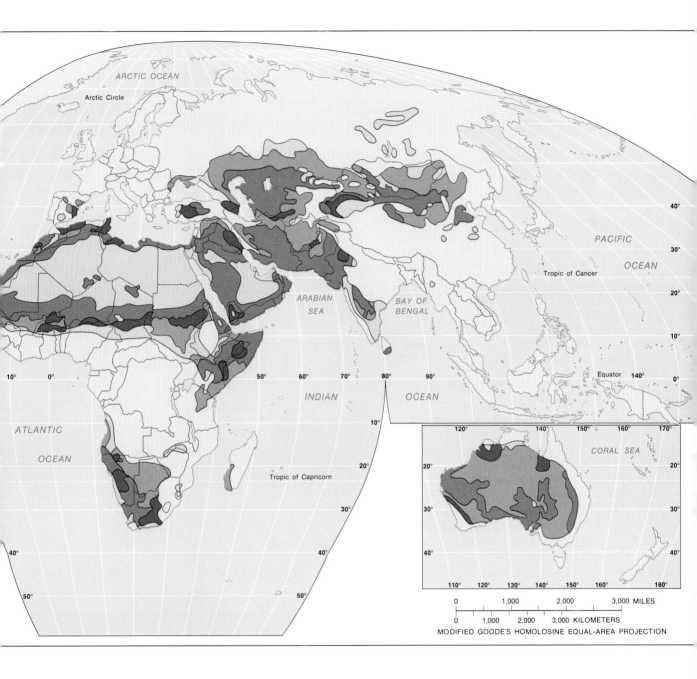

pete only in "cheap labor" areas of manufacturing at a few favored locations. Advantages of cheaper labor costs can be lost in down time without good technicians, or swallowed up in excessive transportation costs.

There has been some progress. Pan African organizations meet and make plans for cooperative programs. Some programs are actually underway. Inter-

national agencies have begun to fund important tropical agricultural research.

Africa South has considerable unused potential. There is vacant land, which, if not highly productive, is usable. Better land, already in crops, can increase productivity. The continent is basically dry, but a reallocation of unused waters could do much to alleviate drought and increase irrigation.

A REALISTIC OVERVIEW | **439**

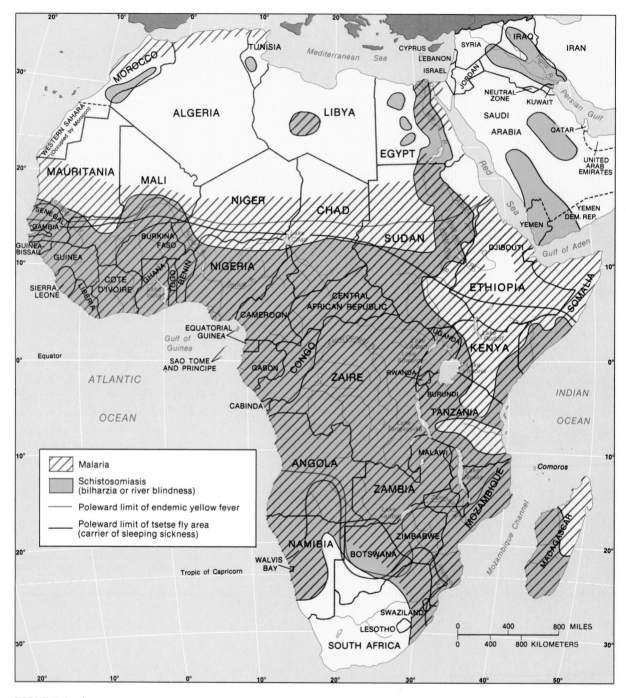

FIGURE 9–4

Endemic Diseases in Africa. Widespread occurrence of diseases and disease-carrying pests have inhibited development over large parts of Africa. The tsetse fly carries parasites that spread sleeping sickness to humans and livestock. Large areas of swamp and seasonally standing water serve as breeding grounds for mosquitoes that carry yellow fever and malaria. River blindness is carried by snails and shellfish that live in sluggish rivers and irrigation ponds. Only the northern deserts and the temperate South are free of these threats to public health. Low population densities in several parts of Africa South are attributable more to these infestations than to an absence of water or to soil infertility.

THE AFRICAN COLONIAL EXPERIENCE

The colonial legacy is still an important influence in the area's economics and politics. Any map of Africa done before 1957 would have shown essentially the same independent states and colonies as a similar map done in 1913. Yet, by 1970, the bulk of the continent was independent (Figure 9–5c). The suddenness with which most of Africa achieved freedom from colonial control was matched only by the speed with which it was placed under that control in the nineteenth century.

The African Slave Trade: The Prelude

Until 1870, European interests in Africa were peripheral, literally and figuratively. Trade with Asia and the Americas was far more profitable. The region's main export from 1500 to 1800 had been people in chains. It is estimated that 30 million people were taken from West Africa alone, and 50 million from the entire continent, during the period of intensive European participation in the slave trade. Slaving both resulted from and encouraged tribal warfare; prisoners were sold to coastal stations maintained by Europeans. Intertribal wars related to the slave trade depopulated whole districts.

Some of the first land claims extending into the interior of West Africa were made on behalf of freed slaves. After Britain abolished its own slave trade (1807), it embarked on a crusade, with the zeal of a repentant sinner, to eradicate the trade entirely. British naval vessels patrolled West Africa's coast, intercepting slave ships, freeing prisoners, and setting them ashore at Freetown in Sierra Leone. In 1816, the American Colonization Society was granted a congressional charter to send freed slaves back to West Africa. The first ex-slaves from the United States arrived at present-day Monrovia in Liberia in 1822. In 1847, Liberia became the first independent republic in Africa.

The African Land Rush

Events in Europe triggered the nineteenth century race for territory in Africa. The creation of a united Germany (1860–71) resulted in German determination to enter the race for tropical colonies. France and Britain also acquired huge new African colonial

territories (Figure 9–5a). The major industrializing powers had decided to develop legitimate trade, ostensibly to compensate both Africans and Europeans for abolishing the slave trade. Real motives were far more practical. Tropical vegetable oils were in high demand in Europe, particularly for manufacturing soap and candles. Native species of oil palms flourished in large parts of Africa. Cotton supplies from America were reduced by the Civil War and the slow, painful Reconstruction. Cotton production in Africa could resolve supply problems in the European textile industry. European powers, lacking crucial raw materials for full self-sufficiency, hoped to find mineral deposits in their colonies.

Africa in 1870 was a temptingly vacant area to Europeans. Their technological superiority in weapons, transport, and communications could make short work of African resistance.

The Political Legacy

The effect of colonial imposition of boundaries by European states has had lasting effects. Many states, especially in West Africa, are unrealistically small. Africa's larger territorial units often have had problems of internal unrest fed by ethnic diversity.

The actions of France in its colonies inadvertently reduced this problem of diversity somewhat. Cultural and political assimilation was the goal; there was no pretense of preparing colonials for self-government. It was assumed that, once introduced to French culture, all colonials would opt for French citizenship in an enlarged France. Basic literacy was emphasized, but was accomplished with a few years schooling exclusively in French. The results of these actions are mixed. Literacy rates in former French colonies are generally higher, and the citizens have a *lingua franca* for communication among themselves and for use in commerce. In some respects, this is an advantage, because language is not a basis for ethnic separatism in former French holdings.

The policies pursued by the British often resulted in greater diversity and larger problems. Using the time-honored technique of divide and rule, the British deliberately pitted one tribe against another. The result was tribal animosities that persisted as internal enmities after independence. These animosities are compounded by class consciousness. Mass literacy, when it occurred, was left to missionaries. The reli-

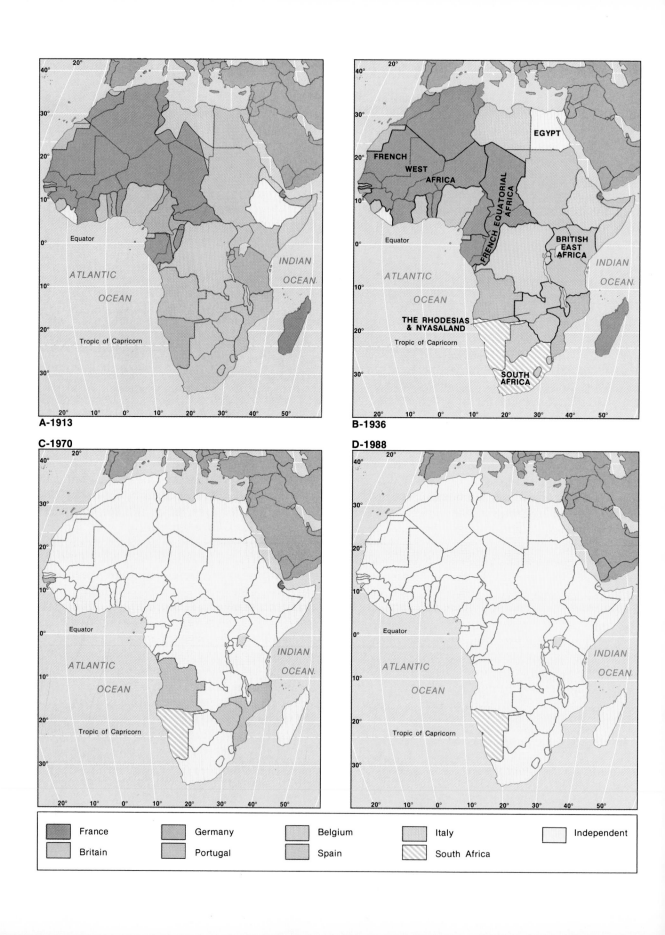

A-1913

B-1936

EGYPT

FRENCH
WEST
AFRICA

FRENCH EQUATORIAL
AFRICA

BRITISH
EAST
AFRICA

THE RHODESIAS
& NYASALAND

SOUTH
AFRICA

C-1970

D-1988

| | France | | Germany | | Belgium | | Italy | | Independent |
| | Britain | | Portugal | | Spain | | South Africa | | |

Developmental Enclave.
Colonial investments were most often concentrated in a few areas. One or two cities, generally capitals or ports, concentrated the bulk of urban development. These cities remain the best developed parts of African countries and continue to attract most new investment, both domestic and foreign. Nairobi, shown here, is typical. Tall, modern skyscrapers and landscaped parks welcome visitors to Kenya's capital. Unseen are the miles of slums and substandard housing that surround this opulent central business district.

gious and linguistic diversity that resulted has complicated matters.

Economic Problems in Common

Africa South has a dual economy: one the precommercial, subsistence agricultural economy, and the other an export-oriented production of plantation and mine. The former employs the bulk of the population yet contributes little to economic growth. The latter, introduced under colonialism, employs relatively few people but provides most of the capital for development. While this is a normal pattern for underdeveloped areas, nowhere is it so startlingly visi-

← FIGURE 9–5
Colonialism in Africa. Seven European powers once controlled virtually all of the African continent. Most of the continent was parceled out among European powers in 1870 at the Conference of Berlin. (a) By 1913, Europeans had removed the last vestiges of Turkish control, and only two independent states remained—Ethiopia and Liberia. (b) In 1936, even Ethiopia was conquered. The number of independent states remained at two as Britain dropped its protectorate over Egypt (1922). Beginning with Ghana in 1957, African colonies received independence. (c) By 1970, only a few remnants of colonial control remained. (d) Currently, Namibia, controlled by South Africa, is the only remaining colonial holding in Africa.

ble as in this region. Each individual economy is geared to the outside; transport systems are separated from those of other states and are almost solely limited to moving export goods (Figure 9–6). High-demand items from the richest mineral deposits or most efficient plantations seem the only commodities shippable because of cost and isolation.

In the realm of tropical agricultural products, these states compete not just with one another, but with other tropical regions as well. A rise in prices brings vast acreage of a single crop into production; the oversupply that results in a few years depresses prices in a perpetual boom-or-bust cycle. Imported goods tend to remain relatively high priced. African economies frequently are overspecialized, and entire economies fluctuate radically with world commodity prices.

Whatever colonial development took place was generally at only a few locations. One or two cities, serving as capital and/or port, concentrate a disproportionate amount of national wealth. They are the only centers served by adequate transportation, so they also draw virtually all new investment. Such cities are the target of virtually all migrants from the countryside and are surrounded by vast and pathetic slums filled with refugees from famine, unemployment, and civil unrest. The developmental disparity between well-developed enclaves, surrounded by a

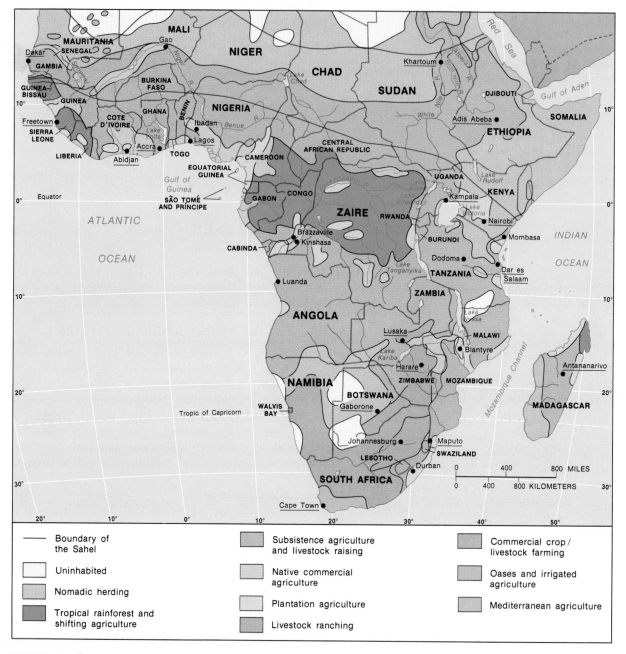

FIGURE 9-6

Land Use Regions in Africa. Few areas of the continent are uninhabited, but much of Africa South has few people because of limited agricultural utility. The drier margins of the north and southwest are devoted to nomadic herding. Vast forests occupy the equatorial core and the monsoon coasts of the extreme west; shifting agriculture is practiced in these forested areas. Most of Africa South is devoted to subsistence farming. Commercial agriculture concentrates in the south and in certain environmentally favored areas organized as plantations during the colonial era. Vast areas are used only for grazing cattle and occasional subsistence plots. The continent can be considered to be underutilized.

primitive economic mass matrix, is nowhere more obvious. Each **development enclave** is more related to the developed world outside than to any other enclave within the state.

The area suffers from a neocolonial status. Domestic economies are unable to generate capital, relying either on whatever capital is supplied by export commodities or on borrowing from the former colonial power. Neocolonialism is best shown in the continued dependence of most African states on their former colonial markets and sources of supply. Even where new outside markets and sources are achieved, there is the tendency for those new sources to interfere in domestic affairs—economically, politically, and even militarily. Even where new investments occur, technical advice, managers, capital equipment, and ultimate markets still tend to be supplied by former colonial powers.

THE PHYSICAL FRAME

Africa South is a huge shield area that rises abruptly above the sea in most places to a relatively high inland plateau (see Figure 9–2). Almost totally within the tropics, the region is characteristically hot except at higher elevations. From its equatorial center toward both north and south, precipitation decreases in total, while it increases in seasonality and unreliability. Orientation in reference to prevailing winds and elevation modifies climates over unusually large areas. Deceptively simple surface and climatic regimes, however, actually contain many exceptions.

Topography

Most of the interior of Africa is a plateau that averages 1,000 to 2,000 feet above sea level north of the equator and somewhat higher south of it. The northern section has been more thoroughly dissected, resulting in rolling hill country over most of the area, with occasional portions that qualify as mountains (see Figure 9–2). The plateau is made of ancient crystalline rock; it apparently has been geologically stable for long periods of time and is a remnant of the ancient and massive continental mass called **Pangaea** (Chapter 1, Figure 1–9).

Along the northern edge of the region are a series of basins stretching west to east near the border of the Sahel with the savanna vegetational zone. They are the basins of the Upper and Middle Niger, Lake Chad, and Bar-el-Ghazal. At the juncture of wet and dry climatic zones, most are partly filled with swamps or salt lakes. Just south of these depressions is a thoroughly dissected plateau country of relatively rugged hills. In the west, these hills are called the Futa Djallon; in the center, the Nigerian plateaus and the Ubangi-Shari upland; and in the east, the Ethiopian Highlands (see Figure 9–2).

The coastal lowland is fairly well developed along the Atlantic and Gulf of Guinea coasts. It varies from 50 to almost 400 miles in width and is an unusually (by African standards) wide strip of fairly level land.

At the hinge, there is a change in coastline orientation from east-west to north-south, marked by a diagonal chain of rugged volcanic mountains, the Adamawa. In the chain's southwestern extension, the peaks are submerged under the sea, forming a series of islands.

South of the equator, the plateau surface is higher, frequently less dissected, and saucerlike. The center consists of two immense sedimentary basins, the Congo (Zaire) Basin, much lower and older, and the higher basin occupied by the upper Zambezi and the Kalahari Desert. Along the Atlantic coast here, the lowland is confined to a narrow coastal strip that rises abruptly to the plateau.

Topography changes equally abruptly along the eastern and southern edges of the continent. Running through the eastern interior is the rift country, a classic proof of continental drift (plate tectonic) theory. Africa is apparently pulling apart, giving rise to a feature known as the Great Rift Valley (Figure 9–7). With its associated bordering highlands, this valley extends from Mozambique to the Gulf of Aden and continues as the Red Sea to the Sinai Peninsula and beyond. At the top of Lake Nyasa, the valley divides into two sections: the western, occupied by a series of lakes, and the eastern, lined with East Africa's great volcanoes. The intervening plateau area is covered with sediments and in part occupied by shallow Lake Victoria. In the Ethiopian section, the highlands spread out to cover most of the country. In the southeast, after disappearing briefly, the rift surfaces again as the Drakensberg Mountains of the South African coast. Two larger fringing lowlands edge the coast of eastern Africa, one extending from Durban to Zanzibar, the other from Zanzibar to the tip of the

FIGURE 9-7

Great Rift Valley. The African continent is gradually splitting apart. The line along which this split is occurring is called the Great Rift Valley. It is actually a long and discontinuous series of valleys occupied in part by large lakes and covered with rich volcanic soils. Most of Africa's active volcanoes and highest mountain peaks are associated with the rift zone. Arrows indicate the direction of movement of pieces of the continental plate.

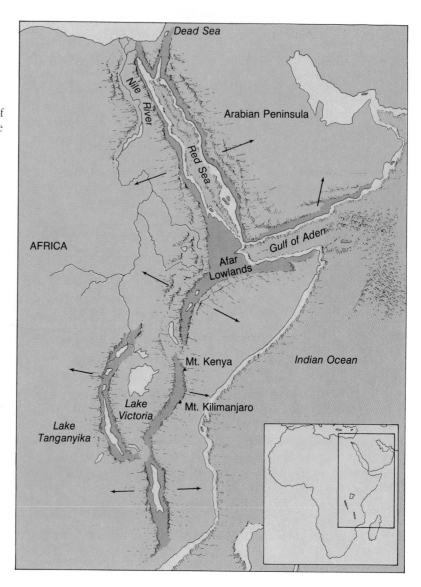

African Horn. The width of these coastal plains varies greatly.

The island of Madagascar is a piece of the African continent that has fully separated from the main continental mass. Its high mountain core is surrounded by coastal lowlands, broad and well developed on the west, extremely narrow on the east.

The Limited Role of the Sea

Over much of the coastline of West Africa, mangrove swamps edge the land, and there are few good har-

bors. Most of the best ports of West Africa today are artificial. The mouths of rivers are often silted and suited only to shallow-draft vessels. Even along the drier shores south of the Zaire River, where mangroves disappear, the proximity of the interior plateau and the lack of good natural harbors hampered early trade and penetration. The best harbor development occurs from Cape Town, South Africa, to Maputo, Mozambique, where the plateaus and mountains reach almost to the shore—areas from which penetration into the interior is exceptionally difficult. The eastern coast had sufficient good harbors, and trade with the Arabs developed early.

The few landlocked African states have come to agreements with surrounding states for the use of ports and transit routes. The volume of trade from most African states is relatively small, so such arrangements are adequate.

Climate: How Tropical Is Tropical Africa?

Little of Africa South extends into the zones of subtropical or temperate climates. However, large parts of southern and eastern Africa are of sufficiently high elevation to experience the cooling influence of altitude. All of the region is warm over most of the year, and only a few areas in South Africa experience frost. Only the West African coast, northern Zaire, and parts of the coast along the African Horn are oppressively hot and humid.

The rainforest climate, with its heat, humidity, and high yearly rainfall totals, occupies a relatively small area from the mouth of the Niger to the mouth of the Zaire and inland to the edge of the Great Rift Valley. Annual rainfall totals of from 60 to 120 inches are common there, and dry periods are really simply less wet (see Chapter 1, Figure 1–15). Similar conditions are found on the coast from Guinea to Abidjan in Côte d'Ivoire. Here, conditions differ in that they are monsoonal, with a pronounced, but brief, dry season. On either side of this narrow belt of almost perpetually hot and rainy climate is a belt of tropical savanna with a warm, dry winter and hot, humid, rainy summer. Rain is brought by the poleward movement of the intertropical convergence. The dry season is longer and more pronounced the farther removed from the equator. On the southern side of the rainforest, the savanna climate penetrates farther poleward. The coasts are dry, but the interior plateau develops convectional storms in summer. Rainfall totals vary from 20 to 80 inches, and rain is more reliable than in the northern savanna. Altitude south and east of Zaire modifies the climate, alleviating some heat discomfort. Monthly temperature averages range from 60° F in the early dry season to the mid-70s in the warmer, rainy summer.

South of Angola, Zambia, and Zimbabwe, the climate becomes steppelike. Rainfall totals fall below 20 inches yearly, and the rainy season is compressed into a few months. The Kalahari Desert is more a semidesert; truly arid conditions are limited to the coast and northwestern portion of the Republic of South Africa. The Republic of South Africa, south of the Tropic of Capricorn, experiences truly temperate conditions. There is a definite winter with some frost danger. Even higher areas, however, are sufficiently close to the sea to experience moderating maritime effects. The Cape of Good Hope area has a Mediterranean regime with dry, hot summers and mild, rainy winters. The eastern highlands and the eastern coast of South Africa have humid subtropical conditions similar to those in the U.S. South.

Along the northern edge of the tropical rainforest, the climate is transitional from savanna to steppe and even to desert conditions in an area of highly irregular precipitation. Rugged mountains in Ethiopia exhibit a **vertical zonation** of climate. At times (and in certain places), monsoonal winds bring moisture in off the Indian Ocean. More commonly, the northward movement of the intertropical convergence brings heavy summer rain. Seasonal and yearly rainfall variability is exceptionally great.

The hot, humid, but almost rainless Danakil coast of Eritrea is classified as desert. It has at least a discontinuous vegetational cover that makes parts of it appear more like semidesert.

Madagascar, an exception, has an east-west zonation of climate. The windward eastern coast has rainforest conditions, the mountains a highland climate, the western coast a savanna climate; even some areas of steppe are found in the extreme southwest.

Vegetation in a Continent of Climatic Uncertainty

The modification of natural vegetation through overgrazing and intensive cropping in some marginal areas has likely changed the original extent of natural vegetation zones. Unlike the Brazilian rainforest, which extends well beyond the normal latitudinal limits of the tropical monsoon and rainforest climates, the African rainforest occupies only the core of those climatic regions.

South of the Zaire basin, in harmony with climates, the zonation is less sharp and the vegetation more varied. Much of the vegetation of wetter Africa South is more forest than savanna. Areas of savanna climate here often exhibit a transitional thorn forest and bush before grasses come to dominate completely in the South African **veld** (temperate steppe). The deserts of most of South Africa and Somalia are more steppelike and grassy than those of the Sahara. Savanna vegetation is like parkland, consisting of a grass matrix with occasional trees. It is typified by acacia trees scattered sparsely over the grassy landscape in the east, scrubby thorn forest in the south, and grass with groves of oil palms or gum trees in the zone of transition along the savanna-Sahel border of the north. As the northern border of the savanna is reached, grasses occur in clumps rather than as a continuous cover; thorny shrub and aromatic bush replace trees.

Drainage and Untapped Hydroelectric Potential

The equatorial core of Africa is the source area for most of that continent's major drainage systems. As rainfall decreases and its seasonality increases, river systems are fewer, smaller, and have less stabilized flow, in a rough approximation of general rainfall patterns. Desert areas have intermittent streams (those that contain water only after storms) and interior drainage. A few permanent streams with their sources in areas of wetter climate flow through the desert as **exotic streams**; they include the Upper Niger and the Nile (see Figure 9–2).

The Zaire's flow is not only enormous but also extremely stable. The Zaire drains a large, shallow depression before descending over a lengthy series of rapids and falls on its way down the escarpment to the sea. With tributaries on both sides of the equator, it receives rain in some part of its 1.5-million-square-mile catchment area at all times. Hydroelectric potential is enormous. Large stretches of the river and its tributaries are navigable, but falls and rapids impede use in both its upper and lower reaches.

The Nile also has its sources in this region, flowing over 4,000 miles, often through desert, on its way to the sea.

Rising in Sierra Leone and Guinea, an area of heavy monsoon rainfall, the Niger first travels north into the Sahel-Sahara borderlands, where it forms a large, swampy inland delta. Much diminished in flow, it winds southeastward back into wetter West African areas, where it is joined by the Benue before it reaches the sea.

The Zambezi has multiple sources and a variable flow. Victoria Falls, much higher than Niagara Falls,

occur where the river plunges from an inland basin through a series of gorges on its way to the sea off Mozambique. The Kariba Gorge below the falls is the site of a huge hydroelectric project.

A series of smaller rivers drain small individual sections of the rest of the region. Noteworthy is the Volta, in West Africa, which has been developed for hydroelectric energy and a major aluminum works in Ghana. The Limpopo forms the northern border of the Republic of South Africa, while the Orange drains its interior. Both are important irrigation sources. The rest of Africa's streams, although unnavigable, could provide local energy sources and valuable irrigation water.

Tropical Soils: An Unsolved Problem

The intense heat and high humidity levels of this region combine to favor rapid breakdown of soil particles and to facilitate chemical reactions. Parent rock material, drainage, slope, original vegetation, and seasonal water regime all significantly affect soil. Because these conditions are so varied in Africa, the pattern of soils is extremely complex. African soils vary widely in fertility; with careful use, some African soils produce crops of high quality and give high yields.

The soils of the equatorial core are generally reddish or yellow-red, indicating that they are acid, high in iron and aluminum oxides, low in calcium and humus, and generally heavily leached. Where the soils are alluvial or result from volcanic material, they can possess a reasonable potential. Plantations occupy the better soils capable of sustained yields. The remainder are used only periodically for shifting cultivation.

Soils of the wet-and-dry tropics are on average somewhat more fertile. The difficulty of using them is associated with water retention, complicated by a concretion, or **hardpan** (a layer of minerals cemented together), beneath their surface. In the wet season, the top soil layer is saturated with water, and the countryside is flooded. During the dry season, water quickly evaporates, and soil becomes baked and bricklike. Heavy machinery can pack it into almost a macadam surface. Fertilization, crop rotation, and fallowing enable some soils of this zone to produce good yields. The best soils of the savannas are chestnut-colored. In areas where they developed under a thick grass cover, they are similar to prairie soils.

The volcanic highlands of Ethiopia, the Great Rift Valleys, and the Adamawa Mountains possess unusually fertile soils. They are capable of constant cropping without serious damage, if erosion is controlled.

The desert soils of the Kalahari and Southwest Africa are calcium rich, if humus poor. Currently used for ranching and nomadic herding, they could produce reasonable yields if carefully irrigated.

THE DISTRIBUTION OF PEOPLE

Africa remains one of the less densely populated continents despite rapid growth over the last 25 years. This low overall density is a reflection of the difficulty of using much of the area. In a region still overwhelmingly agricultural, the general patterns coincide quite literally with known agricultural capabilities.

Unlike North Africa, there are few totally uninhabited areas. The bleakest parts of the southwestern Kalahari dunes and the extremely dry coast of parts of Namibia are the only areas that qualify (Figure 9–8). There are vast areas of extremely low density in the low-grass savannas of East Africa, the desert shrublands of the southwest, and the African Horn. The most infertile portions of the rainforest and savanna lands of the Zaire basin are also only lightly inhabited. The populations of these nearly empty areas subsist on nomadic herding and primitive dry farming in drier zones, or slash-and-burn on a long cycle rotation in the rainforest. They are some of the more thankless natural environments of the continent.

The rich volcanic soils of East Africa's highlands support extremely high densities near the African Great Lakes and reasonably heavy densities in the Ethiopian Highlands. Commercial crop economies of the East and West African coasts also support respectable densities, as do commercially developed portions of Madagascar. Only the high-density areas of Nigeria and the Republic of South Africa reflect urban industrial development. In the rest of the re-

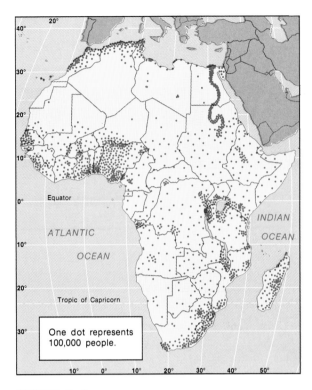

FIGURE 9–8
Population Distribution in Africa. Africa is among the most sparsely populated areas of the world. The deserts of the continent's northern reaches are a virtual population void. The rich volcanic soils associated with the Great Rift Valley appear as population concentrations in Ethiopia, East Africa, and Malawi. Wetter West Africa, with its reliable rainfall and moderately fertile soils, exhibits two concentrations. The Nile Valley and the temperate northern and southern continental extremes concentrate most of the rest of the population. With rare exceptions, the forests and grasslands of the continental interior are lightly populated.

gion, the correlation of population and favorable farming conditions is remarkable.

The largest concentration of high population density is in West Africa, where commercial endeavors were not confined to the coast but were also introduced into the backlands (see Figure 9–8). Plantations and even small native farms there produce commercial crops of irrigated rice, short staple cotton, and the almost ubiquitous peanut. Oil palm groves make it difficult to tell where farms end and

THE PEOPLES OF AFRICA

The world knows little about the early peoples and cultures of Africa South. Painstaking research by Mary and Louis Leakey, noted anthropologists, discovered the oldest known human remains in East Africa. Current theory accepts this area as the original home of humans. Populations did not just diffuse from this area of origin to other continents, but also diffused throughout Africa itself.

Demographers and anthropologists identify five early peoples as inhabitants of the continent of Africa 5,000 years ago: Hamites, Pygmies, Nilotes, Bushmen/Hottentots, and the Negroes of West Africa (essentially the Niger-Congo language group) (Figure Aa). Whereas Hamites are classified as Caucasian (though there has been much mixing with darker-pigmented peoples), the other four are definitely blacks.

With the assumption that East Africa is the home of the earliest humans, it is easy to see the divergence of groups from that area, each to a different environment. The West Africans moved far from the area of origin, occupying the far west of the continent. Their ultimate homeland was the bushlands and monsoon forests of the Atlantic coast. Pygmies were tree-dwelling hunters in the equatorial forests, a limited habitat with limited potential. The Bushmen/Hottentot group spread southward through the scrub of south-central Africa, hunting the plentiful game of the bush. The Nilotes spread westward along the Sahelian grasslands at the contact zone of desert with savanna and part way up and down the Nile. The Hamites seem to have spread along the coasts, north and then westward, separated from the other four groups by Saharan wastes except in the African Horn. At the inception, the groups were probably of approximately equal size (Figure Ab).

The relative demographic success of each group appears to have been related to their environment and choice of lifestyle. The battle between hunter, herder, and farmer over the use of land is as old as the human race. By definition, farming supports the highest population densities, and sedentary agriculture results in the highest birth- and growth rates. Thus, farming peoples were the most prolific, and came to occupy the largest portion of the continent.

The Hamites originally occupied the highlands of Ethiopia. They spread down the Nile and the coast and ultimately occupied the lands of Egypt and North Africa. Wave after wave of Semitic conquest and migration followed. Semites from Arabia crossed the Red Sea and set themselves up as a ruling class in the area of Ethiopia, conquering the Hamite lands on their eastern flank. Finally, the Islamic invasions placed a Semitic people in power over Egypt and the Maghreb, removing the last vestige of Hamite territorial control except in areas of nomadism. Demographically, the Hamites were successful, but the culture and speech were retained by only a few as Semitic speech and culture assimilated urbanites and sedentary farmers among them.

The Nilotic peoples, nomadic herders, spread west along the verdant grasslands to at least the middle of the Sahel. There is linguistic evidence of their interaction with, and perhaps conquest of, areas as far west as today's Guinea. Ruling warrior classes of Nilotes conquered East African farmers of West African stock as far south as Tanzania. As herders, they required larger amounts of land than farmers. Their numbers also grew more slowly. Recent unrest in Uganda is a continuation of the contest between cattle-rearing Nilotes and agrarian Bantu. Only in the last 15 or 20 years have the Bantu farmers of Rwanda and Burundi displaced the Nilotic ruling class of the Tutsi (Watusi). The Masai-Kikuyu conflict in Kenya is a repeat of the same age-old contest.

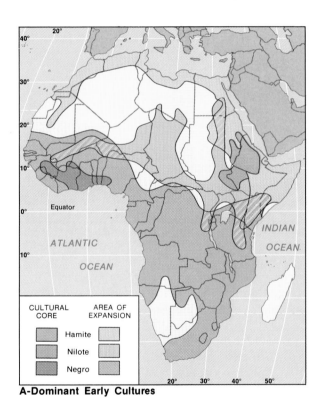

A-Dominant Early Cultures

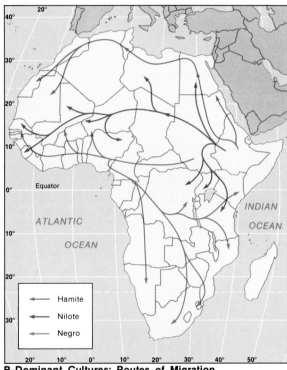

B-Dominant Cultures: Routes of Migration

FIGURE A, pp. 451–452
African Peoples and Migration Patterns. East Africa is considered by many to be the earliest home of humans. From that location, the human race diffused across the earth. (a) Three early cultural groups appear to have peopled most of Africa. (b) Each arose in a different physical environment and spread its control to areas with similar conditions. Two groups were relegated to poorer environments where they form residual populations. (c) Later invasions from outside Africa introduced new elements of language and culture. (d) Africa gave rise to large empires based on agriculture and trade.

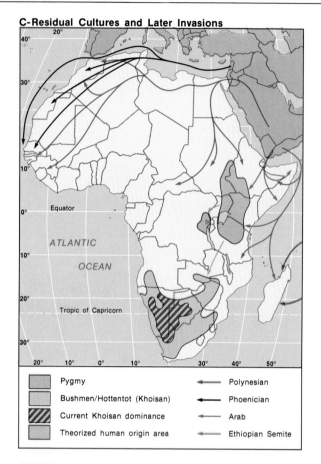

C-Residual Cultures and Later Invasions

Pygmy	Polynesian
Bushmen/Hottentot (Khoisan)	Phoenician
Current Khoisan dominance	Arab
Theorized human origin area	Ethiopian Semite

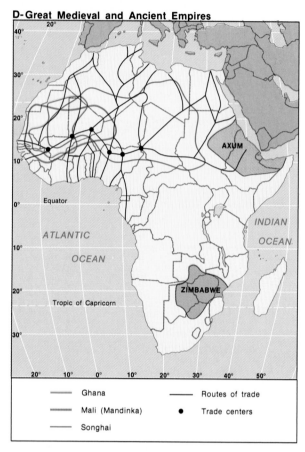

D-Great Medieval and Ancient Empires

Ghana	Routes of trade
Mali (Mandinka)	Trade centers
Songhai	

FIGURE A
continued

Least successful were the Bushmen/Hottentots and the Pygmies. The latter have all but disappeared except in the most isolated forests of Central Africa. The former were gradually forced to the farthest southwest, the inhospitable Kalahari (Figure Ac).

Overall, the West Africans were the most successful. Farmers since about 3000 B.C., the group domesticated yams, several kinds of millet, several root crop varieties, oil palms, certain types of fowl, and perhaps pigs. Sheep and goats diffused to them across the deserts, and contact with the Nilotes brought cattle. This rich and varied diet in a less than hostile environment encouraged proliferation. Most sources agree that about 500 B.C. the Bantu, the easternmost of the West African peoples, spread across the Adamawa Mountains, spilling over into Central Africa. From there, they spread east to the Indian Ocean and south into the farthest reaches of South Africa. Thus, the Bantu language family became the largest, dominating much of Africa South (see Figure 9–1). There are minor exceptions to these trends, but basically the pattern applies. Because of it, there is cultural and linguistic similarity, though not uniformity, over much of Africa South.

forest begins. The peanut crop enriches soils through its ability to fix nitrogen. In the mixed farming economy that prevails, the manure of pigs and cattle and the nitrogen added through peanut cultivation create the possibility of sustained farming and higher population density. Here, too, some commercial-industrial development boosts densities beyond what might have been the case in a totally natural environment.

Few countries are urbanized to any degree, though most boast a large capital city with a disproportionate concentration of the national population total. Where the capital is not also the premier port, national units may contain two important urban centers. Multiple nuclei are present in economies with a heavy emphasis on mineral production, such as Zaire, Zambia, and Zimbabwe, or in areas of dense agricultural settlement with many local market centers.

Though few areas can anticipate supporting densities of a Chinese or European magnitude, most countries are in a position to absorb anticipated heavy population increases by extending agricultural frontiers and diversifying economies. Only Rwanda and Burundi, two small and densely populated East African states with no empty frontier areas, appear to be in serious immediate trouble.

SPLENDORS OF THE PAST

Legends of King Solomon's mines, fabulous ruins in Zimbabwe, fancifully decorated palaces in Kano, and folk tales of the heroic El Haj Omar all hint at the glorious history of Africa's past. It is a history that links the origins of humans to the development of civilization, full of unsolved mysteries and unanswered questions.

There appear to have been distinct centers of advanced civilization in Africa South: East Africa, the inland delta of the Niger River, and the hills of today's Zimbabwe are the three areas for which we have records and architectural evidence of great past civilizations. The most detailed knowledge is available for East Africa, though what is known represents only a fraction of the total picture. The names Ethiopia, Abyssinia, and Axum, all referring to the same general area, appear throughout written history. Tradition places the origin of the Ethiopian Empire in the meeting of King Solomon (of Israel) with the Queen of Sheba, an empire that occupied the southwest corner of Arabia. The empire developed around 500 B.C. at Axum in the northern part of Ethiopia (see Figure Ad, p. 452). Medieval Europe knew only of a fabled Kingdom of Preaster John, an "eastern" Christian monarch who may have been the king of Axum or, equally possibly, the king of a Nestorian Christian kingdom in Middle Asia. One of the Magi of the Christian Christmas tradition has traditionally been depicted as black. His gift, myrrh, used as medicine and for embalming the dead, was a highly valued herb that grows wild on Ethiopia's plains. Little is known of the region's other kingdoms, but advanced societies there left the now-ruined remains of stone-lined wells and irrigation ditches in Kenya, and terraces, marking ancient agriculture practiced on a grand scale, in portions of Kenya and Tanzania.

West Africa sustained a series of medieval empires based on the trans-Saharan trade (see Figure Ad, p. 452). The presence of ivory, ebony, and certain jewels in the holdings of European courts and palaces of all ages indicates an early and sustained trans-African trade. The most impressive writings on these kingdoms flowed from the pen of Leo Africanus, a brilliant poet and scholar who journeyed to the region of the Niger in the sixteenth century.

Pre-dating Africanus by at least 500 years was the empire of Ghana, a mercantile kingdom that probably reached its zenith before A.D. 1000. An inland empire, it never included the modern state of Ghana, its namesake. Ancient Ghana built its fortunes on trade between the Mediterranean and coastal West Africa. Blessed with a surplus of cattle and grain, it traded food, salt, and metal goods southward for rare wood, ivory, diamonds, and gold. It transshipped these items across the desert in return for North African textiles, armaments, and technology. Ghana had a considerable metallurgy and a leather industry that was developed for both local needs and export sales.

Ghana was succeeded by the even larger Mali Empire, which ruled from the thirteenth to the seventeenth centuries. Mali, in turn, was succeeded by the Songhai Empire, with its capital at Gao, a city still in existence. Its focus was farther east, but it may have controlled a territory well to the east of Lake Chad and as far north as modern Libya. It was defeated by the trans-Saharan invasion of Arab and Berber armies with hired European mercenaries in service.

The cities of these empires were large and opulent; some have left impressive ruins despite the ravages of conquering armies.

The reasons for the failure of these empires are complex. Certainly, the conditions that fostered trade did not disappear at once. Here is an area of great crossroads importance, where major east-west routes of the Sahel intersect with a half-dozen north-south routes across the Sahara. Here was the epitome of **complementary** production—three unlike climatic zones (the desert, savanna, and rainforest), with differing production, existing in close proximity to one another, each producing goods the others desired. The great empires were literally in the middle. Trade did not cease. The empires were followed by a series of smaller successor states, each with one (not several) route under its control. These states persisted until the French and British conquests. Trans-Saharan invasions were repeated, even after many states embraced Islam. It was to the advantage of North African states to control *both* terminuses of the trans-Saharan trade. The European invasion was probably more critical to the demise of the empires; the tall, strong, well-fed peoples of the Sahel were highly valued as slaves. There is no doubt that slaves, the militarily vanquished, had been a limited part of ancient trade. The Europeans, however, increased slavery to an unprecedented scale. The coastal peoples of West Africa apparently attacked the kingdoms of the Sahel to obtain slaves at the same time that Moorish forces attacked from the north. Europeans established trading posts in each of the three regions, shipping goods more rapidly and on a mammoth scale by using sea vessels. Past power and glory melted away and were almost forgotten in the process of the European colonial land grab.

In the south of the region, there are large-scale ruins at the site of Zimbabwe—an ancient city from which that modern country's name derives. Ancient mines that once yielded gold, zinc, silver, copper, and iron abound in the surrounding countryside. The city itself contained upwards of a thousand magnificent stone buildings, some of which remain intact. Coins indicate a far-flung trade network that included India and China. The reasons behind Zimbabwe's rise and fall remain unknown.

THE STRUCTURE OF SUBREGIONS

Perhaps more than in any other world region, the subregions reflect a need to subdivide for simple comprehension. There are more than 50 political units in Africa, 42 of them in Africa South. There is only a limited degree of subregional homogeneity, in part because of the extreme diversity of the region itself.

West Africa

West Africa is the most populous part of Africa South, that with which Europeans had their earliest and most sustained contact. This subregion was the prime target area of the infamous slave trade. As various European powers rose to prominence in that trade, they set up a series of fortified supply and trade depots along the Guinea coast. Virtually every usable harbor or river mouth at least temporarily gave birth to such a fort.

Each fort became the nucleus of a colony. Nowhere else in the region is there such a proliferation of countries (Figure 9–9). Average size is quite small by world standards, on the order of individual U.S. states. Only Nigeria is a large unit, encompassing almost one third of the area and over 60 percent of its population. The question suggested by the profusion of so many small states is whether they can survive economically.

Although the area possesses significant mineral wealth, there is a decided paucity of fossil fuels (Figure 9–10). Only Nigeria has a large oil resource and significant coal deposits. While there is some hydroelectric potential, it is not on the grand scale of Zaire's. There are sizable deposits of ferrous metals and some alloys, but with a worldwide steel glut, these resources are in low demand. Nigeria's long-exploited tin deposits are nearing exhaustion. Bauxite abounds, but only Guinea, with its guaranteed East Bloc market, has been able to develop these mineral deposits to its advantage. Ghana, with its huge aluminum-refining capacity, must import considerable bauxite. The area has been poorly prospected; there may be a great deal more mineral resources.

Only where the soils are volcanic, or enriched with alluvium, are they productive. While both en-

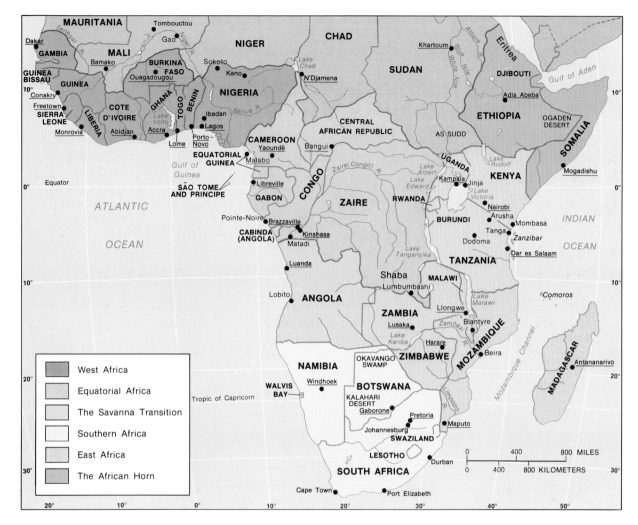

FIGURE 9–9
Subregions and Places of Africa South. Africa has been divided into six fairly homogeneous regions on the basis of physical, cultural, and economic similarities. Populous West Africa contains the largest number of countries. It is a zone of relatively productive crop agriculture. Equatorial Africa is lightly populated and is characterized by a mixture of subsistence farming and large commercial enterprises. The Savanna Transition is literally a zone of physical transition between equatorial rainforests and the temperate south. Southern Africa is composed of the technologically advanced Republic of South Africa and its unofficial economic dependents. Eastern Africa has been divided between a Hamito-Semitic north (the African Horn) and the Bantu-Nilote realm called simply East Africa.

riching occurrences are found in West Africa, they are neither widespread nor typical. The best soils are frequently devoted to tropical export crops; the remaining land is farmed by slash-and-burn farming.

The Ivory Coast has the most well-rounded economy in the region and the highest per capita income (Figure 9–11). Recent drops in world cocoa and cof-fee prices have depressed even this economy. "Oil-rich" Nigeria is not very rich at all, but is the region's second most prosperous economy. Overall, 5 of the world's 10 poorest countries are located here (Table 9–1, p. 458).

Colonially imposed isolation and a history of ad-ministrative separateness have resulted in 13 individ-

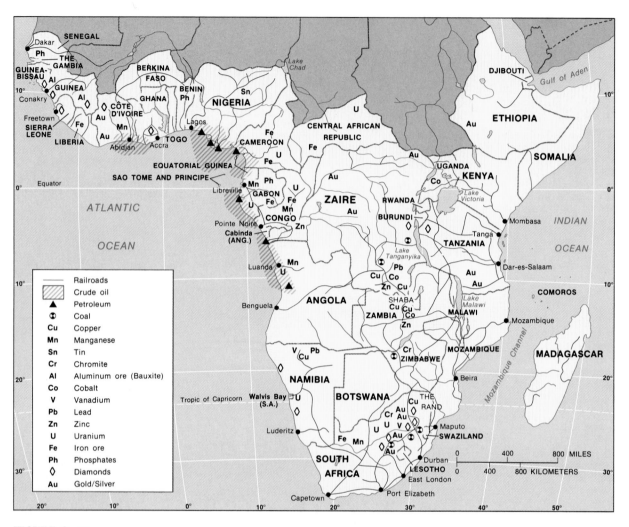

FIGURE 9–10

Mineral Wealth of Africa South. Despite the presence of several world-scale mineral deposits, relatively little is known about the mineral resource base of much of Africa South. South Africa's Rand, the Copperbelt of northern Zambia and southern Zaire, and the oil fields of Nigeria's Niger River delta are the region's most important mineral concentrations. Vast deposits of iron ore, bauxite, and fertilizer minerals are known to exist in both the West African and Equatorial African subregions.

FIGURE 9–11

Social and Economic Characteristics of Africa. The northern and southern extremes of Africa are the continent's best developed portions. Africa South is among the least developed of the world's regions, yet within its confines there is great variety in developmental characteristics among individual states. (a) Per capita income is highest among those states that are mineral rich or characterized by commercial production and activity. (b) Dietary standards follow broadly similar patterns, but drought, government food-pricing policies, and foreign aid result in obvious exceptions. (c) Differing literacy rates can be accounted for by a combination of colonial-era and current government policies regarding education. The lowest rates occur among residents of the Sahel, where nomadism is prominent and economies have been disrupted by drought. (d) Curiously, life expectancies do not correlate strongly with income and dietary standards. Missionary activity and government social welfare policies appear to be the crucial factors.

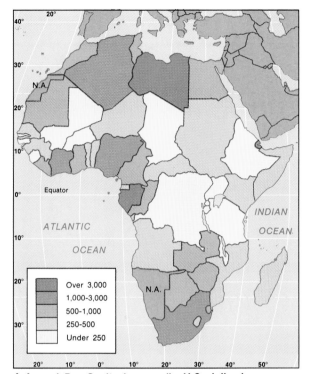

A-Annual Per Capita Income (in U.S. dollars)

	Over 3,000
	1,000-3,000
	500-1,000
	250-500
	Under 250

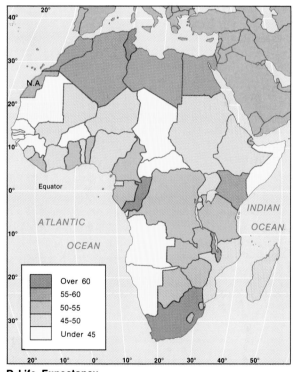

B-Percent of U.N. (FAO) Recommended Daily Caloric Intake

	Over 100%
	90-100
	85-90
	80-85
	Under 80

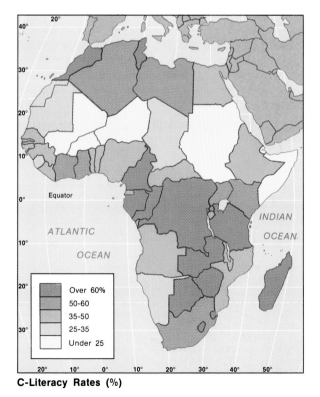

C-Literacy Rates (%)

	Over 60%
	50-60
	35-50
	25-35
	Under 25

D-Life Expectancy

	Over 60
	55-60
	50-55
	45-50
	Under 45

TABLE 9–1
West Africa at an economic glance, 1985.

Country	% Arable Land	Per Capita Income ($)	Per Capita GNP ($)	Inflation Rate (%)	Trade Balance (billions $)	Import-Export Ratio	% Deficit Domestic Budget
Côte d'Ivoire	12	1,100	1,380	5	+0.20	1:1.2	0
Nigeria	34	760	750	20	−1.50	1.2:1	10
Liberia	5	480	420	2	+0.15	1.2:1	3
Guinea	8	395	340	6	+0.10	1:1.2	0
Togo	26	388	330	1	−0.10	2:1	8
Senegal	30	382	480	11	−0.20	1.5:1	4
Benin	16	350	310	7	−0.20	2:1	(no data)
Ghana	12	340	320	129	−0.20	1.2:1	0
Cape Verde*	10	300	360	(no data)	−0.10	18:1	(no data)
Gambia*	28	280	260	22	−0.03	5:4	10
Burkina Faso*	11	180	170	(no data)	(no data)	(no data)	(no data)
Sierra Leone*	25	175	380	75	−0.25	4:1	0
Guinea-Bissau*	8	170	170	(no data)	−0.05	6:1	20

SOURCES: Population Reference Bureau, 1988; World Bank, 1988.
*Among the 10 poorest countries in the world.

ual states. There is little trade among them since they produce virtually the same exports. Economies are generally tied to the former colonial power. There is internal dissension, and frequent governmental changes or military coups are typical. Nigeria's oil and the tourism of the Ivory Coast are bright spots, and Senegal, strategically located, has much potential. Guinea has made great strides in diversifying its economy and has drifted away from the Soviet orbit. The larger states, then, show promise.

Prosperity remains tied to European markets. Arable land can be expanded; crop yields can be increased with the aid of the subregion's mineral fertilizer deposits. There is an attempt to diversify agriculture. Though most of the countries are oil poor, the potential for discovering offshore oil is considerable. Population growth rates are high, but most of the region's states have unsettled frontier areas. There is a strong tradition of commerce and entrepreneurship, if a shortage of ready capital. The future is not well defined, but far from hopeless.

The Corner of Africa. With only 7 million people and despite few natural resources, Senegal has con-

sistently maintained a moderate living standard by exporting agricultural products to France and the European Economic Community. As is typical of the drier fringes of tropical wet-and-dry climates, peanuts are a major crop for export. Tourism is expanding, partly in response to the superb sport fishing available.

The Senegalese themselves are the country's most important resource. They have a long-standing tradition of commercial involvement as tradespeople and entrepreneurs. With a Muslim tradition more than 1,000 years old, many Senegalese are members of Islamic brotherhoods that dominate the nation's business in both rural and urban districts.

Dakar, with over 1 million people, is a modern city of cosmopolitan atmosphere (see Figure 9–8). Broad boulevards, skyscrapers, ornate turn-of-the-century apartment blocks deliberately reminiscent of Paris, mingle with the mosque, minaret, and bazaar so characteristic of the lands of Islam. Western dress is common but appears starkly puritan and subdued alongside the brilliant hues of the more widely worn native dress. Recent drought has dampened national economic growth, but still-busy shops are filled with a variety of imported goods for the affluent urban Senegalese. Actual income in Dakar greatly exceeds regional norms, and official estimates, in a land where black market gains and under the table incomes far exceed official salaries and wages.

Dams are being built to increase irrigated area, and rice acreage is being increased to reduce dependence on imported grain. Relatively stable Senegal, despite its problems, appears to be reasonably successful. It does not suffer the divisive ethnic or political unrest of most West African states. International debt is worrisome and growing, but the government has addressed agricultural problems with research, diversification, irrigation, and modification of pricing structures.

The Gambia, an enclave within Senegal, is a relic of colonial competition. The first British colony in West Africa, it is a narrow strip of territory on both banks of the navigable Gambia River. The international boundary, however, cuts off this superb river from its natural hinterland. The Gambia's only resources are fish and peanuts. A confederation agreement with Senegal, creating Senegambia, allows each state to retain its sovereignty yet now permits joint use of the Gambia River. The Gambia has only about 800,000 people and is among Africa's poorest nations. Gambians are resistant to the confederation, complaining that Senegal reaps all the benefits. Stable Senegal, on the other hand, sees little utility in the underdeveloped Gambia other than the removal of a territorial enclave that hampers communication between its northern and southern portions.

At various times, major portions of West Africa were designated as the Guinea Coast, and colonies called Guinea were differentiated by colonial power (Portuguese, French, Spanish). Two nations of West Africa still use the designation—former French Guinea is simply designated Guinea, while former Portuguese Guinea is designated Guinea-Bissau (after its capital city).

Guinea-Bissau, the remnant of a once extensive Portuguese empire, is one of West Africa's poorest states and one of the 10 poorest in the world (see Table 9–1). On the other hand, geologic exploration has uncovered large deposits of bauxite and phosphates, and there is strong evidence of considerable commercial oil deposits. These incipient riches remain undeveloped. The contrast of potential riches

with actual poverty, so locally characteristic, is a recurrent theme throughout Africa. Guinea-Bissau, unlike most West African states, has no economic umbrella organization, such as the Franc Bloc or the Commonwealth, to aid in development or ensure markets.

The Cape Verde Islands were once politically linked to Guinea-Bissau but opted for a separate state in 1975, two years after the overthrow of Portugal. Like its former partner, the new state is extremely poor, though somewhat better developed. The population is mixed African-Portuguese and was largely assimilated to Portuguese language, culture, and the Roman Catholic faith. An international airport funnels goods and passengers to and from South Africa; the republic has been ostracized by some West African states because of this activity. It must inevitably decide whether its future lies in intraregional cooperation or in its function as a link between Europe and Africa.

Guinea has a varied and impressive stock of natural resources: bauxite, iron ore, diamonds, gold, and a large hydroelectric potential. It was the only French colony to reject membership in the French Community in favor of immediate independence in 1958. France overreacted and withdrew all investments, down to telephones and potted plants. Through its allies, France succeeded in isolating Guinea from the West, forcing it to seek Soviet aid and trade relations. Joint Soviet-Guinean development of bauxite mines has been extremely successful, and Guinea has an alumina plant for first-stage refining before export. The Soviets take almost half of all production.

Huge high-grade iron ore deposits are now exploited, and industrial diamonds are a lucrative mineral export. Mining is the nation's primary endeavor, and mineral products account for 90 percent of exports. Domestic food production is sufficient for Guinea's needs, even allowing for a small, exportable surplus. Exports exceed imports in most years, but borrowing for developmental purposes has led to the familiar large foreign debt. Guinea, however, is in a better position than most to keep up a sustained schedule of payments.

Adherence to strict socialism has waned; independently owned farms are now tolerated, and private entrepreneurship and private traders are encouraged. There has been a shift toward stronger ties with the Arab states, a resumption of relations with France, and increasing interaction with the West. True to traditional form, Guinea pursues an independent course.

Both Liberia and Sierra Leone began as homelands for freed slaves after the end of the slave trade. Sierra Leone was given little assistance. All were expected to fend for themselves in this new, unsettled African homeland, though most lacked farming, hunting, or other survival skills. The original settlers were supplemented later by freed slaves from Nova Scotia and Jamaica (who became the country's social elite and governing force) and some 50,000 slaves released from captured slave ships. The back country was settled comparatively late by Africans arriving from the interior. Thus, an Anglo-African, creole cultural ways prevailed inland.

The history of Liberia is similar. It was founded as a homeland for freed slaves in 1822 by the American Colonization Society. The population descended from this group never exceeded more than 10 percent of the total, yet they ruled as an exclusive elite, dominating the native-born residents of the interior.

Sierra Leone has a natural asset in trade that is underused; Freetown, the capital, has the finest natural harbor in West Africa. Unfortunately, the hinterland is truncated by boundaries created during the colonial era. Sierra Leone, like Guinea, has badly leached, relatively infertile soils over most of its extent, yet it is also mineral rich. It shares Liberia's curious social-cultural makeup, but it has not had the same lucrative economic relationship with Britain that Liberia has had with the United States. The descendants of freedmen speak English and are Christian. The bush dwellers, the overwhelming majority of today's Sierra Leoneans, speak 12 different languages, are dominantly animist or Muslim in faith, and actively seek residence in other countries where work is more plentiful. Despite its diamonds and relatively diverse agriculture, the nation is one of the world's 10 poorest. Social conflicts between the elite of Freetown and the tribal interior complicate internal affairs. Sierra Leone is a country in which literacy, income, and development are all in decline; even a part of the railroad has been dismantled as Sierra Leone apparently retreats into the bush.

Strange as it may seem, Liberia was in some ways disadvantaged by never having been a colony. It lacked the usual colonial investments in transport, mineral exploitation, and plantation development. No sizable Western investment arrived until Firestone Rubber acquired land for the world's largest rubber plantation in the 1920s. Liberia's 12 million high-yield rubber trees provide the second most important export (after iron ore), though recent declines in demand have resulted in the closing of a portion of the plantation. The Liberian economy recorded a second large boost during World War II when American military engineers built a deep-water port at Monrovia in anticipation of a prolonged African campaign. Giant iron mines were opened in the 1950s. Liberia appeared to be well on its way to total development as it maintained strong ties with the United States and a receptive attitude toward investment. The Liberian fleet, a phantom registry operation designed to get around the effects of developed nation insurance, safety, and labor laws, is one of the world's largest and a lucrative source of income.

A coup in 1980 overthrew the traditional ruling elite; little has changed since then. U.S. cultural forces remain strong. Governmental change has not adversely affected the economy, though world market conditions have. Rubber, sugar, and iron prices have declined along with demand. Even though conditions have deteriorated, the country is reasonably prosperous by regional standards. The widespread acceptance of Western norms and culture, the increasing emphasis on education, and the influence of assimilated urban dwellers on their rural family members are helping to build a uniform culture.

The West African Center: Poverty and Prosperity in a Climate of Change. Early European merchant seamen named parts of the West African coast after the anticipated products of trade. Ghana (once called the Gold Coast) and Côte d'Ivoire (literally, the "Ivory Coast") were always considered to be the most valuable commercial holdings in West Africa. These two states are still the most prosperous part of this central portion of the subregion.

Ghana was a British colony with a well-developed commercial agriculture and an educated elite trained in British traditions of democracy. In 1957, it became the first African colony (excepting white-dominated South Africa) to attain independence. Côte d'Ivoire was a French colony with limited commercial development. It received little preparation for the independence it was granted in 1960.

Ghana has a good balance between mineral resources and agriculture. Its status as the world's largest exporter of cocoa, however, has been an asset and a handicap. As with other tropical world exporters of specialty and luxury crops, Ghana had no realistic hope of completely controlling prices. A host of other tropical nations have begun to crop cocoa, and prices are depressed. Cocoa in Ghana, unlike rubber in Liberia, is not solely a plantation crop; a large portion is produced by independent farmers. Diseases have decimated large acreages in both sectors. Crops of corn, rice, plantains, and peanuts are grown for domestic food supplies, but are inadequate for current needs. Small plantings of coconuts, coffee, and rubber have done little to diversify exports.

West Africa's largest hydroelectric project, the Akosambo Dam on the Volta River (see Figure 9–2), was heralded as a great achievement. Nearly half the energy produced is used by a large aluminum works at Tema (see Figure 9–10), but unfortunately, the worldwide aluminum industry has had problems with overcapacity. Power from the Volta Dam is exported to Benin and Togo and also supports light industries in southwestern Ghana.

With a huge surplus of power, a large alumina plant, and a small-farmer class that grows commercial crops, Ghana would seem be in a position to succeed. What went wrong? Kwame Nkrumah, leader of Ghanaian independence, made many of the classic economic mistakes that have come to plague much of the underdeveloped world. In an effort to prove equality with better-developed nations, he invested in huge public works projects that did little to benefit the local economy—stadiums, statues, plazas, and government buildings—and ran up huge foreign debts. Within a decade, the new state was saddled with a cumbersome foreign debt. In an effort to pay debts through increased exports, vast new acreages were cleared and planted in cocoa in the face of falling prices. To avoid internal chaos, cheap food-

stuffs were imported, destroying native agriculture. The dam increased foreign debt further, inundated farmland, and sold its power and water to a relatively low-profit industrial plant. Criticisms of these projects and government actions were met with repression; democracy ended in dictatorship and ultimate military revolt. Experiments with the socialization of agriculture destroyed incentive.

As if this were not enough, falling oil prices have resulted in the expulsion of a million Ghanaians from neighboring Nigeria and the loss of remittances that were once sent to Ghana. Declining cocoa and coffee prices and government forest conservation measures in the Ivory Coast and Cameroon threaten to result in the return home of a half-million more. A 40 percent yearly inflation rate has wrecked savings and played havoc with the manufacturing sector. Ghana now lives increasingly on foreign aid while it undertakes unpopular austerity measures.

Côte d'Ivoire, on the other hand, has such a prosperous economy that upwards of 2 million citizens of other countries live there. Unlike Ghana, it is a diverse agricultural economy. It is the world's largest producer of cocoa and counts itself among the top 10 world producers of coffee. Falling prices for these two commodities have been buffered by increased production of bananas, pineapples, palm oil, citrus, and cotton. With export of over a dozen crops, diversity seems to have ensured stability. New endeavors include winter-grown vegetables for the European market, fiber crops, and even increased acreages of food crops for the neighboring West African market.

Exports exceed imports by 50 percent or more. Per capita income is triple that of Ghana, and per capita GNP is almost 10 times greater. The Ivoirian planter class exceeds 3 million, ensuring a fairly equitable distribution of income. Mineral resources are abundant if not spectacular. There is some oil, and diamond and gold production is up. New iron ore sources await increased demand. Six hydroelectric dams supply the nation's energy demands. There is still room to expand agriculture. Tourism is booming as chic Abidjan draws Europeans, much as the Caribbean draws Americans in winter; the difference is that most of the hotels and shops are domestically owned. There is a good road system, and the country

THE TROPICAL CROP COLLAPSE

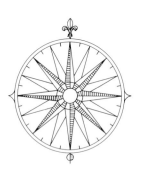

The emphasis on one crop or commodity as the major source of revenue is a familiar problem in underdeveloped economies. Diversity, rather than specialization, seems the route to a balanced economy. The most prosperous and promising African states of the postindependence period were those that specialized in mass-scale production of a single crop. Those crops, and their attendant economies, have since fallen on hard times.

There are two elementary pitfalls in specialization. The most commonly stated one is extreme fluctuation in world market prices. Acreage limitations and storage of surpluses can deal with that problem temporarily. However, changing public tastes and competition from other countries are beyond the control of any one producer. More serious is the question of disease. As plantations and farmers maximize favorable production conditions for a given crop through irrigation, fertilization, and cultivation, they are also providing favorable conditions for predators and diseases that attack that crop. Plant diseases have wiped out vast acreages of cacao in Ghana and Nigeria, following years of near-monopoly in which those two countries produced 40 to 60 percent of the world's crop (Table A). The revival of brewed coffee (as opposed to instant) in the United States, the entry of new coffee producers such as India and the Philippines, and increases in coffee acreage in Central America and Southeast Asia have shaken the once prosperous economy of Uganda.

The situation in Africa South is particularly fragile. Asian tropical economies have always been diverse by comparison, and, belatedly, the lesson of diversity has been learned in Latin America. Sugar, already in oversupply, is not a promising alternative. Palm, peanut, and coconut oil—particularly African specialties—are the victims of the cholesterol scare as millions shift to other vegetable oils for health reasons. Much land

is expanding its hinterland with extensions of its railroad system into Burkina Faso and Mali. Inflation is kept to a minimum, and the country's credit is solid. Unprofitable government-sponsored industries are simply closed. The expulsion of expatriate labor offers a potential safety valve since one in five laborers is foreign.

Côte d'Ivoire has not found all the answers. Strikes are fairly common, and the government is frankly a dictatorship, however gentle. Half the adult population is still illiterate. The country has had to reschedule its foreign debt payments. Nonetheless, long-term prospects are excellent.

Benin, Togo, and Burkina Faso effectively provide the supply of migrants for more prosperous neighbors. It is the essential lack of development in all three that sets migration in motion. Each has mineral resources, but they are largely undeveloped. All carry large trade deficits, are overly dependent on agriculture, and export comparatively few items.

Togo is the least poor of the three. It has developed something of a tourist trade based on safaris.

TABLE A

Tropical crop trends in Africa South.

Country	1971–75	1976–80	1981–84	1985–86
	Coffee (1,000 tons)			
Côte d'Ivoire	243	264	263	185
Uganda	205	140	163	172
Ethiopia	171	187	216	200
Zaire	92	87	80	65
Malagasy	73	79	82	83
Cameroon	80	97	106	128
Kenya	66	85	92	98
	Cacao (1,000 tons)			
Ghana	401	281	196	198
Nigeria	229	168	161	150
Togo	21	14	12	9
Equatorial Guinea	14	7	8	7
Zaire	6	4	5	6
Congo	2	4	3	2
Sierra Leone	6	6	8	9
Côte d'Ivoire	217	324	416	445
Cameroon	110	113	115	120

SOURCES: *United Nations Statistical Yearbook,* various years; World Bank.

in tropical tree crops is unsuited to field crops because of potential erosion. Food crops may alleviate internal food shortages, but governments obtain limited revenue from their production. Windfall profits from tropical crops are at best temporary, and economies with multiple offerings are much better able to weather changing market conditions, price fluctuations, adverse weather, and disease.

Phosphates, long its most important export, have suffered from price fluctuations. Togolese dance groups are well known abroad, and their touring productions earn hard currency.

Benin's economy has suffered heavily in recent years as foreign laborers have been shipped back home from Nigeria and the Cameroons. Once widely employed as civil servants in French-speaking countries of Africa, many Beninois have lost their positions as host countries have trained their own replacements. Local food supplies might be adequate

were it not for the fact that Benin's farmers find better crop prices in Nigeria. Food crosses the unpatrolled borders easily, and the black market flourishes. Benin plans to extend its railroad to Niger, to serve as that landlocked country's port; but Niger also is a poor state with limited development. Large-scale investment in education is a government priority, but to what end is unclear.

Burkina Faso is one of the poorest countries in Africa and the least developed of the three. It is particularly vulnerable as an essentially agricultural state

ABIDJAN: AFRICA'S ANSWER TO PARIS

If commercial agriculture has spread over much of Côte d'Ivoire, industrial and service endeavors concentrate in its capital, Abidjan. The chief port, Abidjan houses one in five Ivoirians in a Paris-like degree of economic and population concentration. It was made into a seaport by dredging a canal across the sandbar that protects its lagoon. Large parts of the sandbar have been converted to tourist beaches backed by luxury hotels and condominiums.

The capital is best described as posh, featuring landscaped boulevards, a replica of the Arc du Triomphe, art galleries, and upscale shops. There is a superficial air of Paris in miniature outside, but a definite tone of native Ivoirian inside, the city's stores and buildings. Textiles and fashions, air conditioners, fine hardwood furniture, elegant shoes, enameled cookware, and wall hangings for sale are all domestically produced in Ivoirian factories and workshops.

Abidjan is not only the national capital, but the regional capital as well. Hundreds of European and American companies occupy office space in an ever-growing echelon of skyscrapers that march along the city's heights. A six-lane highway encircles this city, which boasts a safe, modern water supply, phones that work, and one of Africa's largest and safest airports. Elegant (and spotless) patisseries feature sinful chocolate creations and serve a variety of coffees, at least one of which will satisfy even the most discriminating tastes. There *are* slums, but smaller in number and size, and less intense in their poverty, than those normally found in Third World cities. Thatch-roofed palm wine bistros and impromptu bazaars amid

in the fluctuating rainfall regime of the Sahel. Up to a half-million of its citizens migrate seasonally each year to do harvesting or planting in the Ivory Coast. Renowned in the past for a corrupt and inefficient government, its new name means "land of the incorruptible man." The government has encouraged traditional cooperatives among farmers; a grass roots development has begun to take place through cooperative sharing of tools and technology.

Ethnic diversity plagues all three countries; each has 40 or more recognizable ethnic groups. The inanity of residual colonial borders is strongly in ev-

idence here—two ports with no hinterland, and a hinterland with no port.

Nigeria: Regional Giant. Nigeria is the super-state of Africa. With 98 million people, it has the largest population on the entire continent. Nigeria's size, physical diversity, impressive resource base, and oil revenues make it potentially the most successful African state, *if* it can achieve a united citizenry. The only federal state in the subregion, it has 12 provinces. Although all West African states have some internal problems associated with their multi-

Prosperous Abidjan. *Abidjan is West Africa's most attractive city. A bustling center of commerce and a magnet for European tourists, it is also the largest industrial center of Côte d'Ivoire. Manufacturing is carried on in modern, Western-built plants, but production has the flavor and flair of native Ivoirian crafts. Côte d'Ivoire is West Africa's richest nation. It is rapidly diversifying its economy and plans to become a supplier of industrial goods to all of Africa.*

Abidjan's luxurious surroundings appear more colorful than desperate, more interesting than shocking. Abidjan, like the country it governs, seems to be able to weather economic stress and repeated change without losing its luster, prosperity, or calm. Dakar smells of diesel fuel and peanuts, Lagos of oil and garbage, but Abidjan smells of chocolate, perfume, money, and success.

cultural, multilingual, multitribal status (Figure 9–12), Nigeria alone has suffered a prolonged and devastating civil war.

Nigeria has three distinct regions of high population density (see Figure 9–8). Islam is long established in the Northern savanna zone, and such cities as Kano are recognized cultural centers of Islam. The agriculture, natural vegetation and climate, cultural history, and languages of the North differ sharply from those of the South (see Figure 9–12).

In the South, although the physical factors of climate and vegetation are not markedly different from east to west, cultural differences are critical. While many southerners are of traditional animist religions, the dominant tribe of the Eastern region, the Ibo, has been strongly influenced by Christian missionary activity and has taken advantage of mission schools. Hardworking and eager for more education and wider opportunity, the Ibo impressed the British with their adaptability to administrative work and military service. Northerners and westerners in Nigeria have accused the Ibo of perpetuating themselves in federal positions and the officer corps, or favoring other Ibo at hiring and promotion time.

FIGURE 9-12
***Nigerian Languages and
Unity.*** Fully a dozen major
languages are spoken in Nigeria.
Muslims dominate in the north,
and Christians in the east, while
native religions are strongly
represented in the southwest.
National unity has proven
difficult to achieve in light of
this ethnic-religious diversity.
Violence erupted between
Christians and Muslims in 1987.
Civilian unrest has been kept in
check by military rule.

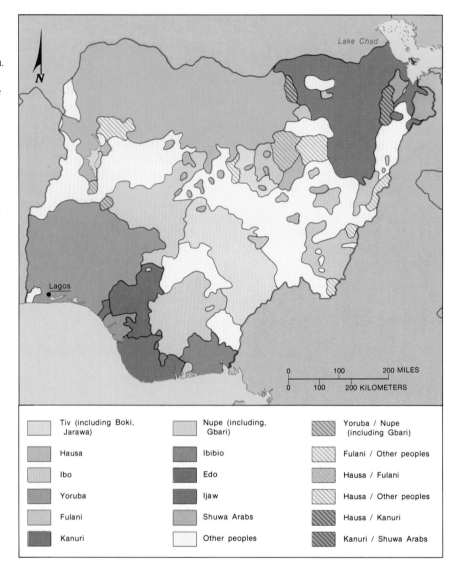

Now that northerners, a majority of the national total, have increased their relative share, the Ibo complain of discrimination.

Oil was first produced in the Niger Delta area (see Figure 9-10). Production costs are high, but the quality is excellent. Oil was a bonanza to what had been Nigeria's poorest region, an area previously dependent on subsistence agriculture and palm oil exports. Many easterners resented having to share this new wealth with the rest of the federation.

In 1967, the Eastern region, containing both the Ibo and the oil, left the federation to form the new state of Biafra. A civil war followed, lasting until 1970. More than 1 million people died, mostly in Biafra, when a federal blockade cut off food aid.

Nigeria has a brilliant future if the bitter legacy of civil war can be overcome. War-related damage to the economy has been erased. Eliminating intertribal and interregional mistrusts engendered by the war will take longer. The "giant of Africa" is taking steps

PROBLEMS OF RAPID URBANIZATION

Most of the world's largest cities are now in developing countries, and they are growing at unprecedented rates. Over 40 percent of the world's people live in cities; by the year 2000, the figure will surpass 50 percent. In 1970, the urban populations of developed countries outnumbered the developing world's urbanites by 30 million. By 1988, the urban populations of the Third World outnumbered those of developed nations by over 500 million.

Africa South is the world's most rapidly urbanizing region. Much of this growth is concentrated in slums and squatter settlements, many of them unserved by basic utilities. In 1960, Nigeria had only two urban centers with more than 500,000 people—Lagos and Ibadan. Twenty-five years later, Nigeria counts 11 "million plus" cities. The urban population of the African continent now is less than that of North America. By the year 2025, Africa is expected to have three times North America's urban population. African cities are currently growing at 8 percent a year, doubling their populations every 10 years. These incredibly rapid growth rates place severe strains on urban infrastructure, housing, employment, education, and health. In Lagos, Nigeria, for example, water is strictly rationed. In some sections of the city, residents must walk long distances to obtain water from scattered pumps that are turned on only for a few early morning hours. Sewerage is nonexistent for 80 percent of Ibadan's households.

Still, the rural-to-urban flow accelerates. Traditionally, cities have offered superior educational opportunities and longer life expectancies for one's children. However, these long-time advantages are the product of previously higher rates of "social benefit" investments in cities rather than in rural areas. It is questionable whether the burgeoning cities of the Third World can continue to provide both better opportunities and better living conditions than rural environments.

to ease tensions and reduce tribal-based discrimination.

Equatorial Africa

Unlike the ethnic and political mosaic of West Africa, the subregion of Equatorial Africa has a degree of homogeneity. It is composed mainly of fairly large states, and the French language and cultural influence provide at least a superficial unity. Climatically, virtually the whole area is either rainforest or wetter savannas; topographically, most of it is within the Zaire (Congo) Basin.

Despite these initial similarities, the region is inhabited by diverse peoples. The infrastructure of roads and railroads is sparse and unconnected. The region contains areas of extremely rich as well as grossly infertile soils. Resource-rich portions exhibit high levels of development yet exist in close prox-

Commercial Lumbering in Equatorial Africa. *Valuable hardwood trees are harvested on a gigantic scale in the tropical rainforests of Gabon. Rich resources, sparse populations, and corporate production are hallmarks of the economies of Equatorial Africa. Oil-rich Gabon is the richest country in the subregion.*

imity to some of the continent's least-developed areas.

This is "corporate" Africa. Huge plantations producing palm oil, coffee, and cocoa from seemingly endless rows of geometrically planted trees; giant mining complexes with huge wastebanks, discolored streams, and brightly lit multistoried cleaning plants; vast logging operations that disfigure the forest with huge scars wrought by macromachines; all attest to the corporate scale and scope of production. Familiar names of European-based conglomerates are emblazoned on signs and equipment, while workers' slogans proclaim the antithesis of capitalism in red banners at the same work site. Seeming contradictions are typical of the realities of Equatorial Africa. The gross production of agricultural and industrial enterprises awes observers with the sheer productivity of the region.

There would seem to be fewer problems here, as well as greater bounty. There is no drought or famine. Refugees are relatively few, and out-migration in search of work is not a necessity. Population pressure has not yet become a problem. Only Zaire appears desperately poor in its income statistics (see Figure 9–11a), but a lively black market and barter economy there make that a doubtful figure. Overall, Equatorial Africa appears rich and fertile with promise.

Cooperation may be the key to unlocking that promise. The region's individual economies are more complementary than competitive. The hydroelectric potential of the Zaire is staggering, and there are large deposits of coal, oil, and gas. The same climatic and chemical forces that combine to impoverish much of the soil concentrate iron and aluminum ores in vast commercial deposits of excellent grade. Rare and base metals of all kinds are mined in multiple deposits, and a full array of tropical export crops contribute to an amazing regional diversity. Coordination of regional investment and production would seem to offer a route to almost unbridled prosperity. But cooperative efforts, agreed to on paper, have not yet been put into effect. For all its success at the moment, the region is still characterized by commodity production and is subject to the unpredictability of world market prices and demand. Regional prosperity, at present, is more apparent than real.

The Zaire Basin. This subregion contains the states of Zaire, Congo, and the Central African Republic, and is the poorer part of Equatorial Africa. Not only is income lower, but international debts are greater and less manageable. These countries have had a fair degree of political turmoil in the recent past and appear to be in a state of economic pause while seeking new direction. All three are located largely within the drainage basin of Africa's greatest river, the Zaire. The Central African Republic stretches north across a low divide to the dry Chad basin, but its orientation is south to the tropical river more than north to the Sahel.

Zaire is the largest state in area in Africa South and ranks second in population. If it were to overcome the shortcomings of its inadequate transport infrastructure and resolve internal dissension among its varied peoples, it could be the African equivalent of Brazil—a large, resource-rich, economically diversified country on the verge of becoming a twenty-first century power. Because Zaire's boundaries are entirely the result of conflicting European colonial ambitions that ignored tribal areas, and because the Belgian administration apparently delayed the preparation of the Zairois for self-government, Zaire has strong potential for further internal conflict. Civil war that quickly followed independence in 1960 was essentially repeated in 1964, 1967, 1977, and 1978. The persistence of these **centrifugal forces** is related to internal disunity and to attempts by the Soviet Union, via its Cuban ally's troops in neighboring Angola, to subvert the Zairian government. This effort has been directed at one of the richest mineralized zones in Africa, Zaire's Copperbelt, in order to deny its minerals to the West.

Enormous plantations are the largest in Africa. About 80 percent of Zaire's agricultural exports —palm oil, rubber, coffee, cocoa, and tea—are produced on plantations. Small African-owned farms contribute most of the cotton and peanuts.

Zaire ranks 1st in world production of industrial diamonds and cobalt, 6th in copper, 8th in manganese, 9th in tin, and 10th in gold. It also produces silver, coal, lead, zinc, uranium, and a variety of rare metals and alloys. Copper prices are down and, as a result, Zaire's per capita GNP has declined precipitously since 1980. Iron ore reserves are large, and there is great potential for discovering oil. Most industry is located either in the Copperbelt's provincial capital of Lubumbashi or in the rapidly growing national capital of Kinshasa.

Most of Zaire's 34 million people are still farmers. Population densities are so light, however, that only 2 percent of the land is cultivated at any time. It is estimated that half the country's acreage could produce crops, so Zaire is among the African states whose potential for food production is hardly touched.

An interesting political reaction to Zaire's cultural fragmentation and thin veneer of European colonial culture is a determined, comprehensive program of "Zairization." The state and river both were to be known as Zaire rather than Congo, and colonial place names were replaced completely. A new flag and national anthem were introduced, and all individuals were legally required to replace "foreign" names with African ones. The Zairians are trying hard to develop a distinctively African, Zairian culture, and with it a loyalty to the idea of a Zairian nation-state. Zaire's fabulous potential is as yet unrealized; the state has the lowest per capita income ($160) of the subregion, and one of the lowest in the world. Unfortunately, the Mobutu regime must be ranked among the most flamboyantly corrupt in the world, illegally diverting many millions better spent on development.

Congo flanks Zaire to the northwest, extending 800 miles inland from the Atlantic. Another sparsely populated area, Congo has only 14 people per square mile. Its 1.8 million people are divided among 75 individual tribes.

Fortunately for Congo, its administrative capital, Brazzaville, became the colonial capital of all of French Equatorial Africa, a colonial amalgam that contained many current states. This position induced large capital investments in the city and its transport

GEOGRAPHIC FACTORS FOR UNITY OR DISUNITY IN THE CONGO BASIN

In July 1960, the former Belgian Congo joined the growing list of newly independent states in Africa. Within days of independence, the country had begun a near collapse into chaos. The economy foundered without direction. Katanga, the southeastern province of the Congo (see Figure 9–9), had declared itself independent, and remained so for several years.

Motivations behind Katangan independence movements are related to the physical, economic, and cultural geography of the Congo basin, and to its history. The Congo state was a product of conflicting European imperial ambitions. Belgian King Leopold hired the explorer Henry Stanley to sail down the Congo River and to establish a territorial claim to the entire basin for a private stock company, the Companie du Congo, headed by the king. The European great powers agreed on the Congo basin becoming, in effect, the world's largest private plantation, because it would act as a buffer state separating conflicting European colonial interests.

To control and exploit this vast territory with the handful of Europeans actually there, the Companie du Congo adopted the familiar colonial tool of "divide and rule." Historical animosities of different linguistic-cultural groups and tribes were taken advantage of by selecting the most aggressive tribes as "native police," better paid than other Congolese, to enforce the will of the Company. Loyalties, then, focused on the tribe and its traditional allies and enemies, rather than on any state. This factor, aided by the failure of the colonial regime to provide higher education for more than a few Congolese, helped lead to civil war and a separate Katangan state.

The Zaire is unique among the world's largest rivers in having rapids and waterfalls near its mouth. The hydroelectric potential is immense, but the barrier to navigation is equally enormous. In developing the transport system of the basin, the Belgians stitched together a patchwork of navigable sections of the river with short rail links around navigational barriers. Such a disjointed system, though cheaper to build, is much more expensive to operate. Each transfer of cargo between different modes of transport (riverboat, railroad, ocean steamer) is a labor-intensive bottleneck.

The Shaba (Katanga) province holds a rich copper deposit and produces uranium (see Figure 9–10). A transcontinental east-west railroad was built to link the Copperbelt with ports on the Atlantic and Indian oceans. This international transport system was far more efficient than using Congo's internal river-rail system. Thus, Katanga could (and did) produce for and trade with the world without the necessity of dealing with the rest of Congo. Its lack of common interests or links with the rest of Zaire is a basic cause of separatist movements.

and communication facilities. As a result, Congo, a country relatively poor in mineral resources and cultivable soils, has prospered. Directly opposite Kinshasa, Brazzaville is the terminus of a railroad from the Atlantic port of Pointe-Noire (see Figure 9–10). Riverboats connect Brazzaville with inland market towns, and the railroad has been extended from Brazzaville to the northern interior. Nonetheless, the north remains practically empty. Some 75 percent of the population is located in the vicinity of Brazzaville and Pointe-Noire, and 60 percent of the population is urban—unusually high for Africa. Pointe-Noire handles Congo's imports and exports and much of the international trade of the Central African Republic, Chad, and Zaire.

Most Congolese are subsistence farmers, producing a wide variety of tropical crops; tropical wood, not crops, is the most important export. Potash mines recently opened in the vicinity of Pointe Noire. There are small deposits of oil and gas, and production exceeds domestic needs.

French interests are dominant, but there is an aggressive campaign to attract other European investment. The construction of a highway to the north is opening lands to colonization, part of a project designed to slow migration to overcrowded cities.

Poorest in both fact and potential is the Central African Republic. Its capital, Bangui, is at the head of navigation upstream from Brazzaville. There are no railroads and, until this century, there were no urban places. Now, under pressure of drought and commodity price declines, towns are growing so rapidly that parts of the countryside have been depopulated. Diamonds account for about 40 percent of exports. Some uranium ore is mined, and it is believed that there are large deposits of iron and a dozen other metals.

The Atlantic Facade. The Atlantic-oriented states are generally in better economic shape than those of the Zaire basin. Gabon is by far the richest state in Africa South, surpassing even the Republic of South Africa. Cameroon, modeling its economy on that of the Ivory Coast, is also quite prosperous. Even the former Spanish and Portuguese possessions, though less well developed, do not rank among the continent's poorest, least developed states. Diversity of agricultural production is the aim of all the states of the subregion. Industrialization has begun to de-

velop at a significant pace. At least for the present, the picture is bright.

Cameroon was originally a German colonial holding. Unlike most other German holdings of the day, it had considerable economic potential, and thus received a disproportionate share of German investment. Rich volcanic soils were cleared and developed for plantations on the slopes of the Adamawa. Because Cameroon was to supply virtually all German needs for tropical goods, there was a great deal of diversity in agriculture from the start.

Unlike many African states, Cameroon processes many of its agricultural commodities. Factories roast coffee and can the ground product for export. Cacao is processed into chocolate in a variety of forms. An aluminum works fully processes domestic bauxite, and small firms make consumer products from rolled aluminum. Boards, veneer, plywood and, now, furniture gradually supplant the export of raw timber.

The impetus to this domestic growth has been oil (see Figure 9–10). Declining oil prices have not depressed this exceptionally diverse economy as sharply as has been the case in many other oil economies. A pragmatic government never used oil revenues as a significant portion of the governmental budget, reserving them instead for capital investment. Dropping oil revenues may affect long-term growth, but they do not interrupt the ongoing economy. Growth continues at an annual rate of 7 percent, well above population growth rates.

There are weaknesses, however. Sewerage, water, and electrical services are inadequate for current needs. Northern Muslims fear dominance by southerners. Smuggling has heightened tensions between Cameroon and its neighbors. There is a territorial dispute between Nigeria and Cameroon over offshore oil (see Figure 9–10). On balance, however, this is one of Africa's more stable and successful economies.

With a per capita income of over $5,000 and a GNP of close to $4 billion, Gabon does not fit the normal patterns of a Third World country. It is atypical of Africa South in many other ways as well. There is little ethnic diversity in this thoroughly modernized, yet very conservative state. The dominant Fang moved into Gabon in the same time period as the first hundred years of U.S. independence. Clever, but not aggressive, they assimilated most of the indige-

LAND ROTATIONAL AGRICULTURE

To an untrained eye, a tropical farm may look somewhat messy and neglected. To a weed-conscious Westerner, **clean tillage** is the norm. Such widely used descriptions as "slash-and-burn" and "primitive farming" have negative connotations. Attempts to supplant native African systems with those of other cultures, however, have often proven disastrous. Traditional systems still give the best results.

The term **land rotational agriculture** is both apt and has a more positive tone. All societies rotate crops and fallow periodically as a means of conserving fertility. In the tropics, land rotational agriculture may be the most practical land use. Generally, the land is held in common by a village. Each year, a family is assigned a plot within an area that is cleared by community efforts. Rotations can be systematic over a 10- to 30-year cycle, or they may be determined by nature with the lushest growth area being the one selected for the next clearing. Trees are killed by girdling; debris is carefully heaped to ensure drying for easy firing. The ashes are a source of potash fertilizer for the upcoming crop. Unburned debris is rapidly infested with termites, which in turn improve conditions for farming by aerating the soil and adding calcium. The soil is hoed rather than plowed; stumps are allowed to remain in place. Both practices minimize erosion. Vegetation is permitted to spring up between rows or hills of crops. Periodically, and never all at once, this green matter is chopped and turned under as a source of nitrogen from rotting vegetation. Phosphorus is supplied by dressings of burned bone, eggshell, or chicken droppings. Thus, all necessary plant nutrients are met within this traditional system.

Intercropping is the usual practice; root crops, corn, beans, melons, and tree crops share common space. Land may be abandoned after one year, or grow successive crops over a two-year cycle. In either case, it is then abandoned to a long fallow—allowing natural restoration of fertility. Most tropical soils are not able to produce more frequently without massive fertilization. It is estimated that the traditional system could support up to 150 persons per square mile in some wetter rainforest areas, and up to 30 per square mile in areas of bush and scrub. Thus, the system is reasonably productive.

nous population to their language and culture. Their native values were, and remain, surprisingly similar to those of the later-conquering French: they are status conscious, frugal, determined to pass on wealth to their children, and attach considerable importance to the land they own. Libreville, the flourishing capital, was founded as a settlement for freed French slaves, but there was little conflict between these newcomers and the sophisticated Fang.

Gabon is rich in oil and is Africa South's second largest producer. Oil profits, spread over a population of only 1.4 million, created almost instant riches. The economy is overwhelmingly dependent on mineral resources, but it has not suffered declines because reserves are large and production costs are very low. It is the world's largest exporter of manganese and has picked up many markets formerly served by South Africa. It is France's chief source of uranium for an expanding nuclear generating industry.

Gabon has few negatives. There is no religious strife, because the country is 75 percent Christian, with no vocal or large identifiable minorities. The population growth rate (1.5 percent) is lower than the world average of 1.7 percent. There is no great public interest in governmental change as progress continues. Health standards are reasonably good, and the facilities at Lambaréné, begun by Albert Schweitzer, have been enlarged and modernized. Gabon serves as a clinic for the entire subregion, and Lambaréné has begun to emerge as a center for research on tropical diseases and new vaccines.

The partially submerged peaks of the Adamawa Mountains form islands offshore that are richly endowed with fertile volcanic soil. Some are a detached part of Equatorial Guinea; the others form the small republic of São Tomé and Principe. Land reform followed independence, but small farms are not as productive as the large plantations of the past. In the midst of rich fishing waters, these two countries seek to supplant food imports with a combination of fish and increased food crop production. Tourism is well established and growing.

The small successor state of Equatorial Guinea was Spain's only colonial holding in Africa South. Whereas Portugal's large African holdings remained poorly developed, Spain's tiny equatorial colony received considerable investment. Despite the exploitative nature of Spain's control, independent Equatorial Guinea is fairly well developed, and its income and living standard place it comfortably within the more prosperous northern subregion of Equatorial Africa.

The African Horn

Ethiopia, Somalia, and Djibouti are included in this subregion. Ethiopia is the oldest continuously independent state in Africa, while Djibouti, a former French colony, is one of the newest. The third unit, Somalia, is a merger of former British and Italian coastal colonial holdings. Ethiopia, plagued by drought and an inefficient, overambitious government, has become one of the most impoverished states in the world. Per capita yearly income is only $140. Djibouti, a ministate the size of New Hampshire, has no natural resources other than a good harbor. Somalia has survived, and indeed developed, by a dependence on foreign aid (first Soviet, now American) and the benevolence of oil-rich Arab states with which it has had a long, historical association.

The subregion is actually one of Africa's poorest, but need not remain so. Whereas the coasts are largely deserts, the highland core is exceptionally well watered. Streams with their sources in this highland irrigate some lands along the Horn's coast and contribute to the flow of the Nile. The water resource has not been fully utilized. Rich volcanic soils prevail over most of Ethiopia, while even those of the surrounding desert can give excellent yields with irrigation. Generally, except in years of drought, the area has produced a crop surplus. Livestock exports to the Arabs form one of the world's longest-established currents of trade. Herbs, gums, and extracts of commercial value have been gathered from the thorny scrub of the semidesert for millennia. Opulent kingdoms rose and fell in this, the fabled land of Punt, when Europe was in its infancy.

The Horn of Africa has always been one of the great crossroads of the world. Here, the ongoing dispute between herders and cultivators over possession of the waters and the land has never been fully resolved. Semite, Hamite, and Bantu peoples have alternately controlled, invading and conquering, retreating or becoming subject, sometimes mingling and intermarrying, but virtually always in conflict. Here the philosophies of Islam clash (and mix) with those of Christianity and native religions.

Ethiopia, the largest state in the Horn, was first described in Western writings by the Greek historian Herodotus. Strategists, anthropologists, and scholars of all types have commonly cited Ethiopia as an example of the significance of topography to defense, cultural isolation, and political independence. The rugged Ethiopian plateau consists of volcanic lava layered on top of sedimentary rocks. It has been split by the trenchlike Great Rift Valley (see Figure 9–7), and the volcanic materials are thought to be associated with the formation of that feature.

Ethiopia the historic state, and Ethiopia the rugged volcanic plateau, have long been synonymous.

The ancient Christian kingdom was long under pressure from Muslim peoples, who surrounded and isolated it in the process of conquest. Originally peopled by Hamites, the present state was constituted when conquering Semites from across the Gulf of Aden invaded and established control.

Ethiopia was (and is) a true empire. It has always contained many different peoples and regions, joined with the original highland core by conquest and held together by the power and authority of the emperor. A newer government now attempts to hold the territory together in the name of Marxist philosophy. The long struggle between Christians and Muslims for political power is related to the balance of religions in the population, about 40 percent Christian and slightly more than 40 percent Muslim, with the remainder adhering to traditional native religions. In the 1974 revolution, the last emperor was deposed, but the Christian church remains extremely powerful. Ethnic groups with strong national ambitions straddle international boundaries here and lie within the control of an alien group. The Horn of Africa is one of the region's least stable areas.

As an outcome of World War II, Ethiopia was awarded coastal Eritrea, a former Italian colony. This province is culturally mixed. The United Nations also gave Eritrea a degree of autonomy that Ethiopia has not been prepared to honor. The continual erosion of local autonomy has resulted in an Eritrean independence movement. Despite central governmental military actions, the guerrilla war continues unabated.

Ethiopia has long had a high level of recognition in the United States. Alone among all the states and peoples of Africa, Ethiopia once successfully repelled European colonial ambitions. In 1935, Italy avenged this humiliation, but the initial sharp defeat of a technologically superior European power gave Ethiopia heroic status. Many black churches in America added the world "Ethiopian" to their titles as a measure of recognition and pride.

The Mengistu regime, after overthrowing the emperor, turned to Soviet bloc aid in confronting the internal insurgency movements. Mengistu proclaimed Africa's first fully Communist state and attempted to force the Marxist-Leninist model on agricultural and industrial production. Food production plummeted. A "red terror" campaign left many of Ethiopia's best-educated people dead or in exile. The ongoing struggle against secession-minded provinces has worsened the severe famine. At one point, 500 children a day were dying of starvation in refugee camps while the government kept grain shipments waiting on board ships, unloading military

Troubled Ethiopia. Drought and starvation have recently plagued this fertile and normally lush highland region of Ethiopia. This village, near Harrar, illustrates the cultural diversity that disrupts national unity. Thatch-roofed circular huts are characteristic of lowland Equatorial Africa. The mosque at the right attests to the local religious preference in a country long ruled by Christians. The villagers are Hamites who speak a variant of the ancient Cushitic tongues.

"VILLAGIZATION" IN ETHIOPIA

Two thirds of Ethiopia's 46 million people live in abject poverty, many at the line of survival. The state's severe food shortages have complex origins but are not aided by the central government's insistence on **villagization** of the rural population. Although it is said to be a voluntary program, foreign observers see it as a forced relocation. The objective of the plan is to move 30 million people from remote, scattered homes to new village centers, ostensibly to provide better access to water, electricity, and clinics. There is no doubt that this plan would facilitate the provision of public services. However, food output has dropped because farmers are forced to spend considerable time "commuting" on foot from their now-distant homes to their fields—time that would be better spent working. Uncooperative families have seen their houses burned. Entire village populations have fled across the border to neighboring states. Refugees leave at a rate of 1,000 a day.

It is believed that the ruling Workers Party may be forcing the relocations for purposes of internal security, reasoning that residents of large villages are more easily controlled and kept under surveillance. On a less negative note, villagization will increase the ability of the government to channel food into starving cities. Whatever the motivations, the resulting rural chaos and reduction of badly needed food harvest are a terrible burden on Ethiopia's poor farmers and consumers.

---

supplies first. Scarce resources have been squandered in building heroic statues and importing luxuries for the new elite.

Ethiopia is one of the earliest centers of domesticated plants, an **agricultural hearth**. Although it is the probable region of origin of three major varieties of coffee trees and produces some of the finest-quality coffees, the value of the crop is reduced by a lack of quality control.

Ethiopia could be self-sufficient in food with a more rational, incentive-based system. More than half the country is good quality grazing land, and cattle exports could have a bright future. Ethiopia's biggest potential natural asset is water. The plateau is the source of the Blue Nile and a dozen other rivers. The country's hydroelectric potential is enormous; irrigation potential is vastly underutilized. Any economic development plans must first consider the

provision of transport. Almost half of all Ethiopians live more than a day's walk from any paved road.

In addition to its potential breakaway province of Eritrea, Ethiopia faces serious problems in its relationships with Somalia and Djibouti. Somalia claims Djibouti as its own territory. The people are largely Somalis, but Ethiopia would lose its only access to a port. Somalia also has waged a continuing war in Ethiopia's Ogaden region in an attempt to incorporate that territory as well (see Figure 9–9).

Somalia has 7 to 8 million people, 60 percent of whom are nomadic or seminomadic pastoralists in a land of desert and thorn scrub. Less than 1 percent of the land is under cultivation. Somalia is a rarity among African states in the homogeneity of its population: 98 percent are ethnic Somalis. There are more camels and cattle than people, and sheep and goats outnumber people three to one.

Other than livestock, exported live and as hides and canned meat, Somalia's basic asset is its strategic 1,700-mile coastline in the area where oil from the Persian Gulf flows either up the Red Sea and through the Suez or around Africa to Europe and North America. This strategic significance is not unknown to the Soviets, who donated a new deep-water port at Berbera, directly south of Aden. Berbera's transformation into a major refueling base for the U.S. fleet is a direct result of Soviet support for Ethiopia and subsequent Somali disenchantment with the Soviets.

Djibouti is a small pocket of former French territory that surrounds a superior harbor, the logical port for much of Ethiopia. Its trade services to Ethiopia are its only resource. The Addis Ababa Railway makes a profit because Ethiopia's own ports are not connected by rail to the capital or the densely settled highlands. Ninety percent of the country is desert wasteland. The upgraded, containerized port facility is East Africa's best. Its political future is in doubt, however, because it is the object of both Ethiopian and Somali territorial claims.

East Africa

The East Africa subregion includes Tanzania, Kenya, Uganda, and the small states of Rwanda and Burundi (see Figure 9–9). There are no mineral-rich areas comparable to the Copperbelt or Nigeria's "oil delta." The five countries of East Africa are primarily agricultural, and all are comparatively poor. Indeed, even the agricultural potential is limited. Large areas of Kenya are semidesert, and vast areas of Tanzania have erosion-prone soils of exceptionally low fertility. More than half the area suffers drought and unpredictable rainfall. Even the rich and well-watered soils of Rwanda and Burundi are so densely populated that there is little or no room for agricultural expansion. Uganda is the only area that seems to contain all the ingredients necessary for a successful agricultural economy; unfortunately, it has been plagued by disruptive civil war.

This is a familiar part of Africa to Americans. Numerous Hollywood movies have been photographed on location here, and Sunday afternoon television has featured the area in many wildlife conservation and safari programs about East African fauna. The temperate highlands of East Africa, lands of virtually

perpetual spring in European perceptions, were deemed suitable for white settlers, and colonization was actively pursued. Crop yields were astounding on virgin soils. Promoters envisioned the emergence of white-run states of exceptional wealth in East Africa. Time, native nationalism, and the truly limited areas of highland suitable to cultivation eclipsed that unrealistic dream.

East Africa is a zone of contacts among major African ethnic groups—the Bantu, Nilotic peoples of the upper Nile, nomadic Hamites (Somalis), and the remnant pygmy groups in Burundi. Tiny Rwanda and Burundi are the least complex ethnically, each with Hutu (Bantu) majorities (89 to 95 percent) and Tutsi (Watusi) minorities. In contrast, Uganda's largest single group, the Baganda, forms only 1/12 the total population, while Tanzania's people are divided among 130 ethnic groups. Kenya is somewhat less heterogeneous, with its largest group, the Kikuyu, forming 20 percent of the population. Remaining white settlers in the highlands and Indian migrants, imported as workers but eventually to emerge as a class of businesspeople, complicate the ethnic picture further.

Islam and Christianity have both been active here. The sultan of Zanzibar established suzerainty over much of the East African coast before European colonial control. In Tanzania, almost a third of the population adheres to Islam, though Christianity dominates in all other states. The particular type of Christianity practiced varies considerably, so religious unity is lacking (Table 9–2).

East Africa, with a fairly high degree of geographic uniformity, is a microcosm of the cultural and societal diversity that affects Africa. Severe internal unrest has characterized four of the five countries in the subregion. At times, civil war has spilled across borders, complicating regional relations.

Recently, Kenya, Uganda, and Tanzania have normalized relations and have begun to plan and undertake joint projects. Unpopular (but necessary) austerity measures are paring trade deficits, while international banks and funds are limiting lending and restructuring runaway debt. There is an uneasy peace in most countries, and even governmental stability in Tanzania. Conditions are improving, but the level of development remains low. Diets remain well below FAO (United Nations Food and Agricultural

TABLE 9-2
Cultural diversity of East Africa.

Country	Religion (%)				
	Catholic	Protestant	Total Christian	Native	Islam
Kenya	(23)	(37)	60	37	3
Uganda	(20)	(30)	50	40	10
Tanzania	(not available)	(not available)	35	28	37
Rwanda	(47)	(8)	55	43	2
Burundi	(54)	(6)	60	38	2

Country	Language and Ethnicity			
	Official Language	Lingua Franca	No. of Other Major Languages Spoken	Largest Ethnic Group (%)
Kenya	English, Swahili	English, Swahili	8	Kikuyu (20)
Uganda	English	English, Swahili	18	Baganda (12)
Tanzania	Swahili	Swahili	30	Masai (4)
Rwanda	French Kinyarwandu	French, Swahili	2	Hutu (90)
Burundi	Kirundi, French	French, Swahili	2	Hutu (85)

SOURCE: *Statemen's Yearbook*, 1981, 1986, 1987.

Organization) food intake requirements, (see Figure 9–11b) and income per capita is among the lowest on the continent.

The two states of Rwanda and Burundi would likely not have emerged as independent countries were it not for the vagaries of colonial politics. Both are part of a remnant of German East Africa that was detached and assigned to Belgium while the bulk of German holdings were mandated to Britain.

Neither of these two nations suffered from the great depopulators of Africa South—the slave trade or the disease-bearing, lowland-infesting tsetse fly. Both are characterized by some of the highest population densities in tropical Africa. With about 95 percent of each country's people in subsistence agriculture, these crowded areas of high relief are being intensively farmed by people who lack the knowledge and tools necessary to minimize erosion. There are frequent famines and wide variations in coffee production (the major export crop). Rwanda has an ideal climate for tea, and tea exports are expected to diversify its one-crop economy.

Both states experienced postindependence conflicts between the Hutu and Tutsi that led to hundreds of thousands of deaths. The Tutsi, a minority in both countries, were the ruling class. Cattle herders of amazing physical stature and strength, the Tutsi enjoyed a privileged economic position. Hutu majorities farmed the land to produce food for the Tutsi and their cattle, while Tutsi cattle were contracted out to Hutu farmers for fattening in return for small amounts of meat. The Hutu people were under obligation to perform personal service for the Tutsi in return for protection. These inequalities eventually led to civil war.

There are brighter spots. The Burundi state breweries have been greatly expanded and now export gross quantities of beer to the mining towns of Zambia and Zaire. Tea and pyrethrum (for insecticides) crops are diversifying agriculture, and Rwanda's government is making a valiant attempt to slow population growth. Both countries seek economic integration with East Africa and with the Central African Economic Community. Economic union will depend largely on their ultimate integration into a planned (but not yet completed) rail network across Africa from Matadi (Zaire) to Mombasa (Kenya) and Dar es Salaam (Tanzania) on the Indian Ocean (see Figure 9–10).

The Republic of Uganda has had so troubled a decade that its per capita income, once the highest in East Africa, has experienced a rapid decline. It suffered one of the most murderous dictatorships in the modern world under the brutal Idi Amin. Economic and social recovery can be expected to be slow after 10 years of Amin and another decade of civil war.

British colonial policy had early forbidden large-scale white land ownership in Uganda, preventing the bitter racial civil war that ultimately plagued Kenya and Zimbabwe. The relatively cooler uplands support a dense population of native farmers. Most of the country lies between 3,000 and 5,000 feet. Shallow Lake Victoria is a major water resource, and a source of the Nile. The great dam at Owens Falls, the outlet of Lake Victoria, generates enough power for a large industrial enclave at Jinja. Kampala, the capital, is linked by rail with the Kenyan port of Mombasa, although relations with Kenya have sometimes deteriorated to the point of closing the border. In short, Uganda has excellent physical resources for agriculture, and a better-than-average economic infrastructure. Its negative economic growth in recent years is entirely attributable to governmental mismanagement and internal dissent.

The largest tribe in Uganda, the Baganda, occupy the fertile and productive shores of Lake Victoria. More than a century ago, the Baganda had organized a good road system and set up a system of chiefs loyal to a "chief of chiefs," the fabled Kabaka of Buganda. This well-established system forced the British to deal with the Kabaka and his chiefs as formidable powers. As independence approached, the Baganda were determined to safeguard their traditional leadership. The hereditary Kabaka of Buganda was elected president, but was overthrown. His successors were not Bagandans and disestablished both the kingdom and the special privileges that the Baganda had enjoyed. Under Idi Amin, about 50,000 Asians, mostly Indians, were abruptly expelled. Unfortunately for the Ugandan economy, the confiscated Indian shops, businesses, and factories did not prosper in unskilled hands.

Recovery since the overthrow of Amin has been erratic. Any rise in prices in the notoriously unstable coffee markets could contribute to recovery, since Uganda is normally the third largest coffee exporter in Africa. On the other hand, large acreages have been diverted from coffee and other commercial crops to promote self-sufficiency in food. Because of 20 years of civil upheaval, almost half the farmland remains unused.

About 4 percent of Kenya's land area is cultivated, a figure that could be doubled. A growth rate of between 4 and 5 percent annually—the world's highest—puts this additional land resource in grim perspective. The highly productive highlands area has been both a positive and a negative factor in the Kenyan economy. The fertile highlands were opened to white (mostly British) settlers. The Kikuyu farmers and Masai cattle herders who occupied the highlands had been decimated by disease. The depopulated highlands looked empty to incoming whites who assumed that their incursion would have little effect on Africans. The "white highlands" coincided with most of the best farmland and became highly productive. Diversified farms produced both tropical commodities and crops usually associated with the midlatitudes. In 1952, the Kikuyu rebelled, and Kenya eventually achieved independence in 1963. The British government bought most of the white-owned farms and redistributed the land to Africans.

Kenya's tourist business has thrived because of Kenya's varied and picturesque scenery, its reputation for political stability (until recent years), and its foresight in setting aside 6 million acres in 10 national parks and four game preserves equipped with luxurious visitor accommodations. In this way, Kenya's wild animal life is converted into an economic asset. There is considerable pressure on the government to convert park and game land to farms, since only 10 percent of Kenya is suitable to cultivation. Stability is threatened by a declining economy, while external relations are threatened by territorial claims advanced sporadically by Uganda and Somalia. Nairobi, an unusually attractive city with an excellent climate and a major international airport, is becoming a convention center.

Tanzania has an unusual, peripheral distribution pattern to its extremely diverse population of 24 million (see Figure 9–8). The high, arid plateaus of the nation's center are sparsely settled compared to the Indian Ocean coast and lakeshores, or the slopes of the volcanic mountains. This pattern of far-scattered population nodes places pressure on a Tanzanian transport network composed largely of railroads and quite dense by African standards (see Figure 9–6).

Tanzania is a result of the voluntary merger of the former British mandate of Tanganyika with Zanzibar (consisting of the islands of Zanzibar and Pemba about 20 miles offshore). These islands once were Arab trade centers involved in the lucrative slave and ivory trades. Zanzibar and Pemba produce most of the world's cloves.

Tanzania is making considerable effort to industrialize, but improvement of social services receives the largest share of government investment. Most industry is located at the port of Dar es Salaam, with smaller nodes in the interior and at Zanzibar City (see Figure 9–9). Tanzania has moved its capital from the old colonial site of Dar es Salaam to the more central location of Dodoma in the sparsely settled plateau, attempting to refocus national attention and development to its interior.

The Savanna Transition

The physical uniformity of much of the Savanna Transition subregion is undeniable. The area is classic African plateau, with only a few areas of dramatic deformation. Only the great gorge of the Zambezi and the tail end of the Rift Valley break its monotonous, seasonally bleak surface. The *siombo*, a dry, thorny woodland, incompletely blankets its surface, thickening along streams, thinning to grasslands and bush where water is less available. The coastal lowlands are narrow and almost harborless on the Atlantic. Only in Mozambique is there a generous, but swampy and malarial lowland. This is tsetse fly country, so cattle are few outside favored locations and often are sickly and thin (see Figure 9–4).

Soils take on a chestnut color, darkening with yearly droppings of humus and the perennial death of grasses in the parching dry season sun. There is no noticeable winter, only drought. The winter season is slightly cooler, but the extremely low humidity allows for the piercing penetration of a merciless midday sun. The dry season is quiet; life is at rest, restricted to the neighborhoods of water holes or

asleep in the meager shade of leafless bush and tree. Only insects move in dull, lethargic flight, their buzz and hum at times deafening in the otherwise silent landscape. There are few people except in favored areas.

The landscape is never lush. Austere in the dry season, brown and nearly lifeless, its angles and planes are simply softened in the wet season with tiny green leaves and a carpet of shimmering grass. There is promise in the seasonal change, almost reflecting the still-unused potential of the land itself. The soils, light and easily eroded, often possess fertility, but their appropriate method of utility remains something of a mystery.

Where crystalline rock is blanketed with sediments, there is coal, and the potential for oil or gas. Metallic minerals are found in abundance from the Zairian border to South Africa along or near the eastern, upturned edges of the plateau surface as it approaches the Great Rift. Copper, lead, and zinc are produced in abundance, but are only a part of the area's riches.

There are six major states of the subregion, many with "new" native names, not yet familiar, replacing past designations that sought to perpetuate the exploits of European explorers.

The whole is an area of transition from tropical rainforest to the deserts of the African southwest or the cooler, subtropical climates of the southeast. It is an area of cultural transition between native, nationalist Equatorial Africa and the white-dominated economy of South Africa. It is undoubtedly the strangest collection of incompatible political bedfellows ever found in one area. Permeating all lands is the sinister intrigue of a white-dominated South Africa, temporarily supporting whichever group will foster its political aims. Guerrilla wars, civil unrest, urban riots, tribal conflict, economic boycott, and a constant shifting of uneasy alliances are characteristic. It is an area in constant, disruptive unrest.

The Marxist Margins. Angola and Mozambique share the heritage of past Portuguese control. In common with most of their subregional neighbors, their independence came quite recently. The Portuguese occupation was not particularly beneficial. In Mozambique, a high exploitative, notoriously corrupt and inefficient private company exercised control; the Portuguese government was relegated to a position of ineffectuality. Primary investment in both units was done more by British, French, and other national interests than by the Portuguese. Portugal itself was poor, so its ability to invest was limited.

Programs of assimilation were widely touted, but assimilated native elements had few rights and enjoyed little economic benefit. The railroads, built and owned by European conglomerates, and the harbors, provided by nature, were the most important assets. Both were utilized to ship the mineral wealth of the interior and South Africa's Rand to the factories and markets of Europe. Transit tolls and taxes were the net gain. The colonies remained largely undeveloped.

In an area lacking natural harbors, the Portuguese early claimed and settled the finest harbors south of the equator. Luanda and Lobito in Angola, and Maputo and Beira in Mozambique, were ports of call on Portugal's trade routes to India in the sixteenth century. Except for Luanda, they are still ports of international significance, serving interior mining and agricultural centers from Zaire's Shaba region to South Africa's Rand when political considerations do not interrupt economic relationships. Maputo in Mozambique is the closest port to the Witwatersrand-Pretoria-Vereeniging area, Southern Africa's sole major industrial area and one of world-class importance (see Figure 9–10).

The long period of terroristic guerrilla warfare that finally overthrew Portuguese domination has never ended. Angola's government long could not control its territory without the assistance of Cuban troops. Oil production is mainly from the tiny enclave of Cabinda just north of the Zaire River where separatists wage a hit-and-run war for independence. The Benguela Railroad has been closed for years because of constant harassment by guerrillas.

Despite all its problems, Angola shows a favorable balance of trade. The Communist government pays its bills, honors contracts, and carries on about 80 percent of its trade with the West. Curiously, Communist Angola's oil has been developed by the United States. Much coffee land, abandoned by the fleeing Portuguese, remains unused—a reservoir of highly usable land for the future. Civil unrest, however, interrupts food production, and malnutrition has become a major problem.

Paradoxes abound in Angola, where Cuban troops protect U.S. oil company pipelines and refineries and the also-Marxist opposition to the government gets aid and seeks recognition from the United States. South Africa supports and arms the opposition, while Angola gives aid, arms, and refuge to Namibian

Copper and Social Welfare in Zambia. *Zambia relies on copper exports for most of its national income. Depressed world demand and lower prices for copper have hurt the economy in recent years. Some copper revenues are used to finance housing and construction of public utilities. Migrants from rural areas flock to Lusaka, the capital, in search of work. Efforts to provide work for the urban unemployed include training in construction. A crew of government trained employees supervises unskilled volunteers in the building of municipally financed cooperative housing.*

troops fighting South Africa. The web of intrigue defies all logic, and the continued survival of the economy is amazing in light of the chaos and conflict.

Mozambique is no more stable than Angola, but it still earns revenue from international railroad traffic. The link with South Africa is particularly strong, with that country sending part of Mozambiquan migrant workers' earnings to the Mozambique government in the form of gold. South Africa maintains a mutually profitable working relationship with a black regime that detests and combats apartheid. If all migrant workers were to return "home" suddenly from South Africa, rural Mozambique would have to support many more people.

Internal problems facing Mozambique are similar to those of Angola, but Mozambique's opposition movement is composed of conservative native elements opposed to Marxism, Christianity, Western influence, capitalism, apartheid, and industrial development at one and the same time. Supported by South African funds, they have little or no apparent public support. Nonetheless, they have been effective in disrupting the economy and have engaged in repeated attacks on the all-important rail lines.

State farms in Mozambique are being abandoned in favor of native communal holdings. Long before the Gorbachev reforms in the USSR, Mozambique purged its public enterprises of corrupt officials and superfluous bureaucrats. Basically, however, communism has not moved the economy forward. Natural disasters have made conditions even worse. Despite its strong opposition to apartheid, Mozambique has had to come to terms with South Africa. As South Africa has gradually withdrawn support from the opposition, it has agreed to prohibit bases of the African National Congress and to expel its activists.

Zambia, Malawi, and Zimbabwe: The Thoroughly Capitalist Interior.

In the late colonial period, Britain attempted to collect these three states into a confederated holding. Britain was well aware of the problems of both landlocked states and individual units too small or too highly specialized for economic survival. The idea was basically sound—creating large, integrated economic units in which leadership would emerge, interests converge, and sound economic development could precede independence. The confederation did not survive. However, recent attempts at joint economic activity in East Africa, as well as in this subset of states, point to the inherent wisdom in the British plan.

Zambia is the largest and least populous of the three states. Rich in mineral resources, much of it is a difficult physical milieu for agriculture. Large areas are covered by soils of exceptional infertility, and the tsetse fly, the carrier of sleeping sickness, is endemic (see Figure 9–4). Because of the state's inadaptability to agriculture, the economy is based largely on mining, and half the population is urban.

Zambian dependence on mining has an unhealthy economic impact in that prosperity depends solely on copper production and prices, both of which have declined in recent years. In an attempt to overcome this dangerously single-faceted dependence, the government has done much to diversify production. Zambia grows more of its own food, and a series of new food-processing plants handle both domestic production and imported cattle and grains. Zambia now refines its imported crude oil, produces fertilizers, and has created a small chemical industry. Unfortunately, efficiency is low in some of these enterprises, requiring subsidies. There have been great improvements in living standards, social services, and medical care. Water supplies, public utilities, and housing are being built and maintained by government-organized cooperative associations.

In the mid-1970s, in opposition to white minority rule in Zimbabwe and South Africa, Zambia deliberately shifted its dependence away from those countries' railways and ports. In one of its first major aid programs in Africa, China built the 1,056-mile Tanzam Railway between Zambia and the Tanzanian port of Dar es Salaam (see Figure 9–10). Railroad operating difficulties and port congestion at Dar es Salaam have forced a reopening of the South African connection.

Malawi's economic development strategy focuses on continuing the significant exports of corn, tobacco, tea, and sugar while developing small consumer industries. Malawi has a fairly consistent, if small, budget surplus. The National Rural Development Program, partly supported by funds from Canada and the European Economic Community (OECD), is a major agricultural settlement and productivity scheme in conjunction with an ambitious road-building program. The country is building a new capital city, Lilongwe, more centrally located than the colonial capital of Blantyre (see Figure 9–9).

Without any significant mineral resources, Malawi's economy is dependent on agriculture—the reverse situation of Zambia. The country's fertile soils and reliable rainfall give good yields. More than 90 percent of the population is rural, and the government actively seeks to improve rural living conditions. Farmers, however, complain that plantation owners receive the bulk of government program benefits. There is no internal unrest, yet progress is painfully slow. The most outstanding governmental success has been the provision of safe and adequate water supplies to well over half the rural population—no mean accomplishment in a continent where access to safe water is a chronic problem for most people.

Zimbabwe was formerly known as Southern Rhodesia. After a decade in rebellion, Rhodesia received its independence as a white minority-run state in 1965; a protracted guerrilla war against the government followed. Eventually, a combination of economic pressures and terrorism forced an end to one of the shortest-lived European-run governments in Africa.

Empire builder Cecil Rhodes obtained mineral concessions from local chiefs here in 1888, and the British South Africa Company was chartered to develop the area. The private company's control was terminated in 1923, and Rhodesia's whites were given a choice of joining South Africa or becoming a part of the confederation mentioned earlier. Southern Rhodesia's white settlers chose to become a separate colony instead. The Zambezi River became more than just a boundary between two colonies; it was the boundary between a colony that whites perceived as theirs and one in which there was implicit recognition of eventual black rule.

When Britain insisted that Rhodesia (Zimbabwe) move toward eventual "one person, one vote" rule, Rhodesia issued a unilateral declaration of independence. Britain did not use force to end the rebellion, instead requesting international economic pressures on the white regime. Economic pressures at first had the opposite effect, with Rhodesian industry booming in the effort to attain self-sufficiency. It is arguable that sanctions produced positive as well as negative effects on Rhodesia's economy since industries flourished. Inflation and the increasing human and financial costs of guerrilla warfare finally resulted in elections in which all adults, regardless of color, could vote.

Zimbabwe has an impressive array of natural resources in a mining-dominated economy. It produces two thirds of the world's chrome ores and mines coal, asbestos, copper, nickel, gold, and iron ore. There is ample power, most supplied by the giant Kariba Dam. The transportation infrastructure is well developed, with a comprehensive rail network and 85,000 miles of roads. There is enormous potential for continuing economic progress.

Unlike the case of East Africa, a majority of the white settlers remain in Zimbabwe. White populations have grown more conciliatory as conflict in South Africa increases. Cooperation is generally viewed to be in the interest of all citizens.

Malagasy: The Atypical Residual. Madagascar is culturally and physically detached from the African continent. The world's fourth largest island, it encompasses the full climatic transition, from rainforest through savanna to desert. Like its African counterparts, it has been subjected to European (French) colonial control and both Arab and European cultural impacts.

The Malagasy Republic is geologically a piece of Africa that split off and drifted away over millennia. Thus physically isolated, it has a flora and fauna of its own that is of great interest to scientists and scholars. The Malagasy language and the dominant people are of Malayo-Polynesian origin. Interestingly, despite its relative proximity to Africa, the island was settled 2,500 years ago by seafarers from Indonesia who ranged halfway around the globe from the South Pacific in one of the world's greatest colonial odysseys. Unlike their Christian South Seas counterparts, or their Muslim brethren in Indonesia and Malaysia, the inhabitants of Madagascar are dominantly adherents of traditional religious beliefs.

French colonial control was brief, about 65 years total. Yet, French influence remains strong, and France is the dominant source of trade and economic aid. French first names dominate in combination with sonorous Polynesian surnames; perhaps 40,000 French colonists remain, and French is widely spoken among urbanites. French occupation was not easily accomplished nor well received. It took France 10 years to consolidate its control, and revolts were frequent. A dozen years after independence, French interests were nationalized, but France is still viewed as something of a model, and the resilient French have renegotiated trade and investment contracts on a new set of terms.

The Malagasy economy has grown slowly since independence; the country remains poor. But in some senses, the country is advantaged; it feeds itself well above FAO standards (see Figure 9–11), and its production is diverse. Arab traders reached the island relatively early, bringing the culture of coffee and cloves. The island produces half the world's supply of vanilla, but that market is depressed by artificial substitutes. Diseases in Tanzania have benefitted local producers of cloves, while pepper and dozens of other spices pick up the slack in markets for sugar. Disastrous tornadoes and intense downpours have recently wreaked havoc with crops.

Over 80 percent of the population farms for a living. Much of the farming is done by slash-and-burn methods (here, particularly destructive in terms of erosion). Despite calamities, there is food for all and for export. The air of desperation that characterizes many African economies is absent.

Southern Africa

Southern Africa is the last stand of colonialism, at least in its more obvious, nineteenth century format. The economies of the "independent" states of southern Africa—Botswana, Lesotho, and Swaziland—are so closely dependent on the Republic of South Africa that these states' freedom of political action is less than total. Within the Republic of South Africa, voting for the national governing body is the privilege of adult whites, a scant 18 percent of the population.

South Africa is a prosperous nation, for some more than others. The program of **apartheid**, an institutionalized form of racial discrimination, applies throughout its economy. Working urban blacks have much lower incomes than whites holding identical jobs. If their living standards exceed those of most Africans, they suffer by comparison with neighbors and fellow workers who are white. Among rural dwellers, particularly those living in the **homelands** (native reserves), there is little difference between

their real income and that of much of the rest of underdeveloped Africa.

As the Republic of South Africa is clearly the dominant economic and military power of this subregion, there are two logical groupings: South Africa and its "fifth province," directly ruled Southwest Africa (Namibia), and the legally independent (but economically dependent) states of Botswana, Lesotho, and Swaziland (see Figure 9–9).

The Land of Apartheid. South Africa is at the same time a developing country and a developed country. In "developing" South Africa, black shepherds may earn $20 a month. In the homelands, the unemployment rate is 25 percent. In some black communities, the infant mortality rate is 30 times higher than in white communities. All these statistics are reminiscent of UDCs. In "developed" South Africa, the most advanced economy on the continent manufactures steel, assembles automobiles, refines oil synthesized from domestic coal, and operates nuclear research facilities. South Africa's population is 8 percent of the region's total, yet it accounts for 33 percent of the region's total output. In urban areas, white South Africans' per capita incomes are 10 times those of blacks; in rural areas, white incomes average closer to 40 or 50 times those of blacks.

South Africa may shortly be involved in total revolution. As reforms are being made and concessions granted, the pressures for change may be intensified rather than diminished by insufficient signs of progress. Most observers agree that South Africa is at the most crucial point in its history.

The resource base of the Republic can be described as fabulous, but with two critical exceptions: a shortage of water and the absence of petroleum. South Africa is one of the world's top 10 producers of minerals; it leads in gold, gem diamonds, antimony, and vanadium and is one of the top 3 producers of asbestos, chrome, industrial diamonds, manganese, uranium, and the platinum group. Its reserves of gold ores are estimated to be two thirds of the world's total. South Africa also has extensive deposits of coal, copper, lead, nickel, and zinc (see Figure 9–10).

Population and Race. South Africa is still a relatively empty land. Its population of 33 million gives it a density of only about 70 persons per square mile,

MAKE IT AFRICAN, FOR AFRICANS

There would seem to be little hope for industrialization in underdeveloped countries not blessed with fuel, raw materials, or proximity to developed-area markets. What of the "poor" selling things to the other "poor" countries? Not promising at first glance, there does appear to be something of a market among UDC consumers. In some cases, rural income exceeds that of urban areas; prosperous farmers in certain African countries do have disposable income. Urban poor work at odd jobs at day or task rates. Because they are accustomed to survival on meager incomes, good days give them cash earnings that are viewed as windfalls and spent on consumer goods. New industrial operations in the Malagasy Republic are geared to just such a market.

Most Africans still lack electricity. To serve this market segment, Malagasy produces batteries by the carload, in all voltages, sizes, and shapes. They are used to adapt television sets, radios, and small appliances to African dwellings without electricity. Two automotive assembly plants produce small trucks, vans, and cars readily convertible to taxis. Granted, most foreign sales are to corporations or governments. Ultimately, as used vehicles, they will enter progressively less affluent levels of society. Factories in Malagasy recondition auto parts and retread tires for just that secondary market. Malagasy's iron sheet plant, operating largely on scrap, produces nails, corrugated roofs (a rural African status symbol as well as a practical building material), and gasoline cans, used mainly for hauling water in the myriad areas lacking a reliable supply. The ubiquitous (and bug-free) foam rubber mattresses are made with Malagasy foam (derived from by-products

comparable to that of the state of Texas. It is very unevenly distributed over a land more than double the size of that American state. Similarly, a "wet" east and a subtropical coast are fairly densely populated, while a dry west is extremely lightly peopled.

Africans constitute 68 percent of the South African population; the largest of many Bantu African ethnic groups are the Zulu and Xhosa, each numbering 5 million. The next largest racial group, the "whites" or Europeans, form 18 percent of the total and are primarily descendants of Dutch, French, English, and German settlers. "Cape Coloureds" are the result of early mixing of indigenous peoples and Europeans. Living mostly in the Cape province, they represent 10

percent of the total and have a separate, official racial status. The group has been enlarged to include the remaining few indigenous Bushmen and Hottentots. Finally, there are the Asians, descendants of both Hindu and Muslim Indians originally brought to Natal to work on sugar plantations.

Race is the key question in this nation; systematic segregation of races acquired legal force and momentum with the triumph of the Afrikaaners, as the Boers prefer to be known, the majority within the white minority. Race became a legal status with far-reaching implications. The government has since pursued the seemingly impossible goals of apartheid: the strict physical separation of the races and of a domestic oil refinery, using imported Arab oil) and sheathed in plastics produced in plants next to the refinery. Metal furniture (termite proof) is a popular item of production, as are metal casement windows (a social step up in urban, self-built squatter homes). Aluminum pots, virtually indestructible and exceptionally cheap, are found throughout mainland bazaars; Malagasy is a major producer with two plants. Treadle-operated sewing machines, no longer made in the technologically triumphant developed world, are another example of adapting to African needs. Without refrigeration in most of Africa, canned and bottled items enjoy high demand. A raft of plants now process and package surplus Malagasy food for export. Cookie and candy plants help to consume the no longer profitable sugar, turning it into profitable, readily marketable items packed in Malagasy-produced tins that have secondary use as household containers and utensils.

A chemical works produces small, and therefore affordable, packages of detergent, bleach, dyes, and household chemicals. Cigarettes (packs of four), chewing tobacco for African miners, two-packs of beer, processed nuts in five-gram packages, cheap naphtha-based soap, twine, and inexpensive work gloves are all adaptations of size, cost, or product preference directly geared to African mainland marketing.

This is a highly practical and quite lucrative beginning to industrialization. Malagasy's success should provide food for thought to both UDCs and international lending and development agencies.

"separate development" economically. The government took South Africa out of the multiracial British Commonwealth, accepted drastically deteriorated relations with Black African states, and was deprived of its vote in the UN General Assembly when it refused to comply with UN rulings on the future of Namibia. This quasi-outcast status of South Africa has strongly influenced the country's economic and military strategy.

The factor that may in fact prevent the ultimate apartheid envisioned is the strong economic interlinkage among South Africans of all races. The continued viability of the South African economy is unthinkable without the participation of the majority of the work force—the blacks. South Africa's Cape Coloureds are highly assimilated, culturally, into the modern urban-industrial societies. Unlike the black South Africans, the Cape Coloured have no stress of detribalization. They are not caught between traditional tribal communities and Europeanized urban industrial cultures. They have no tribal homeland to return to. Despite better-than-Bantu living standards, they are by no means legally equal.

Asians have climbed the economic ladder by dint of hard work and frugal living. Many have generated sufficient capital to open small businesses; others have entered professions. Despite their occasional affluence, Asians are subjected to many restrictions of

HISTORICAL GEOGRAPHY OF SOUTH AFRICA

The Dutch East India Company founded Cape Town in 1652 as a halfway stop on its lucrative trade route to the East Indies. It was thought that a naval base there could control the passage from the Atlantic to the Indian Ocean. Britain seized the Cape in 1795, and Dutch settlers, already established through five generations, resented the take-over. A strong British military presence augmented by migration meant that the Boers (farmers), as the Dutch settlers came to be called, had little option but to submit or leave. They left, beginning in 1836, on the great treks north and east, away from British control, creating two independent and landlocked inland republics (Transvaal and the Orange Free State) (see Figure 9–9).

The discovery of diamonds (1870) at Kimberley, followed by the gold strikes in the Witwatersrand region of the Transvaal (1886), attracted scores of thousands of British immigrants. The accompanying investment in transportation, mining, manufacturing, and urban sectors of the economy seemed like an invasion to the conservative, agrarian Boers. Frictions between the cultures led to two Anglo-Boer Wars, one in 1880–81, and the more famous rematch in 1899–1902. The second war was a particularly bitter one; a scorched earth policy and concentration camps were used by the British during this struggle to pacify Boers and stamp out resistance. The Boer surrender was accomplished by a proposed union of Boer and British holdings—one that incorporated Boer concepts of race relations. It was British policy that only by effectively acknowledging Boer control would a new union be created. Thus, in the long run, the Boers won their war.

apartheid. Because Asians practically monopolize retail trade at the local level, they are often a target of resentment and envy (for their higher standard of living) for South African blacks.

The Economy. South Africa is self-sufficient in food. South African wines have entered world markets with some success, where not boycotted; its produce enters Northern Hemisphere winter markets on a large and lucrative scale. Boer farmers have long prided themselves on raising superior livestock, and animal products are a major export. South Africa operates the continent's largest fertilizer plant and has a modern, scientific, thoroughly mechanized agriculture. A chronic problem for many South African farmers is drought. Many development schemes, domestic and international, involve water storage and transfer.

The diamond- and gold-mining industries, developed in the late nineteenth century, attracted large investments in industrial infrastructure. Great cities such as Johannesburg grew explosively. Manufacturing was already developing and diversifying when World War II produced import shortages. Government policy then emphasized a drive toward industrial self-sufficiency. This policy has become more critical because of the "siege mentality" that resulted from world criticism and increased pressure over racial policies. Manufacturing, spurred by this import replacement strategy, has become the dominant economic sector. Fluctuating currency prices, inflation, stock market crashes, conservative Arab investors, and income tax laws have all stimulated world gold markets. South Africa, whose gold mines are experiencing rapidly increasing operating costs, has benefitted greatly in the process. Increasingly deeper mines now require forced-draft cooling (rock temperatures rise sharply as tunnels probe deeper) and higher energy consumption. It must be noted that South Africa's swift economic development has been based on cheap black labor and generally high prices for gold.

South Africa's labor needs have long been met by importing long-term migrant workers. Well paid by the standards of their own country, migrants are not paid well by developed South African standards. They tend to be used in the most difficult jobs, especially deep mining. Advantages of migrant workers to the South African government are that they cannot settle permanently, do not further tip the racial imbalance against whites, do not consume measurable social welfare benefits, and seldom take an interest in South African politics.

How Stable Is South Africa? To the Afrikaaner, theirs is the "beloved country," the homeland of their ancestors. Afrikaaners have no "home" to go back to, unlike white settlers in Kenya who could return to Britain. They *are* home—they and their people have helped build the country. From their viewpoint, there can be no retreat.

In the long run, even a determined, united, and well-armed white minority may not be able to forestall the evolution of a combination of South Africa's own restive black majority with active guerrilla support from other African states. Economic troubles and internal squabbles in those states, however, make them vulnerable in turn to South African meddling. Though South Africa's neighbors may come to terms temporarily, the Organization of African Unity persists in its condemnation.

Namibia. Namibia is a legacy of the long-defunct German African Empire. It is a thankless land, fronting a barren, desert coast to the Atlantic, backed in turn by hostile cliffs, topped with an arid plateau scrubland stretching off to the horizon. Unclaimed

Role of Migrant Labor in the South African Economy.
The mines and industries of South Africa have labor demands that cannot be fully supplied by even a combination of South African blacks and whites. Additional labor needs are met by using migrant labor from neighboring states and even from as far away as Zaire and Malawi. Wages of migrant workers are lower than those paid to South African black miners. Sometimes, payment is made (in gold) directly to the governments of sending states which in turn issue payment in local currencies to miners' families at home. This system has extended the economic power of South Africa throughout many portions of Southern Africa.

THE BANTUSTANS: NATIVE HOMELANDS

I n an attempt to make the minority a majority, the South African government has conceived a plan of "independence" for native Bantus. It would create, in reality, a series of weak client states economically dependent on South Africa.

Economic development strategies, racial segregation, environmental problems, and political policies are intertwined in the creation of these Bantustans, or tribal homelands (Figure A). The first of the self-governing homelands to be granted "independence" was the Transkei, the Xhosa national homeland. Two others, Venda and Bophuthatswana, have been created since. Eventually, seven areas will have independent status, while still others (smaller groupings) will have self-governance short of independence.

Many Bantu live and work in white areas, though this situation is viewed officially as "temporary." They are registered, by tribe, in their particular homeland, and it is planned to "repatriate" them as soon as it is practical. Blacks must carry a passport document just to leave a Bantustan to go to work in a white area. The technology and capital of "white" South Africa need the labor and markets of "black" South Africa, and vice versa. Physical and geographic segregation will make interaction more difficult.

The "independent" status of homelands has given rise to some interesting cultural and recreational functions. Bophuthatswana's Sun City is a miniature Las Vegas, a few hours' drive from Johannesburg. Activities forbidden in South Africa are legal there. Casino gambling thrives amid interracial mingling; international stars take the opportunity to appear there before nonsegregated audiences. Bophuthatswana's television station can air imported American and European shows that are censored by South African television stations; many white South Africans watch "Bop-TV" as enthusiastically as blacks do.

Many (perhaps a majority) of the people of such "states" as Transkei do not live there permanently; they earn their living in South Africa. Many urban blacks have few, if any, cultural and family ties to their official homelands; their repatriation could amount to exile.

The South African administration has encouraged industrial location within the homelands or at "growth points" on their borders. Few South African industries have selected such locations; the infrastructure of transport, communications, and energy systems tends to encourage urban-suburban South African locations instead. Because countries other than South Africa do not recognize the independence of the homelands, they do not maintain embassies or consulates to aid their own citizens interested in doing business there. The economic future of homelands does not look bright.

South Africa has had to reassess its political-geographic situation in Southern Africa. Its economic interests in its near neighbors, Swaziland,

FIGURE A
South African Homelands.
The policies of apartheid, the
South African system of separate
racial development and
segregation, include the literal
physical separation of the races
into territorial units. The
government has designated 10
such areas, called homelands, as
the future location of all black
residents. Some have already
been declared independent
states by South Africa, but no
other countries recognize this
action. The overwhelming
number of non-whites continues
to live in towns and cities
outside the homelands.

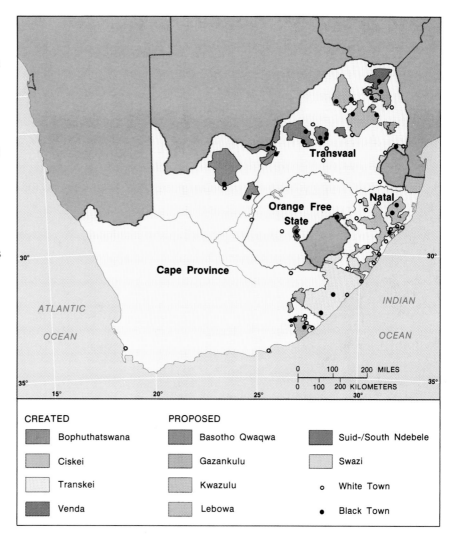

CREATED	PROPOSED	
Bophuthatswana	Basotho Qwaqwa	Suid-/South Ndebele
Ciskei	Gazankulu	Swazi
Transkei	Kwazulu	○ White Town
Venda	Lebowa	● Black Town

Botswana, and Lesotho, and its direct control in Namibia, give it an "outer"
buffer in all areas but one, the direct contact with Mozambique (Figure A).
The "inner" buffer function of the homelands is achieved by placing them
in a very difficult choice situation. If the homelands cooperate with outside
black nationalist guerrilla forces, they can expect denial of access for their
"citizens" to South Africa's economy, or more severe retaliation. If they
refuse to harbor black insurgents, they may bear the brunt of guerrilla
warfare themselves.

by other powers, the land was taken over by Germany in 1885 on the pretext of protecting German missionaries. Little investment was made, as Germany became embroiled in a long and costly war against native elements. World War I followed shortly, and because South African troops already occupied the area, the land was granted to that country as a League of Nations' mandate. Under this arrangement, the people of Southwest Africa were to be prepared for eventual self-governance.

Somehow, this independence never reached fruition. South Africa currently rules the territory as though it were an integral part of that country. Apartheid is officially practiced, and some 100,000 whites, mainly Afrikaners, dominate. No longer an economic wasteland, the area yields a half-billion dollars' worth of alluvial diamonds yearly. Boreholes tapping underground water have enabled limited irrigation and extensive ranching. Rich offshore territorial waters yield a half-million tons of fish annually. There is uranium, a resource important to the South African siege mentality which fears an ultimate racial Armageddon.

For over 20 years, an active guerrilla movement has opposed South African occupation. In 1966, the United Nations revoked the South African mandate, and a deadline for independence was set by the UN in 1975. Plans for a UN-sponsored cease-fire have been put forward periodically. Until recently, South Africa linked Namibian independence to the withdrawal of Cuban troops from neighboring Angola. SWAPO (the Southwest African People's Organization) guerrillas, most African states and churches, and virtually all native Namibians oppose this linkage as irrelevant. Recent Angolan-South African negotiations have proven more positive, and Namibia, it is promised, will obtain independence in the near future.

Botswana, Lesotho, and Swaziland. All three states have similar histories and share a strong economic link with the Republic of South Africa. Each is a ministate in terms of population, with from three quarters to a million and a half people. Each has an extremely rapid growth rate. Because of small size or difficult physical environments, there are serious population pressures.

The Sotho, the people of Lesotho, were placed under a British protectorate in 1868 at their own request, asking for protection against the Boers. The Swazi requested aid against Zulu raiders and were guaranteed independence under British protection. The Tswana (of Botswana), with their traditional territory under Boer invasion, appealed for British protection in 1885. Today's Lesotho is an enclave within South Africa, while Swaziland and Botswana directly border that country. Each is in a customs union with South Africa; there are no tariff barriers among them. Lesotho sends fully half its work forces to South Africa for most of the year; some workers commute daily. Botswana is the most strident of the three in its opposition to South Africa's racial policies. Nonetheless, realistically, it also cooperates economically with South Africa.

Lesotho has the narrowest resource base; water is its most exportable commodity. Near to populous but water-short, semiarid regions of South Africa, mountainous Lesotho has relatively heavy rainfall. It is a poor country where the combination of heavy rainfall and the pressure of population on the limited land have resulted in severe erosion. Tensions have arisen with South Africa, but a working relationship is maintained because Lesotho needs the remittances sent back by migrant workers. Many South Africans visit Maseru (Lesotho's capital) for a weekend of casino gambling and other "forbidden" activities. Water is exported (at a fee) to neighboring South African farms and ranches.

Botswana fares best in resources, with diamonds, copper, nickel, salt, potash, and coal. Its government owns a major portion of each of the mineral reserves and virtually exists on the income that mineral royalties provide. Words such as "sparse" and "dour" describe its monotonous plateau landscape. The huge swamps of the (inland) Okavango delta, and the equally impressive Makarikari salt pan have fascinated scientists for years—giant water features in an area of near desert. It is the Kalahari—the great desert of the south of Africa—that spreads over most of the countryside. In the wet season, a reasonable pasture growth supports large numbers of livestock, but the sere upland plain grants only barest survival during the season of drought. With a density of three people per square mile, it is a truly empty land. Most of the population is huddled along the Zimbabwe-South Africa railroad that clings to the country's eastern frontier. Almost 40 percent of the people live in Gaborone, the capital, within eyeshot of the South African border.

In an effort to commercialize farm production in a country best suited to producing livestock, the government built two huge slaughterhouses and invested widely in boreholes to tap underground water supplies. This investment has spurred increases in the size of herds and the commercial export of beef. Hoof-and-mouth disease in the 1970s, followed by drought in the 1980s, resulted in a decimation of herds. Not a nation to stand idly by, Botswana converted disaster into triumph. The government set up tanneries to process the hides of dead cattle and, in conjunction, private citizens began a series of small shoe factories to use the hides. Diamonds, copper, and nickel account for most of the country's export income. There is export of fertilizer in a nation that can farm only 2 percent of its land but has huge potash deposits. Pastures are greener, healthier, and more able to resist drought here because of this fertilizer supply. Over 50,000 Tswana worked in South Africa as migrant labor until recently. As layoffs occurred and 20,000 workers were sent home, domestic economic functions were able to take up much of the slack. The country is not rich, and many problems remain, not the least of which is dependence on South Africa for imports (80 percent), export markets, and transit.

Swaziland is a ministate of only 6,000 square miles; even by European standards, it is a tiny country. Asbestos, the leading mineral export, has entered a troubled era because of environmental concerns. The mineral economy has now become overwhelmingly dependent on exports of coal. Agriculture is diverse and fairly prosperous; irrigation is widespread. Food exports, particularly sugar, are an important earner of South African currency.

The Swazi government has built a railroad through South Africa (with funds borrowed from South Africa) to the open sea. In return, Swaziland has expelled African National Congress refugees and allowed South African police to sweep through Swazi territory in search of guerrilla fighters. Too small to survive in the face of a more powerful neighbor, too

dependent on South Africa to act independently, the country is attempting to observe neutrality.

THE FUTURE OF AFRICA SOUTH

Knowledge is the key to developing Africa South. The region possesses a physical environment quite unlike that of any other world region. Learning to use that environment to its best advantage is the challenge facing the region's inhabitants and the scientists and technicians associated with international agencies and foreign aid programs. Developed area technologies have often proven unsuitable to African environmental conditions; their application has at times resulted in disaster.

The region, however, is not doomed to perpetual poverty and economic failure. Africa South has a large and varied mineral resource base, a vast untapped hydroelectrical potential, and considerable underutilized land. Population growth is very rapid, but pressures on the land resource are an immediate concern in only a few areas.

Instability of weather and soils is certainly an inescapable fact of the region's physical environment. Recurring droughts have brought starvation and economic disaster to some areas in the last two decades. For the moment, weather and rainfall have returned to normal. Improved usage of available water could help avert such disasters in the future.

Africa South remains a region dependent on commodity export. Declining world prices and demand, with their adverse economic effects, are a temporary situation. Economic diversification could do a great deal to avert repetition of such problems. Multistate cooperative projects can serve as a means of stretching small investment budgets. Improved pricing structures can encourage domestic crop production and help reduce foreign debt.

Internal dissension and intraregional conflict may be the most difficult problems to solve. Education is crucial not only to economic development, but also to the development of sound individual national groupings within the region. Local animosities and tribal affiliations can be eliminated only by the development of national unity. And progress is possible only with peace.

1. What cultural and physical factors justify designating Africa South as a uniform region? Discuss the complexity of African language patterns. How does this complexity affect nation building in the region?

2. Generally describe the topography of Africa. How has topography affected access to the interior? construction of transportation facilities? weather patterns? settlement patterns?

3. How effective has the Sahara been as a barrier to internal trade among the various portions of Africa? Has it had a negative effect on the diffusion of culture? the migration of peoples?

4. Discuss the recent crises in Africa. How and why have government actions affected the production of food crops by indigenous farmers?

5. What is desertification? What human actions seem to intensify the processes and effects of desertification?

6. What diseases are endemic in Africa? How has the incidence of disease and pests affected settlement patterns?

7. Analyze the strengths and weaknesses of the mineral resource base of Africa South. What are the region's major mineral-producing districts?

8. What were the effects of slavery on African populations? What were the major source areas for slaves? What European powers were involved in the slave trade?

9. When did Europeans develop colonial interests in Africa? Which European nations held colonies in Africa? How long did colonialism last? What were its major effects on Africa? When was independence for African states attained?

10. Discuss the major land use patterns in Africa South. Which areas of Africa appear to be underutilized? What is slash-and-burn agriculture? Is it productive? Discuss the problems of applying Western agricultural conditions to farming in the African environment.

11. What are development enclaves? How do they influence internal migration patterns in the countries of Africa South? What is meant by neo-colonialism? Why were colonial capitals usually also development enclaves?

12. How was the Great Rift Valley formed? Relate this physical feature to topographic and settlement patterns in Africa South.

13. Name the five indigenous population groups of Africa. What areas do they currently occupy? Which ones were most successful demographically?

14. Explain the rise and fall of the great empires of the Sahel. Why has it taken Westerners so long to develop a knowledge of these empires?

15. Explain the reasons behind the large number of political units in West Africa. Why does this subregion have such a relatively heavy concentration of people? Which states of the subregion are the most wealthy?

16. What crops are commonly grown in West Africa? Discuss the negative aspects of crop specialization.

17. Discuss the problems of using African soils. What is a hardpan, and how does it limit agricultural utility? What natural forces contribute to fertility of soils in parts of Africa?

18. Which nations of Africa South are among the 10 poorest of the world? Relate life expectancy to dietary standards. Which districts of Africa are comparatively wealthy? the poorest?

19. How effective has black African nationalism been as a unifying factor in this region?

20. Describe some of the physical handicaps to easy navigation of most African rivers.

21. Why is there so little trade among African states?

22. Briefly describe Nigeria's problems of internal unity in terms of its physical, economic, and cultural geography.

23. How did colonial policies and practices facilitate disruptive forces within the newly independent Zaire (formerly the Belgian Congo)?
24. How does the persistence of tribalism threaten the integrity of many African states?
25. How did the site and situation of Ethiopia contribute to its persistence as an officially Christian state in a Muslim matrix? How has a Communist government modified its settlement patterns? its interaction with world powers?
26. How reliable is any attempt at correlating mineral resources and economic development in this region? Why?
27. Discuss the consolidation of Senegal and the Gambia. What are the positive and negative aspects of this union for each country?
28. Contrast the evolution of Sierra Leone and Liberia from their inception to current times.
29. What is the role of European corporate investment in Equatorial Africa?
30. Discuss the problems of Uganda. Why has its economy declined over the last 20 years? How has Tanzania dealt with its ethnic diversity?
31. What is the Copperbelt? Where is it located? Discuss the development of transportation facilities to serve this mineral district.
32. Compare the level of development in the "Marxist Margins" to that of the other countries of the Savanna Transition. Explain why capitalist governments in more developed areas are willing to deal with these Communist economies. How is Angola involved in the dispute over Namibia?
33. How does Madagascar differ from the states of the continental mainland? What adaptations has it made to enhance marketing its production in African markets?
34. Discuss the evolution of apartheid policies in South Africa. How and why have homelands failed? What are the major racial categories given legal status by apartheid?
35. What is the role of migrant labor in the South African work force? How has South Africa extended its political and economic influence to its neighboring states?

SUGGESTED READINGS

Best, Alan, and DeBlij, Harm. *African Survey.* New York: Wiley, 1977.

Boateng, E.A. *A Political Geography of Africa.* Cambridge: Cambridge University Press, 1978.

Christopher, Anthony. *South Africa.* New York: Longman, 1982.

Church, R.J. Harrison. *West Africa: A Study of the Environment and of Man's Use of It.* New York: Longman, 1980.

Church, R.J. Harrison, ed. *Africa and the Islands.* New York: Longman, 1977.

DeBlij, Harm, and Martin, Esmond, eds. *African Perspectives.* New York: Methuen, 1981.

Grove, A.T. *Africa South of the Sahara.* London: Oxford University Press, 1970.

Hance, William. *Black Africa Develops.* Waltham, MA: Crossroads, 1977.

Hance, William. *The Geography of Modern Africa.* New York: Columbia University Press, 1971.

Knight, C. Gregory, and Newman, James, eds. *Contemporary Africa: Geography and Change.* Englewood Cliffs, NJ: Prentice-Hall, 1976.

Morgan, W.T.W. *East Africa: Its People and Resources.* New York: Oxford University Press, 1969.

Morrison, Ronald, ed. *Black Africa: A Comparative Handbook.* New York: Free Press, 1972.

CHAPTER TEN

The Middle East-North Africa

Giant tankers in the Arabian Gulf.

Here is a region as ancient as civilization itself, as important in the time of the pharaohs as it is at present. The secret of its sustained importance lies in its most basic geographic resource: its location at the juncture of Asia, Africa, and Europe. It is a dry area; water, that most critical of all elements, is scarce.

This region was less than promising as a home for early humans because the game, fruits, and seeds that formed the basic food supply of homo sapiens were also scarce. Yet, here were born some of the greatest innovations of humankind, perhaps of the very necessity to survive in a less-than-welcoming environment. The region became one of the world's great agricultural hearths, the place of domestication of many grains and types of livestock. Over time, the cleverness of its early peoples turned adversity into triumph and the barren wastes of nature into bounty with the ingenuity that became the hallmark of humanity. From here, these ideas and practices were diffused to the far reaches of three continents. Innovation continued, in increasingly sophisticated and abstract forms, as the societies of this region grew in number, power, and prosperity. It is the font of Western Civilization.

Not only a center of innovation itself, it came to be receptive to the innovations of others. Since its earliest history, it functioned as a point of exchange. The material goods and wealth of ideas that originated in three continents and many cultures passed through its portals on the way to their ultimate destinations. Much was adopted, changed, and further diffused as a result of this exchange function, enriching the lives and aiding the progress of humankind.

Two distinct cultures appear to have arisen in the area near this continental nexus, each possessing a reliable and permanent source of water. One arose in the fertile plain between the Tigris and Euphrates rivers—a land the ancient Greeks would come to call Mesopotamia, the literal translation of that locational attribute. The other, Egypt, arose along the floodplain of the Nile, a ribbon of green across the sandy desert wastes.

Curiously, neither cultural area appears to have been an agricultural hearth itself. The naturally watered hills north of the Mesopotamian Plain are accounted the original home of that culture's sedentary agriculture, while that of Egypt may have been in the highlands of Ethiopia and East Africa or transmitted from hearths in Southwest Asia. Once each seminal culture had left its mountain homeland, it entered into a milieu where the actions of the group were paramount in the success of a crop, and where the vagaries of nature's rainfall were almost immaterial. Nature provided floods on a seasonal basis, and the rivers were a supplemental source of water as needed. Life developed a reliable rhythm and proceeded in an orderly, predictable fashion. Ultimately, with a sustained food supply and released from climatic uncertainties, the inhabitants could turn their attention from mere survival to other endeavors. Great civilizations were the result.

The wealth of each riverine culture was to sorely tempt each new migration. War is certainly older than civilization, and acquisition is a common human desire. The stability of Egypt, an area reached only with great difficulty except along a tenuous coastal route, was to be less often disturbed. More open to invasion, the lands of Mesopotamia and the Mediterranean hills were to develop multiple kingdoms. Assyrian and Hittite, Philistine and Persian, Phoenician and Jew, were each to develop distinct cultures in this land of abundance, and to war with one another in contests of territorial expansion and power. Not all were Semites, not all were successful, but a series of durable contending cultures came to occupy a portion of this Fertile Crescent and, frequently, attempted to control its entirety.

Invasions were not limited to neighbors. Progressively, they came from farther afield. Alexander's Greece and the Empire of Rome came to subjugate the entire Crescent. Arab, Mongol, and Turkish conquests were to follow. Of all the cultures involved, that of the Arabs was most successful in extending its influence over the entire area. Its religion, Islam, constituted its major force for assimilation. More than a religion, Islam is also a legal system. Both its divine and secular laws were readily adaptable to the various lifestyles of the region's people—urbanite, farmer, or nomadic herder.

WHY "MIDDLE EAST"?

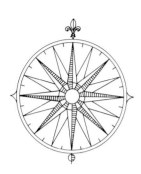

The Middle East (sometimes specifically associated with North Africa) is a "conventional" region to geographers; that is, there is general agreement that this particular part of the world is a distinctive region. If there is a popular consensus that a Middle East-North Africa region exists, there is less agreement on the exact boundaries of that region. The designation Middle East itself, in contrast to the great antiquity of civilizations located there, is quite recent. In the waning years of the nineteenth century, the label "Middle East" was used to designate an intermediate position between the "Near East" and the "Far East." The Near East referred to a part of the Ottoman Empire and encompassed Turkey, the remains of its empire in the Balkans, and the coastlands of the eastern end of the Mediterranean basin. The Far East was then the popular designation for what is now generally termed East and Southeast Asia. Middle East, logically, lay between "Near" and "Far," and originally referred to the area around the Persian Gulf. The Indian subcontinent, then part of the British Empire, was never part of the "Middle East" even though that subcontinent lies between the "Near" and "Far" East.

Near, Middle, and Far East are all, of course, highly ethnocentric designations. They are near, middle, or far east only in reference to Europe and European perceptions of the world. The Near East countries, for example, might just as reasonably be termed the "Middle West" from the viewpoint of India or China.

The regional designation Middle East has gradually extended westward in popular perceptions. Some sources combine all of North Africa (the Sahara and the Mediterranean coast) with the Middle East, stretching eastward to Afghanistan and sometimes even Pakistan. This region is alternatively termed "North Africa-Southwest Asia," which avoids the European bias of relative location. However, the Middle East designation has become so commonly used that it will be applied here despite problems of terminology.

Uniformity in faith brought near-uniformity in language. The Koran was in Arabic, and Arabic was therefore the language of the holy writ. However effective the spread of that faith and its accompanying language, the procession of cultures that preceded and followed the Arabs has also left an imprint on the region.

At times, the area became isolated from Europe by dint of cultural differences, war, or the changing fortunes of trade. Periodically, the innate importance of the region resurfaced, and contacts were resumed. Most recently (until 1920), the region slumbered in poverty and relative isolation under Turkish dominion. It returned to occupy center stage with the conversion of the world to oil as the dominant energy source and the discovery of the world's greatest oil supplies beneath its deserts. The strategic position of the region is perhaps now more important than ever before. The area has actively resumed the role of exchange point between the three continents and has extended its influence virtually around the world. Pivot of history, nexus of continents, superficially uniform but with all the complexity implied in its varied past, the region again emerges as one of the most important in the world.

ROLES OF CULTURE, ENVIRONMENT, AND OIL IN REGIONAL UNIFORMITY

As defined here, the Middle East-North Africa region stretches from Turkey to the Sudan, Iran to Morocco, and even includes the countries of the Sahara-Sahel in Northern Africa. Few regions have such a high degree of uniformity in the perceptions of people living elsewhere. The area is commonly thought of as Arab in speech, Islamic in religion, dry in climate, and rich in oil. Each of these general traits applies to some degree and to major parts of the region, yet none applies totally throughout. Nevertheless, it is on these factors (among others) that regional unity is based.

Arabic is a great unifying influence. Spread widely by the rise of Islam, some classical Arabic is understood by most literate Muslims, wherever they reside. It is widely spoken as a trade language in ports along the Indian Ocean and is used to some extent by the peoples of the Sahel. Its competitors have been far less successful. Turkish, spread into much of the region during the long period of control by the Turkish Empire, remains the majority language only in Turkey. Turkish words, however, have crept into Arabic as spoken along most of the Mediterranean littoral. Iranian is an ancient Indo-European language. Speakers of this tongue periodically have held sway over vast parts of the region. The language is now spoken only over northern areas of the region and has been greatly modified by contact with other cultures. Indeed, it is now written with an Arabic script (as was Turkish before a 1923 decree replaced it with Roman letters). A revised and modernized Hebrew is the official language of Israel. Though a separate language, it is a linguistic relative of Arabic. The languages of later conquerors enjoy some prominence. English is an important second language throughout the region. French is spoken among many urban North Africans and is official in much of the Sahel. Other minority languages are spoken in the Sahara, southern Arabia, the Caucasus fringe, and Iran. Nonetheless, it is Arabic, in all its forms and influence, that dominates.

Islam is the prevalent (and dominant) religion, with the important and obvious exception of Israel, and some sizable Christian minorities in the states of the Middle East. Small Jewish minorities persist outside Israel in Turkey, Egypt, and Iran. Some early predecessors of Islam also persist. Zoroastrianism, once predominant in Iran, remains a minority belief there, and in India, among a class of merchants of ancient origin. Islam has also spread far beyond the bounds of the region. The largest Muslim state is now Indonesia, and important Muslim populations are located in the USSR, Africa South, the Indian subcontinent, and even southeastern Europe and China.

Arid and semiarid climates do prevail, although exceptions occur in two senses: there are small areas of better-watered climate within the region and, of course, there are arid and semiarid climates in many of the world's other regions. Nonetheless, nomadic herding (though of late greatly reduced in importance) and irrigated agriculture, the cultural manifestations of dry environments, alternately prevail over the region's landscape.

Oil is the chief source of current regional wealth. Yet the distribution of oil reserves is not uniform. Some states have none (or very little), whereas others are almost completely dependent economically on their oil production. Saudi Arabia alone accounts for a quarter of the globe's known oil reserves, but the region's two largest states (in population), Turkey and Egypt, have relatively moderate reserves in ratio to population. Nonetheless, oil and oil revenue are now the aegis of cultural change; oil is unquestionably the region's dominant economic influence.

Perhaps the most striking characteristic of the region is the sharply altered perception of it held by outsiders. Eighty years ago, this region was relatively isolated from the main currents of economic and political activity. Now, it ranks as one of the most important centers of both. The frequency of regional visits by American, Soviet, and European leaders, representing political, business, or religious fields, attests to its importance.

The Middle East long has functioned as an important interface area between East and West. To this historic, and continuing, role is now added the interface position between "North" and "South" — between the developed and developing nations. Two of this region's immediate neighbors are highly developed Europe and the USSR; its other two adjacent regions are developing Africa South and South Asia. The Middle East-North Africa truly is a crossroads among the world's regions—in all senses of that term.

The geographic centrality of this region is obvious when one looks at the relationship of the three continents of the Old World—Europe, Asia, and Africa (Figure 10–1). The core of the region, its veritable cultural hearth, lies astride the most probable trade routes among the three continents. Three great monotheistic religions (Judaism, Christianity, and Islam) spread outward from this hearth. Perceptions of the strategic significance of this hearth and the region itself have changed, but for most of recorded history, the region has been considered a key part of the world.

The long struggle of the Christian and Muslim worlds over control of that land called holy by three religions interrupted the important trade routes from the Mediterranean Sea eastward to the Orient. This disruption forced Europeans to refocus on such alternatives as maritime exploration of the Atlantic. In turn, this exploration resulted in both the European "discovery" of the New World and the development of the route around Africa to India. The Mediterranean was no longer the most important and central sea; instead, it became a backwater in the new perceptions of the world. The economic centers of Mediterranean trade, including the Middle East, became relatively less important as a consequence.

This "Columbus syndrome" was followed by two contemporary historical trends linked to each other and to seaward expansionism. European technology and sciences surged forward in the commercial and

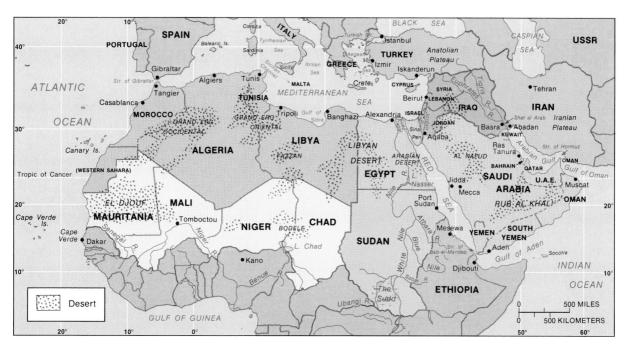

FIGURE 10–1
Strategic Location of the Middle East-North Africa and Its Waterways.
Known as a pivot of history and the home of ancient civilizations, the region draws importance from its location at the juncture of three continents and from its surrounding seas. Trade routes developed early, both overland and on the seas, for the exchange of goods between differing cultures and physical environments. Narrow seas channeled this trade through straits and gulfs whose control offered power and wealth to the controller.

industrial revolutions. Ironically, it was Arab preservation of much of the Greco-Roman world's knowledge of science and geography, together with Arab contributions in those fields and in mathematics, that made it possible for Europeans to achieve these "revolutions." The advances in shipbuilding, navigation, and weaponry that accomplished these revolutions rocketed Europe to the head of the developed world, while relegating the Middle East and North Africa to the underdeveloped world. While Europe prospered and grew in power, the **Dar al Islam** ("House of Islam") seemed to stagnate culturally and technologically.

The significance of the region's core in geopolitical terms was reassessed with the opening of the Suez Canal in 1869, once again placing this region in a vital position astride major trade routes. The discovery (beginning in 1907) of the world's largest deposits of oil completed the perceptual rebound of the region. Once again it had become an area critical to the rest of the world.

The Development of Arab Power and Nationalism

Tradition in both Judaism and Islam links Arab and Jew, both Semitic in language and cultural origin, through a common father, Abraham. Isaac, Abraham's son by his wife, was to sire the nation of Israel; while Ishmael, his son by a concubine, was the traditional father of the Arab nation. The distinction made was less one of legitimacy than one of a separation between the dwellers of the desert (the nomads) and the shepherds of the wetter steppe. This basic difference was further reinforced by the Jewish adherence to monotheism, while the Arabs remained faithful to the traditional gods of nature.

The proto-Arabs were not, however, simply nomadic herders. They developed sophisticated cultures around the wetter margins of the peninsula, and sedentary farmers, as well as cities, were developments that date to at least 3000 B.C. Early Arabs of the peninsular southwest cultivated grains and gathered the gums and herbs that were blended into incense, a major trade commodity because of its wide use in religious ceremonies. The cultures that arose in the Persian Gulf area carried on trade with Persia, Mesopotamia, and the civilizations of the Indus Valley. It was obviously not an isolated area, but it was protected by distance, desert, and the maritime limitations of the era. Even the might of Alexander and Rome never penetrated the Arab cultural hearth. While individual Arab trading kingdoms rose and fell in the peninsula, a brisk business was always carried on between Arabs and the states of the Fertile Crescent.

In Mecca, a town far removed from the urban wealth of Felix Arabia (the southwest), a man of great historical importance was born in A.D. 571. At the age of 40, this man, Muhammad, began to receive revelations from Allah (God) that were to lead to the development of Islam. It was a faith that exacted relatively little from its adherents: the acceptance of one God, daily prayer, alms giving, the observance of a season of fast and, if possible, a pilgrimage to Mecca, the holy site of the revelations. These remain "the Five Pillars"—the central tenets of Islam. Islam recognized the universal presence of God and the equality of all before him, no matter how wealthy or humble. Muhammad became convinced that he was chosen to be God's prophet, and that spreading the word he had received was his divinely ordained mission.

At the time, the Arab community was in a state of upheaval, and many were receptive to the words of the prophet. As Muhammad's following grew, so, naturally, did the forces who opposed him as an enemy of the established faith and power structure. He was soon forced to flee to Medina (Al Medinah), an oasis some 300 miles distant in the relative seclusion of the desert, in order to carry on his work.

In many ways, the Islamic faith forms a continuum of the Judaeo-Christian tradition, incorporating elements of both. Like all major faiths, it developed a lifestyle—a series of guidelines that accompanied its basic precepts. There was a holy day (Friday), and in urban centers there were soon formal places of worship, the mosques. The consumption of pork was forbidden, and ritual circumcision for males was the approved practice. Alcohol and gambling were also forbidden. Polygamy, a native practice, was tolerated, but monogamy was considered more virtuous. The accumulation of wealth was not disallowed but was countered with the dictum of compulsory charity toward the poor. Punishments were prescribed for murder, theft, and other deviant behavior. It was thus

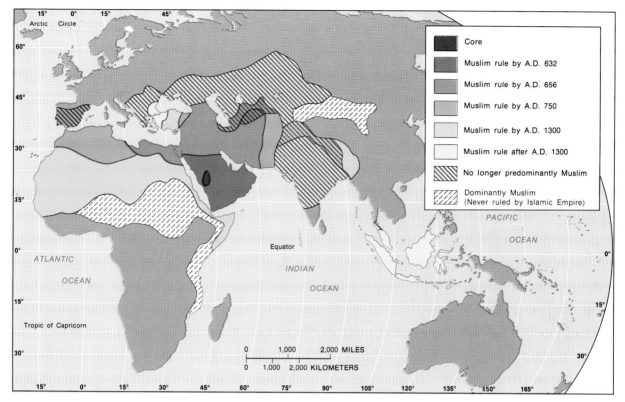

FIGURE 10–2
The Rise and Spread of Islam. Islam, the religion of most of the region's
people, arose in Mecca, a city in Arabia. Muhammad, the founder of this faith, was its
Prophet and teacher. Within a century of his death, the followers of Islam had spread
the faith throughout most of the Middle East and part of North Africa.

a social system as well as a faith, and it addressed the
concerns, problems, and inequalities of the society in
which it arose.

Muhammad's word was spread throughout the
peninsula, gaining adherents among all classes and
tribes. Unified in spirit and full of messianic zeal, the
Arab tribes set forth to spread the faith. Within 200
years, Islam had spread throughout North Africa and
the Middle East. It was carried to China, India, and
Central Asia with the caravans of trade, where it converted large numbers of people. Arab merchants
gained converts overseas in the south and east of Asia
and along the eastern coast of Africa, wherever Arab
dhows sailed in commercial pursuits. Even the Sahara proved no obstacle as Islam spread to the Sahel
and beyond. Adherents were gained even among the

warlike tribes of the Asian interior—Mongols, Tartars, and Turks. It was only in Europe that the faith
met strong resistance. Though its followers conquered and came to rule much of Iberia and the
Balkans, the religion attracted few converts there
(Figure 10–2).

The new religion brought power and wealth to
Arab rulers because it spread not only the faith but
also Arab commerce. A succession of Arab states
arose in its wake, many encompassing the entire
world of Islam as well as non-Islamic lands. Originally ruled from Medina in the religious heartland,
later empires located their capitals nearer the commercial center of activity; the Omayyad Empire ruled
from Damascus and the later Abbasids from Baghdad. The Arabs were not without rivals for power.

The faithful were not limited to Arabs, nor did the faith endow them with a special status. At times, the latent imperialist ambitions of schismatic Persia, wealthy Egypt, and opulent Mesopotamia displaced Arab rulers with those of their own. The dominions of the west (North Africa), successful in their own right and isolated from the core, often formed independent kingdoms. The nomadic Mongols, loosely attached to the faith and bent on the conquest of any who opposed them, succeeded in subduing much of the Fertile Crescent, all of Russia and China, and parts of India. For two or more centuries, they controlled this enormous domain, though they never conquered the Arab core or the western dominions.

The borders of the region coincide loosely with those of the old Arab empires, where the culture and language became early and most firmly entrenched. The latecomer Turks, once the mercenaries of Mongol overlords and Tatar khans, succeeded in again piecing together that imperial Arab framework, along with the holdings of the rump of ancient Rome in the guise of the Byzantine Empire. Neither Mongol nor Turk, however powerful or politically successful, was able to dislodge the Arab language or culture in the process of conquest and rule.

The power of the Ottoman Turks peaked in the seventeenth century, but the death of their empire was prolonged. The decaying Ottoman Empire was gradually but fully replaced by European colonial or indirect rule by 1920. Arab aspirations to self-rule were used by the British and French to topple Turkish rule, but independence was not forthcoming, with European colonial control extended to the area. French colonial control of North Africa, Syria, and Lebanon was matched by British concentration in Egypt, Palestine, Iraq, and the Persian Gulf. Britain maintained its traditional determination to control the passages between all seas and the need to defend the approaches to India. By the 1930s, growing realization of the oil potential of the Persian Gulf lands provided yet another motive. Since 1920, the growing Arab determination to achieve independence from foreign control has formed a unifying focus, although the urgency of individual independence aspirations varied considerably. A resurgent Islam has been another important unifying factor in the region, but individual national political aspirations still run counter to a unified Arab or Muslim state.

THE REGIONAL ECONOMY AT A GLANCE

Most of the region's states would best be described as preindustrial, even though all have planned programs of industrialization (Figure 10–3). Long-term developmental plans are generally geared to reductions in the importation of food and consumer goods, the development of a series of manufactured specialties for regional marketing, primary and intermediate processing of raw materials (rather than direct export of crude, unrefined products), and the development of one or two market specialties for trade outside the region.

There is, in each case, a similarity or pattern in the measures taken to accomplish these goals. The basic first step in all the region's nations is a thorough geologic prospecting. Oil is not the sole resource sought; special attention is paid to the search for groundwater supplies and fertilizer minerals. Basic food supplies obtained by traditional methods at established locations usually have been outdistanced, or at least strained, by rapid population growth. Many of the region's states have large numbers of landless peasants, and land reform is a key goal. But although there is plenty of land, there is insufficient water. New water resources would enable an expansion of acreage and the consequent resettlement of landless farmers in new areas. Fertilizers make possible increased production on existing farms, a measure that in most instances could greatly reduce or eliminate the need for food imports.

A prolific peasantry has continued to produce large families. Shariya (Islamic) law calls for equal division of estates among all sons. Peasant farms have already been reduced in size through inheritance to the point of inefficiency, if not outright inability to survive. Surplus sons now must leave for the city in search of work, where they form a large and growing class of urban unemployed or, at best, eke out a living with occasional jobs. The region's festering urban slums have become warrens of revolution and discontent. Development of small, but numerous consumer goods plants is designed as much to produce jobs for this group of urbanites as to reduce foreign imports.

Oil-based economies have developed refining of crude and, increasingly, the marketing of refined

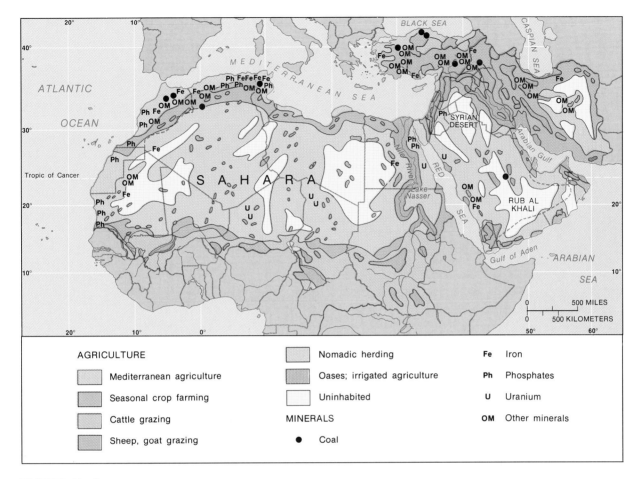

FIGURE 10–3
***Land Use and Selected Mineral Resources in the Middle East-North
Africa.*** Vast areas of the region are devoid of water; therefore, they are also devoid
of permanent settlement and crop agriculture. Only the wetter margins and areas
suitable to irrigation support farming and dense populations. Grazing is the
dominant land use. While oil is the region's most important mineral resource, there
are significant deposits of coal, metal ores, and fertilizer minerals, particularly in
western North Africa, Turkey, and Iran.

products (Figure 10–4). The more advanced (or
wealthier) economies have begun to develop petrochemical industries for the production of coke, plastics, chemical animal feeds, fertilizers, and a host of
other products. Similar intermediate processing can
be applied to other resources produced in the region.

The international marketing of specialized production is less hopeful. There are worldwide gluts of
chemicals, basic food crops, and many consumer

items. Labor is cheap only in those countries that lack
an oil resource. The raw material base of most regional economies is not broad or varied enough to
support many kinds of manufacturing without additional imports. While literacy levels are improving
rapidly, the societal level of mechanical skills is often
quite low.

There has been relatively little international trade
among the region's countries. Highly developed, industrialized countries tend to have more complex

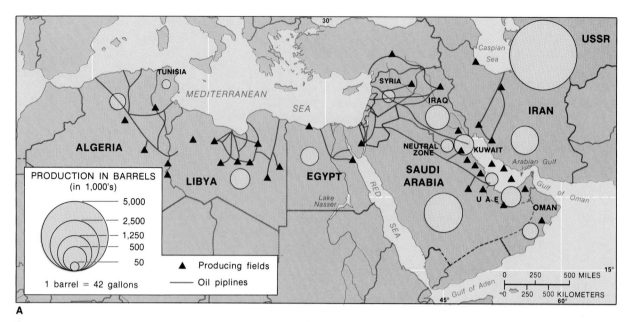

A

B

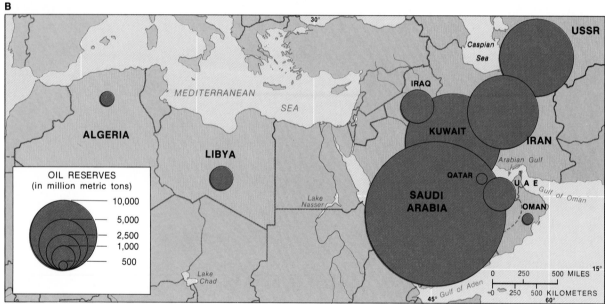

FIGURE 10-4

(a) *Oil Production in the Middle East-North Africa.* The region holds almost half of the all known oil reserves and produces nearly one third of the world's oil. Production is greatest in the states surrounding the Persian (Arabian) Gulf. (b) *Oil Reserves in the Middle East-North Africa.* The oil resource is not uniformly distributed throughout the region. The Gulf states concentrate a disproportionate share. Production is significantly more widespread.

trade relationships, and in this region only Israel is clearly a highly developed country. Substantial and varied resource bases exist in Turkey, Iran, Morocco, Algeria and, after recent discoveries, Saudi Arabia. Other states in the region tend to have a short list of resources, even if there is oil. The distribution of two vital regional resources—water and oil—is extremely uneven. There is a long-established positive correlation between high population density and

freshwater resources. Oil tends to be concentrated where people are scarce (Table 10–1).

Egypt, with the largest national population in the region, has a proven oil reserve of some 292 million metric tons. Tiny Kuwait, with fewer than 2 million people, has reserves of 10,184 billion metric tons (see Figure 10–4). States with enormous ratios of oil wealth to population size make various contributions and subsidies to the large-population, low-oil-

TABLE 10–1
Oil production and prospects for the Middle East-North Africa region.

Country	Pop. (millions)	Pop. Density (persons per sq. mile)	% Pop. Growth	Oil Reserves Million Metric Tons	Oil Reserves World Rank	Oil Production Thousand Metric Tons	Oil Production World Rank	Oil Reserves per Capita (metric tons per capita)
Projected Long-term Exporters with High Reserves per Capita								
Kuwait	1.9	100	2.7	10,184	2	56,712	13	5,360
United Arab Emirates	1.3	16	4.0	4,318	6	73,584	9	3,320
Qatar	0.3	27	3.1	515	18	19,608	20	1,720
Saudi Arabia	11.2	5	3.1	15,911	1	490,800	2	1,420
Libya	4.0	2	3.4	3,719	9	53,856	14	930
Oman	1.2	9	3.5	447	19	15,972	22	370
Iraq	15.5	36	3.3	4,702	5	44,892	15	300
Iran	45.1	27	3.1	6,148	4	65,988	11	140
Major Producers with Lower Reserves and Prospect of Long-Term Reduction in Exports								
Algeria	22.2	9	3.2	1,309	14	36,504	16	58
Syria	10.6	57	3.8	237	25	8,496	31	22
Tunisia	7.2	44	2.4	304	23	5,412	35	42
Limited Producers per Capita								
Bahrein	0.4	695	3.5	37	44	2,624	40	9
Egypt	48.3	47	2.9	292	24	31,800	17	6
Turkey	52.1	64	2.4	30	45	2,388	41	0.5
For Comparison Purposes								
United States	239.5	25	0.7	3,801	8	421,308	3	15
USSR	278.0	12	0.9	7,990	3	609,000	1	28
Nigeria	91.2	103	2.8	1,678	12	71,184	10	18
World Average/Total								
5,200,000,000		36	1.7					

SOURCES: Population growth rates: Population Reference Bureau, 1988; all other entries: George T. Kurian, *The New Book of World Rankings* (New York: Facts on File Publications, 1984.)

resources states. Along with oil money, however, come frequent attempts by the donor to influence the foreign policy of the donee. The wealth of the regional donors (coupled with their development assistance) results in very high proportions of their GNP being spent on foreign aid. For example, whereas the United States spends about one quarter of 1 percent of its GNP on foreign aid, the United Arab Emirates average 10 percent; Saudi Arabia, 6 percent; Qatar, 7 percent, and Kuwait, 3.5 percent.

Arab oil money is invested far beyond the region and tends to be concentrated in building **developmental infrastructure**. For example, Arab subsidies, loans, or gifts are financing airports in Gambia (West Africa); bridges in Taiwan; hydroelectric dams in Ghana, Mali, and Cameroon; fertilizer plants in Pakistan; irrigation networks in Bangladesh; port facilities in Papua New Guinea; and railroads in Congo. Though there is an expected emphasis on extending aid to less fortunate Muslim brethren, Arab development money is also becoming a major factor in some parts of the developing world that have few Muslims. Another form of transferring wealth among Islamic countries is the tendency for oil-rich states to employ workers from poorer states (e.g., Pakistan); these foreign workers then send remittances home.

THE PHYSICAL ENVIRONMENT

The dominant physical unifier in the region is aridity. However, there is considerable variation in the degree of aridity, and certainly in the overall physical landscape. Because desert climate prevails throughout most of the region, the bulk of the population lives in the exceptional zones, either where the climate is wetter, milder, and more hospitable, or where streams and groundwater resources make farming practicable. The physical environment has been important in the formation of the region's culture, and its role is prominent as a limiting factor in the traditional scheme of development.

The Basic Framework of Land and Sea

Higher elevations clearly dominate in the region; the lowlands are few in number and widely scattered.

The northern fringes are demarcated by an impressive series of mountains, extensions of the Alpine and Himalayan systems. The complexity of the topography has been increased by folding and faulting through the ages. This is still a tectonically active area, a crush zone between moving continental masses. Most of the rest of the region consists of ancient, stable continental blocks, sometimes upended into precipitous mountains, sometimes downfaulted into deep, sediment-filled basins. The Great Rift Valley of Africa continues into this region, forming the Red Sea and gulfs of Suez and Aqaba, dividing it almost in half. Occasional remnants of even more ancient mountains rise above the interior plateau in isolated splendor.

Two riverine valleys, the homes of ancient civilizations (the Nile and Tigris-Euphrates), occupy parts of two larger structural basins (Figure 10–5). Other structural basins, lacking water, are simply filled with sediments and salt flats. A series of narrow coastal lowlands edge the Alpine folds in the Persian Gulf and Mediterranean portions of the region, as well as the Red Sea rift. Where such lowlands extend inland, melding with interior basins, the low-utility desert often reaches to the shores. Fault block mountains on either side of the rift form some of the other important heights. The central Saharan highlands sometimes rise to significant heights above the tableland surface, but have little effect on climate. Most other highland areas produce higher rainfall totals. In many cases, they are wet enough to allow dry farming of grains or at least a consistent grazing economy. The mountains, then, constitute a major part of the "exceptional zones."

The sea's importance to the region is far more cultural than environmental. The sea played an important role in bringing portions of this area under Roman control, and its rival, the Phoenician-Punic Empire, was clearly a maritime culture. Nations of the Persian Gulf area were long involved in trade with India, China, and Southeast Asia, and merchants from throughout the Arabian peninsula pursued a brisk trade with East Africa for well over a thousand years. Most of the great Islamic empires, however, were definitely land based, and trade within the region was carried on by overland routes rather than by sea.

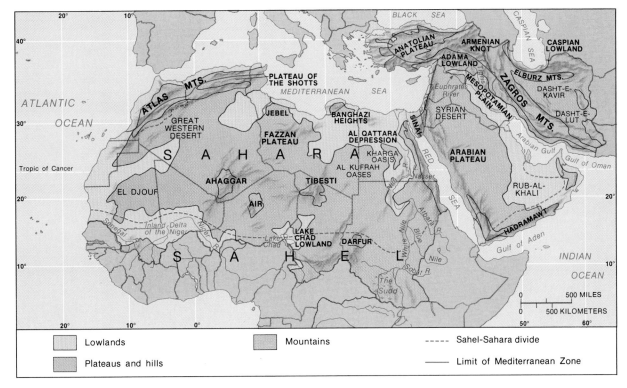

FIGURE 10–5

Physical Features in the Middle East-North Africa. Plateaus and low hills dominate in the region. There are few mountains except at the region's northern margins. The few large lowlands are often the sedimentary basins that contain most of the region's oil. Only the well-watered valleys of the Nile and the Tigris-Euphrates system offer large-scale opportunities for irrigated agriculture. Vast deserts occupy large portions of the region's interior.

The sea has had increasing importance in the last two centuries. As European powers spread colonial control by sea, harbors were improved to facilitate both European trade and military operations. The advent of great regional oil production emphasized the role of the sea, with oil exported primarily via large tankers. Importantly, the region is adjacent to, or includes portions of, major narrows from which ocean commerce can be controlled: the Strait of Gibraltar, the Turkish Straits, the Sicilian Channel, the Suez Canal, the Bab el Mandeb, and the Strait of Hormuz (see Figure 10–1). Each major shift in the direction of world commerce or in the relative importance of commodities carried has focused on one or more of these narrow passages. The very constric-

tion of sea routes in the area has ensured their importance to regional economies.

Climate and Water Resources: The Keys to Survival and Development

The adjectives "hot" and "dry" are unavoidable; they are certainly characteristic. Yet, there are significant differences in the potential utility of various parts of the region. The bulk of the area is in the midlatitudes rather than in the tropics. There is a cooler, sometimes even cold, winter season. Most of the land is definitely desert, but there are large areas of either tropical or temperate grasslands. The areas of Mediterranean climate, with a much higher degree of util-

THE REGION'S OIL AND THE ENERGY CRUNCH

The dominance of the Middle East-North Africa region in world oil markets (62 percent of proven world reserves, 29 percent of world production) is a relatively recent phenomenon. In 1946, the Middle East and North Africa produced less than 1/10 of the world's output; by 1965, it had surpassed the United States as the world's largest producing area. As recently as 1973, Canada and Venezuela supplied 40 percent of U.S. imports, compared to only 15 percent from the Middle East. By 1980, the Middle East's share of the U.S. market for imported oil had risen to 34 percent. After the oil embargo and the energy crisis, it fell to only 13 percent, a startling turnabout. Leading suppliers of U.S. petroleum imports are now, in rank order, Canada, Mexico, Venezuela, Indonesia, Saudi Arabia, and Nigeria. Only the fifth of these top six suppliers to the United States is from this region. Regional production, however, is still a critical factor in West European and Japanese markets (see Table A).

Two physical factors, one geologic and one locational, contribute to the high desirability of Middle East oil. Not only is the area's oil of generally high quality, but the characteristics of underground reservoirs in rock formations permit extremely high rates of production at low costs. In addition, the proximity of most Middle East fields to deep-water ports means that transportation costs are also relatively low. The reasons behind the series of price increases begun in 1973 were neither shortage of supplies nor increases in production costs. Less developed oil economies felt that continued low prices for oil benefitted consuming states at the expense of producing states. Oil economies expected increased prosperity and higher profits; instead, higher world oil prices brought record inflation and serious economic recession in developed countries, and bankruptcy in less developed, oil-importing countries.

The scenario had begun over a decade earlier with the founding of the Organization of Petroleum Exporting Countries (OPEC), in 1960. OPEC's specific purpose was to control the supply and price of the world's oil. The original members were Saudi Arabia, Iran, Iraq, Kuwait, and Venezuela. Additional members recruited to the cartel were Qatar, Indonesia, and Libya (1961); Ecuador and the United Arab Emirates (1967); Algeria (1969); Nigeria (1971); and Gabon (1975). Direct government involvement in setting oil prices began in 1971; the official selling prices for premium-grade crude ("Arabian light") went from $1.34 per barrel in June 1970 to $36.00 per barrel by the end of 1981.

Gains for OPEC members were only temporary. By 1986, the price had dropped to $15 a barrel, as overproduction drove prices lower. Dramatic price drops were the result of several interrelated factors. The sharp price rises of the 1970s produced a strong, largely successful drive to conserve energy among the major importing countries. As oil prices

TABLE A

Percent world oil production by region.

Region		1938	1950	1960	1978	1987
United States and Canada		61.6	51.7	35.6	16.9	18.3
Middle America (mainly Mexico)	Latin America	2.1	2.2	1.6	2.6	5.1
South America		13.1	17.5	16.3	5.4	5.1
Western Europe	Europe	0.3	0.6	1.4	4.0	5.0
Eastern Europe		2.7	1.1	1.5	0.8	0.4
USSR		11.0	7.8	14.5	19.2	23.2
East Asia		0.0	0.0	0.7	3.5	5.3
South Asia		0.0	0.0	0.1	0.5	0.4
Southeast Asia		3.0	2.2	2.2	2.8	2.4
Australia-Oceania		0.0	0.0	0.0	0.8	0.7
Africa South		0.0	0.1	0.3	3.9	3.5
Middle East-North Africa		6.2	16.8	25.8	39.6	30.6
Total world		100.0	100.0	100.0	100.0	100.0

SOURCE: *United Nations Statistical Yearbook*, 1955, 1980, 1989.

rocketed upward, prices for competing fuels also rose. Energy producers in oil-importing states were encouraged not only to prospect for and develop new sources of oil and gas, but also to open new coal mines and to experiment with nontraditional or nonconventional energy sources (e.g., solar energy, windmills, oil shales). Once oil prices began their decline, they were forced increasingly lower by a familiar producer's dilemma. As the unit price (in this case, the price per barrel) begins to drop, producers are tempted to market larger quantities. The hope is to counterbalance lower unit prices with increased production, keeping gross profits at the same level. Increased production, however, drives prices still lower, which often results in still more production.

National economic development plans in OPEC states demanded a certain level of income from oil exports, and it is a brave government that allows recently achieved high standards of living to be reduced sharply. Consequently, OPEC members were drawn into overproduction. The long-term effect of OPEC's use of an international cartel to force prices upward has been the creation of a (temporary) oil glut with serious recessionary problems for major exporters.

ity than the deserts, concentrate much of the region's population in a narrow coastal belt (see Figure 10–3). The eastward extension of this type of climate into Iran is really an **orographic** (mountain-induced) heightening of precipitation, but winter-sown grain production in that country is possible only because of this seasonal rain. With hot, dry summers and mild winters with rain, the Mediterranean climate zone differs little from its counterpart in Europe. Aspect and shelter often affect frost hazards and precipitation totals sufficiently to make individual valleys, piedmonts, or lowlands in the Mediterranean zone more or less usable in accordance with these local climatic controls. The effects are felt far from the Mediterranean climate zone, with the rain in the mountains giving rise to permanent streams that make irrigation possible in drier areas downslope.

Equatorward of the Mediterranean climate zone is a narrow and fluctuating belt of steppe climate; transitional to desert, it also receives its rainfall in winter. This steppe land reaches its greatest extent in the countries of the Middle East, since mountains bordering the rift there increase precipitation enough to promote steppe conditions. In the southwest corner of the Arabian peninsula, the regime is that of dry winter but with sufficient summer monsoonal rain to promote sedentary agriculture, even in unirrigated areas.

In the southern extreme of North Africa, there is another transitional zone from desert to steppe, this time to the wetter equatorial climates. Known as the Sahel, its rains come in summer. Often unreliable and always seasonal, these rains depend on the northward-shifting intertropical convergence (see Chapter 1, Figure 1–15). Whole years may go by without rain, while downpours and floods occur in others. Late rains can be equally disastrous.

The bulk of the region is true desert, with very low rainfall of an unpredictable nature, cold nights, scorching days, and searing winds. The area is located well within the dominance of the belt of subtropical high pressure. Winds are often offshore, so little maritime influence is felt here in either its temperature-moderating or moisture-bringing aspects.

There are few large permanent streams. Vast areas have no surface water at all. Most of the lakes are salty, or at least brackish. Many of the interior basins

have only **intermittent streams**, which flow only right after a rain. The few permanent streams often end in salt lakes or simply disappear into the sands, never reaching the sea. Only some short, swift streams of erratic flow in parts of the Mediterranean climatic zone, or the largest rivers, make the journey to the ocean. The Nile has its sources in the wetter Ethiopian Plateau and the East African Highlands. Even it now barely reaches the sea, because water is extracted en route by thousands of irrigation projects and heightened evaporation. In years of reduced flow, sea waves erode the delta, which is no longer building outward into the Mediterranean.

The highlands of the central Sahara produce a radial drainage pattern—out from the highlands and down to the desert sands. In the rare, wetter years, some of these streams may enter systems that flow south to the Gulf of Guinea. Paintings found in caves in the mountains depicting animals that live in wetter climates are often used as supporting evidence for the claim that Africa's climate is changing. That climatic change has occurred is undeniable. The real question is whether the change is recent and human induced, or ancient and related to the contraction of continental ice sheets.

The Tigris and Euphrates receive both winter rains and summer snowmelt, producing a stable flow. They have few permanent tributaries and combine to form the expanding marshy lowlands of the Shatt al Arab in their lower course. There, date groves are watered by the tides as incoming tidal flow raises the less dense, fresh groundwater levels to the roots of the palms.

Rivers of the wetter areas south of the Sahara touch on, or even enter, the North African portion of this region. The Senegal River, fed by tropical rains, is on the region's southwestern border. The Niger flows north, deep into the Sahara, before flowing southward to the Gulf of Guinea (see Figure 10–5). The large inland delta of the Niger irrigates farms in what would otherwise be empty desert. At the juncture of equatorial and desert climates, Lake Chad is an unusual water feature at the region's southern margin. It fluctuates in size seasonally, and over longer-term wet-and-dry climatic cycles.

Underground drainage is of unusual regional significance. Wells that tap underground streams, **aquifers** fed by different climates in distant areas, and fossil waters (deep groundwater left over from geologic times when wetter surface conditions prevailed) are the only sources of water over vast parts of the region. Two large areas are almost totally without water—the Libyan Desert and the Rub' al Khali of the Arabian peninsula. Smaller basins in Iran and Afghanistan have similar waterless conditions. Large areas of salt desert, with its barren, foreboding surface and treacherous underground salt-slime channels lurking beneath the crust, occupy large parts of interior Iran.

Wood is scarce, even in mountain areas, where human demands exceed supply and reduce the capability for forest regeneration. The vegetational cover in the Mediterranean zone is more scrub and shrub than forest. It thins perceptibly inland to short steppe grasses and bunch grass in semidesert reaches. Even most of the Sahara, however, has some scattered shrubs and patches of grass. Only on the leeward sides of mountains and in several large interior basins does vegetation disappear, leaving a vast empty landscape of shifting sands *(erg)* or barren, rocky wind-swept surfaces called **desert pavement** *(reg)*. Of all the region's states, only Turkey and Iran have fairly large areas of usable forest. Even in those two countries, the forest is under siege. Grass and shrub, not often trees, now mark the wetter, more usable landscape.

THE STRUCTURE OF SUBREGIONS

The first subregion includes Iraq, Syria, Lebanon, Jordan, and Israel, the nations often historically referred to as the Fertile Crescent. This is perhaps the most culturally diverse of the subregions, but history and physical environments provide the uniformity. It is the commercial core of the region. The Arabian Peninsula (including Saudi Arabia, both Yemens, Oman, the United Arab Emirates, Qatar, Bahrein, and Kuwait) forms the second subregion, the Arab culture hearth. The non-Arab states of Turkey, Iran, and Cyprus form the third; Egypt and the Sudan, unified by the waters of the Nile, are treated as a fourth separate subregion. The North African states of Tunisia, Algeria, and Morocco, with their historical linkage as the Maghreb, and nearby Libya, are yet another. The final subregion, the Sahara-Sahel, includes the predominantly Saharan countries of Mauritania, Mali, Niger, and Chad. A portion of each of these states extends into the Sahelian climatic zone. Despite their essential duality, they are the epitome of the Sahel, exhibiting all the cultures, lifestyles, and climates associated with that region of transition (see Figure 10–3).

The Fertile Crescent

The Holy Land of Christians, the homeland of the Jews, the magnificent ruins of Petra, the war-destroyed downtown of Beirut, the temporary camps of Palestinian refugees, and the permanence of ancient Damascus are among the profusion of images linked to this subregion. The designation Fertile Crescent is a popular title given to this area of rich historical background and great commercial tradition. Fertility in this instance refers to the area's relative agricultural productivity when compared to the dour hills north of it and the barren lands to the south. Some of the Crescent's soil is truly fertile, but it is water, really, that makes the difference. It is an area of somewhat wetter climate in general, especially in the west. In its drier eastern portion, fertile soils bring forth crops with irrigation, continuing the green of sown fields across an otherwise barren horizon. It was fertile in the production of civilizations as well, including Babylon, Assyria, Israel, Phoenicia, and many more. It lay between neighbors of great wealth and civilization—Egypt, Greece, Persia, and the merchant kingdoms of Arabia. It pursued an active trade with these prosperous neighbors and, in turn, exchanged technologies and ideas with them. It is the obvious route of passage between the Mediterranean and the seas to the east—the shortest, easiest land connection between the waters, the hospitable area between the difficult mountains and the desert voids. It is the epitome of the crossroads function within a region that has been thus characterized for much of history (Figure 10–6, p. 514).

Here, venerable centers of learning exist in societies where mass literacy is often recent, and sometimes still incomplete. The monuments and holy places of three great religions vie for space and access to their followers, while the faiths themselves quarrel with one another and are rent with division from within. Modern commercial centers rub shoulders with crowded slums.

WATER: THE MOST VALUABLE RESOURCE

An answer to the question, "What is the most valuable resource of this region?" might appear, at first, to be "oil." As valuable as the region's oil reserves are, however, water must take priority. In this, as in all regions, water is essential to life. As would be expected, almost all states of the region strive to develop water resources to the maximum extent possible. Indeed, the ancient empires of this region were termed "hydraulic (water engineering) civilizations" by historian Karl Wittfogel.

The freshwater resources are precious to all the region's people, especially because of their scarcity. Many of the world's countries use irrigation to make agricultural production possible in otherwise inadequately watered lands, to improve outputs in marginally wetter areas, or to grow paddy rice. Some 15 percent of the world's permanent cropland is under irrigation. China alone has over 20 percent of the world's irrigated areas, and 45 percent of China's permanent cropland is irrigated. This entire region's irrigated area is slightly larger than that of the USSR, and significantly larger than that in the United States. Many of the region's states rely heavily on irrigation; Egypt irrigates less than 3 percent of its total land area, but that irrigated area is 100 percent of Egyptian cropland. Israel, the small states of the Gulf, Saudi Arabia, Iran, Iraq, South Yemen, and Lebanon all irrigate at least one quarter of their cropland, in contrast to 11 percent for the United States and 8 percent in the USSR.

Irrigation in itself, however, does not guarantee an agricultural bonanza. Some desert areas are incapable of supporting irrigation agriculture even if water is available. Mismanagement of irrigation projects, inadequate drainage, and salination of soils can contribute to diminishing yields and even abandonment of formerly irrigated lands. Table A indicates that there is no predictable relationship between a high ratio of irrigated land to cropland and the country's capability of increasing food output per capita. It is clear, however, that irrigation is an important part of the agricultural life in this region.

TABLE A
Irrigation and per capita food production trends in the Middle East-North Africa.

Country	Irrigated Land			Index of Food Production per Capita, 1980 (1970 = 100)
	Irrigated Area (1,000 hectares)	% of Total Area	As % of Permanent Cropland	
Algeria	336	0.1	5	80
Bahrein	1	1.6	50	NA
Chad	NA	NA	0	91
Cyprus	94	10.2	22	NA
Egypt	2,850	2.9	100	93
Iran	5,900	3.6	30	112
Iraq	1,730	4.0	32	90
Israel	189	9.3	49	106
Jordan	85	0.9	9	89
Kuwait	1	NA	60	NA
Lebanon	85	8.3	29	83
Libya	140	0.1	11	139
Mali	100	0.1	5	88
Mauritania	NA	NA	4	76
Morocco	500	1.1	6	87
Niger	NA	NA	1	93
Oman	NA	NA	94	NA
Qatar	NA	NA	0	NA
Saudi Arabia	395	0.2	35	69
Sudan	1,700	0.7	14	102
Syria	539	2.9	10	157
Tunisia	140	0.9	4	120
Turkey	2,050	2.7	8	111
United Arab Emirates	5	0.1	36	NA
Yemen (North)	243	1.2	9	94
Yemen (South)	67	0.2	31	103
For Comparison Purposes				
Canada	500	0.1	1	109
China	49,200	5.3	45	116
India	39,090	13.1	23	101
United States	16,697	1.8	11	115
USSR	17,000	0.8	8	108
World	241,035		14	104

SOURCE: United Nations Food and Agricultural Organization, 1987.
NA—not available.

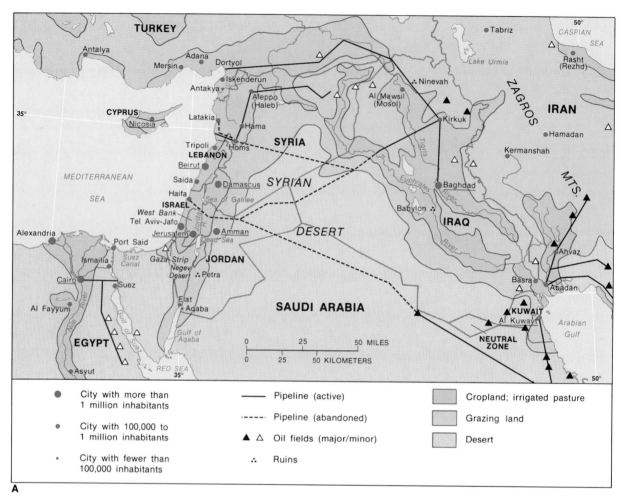

A

FIGURE 10–6

(a) *The Fertile Crescent.* The availability of water creates a crescent of green between the deserts of Arabia and the steppes of Turkey and Iran. This bountiful agricultural region has produced wealth and great civilizations through the centuries. (b) *Israel.* Emigration of Jews from Europe to their traditional homeland in the Middle East resulted in the creation of the state of Israel (1948), displacing Palestinian Arabs in the process. The first of many wars involving Israel and surrounding Arab states resulted in a division of what had been the British mandate of Palestine into an independent Israel and two Arab enclaves. Incorporation of the Gaza Strip, Golan Heights, and the West Bank greatly enlarged the number of Palestinian Arabs residing in Israel. Conflict between these two groups continues to the present. (c) *Lebanon.* Religious diversity has led to almost constant conflict within Lebanon. Conflict among these groups began in 1969, when actions against Israel by Palestinian refugees residing in Lebanon drew Israel and Syria into the conflict. Civil unrest over more than two decades has virtually destroyed the economy.

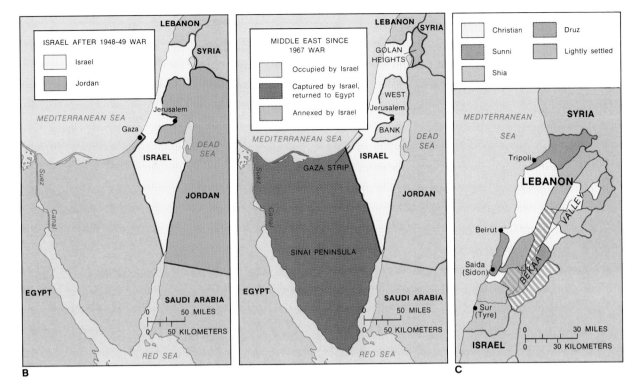

FIGURE 10–6
continued

The area is ancient. Archaeological excavations have recently underlined its age by bringing to light previously unknown civilizations. One was the center of a great Semitic empire 50 centuries ago. Called Ebla by its discoverers, its ruins indicate that it once housed a quarter-million people, an astounding figure for that time. Predecessors may lie beneath its foundations. Younger, and not far distant, is Damascus, often credited as the oldest continuously inhabited city in the world. Tyre, of prosperous ancient Phoenicia, is known to the contemporary world as the Tarabulus (Tripoli) of troubled Lebanon. Jerusalem remains, immutable, through repeated conquests, dispersions of its citizenry, and countless wars.

The seats of power within the area have often changed in stature and location, with always-ambitious rulers presiding over states whose fortunes and domains were either in the process of expansion or suffering the throes of contraction. The capitals, whether of states large or small, were always places of importance. The largest and most splendid of the region's cities were those that served as the administrative centers of great empires. Cities of lesser rank sustained themselves as centers of religion, commerce, or learning. They, too, attained significant size and opulence during periods when the Crescent functioned as an imperial core.

The Legacy of the Recent Past. By the sixteenth century, after the Mongol rulers of Persia had plundered and weakened the Crescent, the Islamic Ottoman Empire extended its control over the area. Centered on Istanbul, that empire had virtually reconstructed the holdings of the older Islamic dynasties. As in the era of Rome, imperial control was centralized outside the Crescent. Though the Turks rebuilt the Crescent's cities and refurbished its ruined countryside, its location and functions were relegated to secondary importance. By the eighteenth century, the Ottoman Empire had weakened. Maintenance became haphazard, administration was decentralized and corrupt, and the processes of economic decay were advancing. By the beginning of the twentieth century, the Fertile Crescent had become one of the poorest, most backward parts of the world.

The ultimate collapse of the empire came in 1918 in a welter of intrigue, greed, military defeat, and rising Arab nationalism. The Crescent did not attain the independence it had hoped and fought for; its territory and people were divided among France and Britain, the victors in World War I. What could have been an independent Arab state became five individual entities under European control. Not officially colonies, they were called **mandates**. Their independence was rescheduled for an undetermined future date.

The modern history and political problems of the Fertile Crescent stem from this long period of domination by a declining imperial Turkey and the policies of France and Britain that held sway during that brief, but crucial interlude of European control. Today, as in the historic past, the area is torn between the promise offered by inclusion in a larger Islamic or Arab entity and narrower local interests. All Arabs in 1918, the population has become split among narrower nationalist groups. The five mandates, under European tutelage, became five independent states. This division has resulted in small states with limited physical diversity and resources. Each controls a portion of the Crescent, but also an additional area of low utility. European rule encouraged cultural diversity, and divergent internal cultures continue to plague the successor states. More important than what Europeans did, is what they did not do. They did not attempt to solve the area's basic problems. Indeed, their inaction led to a compounding and proliferation of difficulties. The complexity and scope of these problems are particularly well illustrated in the cases of Israel, Lebanon, and Jordan.

The Seat of Trouble: A Political Geography of Three States. The problems and fortunes of three individual countries are rarely so intertwined as are those of the small states that anchor the western end of the Crescent. Each is a national crucible in which cultural elements refuse to mix, and only occasionally agree to coexist. Worse than just at odds, the groups have developed intense animosities that lead to outbreaks of civil unrest and, sometimes, outright

war. As if this were not enough, each group maintains separate political parties, receives contributions and aid from outside sources, and maintains separate armies.

The problem itself is rooted in the basic ethnic and religious differences of the people that occupy these territories. Emerging nationalism further compounds the problem. The divisions were often seeded by ancient migrations or the actions of outside forces that exerted temporary control over the region.

The basic cultural division within Lebanon is religious. Traditionally, there has been a balance there between Muslims and Christians (see Figure 10–6c). Christianity in Lebanon is much older than Islam and once encompassed almost the entire population. Islam had limited success in converting the Lebanese. Many of today's Muslims took up residence in the country during the period of Turkish and French control, availing themselves of jobs in the commercial, relatively well-developed Lebanese society. Others were simply included in Lebanon with the formalization of borders in 1921, and again in 1946. The long-standing internal peace of Lebanon was based on a political agreement that recognized an official balance in the population—half Muslim, half Christian. Muslim-Christian detente, however, became uneasy when the Muslims felt that they had become a clear majority. (Out of fear of disturbing the delicate Muslim-Christian religious balance, no census has been taken since 1932.) Serious conflict between the two groups began in 1969.

What is sought is neither independence nor territory. The aim is political control. By tacit agreement and law, the government has always reflected this religious diversity, regardless of political party affiliation. The president was a Maronite Christian, the prime minister a Sunni Muslim and the second-ranking legislative officer a Shia Muslim. All this has changed in recent years in response to changing ethnic makeup. The Christians have the lowest birthrates and the highest rate of emigration. Their numbers and influence have decreased accordingly. Christians now account for only 36 percent of the total population, and they are by no means unified. The Maronites, affiliated with Rome, are the largest group, but there are also Orthodox Lebanese, Armenian Orthodox, Greek Orthodox, and even Protestants. A small, ultraconservative group of Christians has formed an independent military force which has seized control of a part of south Lebanon.

Muslims are not unified either. The community is divided between Shia Muslims (about one third of the total) and a Sunni majority. Some of the Shiites have become involved in Iranian intrigue and are part of the Jihad movement. In addition, there are the Druz, a group considered as heretical by both Christians and Muslims. Their beliefs incorporate some elements of Islam with what is basically native religious beliefs. They have the highest birthrates in Lebanon and, as a result, are increasing their influence. They have two political parties and an agenda of their own.

Four other forces operating in Lebanon further complicate the picture. Only one of these, the Palestinians, is in residence in the country. The Palestinians first entered Lebanon in the aftermath of Israeli independence. Several hundred thousand left Israel, and a like number of Middle Eastern Jews replaced them in an informal but officially tolerated migration that took place between 1949 and 1953. Perhaps 150,000 Palestinians took up residence in Lebanon at the time, about half in UN-sponsored refugee camps. Prior to the oil boom that was shortly forthcoming, prosperous Lebanon was the primary target area for Arab migrants seeking work. Lebanon, therefore, received a disproportionate share of this early wave. Most Palestinians, however, remained in the West Bank area assigned to Jordan (1949) and the Gaza Strip, administered by Egypt (see Figure 10–6b). Numbers increased again after the Six Day War (1967). The final wave came in 1970, when the Palestine Liberation Organization (PLO) transferred its headquarters, leadership, troops, and some supporters to Lebanon under pressure from the beleaguered Jordanian government. The total in Lebanon is about 300,000, or 11 percent of the total population. Numbers are not the cause of the problem. Repeated Palestinian attacks on Israeli territory are what has created the situation.

It is impossible to deal with the Palestinians without discussing the role of Israel, the second of the four forces. A Lebanese government, weary of military attacks and counterattacks on its sovereign territory, attempted to restrain the Palestinians from carrying out raids on Israel in 1969. Raids had been ongoing for 20 years. This restraint, carried out largely by Christian militia, was met with armed re-

sistance in the first direct clash between Palestinians and the forces of their host government. Confrontations between the two became commonplace, culminating in civil war (1975–76). Three Christian factions, aided at times by Israel, fought against a combined force of Palestinians and what have been called "leftist Muslims." Many Muslims had divided loyalties and refused to become directly involved. Israel actually invaded parts of Lebanon in support of the Christian militia. Repeated Israeli invasions took place between 1974 and 1982. Israel finally occupied south Lebanon (1982–85).

The third force is Syria. Syria has never denied its interest in the area, nor its ambitions to combine the two states. In some ways, this is a legacy of French control, since the border between the two at the time was virtually open and migration was unhampered. Even the administration of the two mandates was partially combined. Actual separation came only with Syrian independence. Syria has constantly intrigued in Lebanon, first supporting rebels against the government in 1958. It has occupied portions of Lebanon on several occasions and, in 1976, fought against both Palestinian and Christian militias in an attempt to end the civil war. In 1983, Syria effectively expelled the PLO from Lebanon, supporting a more radical Palestinian faction. In seeking allies among Lebanese groups, Syria formed and supported the Amal militia, a pro-Syrian group of Shia Muslims. Support among Lebanese for union of the two states is uncertain, but Syrian control is factual.

The fourth force is the government of Iran, which has been active among Shia Muslims. Hezballah, the fundamentalist Shia Muslim militia, is its apparent military arm in the area.

Terrorism has stalked Lebanon since 1969 as at least 10 separate militias and twice as many political groups have arisen in support of or opposition to various viewpoints and causes. It is not a simple question of religious and ethnic differences.

The situation in Israel is no less complex, if less disorderly. The forces involved are more clearly defined. The ultimate goal of conflict is distinct, since two national groups claim the same territory as their national state.

The very existence of the state of Israel has been the center of a storm of opposition. It may have seemed, at times, that determination to eradicate Israel was the only unifying force among neighboring

Arab states. The Israel-Palestine question still shows no sign of being permanently resolved to everyone's satisfaction.

Territorial claims commonly are asserted on the basis of discovery, exploration, settlement, economic development, effective administrative control, or historical (but not necessarily continuous) occupation. Israel's claim to its territory is based on antiquity of settlement and a long history of prior occupancy. Both Jews and Muslims, however, claim genealogical ties back to the Prophet Abraham; on this basis, the Arabs view themselves as disinherited from their land. Israel has had many rulers over a long and tumultuous history. The Roman rulers of Palestine (the Roman name for Israel) determined to scatter the Jewish occupants of this highly nationalistic and troublesome (to Rome) area. The **diaspora** (dispersion of the Jews) sent them to all corners of the Roman Empire. Some Jews have always lived in Israel (Palestine), but from A.D. 135 until 1948, the overwhelming majority of the inhabitants of this land were not Jewish. Throughout the centuries following the diaspora, Jews maintained the goal of returning to their ancient homeland. Practical steps toward this goal were finally taken with the Zionist movement of the late nineteenth century. During World War I, the British sought to focus the nationalist ambitions of the Arabs against the Turks, and at the same time tried to influence world opinion by publicly favoring a "Jewish homeland" in Palestine in the now-famous Balfour Declaration. The British government did not actually advocate a Jewish state but rather a more vague "homeland." It also noted that this homeland should not prejudice the rights of the peoples living there already. Throughout the 1920s, Jewish immigration to Palestine slowly grew. Lands, homes, and businesses were purchased, without coercion, from individuals, for individuals. Nazi persecutions in the 1930s enlarged this Palestine-bound flow of European Jews.

The period following World War II is extremely controversial. British, Palestinian, Arab, and Zionist viewpoints concerning this period are widely divergent. The shocking confirmation that Nazi genocidal policies had led to the murder of 6 million Jews (along with millions from other national, ethnic, and religious groups) swung public opinion in Europe and the Americas toward lifting emigration restric-

Iconography of Israel. *For millennia, Jews have mourned the fall of ancient Israel and the destruction of the Temple at Jerusalem, the symbol of Judaic faith and a symbol of the nation. Jews from all over the world come to the Wall to pray and renew their religious commitment. Much of the temple site is currently occupied by a mosque, at a site sacred to Muslims. The city of Jerusalem, and sites and shrines within it, are sacred to three religions. Here a devout Hassidic Jew, a member of an ultraorthodox splinter group within the Judaic faith, meditates at the Wall.*

tions on surviving Jews. (The British, fearing a bloody confrontation between Arabs and Jews in Palestine, had previously sought to strictly control Jewish immigration.) The world realized that an expanding Jewish population in Palestine, fed by immigration, would eventually create a clear majority of Jews over indigenous Arabs. Jewish immigrants were not about to be denied access to their cherished homeland; the ideal of a Jewish state had taken on a new survivalist appeal after the Holocaust. To the Palestinian Arabs, the relaxation of immigration restrictions toward the end of British occupation, plus the unwillingness of many British to enforce their own rules against the refugees, was evidence of unwilling Arab involvement in a European problem. The Arab view was that Jewish immigration was the result of a catastrophic and regrettable eruption of vicious prejudice in Europe. Palestinian Arabs were being deliberately pushed aside to compensate European Jews for a tragedy that had befallen them in Europe, at the hands of Europeans.

Israel, including the West Bank, has a population of 3.2 million Jews and 1.4 million non-Jews, mainly Palestinian Arabs. Since Israel's independence in 1948, the Jewish population has more than quadrupled. Israel's "law of return" stipulates that any Jew can freely enter Israel as an immigrant, and immigrants have contributed over two thirds of this population increase. The Arab minority has tripled since 1948, through high birthrates rather than in-migration. Contributing to the Israelis' sense of insecurity is the fact that Jewish in-migration has slowed considerably, now barely replacing out-migration, while Arab birthrates remain much higher than those of the Israelis.

Recently, the level of Arab civil protest has increased in Israel, and the government's response has been forceful and repressive. World public opinion has tended to be sympathetic toward the Arabs, and criticism of Israel's actions has been widespread. Among the Israelis themselves there is division on the question of how to deal with Arab unrest. There is no doubt that the threat posed by increasing Arab numbers has played a significant role in government decisions to act. There is also no doubt that discontent among Arabs is compounded by this rapid growth as well. PLO recognition of Israel's right to exist entails the creation of a separate Palestinian state, ostensibly on the West Bank and in other portions of today's Israel. Colonization of that area by Israeli settlers in 150 settlements means that there will be less land for the citizens of that projected state or, worse, that Israel has factually, if not publicly, rejected the plan.

In addition to the Palestinians remaining in Israel, there are an equal number (1.4 million) in Jordan, Lebanon, and Syria. Another 300,000 live and work in the oil states of the Gulf, and unknown numbers live scattered throughout the world. Should most or all of these Palestinians return to the projected new homeland, they would form one of the densest populations in the world. They could certainly, and immediately, outnumber the Israelis.

Unlike the Lebanese, the Palestinians had no prior national affiliation. They were simply Arabs, members of an ethnic group, much as Jews were members of an ethnic and religious group. Just as the Holocaust, the exodus from Europe, and the struggle for a homeland created an Israeli nationality, the conflict over Palestine seems to have created a Palestinian nationality. Israelis distinguish between "Arabs" (residents in Israel) and Palestinians (the refugees and military activists outside Israel). Potentially and practically, they are one and the same.

Though not as deep or as troublesome as those in Lebanon, there are cleavages within the Israeli community. These splits are essentially the result of cultural conflict between groups of migrants of different origins. After the diaspora, the Jewish community separated into two widely divergent cultural groups: the Ashkenazim of European background and the Sephardim, whose roots were in Asia and Africa. From the inception of Israel, those of European background controlled the government and wielded the power. The Sephardim were less educated and decidedly less prosperous. Many felt exploited by an economic system that seemed to favor those who came from Europe. The Sephardim were deeply religious, while a majority of Europeans were secularized. Many European Jews were socialists; their Sephardic counterparts were political conservatives.

Recently, the Sephardim have become an active political force. Overtly nationalistic and strongly committed to any programs and policies that they see as beneficial to Israel, they have given their support to the conservative Likud party, effectively removing the Ashkenazi elite from political control. Some Sephardim have risen rapidly in party ranks and have even become cabinet ministers.

The extremely orthodox Jews constitute a much smaller dissident group. These Jews have increased their influence by siding with one party or the other, trading their support for the enactment of laws with religious overtones (dress codes, strict observance of the Sabbath, etc.). They are fundamentalists in the same sense that fundamentalism has arisen elsewhere among the Christian and Islamic communities in the world. Like other fundamentalists, they are often extreme political conservatives.

The political changes in Israel are favored by demographics. The Ashkenazi migrants are a rapidly aging group and now account for only 20 percent of the Israeli total. The Sephardim now hold the balance of power, with 25 percent of the total. They are a young group; most were born after Israel was founded. As a group, their birthrates are more than double those of the Ashkenazim and their descendants. Israel's hope for an end to internal dissension is in the hands of its younger generations. Fully 50 percent of all Israelis were born in Israel.

Outside forces operate in Israel as well. Oil-rich Arab states support Palestinian causes and devise foreign policies that directly affect Israel. Israeli trade and contacts with the distant West are far better developed and more important than those with the region's states. Treatment of Soviet Jews, until recently, was a major obstacle to normalizing relations between the USSR and Israel. The hoped-for loosening of restrictions on migration has already occurred. On the other hand, Soviet Jews more commonly opt for migration to the United States and Canada. The United States, Israel's firmest ally, has great influence on Israeli policies because of the important role U.S. foreign aid grants play in maintaining the economy. On the other hand, the influence of American Jews, once quite powerful, has waned as contributions become progressively less important in the total Israeli budget.

The most telling aspect of Israel's struggle to persist has been the enormous share of budget revenues that must be spent on the military. Continued high defense costs have slowed economic growth to about 2 percent a year. Israel has fought full-scale wars in 1948, 1956, 1967, and 1973; the most recent war was especially expensive. The costs associated with the multiple invasions and occupation of Lebanon, quelling Arab protests, and defending against terrorist attacks and Palestinian raids must all be added to the total. The strains of maintaining approximate parity with Arab military machines fueled by oil revenues threaten the survival of Israel as much as do invasion possibilities.

Israel's internal political stability depends on achieving peace with both neighboring states and with stateless Palestinians. Surrounded by larger, hostile neighbors, Israel remains in a precarious position.

Unlike the rest of the subregion's states, Jordan never existed as a state until the breakup of the Turkish Empire. There are no historical roots with which to identify; in truth, the Jordanian nationality is weakly developed. The wetter, western edge of the country has always been considered a part of the Fertile Crescent. Jordan's most spectacular ancient monument, the ruins of Petra, was built by a society that has long since disappeared and about which little is known. It is no basis for a national mystique in comparison to Masada in Israel or the Phoenician roots of Lebanon.

The leader of Jordan, King Hussein, it can be argued, *is* Jordan. His father, the first king, was from a prominent family in Saudi Arabia. The elder ruler was simply named king by the British during the era of the mandate. Jordan is a new name. The area was originally called Transjordan, a geographic construct that explained its location as "the land beyond the river Jordan."

In addition to its weakly developed nationality, three other factors complicate Jordan's existence: the scarcity of usable land, the threat of Israeli territorial ambitions, and the large Palestinian minority. Eighty-six percent of Jordan is wasteland; only a small fraction of its area is a part of the Fertile Crescent. Water is scarce, and much of the valley of the Jordan River will prove difficult to irrigate and settle because Israel and Lebanon extract most of the river's water before it reaches Jordan.

Israel's claims to the area are far more tenuous than those it applies to its current territory. In ancient times, Israel was the people (the nation), not the state. For much of its history, the nation was divided among the two kingdoms of Judea and Samaria (because of a religious disagreement among Jews of the day). Those two kingdoms coincided with the terri-

JERUSALEM

The personification of Israel is Jerusalem—its history and its present, its problems and its prosperity, its unity and its dissension. It is two cities, not one, both symbolically and in reality. At its heart is the old city, encircled by medieval walls. Within these cramped confines are the shrines and holy places that give the city its special status to three great religions. For Jews, it is the site of the Temple of Solomon, and its later successor the Second Temple, where the faithful from all over the world come to pray and renew their commitment. For Muslims, it houses the Dome of the Rock, the spot from which the prophet Muhammad ascended to heaven. After Mecca and Medina, it is the third most important city to the followers of Islam. For Christians, it is filled with churches built on sites associated with the life and death of Christ. In the manner of the medieval city, it is divided into quarters, in this case, sections for each of the groups to whom it is holy. Within each quarter are groups who represent the dissenting elements of each of those faiths. Even *within* religions, the old city speaks of division.

Surrounding the old city on all sides is the new Jerusalem, built in the 400 years since construction of the old city's walls. Most of it has been built by Israel, for Israelis, since 1948. In new Jerusalem is Yad Vashem (the Holocaust memorial), the Hebrew University, and the modern hotels for tourists who no longer come in great numbers. It is also the home of the Knesset, the Israeli Parliament, because it is the official capital. Fearing antagonisms, however, the embassies of the world's powers still cluster in Tel Aviv some 35 miles to the west.

Both Arabs and Israelis call it home, but they live in separate sections, divided by the cease-fire line drawn in 1949. New Israeli settlements now creep across that line, inspiring fear and hostility among Arabs.

In the center, pressing as near as possible to the old city and the Wailing Wall, are the modest homes and crowded apartments of the ultraorthodox Jews. Dressed somberly, these Jews spend their days in study, prayer, and active campaigns to influence the law and convert the less orthodox to their point of view. In close proximity are the homes of affluent merchants and the successful middle class. Beyond are the rows of functional apartment blocks for the young and the less wealthy, struggling with the massive inflation that has plagued Israel over the last two decades.

It is now Israel's largest city, with a half-million residents. For all its historic importance, it is not the great center of industry and commerce in today's Israel, but the home of its government and the symbol of continuity between the kingdom of old and the modern state. It is a place dominated by memories despite the recency of much of its construction.

tory, more or less, of today's Israel. The control of a portion of Jordan dates back to the time of King David (ca. 1000 B.C.). It was not long, or consistently, a part of ancient Jewish states. Only the most conservative, most nationalistic elements within Israel would put forth a claim to Jordan. Those groups, however, have become more active in the Israeli political arena.

The Palestinian nationals in Jordan have proven to be a major concern. Jordan was awarded the West Bank territory in 1949 and administered the area as an integral part of the Jordanian state until the war of 1967. The population, of course, had even less loyalty to Jordan than the Jordanians themselves. Jordan is also home to the largest number of Palestinian refugees (800,000) of any state in the region. During the occupation of the West Bank, Palestinians were almost equal in number to Jordanians. Efforts by Jordan in 1970 to stem Palestinian raids on Israel from within Jordanian territory met with strong PLO opposition. The unassimilated Palestinians posed a threat to the stability of the state as the heavily armed PLO movement threatened a military takeover and sought public support for their views. Open conflict resulted, and the Palestinians were defeated. Many left for Lebanon in company with the PLO leadership. The exodus resulted in the reduction of tensions and problems within Jordan. The loss of the West Bank and the exodus of some Palestinians reduced the Palestinians' share of the total population to a more manageable 28 percent. Jordan, in support of the PLO plan for an independent Palestine, has relinquished all claims to the West Bank.

The Commercial Tradition Triumphant. Despite the disruption of war and internal turmoil, the economies of Jordan and Israel have continued to grow. It is only in Lebanon that economic collapse has been virtually total. Yet, prior to civil war, Lebanon's economy was one of the soundest in the region. First and foremost, these are urban economies (Table 10–2). The widely studied Israeli **kibbutz**, the irrigation of Jordan's portion of the Dead Sea lowland, and the fabled and fertile Bekaa Valley of Lebanon notwithstanding, urban pursuits dominate overwhelmingly in all three states. Each country has a higher rate of urbanization than does the United States.

None of the economies is mineral rich. All three states are the antithesis of the oil-rich countries in that what few resources are present tend to be dominated by fertilizer minerals. Both Israel and Jordan produce phosphate rock, but only in Jordan is this fertilizer a major export. Both countries also extract potash, salt, and some other chemicals from the Dead Sea. The presence of these fertilizer materials is certainly greatly beneficial to their respective agricultures. Some iron was mined in Lebanon before civil unrest became prominent, and Israel mines some copper, but neither resource is of serious economic benefit. Israel even has a little oil.

Manufacturing, on the other hand, is of considerable importance. Israel has a major clothing and fashion industry. Subcontracting is still done for the New York garment industry, but direct Israeli sales are more common. Israel has developed major chemical industries at Haifa, using imported oil as a raw material, and Tel Aviv is a world-scale pharmaceutical center. As many categories of manufacturing as possible produce for the domestic market to avoid the necessity of imports.

Jordanian industries are much less labor intensive; production runs to chemicals, fertilizers, artificial animal feeds, and light machinery. Unlike the

TABLE 10–2
Economies of Lebanon, Israel, and Jordan.

	% Urban	% Arable	% Agricultural Employment	Per Capita Income ($)	Inflation Rate (%)
Lebanon	83	29	16	1,505	(not available)
Israel	89	21	4	5,995	20
Jordan	75	5	18	2,010	−0.3

SOURCES: *United Nations Statistical Yearbook,* 1988; Population Reference Bureau, 1989.

case in Israel, consumer goods are mainly imported.

Lebanon had an expanding, if small, light industrial base. Even in the face of serious disruption, industrial production managed to grow through 1982.

In all three cases, however, services dominate the economies. Shipping was Lebanon's economic mainstay. As late as 1980, Beirut was handling 2,000 ships a year, and the port of Tripoli (Tarabulus) was a major transshipping center for petroleum and refined products. Lebanese shipping magnates owned vessels registered under the flags of all the major merchant shipping registries—Panama, Liberia, Japan, Italy, Norway, Greece, and the Netherlands. Great investors and speculators, they would buy ships and their cargoes in midocean and even change destinations in billion-dollar business deals. In the Mediterranean cultures, they were viewed as the world's shrewdest merchants and toughest bargainers. Lebanon was the center of the Arab financial world, and Lebanese banks included some of the world's largest. There were over 100 banks in Beirut. Before the collapse, Lebanon earned a third of its income in foreign trade, and 20 percent in banking.

Until 1975, Lebanon's economy was a bright spot in the still largely undeveloped Middle East. The country was a model of harmony and compromise, wherein a multiethnic, multireligious culture existed in peace and relative prosperity. That situation has changed, but business goes on—from Cyprus, Jordan, Switzerland, Canada, and Italy. For Lebanon, it is a tragic decline, but for those who have migrated, business continues. They await the chance to return.

Because of its inability to trade with most of its neighbors, Israel orients its foreign trade largely on Europe. There is a lucrative export business, but unusual services are one of Israel's chief sources of income. The most important of these services is diamond cutting, an old tradition among European Jews. Polished diamonds are the leading export. Diamond cutting in the Western world long has been a Jewish occupation. Many occupations, as well as the ownership of land, were forbidden to Jews in Medieval and Renaissance Europe. Gem cutting was permitted. In a cultural environment often characterized by hostility and prejudice, it was sensible for one to be able to pocket one's stock and flee, carrying in one's brain the necessary "tools" for survival—personal skills. The Israeli government actively encouraged diamond cutting and polishing as a classic high-value-added industry in which transport costs (over whatever distance) remain a tiny fraction of the price of finished goods. Diamonds are mined in Africa, cut in Tel Aviv, and marketed in the West with no significant transport cost.

Other Israeli specialties include research and development. Israel exports its superior agricultural technology and does engineering work of all kinds on contract. Skilled in languages, Israelis do translations of books and films. The restoration and reproduction of art works is a highly exacting specialty for which they are famous. Recently, Israel has begun to develop a computer software specialization, and even does archaeological consulting.

Jordan also has become a service economy, almost bypassing the intermediate stage of industrial development. The country has invested heavily in education. Some 300,000 skilled and highly educated Jordanians work abroad in the nations of the Gulf. Their remittances to family in Jordan amount to over a billion dollars a year. The civil war in Lebanon has resulted in another windfall: offices and investments have left Beirut and relocated in Amman, because Jordan has been of late the most politically stable and least controversial location in the Middle East. Even the Iran-Iraq war brought development and profits as the Jordanian port of Aqaba was enlarged (with Saudi Arabian funds), and Iraq's transit trade passed through Jordan for a time when the port of Basra was closed (see Figure 10–6). Once hemmed in by Israel and Saudi Arabia, Jordan entered into a shrewd territorial trade with Saudi Arabia to gain room for expanding its port.

Tourism has always been a regional specialty. Lebanon once catered to Europeans seeking the exotic and to Arabs seeking a tolerant atmosphere. Jordan has sought to capture that business and has had some success.

Agriculture does play a role in the three economies, even though it is subordinate to other income earners. Even in normal times, all three countries imported food. But clever programs and heavy capital investment have made agriculture a lucrative business.

Israel has a very small portion of its total work force in agriculture (4 percent). Israeli agriculture is practically agricultural science. Its output is astounding, in particular when it is remembered that fully half the national territory is in the Negev Desert.

TEL AVIV-YAFO

Tel Aviv and adjacent Yafo (Jaffa) have grown into one large city of almost a half-million people. An empty beach at the turn of the century, Tel Aviv is new, sleek, and definitely Western. The city served as capital in the early years of Israel, but that function was always a poor fourth after finance, commerce, and industry. The city is Mediterranean in climate and outlook, with beaches and recreation facilities for tourist and resident alike. In 1907, it was a collection of dunes that was purchased by European migrants who were schooled in Zionism, socialist in political outlook, and capitalist in spirit. Here, along deserted shores, these migrants re-created elements of London, New York, Berlin, Paris, and the other cities from which they had come. Waves of later migrants, fleeing Nazi Germany, swelled its ranks in the 1930s, and still later waves of Eastern European Jews came to its welcoming shores in the aftermath of the Holocaust.

To its south is Yafo, the traditional port for Jerusalem, a medieval city grown modern and affluent as the center for the rich citrus district that forms its hinterland. Farther to the north, and not yet fully joined to the other two, is Haifa, the deep-water port, with its much superior harbor. An Arab city in the days of Palestine, Haifa was the European-built outlet for the oil fields of Iraq, to which it was connected by a now-unused pipeline. It is the major industrial center of the country, and smog now obscures the heights of Mt. Carmel which form its backdrop. Collectively, these cities and the towns in between form a conurbation of one and a half million, the largest in Israel.

Unlike Jerusalem, Tel Aviv-Yafo bustles with "new migrants" who seek a living in its atmosphere of commercial growth, rather than fundamentalists seeking salvation or young college graduates seeking government employment. If Jerusalem is the center of learning, Tel Aviv is the center of theater and the arts. Technically oriented, it is a major center of scientific research.

Religion is rarely the topic of discussion, since the ultraorthodox most often avoid Tel Aviv as a place of worldliness and sin. Here the split between Ashkenazim and Sephardim is more keenly felt because the economic gap between the two groups is much wider and more obvious. Here, also, one appreciates the full scope of the diaspora, with Jews who have collected from every corner of the earth in a cosmopolitan mix of incredibly diverse cultures. Yemeni porters, black refugees from Ethiopia, craftsmen from India, intellectuals from Poland, recent émigrés from the Soviet Union, brokers from Argentina, and expatriates from the United States, all Jewish, give the city a flair and flavor unequaled anywhere else in the country. Amid towering skyscrapers and crowded boulevards one senses the proof of the overwhelmingly urban nature of Israel. This is the antithesis of the kibbutz. It is the real power that maintains Israel and generates its wealth.

Production is intense and varied. Citrus dominates agricultural exports, but its relative share is decreasing (despite increasing production) as new specialties are added to the flow of trade.

Israeli agriculture is characterized by very high rates of capital investment and technology. In common with most Mediterranean countries, Israel has a positive balance of trade in agricultural products, importing cheaper grains and animal feeds while exporting higher-priced, high-quality produce and animal products. Israeli farms derive 40 percent of their incomes from exports. Outside wheat and beef, the country produces most of its basic food needs. The predominantly semiarid and arid land is heavily irrigated (49 percent of cultivated area). This intensive irrigation is heavily subsidized by the government, with farmers paying only about one third of the real cost. The Israelis have been markedly successful in increasing their efficiency of water use. The volume of water used in agriculture has stabilized over the last 20 years, whereas irrigated acreage has increased by 18 percent. About half of all irrigation water is used for fruit production, and fruit is its most lucrative agricultural export. Not content to rely on citrus, fruit exports have been diversified, and new agricultural specialties continue to be developed. Seed, breeding stock, bulbs, animal and plant vaccines, nursery stock, and tissue-grafting technology are examples of new products and by-products of Israeli agriculture that are entering export markets.

Considering the semiarid nature of most Israeli agricultural land, it is not surprising that traditionally Mediterranean livestock (goats and sheep) thrive. It is surprising, however, that Israel has a dairy industry that supplies all its domestic needs. Because dairy cattle are acclimated to cool-summer climates, but the European-background culture of many Israelis demands dairy products, Israel produces both through selective breeding and innovative environmental management of dairy herds. Israeli agricultural technology and planning have been so successful that Israel provides technical assistance to developed and underdeveloped countries worldwide.

Communally owned and operated farms in Israel are widely known under the name of their village settlement type, the **kibbutz**. Such a village will house 500 to 1,000 people, many of whom work at tasks that are not directly farm work. Property is jointly owned, and members share the profits of the farm, the crafts produced for sale, and even income from jobs to which many members now commute. The kibbutz is based on the socialist ideals once prominently held by early European migrants.

Without the West Bank, Jordan is a land with limited agricultural utility. Like Israel, it is drilling deep wells in the desert to tap fossil waters and is experimenting with the use of brackish water and sea water for irrigation of certain crops.

Lebanon's agriculture is far more traditional. Olives, grains, and grapes are the normal crops. The combination of war, deforestation, and overgrazing has led to serious problems of soil erosion over the central and northern parts of the country. The fertile Bekaa Valley has been one of the most disrupted parts of the country.

The fortunes of the area are out of kilter with past economic predictions. Jordan's future was once thought bleak, yet Jordan has been transformed into a progressive, reasonably healthy economy. Israel, still the most prosperous economy, has suffered considerably from inflation. The standard of living has fallen, and unpopular austerity measures have had to be taken to restore economic health. Lebanon, once booming, has become an economic shambles.

If peace can be restored, progress would again be possible. All three countries have a tremendous trade imbalance that can be redressed only by service income earnings. Peace would make that remedy more possible.

Syria and Iraq: The Land of Empires. Syria is a proponent of Arab Socialism and one of the more vocal advocates for a united Arab state. Damascus, its capital, was once the seat of a great Arab empire that covered much of this region. It was much larger than most other Arab settlements at the time of independence. As early as the end of World War I, there was a vocal group of Arab scholars and leaders in Damascus who planned a united Arab state (some called it Greater Syria) under Syrian leadership. Since the golden age of Arab culture, Syrians have seen Egypt and Iraq as their possible leadership rivals, and Damascus, Cairo, and Baghdad are still the most influential centers of Arab culture and education.

There are perhaps 12 million Syrians, most of them Arabic in speech and culture. Sunni Muslims form 80 percent of the population, but there is a rich

THE CRADLE OF ALL CIVILIZATION?

The great antiquity of the civilizations of the Middle East-North Africa region has suggested the interesting, but highly debatable hypothesis that all early centers of civilization are the intellectual offspring of this region. If true, it would mean that each recognized early center of the development and advance of civilization was in significant contact with, and derived at least some basic ideas from, the Fertile Crescent (see Figure 10–5).

The "single-source" theory of the diffusion of civilization outward from one center of innovation gains support, if not proof, from the time sequence of the first appearances of agriculture, irrigation, urban centers, and similar revolutionary ideas in various early centers. The spatial diffusion concept discussed briefly in Chapter 1 would postulate that, in a classic diffusion pattern from a center of innovation, places closer to the origin point would have access to the idea or technology first. Places farther away would, presumably, get the news somewhat later.

The Tigris-Euphrates headwaters district is thought to be the earliest center of grain cultivation and agricultural villages, dated to about 10,000 B.C. By 5000 B.C., these key concepts in the development of civilizations had spread throughout the Fertile Crescent. Towns had appeared in the upper Tigris-Euphrates area by 6000 B.C. Later centers developed in the Indus Valley of modern Pakistan, Soviet Central Asia, and even the central Huang He Valley of China by 3500 B.C. New World centers of civilization can be traced to approximately 1500 B.C. in the Mexico-Guatemala and Peru areas.

Thus, the timing of civilization's rise in various centers relative to the Middle East would seem to support the possibility of a single-source origin. But it does not prove it, and most scholars now believe that the key

minority of the Ismaili sect, an active Shiite group, and dozens of religious minorities, including a varied Christian community that may encompass as much as 10 percent of the total.

Syria produces some petroleum, but its production has dropped under the strains of regional disruption. Recent oil and gas finds have ended internal economic woes, since imports no longer are necessary. Syria once depended on transit fees from two major oil pipelines crossing its territory. However, the Tapline from Saudi Arabia to the Mediterranean (at Sidon in Lebanon) and the Iraqi pipeline to Tarabulus are temporarily unused because of tensions over the Iran-Iraq war and disagreements between Syria and the Saudis as well as with Iraq (see Figure 10–4).

The Syrian economy is one of the fastest growing in the region, in part because of increased Soviet aid and generous grants and loans from larger oil-producing states. It is a centrally planned economy in contrast to that of traditionally entrepreneurial Lebanon. In this water-short, unpredictable-rainfall area, major expansion of farmland requires irrigation. Fortunately for Syria, the Euphrates Dam, completed in the mid-1970s, has doubled irrigated land and is enabling Syria to resume food exports despite great population increases. The dam also provides electric power for Syria's ambitious industrialization.

concepts that led to evolving civilization could just as well have been spontaneously developed in more than one center, including several independent centers within this region—termed the "multiple-origin" theory.

The major arguments against the single-source idea lie in the great distances involved and the primitive means of transportation that existed for thousands of years. Still, there are some intriguing bits of evidence and similarities. Why are pottery designs and craftsmanship in Ecuador and Japan highly similar, when both have been radio carbon dated to 3000 B.C.? Why did both ancient India and ancient Mexico use zeroes in their mathematical systems? Is it just an improbable coincidence that the game of Parcheesi, using the same style of board and the same rules, existed in both southwest Asia and Middle America before Columbus?* Perhaps biological evidence is more persuasive. The sweet potato, an American plant in origin, was cultivated in both pre-Columbus Middle America and the Pacific islands of Polynesian culture. Moreover, the plant had the same name in both places.

Is there much doubt that these two cultures were in early contact? Whatever the validity of the single-source or multiple-origin theories of the rise and spread of civilization, there is no doubt that the Middle East-North Africa region was the first source area to appear, if not the sole center of innovation.

---

*George F. Carter, *Man and the Land* (New York: Holt, Rinehart & Winston, 1975), pp. 97–101.

Syria is urbanizing rapidly; Damascus alone contains more than one quarter of the national population. The transport net is most densely developed in the western part of Syria, connecting the capital with Aleppo, the second largest city (population 1 million), and the Mediterranean seaport of Tartus, also a rapidly growing center.

Syria has woven itself the most tangled political web conceivable in pursuing its aims as would-be regional or bloc leader. Its loss of the Golan Heights to Israel was the nadir of its political fortunes (see Figure 10–6b). Since that time, the government of Hafez Assad has asserted itself in every direction. It has become the self-appointed guardian of peace in Leb- anon; no settlement of that country's problems will be possible without Syrian accord. Relations with its neighbors are hardly cordial; pipelines through Syria have been closed, aid and grants have been reduced or eliminated, and diplomatic ties have been broken. Accusations of terrorist involvement have been leveled at Syria by an international community aghast at the Syrian capacity for intrigue. It is certainly an independent foreign policy that Syria pursues, and Syria's actions have led to a reduced level of sympathy and support from other Arab countries. Syria has become isolated, instead of becoming the leader.

The Syrian desert—stony, flat, and almost lifeless—separates Iraq and Syria south of the steppe

DAMASCUS

Damascus thinks of itself as the world's most ancient city. It was never just a trade center and market town, however well developed those functions were, but also a famed center of manufactures and crafts from its earliest history. Damask was the Western name for its richly embossed cloth that commanded a premium price on world markets. Damascene swords were valued for their tempered steel, keen edge, and durability. Spinners and weavers produced luxurious silks of the greatest quality. Books and manuscripts bound in richly tooled leather, brassware of the finest patina, and silver vessels of most intricate design were the products of its craftsmen and workshops through the ages.

The city is both ancient and new. The wealth of old can still be seen in terraced and landscaped courtyards, replete with fountains, caught in a glimpse through ornate and monumental wrought-iron gates. Though the skyline boasts modest skyscrapers, the Omayyad mosque still dominates the urban view. Balconied apartment blocks line the streets of the better sections of town, while Soviet-style worker blocks of prefabricated concrete stretch in lines toward the city's limits. Slums are fewer, and far less wretched, than in many of the region's cities.

The whole is surrounded by the Ghuta oasis, where every usable piece of land is cultivated, giving the city the appearance of being surrounded by a huge and well-kept garden. Viewed from a distance, Damascus equals anything that has been written in its praise. On closer inspection, it shows signs of wear and mild decay. Chipped paint, missing slabs of marble, and crudely patched asphalt streets speak to the strained budget of a country involved in military actions beyond its means and caught up in the rush to industrialize. Somehow gracious despite propaganda slogans emblazoned on cheap cotton banners, it remains one of the marvels of the Middle East. It has endured throughout centuries, as its citizens proudly announce, without falling into ruin or declining into a village. That sense of permanence is pervasive.

Damascus. *Considered by its residents to be the oldest continuously inhabited city in the world, Damascus is surrounded by the lush Ghuta oasis. Since the dawn of antiquity, conquerors have coveted this city and its site. Greeks, Romans, Persians, and Arabs have all ruled the city at one time; each has in some way influenced Syrian culture. This is the interior of the Omayyad Mosque, built when Damascus was the capital of a great Arab empire that bore the same name. Greek, Roman, Byzantine, and Arab elements are visible in its architecture and decoration. It attests to the cosmopolitan nature of Syria and its capital city.*

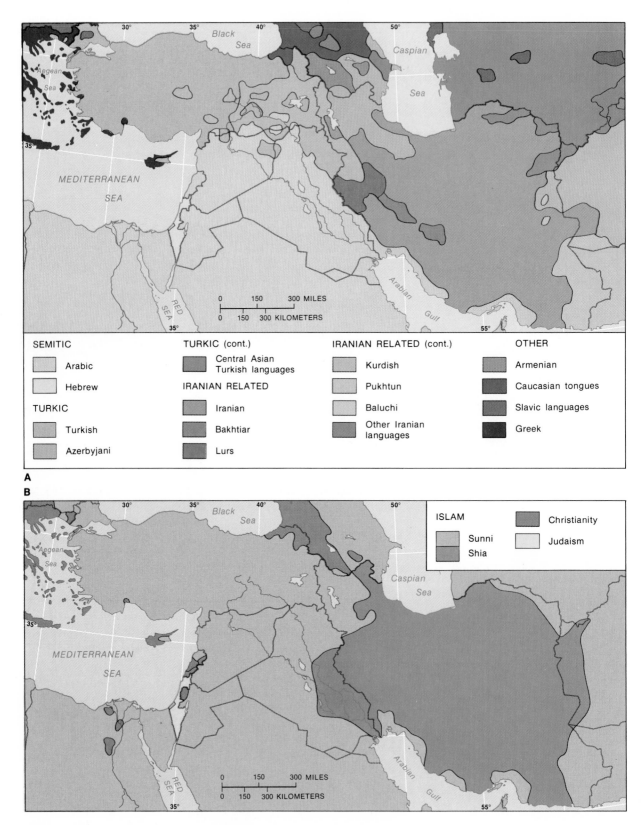

A

B

SEMITIC
- Arabic
- Hebrew

TURKIC
- Turkish
- Azerbyjani

TURKIC (cont.)
- Central Asian Turkish languages

IRANIAN RELATED
- Iranian
- Bakhtiar
- Lurs

IRANIAN RELATED (cont.)
- Kurdish
- Pukhtun
- Baluchi
- Other Iranian languages

OTHER
- Armenian
- Caucasian tongues
- Slavic languages
- Greek

ISLAM
- Sunni
- Shia

- Christianity
- Judaism

lands of the Crescent. If approached by air from the west or south, it is difficult to imagine that this country was perhaps the most coveted land in the Middle East. The desert, however, covers only half its area, and the line between irrigated green and desert brown is as sharply drawn here as it is in the Valley of the Nile. Here, though, the cultivated area is a broad and spacious plain, not a ribbon within the encroaching desert.

The ancient kingdoms of Babylonia and Assyria drew their wealth from this well-watered plain where two rivers combine their flows. Winter rains and summer thaws in the encircling arc of mountains that feed the upper reaches of the Tigris and Euphrates keep water levels deep and flow nearly constant. The oldest permanently used canals in the world siphon off the flow of the rivers into thousands of square miles of carefully cultivated fields. Reed-filled marshes line the depressions that parallel the natural levees of the rivers. Lakes and pools dot the plain in the spring. It is a wet-looking area when crops are in the field. Only at harvest, during plowing time, and in areas in fallow, does it take on the appearance of desert. Yet, islands of unirrigated lands, some quite large, are stark reminders of the actual climate. The whole area smells of mud, the primary

FIGURE 10–7

Language and Religion in the Middle East-North Africa. The region is often thought of as uniformly Arab in language and Islamic in religion. Although these two elements of culture are dominant, there is actually great diversity, particularly in the Middle East portion of the region. (a) Turkic and Iranian tongues dominate in the region's two largest states. Arabic dominates only in Egypt, the nations of the Fertile Crescent, and those of the Arabian Peninsula (its area of origin). (b) The religious picture is simpler. But even within Islam, there are serious doctrinal differences. The conservative Wahabi Muslims of Saudi Arabia contrast sharply with the liberal and highly secular Bekhtashi Muslims of Turkey. The Shiites of Iraq and Iran are regarded as schismatics by the Sunni Muslims, who dominate in most other states. Christian minorities are scattered throughout Egypt and the nations of the Fertile Crescent.

building material of ancient Babylonia, its historic predecessor.

The third portion of Iraq is the mountain and hill country of the north. Some heights reach to 10,000 feet, and all the mountains have a rocky, angular profile. The geologic structure is complex, and both faults and folded structures house deposits of petroleum. Grazing is more common than farming, but fields of winter-sown grain occupy valleys and portions of the piedmont slopes.

Modern, industrial, and often quite Western in appearance, Baghdad, the capital, does not appear to fit into any of the three prevailing landscapes. Seventeen million people dwell in this fortunate land, but there would be room for significantly more if the waters of the rivers were more completely and efficiently utilized. Even in the face of a prolonged and bitter war with Iran, the population has been growing at a rate of over 3 percent annually.

Only three fourths of the Iraqi are Arabs; there is a large (perhaps 20 percent) Kurdish minority in the north and northeast border regions and a small, potentially troublesome Iranian minority (Figure 10–7a). The Kurds are a more or less constant problem for any Iraqi government with their ambitions for a separate state. A 1975 military defeat of insurgent Kurds (the result of withdrawal of Iranian support) may have contained the revolt for a while. The Kurds are linguistically related to the Iranians but are not interested in union with Iran. Independence for all Kurds, who form significant minorities in Iran and Turkey and are also found in Syria and the USSR, has been the ultimate goal since the 1920s.

Potentially more disruptive than the Kurdish minority is the question of religious loyalties. Though 95 percent Muslim, over half the population is Shiite (rather than Sunni), the same sect that dominates in Iran and a cultural residual of long-term control by Iran in centuries past (Figure 10–7b). Sunni means "majority" or "orthodox." The Shiites, or "partisans," believe that Ali (Muhammad's son-in-law) was the Prophet's legitimate successor and that Ali's descendants are the lawful caliphs, or political-religious leaders. The Sunni accept the Omayyad branch of the Prophet's family as the true successors. Iraqi Shiites' loyalty to Iraq could be tested by the fact that Iran's

Ayattolah Khomeini was the self-proclaimed leader of Shiites everywhere. Several important Shiite holy shrines are located in Iraq, and Iran has announced its desire to control the portions of the country that contain them.

Iraq, like Syria, has periodically proclaimed the desirability of a united Arab state. The Iraqi also presume that they could be the obvious leaders of such a unified endeavor. A mildly socialist, frankly military regime has ruled for over 25 years. It is a fairly progressive government with a decidedly modern outlook. Its rejection of Shariya law and many conservative Islamic traditions has made it anathema to fundamentalist Iran.

Oil forms the bulk of Iraq's production and yields the greatest amount of governmental revenues. The reserves are immense (see Table 10–1). While the Kirkuk and Mosul fields of the north still dominate production, important fields in the south, east of Basra, and close to the Iranian border are of increasing importance.

A large proportion of oil revenues seems to be directed toward infrastructural improvement rather than toward agricultural development. Iraq has built a new pipeline from its productive northern Kirkuk fields to Dortyol, Turkey, on the Mediterranean as an alternative to the pipeline through Syria (see Figure 10–4). Pipelines also move oil to a new refinery at Basra, and Iraq has built a major deep-sea terminal for oil exports at Al-Fao on the Gulf.

As in Syria, land reform has taken place, and the worst abuses of the tenancy system have been eliminated. The soil is naturally rich, and crop yields have improved greatly with the spread of improved farming techniques. Iraq has excellent potential for the further expansion of agriculture. It is relatively sparsely populated, with sloping land and available water. Still, agricultural output has lagged behind the birthrate. The major agricultural potential lies in the largely barren Gezira, between the upper Tigris and Euphrates, and north of the southern floodplains. Water management problems are a far greater obstacle there than any lack of water itself.

Iraq has some difficulties with Syria over the division of the waters of the Euphrates. Relations with Turkey have been good, but relations with Iran, which erupted in war in 1980, focus on the continu-

ing problem of the Kurdish minority in both countries and territorial counterclaims. The real causes of animosity may well be politics, history, and religion. The Iranian-Iraqi contest for leadership and dominance in the region and in the affairs of Islam is as old as the two cultures themselves. They are traditional enemies. The religious division of Iraq is a reflection of the long-time control of much of that country by Iran. Fundamentalist Islam versus secular governance and permissiveness is at the root of the problem. After eight years at war and much disruption in the production and shipment of oil, a cease-fire agreement has been reached. Iran and Iraq are in a state of economic exhaustion. A cessation of hostilities was also necessary before a new OPEC agreement could be arrived at.

The war has rearranged the flow of oil to some degree, away from the Gulf and, through new pipelines, to the Mediterranean. Arab solidarity (except for Syria) with Iraq has reshuffled regional alliances. U.S. involvement in the Gulf brought the return of outside intervention after a decade of absence. Terrorism continues. The war and its destruction may have directly involved only the two contending countries, but their ramifications are regionwide.

The Arabian Peninsula

If the Fertile Crescent is thought of as the commercial heart of the Middle East, the Arabian Peninsula is its conscience and soul. It is the homeland of the Prophet Muhammad, where Muslim pilgrims retrace his journey from Mecca to Medina. It is also the homeland of the Wahabi, a strict and orthodox Islamic sect that sees itself as the guardian of the morals of the greater Arab world. It was, until recently, the land of the Bedouin; it is now the land of oil (Figure 10–8).

The compact, blocky peninsula is easily recognized on any map. It is truly a realm of deserts, lightly populated and inhospitable to settlement except at its margins and in favored oases. In the Judaeo-Christian tradition, it was the barren wastes of penitence; the land of exile; the domain of Ishmael, the disinherited son of Abraham.

Great kingdoms arose in its well-watered southwest, where frankincense was produced; that sub-

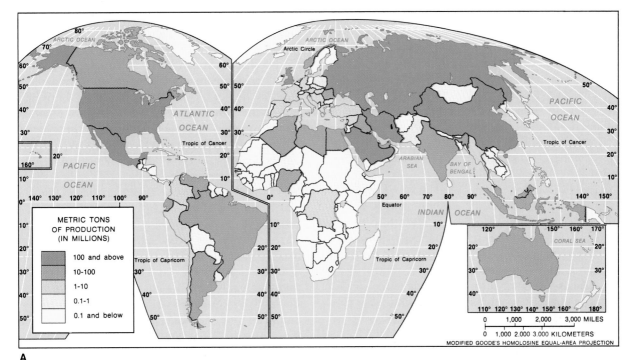

A

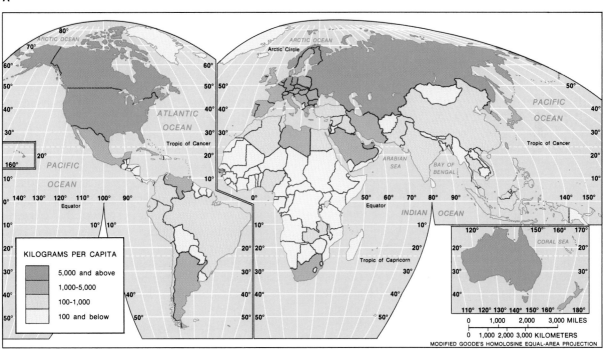

B

FIGURE 10–8

(a) **World Petroleum Production.** Sixty-two of the world's nations produce petroleum, but the 10 largest producers account for 70 percent of the total. Only 2 of those 10 countries, Saudi Arabia and the United Arab Emirates, are located in the region. Collectively, however, the region produces over 30 percent of the world's oil, and it contains about half of known world reserves. (b) **World Energy Consumption.** Oil is the premier energy source at this time. Six countries, all highly developed except for China, consume over two thirds of the world's energy. Of these six, only the United States and the USSR are both major producers and major consumers.

stance became the basis of a wealth-generating trade that flourished for millennia. On its eastern edge, where fishing waters were relatively rich, there arose a series of city-states and small kingdoms that succeeded to wealth as fishermen became sailors and traders. It was always the edges rather than the center that were the sources of wealth and the centers of civilization. The interior was the home of warring nomadic tribes, and grazing was the source of a meager livelihood characterized by the struggle for survival.

The Arabs were unified for the first time, tradition says, by Muhammad in the seventh century A.D. In time, the unity fostered by the Prophet fell apart as a series of outside (non-Arab) conquerors imposed their rule on portions of the peninsula's margins. Powerful local Arab leaders asserted control over crucial oases and ports involved in the Eastern trade. Neither European explorers, nor the grandiose empire builders of the north, nor the Ottoman Turks, the last of the Islamic conquerors, could impose their complete control over the desert interiors.

In 1913, Ibn Saud, a powerful sheikh, overthrew the Turks and, by 1926, had extended his control over most of Arabia. With the discovery of oil in the 1930s, the Arab rulers of this new kingdom acquired a source of wealth much greater than the pearls of the Persian Gulf, the coffee of Yemen, the food crops of the Fertile Crescent, or the incense of the southwest. The possessors of this oil wealth, also controllers of the holy city of Mecca, have exerted increasing influence on the Arab cultural world from their desert domain.

A series of small states exist around the Gulf. Yemen persists, and the once-British Aden region remains as the independent Marxist state of South Yemen. Despite the persistence of these small units, the region is entirely Arab, Islamic, and basically homogeneous in culture.

Despite superficial homogeneity, there are problems. A large number of international boundaries here have not yet been demarcated. When this territorial uncertainty is combined with vast differences in local living standards and strongly polarized international political alignments, the situation is rife with potential for future disputes.

Saudi Arabia: The Dominant Force.

Saudi Arabia, the largest state in this subregion, is approximately the size of the United States east of the Mississippi. However, it has fewer than 15 million people; the population density is truly sparse. Most of the population live in scattered clusters along parts of the coasts, in areas of higher elevation, or in a few oases in the north-central region. The southern interior of the peninsula is called the Rub al Khali, or "empty quarter," due to the scarcity of people and the occurrence there of extensive "seas" of sand dunes (Figure 10–9).

Saudi Arabia contains Islam's two holiest cities, Mecca and Medina. Every Muslim is obliged to make a pilgrimage to Mecca at least once in his lifetime, if possible. With the great leaps in transportation speed and efficiency, relatively large numbers can now make their **haj** (pilgrimage) to the prophet Muhammad's birthplace. The haj draws over 4 million pilgrims annually, generating considerable revenues for service industries in Mecca, Medina (where the Prophet's tomb is located), and Jidda, the Red Sea port that serves Mecca. Saudi Arabia's other claim to world importance is of course its vast reserve of petroleum. With huge reserves, Saudi Arabia can continue to maintain, even expand, production long after many other producers have peaked and are in decline. Some of Saudi Arabia's oil was exported via the Tapline to the Mediterranean, but almost all is now exported from the oil terminal in the Persian Gulf at Ras Tanura (see Figure 10–3). (Arabs would prefer to rename the Persian Gulf the Arabian Gulf.) The Saudi per capita annual income is $12,000, even in the face of declining oil prices.

Although Saudi Arabia is investing its oil wealth in education, economic development, and industrial diversification, 25 percent of the population is still engaged in agriculture. Forty percent of the country is used for grazing, mostly of very low **carrying capacity** (capacity for feeding grazing animals without incurring permanent damage to vegetation and landscape). Only 1.5 percent has been rendered suitable for cultivation, despite huge investments in irrigation.

The Saudis have extended sedentary agriculture at least in part to establish nuclei of controllable population. They have improved and expanded transport with a railway from Ad Damman on the Persian Gulf to the capital (Rijadh), a network of modern roads, and numerous airports. Crop diversification has been encouraged, and problems of poor drainage,

FIGURE 10–9
Land Use and Mineral Resources in the Arabia Peninsula, Iran, and Iraq.
Typical of the entire region, the states that surround the Persian Gulf are largely devoted to livestock herding. Though less than a quarter of the area shown is totally uninhabited, over 80 percent of it is desert. While the area is rich in minerals, particularly petroleum, the agricultural resource is quite limited, with the notable exceptions of Yemen, Iraq, and a portion of Iran. Crop agriculture is being expanded, but most of the nations shown here remain net importers of food.

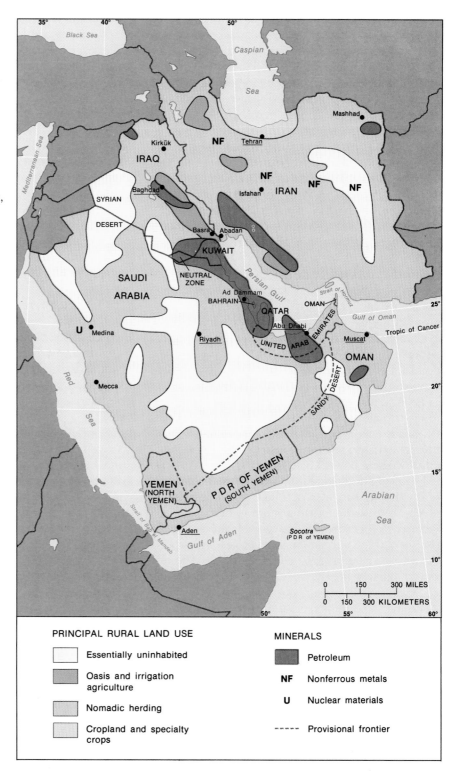

PRINCIPAL RURAL LAND USE

☐ Essentially uninhabited

▨ Oasis and irrigation agriculture

▨ Nomadic herding

▨ Cropland and specialty crops

MINERALS

■ Petroleum

NF Nonferrous metals

U Nuclear materials

----- Provisional frontier

salinity, and erratic water supply have been tackled by American and European agricultural missions.

Government investment in agriculture has continued at a high rate. The traditional pastoral nomadism that long dominated the interior has been under political and economic pressure to change. Nomads are persuaded to become farmers or oil field workers. At the same time, rising prices for meat and hides have encouraged expansion of flocks among the million or so who still follow the nomadic life.

The availability of water at the pumping stations along the Tapline tends to concentrate herds on nearby grazing lands, thus overgrazing them. Overgrazing is also a problem in some formerly remote areas where nomads have switched from camels to trucks. Motorized nomadism moves flocks faster and farther than ever before. Oil revenues encourage the purchase of more animals, and motorization allows each nomad to handle greater numbers of livestock. Overgrazing has so depleted vegetation from some areas that forage is now trucked in. The changes wrought by deep wells, modern roads, and "truck nomads" are forcing shifts even in the traditional economy.

Political and religious conservatism and a strong desire to maintain stability in the critical Persian Gulf dominate Saudi foreign policy. Saudi Arabia has maintained good relations with the United States and other Western powers. The Saudis are in an understandably difficult position in view of their small indigenous population in a country that controls so much wealth and international power in the vital oil industry. To outsiders, there seems an almost frantic pace of development and a strong drive to transform the Saudis into a modern, educated people. The responsibilities of wisely spending the incredible oil-generated wealth are a serious, if enviable, burden. An estimated $25 billion in foreign currency reserves is funding an impressive program of foreign aid and a sizable program of economic development at home. Steeply rising urban demands for water have led to a 5-million-gallon-per-day **desalination** (removing salt from seawater) plant near Jidda and several smaller plants in other towns. Future industrial development will emphasize petrochemicals and more oil refineries. The economic expansion of the last two decades created more jobs than the local population could fill. Temporary workers have been hired from all over the Islamic world to fill the gap, and remittances from workers are an important source of income in dozens of states that are not major oil producers.

The Small States of the Margins. The two contending Yemeni states form the ancient land of Felix Arabia. There is little agreement between them except that there should be only one, unified Yemen. The area is now divided between North Yemen (the Yemen Arab Republic) and South Yemen (the People's Democratic Republic of Yemen). There is no mutually agreeable plan for achieving this unification. The two Yemens share relative poverty; a lack of oil and all other valuable minerals is a major reason why North Yemen's annual per capita income is estimated at only about $475, and that of South Yemen at $300.

In the past, Yemen was part of an important series of trade links between Africa and India. Another source of past prosperity was the considerably wetter climate in the high mountains behind the 40-mile wide coastal strip of desert. These mountains (up to 12,000 feet in elevation) have abundant rainfall and a pleasantly cooler climate than the rest of the peninsula.

North Yemen has been one of the most culturally isolated areas of the world since cessation of traditional trade. It is an isolated mountain fortress surrounded by desert but near the sea. It is both scenic and well watered. There is real potential here for stable agriculture and sustained economic growth. North Yemen, with most of the terraced fields of well-watered crops, was formerly self-sufficient in food. Superior-quality coffee was once the principal source of foreign exchange. Production has fallen, in part due to a severe drought in the mid-1980s, in part because of the increased acreage of Qat, a mildly narcotic herb. Qat is widely used locally, sustaining a mild euphoria in the face of poverty.

South Yemen is a less attractive land and far less suitable for agriculture. The coast is suffocatingly humid, intensely hot and dry, almost without harbors (except Aden), hopelessly infertile, and almost totally barren. Inland, a series of hills rise above the plain like a series of broken steps, culminating in the wetter highlands of Yemen on the west and the confused dunes of the Empty Quarter in the east. The mountains are not green, like those of Yemen; they support a dull scrub of aromatic grasses and dwarf trees, leafless except for a few weeks after the rains. These unimpressive shrubs were the source of frankincense, the trade good of its past glories.

This coast traded primarily with the Horn of Africa and Indonesia. Indeed, thousands of Hadhramis (South Yemeni) migrated to Indonesia in the past. Poverty is worse here than in North Yemen, largely a function of the poorer agricultural possibilities. The British bases, once a source of income and employment, are long gone. The closing of the Suez Canal greatly reduced the strategic significance of the area. Even when the canal reopened, established sea routes around Africa and pipelines continued to handle the bulk of the oil flow.

The capital at Aden, near the Bab el Mandeb (the strategic strait between the Red Sea and the Indian Ocean), was an important trade city and a military base long controlled by the British. After the Suez Canal opened in 1869, Aden became an important coaling station to refuel steamships bound to or from India. As oil replaced coal, a refinery was built at Aden, which still represents the largest source of government income.

South Yemen depends even more heavily on foreign aid; virtually all subsidies are from the Soviet Union and other Marxist states. Its position on the Bab el Mandeb places military units based there minutes away from the oil lifeline of Western Europe.

For twenty-odd years since independence, the government has been militantly Marxist. The Communist party is split into dozens of contentious factions, and the chaotic situation is little different from the old days of tribal warfare and rivalry of local sheikhs.

The ministates of the Gulf (save Oman) have grown rich in oil revenues (see Figure 10–9). They contrast sharply with the Yemeni states in their progress, wealth, and stability. They are the sheikhdoms that, for one reason or another, did not succumb to Saudi pressures or military might. As with the Yemeni states, there are wet and dry parts. The state of Oman, by far the largest, is at least partially usable for crops where monsoon rains and mountain heights occur in conjunction. The other states are best described as environmentally desolate. Like much of South Yemen, the Gulf Coast lacks good harbors. Shoals, limestone reefs, shallow channels, and a linear coast with few embayments made maritime pursuits hazardous and uninviting. The exceptions are Oman (actually beyond the Gulf) and the splendid harbor of Manama on the offshore islands of Bahrein (see Figure 10–9).

Oman is the largest, but not the most prosperous, of the Gulf states. Its importance derives from its control of the Strait of Hormuz at the entrance to the Gulf. Oil is a new source of income here, discovered only in the late 1960s; the traditional source was trade. It was the Sultan of Oman who owned Zanzibar (Tanzania) and dominated the East African trade.

Seven Arab emirates make up the loose confederation known as the United Arab Emirates (UAE). The total population is one and one-half million, but as many as half may be émigré labor from other Muslim countries. The UAE, and their neighbor Qatar, enjoy the highest per capita incomes in the world ($25,000 and $30,000, respectively). Each emirate retains considerable local autonomy. A major political and cultural problem is the flood of foreigners attracted by the oil boom to these rather lightly populated areas. The lack of water will limit development even in these fabulously wealthy states. Abu Dhabi, the largest population group among the seven, has an oil refinery, a natural gas liquefaction plant, and fertilizer and plastics plants under construction. International banking has become a major new function.

Qatar occupies the peninsula jutting into the Gulf from eastern Saudi Arabia. This Arab state is mainly sand, gravel, and barren limestone ridges that receive no rainfall. Summer temperatures surpass 120° F. About the size of Connecticut and Rhode Island combined, the state has a population of 300,000; less than one third are indigenous Qataris; fully 5 percent are Muslims from India and Pakistan, 10 percent are Iranians, and the remainder are Arabs or Africans from other countries of the region. Less than 1 percent of the land is cultivable, but food imports are easily paid for by oil exports. Qatar now has a steel mill with over a half-million-ton capacity, desalinating plants, two petrochemical plants, and a large fertilizer factory, making it one of the most highly industrialized states of the region.

Bahrein, a group of small islands off the Gulf coast of Saudi Arabia to the northwest of Qatar, is an oil-producing state of some 400,000 people crammed into an area of a little better than 200 square miles. It is not as wealthy as Qatar or Kuwait, with oil reserves not expected to last until the end of this century. The per capita income is about $6,500 per year. The population growth rate has declined to 2.7 percent annually. The astonishing subregional growth rates of the 1970s (8 percent for the UAE and 6 percent for Kuwait) have declined with the fortunes of oil. Heavy immigration of people looking for jobs (and also wishing to benefit from the generous social welfare programs for the oil-rich states) has declined. Bahrein has one of the largest oil refineries in the Middle East at Sitrah, where a major oil-loading terminal can accommodate the largest of supertankers.

Kuwait's population of 2 million occupies an area smaller than New Jersey. Free public education up to the university level gives Kuwait an unusually well-educated population for a developing country; the state has a literacy rate of over 80 percent. The per capita GNP of $18,000 is supported by one of the largest proven petroleum reserves in the world, and by abundant natural gas as well. Kuwait has five oil refineries and has constructed large electrical generating, fertilizing, and cement plants. The world's largest water distillation plants are necessary in this state, which has summer temperatures reaching 130° F and averages less than 4 inches of rainfall annually. With little arable land but adequate fresh water from desalination and abundant money for experimentation, Kuwait has pioneered large-scale developments of **hydroponics,** a technique of growing plants in liquids rather than in soil.

Generally, the fortunes of this subregion are tied to oil more than those of any other subregion. It is obvious, however, that industrial diversity and commercial endeavors are developing rapidly as a hedge against overdependence on a single resource. Declining oil prices will tax these economies, but not obliterate them.

Iran, Turkey, and Cyprus: The Non-Arab North

The subregion comprising Iran, Turkey, and Cyprus is culturally marginal in that its people are not Arabs in either language or ethnic heritage. While Iran and Turkey are overwhelmingly Islamic nations, Greek Cypriots form the Christian majority of Cyprus. More European than Middle Eastern in outlook, Cyprus is physically proximate to the regional core. Both Iran (in its older form of Persia) and Turkey once incorporated large parts of the region in powerful empires; both nations were great centers of culture that influenced much of the rest of the region, and everywhere the human landscape is stamped with their cultural imprint. Turkey is relatively homogeneous ethnically, but Iran is quite diverse. Cyprus, torn by a bitter civil war between Greek and Turkish ethnics on the island, has been partitioned.

Iran's large and varied territory has always seemed to hold great potential. It has reasonably fertile soil over sizable, if scattered, areas. The mountains, which virtually ring the entire subregion, serve to temper the heat and extract moisture in an area that otherwise would be largely desert. Higher rainfall gives much of the land the characteristics of steppe, and favored corners are wet enough to be classed as humid. The rains of winter and the melting snows in summer provide sufficient water for irrigation of countless small oases, only collectively providing a sizable cultivable area (11 percent). Iran's seemingly favored circumstances do not yield an easy living. It required human effort to arrange the marriage of waters, often brought long distances, with the scattered pieces of favorable land. By comparison, the valley of the Nile and the plains of Mesopotamia provided much easier environments to organize and improve. Iran is a populous country and has only a little over one acre of arable land for each person, a limitation that seriously restricted the ability of the past government to fulfill promises of land for all farmers.

Following classic development models, the Shah's "bloodless revolution" (begun in 1963) implemented government purchases of land from landlords for redistribution among a landless peasantry. Former landlords were to invest their capital in expanding commercial and industrial enterprises; income from these investments was to compensate for the loss of land rents. The program (and agricultural progress) lagged seriously behind projected targets. Small new farms resulted, and farmers lacked the capital necessary for commercial production. Yields of food declined, necessitating imports. Irrigation works, crucial to production and once centrally managed on huge estates, began to degenerate. War with Iraq and the diversion of manpower and capital to military needs and religious propaganda resulted in a continuation of the decline of irrigation systems under the government of the Ayatollah Khomeini.

Crop-livestock patterns resemble a desert horticulture complementing livestock husbandry on unirrigated grazing land in neighboring districts. Melons, pistachios, almonds, tobacco, apricots, sugar, and tea have the status of basic crops, along with wheat, barley, and rice, the staple grains. Cultivated area actually decreased between 1950 and 1980 at a time of rapid population increase. Grain production, however, grew faster than the population, which increased by 65 percent during the same period. However, these gains were often made at the expense of horticultural crops with potentially greater value. The former Shah encouraged diversion of irrigated land from food crops to industrial crops such as cotton, tobacco, and oil-seeds to supply new factories built under programs for industrialization. Increased meat demands resulted in overgrazing and consequent erosion on marginal dryland pastures. No new progress is evident in food production as Iran continues to import.

Over half of Iran is a dry interior plateau with no permanent streams. Vast areas of salt wastes in the southeast are as barren as the Rub' al Khali of Saudi Arabia (see Figure 10–3). The largest area of potential farmland is the southwestern coastal plain, the area disputed with Iraq. Peopled largely by Arabs rather than Iranians, it is a continuation of the Mesopotamian Plain. Scheduled for huge irrigation schemes (some completed), it was to have redressed the need for food importation. This same land conflict has flared, periodically, over thousands of years among a variety of contending powers that sought control of all or part of the Mesopotamian Plain.

The greatest aid to Iranian development is the state's oil reserve. With over 6 billion tons of proven reserves (almost double the U.S. reserves), Iran has the potential to remain a major oil power through long years of sustained production. There also are commercial deposits of chrome ore, copper, lead, zinc, iron, and manganese, and workable deposits of coal. The location of the oil resource is also crucial. The largest fields are just north of the Persian Gulf and within the disputed area claimed by Iraq.

Despite the fact that Iran was one of the world's earliest exporters of oil (1907), it did not achieve a controlling interest in its own fields until 1973. Oil revenues produced an astounding annual economic growth from the early 1960s through 1973. The strain on the economy of this rapid development was obvious by the late 1970s.

Iran had an uneven development in that the impact of development was felt among relatively small numbers of people and in only a few cities and mineral districts. Large segments of the population were relatively poorer as the gap widened between urban and rural, rich and poor.

Iran has experienced a high population growth rate (3 percent) for several decades; almost half of all Iranians are under the age of 15. The people concentrate in the north and northwest. With over 70 percent of the national territory in mountains or arid plateaus with few inhabitants, densities are exceptionally high in those more favored and heavily populated districts. Almost half the population is urban, an unusually high proportion for a country of Iran's developmental level. Almost 6 million people jam Teheran, the national capital. Teheran, Tabriz, and the Persian Gulf oil-chemical complex are the major industrial centers.

The current rapid growth rate is typical of the second stage of the demographic transition and need not be viewed as a permanent situation. With considerations of wartime manpower viewed as pressing, the current regime encourages continued high (even increasing) birthrates. The ethnic question may yet prove to be more pressing than high growth rates. Until the declaration of the Islamic Republic on April 1, 1979, the official name of the country was the Empire of Iran. It is indeed an empire by definitions of political geography. Its boundaries include many different ethnic groups, several of which aspire to independence (see Figure 10–7a). It is not yet certain to what degree the present Iranian government will be willing to move toward regional autonomy to defuse separatist demands of the Kurds and Azerbyjani.

Only two thirds of Iran's population are ethnic Iranians who speak related Indo-European languages such as Persian, Baluchi, Bakhtiari, and Lur (see Figure 10–7a). Language affinity, however, is no guarantee of national affiliation—Kurdish, too, is an Iranian tongue. Some 25 percent of the population speaks Azerbyjani or another Turkish language. Other ethnic groups with other languages include Arabs in the Tigris-Euphrates lowlands delta area to

THE ISLAMIC JIHAD

The term *Jihad* implies a holy war in the name of Islam. Muhammad once advocated such a war, exhorting his followers in the eighth century to convert their fellow Arabs and to fight against the corrupt practices, tyranny, and injustices of the Arab society of the day. Filled with zeal, they proceeded to conquer the Middle East and North Africa, to spread the religious word, and to extend the Arab domains in the process. The Koran also teaches that all followers of "faiths with a holy book" were to be spared. Thus, Christians and Jews were to be allowed to practice their faiths in peace. Conquest, however, was another matter. That was the political will, not a matter of faith, and the Koran gives no message to cover such things. The memory of the original Jihad is kept alive for modern Muslims in the Koran.

What, then, is the cause of current Islamic militancy? Is this a new Jihad? The militants of today are either born to the poverty and despair experienced by dispossessed rural Muslims in the cities of the region, or are the university-educated elements of Islamic society that have been unable to advance in the stifling economic climate of an underdeveloped country. Militant fundamentalist Muslims are present in all the countries in the region. They generally constitute a small, if vocal, minority, and they often characterize their actions as a Jihad. They have come to power in only one country, Iran, where the excesses and insensitivity of past regimes inadvertently made fertile the ground of religious-political ferment through repression, greed, and gross economic mismanagement.

The success of the fundamentalists in Iran, a country with an overwhelming majority of Shiite Muslims, has tended to create confusion among non-Muslims. Shia beliefs and fundamentalism are often equated, incorrectly, in Western views. Shiite Muslims are a minority presence in most Islamic countries. The Ayatollah Khomeini, a fundamentalist, was their spiritual leader. Many Shiya adherents, however, did not recognize his religious leadership, and most Shiites living outside Iran did not recognize his political leadership. Khomeini was an advocate of Jihadist movements.

the southwest, Armenians in the northwest, Jews, and Assyrians. This complex ethnic and linguistic mix has the potential for disrupting any central government insensitive to their demands. Current policies emphasize religion over nationalism. The death of the Ayatollah Khomeini does not seem to have altered this situation.

Relations with most Arab states, not just Iraq, are decidedly unfriendly. Iran remains in isolation from both its neighbors and the rest of the world. In keeping with its self-proclaimed religious leadership, it justifies war as Jihad, the act of redemption, rather than as an act of territorial expansion.

Although included in the same subregion, the modern, liberal state of Turkey is the antithesis of fundamentalist, traditional Iran. Unlike Iran, Turkey possesses a highly usable land base; its fortunes hinge more on agriculture than on its mineral

His call to holy war was heeded by Shiite fundamentalists in Lebanon, where that group has come to control almost one fifth of the country. The reticence of Sunni Muslims to accept Shiite leadership is probably the reason that the power of the Ayatollah Khomeini did not spread further. Nonetheless, the fundamentalist Jihad movement *has* developed among Sunni Muslims as well. Egypt, the Sudan, Syria, and other nonoil economies in the region have active Jihadist movements that often clash with government forces and demonstrate their opposition publicly.

The dream of the Jihad movement is a series of Islamic republics (for some, a single, united Islamic state), in which Islamic law prevails and the Koran and its teachings become the literal basis for all governance. The group advocates return to traditional dress, the veil for women, strict moral codes, and a strengthening of religious observance. They reject both Western and Communist influence as decadent; they espouse and advocate a return to the traditional lifestyle as opposed to urban-industrial development.

Grinding poverty causes the urban poor to seek new answers, and the leadership of militant Islamic groups claims to have those answers in the form of a return to traditional values and religiousness. The leadership is composed of disaffected intellectuals who often cannot find work suitable to their educational training. They are articulate, intelligent, and understanding; they appreciate the frustrations of the urban poor.

First-generation migrants to the city still have large families. Rural values are retained in tight family households that exist in relative isolation within the unfamiliarity and strangeness of the city. Simple, religious people, unable to (yet yearning for) return to the less complicated country life and the less stressful days of the past, they are prospective converts to the Jihad movement. Scandalized by what they perceive to be the looseness and immorality of the city, the urban poor are caught in an economic trap; they form a bitter majority in many Islamic cities and a potential national majority that could be convinced to join the cause of Jihad.

---

wealth. With 52 million people, it is the region's second most populous nation. The annual growth rate of 2 percent is moderate for the region, if still above the world average. Turks constitute 90 percent of the population, and Kurds are the largest minority. Islam is the religion of 98 percent of the people; as practiced in secularized Turkey, it is a liberal religion. The country has a great variety of physical environments, and there is a varied crop production. All major grains are produced there in quantities sufficient to make the country a major producer by world standards. The coastal lowlands and hills are centers of production for grapes, olives, and figs, the traditional Mediterranean crops. Turkish farms also produce significant amounts of industrial crops: cotton, the famed Turkish tobacco, sugar beets, and tea.

The Turkey of today was the core of the vast Ottoman Empire, a construct that lasted for six centuries. The modern republic, founded in 1923 under the leadership of Kemal Ataturk, determined to turn its back on imperial traditions and to modernize Turkey quickly on the model of European nation-states. Symbolic of this focus on modern nationalism, the capital was moved from Istanbul (the ancient Byzantine Constantinople) to Ankara on the interior Anatolian plateau. As a seaport on the most famous of "narrow seas" between the Black Sea and the Mediterranean, Istanbul was a reminder of internationalist rulers and past empires. Ankara, inland and in the approximate center of the country, was the symbol of a new, inward-looking republic of a nearly homogeneous ethnic nature.

Turkey is experiencing urbanization at such a rapid rate that squatters' shacks spring up on the margins of cities faster than government agencies can move to supply sanitation systems, roads, and other services. Istanbul, still the financial center of the country, houses nearly 5 million in its metropolitan area. Ankara, with over 3 million, grows even more rapidly.

Though not a large mineral producer, Turkey has a rich and varied resource base. Minor oil deposits have been discovered, but the richest potential is offshore in the Aegean Sea. There, Greek islands lie close to Turkey's shores, effectively closing off much of the Aegean to Turkish territorial claims. Relations with Greece have never been good, and the continuing problem of Cyprus is a major sore point. Disputes over the potential Aegean oil fields could worsen an already bad situation.

Turkey started its drive toward industrialization earlier than its regional counterparts. Government action in starting up industries was important in the rapid growth of Ankara, seated in a then-undeveloped area with great mineral and agricultural potential. Private enterprise is now encouraged to promote

Istanbul. *Almost 6 million people crowd the avenues, alleys, and hillsides of Istanbul, the former capital of the old Ottoman Empire. Despite the loss of its capital function to Ankara, Istanbul remains Turkey's largest city, chief port, commercial center, and intellectual hub. For centuries prior to the Turkish conquest, as Constantinople, it had been the capital of the Eastern Roman Empire. Its best known sites are the ancient harbor of the Golden Horn and Haghia Sofia, shown here. Once the mightiest church of Orthodox Christianity, then one of Islam's most important mosques, the structure is now a museum housing the treasures of its ancient and splendid past.*

THE VEIL AS AN INDEX OF ISLAMIC CONSERVATISM

Strict interpretation of Islamic law requires that women conceal their bodies when in public. Religious law does not require a face veil, but the veil has become a strongly entrenched custom in many Islamic societies. Wearing a veil is symbolic of family honor and feminine modesty, and is thought by non-Muslims to symbolize female subordination to males.

The role of women under Islamic law often is misunderstood by non-Muslims. Islam permits a man to have up to four wives at once, and appears to sanction easy divorce by the husband. However, Muhammad actually had reformed previous custom, which placed no limit on the number of wives, nor restricted divorce in any way. His revelations included the right of women to inherit property and to receive alimony if divorced. Relatively few Islamic marriages involve multiple wives, and polygamy often is forbidden by civil law in Islamic states. Custom demands also that widows or divorced women be supported by their sons.

The question of the veil, however, remains controversial for some Islamic societies. Liberal, "modern," secular philosophy leads to diminished importance of the veil, whereas conservative, more religiously oriented societies tend to insist on rigid conformity to its use. Turkey and Egypt abolished the veil in the 1920s. Iran officially abolished the veil in 1935, reflecting a determination to modernize and deemphasize the role of religion in controlling social customs. After the fall of the Shah in 1978, the triumphant conservative religious leadership promptly reinstituted the veil, with severe punishment for noncompliance.

Most Islamic countries, however, have allowed fashion gradually to erode the veil into abandonment. While a full and opaque veil must be worn in Iran, and no veils are permitted in Turkey or Egypt, the relative fullness of the veil and the material of which it is made vary greatly over the rest of the region. In a liberalizing, modernizing society, the small eyeholes in the veil get larger, the veil recedes from below the chin to the lips, and the material gets thinner to the point of near-transparency. A resurgence of more conservative, religious fundamentalism in a society, however, will result in heavier, fuller veils that cover women's faces. The veil literally moves up or down the face in response to prevailing religious-political thought.

new industries and new industrial locations. Government involvement tends to be in high-capital, heavy-industry sectors such as steel, oil, and petrochemicals. The state development bank is an important source of capital; even private industry is strongly influenced by long-range government plans. Turkey has associate membership in the European Common Market, which may soon force changes in many small, uncompetitive Turkish industries.

Turkey has had some serious problems in maintaining a strong, stable, yet democratic government. The choice has seemed to consist of either a strong government or a democratic one. Moslem, yet not Arab, Western leaning, but not fully developed, Turkey reflects its geographical position as a bridge between two continents and a variety of contending philosophies and lifestyles.

Cyprus, the third largest island in the Mediterranean, has been under foreign domination for most of its history. Its strategic position, rather than its productivity, is what has made it so attractive to conquerors. It is, however, quite usable. A fertile central plain lies between the southwestern mountains and the Kyrenia Mountains of the north coast. Nicosia, the capital and largest city, is located in this central plain, along with most of the population of 700,000. The annual growth rate is only 1 percent, reflecting heavy out-migrations occasioned by the long and bitter dispute between Greek and Turkish Cypriots. Cyprus is only 44 miles off the Turkish shores. Greeks and Turks have shared the island for centuries. The Greek ethnics dominate, forming 76 percent of the population. Turks, at 20 percent, are the largest minority.

After millennia of control by different empires, Cyprus passed into the hands of the Crusaders, and then Venice. Turkey seized Cyprus from Venice in 1571. Three centuries later, a weakened Ottoman Empire ceded it to Britain in return for cancellation of an unpaid loan. After 80 years of British colonial status, civil unrest and terror forced Britain to grant independence in 1959. Britain's reluctance to withdraw was based on the certainty that independence would precipitate civil war.

Cyprus's independence was guaranteed by Turkey, Greece, and Britain, and internal actions were subject to the approval of all three. The majority of Cypriots favor *enosis,* that is, union with Greece; Turkey and Turkish Cypriots object. The treaty that granted Cypriot independence specifically forbade union or partition. After 15 years of uneasy independence, Cypriot leaders overthrew the moderate cleric who was president and opted for union with Greece. Predictably, Turkey invaded and literally partitioned the island, occupying about 40 percent of the total territory. Some 200,000 Greeks were forced to flee, while additional Turkish settlers came from the mainland to occupy vacated lands. There is a ceasefire, but all further talks have failed to reach a settlement. In 1983, the Turkish area was declared the Turkish Republic of North Cyprus; the rest remains under Greek Cypriot control and continues as independent. With international help, the economy is fairly buoyant.

Egypt and the Sudan: The Nile Basin

Modern Egypt is a study in contradictions. Its glorious past is overshadowed by the problems of the present. Its large physical size is little help in supporting its population; only 3 percent of the country is arable. Its population of over 52 million has exceeded its ability to produce food; and its growth rate remains high. The large population does not give it the power and influence of its less populous, but oil rich, Middle Eastern neighbors.

The Sudan, in contrast, has not yet fully utilized its portion of the Nile. Poorer and even less developed than Egypt, it has the greater potential of the two states. Historically, Egypt's expansion and cultural influence have ebbed and flowed up the Nile Valley. The Sudan has successfully surmounted all Egyptian claims to political control and could at any time assume the power position in their mutual relationships as ultimate controller of the waters of the Nile.

Egypt is the world's second largest oasis (after Pakistan); 99 percent of Egyptians live in the Nile valley and delta, and 100 percent of its farmland is irrigated. Population, agriculture, urbanization, and general economic development all focus on the Nile. The remaining resources include some iron, phosphates, and copper, and more than enough oil to supply national needs. The reserve is not spectacular but is sufficient to assure Egypt a steady flow of foreign currency as oil exports increase. More critically, it is the energy resource for Egypt's chemical industry—the producer of all the fertilizer used to meet the country's incredibly heavy fertilizer requirements.

Egypt's government recognizes that its central problem is rapid population growth. The population has grown from 10 million to almost 50 million in this century. In the past, population was controlled by natural forces; poor sanitation and poor medical care both led to high death rates, in turn compounded by malnourishment and disease. Famine occurred periodically as fluctuations in the Nile's floods resulted in fluctuations in the amount of land that could be irrigated for crops.

The annual flood of the Nile surged over the banks of the river, contributing a fresh deposit of fertile silt as well as water. Eroded material from the volcanic rocks of Ethiopia, rich in minerals, helped to maintain the highly productive soils despite thousands of years of cultivation. With population on the rise, the obvious solution was to increase the area of irrigated land. British engineers who built the first Aswan Dam did so to stabilize the erratic volume of annual floods. With a major storage reservoir at Aswan, annual variations in the size of the flood could be minimized, reducing the threat of famine. Retention of floodwaters in local basins was replaced by continual flow of irrigation water in long-distance canals. Since completion of that first Aswan Dam, irrigation has been possible year-round. However, the annual addition of a fresh layer of fertile silt has nearly ceased; most of the silt is now at the bottom of the reservoir behind the dam. Irrigated soils have begun to experience problems of salination in their upper layers. Annual floods had formerly dissolved and carried away much of this salt.

The mixed results of the first Aswan Dam have been compounded by the building of the Aswan High Dam a few miles upstream. Built between 1958 and 1970 with Soviet aid, the new dam increased Egypt's land under continual irrigation by 25 percent and enlarged the Sudan's irrigated acreage by 15 percent. In a tropical climate with year-round irrigation, two or even three crops per year are now possible. This multiple cropping and the elimination of the annual input of fertile silt have made it necessary for Egypt to become the world's largest consumer of fertilizer, per cultivated land unit.

The now-stagnant water of the irrigation canals support a variety of snails that act as intermediate hosts in the transmission of trematodes which then infest people, resulting in a disease called schistosomiasis, or bilharzia (river blindness). The environmental impact of the Aswan High Dam reaches far beyond the Nile Valley. The Mediterranean coasts of the Nile Delta, no longer replenished by silt brought down the Nile, are being eroded by waves and currents. Formerly freshwater lagoons behind the coast-

Egypt and the Nile. *This false-color satellite image shows the strong contrast between the stark landscape of the Egyptian desert and the verdant irrigated farmland of the Nile River valley and delta. Egypt is a huge oasis. One hundred percent of its farmland is irrigated, mainly by the Nile. Rapid population growth, compounding an already large base, is rapidly outstripping the ability of the Nile to produce food for Egypt's minions. Population densities exceed 6,000 persons per square mile in the delta area.*

line are being invaded by salt water, killing the freshwater fish that once thrived there. The coastal region is experiencing saltwater intrusion into local water tables. Fisheries in the entire Mediterranean are in decline, as mineral nutrients once contributed by the Nile diminish in volume, decreasing in turn the supply of minute plant and animal life on which the **food chain** is based.

Egypt's population growth rate of 2.6 percent per year will double the population in 25 years, if continued. Although average birthrates have dropped slightly since 1945, those in rural areas and among poor urbanites remain high. In each recent year, about 1 million more Egyptians are added, together with their needs for housing, education, and eventual employment, and their potential as producers and consumers.

The Egyptians are a fairly homogeneous cultural and racial group; there has been no significant migration into Egypt in many centuries. Over 90 percent are Muslims, with a 7 percent Christian minority. Egyptians are extremely conscious of their heritage, and a sound basis exists for Egyptian nationalism (see Figure 10–7).

One of Egypt's main emphases in development is further expansion of irrigated land. Egypt currently produces only about 75 percent of its food needs. Large-scale cotton production began three quarters of a century ago after the first Aswan Dam increased irrigated acreage. Other major crops are Egyptian clover (for animal fodder and as a soil-enriching crop), corn, wheat, rice, vegetables, and sugar cane. The lush delta area has been experiencing an increase in crop yields. With two or three crops per year, this increase might seem to offer the hope of upgrading farmers' living standards—until it is remembered that population densities in parts of the delta reach 6,000 per square mile. Even with very high productivity, poverty will likely prevail.

The 1952 revolution that overthrew the notoriously corrupt monarchy also produced a land reform. At that time, 40 percent of the cultivated land was owned by less than 1 percent of the landlords. More than 70 percent of the farmers owned less than half an acre. In the reform process, no landlord was allowed to retain more than 200 acres, and no single holding was to be less than 5 acres. Although conditions have definitely improved for many farmers, the plight of landless farmers is still serious. It is for social as well as economic production reasons that Egypt is striving so hard to bring new lands under irrigation.

Additional irrigated land is being obtained by three approaches. Areas adjacent to the delta are being irrigated by both Nile water and local groundwater. In the "new valley" area of southern Egypt (west of the Nile and including the large Kharga oasis), fossil water from the great underground reservoirs of the Nubian sandstone is being tapped by wells up to 900 feet deep. Another approach is to use the present supplies of water more efficiently. To this end, open-ditch irrigation canals are being converted to underground pipes to reduce evaporation. Sprinkler irrigation, more efficient than flood irrigation, is gradually being substituted, reducing water consumption per land unit by almost half.

Industrialization is the other emphasis in development. It is hoped that the pressing problems of urban unemployment and the relocation of landless peasants to the city will be solved through the growth of the manufacturing sector. The enormous hydroelectric generation of the Aswan High Dam is used in part to industrialize the mineral-rich Aswan area. A village electrification scheme uses much of the rest of the energy.

Textiles, using locally grown cotton, are produced for a highly competitive world market. They are still the major branch of industry. Over a million tourists visit Egypt annually, generating another large source of income. Millions of people from neighboring countries are attracted to Cairo, the cosmopolitan cultural center and largest city of the entire region. Many Westerners also are attracted by the warm, sunny winters and the fantastic artifacts of one of the world's earliest and greatest civilizations. Suez Canal tolls bring over $500 million a year in government revenues, and a Red Sea-Mediterranean oil pipeline is intended to earn revenues from the many supertankers that cannot use the Suez Canal, even after it has been deepened and widened. Over the last 40 years, Egypt has developed as the media center for the Arab world, producing films, television programs, novels, and magazines for the Arabic language market.

Cairo: A World City. With over 10 million inhabitants in its metropolitan area, and growing faster than many Egyptian planners would like, Cairo is clearly a world city in its cultural leadership, regional influence, and growing international functions. The tragic civil war that has disrupted Beirut's international banking and financial services has presented an opportunity to Cairo as many wealthy Lebanese fled to that city. If Egypt can continue political stability and maintain friendly relations with both the West and the oil-rich Arab states, the potential of Cairo will be even further enhanced.

The Sudan. Over twice the size of Egypt, the Sudan has less than half that country's population. The country extends from the zone of tropical rainforest to the tropical desert. Physically and culturally, the Sudan is two nations; it nearly became two political states. The semiarid and arid northern two thirds of the Sudan contains 16 million people and all the major cities. The northerners are generally Arabic-speaking Muslims who produce most of the country's exports and crops. The 9 million southerners are largely Negro or Nilotic tribes that are animist or Christian in religion, speak a variety of languages, and live in a subsistence economy. Southerners mutinied against the Khartoum administration during the transition to independence. The struggle ended temporarily in 1972, when the south was finally granted autonomy on internal concerns. Civil war resumed in 1981 and continues.

The traditional source of export income has been cotton, grown primarily in the Gezira region between the White Nile and Blue Nile. Agriculture in the Gezira can be greatly expanded; less than half the potentially productive land is currently cultivated. New strains of millet and sorghum, developed by Western agrotechnicians, show promise of stabilizing food supplies in the Sahelian portions of the country. Despite the 5 million or more people affected by its own severe drought in 1985, the Sudan accepted, and tried to feed, some 2 million refugees from Chad, Uganda, Kenya, and Ethiopia.

The promise of the future is oil. For decades, European and American oil corporations have prospected, off and on, in both the Red Sea Hills and the great Sudd (a region of floating vegetation-filled swamps in the south). A combination of southern unrest and falling oil prices periodically halts drilling.

Despite considerable natural wealth, the Sudan remains extremely poor. Here is a country with a reasonably good rail network, a great deal of water, unused land, known commercial deposits of oil in both its south and west, and an active program of family planning in place. These attributes and resources place it in the best possible position to deal with its problems—the same ones that plague almost all the other states in the region. On the other hand, it experiences still other problems: the rainfall variability and drought of the Sahel, and the cultural diversity and extended guerrilla warfare so characteristic of Africa South.

The Sudan holds the answers to the problems of many of its neighbors. The joint Egyptian-Sudanese canal project through the Sudd region would greatly benefit both countries. Half-finished, it awaits a renewal of international funding. The Sudan's vast irrigable lands could make it the granary for Ethiopia, much of Africa South, and the Middle East. Its oil could reduce the energy woes of a dozen nearby states. It is the obvious physical, cultural, and economic link between tropical Africa and the desert lands to the north. The missing catalyst in the formula for success is internal peace.

The Maghreb and Libya

The northwestern corner of Africa, between the sea and the Sahara, is known in the Muslim world as the Gezira al Maghreb, the "Island of the West." In a sense, it is an "island" of more densely populated, somewhat better watered lands, surrounded by the Mediterranean, the Atlantic, and the great desert to the south. Desert and sea meet in Libya, where the coastline of the Mediterranean dips southward along the Tunisian coast (see Figure 10–9).

The Maghreb was always something of a separate world because of its physical isolation. Its various conquerors held it for a time, influenced its culture, and sent colonists. But its peripheral location would ultimately result in a weakening of external ties and a reassertion of independence. It has always been one of the great crossroads of the world, commanding the Atlantic entrance to the Mediterranean and maintaining an often tenuous link through Libya to the Middle East. It was the Maghreb that first introduced and purveyed the products and riches of tropical Africa to the markets of Western civilizations.

Much of the land is rich and produces abundant crops when the meager rains of nature are augmented with irrigation. The Moors of the Maghreb were great engineers whose roads and waterworks were the marvel of their age. Moorish artisans created sumptuous rugs, richly embroidered textiles, and intricate items of brass and precious metals.

When the trade of the Mediterranean was superseded by that of the trans-Atlantic and circum-Africa routes, the Maghreb, like the rest of the Mediterranean lands, suffered a decline. Its importance has since been restored. Tunisia and Morocco host millions of European tourists yearly. Wine, winter produce, and citrus from the region stock the shelves of European stores. Algeria and Libya supply oil and gas to an energy-hungry Europe, and Morocco and Tunisia export the phosphates and fertilizers necessary to sustain the high productivity of European farms. Wealthier than most African states, although still un-

THE SUDD

The Sudd region of the Sudan is one of the world's largest swamps. Here the sluggish White Nile, having tumbled over powerful falls on its way from its sources in the highland lakes of East Africa, spreads over a vast area of perpetually wet, level land. The combination of a high water table and low gradient creates a 50,000-square-mile swamp. Tropical heat and endless water generate enormous amounts of vegetation. Floating islands form from the organic debris, moving, detaching, and reattaching to what might loosely be called the land surface, sodden as it is.

As influential in the total fertility of the Nile as the erosion of the volcanic Ethiopian highlands, the Sudd is the source of nitrogen in the fertile riverine ooze. Though it is cursed for blocking navigation channels and clogging irrigation ditches and hydroelectric intakes, in its rotted form it is the ultimate enricher of soil. Hippopotamuses, fish, flocks of aquatic birds, and hordes of other wildlife eat of the Sudd's plenty, wallow in its cooling waters, and manure it into a primordial soup that ultimately feeds crops downstream on its chemical richness. On the negative side, much of this standing water (24 million cubic yards annually) is lost to evaporation as it lazily sloughs through this area of unpredictably shifting drainage.

For more than a century, there have been plans to canalize a pathway around it, releasing waters to irrigate more land in both Egypt and the Sudan. This radical change in drainage could have serious environmental impacts: destruction of wildlife habitats, lowering of water tables, and even climatic change in the region. The Sudd is in possession of the peoples of the south of the Sudan, who view attempts to drain it as destroying their livelihood and using their resources to enrich the Muslim north. In a political sense, too, the Sudd is a blockage to national unity and development.

---

developed, the Maghreb constitutes a growing market for European manufactured goods.

While currently (and increasingly) within the European economic orbit, the Maghreb is definitely a part of the Arab-Islamic cultural world. Cognizant of their African location and historic ties, the subregion's states are all active and influential members of pan-African organizations. They continue to play an active part in African politics and development. In individual cases, they are the peacemaker, the intriguer, the go-between, the diplomat, the investor —exerting once again the role of broker between

Europe and Africa South, and between Africa South and the Middle East.

The nations of the Maghreb and Libya share some cultural similarities. They are dominantly Arabic in speech and Islamic in religion. Yet all are Hamitic in their roots, and all have sizable unassimilated Berber (Hamite) minorities. Each has a densely populated, well-watered core and vast areas of desert. All but Morocco have large deposits of oil and gas, and all but Libya, not yet fully prospected, have a respectable array of other minerals. Each is relatively highly urbanized, and all are developing rapidly. Though all maintain the same outside contacts, links among them are often grudging and few.

Libya, with its vast oil and gas reserves in relation to its relatively small population of 4.2 million, most nearly resembles a western outlier of the Persian Gulf economies. Ninety percent of the people live on less than 10 percent of the land, mostly along the coast and in the great Kufra oasis of the southeastern portion (Figure 10–10). Independent since 1951, Libya was a poor, underdeveloped state until the dis-

covery of oil in 1959. Production rose quickly; Libya has an income from oil exports that exceeds its development-investment needs.

Oil money is helping to develop the underground water reservoir of the Nubian sandstones. This ancient water reserve, the same one developed in Egypt's "new valley," will support oasis agriculture for several centuries at the maximum planned rate of withdrawal, despite the lack of a natural recharge of the aquifer. The Kufra oasis development will yield

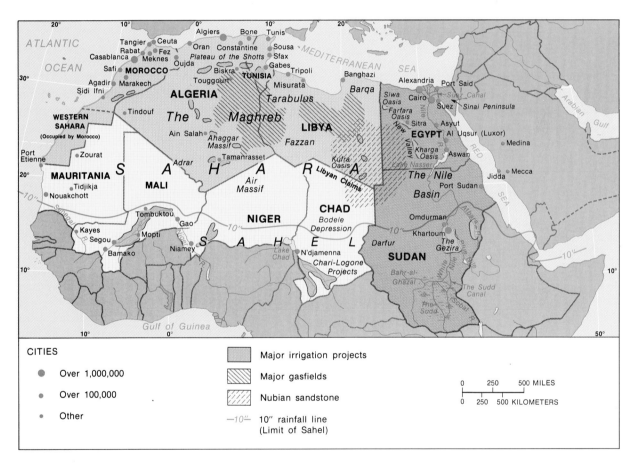

FIGURE 10–10

North Africa and the Sahel. The vast Sahara separates the well-watered portions of the Mediterranean littoral from the savanna and steppe grasslands of the Sahel. Only the Nile River breaches this almost empty desert area, its banks crowded with people in Egypt and the Sudan. The Maghreb and a small part of Libya exhibit the best balance between population and the agricultural resource. The Sahel, quite productive in years of sufficient rainfall, is periodically plagued by drought. The Nile Valley is overcrowded in Egypt and not developed to its full potential in the Sudan.

the food supply needed to meet the future demands of one of the world's fastest growing populations (3 percent per year) (see Table A, p. 513). There is room for further agricultural expansion; farmland could be increased by as much as eightfold. With improved management, grazing lands can support enlarged livestock operations.

Investment, however, has been channeled into industrialization with new establishments manufacturing shoes, textiles, and pharmaceuticals. Though an avowedly socialist state, thousands of small, privately owned food-processing and craft establishments have sprung up in the cities. Fully 20 percent of the population is engaged in manufacturing. Huge sums have been invested in developing highways, airports, and other infrastructure.

Fueled by oil revenues, progress is highly visible. Half of all Libyans are now literate, and a well-educated Libyan technical elite has emerged from graduate schools worldwide. Twice as many people are employed in manufacturing as in oil production. With a large oil reserve, large gas reserves barely tapped to date, a continued surplus of exports over imports, and a respectable reserve of foreign currency and gold, it is not likely that the economy will collapse, though governmental change could occur at any time.

The ambitious, and sometimes irrational Col. Ghadaffi has led Libya into a morass of intrigue involving most of its neighbors, and many Western countries. International intelligence has uncovered planned chemical weapons plants. Libya has armed revolutionary groups in several countries, aided terrorist movements, and engaged in a brief border war with Egypt (1977). It has advanced territorial claims against both Egypt and the Sudan. There is a considerable military outlay as a result of these armed forays, including the contest with the U.S. Navy in the Gulf of Sidra.

The most serious advances, however, have been made against Chad, its neighbor to the south. A multilingual, multireligious state, Chad suffers typical African problems associated with the lack of coincidence between national borders and ethnocultural realities. Muslims in its north have been in active rebellion against the government since independence and have received arms from Libya. Libya has made incursions into northern Chad, claiming to protect Muslim interests and "combat French neo-colonialism."

A more stable picture emerges in neighboring Tunisia, where the economy is better balanced. Tunisia is the subregion's smallest state in area, but the most usable. Fully a third of the land is cultivable, and much of it is quite fertile. Grains and winter vegetable production (the latter destined for European produce markets) vie with vineyards and citrus groves for the fertile, well-watered plains and gentler slopes of the northern third of the country. Olive orchards spread southward across the central portion of Tunisia in a relentless colonization of what had been nomadic grazing lands a generation past. The desert south, in turn, has increased utility as deep water wells are drilled, increasing the livestock carrying capacity there.

There is oil, not in grand amounts, but a comfortable reserve serving domestic needs as well as entering the export market. In addition, there are large phosphate deposits for domestic use and export, and commercial deposits of iron, lead, and zinc. Liquefied natural gas goes to European markets and fuels domestic chemical and plastics industries.

Industries employ 25 percent of the work force. Fine wools, velvets, and traditional brocades are the hard currency earners of a well-developed textile industry. Subcontractors do the needlework for French and Italian firms in the Arab world's largest clothing industry. Yet, the country suffers a trade deficit with machinery and technology imports overshadowing the value of exports. The trade imbalance is redressed by tourism as Europeans flock to Tunisia's clean beaches and exotic bazaars in a climate of winter warmth and receptivity to Westerners.

Tunisia's 7 million people are well fed and reasonably well educated (68 percent literacy rate). Tunisia exudes moderation—moderate growth, moderate prosperity, and moderate politics. With a population growth rate down to a little more than 2 percent, Tunisia is beginning to close the development gap. It is perhaps the most Westernized of the Arab states, and certainly the most accommodating.

Algeria's large size, equivalent to the entire U.S. Midwest, supports a population of 24 million. Fifty-six percent of the population is under the age of 20, giving Algeria a high proportion of dependent youth. The government allocates 12 percent of its total spending to education. It has developed an impressive educational system that has raised literacy to over 60 percent.

Most Algerians live on 15 percent of the land, near the coast. The vast segment of the Sahara held by Algeria has only 1 million inhabitants, while Algiers, the capital, has over 2 million people and grows rapidly through internal migration. Algeria is the part of the Maghreb controlled longest by France (1830–1962). This long colonial domination and the substantial numbers of French ethnics (established there for three or four generations), with their determination to remain, led to a bitter revolt in Algeria that almost precipitated a civil war in France as well.

The highly centralized government of Algeria controls almost every facet of the economy. With much smaller and lower-quality oil reserves, Algerian production is half that of Libya. However, Algeria has the fourth largest proven reserves of natural gas in the world (see Figure 10–10). The Algerians have invested heavily in facilities to liquefy gas for export and for use as a feedstock for domestic industries. A diverse resource base that includes a dozen metal ores, coalfields, and fertilizer minerals augurs well for further economic diversification.

The 4,700-mile highway network includes a paved trans-Saharan highway to the Niger border. Algeria has become a major overland trading partner with the landlocked countries of West Africa because of its highway connections; it is the only state in the subregion to resume actively the traditional trans-Saharan trade.

Agriculture contributed less than 10 percent of the GNP a decade ago, but modernization has improved both the quality and value of production. Algeria has converted the former French estates into huge state farms supporting 4 million people growing grains, potatoes, and citrus. Algeria exports dates, vegetables, and the traditional Mediterranean crops, but it must also import many basic foods for its rapidly expanding population.

Algeria's prime foreign relations problem is its dispute with Morocco over the division of Western (formerly Spanish) Sahara (see Figure 10–10). Algeria wished to become a "two-sea" power by obtaining some frontage on the Atlantic through the territory relinquished by Spain in 1976. Western Sahara also has huge phosphate reserves. Moroccan troops

promptly took the northern part of Western Sahara, splitting the former colony with Mauritania to the south. After Mauritania relinquished its claim, Morocco absorbed the entire area.

Morocco's royal government has been more conservative than any other in the area. It actively supports established regimes in Africa against left-wing rebels. With little oil but a wide variety of other minerals, Morocco is the world's largest exporter of phosphate rock, controlling major reserves of this important fertilizer and chemical raw material. There are substantial deposits of iron ore, manganese, zinc, cobalt, coal, and bauxite (see Figure 10–10).

Morocco's agricultural economy is a blend of tiny subsistence farms and large, modern agribusiness units controlled by a few large landowners. Its population of 25 million does not strain its agricultural resource, and growth rates appear to be moderating. Most people are concentrated in the rich plains of the northwest. Food products, including citrus, fresh and canned vegetables and fruits, fish, and seafood, provide about a quarter of the country's foreign exchange.

Oriental rugs of magnificent design are produced by both native craftsmen and modern factories. Hand-hammered brass and copper jewelry, art objects, and utensils, once the sole domain of highly skilled artisans, are now mass-produced in factories. Parisian-designed clothing is manufactured in the sweatshops of Casablanca and Tangiers, although domestic designers are developing a totally Moroccan garment industry. Leather working, the most ancient of Moroccan craft traditions, thrives in both handicraft and modern forms.

Tourism flourishes, with Morocco offering luxurious accommodations at bargain prices. There is a brisk business in seaside retirement villas for wealthy European nationals. Shoppers, many just weekend tourists, descend upon bazaar stalls, chic shops, merchant quarters, and department stores alike in search of the ultimate bargain. Comfortable, "foreign" yet quite European (French is widely spoken), exotic (but with indoor plumbing), renowned for its cuisine, Morocco has been "discovered" by Europeans and Americans to the tune of several billion dollars yearly.

Involved in the former Spanish Sahara since 1976, Morocco controls Western Sahara's urban centers

The Market at Marrakech. *The Maghreb has been known for its skills and crafts since antiquity. Brassware, leather goods, rugs, and jewelry are the producers of these rich traditions. The pleasant climate, hospitality, and colorful markets filled with bargains have all played a role in developing a brisk tourist trade in Algeria, Tunisia, and Morocco. Culturally a part of the realm of Islam, economically, it is a part of metropolitan Europe. Marrakech is one of five large cities in Morocco.*

and its rich offshore fishing grounds, but not its dissident-dominated deserts. In some ways moderate, its foreign policy has also been decidedly expansionist.

The Sahara-Sahel

All four states of the Sahara-Sahel subregion were a part of the French African Empire. French remains an official language, and France remains the most important trading partner. All four states include large areas of the Sahara, the largest desert on earth, but each has some nondesert (steppe or scrub savanna) territory (see Figure 10–10). The population concentrates in these wetter portions.

Despite recorded temperatures of over 130° F, a paucity of settlements, and places that have recorded no precipitation for at least a decade at a time, the Sahara has been as much a highway as a barrier. It is not an easy or forgiving environment for the uninitiated, but trans-Saharan contacts have existed for many centuries. North Africa can be imagined as a series of approximately parallel belts of climate-vegetation associations trending east-west across the continent. To the south of the desert, the transition is to semiarid, then subhumid, climates, verging on the true savanna. While all four states contain a portion of the desert, each also contains a portion of the land beyond. This climatic transition zone on the southern edges of the Sahara is known as the Sahel. Whereas the poleward fringe of the Sahara near the Mediterranean has a winter "wet" season, the summer "wet" season of the Sahel unfortunately coincides with the season of high heat. Evaporation is high, minimizing the effectiveness of the sparse precipitation. The volume of rains of the Sahel's "wet" season is highly unpredictable from year to year. This variability in precipitation is the prevalent environmental problem for these four states; each state is plagued periodically by destructive, prolonged drought.

The Sahara-Sahel subregion is a major zone of cultural and racial mixing. Berbers and Arabs from the desert and various black African groups from the wetter south have traveled across and settled on the opposite side of the desert. The four states are large in area but relatively sparsely populated (Table 10–3). These nearly empty areas of relatively small economic importance in international trade would appear to be hopelessly isolated. Surfaced north-south roads are only now being completed across the vast desert. Enormous areas are devoid of any permanent settlement.

Yet, this area has a historic importance in the development of cultures. Great black African empires rose in the Sahel zone. Ghana, Malinki (Mali), and Songhai flourished from the twelfth to sixteenth centuries, maintaining trade routes to the Mediterranean and the Near East. They were replaced after 1600 by Arab expansion into the Sahara-Sahel as Arab traders ranged from Spain to Nigeria and from China to Mozambique. Isolation is a relative thing, and trade routes (as well as trading centers) of the Sahel have flourished, periodically, for thousands of years.

Mauritania, named for an ancient Roman province of Africa, has a per capita income of about $460. Meager enough, it is the best figure in the subregion. The other three states rank among the least developed countries in the world. A narrow belt of the Sahel, along the Senegal River, the southwestern border, is productive farmland in wetter climatic cycles. Livestock grazing occupies much of the Sahel area of Mauritania, while the northern desert is barren except for scattered oases. A major iron ore deposit near the border with Moroccan-controlled Western Sahara contributes about three fourths of the country's foreign exchange. The Atlantic waters offshore teem with fish and are now being exploited through better port facilities and processing plants; this is one of the richest of the world's fisheries.

Mali, named for one of the great black kingdoms of the savanna-Sahel, has the lowest per capita income in the region. Exports are primarily livestock products—processed meat and hides. There are known deposits of bauxite, iron ore, copper, and phosphate, but the transport system is inadequate to facilitate development at present.

In the country's interior, the French developed crop farming along the inland delta of the Niger. Rice, peanuts, sorghum, and cotton are grown on

TABLE 10–3
Demographics and economics of the Sahara-Sahel.

Country	Area (1,000 sq. miles)	Population (millions)	Population Density (persons per sq. mile)	Life Expectancy	Adult Literacy Rate (%)	% Farmland	Income per Capita ($)
Mauritania	400.8	2.0	4.9	45	20	0.4	460
Mali	478.8	8.4	17.5	43	18	2.0	160
Niger	489.2	7.0	14.3	44	14	3.0	335
Chad	496.0	4.7	9.5	43	15	2.0	120

SOURCE: World Bank, 1987.

small, freehold, irrigated farms. Crop farmers in the delta have the greatest possibility of economic survival. However, refugees and herders have crowded into farm villages, depleting surpluses that would be sold in normal years. Wind and dust storms are silting canals and reversing years of hard labor.

Niger lost half its livestock in the mid-1970s drought. Before recovery was complete, droughts again ravaged herds in the 1980s. In good years, meat and hides are exported from the drier grazing regions, while peanuts, cotton, and sesame seed are produced in surplus in wetter districts. The economy is sustained, however, by the export of uranium (the fifth largest reserve in the world) to the French equivalent of the Nuclear Regulatory Commission.

Chad reaches further into the better-watered savanna than its Sahara-Sahel neighbors do. It is normally self-sufficient in food (except during disastrous droughts) and exports some peanuts and cotton. In 1981, Libya temporarily occupied Chad and took control of its government. War and civil unrest have disrupted the entire economy.

The Sahel is not without promise, though its immediate prospects may appear bleak. Each of the countries can expand its farmland. Mauritania has joined with Senegal in constructing dams along their common border river. The potential of Mali's inland delta region has been tapped only in minor measure, and Niger is investing heavily in seed and technology to improve production in its dry farming areas of the southern frontier. Better watered overall, Chad has enormous areas of unused land and numerous permanent streams. All four countries have some areas drained by reasonably permanent streams. All have some areas of reasonably fertile soil, and none has exploited its water resource to any great extent. Unfortunately, developmental schemes utilizing irrigation are costly.

All four states have experienced severe problems of internal unrest. They are ethnically diverse in the extreme, despite small populations. The Tuareg, the fabled blue, veiled horsemen of the desert, have engaged in guerrilla warfare in Niger and Chad. The Bantu farmers of Mauritania, ostensibly 20 percent of the population, now claim to be a majority in that country and are demanding a greater role in government. Overwhelmingly Muslim Mali is composed of a dozen major tribes, each with strong community ties and little sense of national belonging. The Islamic north of Chad and its Christian-animist south express strong mutual distrust. An "Arabized" aristocracy in all four states clashes with the sedentary farmers who are their traditional tenants.

There are a few positives. The drought is apparently over, and harvests showed increases in the second half of the 1980s. Senegal, Mauritania, and Mali are engaged in joint water resource planning and sharing that will be mutually beneficial. Mauritania has relinquished its claim to Western Sahara and seeks instead to enlarge and redevelop its Saharan oases. Increased cotton acreage, village agricultural cooperatives, and expanding rice production in the inland delta of the Niger hold the promise of stabilized food supplies and improved farmer income in Mali. Agreements with neighboring states will provide access to ocean ports for landlocked Mali and Niger. Both Niger and Mali actively pursue the trans-Saharan connection as they build new road links to connect with the Algerian network. All four states share in the trade of the Haj, the Muslim pilgrimage to Mecca. There are plans for a trans-Sahelian railroad to follow that traditional pilgrims' route.

Slowly, it would appear that the trade and transit of ancient times is beginning to reappear. Overcoming the problems of nature, however, remains the elusive goal.

REVIEW QUESTIONS

1. Why is it assumed that this region includes an important, early cultural hearth? Which of its countries were the sites of early civilizations?
2. What is meant by the "Columbus syndrome?" What was its effect on trade routes, and thus the economy, of the eastern end of the Mediterranean?
3. Briefly describe the population distribution pattern of this region.

4. What are the major exceptions to the generalization that this region is a Muslim (Islamic) culture area?
5. In what ways is Israeli agriculture distinguished from that of most of the rest of the region?
6. What are the major inflows and outflows of people migrating into or out of Israel? Why does the rising outflow pose a serious long-term threat to Israel?
7. What cultural factors have facilitated the unofficial Israeli-American alliance? Is there any doubt about the future of this alliance?
8. What are the implications of the low spatial correlation of oil and dense populations throughout the region?
9. What are some of the developmental strategies open to oil-rich, lightly populated Persian Gulf states?
10. Why does Greek control over many small islands close to the Turkish coast pose such a serious problem if oil is discovered in the Aegean Sea?
11. What are the underlying causes of the unrest in Lebanon?
12. What is Syria's role in regional politics?
13. What are the basic questions involved in the Iran-Iraq dispute?
14. What are Cyprus's prospects for peaceful economic development in view of its recent civil war?
15. Contrast agriculture and development in Turkey and Iran.
16. What are the strategic considerations in the acquisition of naval and air bases by the superpowers in the seas and straits around the Arabian peninsula?
17. Briefly describe the positive and negative results of Egypt's Aswan High Dam. On balance, has the dam contributed much to the economy?
18. Assess the productive potential of the Sudan.
19. What strategic considerations contributed to Algeria's attempt to acquire part of former Spanish Sahara? to Morocco's attempt?
20. What ethnic, linguistic, and cultural factors are involved in demands for provincial autonomy in Iran?
21. What interactions of precipitation variability and pastoralist's decisions have helped to produce tragic environmental degradation in the Sahel?
22. Speculate on the long-term prospects for stability within the region's states.
23. Contrast the economies of the three Maghreb states.
24. What is the role of migrant labor in the Middle East oil economies?

SUGGESTED READINGS

Beaumont, P.; Blake, G.; and Wagstaff, J. M. *The Middle East: A Geographical Study*. New York: Wiley, 1976.

Brett, M. *Northern Africa: Islam and Modernization*. London: Cass, 1973.

Butzer, K. W. *Early Hydraulic Civilization in Egypt: A Study in Cultural Ecology*. Chicago: University of Chicago Press, 1976.

Clarke, J. I., and Bowen-Jones, H., eds. *Changes and Development in the Middle East: Essays in Honour of W. B. Fisher*. London: Methuen, 1981.

Clawson, M.; Landsberg, H. H.; and Alexander, L. T. *The Agricultural Potential of the Middle East*. New York: Elsevier, 1971.

Cressy, G. *Crossroads: Land and Life in Southwest Asia*. Philadelphia: Lippincott, 1960.

Devlin, J. *Syria: Modern State in an Ancient Land*. Boulder, CO: Westview, 1983.

El Mallah, R. *The Economic Development of the United Arab Emirates*. New York: St. Martin's, 1981.

Fisher, W. B. *The Middle East: A Physical, Social, and Regional Geography,* 7th ed. New York: Methuen, 1981.

Gordon, P. *The Republic of Lebanon: Nation in Jeopardy*. Boulder, CO: Westview, 1983.

Karan, P., and Bladen, W. "Arabic Cities." *Focus* (January-February 1983).

Perkins, K. J. *Tunisia: Crossroads of the Islamic and European Worlds*. Boulder, CO: Westview, 1986.

Waterbury, J. *Hydropolitics of the Nile Valley*. Syracuse, NY: Syracuse University Press, 1979.

Weinbaum, M. G. *Food, Development, and Politics in the Middle East*. Boulder, CO: Westview, 1982.

CHAPTER ELEVEN

Latin America

São Paulo, Brazil.

Latin America is readily distinguished from the United States and Canada on the basis of language and culture. Spanish and Portuguese, both Latin-derived languages, are spoken by the vast majority of the region's people. Even two of the secondary regional languages, French and Italian, are Latin based. Language is not the only element of Mediterranean culture that transferred to the New World. Catholicism came to be the dominant faith, and Iberian (Spanish and Portuguese) landholding systems and architectural styles have imprinted the landscape. Yet, the term *Latin America* is somewhat misleading. Though the impact of this derived culture is great, it is not possible to ignore the enormous influence of indigenous cultures, or the melding of African and Ibero-Indian cultural influences that has taken place over large parts of the region. Pre-Columbian Latin America was relatively densely populated and was organized, in part, into large empires that had developed sophisticated civilizations. European migrants to Latin America constituted an additional cultural force, not a displacement culture as happened in North America.

Though clearly Third World in some ways, Latin America does not share the relative overpopulation of parts of Asia, nor the food shortages of Africa. There is considerable diversity within the Third World, and Latin America potentially is its richest part. It has a huge, not fully tapped agricultural potential, vast mineral wealth and forests, and enormous supplies of water. Some of its individual countries may be poor, but few are destitute. There are huge dichotomies between rich and poor within individual national populations, but commercialization and development are present in both rural and urban areas of the region. Overall, the region is far more urbanized, commercialized, literate, and self-sufficient than much of the rest of the Third World.

Settlement is uneven. Spacious, lightly inhabited interiors of low utility contrast sharply with crowded cities. Infrastructures are well developed in only a few areas, but genuine isolation affects only a small minority of Latins. Food imports are generally more than balanced by a preponderant outflow of commercial crops and raw materials. Latin American countries often exhibit better-balanced economies than many other "developing" countries. The era of the single-dimension economy is past.

High birthrates and rapid natural increase still compound economic problems, but growth rates appear to be leveling off in most of the region's states. Migration from Latin America to the United States (both legal and illegal) relieves temporary population pressure somewhat. Migrants often send part of their earnings home to family members—hard currency that enriches the home economy. Latins residing in North America form a market for specialties produced in their former homelands—a new source of export revenues.

The rate of economic growth remains highly variable. Overheated, rapidly expanding national economies have borrowed to the maximum in their race toward the goal of full development. They often appear to be walking a precarious economic tightrope, vacillating between boom and near disaster, temporarily reeling under the impact of fluctuating world commodity prices. Despite periodic setbacks, a general upward trend is visible.

None of the region's economies is totally dependent on oil or other mineral exports as are some in

TABLE 11–1
Percent urban population by world region.

United States and Canada	75
Europe	73
Australia-Oceania	71
Latin America	67
USSR	65
Middle East-North Africa	47
East Asia	39
Africa South	27
South Asia	25
Southeast Asia	25
World Average	43

SOURCE: Population Reference Bureau.

the Middle East and Africa. Few are totally lacking in energy resources. Except for the small Caribbean islands and some Central American republics, most have a varied resource base. Rapidly advancing technology has made it increasingly possible to use less-skilled labor in factory production. Consequently, industrialization proceeds at a brisk pace.

The old "siesta image" of a relaxed Latin lifestyle is out-of-date. It was an image associated with tradition-bound rural lifestyles in areas in which there was little change. As shown in Table 11–1, two thirds of Latins now live in urban areas. Change there is ongoing. An almost frenetic lifestyle characterizes Latin America's cities. Whether in pursuit of wealth or simple survival, the pace of life is hectic. In the realities of modern Latin America, there is little time for rest; the focus is on the future.

LA INDEPENDENCIA: THE EMERGENCE OF NEW NATIONS

The states of mainland Latin America, except for the Guianas and Belize, achieved independence from colonialism between 1811 and 1826. The more highly valued sugar-producing islands of the Caribbean (except for Hispaniola) gained independence only in this century, many in the last two decades.

Immediately after independence, there were attempts to federate some of these culturally similar areas into a few large states. Most attempts fragmented under the negative pressures of poor land transport and communications and the positive pressures of isolated nodes of settlement focused on a single city, commonly a colonial capital. Gran Colombia, one such example, was held together briefly and primarily by Simon Bolívar's leadership and personality (he was the leading force in Latin independence). Many scattered settlement nodes were consolidated into what became modern Colombia. But even the charismatic Bolívar could not hold together the more distant parts of this original state; they broke away as Venezuela and Ecuador. Brazil, the region's largest state, is the major exception. It has maintained its territorial integrity, and even expanded. Current Brazilian territory includes lands ceded by most neighboring countries during a series of expansionist wars (Figure 11–1). Most generally,

however, as in Africa, the successor states mirror the colonial-era administrative divisions.

The United States strongly supported Latin American independence. The Monroe Doctrine (1823) stated that any attempt by European powers to reimpose colonial control on independent states in the Americas would be viewed as an "unfriendly act" toward the United States. The commonality of interests achieved by these early struggles for independence in the Americas has resulted in a loose hemispheric concept of unity that influences relations of the states of the Americas to this day. Regardless of linguistic and cultural differences between North and Latin America, there is some sense of being American, of being different from Europe.

The Influence of the U.S. Economic Perimeter: The United States as "Big Brother"

Since the early days of the republic, the United States has gradually expanded its economic interests, political-military power, and even its territory toward the Caribbean. Expansion of U.S. interests was first directed toward Mexican territory. After the Spanish-American War (1898), the United States acquired Puerto Rico and began a long and close relationship with Cuba that lasted until the Castro revolution. The Panama Canal Zone was established in 1903, following the "independence" of Panama (actively aided by the presence of the U.S. Navy) from Colombia. The Virgin Islands were purchased from Denmark shortly afterward to help defend approaches to the canal. A naval base was leased from Cuba at Guantanamo Bay to safeguard the Windward Passage into the Caribbean.

To defend its interests in the region against regimes or conditions perceived as unfriendly, the United States sent troops to occupy Nicaragua (1912), Veracruz, Mexico (1914), Haiti (1915), Mexican border areas (1916), and the Dominican Republic (1916). In later years, it supported an abortive invasion of Cuba (1961) and occupied the Dominican Republic a second time (1965). The United States "destabilized" a Communist regime in Guatemala in 1954, dispatched troops to Grenada in 1983, and has sent troops to Honduras in conjunction with support

Guadalajara

30°

UNITED STATES

Louisiana

Florida

ATLANTIC

GULF OF MEXICO

VICEROYALTY

Mexico City

(British)

Santo

(Fr.)

Tropic of Cancer

OCEAN

LINE OF
TORDESILLAS

OF

(Br.)

(Br.)

Domingo

15°

CARIBBEAN

Guatemala

SEA

NEW SPAIN

VICEROYALTY

Caracas

Dutch
Guiana

Fr. Guiana

OF

Bogotá

PACIFIC

Sante Fe

NEW GRANADA

Pará

Maranhão

Piauí

0°

Equator

Quito

Rio Negro

VICEROYALTY

Pernambuco

OCEAN

Lima

OF

Matto Grosso

Cuzco

BRAZIL

Sergipe

Lima

Bahia

VICEROYALTY

OF

PERU

15°

Charcas

Espiritu
Santo

Goiás

Minas

Gerais

Rio de Janeiro

Rio de Janeiro

Tropic of Capricorn

VICEROYALTY

Asuncion

São Paolo

OF

Santa Catarina

30°

LA PLATA

Rio Grande do Sul

Chile

Buenos Aires

Indian

45°

Territory

Falkland
Islands

Borders of Viceroyalties

------ Borders of Audencias (Spanish)
and Capitancies (Portugese)

o   Viceroyalty capitals

Current international boundaries

Line of Tordesillas

Spain

Portugal

France

Britain

Netherlands

Indian Territory

105°   90°   75°   60°   45°

of the Nicaraguan Contra rebels. For the last three decades, the United States has been confronted with an anti-American, aggressively prorevolutionary Cuba.

In recent years, the United States reluctantly arranged for eventual transfer of the Panama Canal to Panama, but in 1989 again asserted its right to protect the canal. The rise of a Marxist government in Nicaragua has caused concern. The United States remains extremely sensitive to political conditions in what it perceives as its "backyard"!

CULTURAL CHARACTERISTICS

Two languages—Portuguese and Spanish—dominate in all the region's states except for the Guianas, Belize, some Caribbean islands, and the Falklands. Widespread use of Spanish represents a powerful cohesive force, facilitating trade and the exchange of technology within the region (Figure 11–2).

Despite the fact that Portuguese is spoken only in Brazil, the language has great regional importance because of the sheer size and population of that country. Brazilians see their version of Portuguese as the virile language of today, overshadowing the original source language.

Indian languages are still widely spoken over significant parts of Latin America (see Figure 11–2). In Mexico, Guatemala, parts of Central America, and the Andean republics (successor states to the empires of the Aztecs, Maya, and Incas), large native populations were never totally assimilated. Half or more of the population in Bolivia and Guatemala is classed as Indian. Such terms as *Indian* and *mestizo* (mixed European-Indian), however, are loosely used. Almost all Mexicans speak Spanish, regardless of ethnocultural origin. On the other extreme, fully 30 percent of Peruvians speak no Spanish at all; half use some Indian language as their primary tongue. Quechua, descended from Incan speech, has official status, but Aymara, the other widely spoken Indian tongue, does not. In Bolivia, where only Spanish is official, over half the population speaks Quechua or Aymara. In Paraguay, on the other hand, 90 percent of the population speaks both Spanish and Guarani, the local Indian dialect; both languages are official.

There are two remaining centers of dominantly Indian speech, centering on the territories of the old Mayan and Incan empires. Their respective cores are Yucatán-Guatemala and the Andean highlands of Peru and Bolivia. Outward from these cores, Indian speech weakens, replaced by Spanish with increasing distance and, seemingly, the relative degree of integration of an area into the national economy. Indian speech virtually disappears in large urban centers (see Figure 11–2).

English is the language of the Caribbean islands longest under British control, such as Jamaica, Barbados, and the Bahamas. Where control varied, as in Dominica and Grenada, French-African Creole speech is spoken in addition to English. Expatriate West Indians, imported as labor for banana plantations on the Caribbean coast of Central America, form English-speaking minorities within those republics. American influence has extended the use of English in Panama and Puerto Rico, though Spanish prevails in both units.

French, however modified to Creole dialect, is spoken in Haiti, French Guiana, Guadeloupe, and Martinique. Guyana and Suriname have adopted English and Dutch, the languages of their former colonial controllers. Asian languages, brought by later migrant labor, have some usage in the Guianas and Trinidad. All these exceptions are minor deviations in the much broader dominance of Spanish and Portuguese speech.

Three highly varied racial-cultural elements have had a strong cultural impact on Latin America: native Indian, African, and European. Though Europeans originally came as conquerors, significant numbers

← **FIGURE 11–1**

Colonial Administration in the New World, circa 1790. Thirteen individual colonies became one country, the United States, on the basis of a common language and a jointly held animosity toward the colonial power. The same essential conditions existed in much of Latin America, but the results were quite different. Brazil emerged as one unified country, but the former Spanish holdings became 18 separate national jurisdictions. Both cultural diversity and physical isolation played a role in the disintegration of the large confederations that originally developed after liberation. Adjustments to borders were made during later wars among regional neighbors, but the similarity of area and borders is still evident.

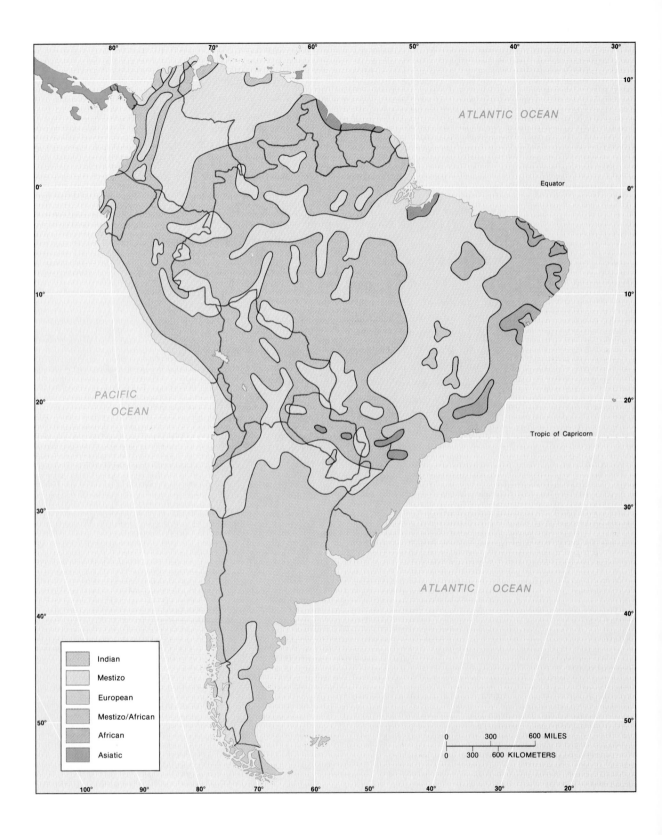

	Indian
	Mestizo
	European
	Mestizo/African
	African
	Asiatic

of settlers followed. The Guianas, the smaller Caribbean islands, the four heavily Indian states of South America, the highlands of northern Central America, and the swampy Caribbean coasts attracted the fewest Europeans (Figure 11–3). A European commercial zone, in which European settlers and commercial interests dominate, includes most of Argentina, Chile, Uruguay, and southern Brazil. Here the population is 80 percent or more of European stock or totally assimilated mestizo.

African racial-cultural input is most strongly marked in areas of early plantation development: northeast-coastal Brazil, the Guianas, the Caribbean islands, and the Central American mainland's Atlantic coast (see Figure 11–2). Large numbers of African laborers were imported to the sugar plantations, and later came to be engaged in the plantation production of rice, cacao, tobacco, cotton, and bananas. The population of Brazil is 15 percent black, and an additional 16 percent are mulattoes (black-white racial mixtures). Great African cultural impact is found in Brazil's art and music. Some regional agricultural methods draw on African cultural influences, regardless of the practitioner's origins. African racial strains and cultural input dominate in most Caribbean islands, particularly those where European powers other than Spain held political control.

Indian influences are most strongly in evidence within the highlands of South America where the indigenous population was numerically greatest and where the most advanced Indian cultures had developed prior to European conquest. The Spanish encountered two advanced cultures (the Incan in Peru and Chibchan in Colombia) and a somewhat less sophisticated one (the Araucanian in Chile) (Figure 11–4, p. 568). The most advanced of these, that of the

FIGURE 11–2
Dominant Cultures and Ethnic Groups of South America. Two languages, Spanish and Portuguese, dominate except in a few small mainland colonies and some islands in the Caribbean. Culture is often a more significant detriment to national unity than are languages. The various admixtures that have developed from a blend of European, African, and Indian racial and cultural elements give the area its distinct regional culture, often simply called Latin or Latin American. In reality, however, it is composed of many highly distinctive subcultures.

Incas, held an empire that extended from their headquarters at Cuzco (Peru) north to Quito (Ecuador) and south through Bolivia to the middle of Chile. They worked metal, engineered irrigation works, created farmland by terracing steep Andean slopes, and domesticated crops and livestock. The Incas controlled a chain of highland basins and valleys interconnected by stone-paved footpaths. The ability of the Incas to hold together this empire with its difficult terrain is attributed to their exceptional administrative talents and feats of engineering. The success of the Spanish against this mighty empire resulted from a combination of superior weapons and a period of internal dissension within the empire.

The Andean Indians display great cultural diversity, probably because of the long isolation of one basin from another. Indian cultural impact elsewhere in South America has been minimal.

In Middle America, cultural inputs vary greatly. However, in areas of strong Indian influence, national cultures have undergone a period of **indigenisimo**, a glorification of Indian roots and an emphasis on pre-Hispanic culture. Mexico provides the finest example. Its flag and other national iconographic symbols emphasize pre-Hispanic themes. The architecture of modern Mexico makes use of Aztec form, style, design, and decorative motif. Murals on the walls of public buildings and Mexico City's splendid subway stations depict themes from Mayan and Aztec civilizations. Archaeology and artifact preservation are prominent academic endeavors. Spanish cultural roots and architectural forms are also strongly in evidence. Yet, it is modern commercial culture that increasingly dominates. Only in rural, southern Mexico are Indian languages and cultural forms still prominent.

The processes of modernization, commercialization, and cultural assimilation dominate Central American urban culture. Indian culture is still strongly in evidence in rural Guatemala and inland Honduras, but has all but disappeared elsewhere. Indians, and all pre-Hispanic cultural remnants, were virtually eradicated in the Caribbean islands during the early years of Spanish occupation.

The landholding system introduced by Europeans developed strong inequities in class and income. The Spanish created and controlled great estates (latifundia) where Indians labored under feudal conditions.

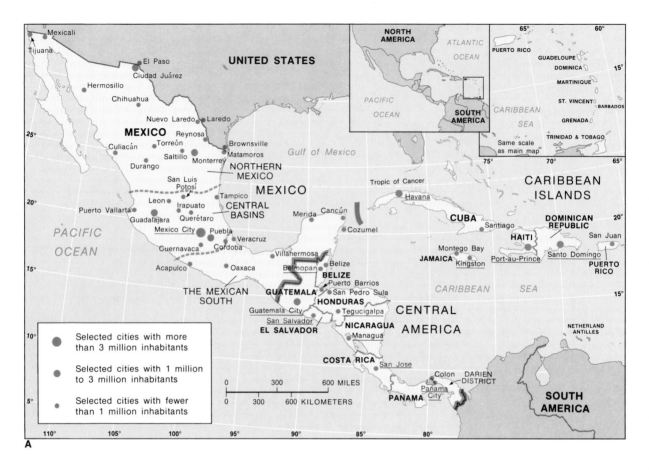

FIGURE 11-3

(a) *Subregions and Places of Middle America.* Middle America is an extremely diverse area, physically as well as culturally. This portion of Latin America is divided into three major sections: Mexico, Central America, and the Caribbean Islands. Mexico, the largest country in Middle America in both area and population, is divided into three distinct subunits. The much smaller Central America is composed of seven individual republics. The islands form the most complex portion, with over 20 separate jurisdictions. (b) *Subregions and Places of South America.* With a few exceptions, the countries of South America are larger than those of Middle America. Brazil, the largest country, covers half the continent. Sheer size and great internal diversity merit the division of Brazil into five subregions. The remaining states are combined into three large subregions: The Caribbean Littoral, the Andean Republics and Paraguay, and the Southern Tier.

They persist as large commercial farms worked by tenants. Curiously, the system may have facilitated the transition to modern, mechanized agriculture. The Portuguese, on the other hand, came to a land nearly devoid of population. Colonists, unlike those of Spanish South America, were rarely drawn from the ranks of nobility or the military. Portuguese settlers with capital sought the rich lands of the north-east for plantations. The other major settlement area was at São Paulo, where poorer Portuguese elements received small grants of land. Known as Paulistas, this group roamed the back country in search of wealth and success. In the process, they colonized much of Brazil.

Late-arriving Europeans have had a great impact on the cultures of southern South America, where

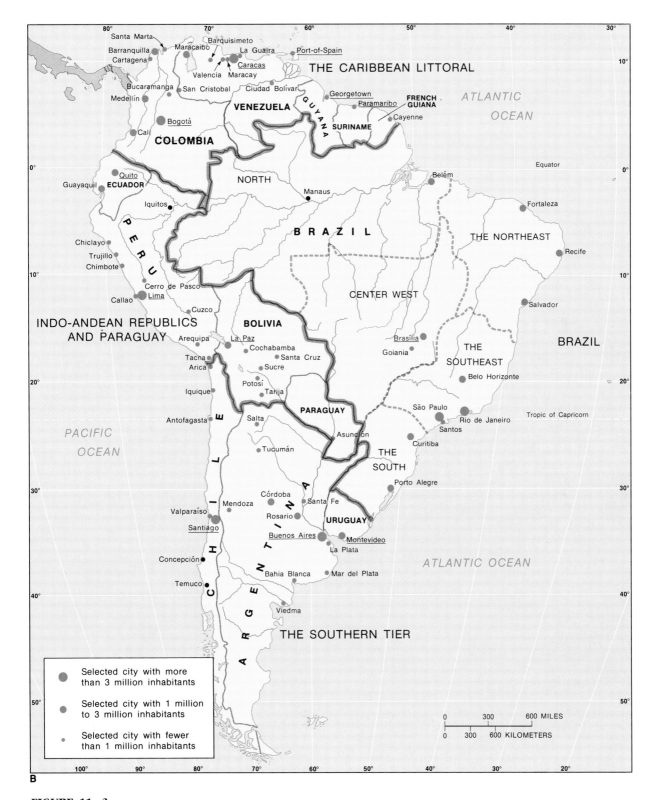

FIGURE 11–3
continued

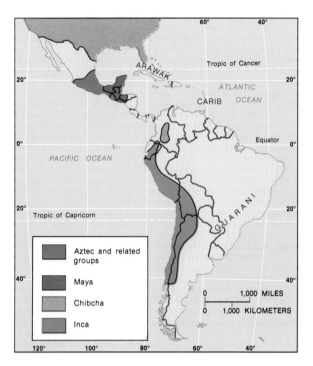

FIGURE 11–4
*Pre-Columbian Cultures and Civilizations of
Latin America.* Unlike in the United States and
Canada, where Indian populations were sparse, some 70
to 100 million inhabitants lived in the area that was to
become Latin America. The Aztecs, Maya, and Incas had
developed advanced cultures, with cities, complex
lifestyles, religions, and monumental works of
architecture and engineering. The Carib, Arawak, and
Guarani Indians appear to have been widely scattered.

Religion

In a very few states, Roman Catholicism is the official
religion. In most of Latin America, the Catholic
church has experienced periods of anticlericalism,
with attendant persecution of both clergy and the
faithful. Church-held lands and property were long
ago nationalized. The Catholic church remains, however, a powerful moral influence. It is a force for
internal nationalism and, increasingly, for economic
reform. It is the voice of social conscience. Activism
on the part of some Roman Catholic clergy through
"liberation theology" has become common in the
last two decades.

Although Roman Catholicism dominates, fundamentalist Protestantism, particularly the Pentecostal
churches, has become important in the region
through missionary activity. Health care and poverty
relief are an integral part of their missionary work.
Protestants form a majority in some former British
Caribbean islands, and a minority in Belize. The Guy-
anas and Trinidad, home to large Asian groups, have
relatively large Hindu and Muslim representations.

Historically, Roman Catholicism has had a lenient
attitude toward the incorporation of native custom
and tradition into church liturgy and observance.
This tolerance was extended to African beliefs as
well. Nonetheless, the religious tenets are essentially
European, as is the rhythm of life, with Sunday and
church holidays as periods free of labor. The reverse
influence, that of Amerindian culture on the church,
is obvious in festivals, architecture, and interior decoration.

The Problem of Land Reform

In virtually all of Latin America, there was class differentiation based on land ownership. In Spanish
America, Indian lands were seized, albeit often from
an Indian ruling class, and distributed to Spanish
settlers. New settler-owners proceeded to operate
their farms on the **latifundium** system, common in
Mediterranean Europe. Land titles included the *en-
comienda,* the right to use the Indian inhabitants of
the lands as labor. The Spanish termed such an estate
a **hacienda**. Workers lived in villages and farmed
small parcels; rents were paid by a share of the crop

they created a zone of surplus grain and meat production for domestic urban and export markets.
Their role often has been to introduce modernization and commercialization into what had been
largely a subsistence pioneer economy. Descendants
of late-arriving Europeans dominate in Uruguay, Argentina, and Chile. In southern Brazil, they have become the bulk of that country's middle-class farmers.
The European cultural input, then, continues, if in a
different form. The mestizo element is frequently
assimilated to European language and some elements of culture. Mestizo is a lifestyle as much as a
biological amalgam, and mestizos form a bridge between native and European cultural elements in
many countries.

or in field labor. The traditional role of the owner was inspector and, socially, that of patron, arbitrating disputes and dispensing gifts. Yields were low, and neither land nor labor were used efficiently.

Independence commonly brought little change. Governments continued to be run and influenced by the landholding class. Those living outside the system often had too little land for subsistence and became hired, temporary labor on the hacienda. The result was poverty for two large classes: the landless peasantry and the subsistence farming community.

This exploitative system has not totally disappeared. Mexico has eliminated less efficient estates, distributing nationalized land to the landless through an ambitious program of land reform. In rapidly developing economies such as Brazil and Venezuela, land has ceased to be the measure of class and wealth. Less exploitative estate systems in Argentina, Brazil's South and Uruguay pay tenants in cash or larger crop shares, enabling former tenants to amass capital and purchase their own lands.

Nonetheless, the problem of the rural, landless poor persists. Migration to the region's cities and colonization of formerly unused areas have lessened pressures but have not completely solved this basic problem; the interrelationship of land, wealth, and power has not been eliminated. In Guatemala, the top one fifth of the population still receives over 60 percent of national income, while the poorest fifth barely subsists on 5 percent of the national total. A mere 8 percent of Bolivians once owned 70 percent of that country's farmland; despite land reform, they still exert a powerful force. In spite of reform, some 5 percent of Mexico's farmers still own no land. In El Salvador, the figure exceeds 20 percent.

In land reform issues, the region's governments are faced with difficult decisions. Production from remaining large estates is efficient; it generates income. Little direct income would be forthcoming from small peasant farms, and virtually none from subsistence plots, were land reform to be carried out to the fullest. The tendency has been to nationalize only unproductive, inefficient estates or portions of large landholdings that are not being used.

Frontier colonization schemes, widely promoted as an alternative to land reform, have had mixed results. The destruction of the rainforests that accompanies colonization has caused worldwide concern.

In Brazil, discouraged homesteaders frequently give up and sell their lands to agents of wealthy landowners who in turn piece together vast new cattle-ranching estates from former peasant holdings. Democracy, quite fragile in the region, will have a difficult time surviving without land reform until such time as the stage of full development is reached.

POPULATION: GROWTH AND MIGRATION

The southern three countries of South America and most Caribbean islands have entered a period of rapidly declining growth rates. Though this change normally accompanies development, sheer population pressure is the apparent cause in the Caribbean. Bolivia and Haiti, with both high birth- and death rates, have not yet entered the period of maximum growth. The signs of a downturn in growth rates are appearing now in Colombia, Venezuela, Mexico, and Brazil. Gradually, rapid population growth will cease to be a regional characteristic, though a regional natural increase of 2.2 percent annually is just slightly below the Third World average (2.4 percent) and still above that of the world (1.7 percent).

Mechanization and modernization of agriculture, a shift from traditional tenancy systems, high rural birthrates, and higher urban wages act as push-pull mechanisms, unleashing massive rural-to-urban movement in Latin America. With a few notable exceptions, shantytowns of the urban poor have arisen around most large cities. These sprawling slums lack most urban services and have become pockets of crime and poverty. Their residents eke out a living through occasional day labor, subsistence gardening on tiny hillside plots, running errands, prostitution, or whatever else earns food or money.

Virtually all South American states have more land that can be colonized and brought into production; the same applies in most of mainland Middle America, if not on the islands (Figure 11–5a). This safety valve may yet contain social unrest and contribute to the eventual elimination of the problems of unemployment and landless peasantries. However, this remaining land is often costly to develop, sometimes requiring irrigation, large-scale drainage, massive

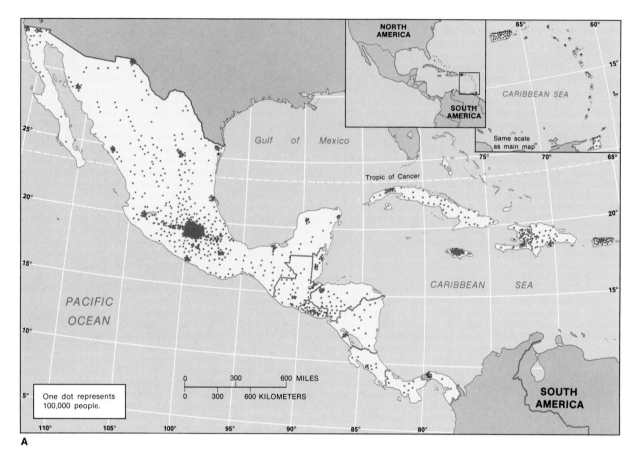

A

FIGURE 11–5

(a) *Population Distribution in Middle America.* In the mainland portion of Middle America, population clusters in highland areas where adequate water and rich soil yield the best agricultural possibilities. The high densities exhibited in central Mexico include large urban centers such as Mexico City. (b) *Population Distribution in South America.* Vast areas of interior South America are virtually empty. The population concentrates on the periphery of the continent.

transportation improvements, and great inputs of fertilizer to become productive. Housing, services, and extensive health hazard control measures are only a few of the additional investments necessary to ensure livability in frontier regions.

There are bright spots. Chile, Venezuela, and Brazil have developed homesteading projects with moderate success. Colombia has extended roads to areas of good soil that lie unused because of poor accessibility. Brazil has developed successful rice-farming settlements by partially draining riverine floodplains. Mexico has irrigated vast areas, for both resettlement and improvement of yields on established farms.

Peru has increased irrigable area by diverting Andean streams from Amazon drainage to the headwaters of the ephemeral desert streams that flow to the Pacific in a series of multipurpose hydraulic projects.

All these schemes have required costly, large-scale investment. For most states, development of the bulk of these unused lands lies in the future. South America, in particular, remains an area in which population and development are clustered around the periphery (Figure 11–5b). The interior is relatively empty; low population densities prevail in most national units. The degree of utility of those remaining empty areas is a question not fully answered.

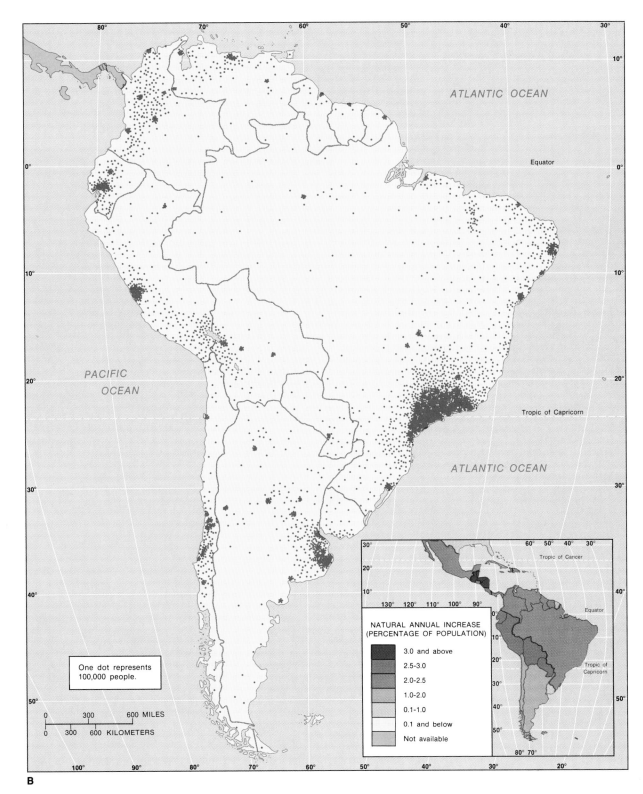

B

FIGURE 11–5
continued

THE PHYSICAL FRAME

Climate is overwhelmingly tropical or subtropical, one mountain chain is dominant, and three huge drainage systems channel most of the region's waters. Deceptively simple at first glance, the physical framework is as amazingly complex as the cultural framework when closely inspected. Population distributions (see Figure 11–5) provide a useful key to understanding the physical framework. The region's larger desert areas—Chile's Atacama, Argentina's rugged deserts of the south, and Mexico's Sonora—are virtual voids except where mountain-fed streams provide irrigation opportunities. At the other extreme, areas of intense, year-round precipitation are equally empty (Figure 11–6). Temperate zones, limited in size, present themselves as densely settled areas at either regional extreme. Throughout, altitude is a significant modifier of climate, and highland areas frequently possess fairly dense populations.

Topography

The topography of Middle America is in many ways an extension of features found in the United States (Figure 11–6a.) Mexico's Baja California is a line of fault block mountains, a continuation, virtually, of southern California's Peninsular Range. The Gulf of California itself is part of the same hill-encircled trough that frames the Imperial Valley of California and the mouth of the Colorado. Most of the northern two thirds of Mexico is a high plateau with basin-and-range features superimposed upon it. The eastern edge, the Sierra Madre Oriental, is a continuation of the eastern ranges of the Rockies. The western edge, the Sierra Madre Occidental, becomes a wide zone of folded and faulted mountains. Its bordering lowland is quite dry and physically isolated from the plateau. The plateau itself becomes higher southward as lava flows and sediments fill in the basins. Densely settled in its southern portion, it terminates abruptly in a transverse volcanic range.

The Gulf Coast Plain stretches from the Texas border to the Yucatán, narrowing from a width of 350 miles in the north to less than 50 miles in the center and again widening southward. The limestone arch that forms the Yucatán is a geologic mirror image of Florida.

North of the 18th parallel, there is a striking re-orientation of trend. The north-south mountain ranges are intersected by an east-west tectonic belt that extends through the Caribbean islands, looping

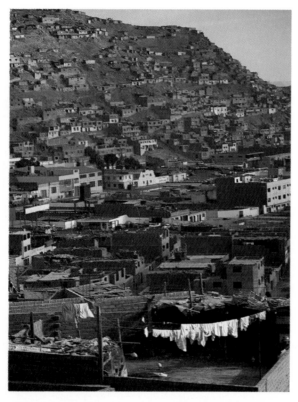

Hillside Squatter Settlements in Latin American Cities. *Rapid population increase in rural areas, mechanization of agriculture, and higher urban wages serve as push-pull mechanisms that have unleashed a massive rural-to-urban movement in Latin America. In the first years after arrival, new arrivals squat on unused land at the edges of the city in makeshift homes that lack utilities and services. After a few years, if they obtain stable employment, conditions grow better. The residents of this barrio outside Lima live under somewhat improved conditions. Homes in the foreground are constructed of sun-dried brick by the residents themselves. However poor, people who live here have a neighborhood water supply and more substantial homes than those later arrivals who live farther up the hill.*

back to become the Andes of South America, the main mountain chain of that continent. These sharp directional changes are associated with two small tectonic plates, the Cocos and Caribbean, caught between the larger Pacific, American, and Nazca plates (see Chapter 1, Figure 1–9). The result is a zone of impressive vulcanism and earthquake activity throughout Middle America.

The Caribbean islands are a classic island arc extending from Cuba through Puerto Rico to Grenada and bordered by deep sea trenches. At the northern edge of the arc, there has been a considerable uplift of ocean bottom. The Bahamas are low coral and limestone islands that rise above this shallow sea platform. Cuba itself is an extension of the east-west trending arc; the Sierra Maestra in eastern Cuba is a continuation of mountains in Guatemala. A second ridge, forming a volcanic spine in Honduras, comes above the sea to form Jamaica. Hispaniola combines the Cuban and Jamaican trends, as does Puerto Rico in a greatly subdued form.

Beyond Puerto Rico, there are two island chains that follow the eastern edge of the Caribbean plate; both are volcanic, and both trend north-south. The outer line of islands is generally smaller, lower, and composed of older volcanics; the inner island chain is the opposite on all counts. The islands of the southern edge of the Caribbean (Trinidad, Barbados) were formed by the drowning of the Venezuelan Andes.

The Andes are a continuation of the subsea mountains that form the Caribbean islands and connect, via a second looping arc, to the mountains of Antarctica. They are the world's second highest mountain range. Though their width is relatively narrow, their height and ruggedness have made them an extremely effective barrier, separating the coastlands from the interior of the same political unit.

The Andes are lowest and least complex in the south, generally a single ridge sculpted by glaciation. Northward of central Chile, they increase in height and split into two ranges; the westernmost is higher and steeper. In Bolivia, the two ranges deviate sharply from their north-south orientation at the bend in South America and diverge to form a high plateau, the Altiplano, between two rows of peaks. High, cool, and inhospitable, much of the plateau is arid and floored with salt flats. In Peru, two ranges become three. There, portions of the basins between these ranges are filled with volcanic debris that has weathered into rich soil (Figure 11–6b).

The Andes change directional trend again at the border of Ecuador, abruptly shifting to run northeast. At this point, they turn inland, allowing the development of a progressively wider coastal lowland. In Colombia, three ridges are widely spaced with intervening valleys, lower and much broader than those of Peru.

There is often a coastal range paralleling the major mountain chain at a lower altitude. It is partially drowned in southern Chile, forming offshore islands. The central Valley of Chile, like the Great Valley of California, is a sediment-filled trough between these coastal ranges and higher, inland mountains.

East of the Andes, there are three major lowlands. The northernmost, the Orinoco lowland, is crescent shaped, broad, and filled with sediment. The Amazon lowland is huge, the largest single feature on the continent. It is triangular in form with the broad, western base against the Andes, tapering to a narrow apex at its mouth. The lowland of the Paraná-Paraguay river system is divided into the dry northern Chaco and the southern Pampas.

The Patagonian Plateau, the Brazilian highlands, and the Guiana Highlands are all ancient crystalline shield areas—old continental block (see Figure 11–6b). In most cases, the land rises abruptly a few miles in from the coast. These steep *serras* (scarps) have made movement into the inland relatively difficult, though they are much lower than the Andes. North of the Paraná-Paraguay river system, the shield reaches its maximum extent as a huge triangle between that river system, the Amazon, and the Atlantic.

The Guiana Highlands are often higher and more rugged than the Brazilian highlands. A very narrow coastal plain fringes the Atlantic in the Guianas.

Climate

Mexico, northward of 20° north latitude, is the largest zone of dry climate found in Latin America. Overall, however, it is wetter than the nearby North American desert. Northern Mexico is more steppe than desert. Most of the rest of Mexico has a savanna (tropical

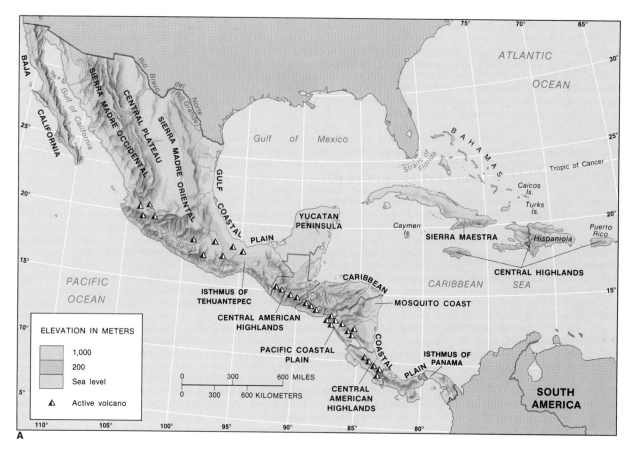

FIGURE 11–6

(a) *Physical Features of Middle America.* Located at the juncture of several continental plates, Middle America is one of the most tectonically active parts of the world. Volcanoes may wreak disaster, but they also provide portions of the area with unusually rich soil. Most of Mexico is a high plateau surface ringed by mountains. The narrow neck of land that connects North and South America contains several sizable lowlands as well as a mountain chain. The Isthmus of Panama has been canalized. Some Caribbean islands are the exposed tops of volcanoes; others are flat, low-lying coral islands. (b) *Physical Features of South America.* Three immense rivers drain most of the continent, forming extensive lowland areas. The Andes, the world's second highest mountain range, occupy the western edge of the continent. Large blocks of ancient crystalline rock, in places covered with sedimentary deposits, form three structural plateaus that occupy the land between the major rivers.

wet-and-dry) climate, though mountain slopes and the highland valleys of southern Mexico can be quite wet. The belt frequented by Caribbean trade winds, the windward eastern coast of Central America, receives year-round rain. The Pacific coast, a leeward coast, receives adequate rainfall only in summer.

The islands from southern Cuba through Puerto Rico receive more frequent rains and greater annual totals on their Atlantic (as opposed to Caribbean) coasts. Smaller islands have a much drier climate and marked winter drought. Violent tropical storms plague many Caribbean islands. The larger islands

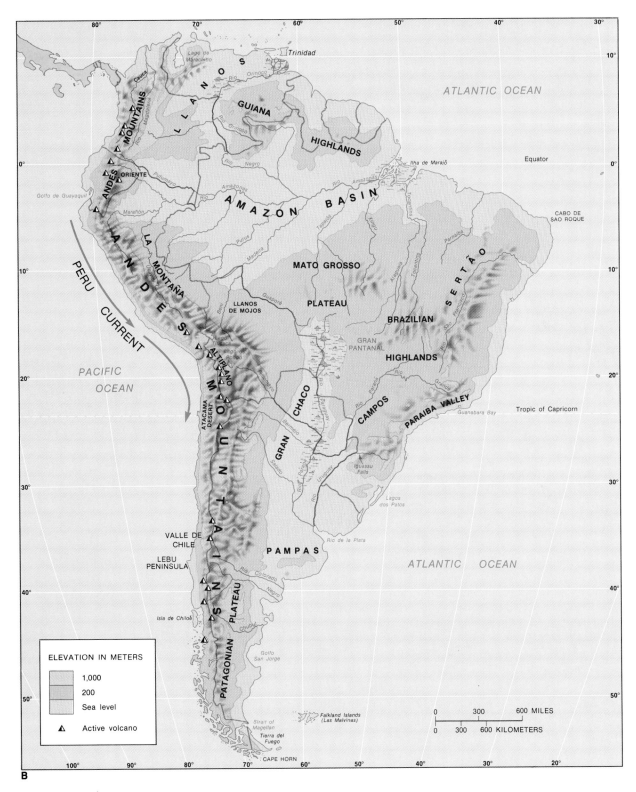

FIGURE 11–6
continued

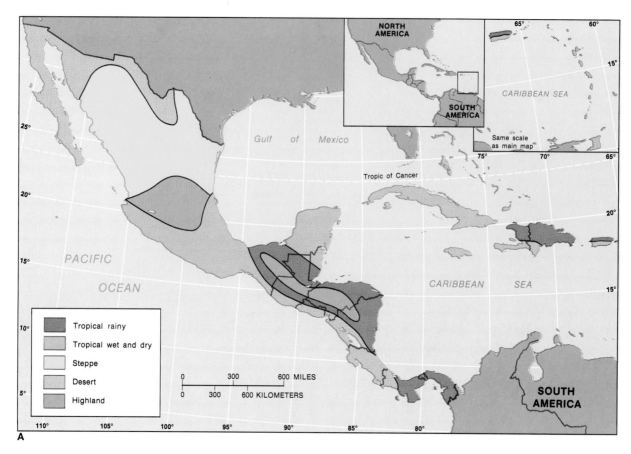

FIGURE 11–7

(a) *Climatic Regions of Middle America.* Except for northern Mexico, Middle America experiences tropical climates. Areas with year-round rain are covered with dense rainforests. Areas with a pronounced dry season are covered with scrub forests and tropical grasses. Except for a few highlands and windward coasts, the Caribbean islands have a tropical wet-and-dry climate that is naturally advantageous to the cultivation of sugarcane. (b) *Climatic Regions of South America.* South America experiences almost every type of climate found on the earth. It contains the world's largest area of tropical rainy climate. Deserts extend in a narrow belt along the Pacific shores from the Peru-Ecuador border to central Chile and beyond the Andes into parts of Argentina. Most of the south-central interior experiences a tropical wet-and-dry climate. The temperate climates of northern Argentina, southern Brazil, and Uruguay support commercial grain production. Climate is intensely modified and influenced by altitude over large areas of South America.

and the Gulf Coast of Mexico have often been referred to collectively as "hurricane alley." Unfortunately, the peak season of occurrence is also the normal period during which crops ripen and harvest takes place. The positive effect of hurricanes is that

they can bring half the year's moisture (Figure 11–7a).

Latin America contains the world's largest zone of the tropical rainforest climate, with its short (or no) dry season, high temperature and humidity levels,

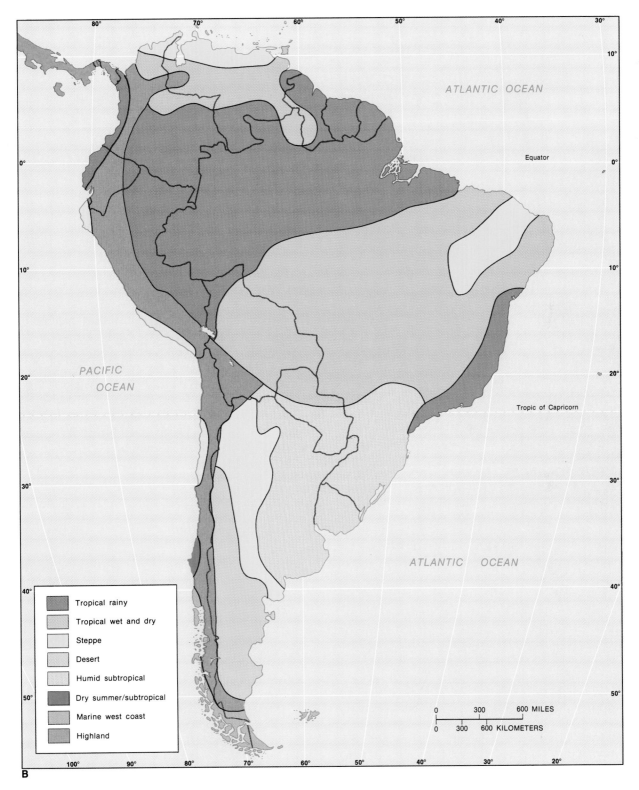

Legend:
- Tropical rainy
- Tropical wet and dry
- Steppe
- Desert
- Humid subtropical
- Dry summer/subtropical
- Marine west coast
- Highland

B

FIGURE 11–7
continued

VERTICAL ZONATION

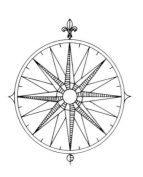

Most of the population of Mexico, mainland Middle America, and the Andean republics live in highland areas where temperatures are modified by altitude. Generally speaking, temperatures decrease at a rate of 2° F to 3.5° F for every 1,000 feet of elevation (Figure A). Though dense populations do occur in some lowlands, both the pre-Columbian Indians and the later European settlers tended to prefer the highland basins.

Each range of elevation has a different kind of climate, which has led to the emergence of different crop patterns and lifestyles in each of the vertical zones (Figure A). The coastal lowland zone (where not desert) from sea level to about 3,000 feet is the **tierra caliente**, or "hot land." It is a zone of tropical crop production and, frequently, plantation agriculture. From elevations of 3,000 to 6,000 feet, the climate is more temperate; the zone is called **tierra templada**. Frost free, yet pleasantly cool, it is a land that can grow corn and wheat well; it is also the zone of coffee cultivation. From 6,000 to 10,000 feet, or even higher near the equator, the high country is known as the **tierra fria**, or "cold land." Potatoes, barley, cabbage, onions, and other cold-tolerant plants are the staple crops. Above this is the **páramos**, a year-round grazing area too cold for tree growth, topped by permanent ice in its highest reaches. The same altitudinal zonations are found in tropical highlands everywhere, though rarely do highlands concentrate such dense populations as they do in Latin America.

and high rainfall totals (see Figures 11–7a and b). The basin of the Amazon, the Guiana Highlands, and some coastlands experience such a climate. Daily temperature variations exceed seasonal average variations, and rain is almost a daily occurrence. Trade winds and sea breezes modify the humidity and **sensible** (human-perceived) **temperature** in coastal areas.

The bulk of Brazil, interior Venezuela, southern Mexico, and the Pacific coast of Central America —indeed more than half the region—experiences a climate that is seasonally dry. The rain-bringing mechanism is the intertropical convergence, which shifts north and south of the equator, following the seasonal migration of the direct rays of the sun. Drought occurs during the respective winter period, though temperatures are hardly wintry. Most of the area receives 30 to 60 inches of rainfall, concentrated in the wet period. These two hemispheric belts of savanna climate have winter (and drought) at opposite periods of the year (see Figure 11–7).

The higher plateau areas of Brazil experience a similar rainfall regime but have cooler temperatures because of altitude. Higher inland areas experience a delightful climate, with temperatures rarely below

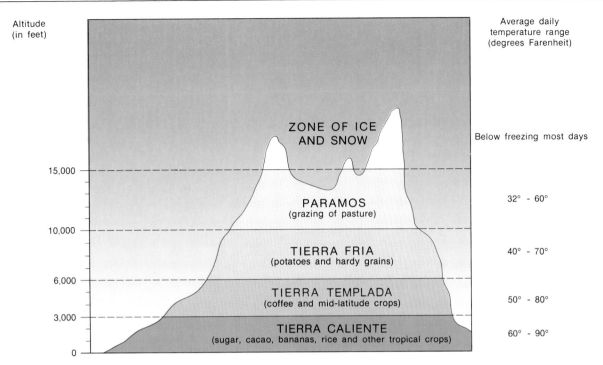

FIGURE A
Vertical Zonation: Climate and Agriculture in the Highlands. In tropical Latin America, far more people live in the highlands than in lowland areas. Temperatures decrease with altitude, so that highland areas offer a variety of climatic conditions. Each of the altitudinal levels has a name and is associated with a characteristic set of crops. An enormous variety of production may occur over a rather limited horizontal area because of this vertical zonation.

freezing. Temperatures at São Paulo, for example, average 64° F with small day-night variation and only a 10° F variation from the coldest to the warmest months. South of Paraná state in Brazil, a zone of temperate climate, much like that of the southeastern United States, occupies the Atlantic side of the continent through the northern third of Argentina. Cold weather occurs but is never intense (see Figure 11–7b).

South and west of the South American temperate areas, the climate grades into steppe and even semi-desert in the belt of prevailing westerly winds on the rainshadow side of the Andes. In western Argentina, conditions are like those of western Texas. In Patagonia, conditions are colder, and it is progressively drier from west to east.

The cold Peru current off the Pacific coast chills the lower layer of air, creating a belt of stable air over an extremely arid desert (the Atacama) that stretches the length of Peru and over the northern third of Chile. South of the Atacama, the climate shades quickly through steppe conditions to a dry-summer subtropical (Mediterranean) climate in central Chile. Winters there are mild and summers hot and rainless. Poleward of 40° south latitude, where Chile becomes a maze of islands and peninsulas, the climate becomes rainy year-round, virtually always cloudy and cool.

Two other dry climatic areas are noteworthy. In the northeast corner of Brazil, inland from the coasts, there is a highly drought-prone area that covers much of the **sertão**, or backland. Rains are capricious, in some years causing intense flooding, in other years totally absent. Similar conditions prevail along the narrow Caribbean coast of Venezuela.

Vegetation

Climatic and vegetational patterns coincide strongly. In Mexico's northwest, the desert landscape exhibits a fairly dense covering. Only in the salt deserts south of the great bend of the Rio Grande (Rio Bravo) country is vegetation truly sparse. Within the Mexican plateau, the vegetational cover is a combination of scrub and steppe grasses. Fringing mountain ranges support dense forests, pine at higher altitudes and in the north, and thorn forests in the south. South of the Tropic of Cancer on the wetter Atlantic side and south of Acapulco on the drier Pacific side, the wetter conditions result in a transition to true tropical rainforest. Drier soil conditions in the northern Yucatán support only scrub.

Within the mountain spine of Central America, rainshadow areas are characterized by a series of scattered tropical grasslands, whereas trees dominate on higher slopes. On the islands, higher peaks and windward-facing slopes support rainforest, while parklike savanna grasslands, with scattered trees, dominate in drier areas. Heavily cultivated, the small sugar islands have little "natural" vegetation remaining.

The **selva**, or rainforest, has an open floor, far less of an obstacle to movement in Latin America than its counterparts in Asia and Africa. The forest grows in three stories whose combined canopies almost exclude sunlight from the floor. Cacao, Brazil nut, and the rubber tree are native and grow in mixed stands, together with valuable tropical hardwoods. In Andean and Central American areas, the tropical rainforest contains commercially valuable species: palms that have uses as fiber, cinchona (which yields quinine for treating malaria), and coca shrubs (whose leaves yield cocaine).

Savanna grasslands have a limited, but valuable, sprinkling of trees. Drought-tolerant trees there are

sources of tanbark, waxes, fibers, resins, dyes, and various chemical raw materials. The *caatinga*, or dry forest, of Northeast Brazil is unique. Carnauba palm (yielding commercial waxes) is its most valuable species, but native tree cottons, rubber-yielding plants, and fleshy cacti are all intermixed in a tangle of shrubs and thorny trees.

The temperate grasslands of South America have all but disappeared due to the extension of cropland. Some, offering pasture to sheep, are still preserved in Patagonia, where the cover is more properly termed bushland.

The temperate forests' species are evergreen broadleaf trees reflecting milder climates. Hardwoods dominate on the Pacific side. Subtropical and temperate forests of the Atlantic yield yerba maté, a tealike drink, and the valuable Paraná pine.

The Atacama Desert, the world's driest, has virtually no vegetation, while desert shrub occupies much of southwestern Argentina.

Soils

Much of tropical Latin America's soil is heavily leached red clay with high levels of iron and aluminum and low amounts of humus. Soils are generally thick but of low fertility, and become exhausted after only a crop or two. Savanna soils are often underlain by hardpans which impede drainage.

Throughout the wet topics, the most fertile soils are generally associated with alluvium or volcanic parent materials. Rich volcanic soils of the Mexican Plateau, Central American Highlands, and parts of the Andes, together with the terra roxa of Brazil and Paraguay, are among Latin America's richest. Alluvial deposits in Colombia and the valleys of eastern Brazil are quite fertile, but those of the Amazon and the Chaco (upper Paraguay) are not universally so. The fertility of alluvium depends on the material from which it is derived.

In large parts of Latin America, soil fertility could prove to be a limitation to agricultural expansion. Some areas of tropical soils are infertile for shallow-rooted field crops yet can produce highly satisfactory yields with tree crops. More important than natural soil patterns may be soil depletion and erosion. Traditional agriculture in many parts of Latin America has proven damaging to soils. Large areas of tropical and subtropical soils can give decent yields with care, rotation, periodic fallowing, and large inputs of fertilizer.

Drainage

There are only three large rivers in all of Middle America: the Balsa, the Rio Bravo del Norte (known to North Americans as the Rio Grande), and the Colorado. Short, swift streams along the Pacific coast of Mexico are used to irrigate areas of lucrative, intensive oasis farming. Much of drier northern Mexico, including the basin of Mexico City, has interior drainage. Mexico City itself occupies an old lake bed. Enormous amounts of water necessary for its rapidly expanding population and industry are drawn from groundwater. An unfortunate side effect is the gradual sinking of the central city.

The seasonally dry, coral (limestone) islands of the Caribbean sometimes lack surface water entirely. A number of islands in the Bahamas remain unsettled because there is no water.

However difficult and costly, a full-scale development of internal water routes is possible in South America. Three great rivers form potential highways into the interior (see Figure 11–6b). Navigation and power are being developed on Venezuela's Orinoco, with its natural 40-foot channel over half its course. The La Plata-Paraná-Paraguay system has achieved a degree of economic importance; a 24-foot channel is maintained to the Argentinian industrial centers of Rosario and Santa Fe. Commercial river vessels ply the waters as far north as Asunción in Paraguay. A joint Brazilian-Paraguayan project has developed part of the tremendous power resource of this river near Iguassú Falls.

The Amazon is the world's largest and most readily navigable stream. Ocean vessels can reach Manaus; deep-draft river steamers can navigate all the way to Iquitos in Peru. The Amazon carries a larger volume of water than any other river.

The Increasing Role of the Sea

The traditional importance of the sea stemmed from colonial-era exports to Europe and the role of the Spanish fleet in defending its colonial territories. Even now, seaborne trade with North America and the connection of both U.S. coasts through the Panama Canal give dominance to the maritime transportation function. Recently, the sea has become an important food source, and recent offshore oil discoveries give it new importance in Mexico, Argentina, and Brazil. Exploration for undersea minerals and increased fishing have induced the region's states to extend territorial borders seaward. The United States (and now, the USSR) maintains military bases in the Caribbean—a continuation of the old defense tradition by successor powers.

Hydroelectric Project at Itaipú. The hydroelectric potential of South America is enormous. Rivers fed by equatorial rains carry huge volumes of water in constant flows. Where these rivers plunge down mountain valleys or fall precipitously over the edge of plateaus, conditions for generating electricity can be ideal. At Itaipú on the Paraná River, Brazil has constructed one of the world's largest hydro power projects. Production is shared with Paraguay since the project is at the border between the two countries.

THE STRUCTURE OF SUBREGIONS

Sheer size and vast economic differences necessitate regional subdivision in Latin America (see Figure 11–3). Groups of similar states can be combined, but Mexico and Brazil are too diverse to be treated as units; they are further subdivided. The Caribbean Islands, grouped into either the larger or smaller categories, share physical and economic characteristics and form a nearly ideal subregion. Central America's states form an equally viable subregion. South America is divided into four subregions: the Caribbean Littoral, the Indo-Andean Republics and Paraguay, the Southern Tier (Chile, Argentina, and Uruguay), and Brazil.

MEXICO, CENTRAL AMERICA, AND THE CARIBBEAN: A MIDDLE AMERICAN OVERVIEW

Middle America is as active politically as it is tectonically. Currently, the most unstable regimes in Middle America are those with low per capita incomes. Nicaragua, El Salvador, and Honduras have the lowest incomes on the Middle American mainland. Honduras, peripherally involved with the conflicts in Nicaragua and El Salvador, is the poorest of the three. Haiti is one of the poorest nations in all Latin America. Grenada, the site of the latest American intervention, has a per capita income of only $500.

Mexico's position (adjacent to rich and expanding U.S. markets), its mineral riches, and its ability to expand both industry and agriculture combine to ensure it the greatest economic potential. Innovative international service offerings by Panama and the Bahamas have generated improved economies in those states but, obviously, not every state can do the same.

Land reform is most likely, politically, where cultivable land remains unused; dissolution of remaining large estates will be difficult to achieve. Mineral development depends on unstable world mineral markets and the vagaries of nature's original distributions. The diversification of agriculture is already well under way in most of Middle America, and industrialization has passed the embryonic stage in a few states. Tourism, regionwide, is well developed, but fickle tourist tastes and a high level of competition may mean that further investment is no guaran-

tee of prosperity. Proximity to North America offers distinct economic advantages, if some potential political disadvantages.

Mexico: The "Superstate" of Middle America

Mexico is the largest state in Middle America, accounting for almost three quarters of the subregion's area and over half its population. It was a dazzlingly wealthy Spanish colony, so rich that earlier Spanish colonies in the islands were partly depopulated and even ignored after Cortes's discoveries. Contemporary Mexico is only two thirds of the state that existed before its collision with a strongly expansionist United States in the 1840s. Current population is over 80 million, and the rate of natural increase (2.5 percent) is still relatively high. In some ways, there is an "expansion" of Mexico today, as Mexicans surge across the border (legally and illegally), often into the very portions of the United States that were once a part of Mexico.

In an important sense, there are "two Mexico's," that of the rich and that of the poor. Although this disparity persists, Mexico's economy has grown at a yearly rate of 5 to 10 percent or more for most of the last 50 years, and a middle class is rapidly emerging. Mexico has achieved a 100 percent increase in industrial output per capita in a single generation, a remarkable feat in the face of rapid population growth. The cumulative effect of this healthy growth rate places Mexico among the few so-called developing nations that are unquestionably developing.

Stability and investment are the keys to Mexican success. Mexico has been governed by one political party, backed by popular consensus (at least until recently), since 1928. In the last 40 years, the percentage of the work force employed in manufacturing has doubled as the government builds new industries itself and fosters private investment as well. Mexicans save more than any other Latin Americans, financing a significant share of industrial growth and general expansion in the process. Foreign manufacturing investment has more than doubled.

In 1910, before reform, 3 percent of Mexicans held 90 percent of the farmland. Over half of all the arable land in Mexico has been redistributed since 1917. New agricultural colonies have been developed in both the dry North and the wet South as

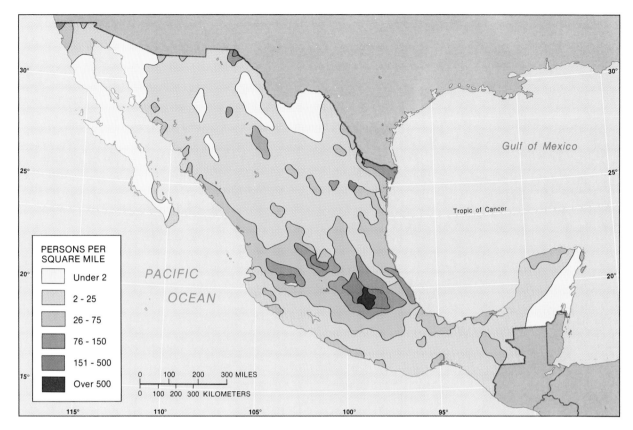

FIGURE 11–8
Distribution of Population in Mexico. With 80 million people, Mexico
concentrates most of the Middle American population. Vast empty and lightly settled
areas of the North will likely remain that way because water is in short supply. The
dense population of the Central Basins represents the concentration of major urban
centers. Mexico City, the world's largest urban place, has at least 18 million
inhabitants. Veracruz, the country's chief port, and Guadalajara, its second largest
city, are also located in this subregion.

irrigation, flood control, and drainage extend frontiers. Despite land reform, tens of millions of rural
émigrés have crowded into the nation's cities.

By 1980, over 70 percent of all Mexicans lived in
cities (Figure 11–8). Many, however, dwell in makeshift houses without sewerage or water in hillside
barrios. Hidden from view by burgeoning construction, they await their legacy from Mexican growth.

As impressive as all the gains are, they are still
unevenly distributed. Industrial, urban, rising middle-class Mexico differs sharply from the poorly educated and underfed Mexico that remains in rural
areas or clings to life in squalid slums. A million

farmers and their families still own no land. A similar
number hold only enough land for the barest survival. In cities, it may take a full generation for migrants to attain stable employment and a place in
public housing.

Mexico has almost achieved self-sufficiency in basic foodstuffs, and exports of commercial crops, produce, and meat have increased. The discovery of vast
oil and gas reserves initiated the most recent boom;
declining oil prices have now triggered rampant inflation, temporarily dampening prosperity. Huge foreign debts, incurred to develop the national economy, hamper real growth as income is diverted to

MAINLAND VERSUS RIMLAND

The Aztec civilization that developed in the steppes of highland Mexico was based on irrigated agriculture. It incorporated areas much larger than most European states of the day. The Mayan culture arose in the tropical rainforest and extended its control to tropical grasslands, ultimately controlling much of southern Mexico and northern Central America. Both societies developed sophisticated architecture, large urban centers, and highly specialized agriculture. They organized religions and developed far-flung transportation networks. Each produced refined mathematicians and astronomers.

Mainland Middle America, then, was densely populated and well developed prior to European intrusion. It was one of the world's great agricultural hearths (sites of sedentary agricultural and plant domestication). This situation sharply differentiated mainland Middle America from the islands of the region. The islands had a more primitive culture, still based largely on hunting, fishing, and gathering and concerned only with simple subsistence.

Of all their domains, the Spaniards preferred those of Mexico and Peru, the sources of precious metals. Yet the dense populations of those areas precluded heavy Spanish settlement, and their physical environments were not conducive to the crop that Europeans saw as most lucrative—sugar.

Following the example of the Portuguese in Northeast Brazil, an area of broadly similar climate, the Spanish (and later, other European states) proceeded to clear the Caribbean Islands and to establish Portuguese-style sugar plantations. Disease, hard labor, and a policy of exterminating resistors led to the decimation of native populations. Over time, the organization of the islands for plantation production and the replacement of Indian labor with slaves from Africa sharply differentiated the two areas. John Augelli, a specialist in Latin American geography, noted the distinction, designating them as "mainland" and "rimland" (Figure A). The nineteenth century extension of the plantation economy to the Atlantic coast of the mainland (complete with African-American labor) has effectively extended the rimland to that area. The Indian-Mestizo-Hispanic mainland remains the successor of Indian cultures. It is devoted to highland agriculture and ranching. The rimland economy is based largely on sugar and bananas. African population and cultural influences, combined with a variety of languages and cultures derived from historical control by various European states, blend into a unique Creole subculture.

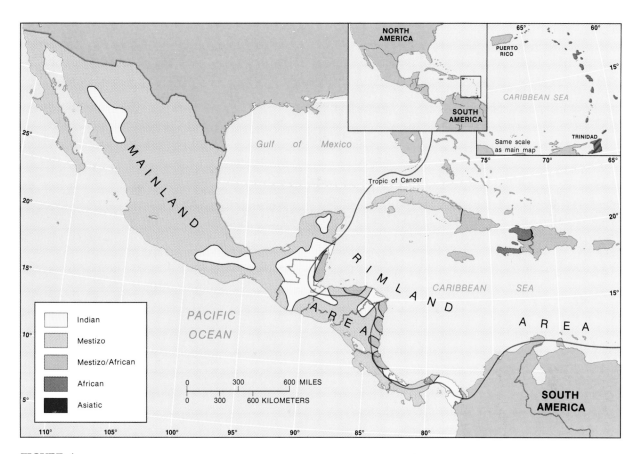

FIGURE A

Mainland and Rimland in Middle America. There is a sharp contrast
between the cultures of the mainland and those of the islands and Caribbean coast.
The mainland area was occupied by a relatively dense population at the time of the
Spanish conquest, while the islands and Caribbean coast were sparsely populated.
The mainland area was devoted largely to mining, grain production, and cattle
grazing. Labor was performed by Indians. The latter area was dominated by sugar
production, and labor was performed by slaves imported from Africa. Spain retained
control of the mainland until independence. The result is a mainland dominated by
Spanish and Indian culture, including the Spanish language. The population is
dominantly mestizo and Indian. The rimland came to be dominated by other
European powers. A variety of languages are spoken, blacks dominate, and a Creole
culture with strong African inputs is characteristic.

pay interest on the debt. Still, accomplishments outweigh problems.

Northern Mexico: South of the Border. The rapidly developing North is as different from the core of Mexico as it is from the neighboring United States. This amalgam of desert and semidesert, mountain and basin, oasis and grassland, includes well over half the country but only 30 percent of the Mexican population. Large areas are sparsely populated. The land between oases is poor grazing country, at times desolate. Large border cities, such as Juarez, Tijuana, and Mexicali, concentrate much of its population. The Baja Peninsula remains virtually empty.

Its central portion is a land of mines and ranches. Its cities—Durango, Chihuahua, and Torreón—bear names more familiar to Americans than many of Mexico's larger cities to the south. This portion of the North is highly mineralized and contains many of Mexico's most important metal ore and fuel deposits, including most of its coal, gas, iron, and copper. Mining has become proportionately less important to the regional economy as manufacturing and irrigated agriculture expand at more rapid rates.

It is the border itself that gives the subregion its lively economy and special character. Over 1,900 miles in length, it is often loosely patrolled and poorly marked. In places, it crosses sandy wastes and rocky hills in straight lines; in others, it follows river courses long abandoned by the shifting bed of the Rio Grande. Sizable towns mark its official crossing points, but holes in fences attest to heavy unofficial traffic. Often viewed as a precipitous divide between the wealth of the First World and the poverty of the Third, the border today is really a transition zone between Mexican and U.S. cultures and economies.

Border cities generally exist in pairs; most often, the Mexican one is larger in size. Contrasts are rife. Prosperous San Diego sprawls its spacious dwellings over carefully landscaped acreage, while Tijuana's cramped quarters more often huddle on patches of foot-trampled barren earth. Yet such stark comparisons can be overdrawn. Tijuana, too, boasts a bustling downtown, and it also has its mansion districts. Something of the character of both, some would say, is revealed in their respective names. The wry Tijuana (loosely, "Aunt Jane") suggests that city's rollicking nightlife, whereas the American neighbor, named after a saint, is certainly far more staid. Yet both are booming centers of business and industry, in some measure relying on each other and benefitting from their location near the border.

While day trip tourism, shopping, and entertainment are the functions for which these border cities were originally noted, the cities all have become primarily large manufacturing centers. These are the centers of Mexico's "export only" and "in transit" industries. In the former, foreign investors are allowed to produce semiprocessed goods, using Mexican raw materials, for final assembly or finishing in the United States ("in-bond plants") or to assemble finished products in so-called free-trade zones. In either case, the industry uses cheaper Mexican labor in labor-intensive functions. Products enter the United States duty free or with a small value-added tax. In the latter case, goods are processed in transit on their way from one portion of the United States to another in interim facilities in Mexico. Urban growth here now rivals that of the capital. Juarez and Tijuana both have a million inhabitants.

Manufacturing is not limited to border towns. One of the earliest industrial ventures of modern Mexico was the steel complex at Monterrey. Monterrey has grown to become Mexico's second largest industrial center, diversifying its original base with alloy smelting, engineering, chemicals, glass, and clay products—the normal "partner industries" of a coal and steel economy.

The general profusion of metal ore mines in the subregion led to major smelting complexes at a dozen towns. New phases of mineral industry are based on the large gas fields near Reynosa in the Rio Grande Valley (Figure 11–9); they supply both energy and raw materials for the expanding chemical sector.

Agriculture is concentrated in the lower Rio Grande Valley and a string of oases along the Gulf of California. The Laguna district, near Torreón, was once Mexico's most prosperous farming area and home to its most successful ejido projects. (The **ejido** is a communal farm settlement developed to farm cooperatively land that was taken away from estate landlords and redistributed to peasants. Similar to cooperatives, its purpose was to return to pre-Spanish communal village land tenure systems.)

The North's future is exceptionally bright. The district is well served with rail and highway transporta-

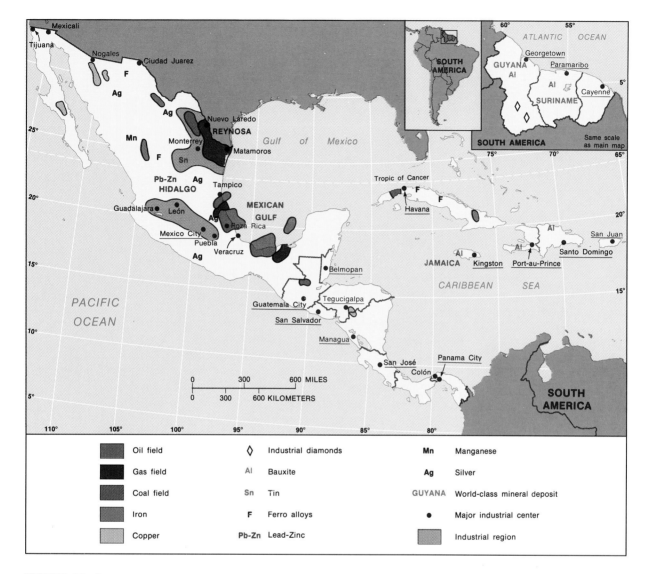

FIGURE 11–9
Minerals in the Economy of Middle America. Mexico has long been the chief
mineral-producing area of Middle America. Gold, silver, and copper were the chief
sources of that country's wealth. The gas fields of northern Mexico and the oil fields
of its Gulf Coast are the country's new sources of mineral wealth. Cuba has
important deposits of iron and nickel. Bauxite deposits (aluminum ore) are
widespread throughout the larger Caribbean islands and the countries collectively
called the Guianas.

THE BORDER CITIES OF MEXICO

By definition, borders separate different jurisdictions, governed by different laws. By extension, borders separate different cultures, with different values and economies. Internally, Americans can cross state borders in pursuit of cigarettes, liquor, or even major appliances at cheaper prices, taking advantage of differing tax laws. International borders are little different in this respect, if more complex because of currency exchange rates. Borders act as a divide, but they also encourage trade because of the essentially different opportunities offered on either side.

The U.S.-Mexican border does a brisk business in day trip trade because of its relative openness and the strong economic differential between the two countries. In most cases, both paired U.S. and Mexican border cities exhibit a Hispanic majority. The cultural attributes of landscape and lifestyle observed in both towns contain elements of both cultures. Bilingualism is the norm, regardless of the ethnic heritage of city dwellers.

Today's Mexican border town offers considerable economic opportunity. As in the case of traditional foreign migration to the United States, there is an "immigrant's ladder"—a route to success. A typical female job progression would be (manufacturer's discount) coupon sorter, secretary, word processor, then computer operator. They are all labor-intensive tasks that can avail themselves of cheaper Mexican labor costs without violating tariff or goods quota legislation enacted by either country. Words, signals, electronic impulses, or symbolic paper—rather than goods—cross the border readily. A typical male progression would be truck unloader, forklift operator, taxi driver, then mechanic, indicative of the massive transportation function associated with an economically active border.

There are hundreds of millions of legal border crossings yearly for a variety of reasons, including shopping, education, recreation, and work. Mexicans daily cross to work in American factories, retaining residences in Mexico where living costs are cheaper. "Twin cities" often have twin industrial plants, shuttling subassemblies between them for ultimate sale in both jurisdictions, employing both Mexicans and U.S. citizens. These plants attract skilled labor but also offer training for the unskilled. A devalued peso draws floods of American shoppers and creates retail and service employment opportunities on a grand scale.

Residents often have friends and family on both sides of the border. Generally, the standard of living in a Mexican border town is higher than the Mexican national average, and much higher than that characteristic of the Mexican countryside. Such cities offer opportunity and excitement to both Americans and Mexicans. Emporiums for the goods of both countries, they are New World bazaars, bustling with traffic and brimming with industry and entrepreneurship.

Tijuana. *The U.S.-Mexican border cities are now the fastest growing urban centers in the country. The maquiliadoras, foreign-owned plants that employ Mexican labor, are one of the causes of this extremely rapid growth. Cheaper labor costs are the attraction to investors. The other significant factor is tourism. Tijuana, with well over a million inhabitants, is the largest of these border cities.*

tion facilities; there is a growing regional pool of skilled workers. Adjacent to some of the most rapidly expanding market areas in the United States, it is progressive, modern, mechanized, and growing.

Mexico's Central Basins. The most densely populated and intensely used portion of Mexico, the Central Basins support half of the country's population (see Figure 11–3a). The dominant area of Mexico, even in precolonial times, it is a district characterized by rampant growth and serious problems. The mineral deposits that first stimulated growth here are not exhausted, though gold and silver production is declining. The agricultural base, although rich, is in some senses limited by rural overpopulation, in others by the coolness of the highland climate and the paucity of water.

The ejido system and the government program of land reform have been most thoroughly carried out here, fostering political stability. Diversification of agriculture is in progress, even if the pace lags behind regional needs. Irrigation expands, but the region's growing cities compete for a meager water resource. Greatest growth and development are evident in the industrial sector. Central Mexico is served by a well-developed network of highways and railroads, and it is Mexico's largest single market.

Mexico City, the national capital, is perhaps now the largest city in the world. Conservative estimates place its population at over 18 million. As in other great metropolitan centers, a large part of Mexico City's industry produces consumer goods for the city market itself (alcoholic beverages, clothing, and processed food are typical). Production varies in sophistication from a step above native crafts through assembly of automobiles and household appliances to sophisticated production such as machine tools and medical devices.

The rapid growth of Mexico City has multiplied the city's problems as well as its living standard. Air pollution reaches disastrously high levels as congestion approaches a perpetual traffic jam. High costs of

Mexico City. *The Plaza de La Reforma is the heart of Mexico City, and Mexico City is the very heart of the country. The rapid growth of Mexico City has multiplied the city's problems. Air pollution reaches disastrously high levels, and congestion approaches a near-perpetual traffic jam despite the building of a large and efficient subway system to relieve traffic pressures. The site of the city is prone to earthquakes. Withdrawal of groundwater to meet urban and industrial demands has caused parts of the city to sink. Recent legislation designed to discourage development and investment in Mexico City has had only limited success.*

living result in high wage structures and inflation, and slums engulf the city on several sides. The Mexican government has placed restrictions on further expansion. No new plants are allowed to build in the city itself, and plant expansions are strictly controlled. New satellites are developing, but not sufficiently far away to alleviate the problems of congestion and pollution.

The largest nearby central city is Puebla, now with a population of over 800,000. Not just a local market center for its agricultural hinterland, it is a major textile center. Puebla is located between the Gulf Coast oil fields and the capital city on the main routes to the port of Veracruz.

Mexico's second largest city, with a population of over 3 million, Guadalajara draws its importance from its superb agricultural hinterland, a relatively level land covered with rich volcanic soil. Traditional subsistence crops in small fields mingle with commercial production such as citrus and agave, the raw material for pulque and tequila. More traditional than the capital, it retains more of older Mexican culture despite its size and recent growth.

The Mexican South. The Mexican South is ethnically different from the rest of Mexico; it is peopled largely by pure-blooded Indians. Until recently, it was peripheral to the economy. Still overwhelmingly rural, it contains large tropical crop plantations, though most of the populace remain small subsistence farmers. It has recently expanded its economy on the basis of oil and gas and diversification of farming. In some ways, it is a frontier, particularly in rainforest areas where government-sponsored colonists from the overcrowded Central Basins clear land for new farms.

The most developed portion of the region is the Gulf Coast, with Mexico's most important port, Veracruz, serving as entrepôt for both the capital and the oil fields. There are new finds of oil and gas both on- and offshore (see Figure 11–9a). Oil and gas deposits continue in an almost unbroken arc southward from the U.S. border to the Yucatán. There are also reserves in Chiapas on the Pacific side. The government has invested in multipurpose water projects throughout the South to provide power, control floods, increase irrigation, and eliminate malaria. They are the basis for massive colonization and development schemes planned for this heretofore

sparsely populated territory. Petrochemicals are the leading industry, providing an expanded industrial base, but sullying the air and water with effluent and smoke.

The region includes Acapulco, Mexico's premier tourist center, a resort of world-scale importance. Oaxaca, the penultimate Indian cultural center, specializes in jewelry and native crafts. It is a favorite center for domestic tourism. Atlantic coast tourist centers, especially Cancun and Cozumel, tap the tourist market potential of the eastern United States. These newer resorts blend modern amenities with traditional cuisine, seashore recreation, and nearby Mayan archaeological sites, in an extremely successful commercial venture.

Central America

The seven countries of Central America, south of Mexico and north of Colombia, exhibit a degree of cultural diversity worthy of a larger region. Five of these countries, four of which are members of the Central American Common Market (CACM), regard themselves as truly Central American. Belize (formerly, British Honduras), dominantly black and English speaking, is singularly different. Panama did not become independent of Colombia until 1903, following Colombia's rejection of a treaty enabling the United States to build a canal. The "five" were all part of a Federal Republic of Central America from 1823 until 1838. Their ultimate fragmentation reflects historically separate settlement cores within the rugged mountains (Figure 11–10). The historic federation remains a political ideal of some area leaders.

Unlike Mexico, this has not been an area noted for political stability. Salvador is torn by civil strife, and Nicaragua has adopted a Marxist developmental program that is viewed as a threat by many area governments. Panama has refused to associate itself with the five in any of their sporadic attempts at confederation, or in their common market (Table 11–2). Belize, possessing an ethnic flavor more Caribbean than Central American, is claimed by Guatemala. There is little commonality of interest and little in the way of economic complementarity among the countries of Central America.

Coffee is now the most important crop in most of the region's economies, providing 30 to 70 percent of each country's national export earnings among

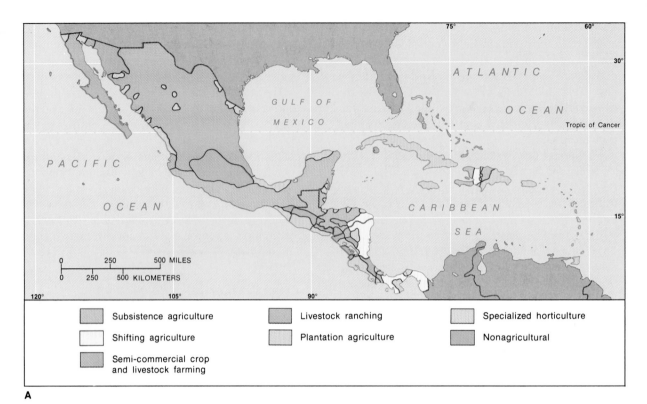

Subsistence agriculture	Livestock ranching	Specialized horticulture
Shifting agriculture	Plantation agriculture	Nonagricultural
Semi-commercial crop and livestock farming		

A

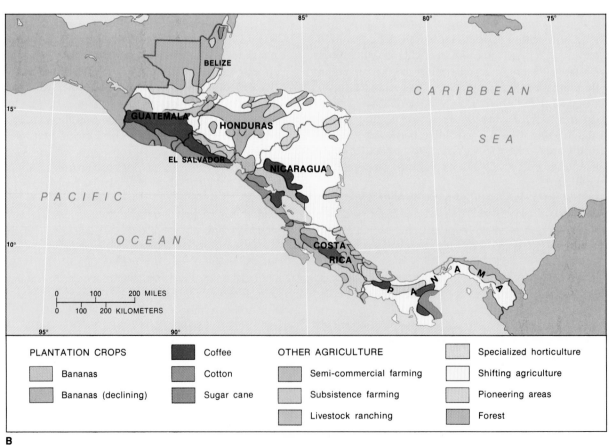

PLANTATION CROPS
- Bananas
- Bananas (declining)
- Coffee
- Cotton
- Sugar cane

OTHER AGRICULTURE
- Semi-commercial farming
- Subsistence farming
- Livestock ranching
- Specialized horticulture
- Shifting agriculture
- Pioneering areas
- Forest

B

TABLE 11-2
Comparative data for Central America, 1987.

Country	Population (millions)	Pop. Growth Rate (per thousand)	Pop. Density (per sq. mile)	Per Capita Income ($)	Major Ethnic Groups	Language	Economic Organization
Belize	0.178	2.7	19	1,140	African	English	Caricom
Costa Rica	2.84	2.7	140	2,250	Spanish, Mestizo	Spanish	CACM
El Salvador	5.30	2.6	642	810	Mestizo	Spanish	CACM
Guatemala	8.57	3.0	205	1,720	Indian	Spanish	CACM
Honduras	4.70	3.1	109	620	Mestizo, Indian, African	Spanish	
Nicaragua	3.52	3.4	65	800	Mestizo, African	Spanish	CACM
Panama	2.31	2.0	71	2,070	Mestizo, African, Indian	Spanish, English	

SOURCES: Inter-American Bank; Population Reference Bureau, 1988.
CACM—Central American Common Market; Caricom—Caribbean Community.

CACM members. Bananas, once the premier crop, have become proportionately less important as other export crops have developed.

Caribbean coastal banana plantations prospered from 1900 to about 1930. Since then, production has dropped rapidly because of disease, causing abandonment of many thousands of acres. The Central American banana industry then moved across the mountains to the Pacific coastal lowlands. Ultimately, disease spread to Pacific coast plantations. Experimentation with disease-resistant varieties of bananas has benefitted both flanks of the mountain spine, but bananas have never recovered their former regional supremacy (Figure 11–10b). Foreign-owned fruit companies, once the dominant planters, now concentrate on marketing, providing technical assistance, and financing local growers, often their former tenants.

← **FIGURE 11-10**

(a) *Agriculture in Middle America.* Cattle grazing dominates in the Mexican North. Wetter tropical areas such as the coastal lowlands of Central America and southern Mexico are used dominantly for shifting agriculture. Subsistence farming dominates in areas with a high incidence of unassimilated Indian population and in areas characterized by isolation or rough topography. Plantation agriculture dominates in most of the Caribbean islands, and in parts of coastal Mexico.

(b) *Agriculture in Central America.* Historically, bananas were the most important commercial crop in Central America; however, cotton and sugarcane have increased in importance, and coffee has become the dominant crop.

Individual Economies: The Question of Diversification or Cooperation. Guatemala is the region's most populous state (see Table 11–2). The population is young, and it is projected to double in 20 years. More than half the population is Indian, and many are landless peasants. The economic plight of Indians is a social problem with explosive potential. Pioneering farm areas in the pest-ridden eastern provinces, flight to Mexico, or migration to the capital are the alternatives to rural poverty. The rapid growth of Guatemala City (now with over 1.5 million inhabitants) is closely related to the exodus of landless laborers from the countryside.

MEXICAN OIL

To understand the Mexicans' intense nationalism concerning their oil resources, it is necessary to consider the history of the oil industry there. International (primarily American) oil companies initially found and developed Mexican oil. They were later accused of "skimming" the oil resource, pumping wells for maximum immediate production rather than managing them for long-term sustained yields. Companies were reluctant to hire Mexicans above the laborer level, and no Mexican had executive status before nationalization. Oil companies were accused of deliberately cutting production to threaten a government that had come to rely heavily on oil revenues (Table A).

Finally, in 1938, the oil industry of Mexico was nationalized. The response of the international oil companies to nationalization was to press for immediate payment in cash for nationalized properties, while claiming ownership of all oil still in the ground. Tankers delivering Mexican oil to foreign ports were served with legal papers claiming that their cargo had been illegally seized from its rightful owners. Chemicals and equipment vital to the oil industry could not be sold to Mexico without risking a boycott from the most important customers—the international oil corporations.

The legacy of this bitter struggle has been a nationalized Mexican oil industry (Petroleos Mexicanos, or Pemex). Mexicans are determined to be hard bargainers in dealing with the United States and Europe in oil and gas sales, because this means dealing with many of the same companies that caused past problems. Mexico's oil and natural gas reserves are in the same class as those of the USSR and Saudi Arabia, and the country's pace of economic development may rest on an increasing output of oil and gas.

Guatemala is the most industrialized of the area's economies. It has been a major beneficiary of CACM investment, largely because of its sizable population. The jungles of its north, the Petén, hide the remnants of ancient Mayan cities. Once evidently prosperous, the region was abandoned in pre-Columbian times. Planned national development hinges on recolonization of the area, and the development of tourism based on the ruins of its Mayan past.

Neighboring Honduras is Central America's poorest country. Most of the Honduran coast fronts on the Caribbean, orienting its trade, interests, and outlook toward the eastern United States. Almost 90 percent

mestizo, Honduras is culturally quite uniform. If any Central American country truly merited the designation "banana republic," it was Honduras in the 1920s, when a third of the world's bananas were produced there. When disease drastically cut production, the United Fruit Company (an American corporation) gave abandoned banana lands to farmers.

Honduras fought a brief, bloody war with El Salvador in 1969. Salvadorans, whose nation's population density is more than five times the regional average (see Table 11–2) had been illegally entering Honduras and squatting on unused lands. Honduras's use as a staging center for Contra activity against

TABLE A

Petroleum production in Mexico.

Year	Quantity (million barrels)	Comments
1910	3.6	
1917	55	
1921	193	Mexico produced nearly 1/4 of world supply
1932	33	
	—	Oil industry nationalized 1938
1950	67	
1963	125	
	—	Contracts with foreign oil companies ended 1969
1970	177	
1975	294	
1979	483	
1983	931	Mexico now the 2d largest source of petroleum imports into United States
1984	1,176	
1987	1,170	

SOURCES: U.S. Department of State, *Background Notes—Mexico* (Washington, DC: Author, 1982); *United Nations Statistical Yearbook,* 1988.

the Sandinista regime in Nicaragua has brought it large amounts of military and economic aid—both direct and indirect. American industrial investors have been encouraged to locate factories in Honduras, and the increased economic activity has produced some benefits. The country's dependence on the United States has caused its neighbors to label it "America's largest aircraft carrier."

El Salvador, the second smallest state in Central America, lacks a two-ocean frontage. Its high population density is more typical of the Caribbean Islands than Central America. El Salvador has had to cope with being on the "wrong side" of Central America in terms of accessibility to its major markets (the eastern United States and Europe). In Central America, El Salvador alone lacks unsettled, cultivable lands to act as a safety valve for its rapidly expanding population (see Figure 11–10b). Half of Salvadoran farms are less than five acres, and the gap between rich and poor has fueled guerrilla-style revolution. Coffee and cotton, which provide over half of the export income, take the best lands; food production is grossly inadequate. Soils are heavily eroded as the desperate need for more food production has pushed cultivation onto steep slopes exposed to heavy seasonal rains.

As with other Central American states, the wide fluctuations in world coffee markets have had serious impacts on the balance of trade, unemployment, and government revenues. All the region's states are in the uncomfortable position of relying heavily on coffee exports to provide tax revenue and investment funds with which to diversify economies and ultimately decrease dependence on coffee.

Belize is culturally and economically atypical of mainland Middle America. It has by far the lowest population density of the seven republics. The offi-cial language is English, and many of its people are of African or part-African stock. Nearly a third live in Belize City, the former capital and main seaport. Europeans were first attracted to Belize by the commercial possibilities of mahogany and tropical cedar. English-speaking blacks from the Caribbean islands were imported by the British as forestry labor.

Belize has decided to move its capital from the hurricane-damaged colonial center at Belize City to a new, more central location at Belmopan (see Figure 11–3a). With the best tropical hardwoods virtually

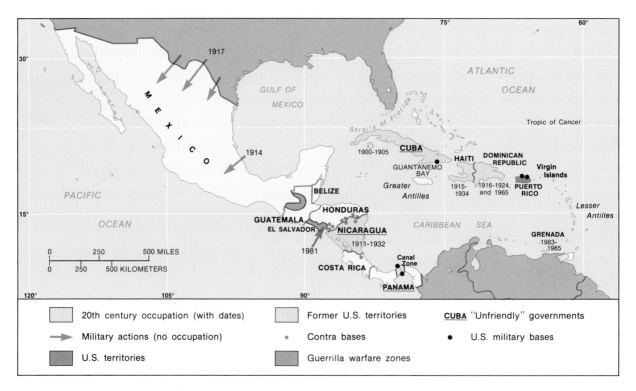

FIGURE 11–11

U.S. Military Intervention in Middle America. The respective levels of development in the Americas diverged rapidly in the postindependence era. A young, expansionist United States quickly came into conflict with Mexico over territorial control of Texas, California, and the lands in between. American troops repeatedly invaded and occupied parts of Middle America during the first four decades of the twentieth century. The United States actively developed military bases in the area to protect the crucial Panama Canal. Military intervention has continued, but with less frequency. The canal ownership has been transferred to Panama but relations between the two countries continue to be strained. Anti-American regimes are in control in Nicaragua and Cuba. The past American record of intervention still clouds regional relations.

gone, forestry emphasis has shifted to pine lumber for the wood-short Caribbean islands. Agriculture is poorly developed, and towns rely on imported food. The rich fisheries offshore have developed rapidly, though shellfisheries have already expanded beyond the natural replenishment limits.

Nicaragua has the unfortunate distinction of having been repeatedly occupied militarily and administered directly by the United States (the last time, from 1926 to 1933) (Figure 11–11). Early American interest in Nicaragua was related to the great Nicaraguan Rift, a trenchlike valley across the highlands of the narrowing isthmus that provided the second best routeway across Central America (see Figure 11–6a). A Nicaraguan canal has been discussed sporadically for more than a century. Unfortunately, the same geologic forces that created the rift provide frequent earthquakes and volcanic eruptions.

The extremely uneven distribution of land and wealth provided considerable popular support for the Sandinista guerrilla war against a corrupt dictatorship. The government gradually lost all support, and the Marxist Sandinista regime came to power in 1979. Relations with the United States have been strained ever since. Reforms under the Sandinista regime have been slow.

Nicaragua is one of the least industrialized Central American states, with its industry concentrated in the capital. Most Soviet bloc investment has been in ports and infrastructure, and most aid has come in the form of armaments. The country remains poorly developed; its present economy is neither healthy nor expanding.

Costa Rica, in sharp contrast to Nicaragua, has had a history of peace and prosperity. About 80 percent of its people claim to be unmixed descendants of Spanish colonists, though incidence of mestizos is certainly higher. Because it was located outside the dominant Meso-American Indian culture area, its scattered Indians were early absorbed or disappeared. The typical colonial emphasis on huge estates owned by Spanish and worked by Indians thus did not materialize here. Instead, the country has a tradition of small holdings owned and operated by individual farmers. Costa Rica has the most evenly distributed national income, one of the highest per capita GDPs, a high literacy rate, and the strongest tradition of democratically elected governments in all of Central America. Prosperity has been main-

Banking in Panama. *Unlike most of its Central American neighbors, Panama depends more on services than on agriculture for most of its national income. It also does a large business in the registry of ships. Banking and financial functions are rapidly becoming its chief source of income. Panama uses the U.S. dollar as official currency. This, in combination with favorable banking laws, numbered accounts, excellent communications facilities, and high interest rates, has resulted in a multiplication of the number of banks licensed and located there.*

tained by pioneering new farming areas on the Caribbean and Pacific flanks of the central core (see Figure 11–10b).

Panama holds itself apart from Central American politics and economic cooperation schemes. Until 1903, it was a province of Colombia. To ensure Panamanian independence (and control of the canal site), the U.S. Navy interposed its ships between Panama and the Colombian navy during Panama's separatist revolt. Panama's greatest resource has always been its location at the narrowest portion of the isthmus connecting North and South America. It has profited from servicing the resulting transit trade. The canal, although too narrow to admit supertankers, is still viable. Its unavailability would still cost American consumers heavily. The Panama Canal Treaty (1978) provides a new basis for U.S.-Panamanian cooperation on the canal (see Figure 11–11). The new treaty reaffirms the United States' primary responsibility for the defense of the canal until the year 2000, though American jurisdiction in the Canal Zone has ended.

Culturally, Panama is two countries. The sophisticated urban populations of Panama City and Colón contrast sharply with the provincial, agricultural populace of the Pacific lowlands. Bananas, pineapples, cocoa, and sugar are grown for export, but the real props to the Panamanian economy are oil refining, canal revenues, international banking, and liberal laws governing the registration of merchant ships. The Panamanian flag flies over one of the world's largest merchant fleets because foreign owners prefer Panama's less stringent safety and labor laws, its low taxes, and minimal insurance requirements. Panama uses the U.S. dollar as currency; this, in combination with its favorable banking laws, excellent communications facilities, and absence of exchange controls, has multiplied the number of banks licensed there.

Panama in many ways reflects the totality of problems facing Central America. Land pressures, even in this heavily urban country, have resulted in destruction of the rainforests and infringement on Indian tribal lands. The *campesinos,* farmers with holdings too small for anything but the barest subsistence, have begun to clear lands in the hills. The erosion that results as vegetational cover is removed has led to intense flooding in older established lowland farming areas. Much of the newly cleared land is only fleetingly fertile, and is shortly abandoned in favor of new clearings.

The Caribbean Islands: The Sugar Syndrome

The Caribbean Islands are the most densely populated portion of Latin America. They were highly contested prizes in a series of European wars as each European power sought to acquire its own source of sugar. They were so highly valued at one point that France relinquished control of Canada to England in order to keep its sugar islands!

Overpopulation is a direct outgrowth of sugar production. To meet the work force needs associated with this labor-intensive crop, Europeans imported large numbers of African slaves. Island economies achieved 100 percent of their labor needs in the late eighteenth century; succeeding population growth has been automatically surplus. The traditional route to survival has been emigration. During the nineteenth century, emigration was possible because of increased labor demands elsewhere in the Caribbean—the logging of mahogany, the banana boom, and the building of the Panama Canal. Because these were either English or American investments, residents of the then-British colonies benefitted the most. The only recent labor demands were associated with the building of oil refineries in Trinidad and the Netherlands Antilles. Pressures have intensified as opportunities for emigration decrease.

Out-migration to France, the United States, and the Netherlands has benefitted only the islands controlled by those countries. Restrictive migration laws have limited flows from former British holdings to the United Kingdom or the Commonwealth. Long-independent Haiti has no outlet, and suffers the consequences most intensely. Trinidad and Barbados have oil resources to provide capital and job alternatives. The formerly Spanish-held areas, never as fully developed or as densely settled, have generally experienced a lower degree of pressure.

The Greater Antilles: Cuba, Jamaica, Puerto Rico, and Hispaniola. The larger islands cf the Caribbean were among the first and richest colonies of Spain. Only later were they neglected in favor of the fabulous mineral wealth of Mexico and Peru. Their Amerindian population (about 1 million before Columbus) had been largely decimated; Jamaica, the smallest island, was thoroughly converted to a plantation economy when it became a British possession. It shares the African cultural heritage, and the economic problems, so characteristic of the smaller sugar islands. France followed suit in its possession of the western part of the larger island of Hispaniola. Spain's pursuit of sugar wealth was belated. Its holdings were more often peopled by small farmers from Spain. Most African-Americans came there as freedmen from the other islands, and not until the nineteenth century. Consequently, the populations of those areas are more racially mixed than those of other formerly European-controlled islands.

Haiti was the first to attain independence, after a slave revolt in 1804. It was followed by the Dominican Republic. Cuba and Puerto Rico, unable to join the mainland of Latin America in freedom from Spain in the 1820s, were finally freed as a result of the Spanish-American War (1898) (see Figure 11–11). Cuba became a theoretically independent republic, but with strong economic ties to the United States; Puerto Rico became a U.S. territory. These political-economic associations have been extremely important in the development of modern Cuba and Puerto Rico. Jamaica is the most recent of these larger islands to attain independence.

Cuba is by far the largest of the Caribbean islands. Its strategic location effectively separates the Gulf of Mexico from the Caribbean. To its southeast lies the Windward Passage, a narrow strait providing entrance to the Caribbean; the Straits of Florida separate it from the United States (see Figure 11–11).

Of Cuba's 10 million people, about 40 percent are black or mulatto (mixed Caucasian and black); the rest are mestizo or white. Cuba has the lowest population density of the larger Caribbean islands and could be the most prosperous with its mineral reserves of nickel and iron, relatively good ratio of cultivable land per person, and accessibility to the American market. Cuba languished under Spanish control as that nation's power declined during the nineteenth century. Bereft of its status as supplier to an empire, it produced sugar only for the limited market of Spain. The climate and terrain of Cuba were nearly ideal for sugarcane. Cuba had large areas of gently rolling to level land and a tropical wet-and-dry climate. (Heavy rains are necessary to rapid cane growth, and the dry season is necessary for harvesting.) This natural suitability did not go unnoticed by American investors.

American influence increased rapidly after 1898. Held at first as a protectorate of the United States, Cuba was granted independence in 1902 with the proviso that the United States had the right to intervene to "preserve stability." American capital flowed in to boost sugar production, and traditional plantations were replaced by foreign-owned corporations.

The combination of proximity to American markets, heavy American investment in sugar plantations, and U.S. paternalism toward Cuba resulted in Cuba having 95 percent of the U.S. sugar import quota at a guaranteed price far above the world market level.

The tragedy of the Castro revolution was that it did not eliminate the large corporate landholdings worked by landless peasants, but rather reinforced their fundamental effects on the economy. Luckless Cuba has never been free of economic colonialism. First a part of the Spanish Empire, then subjected to exploitative "economic colonialism" imposed by the American-Cuban sugar connection, Castro's Cuba is now caught in another exploitative relationship, this time with the Soviet Union. Instead of distributing land to peasants, Castro simply has nationalized plantations, keeping them intact as "state farms."

Cuba must import many of the basic foods because its best land is in sugar. A major trade deficit has given the country a large per capita foreign debt (owed the USSR, which accounts for two thirds of Cuba's trade). Sugar prices remain low in response to both a world glut and selling to the USSR, the world's largest (beet) sugar producer. Just as Cuba's neighbors in Latin America scrambled for its U.S. sugar market, they also eagerly picked up shares of Cuba's once lucrative U.S. tourist trade.

Noteworthy exceptions to this otherwise glum picture are vastly improved health and educational standards. Literacy rates have doubled to almost 90 percent, and life expectancy has risen by 20 years. Migration of refugees and their family members to the United States has reduced population pressure *and* potential opposition.

After 30 years, the revolution has stagnated. The anticipated benefits of Soviet alliance have not materialized. The Soviets, with food supply problems of their own, cannot meet Cuba's needs for imported grain. Low prices paid for Cuban sugar under CMEA pricing agreements barely cover production costs. Tobacco, Cuba's second most important export crop, is in low demand in Soviet and East European markets. Tourism suffers under Soviet currency and travel restrictions. Generally, Cuba has had to be responsible for solving its own problems.

Jamaica's history and culture are similar to those of the other formerly British islands. Its people are overwhelmingly of African origin. It is an island of great beauty and superb climate with rich soils, but far too many people to support. There are serious problems confronting the government, and no easy solutions. High crime rates, narcotics, sprawling shantytowns, and undereducation are as much a part of this island in the sun as are its lovely beaches and fabled hospitality. Unlike the sensuous salsa of Puerto Rico, reggae, Jamaica's world-renowned form of music, speaks to poverty, problems, and survival. It is perhaps the ultimate expression of national feelings.

The land is beautiful but much of the island is unusable. Karst hills cover over half the island and are honeycombed with underground drainage, caverns, and sinkholes. This is the so-called pocket country, the lair of pirates in the days of buccaneering, and home to smugglers today. Inhospitable to agriculture, the hills hold huge reserves of bauxite in their shallow hollows. Ore is easily produced from these limestone "pouches" by cheap open-pit methods. Reserves will last 100 years at current rates of exploitation. Much of the ore receives first-stage refining (alumina) before export.

Sugar, once guaranteed a duty-free market in Britain, has declined in importance. Marijuana production, technically illegal, supports many farm families. Tourism is booming. It is the brightest side of the economy. Air Jamaica connects the island with many U.S. cities to increase both tourism and the share of the tourist dollar retained by Jamaica.

Puerto Rico has a population density of over 1,000 persons per square mile. This high population density in a mountainous island with severe soil erosion and no important minerals led to its unfortunate nickname, "poorhouse of the Caribbean," in the 1930s. The first 40 years of association with America had done little to improve its quality of life. Times have changed. Puerto Rico has achieved one the highest standards of living in Middle America, largely because of its special relationship with the United States. Its citizens enjoy many of the benefits of U.S. citizenship, and there are no tariff or immigration barriers into the United States. Political stability, despite a small but active minority of *independistas,* attracts U.S. investment. Tax advantages and lower wages than on the "mainland" have brought in many industries: garment, electronics, pharmaceuticals, and component assembly. Although income is much lower than the U.S. average, it is very high when compared to those of the rest of the region. The economy is diverse. Since the development of man-

ufacturing, agriculture employs fewer people than any other sector.

A wide range of climate and soil types supports a variety of tropical crops (led by sugarcane, but also including tobacco, pineapples, and coffee). These crops enter the lucrative U.S. market on a preferential basis. Unrestricted access and cheap air fares to the mainland have reduced the pressures of overcrowding. Half of all Puerto Ricans live on the mainland. There is frequent movement back and forth between the island and the big cities of the eastern United States. American popular culture and ideas have reached virtually every village. Still, Puerto Rico is not "New York in the sunshine"; its people are proud of their Spanish heritage and determined that the island will retain its distinctive culture.

Hispaniola, the second largest of the Greater Antilles, contains two countries. The Dominican Republic shares the island, somewhat uneasily, with Haiti. Their mutual boundary is also a sharp cultural-ethnic divide. The Dominican Republic's per capita income is over four times that of Haiti's; its population density is little more than half, and its quality of life is far superior. The border with Haiti has always been tightly controlled to prevent Haitian encroachment into the sparsely settled western mountains. The Dominicans regard this area as a land reserve for the absorption of future population increases.

Traditionally, water supply, rather than fertility, was the measure of suitability for farming in the Dominican Republic. Cropland has increased substantially with irrigation. Sugar, cocoa, and coffee are the primary export crops, and the country is now self-sufficient in basic foodstuffs. The Republic was a major beneficiary of Cuba's loss of preferred access to U.S. tobacco and sugar markets.

The population is overwhelmingly mulatto, with many blacks tracing to an earlier Haitian military occupation. The culture is definitely Hispanic. The Dominican Republic's highly favorable ratio of land to people is changing for the worse as population increases, though at a slightly lower rate than the regional average. Varied exports, industrialization, and a growing tourist and convention trade keep the economy healthy and reduce the dependence on agriculture.

Neighboring Haiti is the hemisphere's poorest country. Its infant mortality rate is the region's highest, and its outlook for the future the least encouraging. Almost 80 percent of Haiti is mountainous, and the frantic search for firewood, cultivable land, and livelihood has contributed to horrendous erosion in steep lands with very heavy rainfall. Haiti's population growth rate has dropped slightly, but birthrates remain high. Overpopulation is extreme; Haitians are not able to feed themselves on an acceptable level of quantity or quality. There is no agricultural frontier; the 75 percent of the population dependent on agriculture have used even the unusable land. Soil quality and quantity degenerate as Haiti's land resource grows poorer with time and misuse.

This unfortunate land was once the pride of the French Empire, the richest prize of the Caribbean. But its fabulous wealth was concentrated in the hands of a few wealthy landowners, and won at the expense of the many enslaved on sugar plantations. Pressed beyond human endurance, the slave population revolted and dispossessed the French, administering a stunning defeat.

Haiti's landscape is dotted with the ruins of eighteenth century French sugar mills and estates destroyed as symbols of slavery. The practical consequence of this destruction was, however, the loss of infrastructure as Haitians reverted from commercial production to subsistence. Haiti quickly moved from the Caribbean's richest colony to its poorest republic as peasant plots in the hills replaced plantations. Sugar and rice grow on poorly maintained irrigated lowlands. Coffee trees cling to patches of land on precipitous mountain slopes. Food grows wherever the land will produce a crop. Everywhere there are too many people.

Vestiges of French culture remain. French is the official language, but only 10 percent of Haitians speak it. Creole (a mixture of French, Spanish, English, and traces of African and Amerindian languages) is the majority language. A small, disproportionately wealthy mulatto elite proudly adheres to European cultural values. The bulk of the population, nominally Catholic, follows a unique mixture of Christianity and traditional African beliefs. Vodoun (voodoo), with its rituals, dance, and ability to cast and remove spells, is still an active force in rural areas and city shantytowns.

Haiti's chances of improving its citizens' quality of life would seem to lie in increased industrialization. Light industrial development is encouraged by the absence of controls on foreign capital flows and the "repatriation" of profits. Port-au-Prince, the capital, has attracted some textile, garment, and assembly industries, but far too few. Tourism, the Caribbean panacea, develops very slowly because of domestic strife and, now, the fear of AIDS. Nearly constant political instability has discouraged potential investors. The populace suffers repression and exploitation as the elite garners whatever wealth remains. Desperate Haitians flee to any refuge, living in the shadows as illegal migrants, working on whatever tasks hold out the hope of survival.

The Lesser Antilles. The smaller islands of the Caribbean lie in an arc between Puerto Rico and South America. To this scattered chain are added the Bahamas to the southeast of Florida, the Turks and Caicos islands, and the Cayman Islands south of Cuba (see Figure 11–6a). The islands are extremely varied in physical structure, ranging from the low, flat Bahamas to the volcanic peaks of Dominica and Martinique. Only their colonial histories and cultures provide a semblance of regional unity, and even here, the picture is one of a complex mosaic in which a few cultural-economic themes predominate.

A revolution in the settlement and economy of the islands was precipitated by the rapid spread of sugarcane cultivation. The fierce and tenacious Caribs, the original inhabitants, had virtually disappeared under the European onslaught; European labor was both unwilling and expensive. The lucrative European markets for cane sugar inspired large-scale importation of West Africans, producing the prevalently black ethnic makeup of all the smaller islands except Trinidad. There, blacks (45 percent) are almost matched by East Indians (40 percent) brought in as "contract" (indentured) plantation labor after slavery ended.

The "sugar islands" became enormously valuable to the Europeans. From the start, they were one-crop economies. The notable exceptions to this scenario were not suitable to sugar production. Export crop diversification has been relatively slow. Bananas became important after disease racked mainland plantations. Land is scarce on all the islands, and no crop can guarantee a livelihood for all.

Industry throughout the islands is based on crop processing (rum, fermented from sugar, is a regional specialty) or the refining and transshipment of oil products. Tourism is an economic mainstay, and island governments create receptive political climates for international banking. There are few natural resources. The long-term persistence of direct colonial control helped to encourage European and American investments in oil refineries to process Venezuelan (and, lately, African and Middle East) crude. Much of the fuel oil supply for the northeastern United States is processed here.

Trinidad, by virtue of its large size and diverse economy, is an exception to the generalizations applicable to the Lesser Antilles. Geologically, it is a detached fragment of South America. Geologic similarity to nearby Venezuela is reflected in oil, gas, and asphalt reserves. Offshore discoveries in recent years have improved reserves, but Trinidad's refineries now rely on imported crude for half their production.

Trinidad and Tobago (a smaller island to the northeast) are associated politically but are quite different ethnically and economically. Tobago is a classic plantation economy and is ethnically African. Trinidad's East Indian population element is unique in the islands. Trinidadian plantations produce coffee, cacao, citrus, and bananas as well as sugar, but the economy is dominantly industrial. This situation is fortunate for Tobagans, who have unrestricted migrational access to more prosperous, diversified Trinidad.

A curious cultural outgrowth of Trinidad's massive oil-refining industry has been the steel drum band. The readily available old oil drum found many uses in Trinidadian homes—washtub, cooking utensil, water storage facility and, ultimately, musical instrument. Trinidad is the originator of calypso music, with its steel drum instruments and characteristic sound. This original Caribbean musical innovation has since proliferated throughout the islands.

The Bahamas (and their neighboring Cayman Islands) have found prosperity as "offshore" banking areas. Banking laws there guarantee secrecy of financial transactions, ensuring in return a certain amount of government revenue and white collar employment. The apparent economic success of the Bahamas is directly related to their proximity to the United States. A more than billion dollar tourist industry contributes 70 percent of the country's GNP (Table 11–3). After Castro's rise to power in Cuba, Americans sought a tropical replacement for Havana.

The Bahamian casinos proved to be an additional attraction to Americans with only Nevada (and later, Atlantic City) as alternatives.

With a 25 percent unemployment rate, the economy does have weak spots. The huge refining industry is severely depressed. Drug trafficking has become a major issue of contention between the United States and the Bahamas. Per capita income has declined 15 percent since 1986. Shack settlements have sprung up on beaches and in wooded areas, where they house over 20,000 illegal Haitian migrants and an equal number of illegal aliens from other Caribbean islands.

The French dependencies of Guadeloupe and Martinique are somewhat poorer than Trinidad and the Bahamas, but they evidently seem enormously prosperous to their less-fortunate immediate neighbors. Thousands from Dominica, St. Lucia, St. Vincent, and other Caribbean islands migrate illegally to Guadeloupe and Martinique in search of work.

Both the islands are officially a part of France and, therefore, eligible for all manner of government subsidies. Technically a part of the European community, they have received a host of multinational investments in business and industry.

Slightly more exotic because of their thoroughly French-influenced culture, architecture, and landscape, they are a favored stop on cruises. Modern, large airports and superior harbors increase these islands' accessibility.

Barbados is another bright spot in the regional economic fabric. It is still a sugar island, quite literally, since this staple remains its chief crop. Barbadan farmers, however, have proven to be more efficient and competitive than most of their neighbors. The discovery and production of offshore oil has relieved the economy of what was once its largest, most costly import. Japanese and South Korean investments have enlarged industrial production, and tourism increases rapidly in what Barbadans correctly boast of as a clean and safe environment.

The former Netherlands Antilles (Curaçao, Bonnaire, and Aruba, among others) are tourist meccas for Europeans, wealthy Latin Americans, and North Americans alike. Neat and prosperous, with classically Dutch architecture adding charm and interest, they are well established on the tourist circuit. Their limited rainfall has always hampered agriculture, and sugar was never important. Oil refining, largely Venezuelan crude, has been the economic mainstay since 1920.

The American Virgin Islands subsist almost entirely on a lucrative tourist trade (see Table 11–3). The neighboring British Virgin Islands perform as a nearby supplement, but with considerably less economic success. Population pressures in the American sector have been relieved by out-migration, while labor flows from the British islands supply seasonal demand peaks there.

The remaining islands live at a considerably lower standard. Sugar is no longer the chief crop or source of income on most islands because oversupply and lowered prices have rendered island production uncompetitive. The new boom crop in many of the islands is bananas as both American and European markets for that fruit increase.

Perhaps the most poignant example of the sugar island syndrome is found in Grenada. In all the islands, governments are hard pressed to reconcile long-term economic investment needs with public demands for badly needed social investment in health care, education, and housing. The periodic coups and demonstrations that wrack the islands often stem from this basic conflict in investment priority aims. Emphasis on human welfare projects in Grenada simply raised costs without generating income, if it greatly benefitted average Grenadans by addressing their most pressing human needs. With huge debts and little available credit, the government in power gratefully accepted Cuban aid to build an enlarged airport. The government announced that the airport was to enhance tourism, but the United States viewed it as a potential enemy military base.

The assassination of the then head of government by domestic, pro-Castro military leaders led to American intervention. During the occupation, the airport was indeed built. Most troops have been withdrawn. The invasion itself was not universally approved nor disapproved of, and public opinion remains divided.

Economic opportunities are quite limited in the generally crowded little islands. Occasional hurricanes, stiff competition in foreign markets for the islands' agricultural products, and the low-pay status of most tourist industry service jobs present a discouraging picture for economic expansion. Life may appear carefree to the casual tourist, but social-economic tensions are surfacing with increasing vehemence.

South America: Poised for Takeoff

South America, though by no means simple in its cultural makeup, is less complex than Middle America. The countries are larger, and their economies are frequently more diverse. Population pressures are much less strongly felt, and there are reservoirs of unused land. Increasingly, a diverse South America sees advantages in regional cooperation and trade. Industrialization proceeds rapidly, but heavy foreign debt and domestic strife are serious threats to continued progress. An oil-rich Venezuela is the continent's richest state, and only the economies of Peru, Bolivia, and Guyana are desperately poor. The continent's giants, Argentina and Brazil, have strong international ambitions. Drug production and trade muddy the relations of Colombia and some Andean countries with the United States; tensions ebb and flow.

Urbanization has reached unusual proportions in South America, where already huge cities continue to grow rapidly. Urban slums, congestion, and pollution present planners and politicians with enormous problems. Population growth rates are dropping, but in most cases still exceed world averages. Incomes have been rising in most countries, and a strong middle class has developed in many states. Progress and development, although uneven, are a consistent and positive theme in most South American countries. The general economic trends are encouraging.

The Caribbean Littoral: Colombia, Venezuela, the Guianas, and Suriname. The nations of South America that border the Caribbean are a transition zone between Middle and South America. The Guianas began their colonial existence as mainland extensions of the Caribbean Islands' sugar plantations, and early settlements in Colombia and Venezuela were tied to sugar. This region contains all the ethnic elements, and many of the important physical environments, prevalent in the total Latin American picture. It also has all the problems and positive attributes of Latin America. It is both rich and poor, historically and currently representative, long established and newly independent in its various portions.

TABLE 11–3
The Lesser Antilles at a glance.

Unit	Area (sq. miles)	Pop. (thousands) 1940	Pop. (thousands) 1989	% Urban	Pop. Density (per sq. mile)	% Literacy	% Arable Land
Bahamas	5,380	80	247	75	45	93	2
Barbados	168	200	256	42	1,542	99	76
Trinidad and Tobago	1,980	610	1,261	21	636	95	30
Windward Islands							
Dominica	290	51	80	30	262	80	23
Grenada	133	88	87	22	654	95	41
St. Lucia	238	69	128	49	537	78	28
St. Vincent and the Grenadines	150	58	112	30	746	85	50
Leeward Islands							
Anguilla	35	6	7	N.A.	206	81	N.A.
Antigua and Barbuda	171	35	69	34	403	90	18
Br. Virgin Islands	59	6	12	36	203	86	N.A.
Montserrat	32	14	12	N.A.	363	93	21
St. Kitts-Nevis	139	32	55	52	396	80	26
The French Group							
Guadeloupe	667	312	335	39	502	92	68
Martinique	425	241	329	58	774	94	37
Netherlands Group							
Aruba	76	28	68	N.A.	894	92	N.A.
Netherlands Antilles	310	72	202	63	652	90	N.A.
U.S. Virgin Islands	133	25	118	39	887	N.A.	N.A.

SOURCES: *Statesmen's Yearbook, 1987–88;* World Almanac, 1989; World Bank; Preston James & C. W. Minkel, 5th ed. 1986, [Wiley]; and *Statistical Yearbook,* World Bank, 1988.
N.A.—not available

Income per Capita ($)	Agricultural Employment (%)	% Natural Increase	Major Crops	Tourism (receipts in millions $ or numbers)
7,598	5	1.8	Fresh vegetables	$ 1,100
3,040	12	0.7	Sugar	$ 326
6,800	10	2.0	Sugar, cacao	N.A.
1,034	40	1.3	Bananas, citrus	23,000
800	33	2.0	Nutmeg, bananas	$ 26
1,220	36	2.4	Bananas	$ 69
920	30	2.3	Bananas, arrowroot	$ 27
N.A.	N.A.	N.A.	N.A.	N.A.
1,980	N.A.	2.6	Cotton	167,000
N.A.	N.A.	N.A.	N.A.	450,000
N.A.	N.A.	N.A.	N.A.	N.A.
1,250	48	2.9	Sugar	42,000
3,960	N.A.	0.8	Sugar, bananas	390,000
5,040	N.A.	0.7	Bananas, sugar	1,500,000
N.A.	12	N.A.	Corn ⎫	1,600,000
N.A.	8	N.A.	Corn, poultry ⎬	
7,865	N.A.	1.2	Fruits, vegetables, ⎫ flowers ⎬	1,200,000 $ 377

TOURISM: A DECEPTIVE INDUSTRY?

Tourism is virtually the only industry in many of the Lesser Antilles. Tourist investments have been deliberately encouraged by island economies seeking diversification and new sources of employment. Tourism, though, is a highly footloose economic phenomenon. Civil disturbances, especially those with racial overtones, can cause dramatic drops in tourism. Strikes at ports, airports, or hotels can quickly redirect tourists to other destinations, with long-term effects on "revisit" trade.

Tourism can be a deceptive industry for developing countries, since much of the apparent income goes right back to the developed countries that provide the tourists. Visitors place high demands on scarce freshwater supplies and transport facilities, yet they are there only on a highly seasonal basis. Locally owned, smaller inns and guesthouses tend to receive only the overflow when big hotels are filled. Developed countries' citizens own the hotel chains, car rental companies, airlines, cruise ships, and casinos. Tourists' tastes tend toward prime meats, frozen vegetables, and alcoholic drinks, which are mainly imported. Islanders are the seasonally employed, filling low-paying service occupations (waiters, maids, bartenders, taxi drivers, "beach boys"), but are not the managers. While seasonally unemployed, islanders can enjoy the beaches not restricted to hotel guests, if the water hasn't been too polluted by the concentration of large hotels.

It is the ultimate crossroads of Latin America, in the sum of its parts, a composite of the whole. It is that portion of South America most directly tied to the United States as a supplier of raw materials and tropical products.

Colombia is one of South America's most populous units. Its natural resource base, though large and varied, has not provided the economic bonanza that oil and iron ore have for neighboring Venezuela. Once a net exporter of oil, Colombia has increased domestic consumption through industrialization to the point of becoming a net importer. The country also has large coal reserves. New large-scale investments at El Carejo in the Guajira Peninsula will significantly increase production. Ninety-five percent of the world's emeralds are mined in Colombia. The government controls the largest mines, but many emeralds appear in "uncontrolled markets." Colombia is also a large gold producer (Figure 11–12).

Official statistics on foreign-exchange earnings are underestimated because of Colombia's large-scale production and shipment of illegal drugs to American markets. This marijuana-cocaine connection to North America has the grave potential for destabilizing the government through the huge possibility for bribery and corruption associated with illicit profits.

Colombia has been plagued by political unrest associated with the large-scale rural-to-urban migration. This process has created 23 cities of over 100,000 people (including the capital, Bogotá, at 5 million, and three other "million" cities). Urban growth is recent, rampant, and not easily absorbed. The high proportion of young people in the population has added demands on educational and social

Caribbean Tourism.
Jamaica's economy is varied; tourism is only one facet. Jamaica owns its own airline, and some hotels are owned by Jamaicans. With a long-standing tradition in the tourist trade, many Jamaicans are trained professionals, holding higher-paid jobs as managers, head waiters, chefs, and directors.

services. Any Colombian government must fulfill rising expectations, maintain order, and balance a precarious budget—a set of heavy demands.

Officially, half of Colombia's foreign exchange is earned by coffee; Colombia is the world's second largest producer after Brazil. Renowned for its quality, coffee is produced largely on small farms rather than on estates. Any fluctuation in world coffee prices adversely affects the economy. Colombia has encouraged a drive toward export diversification to minimize dependence on coffee.

If it grows in South America, Colombia is a producer: cocoa, sugarcane, bananas, tobacco, coffee, cattle, citrus, corn, wheat, potatoes, pineapples, and every vegetable imaginable. Agricultural colleges have special courses of study for "cold-climate" and "warm-climate" farming, since altitude supplies a

ready milieu for both. There is great potential for expansion in agriculture, with much fertile soil still uncultivated. With a declining annual growth rate (2 percent), Colombia can provide for more people.

Among the early migrants to Colombia were the Antioqueños, descended from Christianized Spanish Jews and Moors, who settled near Medellín. Skilled in crafts, they began the country's first industries. Today, Medellín is a center of diverse and expanding industry and ranks as the second most important Colombian city. It is also the home of the country's infamous drug cartel. Bogotá, the capital and cultural center, has also attracted considerable foreign investment.

With its great mineral wealth, Venezuela has by far the highest GNP per capita in South America. A founding member of OPEC, it nationalized its oil

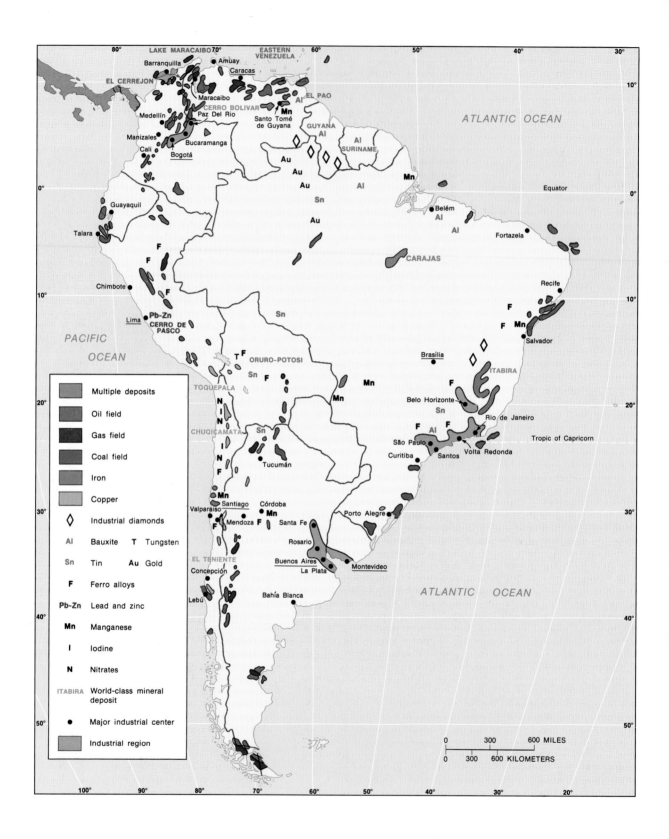

Multiple deposits

Oil field

Gas field

Coal field

Iron

Copper

◇ Industrial diamonds

Al Bauxite   T Tungsten

Sn Tin   Au Gold

F Ferro alloys

Pb-Zn Lead and zinc

Mn Manganese

I Iodine

N Nitrates

ITABIRA World-class mineral deposit

• Major industrial center

Industrial region

Venezuela and Oil. Oil has been the driving force behind the rapidly expanding Venezuelan economy, and the country's chief source of wealth. Venezuela enjoys the highest per capita GNP in South America. Oil production peaked in 1970, as the long-producing fields under Lake Maracaibo, shown here, began to show signs of depletion. Recent drilling at deeper levels under the lake has revealed vast new reserves.

industry in 1976, placing it in the hands of a government corporation. Oil accounts for 80 percent of the country's foreign exchange earnings and two thirds of government revenue. Recent drilling (at much deeper levels) under Lake Maracaibo, together with exploration of the "Orinoco Tar Belt," has discovered what may be the world's largest deposit of "heavy oil" (thick, almost tarlike oil that cannot be retrieved through traditional oil wells). Steam injection, an expensive process, may be necessary to recover this oil; high prices are needed to justify costs. Venezuela is also a major producer of iron ore. Hydroelectric power potential is immense; Guri Dam

FIGURE 11-12
Mineral Resources in South America. Gold and silver were the chief minerals sought by the Spanish conquistadores. Though still important, they are vastly overshadowed by the production of other minerals. Bolivian tin, Chilean copper, and Venezuelan iron ore are all of much greater importance in today's world. The continent is exceptionally rich in metal ores. Bauxite is produced commercially in seven of the continent's countries. With the continent once thought to be lacking in fossil fuels, oil has become a major export from Ecuador, Peru, and Venezuela. Colombia is developing its coal reserves on the Guajira Peninsula. Both Chile and Peru produce coal in excess of national demands.

generates 9 million kilowatts, ranking it above all but a handful of power projects in the world.

Part of the explanation for Venezuela's impressive prosperity (though incomes *are* extremely unequal) is the relatively low ratio of people to this fabulous resource base. Venezuela has a population density lower than the regional average, but a relatively rapid growth rate (2.7 percent). Most Venezuelans live in the Andes basins and along the Caribbean coast (see Figure 11–5). Although about half of Venezuela lies south and east of the Orinoco River, that area contains only 5 percent of the people. A new $4-billion city, Ciudad Guyana, is being completed 300 miles southeast of Caracas to spur development in the "empty South."

Among other developmental thrusts, Venezuela has set about developing the Llanos. Cattle ranching is the major form of agriculture there, but rice has been introduced in the Orinoco Delta. Large areas are being colonized in conjunction with a land reform program. Ultimately, 100 million acres are to be settled (Figure 11–13).

The three smaller units on the northeast coast of South America remained colonies long after the dissolution of the Spanish and Portuguese Latin American empires. From west to east, they were controlled by the British, Dutch, and French. Population densi-

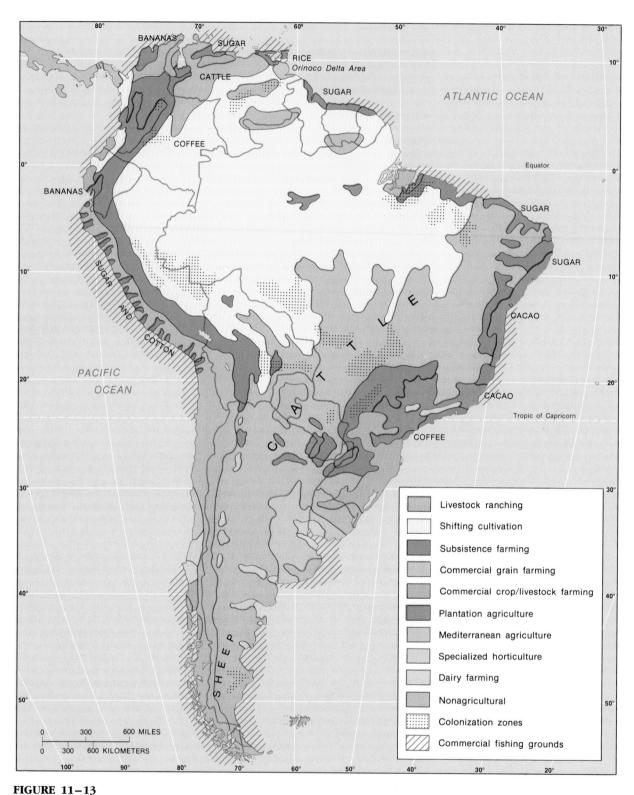

FIGURE 11–13

Agriculture in South America. Extensive forms of agriculture such as cattle grazing and shifting cultivation dominate over more than 80 percent of the continent. Increasingly, the countries of South America are emphasizing self-sufficiency in foodstuffs. Agricultural production, while locally specialized, is becoming much more diverse.

CARACAS: AN EXAMPLE OF THIRD-WORLD URBAN GROWTH

The familiar hillside slums, chronic housing shortages, utility supply problems, and traffic jams so characteristic of Latin American cities of today are perhaps only the beginning of a potential urban crisis that will face all Third World nations. Current population projections estimate that the world will add 3 billion people by the year 2025. Most of that growth, both absolute and relative, will occur within Third World nations. Most of their growth, in turn, will take place in urban rather than rural areas, dominantly in a few primate cities.

Since 1950, the population of Venezuela has grown almost 400 percent, while that of Caracas has increased over 1,000 percent. In 1950, Caracas and La Guaira, its port, together accounted for a mere 8 percent of the national total; today, that relative figure has long since surpassed 25 percent. By 2025, metropolitan Caracas could house over 10 million people, more than one-third the national total. Most of these people will be crammed into a narrow, 15-mile-long valley, an enclosed highland basin subject to temperature inversions and, consequently, intense smog.

Most of the nation's industry concentrates here; there is no better, more economical location than this capital city. It is the largest, most affluent national market, well connected with all other markets and located near a major port. Yearly, it attracts a stream of rural migrants, a seemingly inexhaustible supply of cheap labor. Some 3 million illegal foreign migrants swell the ranks of those seeking work in Venezuela, generally in the capital, where the largest share ultimately seek jobs and the wages are highest.

Migrant shantytowns are repeatedly bulldozed to make way for new, higher-value construction, only to be replaced by new and larger barrios along the ever-expanding city fringes. More than half the nation's 3 million cars congest the capital's streets and pollute its air. Recently, a subway system has been built in an attempt to ease congestion and reduce pollution.

The rampant growth of Third World cities is at odds with the realities of tight national and municipal budgets. Public demands for better living standards translate readily into political pressures for diversion of funds into social (as opposed to primarily economic) investments. Recent riots in Caracas were a manifestation of just such demands.

If the problems can be solved, the solutions seem more likely to be provided by an affluent Caracas than by other, less prosperous cities. If uncontrolled urban growth presents serious problems there, what will be the scope of these problems in less affluent nations?

ties are extremely light, and total population is less than 1.5 million.

In several economic and cultural senses, the Guianas have more in common with the Caribbean Islands than with mainland South America. They were viewed by colonial powers as extensions of their Caribbean sugar island holdings. Because all the Guianas were originally sparsely inhabited by Amerindians, the colonial nations imported labor for sugar (export) and rice (subsistence) production. East Indians, who came as indentured servants, form the majority in Guyana, the largest single group (37 percent) in Suriname, and a small minority in French Guiana. Africans, imported as plantation labor, form the next largest ethnic group.

Sugar remains an important export crop of Guyana and French Guiana but is only a minor export of Suriname. Suriname's bauxite deposits rank among the richest in the world; the country mines, processes, and exports both bauxite and alumina. It also manufactures aluminum using local hydroelectric energy. Guyana also mines bauxite for export and plans a huge hydroelectric project to produce higher-valued aluminum for export (see Figure 11–13). Forest product industries have been expanded significantly in Suriname and French Guiana.

The Indo-Andean Republics and Paraguay.
While Amerindian cultural influences are present to some degree throughout Latin America, they are particularly strongly represented in the areas that were dominated by the great New World civilizations that held sway at the time of the Spanish conquest. Two of these, the Chibchan and the Incan, were developed in South America (see Figure 11–4). The most powerful and impressive was that of the Incas, who had extended their control over 2,000 miles from northern Ecuador through northern Argentina and central Chile. Unlike the great states and empires of Europe that arose primarily in lowlands or river valleys, the Incas created an enormous, land-based empire composed of relatively isolated mountain basins and scattered, irrigated desert valleys. It was held together by a remarkable system of roads and paths. To utilize this harsh physical environment, the Incas developed elaborate terracing and irrigation systems. They domesticated livestock from among the sparse fauna—llamas, alpacas, ducks, guinea pigs, and dogs. These peoples originally domesticated potatoes, corn, and several kinds of beans, their principal gifts to the rest of the world.

The conqueror Pizarro arrived with fortuitous timing (just after an internal power struggle) and subdued this huge empire over a decade. The stunning output of gold and silver made Peru the most valued Spanish colony in South America, and Lima the most splendid colonial capital. Spanish colonial policy was directed at safeguarding this magnificent prize and the routeways to it. Spanish Peru's trade flowed on either of two routes: south to the Atlantic via tributaries of the Paraná system, through Asunción and Buenos Aires; or northeastward by ship to Panama, then by portage across the peninsula to an Atlantic port, and on to Spain.

The outstanding features of the four countries in the subregion are poverty, Indian ethnocultural dominance in the population, and a low level of development. All but lowland Paraguay were historically a part of the Incan Empire. The addition of Paraguay to the subregion is based on its ethnic and economic similarities to the other three states (Table 11–4).

TABLE 11–4
Selected comparative data for the Indo-Andean Republics and Paraguay.

Country	% American Indian	% Mestizo	% Urban	% Growth Rate	% of Labor Force in Agriculture	Per Capita Income ($)
Bolivia	55	30	45	2.2	47	570
Ecuador	25	55	45	2.7	40	1,450
Peru	35	49	40	2.5	40	660
Paraguay	41	54	65	2.9	34	1,620

SOURCES: Inter-American Development Bank; Population Reference Bureau, 1988.

The Incas. *The Incas created an advanced civilization in one of the world's most difficult and unlikely physical areas. Steep mountain slopes were terraced into farmable plots of level land. Large granaries, shown in the ruins at the left, stored surplus production. A vast network of roads was created to ensure adequate food supplies for all and constant communication among the far-flung settlements of an empire that stretched from southern Colombia, through the Andean countries, to central Chile. Irrigation works were erected on a vast scale to ensure adequate water supplies. Production was enormous, despite the absence of the plow and draft animals.*

Ecuador, on the northern fringes of the Indian culture zone, has three distinct physical-cultural sub-regions, which Ecuadorans call *costa, sierra,* and *oriente*. The sierra, the Ecuadoran portion of the Andes, contains the capital, Quito, but the costa is the country's most populous and productive zone. The oriente ("the east") is the empty zone, the land anticipated to house the overflow from the sierra and to provide new mineral wealth for economic diversification.

The cool sierra belies the country's equatorial location. ("Equator" is the literal translation of its name.) The population there concentrates in 10 crowded basins, generally at elevations of over 10,000 feet. A patchwork of small fields uses every piece of available land; farms just barely support the farm families that rely on them. Here the bulk of the population lives at subsistence levels. Fields produce midlatitude or tierra fria crops, a response to the cooling effects of altitude. Corn, wheat, potatoes, and pasture are the predominant land uses. Birthrates are high, land is expensive, and developmental levels are low. There is little future there.

The lush costa is the antithesis of the sierra. Climate is warm but bearable, and the soils are richer than those of the sierra. Agriculture is commercial and highly productive; its people are comfortable though hardly wealthy. There are plantations, but production comes mainly from small and medium-sized farms. The lowlands around the Gulf of Guayaquil support a rich and diverse agriculture. About a third of the nation's population live here, an amalgam of European, mestizo, and some black stock. Race has little bearing on income; all are relatively prosperous. The costa yields the world's largest export crop of bananas and generous bounties of rice and cacao. Guayaquil, the district's port, is half again as large as Quito.

Residents of the crowded sierra show some willingness to move to the costa, but migration to Oriente province has been too limited to develop fully that area's potential for food and export crop production. Jivaro Indians, still pursuing a hunting and gathering existence, successfully repulsed early organized colonization; their past reputation as headhunters apparently is still a discouraging factor.

The plight of the overcrowded highlands is difficult to solve. Family and community ties are strong, and residents leave only when existence becomes unbearable. The land, although not infertile, is too cool for lucrative cash crops. Government efforts to raise the living standards of the sierra's people concentrate on industrial development and crop diversification.

Ecuador has a varied resource base, but oil is its most important commodity. Important oil fields are found on the Pacific coast and in Oriente, Ecuador's slice of Amazonia. Pipelines transport oil from Oriente to the coast. Total reserves are estimated at over 2 billion barrels, a moderate-sized reserve that allows for export (see Figure 11–12).

Ecuador has a comparatively high annual growth rate; however, it is declining with development. Growth is steady, if not spectacular. Ecuador is no longer abjectly poor, although it is far from as rich as Venezuela. Like Paraguay, at the other regional extreme, it has made considerable progress.

Home of the original Incan culture and exceptionally wealthy in minerals, Peru is economically the most promising of the four states of the Indo-Andean subregion. It is by far the largest and most populous. A series of moderately large oil fields are located along Peru's coastal plain, and large new fields began production in the Amazon Basin during the last de-cade. A trans-Andes pipeline is completed, and new exploratory drilling is taking place offshore. Large deposits of high-grade iron ore, fortunately not far from sea transport, are used for export and a growing domestic steel industry. Peru has important reserves of most common and precious metals, as well as a series of alloys and rare metals (see Figure 11–12). The recent oil glut and weak metals markets have depressed Peru's mineral-based economy.

The cold Peru current, sweeping close inshore and transporting mineral nutrients to support the base of the oceanic food chain, sustains one of the world's largest and most prolific fisheries. The almost rainless coastal climate allows deep layers of bird **guano** to develop. This rich source of nitrate, easily acquired, was important in the munitions industries of the world before chemical processes for atmospheric fixation of nitrogen were developed. Guano is still highly desirable as fertilizer, and a government monopoly regulates production. During the 1970s and early 1980s, the fish catch plummeted when the current changed location (during the famous weather phenomenon called El Niño); overfishing has also contributed to the decline.

The historical core of the country is found in the areas of cooler climate at high altitude, the land of the Incas. Peruvians were quicker to develop their Pacific lowlands than were Ecuadorans, even though that coast is arid. The narrow coastal plain of Peru south of Chiclayo contains a series of long, narrow valleys reaching back into the Andes at right angles to the coastline (see Figure 11–3b). This pattern translates into a series of irrigation-based ribbons of cultivation and settlement, separated by steep mountains or patches of narrow desert coastal plain. A series of "miniature Egypts" produce cotton, sugar, rice, and basic foods in these coastal oases. A pipeline from the interior now transports water from the Amazon over the Andes to the headwaters of some of the coastal rivers, replenishing water supplies and allowing for an expansion of irrigation.

Lima, the capital, is approximately in the center of Peru's coastal plain. With its nearby port of Callao, the metropolitan area contains 5 million people, about one fourth of the country's total population. Along with its administrative functions, Lima functions as a manufacturer of consumer goods. It is a major South American air transport node.

The highland basins are almost purely Indian in culture. As in Ecuador, farms are small and the population exists at a subsistence level. The climate is cool and the air is rare; breathing is a major task for those unacclimated to such extreme heights. Cuzco, the ancient Incan capital, is the commercial center. The fabled Cerro de Pasco mines still produce; mining is the basis of existence for those who do not farm.

Peru enjoyed brief prosperity in the 1970s when the demand (and price) for its minerals provided a lucrative income. At the time, the government, rich on oil revenues, invested in a series of developmental schemes that involved Peru's Amazonian interior. Two Trans-Andean Highways were completed, and the north-south highway along the Andean margins was begun. These highways were built to encourage colonization of the interior and to link the country's three zones into an integrated national economy.

Peru has many of the requisites for development in place. There is obviously a rich metal mineral base, plentiful fuel, a fairly dense population, something of an infrastructure, and a commercial agriculture. What seems to have gone wrong? Peru, like Mexico, invested heavily in developmental projects. Oil revenues were to pay the bills. Depressed oil prices left Peru without the income necessary to enable the sustained payment of its huge debts. A rush to implement land reform replaced rich landowners, who might have weathered the economic storm, with small, poor farmers, who could not. Half-finished projects remain uncompleted in the face of austerity and severe economic stress. International economic credit was tightened when all of Peru's economic endeavors collapsed at once. The predictable results were civil unrest, guerrilla warfare, repeated coups, and further economic retrenchment. Leftist guerrilla activity in the south has proven impossible to halt. Half the urban population and almost 20 percent of rural dwellers are unemployed.

Peru is slowly mending its economy. Irrigation is cautiously expanding. Small farms have been organized into cooperatives similar to the Mexican ejidos. Education introduces new techniques and conservation to farm families. The eastern slopes of the Andes and the Brazilian borderland, the so-called La Montaña district, are slowly being colonized by surplus farm labor from the highlands. Current programs proceed more cautiously. Perhaps too much had been attempted too soon. Certainly, the timing was wrong.

Bolivia is literally the top of the Andean world. Most of its population lives on the eastern, wetter side of the altiplano, a high, wide plateau between the two major ranges of the Andes. The country itself is relatively large, about twice the size of France. Most of it is of apparently low utility. The landscape of the altiplano is bleak, in many ways similar to that of the Tibetan plateau, with large salt lakes, sparse pasture, and a desertic absence of rain. A few favored basins and depressions in and around the mountains along its eastern edge concentrate the bulk of Bolivia's crop agriculture and people.

Settlement is gradually descending the eastern slopes of the Andes as peasants leave crowded basins in search of new land. A series of deep valleys there, paralleling the Andes, offer ribbons of fertile soil along permanent streams. The climate is warmer and more moderate, and a greater variety of crops can be grown. Though the eastern slopes are increasing their share of total national population, their potential for absorbing excess highland population is limited. The amount of usable land is small, and isolation poses a serious problem. Only a few commodities produced are valuable enough to bear the cost of shipment to markets, and agriculture is often mere subsistence. The bulk of the area beyond these front ranges of the Andes, fully one half of the country, consists largely of inaccessible rainforest, pest-ridden grasslands, or the raw scrublands of the Chaco.

Almost half of the nation's farmers barely attain subsistence on small individual plots or communal Indian lands. Countless others are tenants on the estates of a small but wealthy landed gentry. Miners, despite unionization, earn a living that is only marginally better. The bulk of the rural population remains extremely poor. A small upper class still controls most of the national wealth.

The cultural cleavages between white, mestizo, and Indian are not as great as in Peru. Indian languages have long been fully official and are understood by virtually everyone. The government is attuned to Indian culture and needs. A dual society persists as the culture of whites and mestizos coexists, rather than mixes, with that of the Indians. Pov-

THE GEOGRAPHY OF COCAINE

Cocaine is extracted from the leaves of a shrub native to Andean South America; its use is ancient. The coca plant grows wild, though it is now actively cultivated. Mountain Indians viewed it as medicinal and used it as a mild stimulant to soften life in a harsh environment. Chewing a few leaves, as is the Indian habit, produced a mild euphoria. Its use as a high-powered painkiller and a numbing agent developed in the nineteenth century, when Europeans began to refine it (increasing its potency) for anesthetic purposes.

Production concentrates in the most isolated areas possible—the valleys of the Andes and the interiors of Peru, Bolivia, and Ecuador. It takes 1,200 pounds of leaves to make a kilo (2.2 pounds) of cocaine. Primary refining takes place in native stills, reducing bulk, increasing value, and making transportation, even on foot, economical. Reaching and destroying such isolated production sites would be an almost impossible task. Thousands of clearings in the tropical forest can act as primitive airfields for transportation by plane. At any point along the trail, shipments can be diverted to any one of several dozen small ports for transshipment on one of thousands of fishing boats that can rendezvous with ships at sea. Change from one form of transportation to another can occur at many hundreds of locations. The complexity of the transportation and the various permutations possible are staggering.

Native growers and foot messengers are relatively poorly paid for their endeavors. Yet, a small crop of cocaine brings much more income than acres of grain. Many subsistence farmers receive all or most of their cash income from cocaine.

Indicting and convicting any individual or group will not halt operations. Matching the brokerage and production units of South America is an equally complex distribution system within the United States. Where should enforcement begin? Where would it be most effective? Basically, cutting off supplies is virtually impossible and, in any case, extremely costly. Cuts in consumption are the most logical approach in view of the geographic realities of the production and shipment system.

erty is of the proportions found in Guyana. A quarter of all children die before reaching the age of five. Exploitation, however, once at a level equivalent to the worst found anywhere in traditional Latin America, has eased with sporadic social reforms.

During the Spanish colonial epoch, Bolivia was wealthy, second in importance only to Peru. In both cases, the importance was based on minerals as gold flowed from Peru and even larger flows of silver left Bolivia destined for Imperial Spain. Copper, known in Bolivia since ancient times, occurred there in its "native" or pure metallic form. The center of mining, then as now, was the Potosi district. Silver has declined in importance, yielding pride of place to tin and tungsten. Mining continues despite unusually high costs of processing and transporting ore within and from mountainous, landlocked Bolivia. Recent economic distress has resulted largely from Bolivia's continued dependence on mining.

In another age, the position of Bolivia was pivotal. Highly valued gold and silver and the purest of copper could profitably make the arduous journey down from the altiplano on mule, llama, or foot to ports on the Pacific or to the rivers that led to Buenos Aires and the Atlantic. There were no political barriers to movement. Today, individual national jurisdictions complicate movement, and the base metals produced are lower in value. Bolivia does have rail connections to Arica and Antofagasta, two ports on the Pacific, which unfortunately are located in Chile. Bolivia was much larger at the time of its initial independence (Figure 11–14); it had its own corridor to the sea. In the disastrous War of the Pacific (1879–83), Chile seized Bolivia's port as well as its nitrate deposits. During the rubber boom that followed, Bolivia lost its most productive rubber territory to Brazil and, with it, access to a navigable tributary of the Amazon. In the war of 1932–35, Paraguay pushed its territory deeper into the Gran Chaco at Bolivia's expense, depriving it of a potential oil reserve and effectively blocking Bolivian access to the traditional route out via Argentina. In each case, Bolivia lost territories that it had not effectively settled. Undoubtedly, Bolivian development and progress have been hampered by its difficulty of access to the outside world.

Bolivia currently imports food that it could produce domestically. At present, only 3 percent of the land is devoted to agriculture. The mountain basins of La Paz, Cochabamba, Sucre, and Tarija are filled to capacity. If there is to be expansion of food production, it must be through development of the interior since most of the altiplano is barren. The interior is not always inviting. Rainforests cover exceptionally poor soils in much of the northeast, while the east-central interior lowland is a poor clay plain. Cocaine is the major frontier crop of the outlands, where it is grown on thousands of small plots and a few major plantations. Though the Bolivian government has been cooperative, and has even invited U.S. troops to assist in the cleanup, the illicit trade persists. Cocaine, like gold before it, is sufficiently valuable to

overcome the shortcomings of distance and inadequate transportation.

Paraguay's 4.2 million people occupy a country with little known mineral wealth but ample agricultural potential. The population is the smallest of any Spanish-speaking South American country, a fact related to past disastrous wars.

In 1865, Paraguay began a war designed to provide it with an ocean port. Brazil, Uruguay, and Argentina joined in alliance against Paraguay and fought for five years until an exhausted Paraguay surrendered. Paraguay lost nearly half its original population, finishing the war with only 20,000 males. Paraguay's second adventure in territorial expansion —the Gran Chaco War with Bolivia in 1932–35— was more successful (see Figure 11–14). With extensive water engineering works, the Gran Chaco could become a reasonably prosperous agricultural district. At present, none of the four nations controlling portions of it (Paraguay, Bolivia, Brazil, Argentina) have effectively settled their Chaco lands and brought them within their respective national economies.

Landlocked Paraguay's access to the world has been complicated by the fact that the country's rail connections to Argentina cross the maze of shifting channels of the Paraná River via ferries rather than via bridges (Figure 11–15, p. 620). Costs for shipment to the port of Buenos Aires are high. Bridged rail routes eastward encourage traffic to flow through Brazil instead. Forest products, tobacco, meat, and live cattle are major exports in a land of surplus food.

FIGURE 11–14, pp. 618–619   →
Territorial Changes in South America. A series of wars and border conflicts have altered the outlines of many South American countries. Brazil has been a rather consistent winner in its territorial conflicts. Beginning as a series of coastal settlements, it has extended its control to most of the Amazon Basin, and ultimately added pieces of territory once claimed by each of its neighbors. It controls virtually one-half the continent. Luckless, landlocked Bolivia, a consistent loser, is negotiating for a territorial exchange with Peru and Chile that would return its corridor to the sea. Venezuela and Guyana dispute their common border, as do Ecuador and Peru. The contest between Britain and Argentina over the Falklands was the latest territorial war.

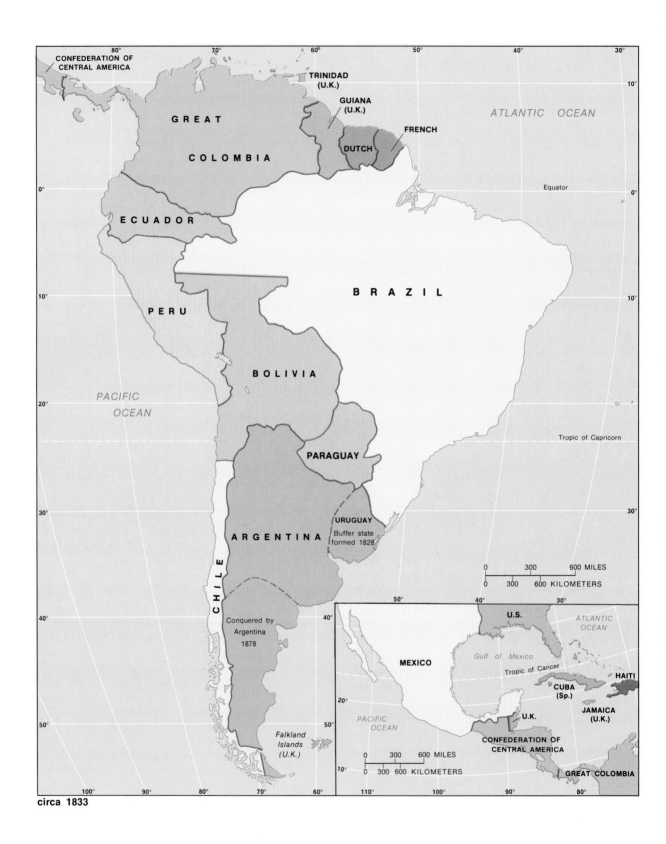

CONFEDERATION OF
CENTRAL AMERICA

TRINIDAD
(U.K.)

GUIANA
(U.K.)

FRENCH

DUTCH

ATLANTIC OCEAN

G R E A T

C O L O M B I A

Equator

E C U A D O R

B R A Z I L

P E R U

PACIFIC
OCEAN

B O L I V I A

Tropic of Capricorn

PARAGUAY

URUGUAY
Buffer state
formed 1828

A R G E N T I N A

C
H
I
L
E

Conquered by
Argentina
1878

Falkland
Islands
(U.K.)

300   600 MILES

300   600 KILOMETERS

U.S.

ATLANTIC
OCEAN

MEXICO

Gulf of Mexico

Tropic of Cancer

HAITI

CUBA
(Sp.)

PACIFIC
OCEAN

U.K.

JAMAICA
(U.K.)

CONFEDERATION OF
CENTRAL AMERICA

GREAT COLOMBIA

300   600 MILES

300   600 KILOMETERS

circa 1833

618 | LATIN AMERICA

Territorial Transfers and Border Disputes

COUNTRIES WITH
TERRITORIAL LOSSES
(with dates)

——— Bolivia

——— Colombia

——— Ecuador

——— Paraguay

——— Uruguay

——— Venezuela

Currently disputed
areas

Panama
independent
1903

VENEZUELA

COLOMBIA

GUYANA

SURINAM

FRENCH
GUIANA

ATLANTIC OCEAN

Equator

ECUADOR

1904

1907

1859

1904

1936-42

1902

1867

PERU

BRAZIL

1902

1907

BOLIVIA

1870

PACIFIC
OCEAN

1938

PARAGUAY

1895

1830

1883-94

1870

1851

ARGENTINA

Tropic of Capricorn

URUGUAY

0   300   600 MILES

0   300   600 KILOMETERS

CHILE

Border settled 1881

Indian
territory

annexed
1879

Falkland
Islands
(U.K.)

ATLANTIC
OCEAN

Gulf of Mexico

Tropic of Cancer

PACIFIC
OCEAN

HONDURAS

Mosquito
Territory

NICARAGUA

0   300   600 MILES

0   300 600 KILOMETERS

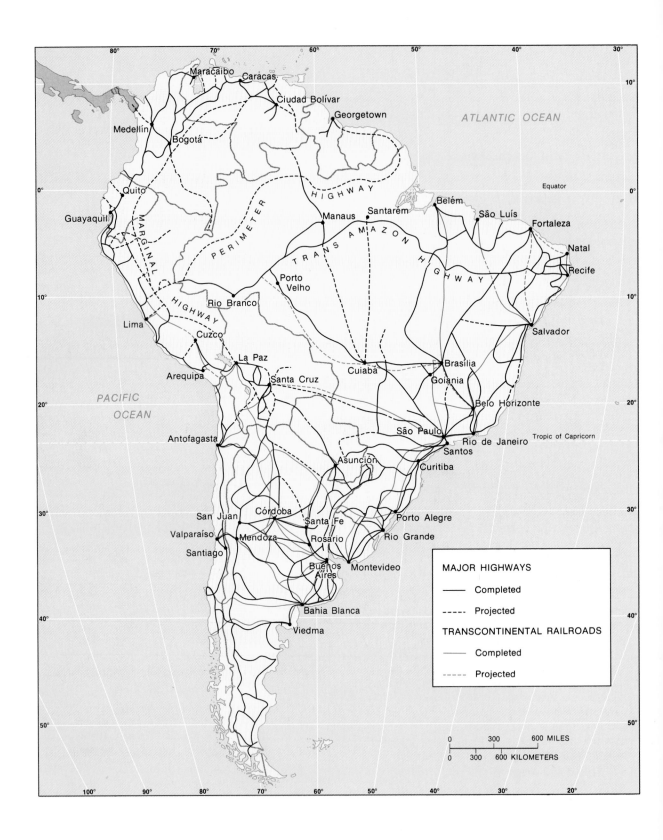

MAJOR HIGHWAYS

— Completed

- - - Projected

TRANSCONTINENTAL RAILROADS

— Completed

- - - Projected

Hydroelectric development (the huge, joint project with Brazil near the Iguassú Falls on the Paraná River) has dramatically decreased the need for oil imports. Surplus power is exported to Brazil and Argentina, earning foreign exchange.

Position is now Paraguay's advantage as Argentina develops its north and Brazil its interior west. Development and a positive change in economic fortunes come to Paraguay less as a result of what *it* does than as a result of the actions of its neighbor states.

The Southern Tier: Chile, Argentina, and Uruguay. These three countries are often treated as a unit because of their relatively high degree of development, temperate climate, and strong European cultural influences. All three economies have had strong historical links to Western Europe, remaining largely outside the sphere of American influence until the last 40 years. Collectively, these countries were a major destination for European migrants after 1920.

Though most Chileans have some Indian blood, and some Indian culture areas still exist in Chile and Argentina, these three countries have the most completely European populace and culture of all Latin American countries. All three have undergone severe economic readjustments during the last three decades. Industrialization took place early in all three, though it has not been characterized by even growth, either regionally or over time. Their farm economies all moved away from wheat, wool, and cattle for the export market and toward industrial crops to meet the needs of domestic and foreign markets. All have

a reasonably high standard of living, but they no longer lead the region.

There is great unrealized potential, and, with great strides taking place elsewhere in Latin America, the next few years may be crucial in determining whether these countries lead or are left behind.

Chile spans 2,650 miles. A collection of basins, islands, peninsulas, and oases, it would appear, superficially, to be the least likely place to achieve national unity. Yet, Chilean nationalism is well developed, and a relatively dense network of roads, railways, water, and air routes has created a reasonably well integrated unit (see Figure 11–15).

The population is most heavily concentrated in the central zone of Mediterranean climate, the fertile and moderately well watered Valle de Chile. This zone serves as Chile's single well-outlined and prosperous core. To the north, the desert zone is not well suited to irrigated agriculture, but has an economy based on mining, fishing, and industry. South of the core is a zone of cooler, wetter, less fertile land that has a reasonably high agricultural potential. Containing a coal and iron resource, suited to dairying, and rich in timber, the southern portion was a pioneer zone of advancing settlement frontier until 1973. The fiorded southern coast is similar in both climate and appearance to Norway and British Columbia. Agricultural potential there exists in scattered valleys, on the leeward sides of islands, and is, in all cases, relatively isolated from the core. The exploitation of this territory's hydroelectric potential and rich timber resources is just beginning.

Normally, copper represents 30 to 50 percent of all Chilean exports by value. Expanding production of other minerals ensures Chile's self-sufficiency in most mineral needs. Coal deposits of moderate grade and rich gas fields provide a more than adequate fuel base for intensive industrial development. Industry is comparatively well developed. The country has one of South America's few integrated steel mills, which operates entirely on domestic resources. With 83 percent of the population living in urban areas, and one of the highest urbanization rates in the world, Chile has developed a large range of consumer goods industries geared to both foreign and domestic markets.

Despite 50 years of attempted land reform, agriculture is still dominated by large, underproductive estates, which have survived since colonial times. In

← **FIGURE 11–15**

Transportation in South America. Once-isolated interiors are gradually being linked to the core areas of their respective national jurisdictions. Brazil's ambitious national highway system is nearly completed. Marginal highways projected for the eastern slopes of the Andes from Venezuela through Bolivia are in varying stages of construction. Most of the western perimeter countries have completed at least some trans-Andean road projects. Chile has integrated its most important settlements with both road and rail facilities. Well-settled portions of the Atlantic coast from Argentina's Pampas to the Northeast of Brazil boast a fairly dense rail network. Brazil has extended trunk rail lines to Bolivia, Uruguay, and Paraguay in an attempt to divert trade of those countries to its ports.

a climate similar to that of California, an increasingly wide variety of crops are produced in that district, though large areas are still devoted to grazing, dry farming of wheat, and other forms of low-intensivity agriculture. Recently, there has been an increase in produce sales to the United States and Europe, utilizing the opposition of the Southern Hemisphere's seasons to an economic advantage. A recent "poisoned fruit" scare temporarily halted exports.

Chile carries a heavy yearly trade deficit. The economy stagnates while the population continues to grow, although at a moderately low rate. Relatively well developed already, and possessed of a magnificent resource base, Chile has suffered a decline in regional importance and income.

Although Argentina suffers the same economic slowdown, its situation appears brighter. Argentina encompasses over 1 million square miles and has 32 million people. Exceptionally diverse, this most Europe-oriented of the local economies seems to have rediscovered its Latin American origins.

Argentina came to be the most prosperous state in South America by the close of the nineteenth century. (Oil-rich Venezuela has now surpassed Argentina.) The greatest resource of the country is its rich soil. An immense proportion of Argentina is usable for agriculture (40 percent is in pasture; 20 percent is in cropland). It is the fourth largest producer of cattle in the world, and the largest meat exporter. It is a major producer and exporter of wheat, corn, soybeans, wool, mutton, and pork. When the United States embargoed grain sales to the Soviet Union, Argentina filled the gap in the grain trade.

In addition to its rich agricultural base, Argentina possesses a sizable mineral resource, only now beginning to be exploited on a large scale. Oil, long a crucial import, is now more than sufficient for the domestic market; it comes largely from fields around Comodoro Rivadavia in Patagonia and Tierra del Fuego in the far south. A vast hydroelectric potential is being harnessed.

The population is highly urbanized; almost one-third live in metropolitan Buenos Aires, making it one of the world's largest cities. That city is Argentina's largest and most diverse manufacturing center, producing a full range of consumer goods and machinery in addition to processing agricultural production.

Almost one fifth of Argentina and all of Uruguay are in the **Pampas,** an area of prairie and steppe lands of incredible fertility (Figure 11–16). The Pampas has a flat to gently rolling terrain, originally carpeted in thick, rich grass that stretched as far as the eye could see. It is considerably flatter than the Great Plains of North America, an area with which it is often compared. The similarities of the two grasslands in appearance and settlement history are amazing. In both cases, the climatic transition is drier to the west, wetter to the east, and agricultural products tend to flow eastward to centers for processing, distribution, and consumption. However, whereas the U.S. and Canadian Plains together have many east-west major transport routes and many medium-sized cities engaged in processing and marketing, the grasslands of Argentina and neighboring Uruguay have but one major focus each (see Figure 11–16).

In both areas, there was rapid economic change related to changes in the world economy and in agricultural and transport technology. British capital helped create a dense rail net on the Argentine Pampas, clearly focusing on Buenos Aires (see Figure 11–16). The lines built in Uruguay focused their Pampas on its capital, Montevideo. During the second half of the nineteenth century, a technological revolution in farm machinery, transport, commercial canning, and refrigeration set into motion a series of changes that drastically altered land use, production, and lifestyle in the Pampas. Purebred cattle grazing on cultivated alfalfa pastures replaced scrub cattle subsisting on uncultivated, coarse native grasses. Commercial wheat farming then partially displaced cattle on the wetter fringes. Corn later displaced wheat in areas closer to Buenos Aires and in the wetter north; it was then partially replaced by still other crops as land use competition from market-oriented fruit and vegetable production, dairying, and industrial crops proved more profitable. The resulting pattern is roughly one of concentric circles of different land uses outward from Buenos Aires (see Figures 11–12 and 11–16). In Uruguay, where sheep are more important than cattle, the evolution of the landscape has proceeded more slowly.

North of the Pampas proper is a much wetter (though nonetheless fertile) area designated Entre Ríos ("the land between the rivers"). Corn, and now rice, are the major grains. Rosario (population 1.6

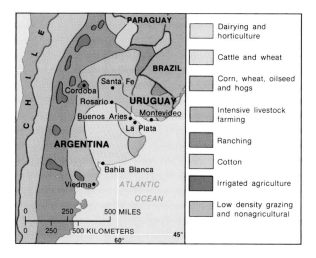

FIGURE 11–16

The Pampas. Argentina and Uruguay were of little interest to the early Spanish conquerors and settlers. Deficient in the sought-after gold and silver, they possessed one of the world's greatest agricultural resources in the form of subtropical grasslands endowed with fertile soils. Early settlers grazed longhorn cattle and sheep on the rich natural pastures, exporting low-quality meat and wool to the markets of Europe. In the intervening years, agricultural production has intensified. Farm production is in increasing harmony with the potential of the environment.

million) and Sante Fe (population 535,000) are the large cities that dominate the area. Both are large industrial centers with diverse production. Rosario has a sizable steel industry.

In the northernmost portion, there are rich mineral-bearing districts in the Andes and the high plateau. Tucumán, the most densely populated rural district in Argentina, is a major sugarcane producer. The city of Tucumán has been designated as a growth pole by Argentine planners in an effort to decentralize function and population from the single dominant center of Buenos Aires.

A major oasis center is Mendoza, an area of semi-desert with winter rains and mild winter temperatures. The extension of irrigation there has created vast orchards and vineyards in a countryside of Mediterranean flavor. Most Argentine wines, olives, and major amounts of peaches, cherries, plums, and apricots are produced here. Well tended and picturesque, the district is a favorite resort for Argentine, Chilean, and European tourists. The city of Mendoza is also industrializing rapidly, and the whole district constitutes one of Argentina's most rapid growth areas.

Patagonia, the country's last frontier, is a vast triangular plateau. It is mainly desert or steppe, and grass is the dominant vegetation. Sheep herding is the principal function, with the cool climate producing a thick coat of wool. The government has targeted this area for rapid in-migration. Argentina's growing fishing industry is based here, and the country's largest coal, oil, and gas reserves are all located in Patagonia. However, distance and poor connections with the core area limit the frontier's potential.

Uruguay is physically an extension of the Pampas on the other side of the Plata estuary. Culturally, it is a transition area between Argentina and Brazil. Like Argentina, it has a population largely of European origin. Though virtually everyone speaks Spanish, fully a quarter of the population is of Italian descent, and almost 40 percent of the total is of non-Spanish ethnic origin. While its wealth is almost entirely based on agriculture, only 16 percent of the work force is directly employed in farming. Industry, like that of Argentina, consists of food processing and production of consumer goods for the domestic market. The economy of Uruguay is even more centralized in the capital city. Some 85 percent of the population is urban, and almost 60 percent (1.7 million) of the country's people live in metropolitan Montevideo.

Uruguay has an excellent supply of water and a temperate climate in which frost occurs only about once a decade. The soil is exceptionally fertile and gave rise to a natural cover of rich, nutritious grasses. Stock raising is the most important occupation in a land with 10 times as many stock animals as people. Sheep rearing is most important, though cattle raising is rapidly overtaking it.

Uruguay is prosperous, with a yearly per capita income higher than that of Argentina or Chile. It has a high rate of literacy, a large middle class, a fairly equitable distribution of wealth, and a reasonably high standard of living. Meat and other basic foods are so cheap that the per capita income of $2,100 is deceptively low in terms of actual purchasing power.

With a 60 percent yearly inflation rate in the 1970s and a rapidly aging industrial plant in its food-

Patagonia. *Patagonia remains Argentina's last frontier. It is mainly a plateau, at times steppe or desert, and grass is the dominant vegetation. Sheep herding is the principal function, though the area could produce grains in its wetter and warmer northern reaches. The tenant farmer shown here will cultivate grain (oats or barley) after plowing up the overgrown pasture. He will receive the bulk of the proceeds from the sale of the crops in return for his labor. After three to five years of cultivation, he will seed alfalfa and move to another location. The renewed alfalfa pasture land will graze high-quality merino sheep for seven years.*

processing sector, Uruguay has experienced some economic problems. A costly program of public health care, pensions, guaranteed minimum income, and other social benefits has often been blamed for economic chaos. Formerly, one worker out of every four was a government employee in the world's most cumbersome bureaucracy; the ratio has been reduced to less than one in five in recent efficiency moves. The greatest single cost has been that of importing petroleum. The antidote to the energy problem has been a series of hydroelectric dams built to create huge reservoirs. In the relatively level terrain, much excellent land has been inundated with water in the process. Land reform moves ahead as inefficient estates are nationalized, subdivided, and resettled by farm families. Millions of fruit trees have been planted, and oranges and orange juice are a new lucrative export.

With few raw materials at hand and a small domestic market, Uruguay's industrialization has been severely handicapped. The highly successful wool and cotton textile industries, both using domestically produced raw materials, are major exporters.

Tourism is expanding rapidly. With its pristine beaches, lovely climate, and excellent hotels, Uruguay is attracting droves of tourists from Argentina and Europe.

Brazil: The Giant Awakens. World observers have coined such terms as "sleeping giant" in reference to Brazil's enormous economic potential; critics have noted, somewhat sourly, that Brazil is perpetually "poised for takeoff, but never gets off the ground." As a nation, Brazil has demonstrated both genius and unpredictability in its race to achieve full development and recognition as a world power. Brazilians are convinced of their destiny, and undeniable progress has been made toward achieving those goals during the last 35 years. Words such as "flamboyant," "innovative," and "dynamic" are the adjectives most widely applied to Brazil; they are not the usual words used to describe Third World nations.

Detractors and doubters emphasize Brazil's problems: unmanageable foreign and domestic debt, repeated financial crises, serious inflation, devalued currency, and exploitative development (with dire environmental consequences). A brief mental review of the first century of American history, however, reveals the identical problems. From the positive side of the ledger comes wealth of natural resources,

THE STRUGGLE OVER THE FALKLANDS

The islands known to the British as the Falklands, and which Argentina calls Las Malvinas, constitute some 200 islands and rocky pinnacles in the South Atlantic, totaling 4,700 square miles of land. The total population is only about 1,800. The remote, treeless, windswept islands seemed like a cold, desolate desert to early explorers. Apparently, no human habitation preceded European discovery, which is of uncertain and controversial date. The first recorded landing is credited to an English sea captain in 1690. No country seemed overly excited at the prospect of colonizing the barren islands. The French established a small settlement in 1765, and sold the colony to Spain in 1766. The British, also in 1766, established a small force on another island. Britain and Spain achieved a standoff in a vaguely worded treaty in 1770, and Britain abandoned occupancy in 1774. Argentina claimed the islands in 1820 and sent settlers; Britain took possession by military force in 1833. Britain exhibited little official interest in the islands, or their inhabitants, who were, and are, mostly employees of the royally chartered Falkland Islands Company. The company owns about half the total land area. Prior to the Argentine invasion in 1982, the only regular civil air service was provided by Argentina.

The Argentine invasion, British counterinvasion, and associated war at sea were costly in losses of ships, aircraft, and service personnel. The costs of continuing defense will be a severe strain on the United Kingdom's economy. Why did Argentina attempt to impose control over people, 97 percent of whom are of British heritage and wished to remain associated with Britain? Why did Britain fight so vigorously to retain control over half a million sheep and a handful of colonists nearly 8,000 miles away? After all, Dr. Samuel Johnson, in a 1770 pamphlet, termed the islands "a bleak and gloomy solitude . . . where a garrison must be kept in a state that contemplates with envy the exiles of Siberia; of which the expense will be perpetual and the use only occasional." Many would accept this as a modern judgment as well. In 1982, the Argentine writer J. L. Borges compared the war to "two bald men fighting over a comb."

The answer involves competing national prides but ultimately may focus on the adjacent seas and seabed rather than on the islands themselves. As the technology of undersea gas and oil exploitation evolves, there will be a scramble for national claims and control over the ocean floors beyond the continental shelves. The controversial "lake" approach would divide the right to exploit minerals on or under the seabed, measuring "boundaries" halfway between seacoasts, or islands. The Falklands, if retained by Britain, take on new significance as basepoints for assigning Britain potential control over a major portion of the South Atlantic. Many geologists believe that the seabed between Argentina and the Falklands could contain a major oil field whose future ownership may be the major point of the Falklands dispute.

territorial size, available land for settlement expansion, ambition and drive. Again, the similarities with the United States are astounding. The prerequisites for success are all in place.

Brazil is enormous; it is roughly equivalent in size to the United States, Canada, or China. The economy is now multidimensional after a long flirtation with one boom commodity or crop after another. The capital has been moved to the interior in a bold step to develop the hinterland and expand the frontier of development. Enormous sums have been spent on infrastructure to integrate the diverse regional components into a unified economic whole. The country never lacked a cultural unity; the self-cognizance of its people as Brazilians, of its culture as definitely and uniquely Brazilian, is undisputed.

Despite the strong national cultural unity, sheer size and geographic diversity preclude dealing with the country as a unit. Here, Brazil has been divided into five subregions, each with a distinctive character and a particular set of problems (Figure 11–17). The Northeast—the earliest settled and currently the poorest—encompasses the familiar "bulge" of Brazil from the edge of the Amazon floodplain to the hill lands south of Salvador. It is plagued with overpopulation, a disastrously fluctuating climate, and an obsolete economy. It is the part of Brazil that is characteristically and unequivocally Third World. The Southeast is the country's economic core. It is dynamic, growing, increasingly industrialized, yet still expanding. It is also a prodigious agricultural region and source of raw materials. The South is temperate Brazil, overwhelmingly geared to the production of food for the domestic economy. It has a modern, well-mechanized agriculture and contains Brazil's chief sources of coal. Active pioneering continues as settlement spreads to its remaining fringes, though the economy is already in transition to the industrial phase. It is a fat and generous land with a strong European influence stamped on its landscape. The interior of Brazil is a pioneer region. Climatically, it consists of a rainforest north and savanna south. More important, the southern part of the interior (officially designated the Center West) is the region of active, rapid colonization, whereas the northern portion (the North) remains more an area of unrealized potential. The Center West is centered on the new capital, Brasilia, and constitutes a zone of intense development and massive investment. It is the land of the future. The North, much more difficult to utilize, is still the land of boom and bust, the final frontier.

The Northeast: Third World in Fact. Brazilians concede that the Southeast is the "heart" of Brazil, but consider the Northeast its "soul." The Northeast is the source area for much of Brazilian music and literature. Here is where African influences are strongest, the environment most taxing and insidiously unpredictable, and poverty most visible and grinding. In some ways, it is a residual of the colonial era, out of phase with modern Brazil. Yet, it is a physically beautiful area and warmly hospitable. It was not only the site of the first settlement, but also that of the first Brazilian boom—sugarcane. It was only lightly populated at the inception of development, and the labor force consisted of slaves imported from Africa, a migration that persisted until the 1870s. The sugar boom barely lasted a century as the evolution of Britain, France, and others into colonial powers (in the eighteenth century) brought the Portuguese monopoly to an end. Lesser booms in cacao, tobacco, and cotton followed, each bringing temporary prosperity and ultimately ending in decline as more efficient production began elsewhere. Though successive booms were shorter lived, each brought a new wave of settlers. The better-watered coastal areas continued in production, but expansions into the drier back country, the **sertão**, often never reached full potential before drought brought an untimely end to boom.

Depressed for the last hundred years, the Northeast was long the most important section of Brazil. In an effort to stem economic decay and to revive the area, the government has begun an intensive campaign to industrialize cities and to improve agriculture through irrigation, fertilization, and capital investment. Per capita income is less than half the national average; economic redevelopment is the most pressing need. The world's second largest petrochemical complex has been built outside Salvador in Bahia state. A capital-intensive type of industry, however, it requires few employees. The choice of the site was not based solely on economic need, since Brazil's largest producing oil field is located in Bahia state and promising new fields have been discovered offshore of the Northeast (see Figure 11–12). Other large investments include several hy-

FIGURE 11–17
The Subregions of Brazil.
Brazil, with an area of 3 million square miles, is one of the largest of the world's countries. Settlement was long confined to the states bordering the Atlantic from Ceará in the northeast to the Uruguayan border. The interior was divided among a sparsely populated backland, called the sertão, the almost empty rainforests of the selva, and a pestiferous zone of wet-and-dry climate that alternated between scrubby deciduous forest and rough grassland. The coastal provinces are divided into three subregions: the Northeast, a zone of poverty and plantation economy; the Southeast, Brazil's burgeoning industrial area; and the South, a subregion devoted to temperate crops and production of domestic food supplies. The interior is divided between the North, the still lightly populated selva, and the Center West, containing the capital and the agricultural frontier.

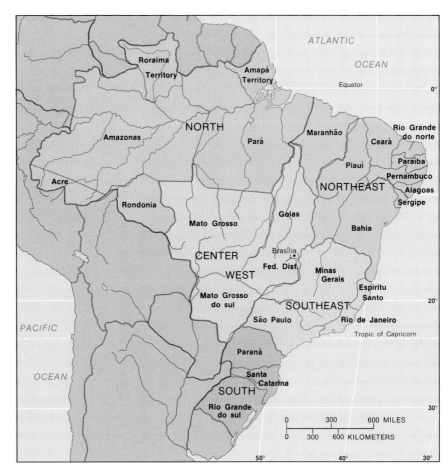

droelectric projects, major port improvements, large-scale irrigation schemes, and huge cement plants. The Northeast possesses perhaps the best developed highway transportation network in the country. It is now fully linked to the road network of the Southeast, to Brasilia, and to the rapidly growing Center West.

Despite this intensive investment, the Northeast lags on almost all fronts. Birth- and death rates are the highest in Brazil. Much of its urban growth results from the flight of farmers escaping drought and disaster in rural backlands. Irrigation projects can locally ameliorate the effects of drought but are extremely costly to construct. The bulk of the sertão is still not irrigated. Northeast Brazil is no longer stagnant, but its problems remain immense.

The Southeast: Heartland and Headquarters. Exciting and urbane Rio is not Brazil's largest city, but it is certainly its most magnificent. Its less sparkling but much larger rival, São Paulo, supplies the financial backing for Brazil's rapid growth. Rio is Brazil's welcoming facade—elegant, fashion conscious, full of artistic and architectural brilliance (if bordered by pathetic slums), and quite comfortable in its casual and unhurried lifestyle. São Paulo is frenetic, success oriented, and highly commercial; it is Brazil's money machine. Paulistas often characterize their city as "the brains and the work" that supports Brazil, and Rio as Brazil's seat of "relaxation and fun." Both are backed by well-developed and productive hinterlands that contribute export crops, raw materials, foodstuffs, and manufactures to these two urban giants that dominate almost every aspect of Brazilian life.

The Southeast contains over 40 percent of Brazil's population, most of them dwelling in its three huge urban complexes. São Paulo, with over 15 million people in the central city itself, may yet become the

BRASILIA

No one believed it would work. Since the 1920s, Brazil had had a zone reserved in the interior for the purpose of constructing a new federal capital. It was to serve as a new national focus, associated with the development of Brazil itself, away from the coast with its European ties. Opponents pointed to the immense expense of construction in a lightly settled and poorly developed portion of the country. Advocates defended it as a necessary investment, without which the riches of the interior would remain untapped.

President Juscelino Kubitschek, an exceptional leader by any standard, simply decreed in 1956 that the long-planned capital be built. Ground was broken almost immediately; plans were fully drawn by 1958, and the capital was inaugurated and operative in 1960, all within his five years as president.

Brasilia was to be ultramodern (reflecting Brazil's break with its past and commitment to the future). It was to act as a growth pole, an attraction for both developmental investment and migration to the interior. Brazil has a long tradition of architectural excellence; the new capital, it was decided, should reflect that tradition. Contests for best design were held for the total plan and for individual structures. In keeping with the modernity theme, the city was designed in the shape of an airplane. Located in the headwaters of a dozen streams that were easily impounded, the otherwise unimposing physical site was enhanced by a series of splendid artificial lakes. Wide, landscaped boulevards were to connect each section of the city, and parkland was to provide greenery and recreation in every neighborhood.

The start was slow; virtually everything, from worker housing through food, cement, and construction equipment, had to be flown in at great expense. It appeared from afar as though a city from the future (or outer space) had been dumped in the middle of the tropical forest. Up close, Brasilia appeared rather raw, barren, and decidedly empty in its early years.

From the first, there was resistance to moving to the new city. Offices often remained in Rio, sending a few staff to Brasilia as a token presence. When employees were physically transferred there as office space became available, most left for Rio on Thursdays, retaining residence in the old capital. Government decrees prescribed firing for employees who did not move families there; in turn, employees protested that amenities were not available in Brasilia. Foreign governments who maintained their embassies in Rio were threatened with a break in diplomatic relations. Somewhat grudgingly, the world's nations complied; after all, Brazil was a large country and, frequently, an important commercial customer.

Growth proceeded, but many problems plagued the project. The design had assumed public transportation; there was no place to park cars, either at work or at home. Brownouts were frequent because local power supplies were inadequate; shrubbery shriveled in the merciless sun, and lawns refused to grow in the baked lateritic clays. Buses did not run since scarce fuel supplies were allocated to the construction industry on a priority basis. People climbed up 10 floors to barren apartments (there were no

Brasilia. *Resembling a surrealist dream or a futuristic, science fiction scene, Brasilia rises amid the wild forests of Brazil's interior almost without warning. Brasilia was created during the last 30 years on a raw, uninhabited site, far from the thickly settled areas of the coast. It was built with a purpose: the redirection of Brazil's energies to the development of its vast, underutilized interior. Brasilia was designed by some of the world's most outstanding architects to symbolize this break with the traditions of the past.*

furniture stores) rather than hazard being stuck in a suffocatingly hot elevator during outages.

For those who were not originally included in the plan, the city did become an instant attraction. The rural poor flocked to Brasilia in great numbers. Despite government efforts, shantytowns arose in peripheral locations. Despite zoning, farmers cleared land within the city limits and commenced production. Brothels and bistros opened shop on vacant lots and at the end of bus routes. Those who were "planned for" often did not like the plans. Some middle-class Brazilians balked at living in huge apartment blocks which they nicknamed "filing cabinets," erecting private, single-family homes on vacant ground to which they had no clear title. Plans had to be modified accordingly; Brazilians are nothing if not independent.

By 1970, a decade after its official opening, the city had exceeded a half-million inhabitants and outgrown its plan. By 1990, the population was nearly 2 million. The project had succeeded beyond anyone's dreams. Problems still abound, but now most streets are paved, electrical service is reliable, and the water problems have been solved. Food is plentiful, and prices for everything (except housing) are within national norms. Until recently, Brasilia was purely a governmental center. Light industry is now proliferating in carefully zoned districts; services have become available. The raw atmosphere has disappeared as carefully tended greenery (now chosen from species adaptable to the environment) replaces barren earth. It is now a capital city that inspires admiration and respect, if still a little stark for Brazilian tastes.

More important, Brasilia has functioned as the anticipated growth pole. Fully (and fairly directly) connected to all the country's regions, it can function efficiently as a national capital. New commercial farming districts have opened throughout the central highlands of Goiás. It is still a lightly peopled area, but certainly not a wilderness any longer.

world's largest metropolitan area. Coffee supplied its original wealth, but industry is its life's blood now. Literally thousands of industrial plants of all sizes and representing all types of production support its staggering rate of growth. The problems promoted by this rapid rate of industrialization are legion, but the city largely ignores them. It shrugs off its daily smog and even takes a perverse pride in holding the world's record for the largest and longest (in duration) traffic jam.

Santos, São Paulo's outport, is not a good natural harbor, requiring constant deepening and expansion to keep pace with the volume of São Paulo's production and trade. A half-dozen major cities (population over 500,000) inland from São Paulo bristle with new industrial investment, absorbing the overflow of both population and investment from this colossus of Brazilian cities.

Rio has passed the 10 million mark as it spreads almost indiscriminately over hill and valley in search of room at its congested site. Explorers, mistaking the elongated bay for a river, named it Rio de Janeiro ("river of January") to mark the month of its discovery. Most of coastal Brazil is flanked by a narrow coastal plain that rises abruptly in a steep scarpland to the edge of the Brazilian plateau. There are few good harbors; Rio and Salvador are the major exceptions. It is the plateau beyond, rather than the coastal plain, that made Rio wealthy. Gold from the interior was shipped out through the harbor. The thick and fertile soil of the land beyond the pass was to sustain its growth long after the gold was gone.

As with São Paulo, coffee supplanted gold as the source of local wealth, but minerals never totally lost their importance. Rio's splendid harbor, the largest and best in the country, served to export both. The city has always placed a priority on good living and exudes an atmosphere of excitement. Though it has lost the capital function, it remains "the center of the action," retaining its role as the dominant center for media, entertainment, tourism, fashion, and the arts.

Rio is also a huge industrial center. South America's largest shipyards, actively competitive with those of Europe and Japan, are among its major employers. Giant factories, refineries, and chemical complexes ring the outer edges of the bay (which they threaten with pollution), and over a million workers crowd

into its garment shops and shoe plants every day. Add to this the miles of beaches in a perpetual summer climate, and the attraction of Rio for millions of migrants from the interior ceases to mystify.

Between these two huge cities is the growing urban-industrial district of the Paraíba Valley which will soon physically link the two. The valley was the first area of coffee production; now this land is filled with lucrative citrus groves and rice paddies. As southern California and Florida urbanize, their fabled groves have gone to development, and Brazil now dominates world orange and orange juice production as concentrate leaves Rio in a fleet of tankers for world markets. The huge steel mill at Volta Redonda, efficient and ever-expanding, is the prime industrial objective of the valley, and dozens of machinery and transportation equipment plants now process its steel into higher-value items.

There is a third major industrial center in the Southeast, in the state of Minas Gerais ("general mines"). Belo Horizonte, a city of 3 million people, was founded as the state capital 100 years ago. It is also a mining and industrial town experiencing its seventh mineral boom. The district of Itabira nearby is one of the largest known deposits of high-grade iron ore (over 65 percent iron content). The iron reserves are virtually inexhaustible. Steel mills, manganese reduction plants, metal smelters (other ores also abound), and foundries are scattered through the district. Vast mineral wealth provides a solid base for continued industrial growth.

The South: The Land of Plenty. The South differs strongly in culture and history from the rest of Brazil. It is the most thoroughly Europeanized of Brazil's regions because of original settler origin and the strong presence of recent migrants in the total population. This was the last frontier of coastal Brazil, settled largely since 1860 and still clearing its last acreages. Not well served by transportation until recently, it originally developed much like neighboring Argentina and Uruguay—a land of ranching that turned to crop farming comparatively recently.

The major ethnic groups settling the area were Italians, Germans, and Slavs; most came as pioneer farmers during the late nineteenth and early twentieth centuries under Brazilian government-sponsored settlement programs. The Germans and Swiss produced a mixture of traditional European and American crops, adopting corn and beans but devoting considerable space to the rye-potato-pig and dairy culture combination that was their familiar European standard.

Descendants of Italian migrants farmed the slopes, developing orchards and vineyards, producing wine for Brazilian and export markets and hearty durum wheat for traditional pastas which now enjoy widespread popularity in Brazil.

Slavs of Polish, Russian, Ukrainian, and Serbian origin cluster mainly in Paraná state. They often have much larger farms and have entered into coffee and other cash crop production in addition to large-scale dairying and beef production (see Figure 11–13). The landscape exhibits many signs of this non-Iberian ethnic input. Half-timbered houses, onion-domed churches, blue-stuccoed cottages, concrete barns, and pious roadside chapels are all cultural elements of Central Europe transported to subtropical surroundings.

Brazilians of other regions stereotype the South as hurried, disciplined, hard working, and a little dull, but the general optimism and willingness to take a gamble are as evident here as anywhere in Brazil. Loyalties to the country are strong (as is assimilation), so the different ethnicities do not imply any separatism. Life in Brazil has brought prosperity, and periodic economic crises have had little adverse effect. In the words of Southerners themselves, life in Brazil is "a kind of beloved chaos, where opportunity far outweighs adversity." Stress is considered unnecessary and in bad taste. The subregion is Brazil's second richest, and income is well distributed. Pôrto Alegre, with nearly 2 million people, is Brazil's sixth largest city and functions as one of its regional capitals.

Rio Grande do Sul, the southernmost state, is transitional in culture and function to the Pampas yet is a cultural microcosm of Brazil. Varig, the giant of Brazil's airlines is based here and typifies the entrepreneurial tenor of the South which hides deceptively behind the bounty-inspired outward calm. Itaipú, the giant hydroelectric project on its western borders (and funded in part by local private capital), exhibits the region's commitment to the Brazilian dream.

Paraná, the northernmost state, is caught up in the race for development. The expansion of industry from the Southeast is as readily visible as the expansion of agricultural frontiers. In-migration boosts population growth to a rate of 7 to 10 percent yearly in a fast-paced economy. Coffee plantations spread out over hills that were forested only a decade or two ago. This state is the new center of national coffee production. Yields are high, and quality is excellent as Brazil seeks higher returns on its shrinking coffee crop. Frosts are a costly hazard; this area is the physical limit of expansion for coffee.

The Center West and North: The Final Frontiers. Much as Horace Greeley once advised American youth to "Go West," so does the Brazilian government. Brazil has always envisioned its interior as a land of untapped wealth and unlimited opportunity. It is really two regions, the tropical selva (rainforest) of the north and the scrub and savanna lands of the south. Neither region is easily put to conventional use, and the "promise" is not guaranteeable. Nor is this the first time that a boom has hit the interior.

The North encompasses 40 percent of Brazil's territory, yet, with only 7 million people, it remains virtually empty. Forests range from magnificent stands of timber to stunted bush in response to variable soils and local climatic conditions. It is neither the "green hell" of some writers nor the "green mansions" of another. The humidity level is perpetually steamy, and rain is virtually a daily occurrence, if brief in duration. Diurnal and seasonal temperatures vary little; monotony is as universal as humidity. The forest floors are virtually clear of vegetation, and an atmosphere of dusk prevails as the canopy of leaves nearly shuts out the sun. It is largely silent unless birds are disturbed by human intrusion or a troop of monkeys passes through. Mosquitoes abound, and insect life is profuse and varied; there is a faint odor of vegetational rot. Piranhas and boa constrictors do exist, but they are far from an omnipresent danger. The air of mystery is a product of the human mind; the fascination experienced by visitors is understandable.

Since first observed by Europeans, the selva has been thought to be incredibly rich. Generations

RONDÔNIA: FRONTIER OF PROMISE OR DISASTER?

The Amazon River basin contains the largest remaining area of rainforest in the world, and Brazil's share of it dwarfs those of neighboring states. Scientists worldwide have come to understand the role of the world's rainforests in weather and climate and in the preservation of a breathable atmosphere. Conservationists see it as the last of a unique environment, one that houses plant and animal species not yet known and with potential value to humanity that may be lost forever in the process of deforestation. Brazil sees it as an area of unused land that offers promise as a relief for its rural overpopulation and a springboard to total development.

Lumbering is not the cause of the destruction of the Amazon rainforest. Some 70 percent of the world's tropical hardwood lumber comes from Southeast Asia, where pressures of population and commercial exploitation are much greater. Brazilian pressures on the rainforest come mainly from agricultural colonization.

At the peak of the land rush, 150,000 settlers a year moved to Rondônia. What had begun as a trickle of migrants in the 1970s turned into a torrent after the paving of the highway in 1984. The flow has since diminished, and with the best land gone, migrants are pushing on to the neighboring state of Acre.

Though undoubtedly destructive, this is not the end of the rainforest. The Rondônia scheme may have resulted in the sacrifice of 3 percent of the total Brazilian rainforest, much less than that destroyed in Asia and Africa through lumbering alone. Rondônia, and neighboring Acre, like the state of Goiás, do have areas of fertile soil suited to sustained farming. In truth, most of the rainforest, however, has highly infertile soil and is not likely to be released to colonization. Malaria and other tropical diseases are rife. There is no electricity or running water. There are insufficient schools and medical services. It is difficult to imagine the factory workers of São Paulo or members of the urbane business community of Rio packing up and moving to Rondônia. Even the urban poor have little incentive to leave the **favellas** and return to the backland they left before.

Who migrates to the rainforest colonies? Most are from the ever-dwindling Brazilian farm population, which is statistically more likely to migrate to the city than to the wilds of the rainforest.

Preston James, a noted Latin American geographer in the United States, generated the concept of the "hollow frontier" in his observations on Brazil. He noted the tendency for Brazilians to move in and develop an area, only to abandon it after a short time, having exploited it to the maximum in the process. The observation was correct in its time, but that time is past. A largely urban, better-educated Brazil no longer equates with a historical Brazil where 80 percent were farmers and the level of learning and opportunities were much lower.

There are genuine concerns. Cattle barons may buy up the land of discouraged farmers and attempt to replace farms with cattle ranches. The destruction of the rainforests could indeed have adverse environmental effects, but Brazilians see a parallel between the development of North America and Brazil, pointing to our agricultural wealth and the role of our frontier in attaining total development.

The few remaining Indians who occupy portions of Rondônia do have rights, and the Brazilian government has recently shown increased interest in protecting those rights. The Indians have developed unlikely allies among the professional gatherers of rubber, Brazil nuts, and other forest commodities who also oppose rural colonization. The World Bank and other funding and lending agencies, on which Brazil relies for credit, have insisted on a stringent plan that conserves areas of forest, preserves Indian holdings, and provides a conservational framework for settlement. Brazil has followed these guidelines within reason. Fears are blown out of proportion, as even the undisciplined rush of squatter farmers out to the highways of the northeastern interior (in the 1970s) really led to clearings that are rather widely spaced. There was not any wholesale destruction of rainforest then.

While the developed world sermonizes about conservation and raises the specter of world environmental disaster, the less developed countries counter with questions about the morality of a world banking system that sustains poverty and, in effect, forces such answers as colonization schemes. Demands from the developed world for lumber are a more serious cause of rainforest destruction, overall, than a half-million Brazilian peasants with axes and hoes. There is the serious question of sovereignty, the right of Brazil to pursue its own domestic policies without outside interference. The problems involved here are complex, and the answers are not clear. What is apparent, however, is the need for a reasoned, balanced response.

passed before science fully understood how so poor a soil could produce so rich a forest. In the meantime, many cleared the land in anticipation of rich yields that never developed and, after abandonment, the forest reclaimed the clearing. Little truly virgin land exists; humans have been virtually everywhere within the selva in search of rubber, lumber, minerals, or farmland.

The Amazon was the selva's only transportation route until the dramatic development of the Trans-Amazon Highway with its tributary roads (see Figure 11–15). The highway is now connected with the economic heartland of Brazil, but road maintenance is a major and continuing problem.

A brief rubber boom occurred at the turn of the century; rubber trees are native to the area, and the gathering of wild rubber was a major and lucrative source of employment. Rubber plantations developed by the British and Dutch in Southeast Asia soon eclipsed the boom. Attempts to establish rubber plantations in Brazil were dogged by isolation and the shortage of labor. Isolation has now been minimized, but the labor shortage remains. Synthetic rubber, technically superior for many uses, has depressed chances of revival. Medicinal roots and herbs, nuts (e.g., Brazil nuts), dyewoods, and some commercially valuable fibers are still gathered, enabling local people to supplement subsistence farming with a meager cash income. Slash-and-burn clearings are still observable.

Manaus, replete with its diamond-embellished, green-domed opera house and other relics of the rubber boom has begun to grow anew under the impetus of a new highway-inspired boom and a nationwide retail function. Its population approaches a million as new investments in gold, timber, and agriculture enter the region. This time, the boom may be of longer duration; it is multifaceted, and failure of one commodity or endeavor will no longer result in regional economic collapse. In the 1970s, Manaus was designated a duty-free port to stimulate the local economy. The results have far exceeded expectations, with millions of Brazilians yearly descending on its markets to purchase imported appliances, luxury goods, and consumer items. Virtually a legalized black market, it is a world-class emporium with the atmosphere of a bazaar.

Mineral resources are evidently plentiful and rapidly coming to light in the North as geologic prospecting inevitably accompanies settlement. Gold, the persistent temptation that has lured people to the interior since the discovery of Brazil, has touched off the most recent boom.

The mining camps of today are as tawdry, dangerous, lively, and grimy as those of the California '49ers. Prices there are inflated beyond belief, theft and murder are rampant, alcohol and drugs abound, and very few grow rich. The glamour is in the gold itself, not the camps or their lifestyle.

Other mineral booms are less spectacular draws but far more sustainable. Geologists have uncovered the world's largest known body of high-grade iron ore at Carajas in eastern Pará state; ores range from 60 to 80 percent iron content, an incredible richness. Depressed international iron ore markets and current Brazilian oversupply have not dampened enthusiasm. Brazil will likely dominate world steel production in the twenty-first century as the cheapest producer and supplier. Known bauxite reserves are immense, and prospecting is only beginning. Large deposits of manganese are already being exploited from huge reserves in Amapá territory. Industrial diamonds are recovered in placer operations. There is oil. Original finds were small, but two new fields near the Peruvian border appear promising.

Despite the obvious mineral wealth, agriculture may offer a more stable promise for the future. Brazilian agroscience has developed disease-resistant cattle breeds; uncovered native, soil-enriching legumes and grasses; and otherwise made agriculture in the area more promising. The so-called Great Carajas scheme, underway in eastern and southern Pará state, has cleared forest and scrub for massive cattle ranches. To reduce environmental degradation (and to diversify the regional economy), it envisions a combination of sustained forestry operation, rotational crop farming on mature pasture lands, the already mentioned iron mining, and a series of hydroelectric projects on some of the Amazon's major tributaries. This grand-scale effort in megaregional planning is intended to alleviate the population pressure and poverty of the Northeast, supplying a source of employment and migrational opportunity, as well as serving as a means of developing the interior.

The Center West is apparently a more fortuitous developmental region. It now houses almost 8 percent of Brazil's population and is growing at a rate of 8 to 10 percent yearly. Brasilia, despite a slow beginning, has proven to be a qualified success. The improved transportation necessitated by Brasilia's creation, and the large regional market created by the capital itself, have combined to reverse the fortunes of the area, eliminating its isolation and enabling local sales of farm production.

Brasilia has grown more rapidly than expected, as thousands of migrants from the Northeast and the sertão throng to its environs. The official, planned city contains only half the population, with temporary barracks for construction workers, hastily constructed slums, and (sometimes self-built) single homes housing the rest. Planned for a population of 500,000, it now contains at least three times that number.

The farmlands of southeastern Goiás state are a virtual continuation of the fertile Brazilian Highlands. Coffee production is of increasing importance, but wheat and corn for the capital market are also major crops. Market gardening has been developed by squatters, while cattle rearing spreads over the savannas of the southern half of the province. The capital generates industrial demands that are best met by nearby districts; thus food-processing, clothing, and printing plants have already sprung up in what promises to become a new light industrial base. The railroad has been extended to Brasilia and will ultimately reach Belém at the mouth of the Amazon. New highways already reach each of the country's major districts. This intensification of the transportation network enhances the possibilities for further development.

The higher areas of eastern Mato Grosso do Sul are gradually coming under the plow as the agricultural frontier spills over from neighboring Paraná and São Paulo states (see Figure 11–17).

CONCLUSIONS

The traditional solidarity of the Western Hemisphere was once the keystone of American foreign policy. Rising differences in the degree of development and standard of living between the United States and Latin America slowly introduced a note of discord into what had been the most harmonious of regional relationships. Hemispheric relationships became cool, if not overtly unfriendly, for most of this century.

As growth and development proceed in this region, there is a gradual but marked change in attitudes. There is a renewed self-confidence on the part of some Latin American nations. Latin America looms large as a customer for American goods; on both sides, there has been an increased willingness to trade. Discussion between the United States and Latin American states is now more frank and open. There are many new joint projects and attempts to find joint solutions to problems. Even the enormous Latin country debts, largely owed to U.S. banks, have of necessity brought the region's nations closer together.

Europe is admired, but still distrusted, by most Latin American states. Relationships with the USSR are less than cordial in most cases. Relationships with the rest of the world, except Japan, are rather distant because of a lack of interaction. For the moment, internal development is deemed more crucial than foreign contacts; intraregional relationships are being explored, and hemispheric concepts are being revived. Those Latin American countries that have attained a fairly high degree of development appear to be ready to embark on the road to higher goals. Mexico, Colombia, Venezuela, Argentina, and Brazil, as well as some smaller countries, aspire to entry into the First World. Brazil makes no secret of its desire to attain superpower status. The next few decades will reveal whether these goals will be achieved.

1. What is the origin of the designation Latin America? How accurate is it as a descriptive term for the region?
2. Contrast the conventional image of Latin America with the realities of modern Latin America.
3. How has the role of the United States changed in inter-American (hemispheric) relations?
4. List the common cultural characteristics of Latin America. What are the major exceptions to language uniformity? Where are Indian influences greatest? How was the influence of Indian culture affected by pre-Columbian settlement patterns?
5. Discuss the problem of land reform in Latin America. What countries have developed a successful program of land reform?
6. What are push-pull mechanisms? What has been their role in rural-to-urban migration in the region? What possibilities exist for colonization of new lands in Latin America?
7. Discuss the distribution of climates in Latin America. How usable are the region's rainforest areas? its tropical grasslands?
8. How does altitude affect the climates of Latin America? How is this altitudinal effect reflected in the region's crop patterns?
9. Contrast the mainland and rimland portions of Middle America. What were the mechanisms by which rimland cultural elements were spread to Central America?
10. Contrast the level of development and dominant economic features in each of Mexico's three subregions.
11. What economic functions have created boom conditions in the Mexican cities located along the U.S. border?
12. Discuss the past and present importance of mineral resources in the Mexican economy.
13. What seven countries constitute Central America? Which of these states are the richest? On what products and factors is their prosperity based? Which among them have the highest growth rates?
14. What are the prominent crops of Central America? What changes have occurred in banana production in Central America?
15. Discuss the interregional and international tensions that exist in Central America.
16. What are the bases of Panama's prosperity? How was Panama chosen, from among several sites, as the site of a major canal?
17. What have been the ramifications of centuries of sugar production for the Caribbean islands? How has the subregional economy diversified in recent years?
18. Contrast the economies and cultures of Haiti and the Dominican Republic.
19. What have been the positive and negative results of Cuba's association with the USSR?
20. Discuss the advantages and disadvantages of tourism as a replacement for sugar production in the economies of the Lesser Antilles. Which of these islands experience the greatest relative prosperity? Why?
21. Compare the economies and relative prosperity or poverty of Puerto Rico and Jamaica.
22. Why has it been so difficult to unify Colombians into one nation and economy? Analyze Colombia's mineral resource base. What is the country's agricultural potential?

23. Discuss Venezuela's attempts to diversify its economy and to develop its interior. Compare urbanization and urban settlement patterns in Colombia and Venezuela.
24. Why has it proven so difficult to eradicate the cocaine trade?
25. In what ways are the three countries forming the Guianas atypical of South America? Compare the fortunes and problems of the three.
26. Why is Peru potentially the richest of the Andean countries? What are its major internal regions? Why has its economy declined in recent years?
27. Contrast highland and lowland Ecuador in terms of cultural and economic characteristics.
28. Explain the role of isolation in limiting development of the Bolivian economy.
29. What are the major regional subdivisions of Chile? Compare the production of each of these subdivisions.
30. Trace the development of Argentine agriculture from settlement to the present. What is the role of Buenos Aires in the Argentine economy?
31. What are Uruguay's major economic problems? What steps have been taken to restore prosperity?
32. Why has Northeast Brazil become the poorest part of that country? Contrast its cultural and economic makeup to that of the rest of Brazil. What steps have been taken to improve its situation?
33. Compare Rio and São Paulo, the urban giants of Brazil.
34. Discuss the role of various commodity booms in the development of Brazil's economy.
35. Analyze the productive potential of Brazil's interior subregions.

SUGGESTED READINGS

Blakemore, Harold, and Smith, Clifford T., eds. *Latin America: Geographical Perspectives*. New York: Methuen, 1983.

Blouet, Brian W., and Blouet, Olwyn M., eds. *Latin America: An Introductory Survey*. New York: Wiley, 1982.

Brand, Donald W. *Mexico: Land of Sunshine and Shadow*. Princeton, NJ: Van Nostrand, 1966.

Cole, John P. *Latin America: An Economic and Social Geography*. London: Butterworth, 1975.

James, Preston, and Minkel, C. W. *Latin America*, 5th ed. New York: Wiley, 1986.

Johnson, John J. *Continuity and Change in Latin America*. Stanford, CA: Stanford University Press, 1964.

Lowenthal, David. *West Indian Societies*. New York: Oxford University Press, 1972.

Meggars, Betty J. *Amazonia: Man and Culture in a Counterfeit Paradise*. Chicago: Aldine-Atherton, 1971.

Morris, Arthur S. *Latin America: Economic Development and Regional Differentiation*. Totowa, NJ: Barnes & Noble, 1981.

Pico, Rafael. *The Geography of Puerto Rico*. Chicago: Aldine, 1974.

Sauer, Carl O. *The Early Spanish Man*. Berkeley, CA: University of California Press, 1969.

Webb, Kempton E. *Geography of Latin America: A Regional Analysis*. Englewood Cliffs, NJ: Prentice-Hall, 1972.

West, Robert C., and Augelli, John P. *Middle America: Its Lands and Peoples*, 2d ed. Englewood Cliffs, NJ: Prentice-Hall, 1976.

Appendix
RESOURCE MATERIAL

SUGGESTED ATLAS REFERENCES

Goode's World Atlas. Chicago: Rand McNally, 1983.

Hammond Medallion World Atlas. Maplewood, NJ: Hammond, 1984.

National Geographic Atlas of the World. Washington, DC: National Geographic Society, 1981.

The New International Atlas. Chicago: Rand McNally, 1984.

The Times Atlas of the World. London: Times Books/John Bartholomew & Son, 1981.

The World Atlas (English edition of *Atlas Mira*). Moscow: Glavnoe Upravlenie Geodezii Kartografii, 1967.

SUGGESTED STATISTICAL AND BIBLIOGRAPHICAL REFERENCES

Harris, Chauncy, ed. *A Geographical Bibliography for American Libraries*. Washington, DC: Association of American Geographers, 1985.

Statesman's Year-Book: Statistical and Historical Annual of the States of the World. London: Macmillan. (annual)

United Nations Statistical Office. *Demographic Yearbook*. New York: United Nations Publications. (annual)

Glossary

Aboriginal people The earliest settlers of a place, or those inhabitants, usually of a low level of development, in place at the time of "discovery" by more advanced cultures.

Acculturation The adoption, or modification to some degree, by people of one culture, of behavior patterns, language, religion, technology, or lifestyle of another culture.

Advance capital *See* March capital.

Agglomeration A cluster or concentration of population or economic activities. It most often refers to a group of settlements in an area of high population density or to a series of factories or businesses.

Agrarian Pertaining to agriculture, including the use of the land for crops or livestock or anything related to farmers, farm production, or rural societies.

Agribusiness The name given to large-scale, highly mechanized farming in the United States, commonly controlled by corporations.

Agricultural hearth An area of original domestication or development of a crop or form of livestock.

Agricultural revolution The major advance in stabilizing the human food supply by selecting, planting, and cultivating crops and selectively breeding and herding animals.

Agriculture The use of land and other elements of nature (particularly water) for the production of crops and livestock; the modification or selective use of the natural environment for food production.

Alloy Any of several metallic minerals that readily combine with other metals in the smelting process. Alloys generally impart special properties to the combined metal. Most ferroalloys contribute to the strength or durability of steel.

Alluvium Unconsolidated sediment deposited by a stream.

Altiplano A basin, valley, or plateau at high elevation, generally near or above the climatic limit for the growth of trees. The most common land use in such areas is grazing. It is a Spanish language term literally meaning a high, level area.

Apartheid The official policy of "separate development" of racial groups, including residential segregation, of the government of South Africa.

Aquifer Rock or soil through which groundwater moves easily.

Arable Suited to cultivation and the production of crops.

Aryan Referring to a group of languages, people, and cultural practices that arose in the steppes of Asia and spread to Europe, northern South Asia, and parts of southwestern Asia. *Indo-European* or *Indo-Aryan* are terms that designate the same peoples and languages.

Assimilado In Portuguese colonial Africa, the status of "assimilated" to Portuguese culture attainable by Africans; citizenship by aspiration to the ruler's culture.

Assimilation pressures "Melting pot" pressures on a minority to minimize or abandon their distinctive cultural traits, including their religious practices, dress, and food preferences.

Atoll A low-profile island of coral taking a roughly broken-circle form enclosing a lagoon. It is thought to be associated with volcanic subsidence (Darwin's theory).

Autarky The achievement of self-sufficiency in production of all raw materials and manufactures. Theoretically, a state in autarky would be invulnerable to interruptions or dislocations in international trade.

Balkanization The political fragmentation of an area into mány small units based on a mosaic of ethnic, linguistic, historical, and/or religious associations.

Banana republic A derogatory name once collectively applied to Central American republics; it implied dictatorship, corrupt government practices, and military control.

Bantustan In South Africa, a theoretically independent state composed of a "tribal homeland" imbedded within South Africa. No state other than South Africa recognizes the independent status of such states. *See also* Homeland.

Barrier island A low, elongate ridge of sand that parallels the coast, forming beaches.

Barrio A name used to imply an ethnic concentration of peoples of Hispanic language and culture within North American cities. In Latin America, it may be used as a synonym for slum. Literally, a district within a city.

Basement complex In geology, the crystalline (igneous) rock of the crust underlying later veneers of sediments or recent volcanics. Ancient continental rock materials from the earliest geologic times; extremely resistant to erosion except in humid tropical areas.

Bauxite The ore from which aluminum metal is derived. The name is derived from the place in France where it was first discovered.

Beaches *See* Barrier island.

Beneficiate In mining, the process of enriching an ore by concentrating the economic mineral. This then enables longer-distance shipment of ores. By extension, the enrichment of a material or activity to make it more viable.

Birthrate Usually expressed as number of live births per thousand people per year.

Bloc A group of countries that are combined for a specific economic or political purpose. Members act collectively for a common interest or goal.

Break-of-bulk function Performed by ports and some inland transport centers where goods are transshipped to other types of carriers.

Buddhism A religion that began in India in the sixth century B.C. and spread to much of the rest of southern and eastern Asia. It has exerted a strong influence on the cultures of China, Japan, and many countries of Southeast Asia.

Buffer In political geography, a political unit that functions, by design or accident, to separate two potential belligerents. It is intended to reduce tensions.

Caliche A crust of mineral salts in soil, left behind by evaporation of water. It occurs where evaporation potential greatly exceeds precipitation.

Capital-intensive Industry characterized by very high investment in plant and equipment per employee (e.g., oil refining). The same applies to highly mechanized agriculture.

Carrying capacity A measure of the capability of pasture to "carry" or support livestock without suffering any permanent damage or deterioration. By extension, a measure of the ability of any areal resource base to carry any human use without deterioration.

Cartel An organization for the purpose of monopoly or protection; an economic mechanism composed of several companies or countries that use joint efforts to eliminate competition and control supply or price. OPEC is the classic case at the current time.

Cartography The art and science of making maps, an important and characteristic technique of the field of geography.

Cash crop A highly specialized crop grown for the purpose of earning cash income on an otherwise subsistence farm.

Caste A class grouping in the Hindu religion and society based on occupation. Caste reflects social status and has broad economic implications in India.

Causal relationship A cause-and-effect relationship, often implied, but not necessarily proven, by a spatial correlation.

Central business district (CBD) The commercial heart of the city, characterized by the retail, wholesale, and office functions and the highest land values.

Central place Place in which goods and services are made available to surrounding hinterland populations. Central places are central to their trade area and are part of a hierarchy of such places.

Centrifugal force In industrial location, a force that favors movement outward from the center of production. In political geography, a force that tends to divide a country and reduce national unity, leading to civil or political conflicts and perhaps even ultimate separation.

Centripetal force In location, a force that favors concentration at the established center of production (e.g., cheaper transport to distant markets). In political geography, a cultural force that tends to unite a country.

Changing interactions The phenomena and study of the changing interrelationships—economic, political, cultural, and strategic—among and within world regions.

Chernozem A Russian term, now in general use, for "black earth" (humus) and mineral-rich prairie soils capable of supporting heavy crop yields when supplied with adequate moisture.

Circulation In political geography, the set of factors such as transport and communications facilities that facilitate strong interdependence among parts of a national unit. Strong circulation favors unity.

City-state A state whose territory is limited to the city and its hinterland. City-states were common in ancient Greece. Singapore and Hong Kong are current examples.

Clean tillage A Western agricultural practice that implies empty spaces between rows or hills of crops. Soil is hoed or plowed to keep it free of vegetation other than the specific crop being grown.

Client state A state in a subordinate, dependent relationship with a more powerful state.

Climate A description of aggregate weather conditions; the sum of all statistical weather information that helps describe a place or region; average annual weather conditions in a place or, categorically, over an area.

Closed city A city to which further migration is restricted by government; common in Communist societies.

CMEA *See* Comecon.

Collective farm Communist area farm theoretically owned by members but actually owned and managed by the government. Workers are paid in proportion to the economic success (or failure) of the farm. Recent liberalizations have granted more actual decision-making powers to workers.

Collectivization In Communist states, the process of forming collective farms, usually replacing private farms.

Colony A territory directly under the control of another state. Colonies were usually founded or conquered by that controlling state, which then exercised virtually full control of both domestic and foreign affairs.

Comecon (CMEA) The Communist counterpart of Western Europe's European Economic Community. Full members are the USSR, Poland, East Germany, Czechoslovakia, Hungary, Romania, and Bulgaria. Yugoslavia is an associate member.

Commercialization The process of moving from a subsistence economy toward an exchange, cash-based economy, with implied individual specialization of production.

Commercial revolution The five centuries in Europe prior to the industrial revolution, in which the techniques, technology, and systems of long-distance trade and finance were established, providing the groundwork for industrialization.

Commodities In international trade, such items as export crops, mineral ores, and raw materials; useful products of farm, mine, and forest, not fully processed (but of reasonable value and utility), that enter into trade.

Common market Any grouping of sovereign states agreeing to eliminate tariff and regulatory barriers to international flows of materials, labor, capital, energy, and so on. Specifically, the European Economic Community, the original common market.

Commune Communal living center, in mainland China in the 1950s, designed to destroy old traditions and group people together for more efficiency. The role of communes has been reassessed in the last two decades.

Communist bloc The group of Communist countries that normally follow the political leadership of the Soviet Union (i.e., Eastern Europe, Cuba, North Korea, the Mongolian People's Republic).

Complementarity The situation in which two or more areas produce different goods, resulting in flows (movement) for the purpose of exchange. Farm production and factory production are complementary, assuring movement in both directions between urban and rural areas.

Condominium In its international political sense, the joint control of an area by two countries. Egypt and the United Kingdom once shared political control of the Sudan. In its more general usage, joint control, ownership, tenancy, management, or occupancy of territory or property.

Continental drift The imperceptibly slow movement of continents, or parts of continents associated with plate tectonics. *See also* Tectonic plate.

Continental drift theory A theory that originally proposed that the continents moved around the earth's surface. It has been replaced by the plate tectonic theory, a refinement of the original thesis now essentially proven by geophysical observation.

Continental glaciation Referring to the scale and results of glacial activity that took place during the Pleistocene geologic epoch. At that time, huge ice sheets enveloped much of the area within the northern latitudes of the earth. The entire Pleistocene epoch encompassed some 3 million years, during which four or more ice sheets modified the soil and topography of large areas of Europe, North America, and the USSR.

Continental glacier A massive accumulation of ice that covered extensive land areas. Flow of the ice was on a continental scale. The ice usually was not controlled by the underlying topography, resulting in large areas being leveled or filled.

Continentality In climate, the set of climatic influences of large landmasses, heating and cooling faster than sea surfaces, leading to more extreme temperatures in landmass interiors than over or adjacent to seas.

Continental shelf The gently sloping submerged portion of the continental margin extending from the shoreline to the continental slope; the shallow suboceanic land surface adjacent to the continental land surface.

Conurbation The term used to designate a coalescing of urban areas through growth at city margins. Essentially the same phenomenon observed in megalopolises on a smaller scale. A term first developed by British geographers.

Convention A widely accepted rule of behavior. In making maps, certain rules of color, balance, and position are widely followed and have become standard; these rules are known as cartographic conventions.

Core The heartland of a state, an area that is the center of political and economic power and from which flow political direction and economic development. It is usually relatively densely settled.

Correlation A geographic association of two or more factors. It may suggest a cause-and-effect relationship between the factors, but this relationship must be investigated cautiously.

Corridor functions A set of transport services associated with a heavily used routeway. Often related to physical factors such as mountain passes or river valleys, these routeways channel a preponderance of total traffic.

Cottage industry Any complete production or phase of a manufacturing or processing sequence (e.g., assembly) that is carried out in the worker's residence rather than in a factory.

Coup A popular or military uprising to overturn the government in power.

Crop agriculture The form of agriculture devoted exclusively or overwhelmingly to the production of food crops derived from vegetational forms.

Crop rotation A cycle of changing agricultural land use for the purpose of maintaining fertility. Crops are chosen for their selective use or addition to the soil of certain chemical elements (nutrients).

Crust The very thin, outermost layer of the earth.

Cultural crossroads A part of the world, as in East Europe or the Middle East, that has experienced many diverse cultural contacts, invasions, and currents of history.

Cultural determinism The school of thought that emphasizes human abilities to overcome or dominate the forces of nature; the opposite of environmental determinism. *See also* Possibilism.

Cultural geography The study of the distributional patterns of human culture and the reasons for observed patterns.

Cultural hearth The center of origin of a culture or an item of culture; the area of origin of an idea or technology that subsequently diffused to other areas and peoples.

Cultural landscape The whole complex of created landscape that a culture superimposes on the base of the physical landscape. It is a distinctive, artificial landscape.

Cultural region An area that exhibits some degree of homogeneity in culture or cultural traits and artifacts.

Culture The total complex of learned and inherited lifestyle, including material items such as technology, architecture, and clothing and such abstracts and behavior patterns as language, law, and religion.

Dar al Islam The "House of Islam," or traditional extent of Islamic religion and culture—the Arabian Peninsula, North Africa, the Near East, Turkey, Iraq, and Iran.

Death rate Usually expressed as number of deaths per thousand people per year.

Decentralization The policy and/or process of spatially dispersing any function or activity from a former concentration.

Deciduous Vegetation that loses leaves during cold or dry seasons as a means of conserving moisture.

Delta An accumulation of sediment formed where a stream enters a lake or ocean.

Demographic transition A model of population growth (birth- and death rates) that characterizes a society moving through and completing the industrial revolution and associated health care revolution.

Demographic winter The anticipated future situation in aging industrial nations, in which the combination of already low and falling birthrates with a high concentration of elderly leads to a numerical decline in the nation's population.

Demography The study of population, including the characteristics and composition of a population, growth trends, and predictions concerning anticipated changes.

Density A description of the relationship between the number of any phenomenon and the size of the area in which it is distributed. A large number in a small area is a high density.

Deposition The processes of laying down (depositing) water or wind-transported materials that have been eroded at other locations.

Desalination The desalting of seawater to produce potable water; the flushing out of salt concentrates from long-term irrigated soils.

Desert An area of very sparse vegetation in which evaporation potential greatly exceeds precipitation. Deserts exhibit daily extremes of heat and cold and a low incidence of surface water.

Desertification The process of human-induced microclimatic changes that expand the desert at the expense of its subhumid margins. Environmental mismanagement (e.g., destruction of forests, overgrazing) contributes to desertification of formerly productive lands.

Desert pavement A layer of coarse pebbles and gravel created when wind removed the finer material.

Detribalization The movement of tribal members into an urban-industrial society to the point of losing touch with the remaining tribal groups and discontinuing traditional practices. Incomplete acculturation into the modern society may mean a cultural limbo status for a time.

Developing area An optimistic term for a state or region that has not yet attained full development or industrialization. Continuing advance is implied but may not occur.

Development A general level of technology, economic sophistication, and standard of living; an index of economic achievement.

Developmental differential A strong and obvious difference in developmental level, whether within or between states.

Developmental gap The ominous trend in which rich, industrial countries grow richer faster than poor countries become less poor. The gap between rich and poor is growing wider rather than narrower.

Developmental infrastructure The complex of transport, communications, energy development and transmission, and basic production facilities that must be established in order for development to proceed.

Developmental strategy The conscious plan for directing economic development based on selective investment and exploitation to speed overall development.

Development enclave A zone of relatively modern industrialization within predominantly underdeveloped countries.

Dhow Traditional Arab sailing vessel that carried the commerce of early merchants from Arabia and the Islamic empires of the past on the Mediterranean and Red seas, the Indian Ocean, and in the Persian (Arabian) Gulf.

Diaspora The deliberate dispersal of the Jewish population of Palestine under the Romans, following the destruction of Jerusalem in A.D. 70 (other sources list it as A.D. 135).

Diastrophism The collective processes of landform creation through the forces of tectonic upheaval.

Diffusion The outward spread, not necessarily in proportion to distance and time, of an idea, an innovation, or a piece of material culture. There are both physical and cultural barriers to the spread and adoption of new ideas or material culture.

Direct water power The use of falling water to turn a wheel, creating mechanical energy in the process. Water is used directly rather than first being converted to another source of energy (e.g., hydroelectricity).

Distributional pattern The spatial incidence of any variable forms a distributional pattern, which may provide clues to causal relationships when compared to the distributional pattern of other phenomena.

Diurnal The daily range in temperature between the lowest and highest readings; daily.

Doubling time The time required for any population to double in size.

Drainage In agriculture, the removal of excess water from swamps or other wetland areas to make them more suitable to crop production. It is accomplished by lining fields with tiles placed beneath the topsoil.

Drainage basin The land area that contributes water to a stream.

Drip irrigation A highly efficient method of irrigation that carefully controls the amount of water received by any plant. It conserves water by making the most efficient use of that life-sustaining liquid. It was first developed on a commercial scale in Israel.

Drive to maturity One of Rostow's stages of economic development, in which the forces for increased production and development begin to gather speed and proliferate. Factories multiply, diversify, and spread to new areas as the stage unfolds.

Earth-sun relationships The seasonally changing location of the direct rays of the sun that results from the tilting of the earth 23½° away from the vertical.

Economic colonialism The economic control of an area by a larger, outside power. It implies the indirect control of certain political functions through exerting economic pressures or the control of certain segments of an economy.

Economies of scale Savings that occur from producing and consuming in large amounts. The costs are reduced as the scale of the enterprise increases.

Ecumene Originally described the (known) inhabited world; now used more specifically to delimit the area populated by more than two people per square mile.

Ecumene triangle In the USSR, the triangle that approximates the most populated, most productive, most industrialized part of the country; in general terms, Leningrad to Novosibirsk to Odessa to Leningrad.

Ejido A Mexican land reform settlement in which land is managed collectively and education and technology are used to improve the efficiency of production.

Empire A strongly centralized state, usually emphasizing obedience to the will of a hereditary ruler over a polycultural political unit; a collection of nationalities governed by one ruler and located in one jurisdiction.

Enclave A piece of foreign-controlled territory within a state's boundaries.

Entrepôt A transport center and exchange point, almost always a seaport, that serves as a trading center for another area, sometimes across international boundaries.

Environmental degradation The deterioration in the quality of the environment that occurs through erosion, pollution, and general mismanagement or rapacious management.

Environmental determinism The belief that human cultures are strongly influenced and molded by forces of the physical environment, primarily climate.

Environmentalism The appreciation and concern for the conservation and protection of the total physical environment; not to be confused with environmental determinism.

Equator The name given to a line of latitude that bisects the earth into two equal parts, midway between the pôles. It is the 0° latitude line.

Erosion The incorporation and transportation of soil and rock material by a mobile agent, such as water, wind, or ice.

Estancia Literally, "estate;" essentially the same as *hacienda*.

Ethnic federalism The recognition of the nationalistic nature of ethnic homelands and historic states by granting some degree of local government or provincial status within the structure of a federal state. Examples are Yugoslavia and the USSR. *See also* Federal state; National minority.

Ethnicity The fact and self-consciousness of membership and participation in a distinctive cultural group which may or may not have any racial association or distinction.

Ethnic minority A culturally distinct group within a larger, politically dominant culture but without ambitions for political independence.

Ethnic region A sizable territory occupied predominantly or exclusively by an ethnic group.

Ethnocentrism The tendency for any nationality or ethnic group to view the world as it relates directly to them. Implicitly or explicitly, there is an assumption that one's own homeland is the center of the world.

Eurasia The great landmass customarily separated into the continents of Asia and Europe.

European Economic Community Community founded in 1958 by France, West Germany, the Netherlands, Belgium, Luxembourg, and Italy, and joined later by the United Kingdom, Eire, Denmark, Spain, Portugal, and Greece. The community's members have no tariffs or restrictions on interchanges of goods, raw materials, capital, or labor.

Exclave A piece of one state's territory physically separate from and surrounded by the territory of another state. An exclave of one state would be an enclave within another state.

Exotic stream A permanent stream flowing through an area whose climate does not support permanent streams. The exotic ("out-of-place") stream exists due to water sources located in a more humid climate.

Extensive agriculture A land use system in which relatively little time and effort (or machinery use, fertilizer, etc.) are invested per land unit. Productivity per land unit is comparatively low but is usually compensated for by large land units.

Exurbia The name given to the lowest-density fringes of suburbia; the far-flung outer limits of commuting.

Fall line In the eastern United States, the physiographic boundary between the Piedmont of the Appalachians and the Coastal Plain, marked by falls or rapids where streams leave the resistant rock of the Piedmont for the softer sediments of the Coastal Plain.

Fault A break in a rock mass along which movement has occurred.

Favellas The slums created by rural migrants to the cities of Brazil. They usually occupy the city margins or physical sites that are difficult to build upon.

Fazendas Large estates producing sugar, coffee, or other export crops in Brazil. Like haciendas in Spanish-speaking Latin America, they have villages to house workers, and labor-landlord contractual obligations are similar. Unlike the hacienda, in which the peons (tenants) were tied to the land by perpetual debt, laborers on fazendas were (and are) free to leave after the contract was completed.

Federal state Any country in which the governmental function is divided between two levels of jurisdiction. A certain degree of autonomy and difference is allowed at the lower level of government. Countries that are large in area, ethnically diverse, or composed of subunits with a prior history of separate governance generally tend to use a federal system.

Fertile Crescent The traditional term identifying the subregion of the Near East that extends from present-day Israel, Lebanon, and Syria's better-watered Mediterranean-type climate to the great oasis of the Tigris-Euphrates valleys. The contrast to less favored deserts and rugged mountains elsewhere led to the designation fertile.

Fertility The measure of the ability of a soil to produce crops; the number and amount of chemical nutrients found in a given soil. Soils that are capable of sustained crop production without the addition of chemicals are deemed fertile.

Feudalism A social system that evolved in medieval Europe. Under feudalism, land was owned by a nobility and farmed by a class of landless peasants. Nobles were to defend the peasants in return for the peasants' work; in fact, it became a highly exploitative system.

Finca A small or moderate-sized estate for the production of export and/or commercial cash crops; common in Colombia, Venezuela, and parts of Central America. Either owners, tenants, or hired workers or a combination of them may provide labor. It is the size that distinguishes these holdings from other types (e.g., the much larger *hacienda*).

Fiord A U-shaped, glaciated mountain valley now partially inundated by the sea; specifically, such deep, steep-sided arms of the sea along the Norwegian coast.

Floodplain The low-lying area of nearly level land that occurs close to any river or stream. It is the area subject to flooding.

Flow Any movement, in any direction, along a link (route) or network. Flows may consist of people, commodities, or even electrical energy or telephone calls.

Flow resource Resource that with prudent management will continue as a resource forever. Forests, hydropower, and fisheries could (and should) be managed as flow or renewable resources.

Food chain The system of predator-prey relationships among lifeforms within an ecosystem. Almost all food chains are based on plants consumed by animals that are in turn eaten by other animals, which also might be preyed upon.

Footloose industry An industry that is not tied to any particular location(s) through dependence on a source of raw material, energy, or access to water or rail transport, and therefore can move about in response to other, less traditional locational considerations.

Formal region There is no dominant center implied in a formal region. It may be uniform concerning a specific criterion or at least exhibit an acceptable degree of uniformity.

Fossil fuel A source of thermal energy resulting from the fossilization of organic material that flourished in past geologic eras—coal, lignite, natural gas, or petroleum.

Fossil water Water from underground reserves that is the result of more humid conditions in past eras. The contemporary climate is not capable of producing or recharging this underground supply, which is a fossil of different conditions in the past.

Freehold Private property, owned outright, usually by the present occupant. In farming, freeholds are privately held and managed units.

Free-trade zone A legally designated area within which imported materials may be processed, stored, and re-exported without payment of tariffs or other taxes on import-export.

Friction of space The cost and/or inconvenience of overcoming distance. Decreases in time-distance or cost reduce friction of space.

Frontier In political geography, more a zone than a line; an indefinite area of political control.

Frontier crop A crop produced on settlement frontiers because of its ready market and storageability. Wheat is a common frontier crop.

Frontier mentality The attitude, alleged to be common among frontier pioneers, that resources are virtually limitless and thus need not be conserved.

Frost Belt The popular designation given to those areas of the United States with a marked winter (coined during the energy crisis of the 1970s).

Functional classification of cities A categorization of cities by their basic economic function—mining, administrative, resort, and so on. An alternative is classification by size.

Functional region Sometimes called a nodal region, a region with a functional relationship with a node or center. A seaport's hinterland would be a type of functional region.

Functional zonation The tendency for the gross land use pattern of cities to show grouping together of similar functions—office, light industry, upper-class residential, and so on.

Fundamentalism The emphasis on basic values and the literal interpretation of scriptures that accompanies religious revival in any major religion. Strict moral behavior is expected of members of the faith; fundamentalist groups may be militant at times, seeking to influence or even take over the government function.

Gastarbeiter Literally, "guest worker;" a German term for international migrant labor.

General farming Unspecialized agriculture with a wide variety of crops and livestock produced on every unit.

Geodysic location The location of a site, settlement, or area given in terms of latitude and longitude.

Geographical perspective The viewpoint that seeks to place any phenomenon or pattern within a geographic context; that is, seeking spatial relationships.

Geographic context *See* Geographical perspective; Spatial correlation.

Geographic explanation An explanation of a distributional pattern or phenomenon considering a wide range of potential influence from the physical and cultural sets of geographic or spatial phenomena.

Geography   The science of distributions. Geography is concerned with spatial variations in any physical or cultural phenomena.

Geopolitics   The field of study that focuses on the relationships between the political structure of the state and all relevant aspects of the geography of the state and its neighbors.

Geothermal power   Energy generated by using the internal heat of the planet. Geothermal energy is tapped on a very limited scale because suitable geological conditions are highly localized.

Ghetto   The voluntary or involuntary residential segregation of a racial, ethnic, or religious minority. Ghettos are almost always unofficial, but they have been legislated at times.

Glacier   A thick mass of ice originating on land from the compaction and recrystallization of snow that shows evidence of past or present flow.

Glasnost   The policy of "openness" (as opposed to secretiveness) advocated by Soviet Premier Mikhail Gorbachev.

Global interaction   As a large-scale phenomenon, a characteristic of the modern world with its global interchanges of raw materials, technology, people, energy, capital, manufactured goods, information, and culture.

Gondwanaland   The southern portion of Pangaea, consisting of South America, Africa, Australia, India, and Antarctica.

Great Leap Forward   The ill-considered plan for accelerated economic development in the People's Republic of China during the 1950s.

Greenbelt   A buffer zone of open space used to separate different functional areas in planned communities.

Green Revolution   A description of the recent, rapid strides in agricultural science in selective plant breeding, cultivation, and fertilization techniques that significantly increase food crop yields per land unit. The productivity increases, however, also rely on large inputs of energy and fertilizer.

Grid   A network of lines that cross at right angles. Streets of cities may exhibit a grid pattern where they cross consistently at right angles. The network of imaginary lines across the earth's surface—the lines of longitude and latitude—are called the earth's grid.

Gross national product (GNP)   A popular measure of the level of industrialization and economic development. It is the total monetary value of all goods and services produced per year, commonly expressed per capita.

Groundwater   Water stored underground in soil and/or rock. It is water left from precipitation that has been absorbed into the porous rock or soil. Wells are used to tap groundwater supplies.

Growth pole   In developing nations, a city or economic region selected for planned investment and growth which, it is hoped, will stimulate general regional growth by attracting other economic activities.

Growth rate   The product of additions (births and in-migrants) and subtractions (deaths and out-migrants) in a population per thousand per year.

Guano   A generic term for dried bird manure, highly valued as a fertilizer for its high nitrogen content.

Guerrilla warfare   War carried on under clandestine circumstances by dissident elements seeking to overthrow the government in power.

Hacienda   The name given to a class of large estates common in Spanish-speaking Latin America. Owners were a class of landed gentry. Tenants were attached to the land by contract; in return for their labor, they received a dwelling and subsistence. The landlord had certain traditional obligations to the tenants and their families.

Haj   The pilgrimage of a Muslim to the holy city of Mecca. Every Muslim is to make this pilgrimage at least once in a lifetime, if at all possible.

Hardpan formation   The development of a crust or concretion of minerals within soil. This hardpan hinders drainage and the development of root systems and generally lowers the agricultural utility of soil.

Harijan   Literally, "beloved ones"; the name assigned by Gandhi to the lowest castes of Indian culture in an attempt to readjust the Indian public attitude toward those of lowest caste. Unfortunately, the term has become an epithet and a derogatory term, the opposite of that which Gandhi intended.

Hearth   The source area (in agriculture, culture, etc.) of any innovation; the area of origin from which an idea, technique, artifact, crop, or good is diffused to other areas.

Heartland   The most important, concentrated area of agricultural, mining, and/or industrial production within a state; most probably, the most heavily populated region.

Heartland theory   A theory in political geography, developed by Halford Mackinder, stating that whatever power controls the great core of the Eurasian landmass potentially holds the key to world domination.

Hidden export   The providing of a service rather than the exporting of tangible goods for earning foreign exchange. Tourism, insurance and financial services, transport services, rental of facilities, and ships' crews could all represent hidden exports.

High-value-added   A product or industry in which the total value of the finished goods is much higher than that of its constituant raw materials; a measure of the degree of industrialization.

Hinge function   A function resulting from a location between two complementary or unlike physical areas or between two different cultural areas. These locations with hinge functions are centers of trade and interchange.

Hinterland   The surrounding or even distant area that has a close economic relationship with the city or region whose hinterland it constitutes; commonly, a raw materials supply and/or market zone.

Homeland   The current, perhaps more palatable, and socially acceptable name given to separate black states within South Africa. Formerly called Bantustans, homelands are pseudoindependent areas, client states of South Africa. The aim of creating homelands is separate development and racial segregation for the country's blacks under the program of apartheid. *See also* Apartheid.

Horn   A pyramidlike peak formed by glacial action in three or more cirques surrounding a mountain summit; the sharp, triangular peak characteristic of certain heavily glaciated mountains.

Human geography   *See* Cultural geography.

Humus   Organic matter in soil produced by the decomposition of plants and animals, generally considered an enhancement of fertility.

Hydraulic civilization   Civilization based on elaborate water engineering systems to irrigate desert land, as in the Nile Valley, the Tigris-Euphrates region, or central Mexico.

Hydraulic mining   The sluicing of river sands or gravels (alluvium) with high-pressure jets of water to separate out denser particles of mineral ores.

Hydroelectric   Referring to electricity generated by falling water. Spillways from dams provide the energy source to convert mechanical to electrical energy by turning turbines.

Hydrologic cycle   The endless cycle of evaporation and precipitation by which water is evaporated from ocean (major) and land (minor) surfaces, enters the atmosphere, is transported within the atmosphere, cooled, condensed, precipitated, stored temporarily in snowfields, groundwater, or reservoirs, and ultimately runs off through surface streams to the sea.

Hydroponics   The science of growing plants with their root systems immersed in, or flushed by, nutrient-laden water rather than in soil.

Iconography   In political geography, the set of shared beliefs, heroes, myths, sense of destiny, and interpretation of history that is a basis for national unity. An icon is a symbol of belief and faith.

Ideograph   A symbol that represents a specific idea or object, as in Chinese ideographs, rather than the sounds of spoken language, as in the Latin alphabet.

Imperialism   The policy or act of creating a political or economic empire by subjugating areas beyond a country's borders.

Import replacement   The strategy of fostering domestic production of consumer goods to reduce imports in a developing economy.

Indigenisimo   In Latin America, the resurgence of interest and pride in native Amerindian culture and ancestry.

Industrial inertia   The tendency for a plant, once established, to remain in that location even as other locational factors become less favorable. The costs of abandonment or moving argue for continuation at the established location.

Industrial revolution   The eighteenth century (and continuing) period of rapid advance in use of inanimate power, more complex machinery, and the factory system to revolutionize the scale and efficiency of industry, transport, communications, agriculture, and mining. The antecedents of this revolution developed over centuries in many parts of the world.

Infant mortality rate   A statistical measure of children, other than the stillborn, who do not survive to their first birthday. This rate is often used as a key indicator of development level.

Infrastructure (economic)   The transportation networks (roads, rails, etc.), capital equipment, housing stock, and other physical structures found within an area; the internal collection of structures, facilities, and phenomena with capital value in any economy.

Innovation   The creation of a new idea or technique, new technology or modification of technology, or extension or new application of existing concepts. The spread of the innovation over space is diffusion. *See also* Diffusion.

Integrated economy   An economy in which primary, secondary, and tertiary sectors of the economy are at approximately the same level of development and interrelated within the domestic economy.

Intensive agriculture   The intensive use of land for production. Large amounts of labor (and capital equipment, fertilizer, etc.) are applied per land unit. Productivity is very high, but so is investment of time and energy.

Intercropping   *See* Interculture.

Interculture   The agricultural practice of growing two or more different crops on the same plot at the same time.

Interdependence   Mutual dependence; relying on one another. In the global geographic sense, the dependence of the world's nations on one another for a collective food supply, sources of raw materials and fuel, and efficient production of manufactured goods. In the environmental sense, the concept that the actions of any one or group of nations have repercussions for all others.

Interior drainage A phenomenon, common in deserts, in which drainage systems do not reach the sea due to low precipitation and high evaporation; streams end in salt lakes or disappear in valleys with no outlets.

Intermittent stream A stream with seasonal or occasional flow; a stream that flows only after a rainstorm. Intermittent streams are often the sole drainage feature in desert areas.

Intermontane "Between the mountains"; usually refers to a valley, area, or basin between highlands.

Interruption (of projections) A method of cartography that maximizes land area and minimizes distortion on projections of the world by eliminating large portions of the oceans. Because most geographic phenomena occur on land, this is an acceptable form of distortion for showing distributions at a world scale.

Intertropical convergence (ITC) Sometimes called the equatorial low, the zone in which air circulating equatorward from each hemisphere's subtropical high pressure converges and then rises vertically, producing heavy precipitation.

Invisible export Any of the services involved in international trade relationships. *See also* Hidden export.

Invisible income Income earned through services rather than through the production and exchange of commodities and goods. Marine insurance, income from bank transactions, monies spent by tourists, interest payments, transshipment or cargo fees, and royalties are all forms of invisible income.

Irredenta Literally, "unredeemed territory"; territory claimed by a state on the basis of its cultural, ethnic, or linguistic makeup, but still controlled by another state.

Irredentism The cultural inclusion of an area outside a country's political jurisdiction. It may include planned or active attempts to annex that cultural group and the land it occupies.

Irrigation The use of water to produce (or increase the production of) crops. The water so applied is a substitute for (or supplement to) rainfall.

Islam Literally, "submission to God;" the official name for one of the great religions of the world (over 900 million adherents). This religious philosophy originated in Arabia and was spread from there throughout northern Africa and over much of southern and western Asia. It is monotheistic (accepts only one God); the holy book of Islam is called the Koran. The religion represents a later continuation of the Judaeo-Christian religious tradition. *See also* Muslim.

Island arc A group of volcanic islands formed by the subduction and partial melting of oceanic lithosphere. Japan, the Aleutians, and the small Caribbean islands are examples.

Isthmus A narrow section of land connecting larger landmasses and situated between water bodies.

Jet stream Swift (120 to 240-kilometer-per-hour), high-altitude winds that move weather systems along.

Jihad A holy war in the name of Islam. Currently a title appended to the name of several groups involved in terrorist acts and/or the spread of fundamentalist influence or goals.

Journey to work The distance traveled (and the trip itself) in commuting to and from work each day.

Jute A strong fiber used in the production of burlap bags, carpet backing, and rope. It is widely produced in India and Bangladesh and is valued because of its durability and resistance to rot.

Karst A topography consisting of numerous depressions called sinkholes, caves, and so on associated with limestone bedrock.

Kibbutz In Israel, a communally owned and operated settlement, often a farm settlement but also likely producing some craft or even industrial goods.

Kombinat In Soviet economic planning, the "pairing" of two distant industrial complexes based on complementary raw materials. For example, iron ore is carried from one area to a coal-based "partner," and coal is carried on the return trip.

Koran The holy scriptures (or book) of Islam. It contains an exposition of the religious and moral code of Islam.

Labor-intensive Endeavor in which labor costs are an unusually high component of the total production cost of the finished product or service. Education is an example.

Landlocked Any state not having direct access to seaborne trade through ports within its national territory. Examples are Austria, Uganda, and Nepal.

Land reform The division of large, landlord-owned estates into smaller, individually owned units. It is generally done under the auspices of government.

Land rotational agriculture Tropical agricultural system that preserves soil fertility through long periods of fallowing after brief cultivation. *See also* Slash-and-burn agriculture.

Landscape The sum total of the physical and cultural environments in an area. Landscapes reflect nature, history, and culture.

Land survey system The mathematical basis for surveying property and political boundaries in any area, as when former Indian lands were surveyed prior to large-scale non-Indian settlement in the United States.

Land tenure A form of land control and ownership. The various forms of land tenure include freehold (private ownership by the farmer), tenancy, plantations, rented lands, and the like.

Land use A fundamental concern of geographers; the type of economic use to which land is put. In the context of planning, the basic concern of planning.

Land use inventory A complete map of existing land use within the planning agencies' jurisdiction.

Language The total of words and systems of combining words to give expression to ideas, feelings, perceptions, or desires.

Language diffusion The process of geographic spreading and intermingling of a language or some of its vocabulary and usage into the domain of another language.

Language family Group of languages related through evolution from a common origin language and language hearth.

Laterite A soil condition, common in the tropics, in which water-soluble minerals are leached out by abundant water, leaving mostly iron and aluminum oxides, which, exposed to the atmosphere by erosion, can form an almost impenetrable layer.

Laterization The leaching process that leaves a residue of iron and aluminum oxides in tropical soils.

Latifundium An estate owned by landlords, often absentee, who hire day labor; characteristic of the Mediterranean, parts of the Middle East, and much of Latin America.

Leaching A natural process in which plant nutrients are rinsed from the top layers of the soil to deeper levels, making it difficult for roots of crop plants to reach a food supply. It occurs in areas of heavy precipitation.

Legumes A class of plants that enrich soil by adding nitrogen through root nodules and preserve it with a "fixing" process that effectively keeps soil nitrogen from being dissolved.

Leisureopolis A leisure- and recreation-oriented linear strip of urbanization in which development is aligned with, and based on, a natural recreation amenity such as a shoreline.

Less industrialized country (LIC) A country yet to move fully into the industrialization process. Examples are Laos, Angola, and the Sudan.

Lignite A low-quality (low-thermal-content), usually brownish coal that burns with a great deal of pollution.

Lingua franca Any mutually understood language of commerce developed from a mixture of languages and different from the native language of all its users; a language of convenience.

Link A connection (by whatever mode) between two settlements or crossroads (nodes); any line of or facility for transportation.

Lithosphere The rigid outer layer of the earth, including the crust and upper mantle; the "rock" layer.

Littoral The coast; the land adjacent to the sea.

Llanos Literally, "grassy plains;" the name given to areas of tropical natural grasslands in parts of Latin America.

Locational analysis A set of theories and principles in geography (and industry) to aid in reaching rational decisions for the location of an investment.

Location strategy The practical application of geographic analysis; the system of identifying the point of best potential for an economic enterprise, maximizing profit and minimizing costs in the process.

Loess Fine-grained materials transported and deposited by wind. Frequently of glacial origin, this windborne material can become the basis of exceptionally productive soil.

Machine tool A machine that makes other machines; metal-fabricating and -shaping machinery; capital equipment. A machine tool industry is at the apex of industrial development.

Maghreb Literally, "the western island"; an Arabic (and cultural, geographic) term used to refer to the Islamic states of North Africa that lie to the west of the Gulf of Sidra in Libya.

Magma A body of molten rock found at depth, including any dissolved gases and crystals.

Magnetic anomaly An apparent concentration of metallic (ferrous) ores, indicating the desirability of further exploration, detected by ground exploration or satellite scan.

Mainland-rimland The basic cultural region division in Middle America, proposed by geographers John Augelli and Robert West.

Malthusian Referring to the theories and viewpoint of Thomas Malthus, who wrote a treatise on population. Malthus stated that population growth was outpacing the ability of the earth to produce food. Those who currently support similar positions are called neo-Malthusians.

Malthusian theory of population growth Theory that states that a human population will expand geometrically (2, 4, 8, 16, 32, . . .) much faster than the food supply will increase, leading to widespread famine disaster about every 25 years.

Mandate A form of political control in which a colonial area is to be tutored and otherwise prepared for eventual independence.

Manufacturing The processing of a raw material into a finished product through the application of power, labor, and capital equipment. It generally takes place in a structure, produces something that is not a service or a maintenance activity, and utilizes some sort of inanimate energy in the process. *Industry* is a more general term; manufacturing is a type of industry.

Map interpretation Analysis aided by the use of a map that goes beyond simple map reading; interpreting features and functions not directly shown on the map from data that are present on the map. Through skillful map interpretation, certain elements of culture, subsurface geology, and even accurate slope measurements can be derived from a map that does not directly display them.

Map legend A written and graphic key that explains map symbols and aids in the reading of the map.

Map projection A system whereby the amount, location, and type of distortion are controlled in the process of reducing the three-dimensional world to a two-dimensional map.

March (advance) capital A capital deliberately located in a position close to an advancing or contested frontier rather than in a more central, secure location; a capital located for close communication with military forces.

Maritime (effect) Pertaining to the sea or the use of the sea. In the climatic sense, the spreading of moderating effects to weather conditions and climates over the land surface. The reduction of temperature extremes that is characteristic over water can be extended to adjacent land areas in the absence of such blocking phenomena as mountains.

Maritime orientation The psychological, economic, and cultural emphasis on seaborne opportunities and contacts.

Market The region or point at which a commodity can be sold; the place of commercial exchange of one's goods or services.

Market-set prices Prices for crops and commodities that result from consumer demand rather than from government action or dictate.

Material culture The complex of tangible items produced in a cultural tradition, reflecting the values, lifestyles, and technology of a society. Culture is not necessarily material, but material items can be cultural expressions.

Material culture complexes The various items of material culture associated with a specific cultural group.

Maturity In Rostow's stages of economic growth, the stage by which all major sectors of the economy have been modernized through application of the principles of the industrial revolution.

Mechanization of agriculture The application of machines and capital to agriculture in order to save labor and time. Tractors are substituted for draft animals, and machines do labor formerly performed by human hands in combination with simple tools.

Megalopolis The term assigned by geographer Jean Gottmann to an urbanizing region in which two or more great cities are growing toward one another along interconnecting transport routes; originally, the urban complex of the northeastern United States.

Megalopolitan Characteristic of, or potentially resembling, the original, northeastern U.S. seaboard Megalopolis.

Melanesia Region of (racially) black islands in the South Pacific.

Melanin The dark skin pigment present in widely varying degree in all humans except albinos.

Melting pot concept The assumption, not necessarily true, that immigrants to America would merge into a homogeneous American culture, abandoning their distinctive ethnic cultures.

Mestizo In Latin America, a person of mixed, usually European and Amerindian, racial heritage.

Metropolis A city great in size, variety of functions, and extraregional importance.

Metropolitan area The contiguous built-up area of a city; a city plus its suburbs; a zone of influence and the commuting area for a large city.

Microclimate Relatively small variations in temperature and humidity, located within varying vegetation environments of small areas. It is often changed critically by human mismanagement or human modification.

Micronesia A region of tiny islands in the Pacific Ocean.

Migrant A person who temporarily or permanently changes residence, frequently for economic opportunity.

Migration The temporary or permanent movement of people over a significant distance.

Mineral In geology, naturally occurring, inorganic crystalline material with a unique chemical composition. In geography, a minable resource that yields a fuel, metal, or another raw material for commercial use or manufacturing.

Ministate An independent political unit of relatively very minor territorial scale. Examples are Andorra, San Marino, and Monaco.

Mirror image colony Similar in some aspects of physical geography and having a relatively small aboriginal population, a colony that was seen as a potential replica of the "parent" society.

Mixed economy An economy in which there are some elements of both socialistic (government ownership of the means of production) and free-enterprise types of control.

Mode Form or means of transportation or type of carrier. Rail, highway, and pipelines are three types of transportation modes.

Model An abstract simplification of reality designed to eliminate peripheral or unrelated phenomena in order to focus study on basic factors and relationships.

Monetary self-sufficiency A system of balancing food import costs with exports of higher-priced, higher-quality food products of a specialized type.

Monotheistic The belief in one god; religions based on that premise.

Monsoon The seasonal reversal of winds, onshore in summer, offshore in winter, associated with the presence of temperature-induced pressure systems over land and water surfaces. Its most dramatic effects are felt over southern and eastern Asia.

Monsoonal Pertaining to the effects of, or having some characteristics of, the monsoon phenomenon.

Moraine Ridge, hill, or veneer of glacially deposited materials, characteristically of unsorted debris.

Mulatto A person of mixed black and white racial ancestry, implying also a mixture of African and European cultural values and practices in some parts of the world.

Multicropping The farming practice of growing two or more crops in succession on the same piece of ground. For example, winter-sown wheat (actually planted in the fall) is harvested in June and will be followed by a crop of barley.

Muslim (Moslem) A follower of the Islamic faith and the teachings of its prophet, Mohammed (Muhammad).

Nation A cultural unit; a group of people tied and identified by their cultural distinctiveness. A nation may or may not constitute a state as well, but the desire for an independent state is implicit.

Nationalism The cultural complex of loyalties based on commonalities of interest and experience. Loyalty must be to the larger political unit rather than to tribal or subregional authorities.

National minority An ethnic, linguistic, or cultural minority controlled by another state, but with ambitions for political independence.

Nation-state A happy coincidence, territorially, of a nation (a culturally distinctive group) and a state. The national territory may still have some unassimilated minorities, however.

Natural boundary A boundary that represents the concept that rugged terrain, mountain streams, deserts, and the like make the least troublesome boundaries and are therefore "natural" ones.

Natural increase rate The rate of increase in a population exclusive of migration; the product of the crude death rate subtracted from the crude birthrate.

Natural resources A collective name given to the numerous fuels, minerals, agricultural products, or vegetational materials that can be used to benefit or develop a society. Frequently, they can be used as raw materials in manufacturing or processing. All are products of the natural environment.

Natural (native) vegetation The vegetation (plant life) of an area as it evolved in an undisturbed physical environment, in harmony with particular soil and climatic conditions found in an area. Natural vegetation has been removed over vast areas of the earth.

Neocolonial Relating to the policies of industrial states to dominate the economies of independent states that were colonies and thus limit the freedom of action of these former colonies.

Net migration The balance between in-migration and out-migration in any area. If arrivals exceed departures, there is a positive net migration; if the reverse is true, there is a negative net migration.

Network The system of transportation and interconnection found within an area. It consists of nodes (settlements or points of activity) and links (the routes of movements that occur between nodes).

New Economic Policy (N.E.P.) The "new economic policy" of Lenin; a temporary return to limited individual ownership and enterprise that occurred during the early years of Communist control in the USSR.

Newly industrializing country (NIC) A country that is clearly moving through the industrializing process, characterized by rapid expansion of industrial output. Examples are South Korea, Brazil, and Mexico.

New town In planners' definitions, a community with a complete set of urban functions (residential, recreational, retailing, administrative, manufacturing, etc.) that was totally designed and planned prior to construction.

Nodal region *See* Functional region.

Node An urban center; a settlement: the central functioning point of a nodal or functional region. In transportation, any settlement or crossroads.

Nomadic herding Primitive livestock herding in which mobile people follow (and guide) the animals in search of pasture; generally confined to arid or semiarid areas.

Nomenklatura In the Soviet Union, those on the secret roster of high officials, military officers, and other leaders entitled to special privileges.

Nonalignment The political stance of states refusing to be identified as consistent supporters or allies of either of the contending superpowers seeking global influence.

Norden The northern tier of European countries—Norway, Sweden, Finland, Denmark, and Iceland; the Fenno-Scandinavian culture area.

Oasis An area within a desert in which water is available for the sustenance of life, including agriculture. The water source may be a spring, well, or river, and the area may be large or small.

Occidental A term used to refer to Western areas and cultural practices. It is the opposite of oriental. Both these terms are used to imply strong differences between the cultural attributes and lifestyles of Asia and

Europe and the areas that came under their respective influences.

Ocean currents The relatively swiftly moving surface waters of parts of the ocean given direction and motion by prevailing atmospheric circulation patterns and winds. Currents serve to exchange heat across the water surface of the planet; they may indirectly affect precipitation and temperature conditions over adjacent land surfaces.

Ocean island An island that is relatively remote from all continents and is geologically not part of a continental mountain system but rather results from tectonic and/or coral activity.

Ocean trench A huge, arclike deep trench in the ocean floor just off a continent-fringing island arc, associated with tectonic plates in collision.

Offshore banking A term broadly applied to international banking as developed in Third World nations. "Offshore" implies a location beyond the shores of developed areas, with their strict banking practices. Numbered (anonymous) accounts, gross currency exchanges, gold sales, and corporate ownership transfers are common transactions. The avoidance of income tax or other tax payments is their main attraction to customers.

Oil bridge The term applied to the continuous shuttle of oil tankers between the Persian Gulf and Japan.

Oil patch A popular designation for the oil-producing U.S. states in sharp economic decline as a result of the world oil glut. Examples are Texas, Oklahoma, and Louisiana.

Older industrialized country (OIC) A country characterized by a mature industrial economy, with a marked shift toward more emphasis on high-tech industries and services. Examples are the United States, the United Kingdom, and Japan.

Olduvai Gorge A site in Tanzania, East Africa, where the oldest yet discovered physical remains of humanlike creatures or "early man" have been found.

OPEC *See* Organization of Petroleum Exporting Countries.

Ore A variable, economically determined classification of any rock or soil material with a sufficiently high concentration of a desired mineral to warrant its mining and processing.

Organization of Petroleum Exporting Countries (OPEC) An organization that attempts to control world oil prices through limiting production. Its members all produce a surplus of oil for export.

Oriental A term referring to the East and things Eastern. It is a European perspective and was used to refer to the areas, cultures, and cultural practices found to the east of that continent.

Oriental intensive agriculture The East Asian culture area's characteristic heavy inputs of labor and organic fertilizer to produce multiple crops intercultivated with wet rice.

Orographic precipitation Precipitation induced by an air mass being forced to rise over a mountain barrier, decreasing its temperature; intensification of precipitation caused by passage of clouds over a topographic high point.

Overcapacity Physical means of production (e.g., in steel mills) that are, at least temporarily, surplus to demand.

Overdeveloped economy An economy that is heavily dependent on imported industrial resources, even imported foods, and is out of balance with the local resource base.

Overdevelopment A situation in which a country or region is heavily dependent on imported resources, including food; the condition of an intensively developed area being out of balance with the local resource base.

Overgrazing The practice of allowing too many domestic animals to graze a specific area, resulting in permanent damage to the vegetation and soil of that area.

Overheated economy An economy that has outstripped domestic capital and demand in its attempt to attain increased levels of production and development.

Overpopulation Usually defined as a situation in which a country or region is consistently unable to feed all of its people adequately. Rather than being a specific ratio of people to area, the precise definition varies, depending on the level of economic development.

Pacific ring of fire The general area of tectonic (earthquake) activity and vulcanism around the edge of the Pacific basin that is associated with plate tectonics.

Paddy An enclosed basin for the growing of rice; irrigated rice culture. The rice plant is grown in standing water; drainage takes place shortly before harvest.

Pampas The lush natural grasslands of temperate South America, an area of extremely fertile soil and highly productive agriculture. A part of Argentina and all of Uruguay are the areas traditionally associated with and included in the Pampas.

Pangaea The proposed supercontinent which 200 million years ago began to break apart and form the present landmasses.

Páramos In Latin America, that part of vertical zonation agriculture that is too cool for crops but used for grazing; the high-altitude area devoted to permanent pasture and few, if any, crops.

Partition In the political geographic sense, the division or separation of the people and national territory into separate political entities, as in the division of Germany into two political entities.

Passive resistance A series of protest and civil disobedience techniques developed by Gandhi in his campaign for the independence of India; resistance to authority through civil disobedience without violence or active fighting.

Pattern A degree of regularity in a spatial distribution. Pattern is repeated (or repeatable) and predictable to some extent.

Pax Britannica The historical period, approximately 1815 to 1914, during which Britain's paramount power acted to restrain major wars among great powers.

Peasants Those who work the land as farmers. In a traditional society, they are an economic underclass. When they are tenants or sharecroppers on the lands owned by others, they are referred to as landless peasants.

Peninsula A body of land almost surrounded by water but joined to a larger land area; "almost" insular or islandlike.

Peonage A system of land tenure common in some parts of Latin America. Peasants were tied to the land by being permanently in debt to the landlord.

Per capita Literally, "per person"; term used in measures such as income, debt, and production; the value of such variables for each person.

Perception The phenomenon of different people selectively observing and interpreting different items in the same environment or section of reality. Varying cultural backgrounds may lead to very different conclusions and assumptions about a landscape.

Periphery The "edge," fringe, or frontier, as opposed to the core or the center.

Permafrost The permanently frozen subsoil of arctic environments that hinders drainage and complicates construction of buildings and roads. Solar energy in these areas is sufficient to thaw seasonally the surface or topsoil, but not the soil or rock underneath. Permafrost presents special difficulties for the building and maintenance of structures and transportation facilities.

Phosphate A mineral containing the element phosphorus, which is necessary to plant growth, particularly the development of roots, bulbs, and tubers; a fertilizer made of these minerals.

Photosynthesis The process by which green plants use chlorophyll to convert solar energy into organic molecules, and thus begin the chain of food which supports virtually all forms of life.

Physical geography The portion of geography that emphasizes the study of the physical environment; it is concerned with topography, soils, weather, climate, natural vegetation, and water features.

Physiographic Having to do with the physical geography of an area, particularly the landforms. A physiographic region exhibits a similarity or uniformity of such factors as climates, soils, and landforms. Physiographic maps show primarily drainage and elevation.

Physiological density The density of human population relative to cultivated area rather than to total area.

Planetary engineering Any massive-scale engineering project with the potential for significantly affecting the environment and ecology of large areas. The Siberian River Reversal Scheme is an example.

Planned economy A centrally planned economy with either absolute government control or majority involvement. All construction, expansion of plant, and production are planned by a central governmental agency.

Planning As used by professional planners, the comprehensive, long-range planning for orderly growth, efficient and nondisruptive land use, intelligent use of resources, and development of an attractive and healthful living environment for all.

Planning inventory An inventory or comprehensive study of existing land use patterns and recent trends—a necessary first step in the planning process.

Plantation Huge farm specializing in one or two crops for export, generally found in tropical or subtropical areas. Luxury items or industrial crops dominate, and labor is hired on a contract basis (most plantations are in labor-short areas).

Plantation economy An economy in which an important or leading sector of the economy is represented by large landholdings using hired labor to produce commercial crops for export.

Plant obsolescence The process of a plant (building, machinery, and associated facilities for production) becoming less economic to operate in competition with newer plants using the latest technology.

Plebiscite A poll of the citizenry on a possible change in political affiliation to another state; determination of national political control by referendum.

Political ideologies Ideas, ideals, theories, and practices that characterize political philosophies and governments.

Polynesia Literally, "many islands"; region of the Pacific basin; the area of larger islands in the western Pacific basin.

Popular culture That which reflects mass production, mass media, and commercialized arts and entertainments. Unlike folk culture, it has no traditional basis.

Population cohort Group of similar-aged people, usually in five-year intervals. *See also* Population pyramid.

Population explosion The term describing the unprecedented sharp, sustained increase in world population growth rates over the past three centuries, related primarily to declining death rates.

Population pressure The perception that a country or region has difficulty in adequately feeding the human population. There is no density figure that defines this pressure; rather, it is related to economic development and resource utilization as much as to physical density.

Population projections   Systems of predicting the size and characteristics of a population, based on analysis of current trends such as birth- and death rates and migration factors.

Population pyramid   A graph used to display the age and sex characteristics of a population. A "normal" population graph has an approximately pyramidal shape, since there are large numbers of youth compared to the number of the elderly. Age groups are shown at five-year intervals.

Possibilism   The philosophical view that humans take an active rather than passive role in choice of cultural development in a physical landscape; their choice is influenced more by culture than by physical environment.

Potash   The potassium necessary to plant growth, particularly the development of leaves, flowers, and fruits. Potash is obtained in mineral form and by burning vegetation in closed containers (pot ash).

Power vacuum   A temporary situation in which a particular area has no political or military power beyond its territory, and there is no involvement of outside powers.

Prairie   A grassland area noted for its rich soil. Prairies are not natural grasslands but develop in response to repeated burnings by humans (or natural forces) that prevent tree growth and preserve the soil-enriching grass cover.

Preindustrial   The economic and technological condition of not having significantly begun the process of industrialization.

Primary employment   Employment in producing or extracting raw materials, as in mining, forestry, agriculture, or fishing.

Primate city   A city that occupies the pinnacle of the urban hierarchy in a country or major region. It has the greatest array of specialized goods and services and, in some senses, a nationwide hinterland.

Private plot   In Communist states, a small land unit from which a farmer can sell produce of any kind on free markets and keep the proceeds, which he cannot do in collective production. Much of fresh fruits, vegetables, and eggs of Communist states are from such limited "private" production plots.

Privatization   In the People's Republic of China, the policy of encouraging a controlled amount of private enterprise in agriculture, crafts, and services. The term also applies to the denationalization of some European industrial and transportation objectives.

Probabilism   Theory which asserts that if one knew all the physical environmental constraints and cultural values of a society, one could predict the cultural development that would follow. Probabilism is disavowed as an essentially racist-supremacist philosophy.

Protectionist tariff   An import tax imposed specifically to give market advantages to domestic sources.

Protectorate   A political unit with less sovereignty than a true state but more than a colony. Typically, foreign policy and defense are controlled by the "protector," but domestic policy is only generally guided and not administered from outside.

Push-pull mechanisms   Negative and positive location factors. A "push" mechanism favors an activity moving out to a more attractive location; a "pull" is a positive reason attracting an activity to it. The term can also be applied to domestic and international migration.

Quaternary employment   In the traditional threefold division of economic activity, "primary" is obtaining raw materials, "secondary" is manufacturing, and "tertiary" is all services. Jean Gottmann's "quaternary" category segregates services, largely intangible, performed by highly specialized, highly educated or trained people.

Race   An emotionally loaded term that describes differentiating physical characteristics of major subdivisions of the human race, but that can also refer to perceived cultural and behavioral traits.

Radial drainage   A system of streams running in all directions away from a central elevated structure, emanating from a central point.

Rainforest   A three-tiered forest found in areas of hot, wet tropical climate. The number of individual species found in this vegetational association is astounding. The rainforest's role in preserving the earth's atmosphere and generating rainfall is only now coming to light. The destruction of the world's rainforests for the extension of agriculture and through lumbering has become a serious concern. *See also* Selva.

Rainshadow   A dry area on the lee side of a mountain or mountain range.

Ranching   A commercial version of livestock production in which machinery and a variety of techniques are used to lower costs, reduce labor needs, and increase the amount and efficiency of production.

Rationalization of holdings   The trading of one compact plot for the many scattered strips of land that may be in one ownership due to relics of feudal land systems, inheritance, and dowry.

Raw materials   Naturally occurring materials that must enter some form of processing or manufacturing before final consumption or utility.

Region An areal unit smaller than the globe, defined by criteria selected for degree of relative homogeneity (e.g., a climate region); a region involving some degree of functional interaction (e.g., a port hinterland or a metropolitan commuting region).

Regional approach *See* Regional geography.

Regional capital In large areal units, such as Canada or the United States, a city that functions as an economic, administrative, and cultural "capital" even if it is not also a provincial or state political capital. U.S. examples include Atlanta and Los Angeles.

Regional geography The study of all factors of the total environment—cultural and physical—that have significance to the understanding of that subdivision of the earth's surface. The focus is on an area and all geographic variables interrelated there.

Regional growth pole *See* Growth pole.

Regional interdependence A strong degree of mutual economic benefit tying together various parts of a region.

Relict population A small survivor of a formerly more dense population, for example, of a small remnant population of a once widespread plant or animal species.

Relief The difference in elevation over an area. The difference between the elevation of the highest and lowest points within an area is called local relief.

Religion A system and body of faith and worship, including a set of values and behavioral standards; an ethical code followed in relation to other people and to the Supreme Being.

Religious diffusion The outward spreading and subsequent acceptance or modification of a religion from its center of origin.

Remote sensing Viewing an object or a portion of the earth from a distance with the aid of a photographic or other device. Satellite images and air photos are the most widely used products of remote sensing.

Residual minority Any group of some ethnic, racial, or cultural distinction and identity, once a majority, now outnumbered by later-arriving people of different cultural, ethnic, or racial makeup.

Rump state The small residual left from the defeat or collapse of a once much larger political unit, as in the end of an empire. Austria and Hungary are rump states of a once much larger empire.

Rural The opposite of urban; pertaining to the agricultural landscape; generally refers to circumstances surrounding a dwelling place; countryside.

Russification In the Soviet Union, as in tsarist Russia, the policy of encouraging, or forcing, the cultural assimilation of non-Russian ethnics into the majority culture.

Rust Belt A popular designation for states and areas in which there is strong industrial decline, particularly in the production areas of metallurgy, metal engineering, heavy machinery, and transportation equipment. With exceptions, the northeastern United States.

Sahel The transition zone between the Sahara Desert and the wetter lands to the south in Africa. It is a region of transition with unreliable precipitation and a high frequency of drought.

Salination The process in which soluble minerals become poisonously concentrated in the top layer of desert soils by surface evaporation of irrigation water. It can be avoided or slowed by "flushing" and drainage systems.

Salinity The proportion of dissolved salts to pure water, usually expressed in parts per thousand (0/00). The designation may also be applied to soil.

Salt flat A white crust on the ground produced when water evaporates and leaves its dissolved materials behind; large areas of salt-encrusted soil, often old lake bed.

Satellite Although theoretically sovereign, a state that in fact has foreign policy and general policy guidelines, at least for some domestic affairs, imposed from without. The term is popularly applied to East European Communist countries dominated by the USSR.

Savanna A name given to a type of tropical vegetation that is often described as parkland since it is a mixture of trees and grass. The climatic designation "tropical wet-and-dry" is a broader term that encompasses a greater variety of vegetation.

Scale A uniform method of reduction in size. On a map, the degree of reduction that has taken place is expressed as map scale. It is the ratio of distance or area in reality to distance or area on the map. It can be expressed as a ratio, as a fraction, in words, or in a horizontal linear bar graph.

Scarp A cliff or steep slope. It is always composed of material resistant to erosion and may be the edge of a plateau or shield. Also known as an escarpment.

Scientific revolution The practical application of scientific principles to improving production efficiency in all sectors of modern economies, from agriculture to transport to manufacturing.

Scrub A generic term used to define a landscape covered with dwarf trees and bushes. More than grass, less than forest, it is often a stunted version of the latter. It is found in areas of dry or cold climatic extremes or of abandoned farmland.

Secondary employment Employment in processing or manufacturing raw materials, increasing their utility and transforming their nature.

Sector theory Theory which asserts that land use patterns of functional zones will assume wedge-shaped sectors outward from the central business district. This theory modified the concentric zone theory of urban land use.

Sedentary (agriculture)   Permanent residence in a given area for the purpose of gaining a livelihood. The term generally applies to crop farming by an established population in a well-defined settlement; it is the opposite of nomadic herding.

Sediment   Unconsolidated particles created by the weathering and erosion of rock, by chemical precipitation from solution in water, or from the secretions of organisms, and transported by water, wind, or glaciers.

Sedimentary rock   Rock formed from the weathered products of preexisting rocks that have been transported, deposited, and lithified (converted to rock).

Selva   The name given to the rainforest of the Amazon. Selva occurs in areas of perpetual high temperatures and humidity that experience little or no dry season. It is the archetypical tropical rainforest of broadleaf evergreens.

Sensible temperature   The perceived temperature affecting people; a result of temperature, humidity, and wind speed factors.

Sertão   The backlands of Brazil, particularly those of the northeastern part of the country. They are associated with recurring natural climatic disaster and consequent poverty and out-migration.

Settlement nucleus   Node of greater density of settlement within a less dense matrix.

Settlement pattern   The physical arrangement of structures, roads, and other major created landscape that reflects cultural values, level of technology, population density, and livelihood.

Shantytown   Group of makeshift housing constructed by the poor inhabitants, usually recent in-migrants, on the fringes of rapidly expanding cities of the developing world.

Sharecropping   An economic-agricultural system in which land is rented from a large landholder and paid for by a prearranged portion of the crop produced (or a percentage of cash receipts from the sale of a crop).

Shatterbelt   The territory, commonly fragmented politically, that lies between more powerful, expansionist cultures. Frequently, a politically unstable area. Eastern Europe and Southeast Asia are examples.

Shatter zone   *See* Shatterbelt.

Shia   The smaller of the two major divisions within Islam. The division arose over questions of who were the legitimate heirs or successors to Muhammad. Shiite Muslims dominate in Iran and Yemen and are a plurality in Iraq. Elsewhere, the Shia followers are scattered minorities among a Sunni majority.

Shifting agriculture   *See* Land rotational agriculture; Shifting cultivation; Slash-and-burn-agriculture.

Shifting cultivation   The land use system that evolved in areas of marginally fertile soils, commonly in the wet tropics, in which the farmers clear, plant, harvest, and then move onto a fresh patch of land after a season or two, using the land in a long-term rotation rather than planting the same land year after year.

Shotgun house   A folk style of narrow homes with one room behind the other, common in the American South; may represent a West African style transmitted via Haiti and New Orleans.

Siberian River Reversal Scheme   The huge water engineering project, partially complete, to partially "reverse" the northward flow of Siberian rivers (Ob, Yenisei, Lena) in order to increase irrigation in Soviet Central Asia, generate power, and increase navigability of several rivers.

Sierra   A name frequently appearing on maps that designates mountains or highland complexes; a Spanish term widely used in areas that were once under Spanish control. *Serra* is the Portuguese (and Brazilian) variation.

Sikhs   A religious group in India that combines elements of Hindu and Islamic beliefs and arose at a time of serious pressure against those who did not accept Islam, the official state religion. Sikhs are now a dissident element within India and are seeking autonomy or independence.

Silvaculture   The growing of trees as a long-term crop through selective breeding, planting, and culling of poor specimens or unwanted, competing species; professional or commercial forestry.

Single-source theory   The idea that all civilizations stem from one early source area or innovator.

Site   The most local and immediate scale of location; the narrowest definition of location, it refers to the specifics of the immediate surroundings.

Situation   A broader definition of location than site; location relative to a larger region or country or to the world.

Slash-and-burn agriculture   A land rotational agricultural system in which trees of the forest are girdled and burned to clear the land and add fertility to the soil. A year or two of cropping follows, after which the land is abandoned and another clearing is made at a different location. In this manner, land is fallowed to allow the return of fertility. *See also* Land rotational agriculture.

Snow line   The lower limit of perennial snow in a mountain area.

Soil   A combination of mineral and organic matter, water, and air; that portion of the decomposed rock that supports plant growth.

Soil horizon   A layer of soil that has identifiable characteristics produced by chemical weathering and other soil-forming processes. Most soils have three horizons, designated A, B, and C, from top to bottom.

Sovereignty The concept of totally independent right and power to govern a territory and pursue foreign policy without interference from outside interests.

Soviet model The economic development model in which producer goods, such as steel mills and truck factories, receive priority over consumer goods. It could also be called the "heavy-industry-first" model.

Spatial Spread over area or space.

Spatial correlation An obvious relationship between two or more variables through space. The presence or absence of one variable is related, apparently but not necessarily, to the presence or absence of another variable.

Spatial interaction Interchange of materials, people, culture, or other economic and political forces, not necessarily tangible, across space or territory.

State An independent, sovereign unit with an established territory and a permanent population.

State farm Associated with Communist economies, a large, state-owned farm whose labor is paid wages in proportion to hours spent at work at a job category assigned a particular hourly rate, much as in a factory. State farms often have production of seed, experimentation, and breeding as their major functions.

Steppe Temperate natural grasslands and the semiarid or subhumid climate that supports a vegetational cover of grass and is too dry to support the growth of trees. The name is Russian in origin and originally applied to the treeless plains of the south of the USSR.

Strait A narrow, constricted body of water connecting two larger bodies of water. These narrow waterways have extreme military and commercial significance, since control of nearby land areas can protect (or interfere with) the passage of ships.

Stream A general term to denote the flow of water within any natural channel. Thus, a small creek and a large river are both streams.

Stream capture The "capture" by a headward-eroding stream of part of the headwaters of another stream system by providing a steeper gradient.

Stream piracy The diversion of the drainage of one stream resulting from the headward erosion of another stream.

Strip mining *See* Surface mining.

Subculture A subdivision of a culture; a group of people within a culture who share many basics of the majority culture, yet who also have distinctive cultural traits, beliefs, and material culture.

Subregion A smaller division of a region, again defined by selection of a degree of similarity on a specific economic or cultural relationship.

Subsistence farming A type of farming in which the farm family is essentially self-sufficient and in which little surplus is marketed. It is the least advanced farm economy and the opposite of commercial farming.

Subsoil A term applied to the B horizon of a soil profile; the soil beneath the surface or topsoil. *See also* Soil; Soil horizon.

Suburbia The lower-density fringes of a central urban place; in the United States, "suburbia" contains more people than do central cities, and many urban functions are carried out in suburbia. Suburbia also refers to smaller, individual political jurisdictions outside the central city.

Successor state A new state that arises from the re-forming of political units following the collapse of an empire or very large state. Zaire and other African states, formerly colonies, are successor states.

Sun Belt Generally, any area of comparatively warm, sunny climate that functions as a resort and retirement area attracting new migrants in search of a superior lifestyle; originally, the southern United States.

Sun Belt tourism Domestic or international tourism motivated by a warmer, sunnier climate and related amenities rather than by cultural considerations.

Sunni The larger of the two major divisions within Islam; the orthodox division of Islam which accepts traditional law as equal to the teachings of the Koran.

Superpowers The largest, most powerful states; those exerting most power in international affairs and having the greatest military and economic forces; usually the United States and the USSR.

Supply-generated flow The movement of goods to markets in response to production of a surplus. For example, the production of surplus of citrus produced in California generates flows to other parts of the United States.

Surface mining The form of mining in which (rather than using underground tunnels to retrieve the mineral), the surface and covering materials are removed, exposing the mineral at the new surface.

Synthesis The development of material of identical or highly similar properties by processing and manufacturing an artificial substitute. The product is called a synthetic.

System A group of phenomena that are associated by some form of interaction.

Systematic geography Geography that focuses on the spatial patterns of a specific physical, economic, or cultural factor. Agricultural or transportation geography represent systematic organization of spatial patterns.

Taiga A term for the slow-growing boreal forest; the transition from midlatitude forest to tundra.

Takeoff In Rostow's stages of economic growth, the stage of high reinvestment in expanding enterprises, coupled with proliferation of the principles of scientific research and mass production in major sectors of the economy.

Technology Applied science; the collection of practices and skills that advance industry and personal convenience in a society; the application of science and scientific technique to production, increasing efficiency in the process.

Tectonic activity Mountain-building or crustal deformation activity, as in earthquakes; associated with tectonic plate movement.

Tectonic plate A huge plate of crustal material, of which there are an unknown number, that is in motion and colliding with other plates, resulting in "drifting continents," vulcanism, and earthquakes.

Tenant farming An economic situation in which lands are rented by farmers. Tenant farming is common worldwide but is particularly widespread in less developed economies.

Terai The name given to the lower slopes and near-tropical climatic areas of the Himalayan foreland.

Terrace A flat, benchlike structure produced by a stream, which was left elevated as the stream cut downward; a manmade structure with similar appearance.

Terracing The process of creating level land in areas of steep slope. Walls and earthworks are used to create "steps" on hillsides. It is a method of extending farmland in mountain environments, and aids in the reduction of erosion.

Terra roxa A name given to highly fertile red soils in Brazil that formed from the decomposition of mineral-rich metamorphic rock of the Brazilian Shield. Such soils were particularly highly valued for the production of sugar and coffee.

Texture The size, shape, and distribution of the particles that collectively constitute a rock or soil.

Thematic map Any map designed for a special purpose and showing distributions not commonly found on general reference maps.

Theocracy A state ruled or controlled by a religious leader and in which religious law is treated as civil law.

Third World Not part of the Western industrial world or the Communist bloc; the developing countries, especially those politically nonaligned.

Thorn forest A deciduous forest that loses its leaves in response to a dry season, rather than to a cold season. As a means of reducing evaporation of scarce water, trees lose their leaves. Thorn forests are the characteristic vegetation in some of the drier savanna climate areas that experience a long and pronounced annual dry season.

Three-field system A rotational system of farmland management, common in feudal Europe, in which each farmer worked three plots, alternating crops, pasture, and fallow.

Threshold In studies of the geography of service centers and service functions, the minimum number of potential customers in the trade area necessary to support a given function.

Tierra caliente In Middle and South America, the "hot land" or tropical zone at the lowest elevation in vertical zonation.

Tierra fria The "cold land" at higher elevations in vertically zoned agricultural lands; typical crops include potatoes.

Tierra templada The "temperate land" in vertical zonation; crops include grains and midlatitude fruits and vegetables.

Time-distance Distance as measured in the amount of time consumed in travel between two points (e.g., a half-hour walk, a 20-minute drive).

Time-place utility Part of the set of variables that determine whether a particular material is of economic use. Uranium would have lacked time-place utility to pre-Columbian Indians.

Topography The term used to describe the elevation and landforms in a landscape; almost literally, the "ups and downs" on the surface of the earth. It also refers to the study of landforms.

Topsoil The uppermost layer of the soil, at times rich in humus and generally more fertile than the subsoil beneath it. *See also* Soil; Subsoil.

Trade balance The difference between imports and exports in a country's international trade. A surplus of exports creates a positive balance of payments (an inflow of cash); the reverse is a negative balance of trade (an outflow of cash).

Traditional society A society in which land is still the basis of power and status. In an economy dominated by agriculture, wealth is concentrated in the hands of a minority of large landholders; capital is invested in luxuries and conspicuous consumption by members of that group, who also control the government.

Traffic shadow A phenomenon in air passenger traffic, though applicable to other modes of public transport. The largest airports will attract more traffic from a larger hinterland, overshadowing nearby airports.

Transhumance The seasonal shift of grazing animals from one place to another to take advantage of seasonal differences in carrying capacity and forage availability; common in summer pasturing of animals in mountain meadows that are unusable in winter.

Transport infrastructure The physical facilities for modern transport that support economic development. Examples are port facilities, airports, roads, and railroads.

Tribute territory In imperial systems, such as that of the ancient Chinese Empire, a territory enjoying some degree of autonomy so long as it acknowledged the "overlord" by sending tribute.

Troposphere The lowermost layer of the atmosphere, generally characterized by a decrease in temperature with height.

Tundra Both a climate type and vegetation term; the area of the subpolar zone and in high mountains in which the brevity and relatively low temperatures of the summer season limit plant growth to slow-growing grasses, sedges, mosses, and the like. No month is free of frost.

Underdeveloped country (UDC) A country that does not meet a set of developmental norms; a country that exhibits lower living standards, incomes, and nutrition. Normally, underdeveloped countries are dominated by agricultural or other primary production. They are in the early stages of the Rostow scheme of development.

Underemployment Part-time or occasional employment common in underdeveloped areas and countries with a surplus of labor in relation to available jobs.

Unitary state The opposite of a federal state; all government power is concentrated in one (the national) level.

Untouchables Members of the lowest castes in Hindu society.

Urban area A city or other densely populated area that exhibits the traits of a city, including a certain style of life and the absence of farming as an occupation.

Urban explosion The unprecedented rapid growth of cities associated with the industrial revolution.

Urbanization The process of becoming more like a city; the shift in a population from rural to city dwellings and to nonfarm occupations; the expansion of a city through the absorption of surrounding areas.

Use pressure The burden of large numbers of users or visitors who inevitably, if unwittingly, deteriorate a facility or landscape by sheer numbers.

Value added A measure of the difference in cost of raw materials and price of the finished goods. It is a result of the skill of labor and the use of capital equipment.

Veld The name given to the South African steppe or temperate grasslands. Unlike the savannas, the veld is subject to seasonal frost. In wetter areas, it contains copses of trees; in drier areas, herbaceous shrub is intermixed with grasses.

Vertical zonation The altitude-induced changes in climate (temperature) that enable cultivation of different crops at different altitudes.

Villagization (campaign) A resettlement scheme in rural Ethiopia that proposes to forcibly collect scattered farm families now living in traditional homesteads into villages.

Virgin Lands Program An effort, between 1954 and 1959, to increase Soviet cropland by about 20 percent. The Virgin Lands, mostly in Kazakh SSR, were too dry to have been cultivated before. The program was not a success.

Vital rates The two rates, birth- and death, that are used in conjunction with migration data to depict changes in, and prospects for, a population.

Wadi An Arabic term for intermittent streams; streams that flow only after a rain; the valley carved (and occupied) by an intermittent stream; a dry streambed. Dams in some wadis impound rainwater for crop irrigation.

Waldsterben Literally, "forest death"; the death of trees attributed to direct or indirect effects of environmental pollution; a German term widely used in Europe to describe this phenomenon.

Water table The level of groundwater in soil or rock beneath the surface; pertaining to the water supply that exists beneath the earth's surface in any given area.

Weather The state of the atmosphere at any given time.

Weathering The disintegration and decomposition of rock at or near the surface of the earth through the aegis of chemical or physical forces or both.

World city A city of international significance, politically, economically, and/or culturally. Tokyo and Paris are examples.

World language A language that is widespread in use due to its commercial, scientific, and academic utility. World languages are commonly multinational in use (e.g., French, English, Spanish).

Xerophyte A plant highly tolerant of drought; typical plant found in a desert area with adaptations for water storage and/or the reduction of evaporation.

Zaibatsu Very large corporations in Japan; they tend to be conglomerates, that is, they are engaged in many, unrelated business activities.

CREDITS

Chapter One

Chapter opening photo: Map of the World by Wellem Blaeu.

p. 14 Courtesy Rosensteil School of Marine and Atmospheric Science, University of Miami.

p. 25 Copyright © Wouter Loot-Gregoire/Valan Photos.

p. 29 Courtesy of Wade R. Currier/Geovisuals.

p. 63 Debby Rogow/Photo Edit.

Chapter Two

Chapter opening photo: National Park Service, Statue of Liberty National Monument.

p. 105 Produced from USAF DMSP (Defense Meteorological Satellite Program) film transparencies archived for NOAA/NESDIS at the University of Colorado, CIRES/National Snow and Ice Data Center, Campus Box 449, Boulder, CO, U.S.A. 80309.

p. 111 by Gail Meese/Merrill Publishing.

p. 115 by Charles Stansfield, Jr.

p. 116 Copyright © by Robert Perron.

p. 121 Copyright © by Ron Sanford.

p. 125 by Larry Lee/Chevron/Mobil.

p. 131 by Jean-Marie Jro/Valan Photos.

p. 133 by Marcel/Photo Edit.

Chapter Three

Chapter opening photo: Copyright © by Ron Sanford.

p. 154 by Ed Wojtas.

pp. 157, 161, 185, 186 Copyright © by Ron Sanford.

p. 165 by Alan Oddie/Photo Edit.

p. 171 by Tony Freeman/Photo Edit.

p. 177 Copyright © Paul Conklin/Photo Edit.

Chapter Four

Chapter opening photo: Ric Ergenbright Photography.

p. 206 by Soyfoto/Eastfoto.

pp. 217, 223, 224, 233, 235, 240 by Gwendolyn Stewart.

pp. 242, 246, Copyright © Wolfgang Kahler.

Chapter Five

Chapter opening photo: Copyright © Wolfgang Kahler.

p. 268 by N. Mendham/Valan Photos.

p. 274 by D. Krause.

p. 276 Copyright © Wolfgang Kahler.

Chapter Six

Chapter opening photo: Jerome Wyckoff.

pp. 282, 299, 301, 306, 310, 317 Copyright © Mike Yamashita.

pp. 288, 313 Copyright © Wolfgang Kahler.

p. 335 Copyright © Galen Rowell/Mountain Light Photography.

Chapter Seven

Chapter opening photo: Copyright © Larry Tackett/Tom Stack & Associates.

pp. 367, 384, 390 Copyright © Tim Gibson/Envision.

p. 372 Copyright © Galen Rowell/Mountain Light Photography.

p. 375 Copyright © Carl Purcell.

p. 381 Copyright © Paolo Koch/Photo Researchers, Inc.

Chapter Eight

Chapter opening photo: Copyright © Wolfgang Kahler.

p. 403 Copyright © George Holton/Photo Researchers, Inc.

p. 421 Copyright © Wolfgang Kahler.

p. 425 Copyright © Michael Yamashita.

Chapter Nine

Chapter opening photo: Copyright © David Waters/Envision.

pp. 443, 465, 468, 480 by World Bank Photos.

p. 474 Copyright © Tim Gibson/Envision.

p. 487 Copyright © Blair Seitz.

Chapter Ten

Chapter opening photo: James Karageorge/Chevron Corp.

p. 518 Copyright © Carl Purcell.

p. 529 by Alan Oddie/Photo Edit.

p. 542, 552 Copyright © Wolfgang Kahler.

p. 545 by NASA.

Chapter Eleven

Chapter opening photo: Copyright © Robert Holmes.

pp. 572, 597, 607 Copyright © Carl Purcell.

p. 581 Copyright © Noel R. Kemp/Photo Researchers, Inc.

p. 589 by Tony Freeman/Photo Edit.

p. 590 Copyright © Byron Augustin/Tom Stack & Associates.

p. 609 Copyright © George Gerster/Photo Researchers, Inc.

p. 613 by Kent Legrow.

p. 624 Copyright © Galen Rowell/Mountain Light Photography.

p. 629 Copyright © Don Rutledge/Tom Stack & Associates.

Index